Lecture Notes in Bioinformatics 10209

Subseries of Lecture Notes in Computer Science

More information about this series at http://www.springer.com/series/5381

Ignacio Rojas · Francisco Ortuño (Eds.)

Bioinformatics and Biomedical Engineering

5th International Work-Conference, IWBBIO 2017
Granada, Spain, April 26–28, 2017
Proceedings, Part II

 Springer

Editors
Ignacio Rojas
Universidad de Granada
Granada
Spain

Francisco Ortuño
Universidad de Granada
Granada
Spain

ISSN 0302-9743   ISSN 1611-3349   (electronic)
Lecture Notes in Bioinformatics
ISBN 978-3-319-56153-0   ISBN 978-3-319-56154-7   (eBook)
DOI 10.1007/978-3-319-56154-7

Library of Congress Control Number: 2017935843

LNCS Sublibrary: SL8 – Bioinformatics

Printed on acid-free paper

This Springer imprint is published by Springer Nature
The registered company is Springer International Publishing AG
The registered company address is: Gewerbestrasse 11, 6330 Cham, Switzerland

Preface

We are proud to present the set of final accepted full papers for the third edition of the IWBBIO conference "International Work-Conference on Bioinformatics and Biomedical Engineering" held in Granada (Spain) during April 26–28, 2017.

The IWBBIO 2017 (International Work-Conference on Bioinformatics and Biomedical Engineering) seeks to provide a discussion forum for scientists, engineers, educators, and students about the latest ideas and realizations in the foundations, theory, models, and applications for interdisciplinary and multidisciplinary research encompassing the disciplines of computer science, mathematics, statistics, biology, bioinformatics, and biomedicine.

The aim of IWBBIO is to create a friendly environment that could lead to the establishment or strengthening of scientific collaborations and exchanges among attendees, and, therefore, IWBBIO 2017 solicited high-quality original research papers (including significant work in progress) on any aspect of bioinformatics, biomedicine, and biomedical engineering.

The following topics were especially encouraged: new computational techniques and methods in machine learning; data mining; text analysis; pattern recognition; data integration; genomics and evolution; next-generation sequencing data; protein and RNA structure; protein function and proteomics; medical informatics and translational bioinformatics; computational systems biology; modelling and simulation and their application in the life science domain, biomedicine, and biomedical engineering. The list of topics in the successive call for papers also evolved, resulting in the following list for the present edition:

1. **Computational proteomics**. Analysis of protein–protein interactions. Protein structure modelling. Analysis of protein functionality. Quantitative proteomics and PTMs. Clinical proteomics. Protein annotation. Data mining in proteomics.
2. **Next-generation sequencing and sequence analysis**. De novo sequencing, re-sequencing, and assembly. Expression estimation. Alternative splicing discovery. Pathway analysis. Chip-seq and RNA-Seq analysis. Metagenomics. SNPs prediction.
3. **High performance in bioinformatics**. Parallelization for biomedical analysis. Biomedical and biological databases. Data mining and biological text processing. Large-scale biomedical data integration. Biological and medical ontologies. Novel architecture and technologies (GPU, P2P, Grid etc.) for bioinformatics.
4. **Biomedicine**. Biomedical Computing. Personalized medicine. Nanomedicine. Medical education. Collaborative medicine. Biomedical signal analysis. Biomedicine in industry and society. Electrotherapy and radiotherapy.
5. **Biomedical engineering**. Computer-assisted surgery. Therapeutic engineering. Interactive 3D modelling. Clinical engineering. Telemedicine. Biosensors and data acquisition. Intelligent instrumentation. Patient monitoring. Biomedical robotics. Bio-nanotechnology. Genetic engineering.

6. **Computational systems for modelling biological processes**. Inference of biological networks. Machine learning in bioinformatics. Classification for biomedical data. Microarray data analysis. Simulation and visualization of biological systems. Molecular evolution and phylogenetic modelling.
7. **Health care and diseases**. Computational support for clinical decisions. Image visualization and signal analysis. Disease control and diagnosis. Genome-phenome analysis. Biomarker identification. Drug design. Computational immunology.
8. **E-health**. E-health technology and devices. E-health information processing. Telemedicine/e-health application and services. Medical image processing. Video techniques for medical images. Integration of classical medicine and e-health.

After a careful peer-review and evaluation process (each submission was reviewed by at least two, and on average 3.2, Program Committee members or additional reviewers), 120 papers were accepted for oral, poster, or virtual presentation, according to the recommendations of the reviewers and the authors' preferences, and to be included in the LNBI proceedings.

During IWBBIO 2017, several Special Sessions were carried out. Special Sessions are a very useful tool to complement the regular program with new and emerging topics of particular interest for the participating community. Special Sessions that emphasize multi-disciplinary and transversal aspects as well as cutting-edge topics were especially encouraged and welcomed, and in IWBBIO 2017 they were the following:

– **SS1: Advances in Computational Intelligence for Critical Care**

Decision-making in health care in clinical environments is often made on the basis of multiple parameters and in the context of patient presentation, which includes the setting and the specific conditions related to the reason for admission and the procedures involved. The data used in clinical decision-making may originate from manifold sources and at multiple scales: devices in and around the patient, laboratory, blood tests, omics analyses, medical images, and ancillary information available both prior to and during the hospitalization.

Arguably, one of the most data-dependent clinical environments is the intensive care unit (ICU). The ICU environment cares for acutely ill patients. Many patients within ICU environments, and particularly surgical intensive care units (SICU), are technologically dependent on the life-sustaining devices that surround them. Some of these patients are indeed dependent for their very survival on technologies such as infusion pumps, mechanical ventilators, catheters and so on. Beyond treatment, assessment of prognosis in critical care and patient stratification combining different data sources are extremely important in a patient-centric environment.

With the advent and quick uptake of omics technologies in critical care, the use of data-based approaches for assistance in diagnosis and prognosis becomes paramount. New approaches to data analysis are thus required, and some of the most interesting ones currently stem from the fields of computational intelligence (CI) and machine learning (ML). This session is particularly interested in the proposal of novel CI and ML approaches and in the discussion of the challenges for the application of the existing ones to problems in critical care.

Topics that are of interest to this session include (but are not necessarily limited to):

- Novel applications of existing CI and ML and advanced statistical methods in critical care
- Novel CI and ML techniques for critical care
- CI and ML-based methods to improve model interpretability in a critical care context, including data/model visualization techniques
- Novel CI and ML techniques for dealing with non-structured and heterogeneous data formats in critical care

Organizers: Dr. Alfredo Vellido, PhD, Department of Computer Science, Universitat Politécnica de Catalunya, BarcelonaTECH (UPC), Barcelona (Spain). Dr. Vicent Ribas, eHealth Department, EURECAT Technology Centre of Catalonia, Barcelona, Barcelona (Spain).

- **SS2: Time-Lapse Experiments and Multivariate Biostatistics**
Biological samples are evolving in time, phases, periods, behavior. To be able to understand the dynamics, we need to perform time lapse experiments. Today's technique and measurement devices allow us to monitor numerous parameters in semi-controled environments during the experiment. The increase of measured data is enormous. The interpretation requires both qualitative and quantitative analysis. There are useful methods of biostatistics, multivariate data analysis, and artificial intelligence, namely, neural networks, genetic algorithms, and agent-based modeling, respectively.

In this special section we will provide a discussion on broad examples from time-lapse experimental design through information and data acquisition, using methods of bioinformatics, biophysics, biostatistics, and artificial inteligence. The aim of this section is to present the possible increase in data interpretation and related methods.

Organizer: Dr. Jan Urban, Laboratory of Signal and Image Processing, Institute of Complex Systems, Faculty of Fisheries and Protection of Waters, University of South Bohemia.

- **SS3: Half-Day GATB Tutorial. The Genome Analysis Toolbox with de Bruijn Graph**
The GATB programming day is an educational event organized by the GATB team. This free event is open to everyone who is familiar in C++ programming and wants to learn how to create NGS data analysis software.

The tutorial has a focus on the high-performance GATB-Core library and is taught by the developers of this library.

During this half-day tutorial, some of the following topics are explored:

- A theoretical introduction to GATB: the basic concepts.
- GATB-Core practical coding session 1: I/O operations on reads files.
- The GATB de Bruijn graph API.
- GATB-Core practical coding session 2: k-mer and graph APIs in action.
- GATB-Core practical coding session 3: writing a short read corrector tool.
- Q and A session: obtain answers from GATB experts.

Organizers: Dr. Dominique Lavenier, GenScale Team Leader, Inria/IRISA, Campus de Beaulieu, Rennes, France.
Dr. Patrick Durand, Inria, Genscale Team, Campus de Beaulieu, Rennes, France.

- **SS4: Medical Planning: Operating Theater Design and Its Impact on Cost, Area, and Workflow**

  The design of operating rooms is one of the most complicated tasks of hospital design because of its characteristics and requirements. Patients, staff, and tools should have determined passes through the operating suite. Many hospitals assume that the operating suite is the most important unit in the hospital because of its high–revenue. Arch design of these suites is a very critical point in solving an optimized problem in spaces, workflow of clean, dirty, and patient in/out in addition to staff together with their relations with adjacent departments. In this session, we illustrate the most common designs of operating suites and select the most suitable one to satisfy the effectiveness of the operating suite, maximizing throughput, minimizing the costs, and decreasing the required spaces related to available resources/possibilities. The design should comply with country guidelines, infection control rules, occupational safety and health, and satisfy the maximum benefits for patients and staff. A comparative study was performed on 15 hospitals and it recorded that the single input–output technique is the best design.

  Motivation and objectives for the session: Operating suite design is a very critical task owing to its impact. Biomedical engineers should participate in the design and review the workflow and available functions.

  Organizer: Dr. Khaled El-Sayed, Department of Electrical and Medical Engineering at Benha University, Egypt

- **SS5: Challenges Representing Large-Scale Biological Data**

  Visualization models have been shown to be remarkably important in the interpretation of datasets across many fields of study. In the context of bioinformatics and computational biology various tools have been proposed to visualize molecular data and help understand how biological systems work. Despite that, several challenges still persist when faced with complex and dynamic data and major advances are required to correctly manage the multiple dimensions of the data.

  The aim of this special session is to bring together researchers to present recent and ongoing research activities related to advances in visualization techniques and tools, focused on any major molecular biology problem, with the aim of allowing for the exchange and sharing of proposed ideas and experiences with novel visualization metaphors.

  Topics of interest include:

  - Genome and sequence data
  - Omics data (transcriptomics, proteomics, metabolomics)
  - Biological networks and pathways
  - Time-series data
  - Biomedical ontologies
  - Macromolecular complexes
  - Phylogenetic data

- Biomarker discovery
- Integration of image and omics data for systems biology
- Modeling and simulation of dynamical processes

Organizer: Prof. Joel P. Arrais FCTUC – University of Coimbra, Portugal.

– **SS6: Omics of Space Travelled Microbes – Bioinformatics and Biomedical Aspects**

The National Research Council (NRC) Committee for the Decadal Survey on Biological and Physical Sciences in Space reported that "microbial species that are uncommon, or that have significantly increased or decreased in number, can be studied in a "microbial observatory" on the International Space Station (ISS)." As part of the microbial observatory effort the NRC decadal survey committee suggested that NASA should: "(a) capitalize on the technological maturity, low cost, and speed of genomic analyses and the rapid generation time of microbes to monitor the evolution of microbial genomic changes in response to the selective pressures present in the spaceflight environment; (b) study changes in microbial populations from the skin and feces of the astronauts, plant and plant growth media, and environmental samples taken from surfaces and the atmosphere of the ISS; and (c) establish an experimental program targeted at understanding the influence of the spaceflight environment on defined microbial populations."

The proposed session discusses state-of-the-art molecular techniques, bioinformatics tools, and their benefit in answering the astronauts and others who live in closed systems.

Organizer: Dr. Kasthuri Venkateswaran, Senior Research Scientist, California Institute of Technology, Jet Propulsion Laboratory, Biotechnology and Planetary Protection Group, Pasadena, CA

– **SS7: Data-Driven Biology – New Tools, Techniques, and Resources**

Advances in sequencing techniques have accelerated data generation at diverse regulatory levels in an unprecedented way. The challenge now is to integrate these data to understand regulation at a systems level. As the sequencing technologies evolve, new tools and resources follow, revealing new aspects of complex biological systems.

This special session brings together experts from computational biology and machine learning to present recent advances in the development of new tools and resources using next-generation sequencing data including novel emerging fields such as single-cell transcriptomics. The session features an invited speaker and three/four short talks. To promote emerging leaders of the field, we select invited speakers who have gained their independence in recent years.

Organizer: Dr. Joshi Anagha, Division of Developmental Biology at the Roslin Institute, University of Edinburgh

– **SS8: Smart Sensor and Sensor-Network Architectures**

There is a significant demand for tools and services supporting rehabilitation, well-being and healthy life styles while reducing the level of intrusiveness as well as increasing real-time available and reliable results. For example, self-monitoring applications need to be improved to move beyond tracking exercise habits and capture a more comprehensive digital footprint of human behavior. This session focuses on primary parameter capturing devices and networks demonstrating advances in sensor development including a customized algorithmic shell research to support diagnostic decisions. Target domains are, for example, continuous differentiating between mental and physical stress, blood pressure monitoring, sleep quality monitoring, HRV etc.

Organizers: Prof. Dr. Natividad Martinez, Internet of Things Laboratory, Reutlingen University, Germany.
Prof. Dr. Juan Antonio Ortega, University of Seville, Spain.
Prof. Dr. Ralf Seepold, Ubiquitous Computing Lab, HTWG Konstanz, Germany.

– **SS9: High-Throughput Bioinformatic Tools for Genomics**
Genomics is concerned with the sequencing and analysis of an organism's genome. It is involved in the understanding of how every single gene can affect the entire genome. This goal is mainly afforded using the current, cost-effective, high-throughput sequencing technologies. These technologies produce a huge amount of data that usually require high-performance computing solutions and opens new ways for the study of genomics, but also transcriptomics, gene expression, and systems biology, among others. The continuous improvements and broader applications on sequencing technologies is producing an ongoing demand for improved high-throughput bioinformatics tools.

Therefore, we invite authors to submit original research, new tools or pipelines, or their update, and review articles on topics helping in the study of genomics in the wider sense, such as (but not limited to):

- Tools for data pre-processing (quality control and filtering)
- Tools for sequence mapping
- Tools for de novo assembly
- Tools for quality check of sequence assembling
- Tools for the comparison of two read libraries without an external reference
- Tools for genomic variants (such as variant calling or variant annotation)
- Tools for functional annotation: identification of domains, orthologues, genetic markers, controlled vocabulary (GO, KEGG, InterPro)
- Tools for biological enrichment in non-model organisms
- Tools for gene expression studies
- Tools for Chip-Seq data
- Tools for "big-data" analyses
- Tools for handling and editing complex workflows and pipelines
- Databases for bioinformatics

Organizers: Prof. M. Gonzalo Claros, Department of Molecular Biology and Biochemistry, University of Málaga, Spain.
Dr. Javier Pérez Florido, Bioinformatics Research Area, Fundación Progreso y Salud, Seville, Spain.

- **SS10: Systems Biology Approaches to Decipher Long Noncoding RNA–Protein Associations**
Long noncoding RNAs (lncRNAs) make up large amounts of the RNA and total genomic repertoire. Studies on the functional characterization of lncRNAs have resulted in data on interactions with their RNA peers, DNA, or proteins. Although there has been an increase in evidence on the link between lncRNAs and diverse human diseases, there is a dearth of lncRNA–protein association studies. Additionally, existing methods do not provide theories about the possible molecular causes of such associations linking to diseases. How such regulatory interactions between classes of lncRNAs and proteins would have a significant influence on the organism and disease remains a challenge. A good number of bioinformatics approaches have arisen in the recent past exploring these challenges. The idea behind this session is to bring together the wide gamut of researchers who have worked on these methods across different organisms.
The following are the sub-topics of the proposed session, which we would like to call for papers.

- LncRNAs in genomes: annotation and curation
- LncRNA–protein interactions leading to important diseases: systems Biology approaches
- Identifying lncRNAs with respect to their mechanism and dysregulation in diseases
- LncRNA databases and webservers
- Machine learning approaches and prediction servers

Organizer: Prashanth Suravajhala, PhD, Department of Biotechnology and Bioinformatics, Birla Institute of Scientific Research, India

- **SS11: Gamified Rehabilitation for Disabled People**
Gamification is a hot topic in many areas as it aims at motivating people to do things driven by different innate needs like the wish to accomplish tasks, to compete against others, or to gain something. These and other motivators are efficiently applied in computer games and could be extraordinarily useful in ensuring that patients perform their daily exercises regularly and have fun.
The idea of exergames (exercise games) is not new; the literature reveals that much effort has already been made and with great success. Nevertheless, most applications have been developed for special problems or diseases (i.e., stroke, parkinson, cerebral palsy etc.) and are not generally applicable. In general, people suffering from severe disabilities and chronic diseases are rarely addressed as a target group. Also, the focus is generally set on the medical achievements, which is correct, but the next step would be to enhance the fun factor because no tool is of much use if the patient is not using it because of boredom or demotivation.
The objective of this special session is to gather new ideas about the combination of need and fun, i.e., find ways to create exercise platforms that fit everybody's needs, provide access to the therapist for monitoring and configuration, while the patients benefit physically and mentally when having a good time.
Target groups would be people of:

- All ages, while focusing on younger people, who can be involved more easily but are less addressed in the literature
- All diseases, while focusing on chronic illness and severe disabilities (e.g., muscle dystrophies and atrophies)

The contributions should show advances in at least one of the following areas:

- Adaptability to users with all kinds of problems (e.g., possibility to configure the limbs used to play or playing with facial movements, wheelchair and standing modes, coping with muscle weaknesses etc.)
- Implementation of gaming techniques and special motivators
- Physical or mental exercises, aimed at rehabilitation or daily practice
- Understanding the users, awareness of their level of motivation, fatigue or progress and react accordingly

Organizer: Dr. Martina Eckert, Associate Professor at University of Madrid, Spain.

- **SS12: Modelling of Glucose Dynamics for Diabetes**
  Diabetes is the eighth most common cause of death, while its treatment relies on technology to process continuously measured glucose levels.

Organizer: Dr. Tomas Koutny, Faculty of Applied Sciences, University of West Bohemia

- **SS13: Biological Network Analysis in Multi-omics Data Integration**
  In many biological applications, multiple data types may be produced to determine the genetics, epigenetics, and microbiome affecting gene regulation and metabolism. Although producing multiple data types should provide a more complete description of the processes under study due to multiple factors such as study design (synchronization of data production, number of samples, varying conditions), the analysis may leave more unfulfilled promises than synergy expected from the wealth of data.
  In this session, some of the following challenges are addressed:

- How to conduct meaningful meta-analysis on historical data.
- How to use biological knowledge (represented in reproducible and interoperable manner) in the analysis of large and sparse data sets more effectively.
- How to fill the gap between hypothesis-driven mechanistic studies, e.g., applying modelling to very well studied biochemical processes and data-driven hypothesis-free approaches. How omics data can help.
- Beyond meta-transcriptomics and metagenomics: integration and interpretation of microbiome and host data.

We would like to bring together communities concerned with these topics to present state-of-the-art and current cutting-edge developments, preferably work under construction or published within the past year.

- Objective 1: presentation and discussion of newest methods
- Objective 2: round-table discussions on the topics highlighted above and other related topics suggested by session participants

An additional topic that does not fit the proposed session but that I would love to see addressed is: How to improve open access to the data that is not next-generation sequencing (e.g., metabolomics, proteomics, plant phenotyping). For this an active participation of journal editors would be necessary to discuss opportunities to change journal publication policies.

Organizer: Dr. Wiktor Jurkowski, Jurkowski Group, Earlham Institute, Norwich Research Park, UK

– **SS14: Oncological Big Data and New Mathematical Tools**
Current scientific methods produce various omics data sets covering many cellular functions. However, these data sets are commonly processed separately owing to limited ways in how to connect different omics data together for a meaningful analysis. Moreover, it is currently a problem to integrate such data into mathematical models. We are entering the new era of biological research where the main problem is not to obtain the data but to process and analyze them. In this regard, a strong mathematical approach can be very effective (see J. Gunawardena's essay "Models in biology: accurate descriptions of our pathetic thinking," BMC Biology 2014, 12:29).

In this Special Session we focus on big data (omics and biological pathways) related to oncological research. General biological processes that are relevant to cancer can also be studied. Mathematical tools basically mean statistical learning (data mining, inference, prediction), modeling, and simulation. We want to place a special emphasis on causality. Closely tied to mathematical tools, efficient computational tools can be considered.

Organizers: Dr. Gregorio Rubio, Instituto de Matematica Multidisciplinar, Universitat Politecnica de Valencia, Valencia, Spain. Dr. Rafael Villanueva, Instituto de Matematica Multidisciplinar, Universitat Politecnica de Valencia, Valencia, Spain.

In this edition of IWBBIO, we were honored to have the following invited speakers:

1. Prof. Roderic Guigo, Coordinator of Bioinformatics and Genomics at Centre de Regulacio Genomica (CRG). Head of the Computational Biology of RNA Processing Group. Universitat Pompeu Fabra, Barcelona, Spain
2. Prof. Joaquin Dopazo, Director of the Computational Genomics Department, Centro de Investigación Príncipe Felipe- CIPF, Valencia, Spain
3. Prof. Jose Antonio Lorente, Director of Centre for Genomics and Oncological Research (GENYO). Professor of Legal and Forensic Medicine, University of Granada, Spain

It is important to note, that for the sake of consistency and readability of the book, the presented papers are classified under 16 chapters. The organization of the papers is in two volumes arranged following the topics list included in the call for papers. The first volume (LNBI 10208), entitled "Advances in Computational Intelligence: Part I" is divided into seven main parts and includes the contributions on:

1. Advances in computational intelligence for critical care
2. Bioinformatics for health care and diseases
3. Biomedical engineering
4. Biomedical image analysis
5. Biomedical signal analysis
6. Biomedicine
7. Challenges representing large-scale biological data

The second volume (LNBI 10209), entitled "Advances in Computational Intelligence: Part II" is divided into nine main parts and includes the contributions on:

1. Computational genomics
2. Computational proteomics
3. Computational systems for modelling biological processes
4. Data-driven biology: new tools, techniques, and resources
5. E-health
6. High-throughput bioinformatic tools for genomics
7. Oncological big data and new mathematical tools
8. Smart Sensor and sensor-network architectures
9. Time-lapse experiments and multivariate biostatistics

This fifth edition of IWBBIO was organized by the Universidad de Granada together with the Spanish Chapter of the IEEE Computational Intelligence Society. We wish to thank to our main sponsor and the Faculty of Science, Department of Computer Architecture and Computer Technology and CITIC-UGR, from the University of Granada, for their support and grants. We also wish to thank the Editors-in-Chief of different international journals for their interest in having special issues with the best papers of IWBBIO.

We would also like to express our gratitude to the members of the different committees for their support, collaboration, and good work. We especially thank the local Organizing Committee, Program Committee, the reviewers, and special session organizers. Finally, we want to thank Springer, and especially Alfred Hofmann and Anna Kramer, for their continuous support and cooperation.

April 2017   Ignacio Rojas
                                                                Francisco Ortuño

The original version of the book was revised:
For detailed information please see Erratum.
The Erratum to this chapter is available at
https://doi.org/10.1007/978-3-319-56154-7_65

Organization

Steering Committee

Miguel A. Andrade	University of Mainz, Germany
Hesham H. Ali	University of Nebraska, USA
Oresti Baños	University of Twente, the Netherlands
Alfredo Benso	Politecnico di Torino, Italy
Giorgio Buttazzo	Superior School Sant'Anna, Italy
Mario Cannataro	University Magna Graecia of Catanzaro, Italy
Jose María Carazo	Spanish National Center for Biotechnology (CNB), Spain
Jose M. Cecilia	Universidad Católica San Antonio de Murcia, Spain
M. Gonzalo Claros	University of Malaga, Spain
Joaquin Dopazo	Research Center Principe Felipe, Spain
Werner Dubitzky	University of Ulster, UK
Afshin Fassihi	Universidad Católica San Antonio de Murcia, Spain
Jean-Fred Fontaine	University of Mainz, Germany
Humberto Gonzalez	University of the Basque Country, Spain
Concettina Guerra	College of Computing, Georgia Tech, USA
Andy Jenkinson	Karolinska Institute, Sweden
Craig E. Kapfer	Reutlingen University, Germany
Narsis Aftab Kiani	European Bioinformatics Institute, UK
Natividad Martinez	Reutlingen University, Germany
Marco Masseroli	Politechnical University of Milan, Italy
Federico Moran	Complutense University of Madrid, Spain
Cristian R. Munteanu	University of A Coruña, Spain
Jorge A. Naranjo	New York University, Abu Dhabi
Michael Ng	Hong Kong Baptist University, SAR China
Jose L. Oliver	University of Granada, Spain
Juan Antonio Ortega	University of Seville, Spain
Julio Ortega	University of Granada, Spain
Alejandro Pazos	University of A Coruña, Spain
Javier Perez Florido	Genomics and Bioinformatics Platform of Andalusia, Spain
Violeta I. Pérez Nueno	Inria Nancy Grand Est, LORIA, France
Horacio Pérez-Sánchez	Universidad Católica San Antonio de Murcia, Spain
Alberto Policriti	University of Udine, Italy
Omer F. Rana	Cardiff University, UK
M. Francesca Romano	Superior School Sant'Anna, Italy
Yvan Saeys	VIB - Ghent University, Belgium
Vicky Schneider	The Genome Analysis Centre, UK
Ralf Seepold	HTWG Konstanz, Germany

Mohammad Soruri	University of Birjand, Iran
Yoshiyuki Suzuki	Tokyo Metropolitan Institute of Medical Science, Japan
Oswaldo Trelles	University of Malaga, Spain
Renato Umeton	CytoSolve Inc., USA
Jan Urban	University of South Bohemia, Czech Republic
Alfredo Vellido	Polytechnical University of Catalonia, Spain

Program Committee

Jesus S. Aguilar
Carlos Alberola
Hisham Al-Mubaid
Rui Carlos Alves
Yuan An
Georgios Anagnostopoulos
Eduardo Andrés León
Antonia Aránega
Saúl Ares
Masanori Arita
Ruben Armañanzas
Joel P. Arrais
Patrizio Arrigo
O. Bamidele Awojoyogbe
Jaume Bacardit
Hazem Bahig
Pedro Ballester
Graham Balls
Ugo Bastolla
Sidahmed Benabderrahmane
Steffanny A. Bennett
Daniel Berrar
Mahua Bhattcharya
Concha Bielza
Armando Blanco
Ignacio Blanquer
Paola Bonizzoni
Christina Boucher
Hacene Boukari
David Breen
Fiona Browne
Dongbo Bu
Jeremy Buhler
Keith C.C.
Gabriel Caffarena
Anna Cai

Carlos Cano
Rita Casadio
Daniel Castillo
Ting-Fung Chan
Nagasuma Chandra
Kun-Mao Chao
Bernard Chen
Bolin Chen
Brian Chen
Chuming Chen
Jie Chen
Yuehui Chen
Jianlin Cheng
Shuai Cheng
I-Jen Chiang
Jung-Hsien Chiang
Young-Rae Cho
Justin Choi
Darrell Conklin
Clare Coveney
Aedin Culhane
Miguel Damas
Antoine Danchin
Bhaskar DasGupta
Ricardo De Matos
Guillermo de la Calle
Javier De Las Rivas
Fei Deng
Marie-Dominique Devignes
Ramón Diaz-Uriarte
Julie Dickerson
Ye Duan
Beatrice Duval
Saso Dzeroski
Khaled El-Sayed
Mamdoh Elsheshengy

Christian Exposito
Weixing Feng
Jose Jesús Fernandez
Gionata Fragomeni
Xiaoyong Fu
Juan Manuel Galvez
Alexandre G.de Brevern
Pugalenthi Ganesan
Jean Gao
Rodolfo Garcia
Mark Gerstein
Razvan Ghinea
Daniel Gonzalez Peña
Dianjing Guo
Jun-tao Guo
Maozu Guo
Christophe Guyeux
Michael Hackenberg
Michiaki Hamada
Xiyi Hang
Jin-Kao Hao
Nurit Haspel
Morihiro Hayashida
Jieyue He
Pheng-Ann Heng
Luis Javier Herrera
Pietro Hiram
Lynette Hirschman
Ralf Hofestadt
Vasant Honavar
Wen-Lian Hsu
Jun Hu
Xiaohua Hu
Jun Huan
Chun-Hsi Huang
Heng Huang
Jimmy Huang
Jingshan Huang
Seiya Imoto
Anthony J. Kusalik
Yanqing Ji
Xingpeng Jiang
Mingon Kang
Dong-Chul Kim
Dongsup Kim
Hyunsoo Kim

Sun Kim
Kengo Kinoshita
Ekaterina Kldiashvili
Jun Kong
Tomas Koutny
Natalio Krasnogor
Abhay Krishan
Marija Krstic-Demonacos
Sajeesh Kumar
Lukasz Kurgan
Stephen Kwok-Wing
Istvan Ladunga
T.W. Lam
Pedro Larrañaga
Dominique Lavenier
Jose Luis Lavin
Doheon Lee
Xiujuan Lei
André Leier
Kwong-Sak Leung
Chen Li
Dingcheng Li
Jing Li
Jinyan Li
Min Li
Xiaoli Li
Yanpeng Li
Li Liao
Hongfei Lin
Hongfang Liu
Jinze Liu
Xiaowen Liu
Xiong Liu
Zhenqiu Liu
Zhi-Ping Liu
Rémi Longuespée
Miguel Angel Lopez Gordo
Ernesto Lowy
Jose Luis
Suryani Lukman
Feng luo
Bernard M.E.
Bin Ma
Qin Ma
Malika Mahoui
Alberto Maria

Tatiana Marquez-Lago
Keith Marsolo
Francisco Martinez Alvarez
Osamu Maruyama
Tatiana Maximova
Roderik Melnik
Jordi Mestres
Hussain Michelle
Ananda Mondal
Antonio Morreale
Walter N. Moss
Maurice Mulvenna
Enrique Muro
Radhakrishnan Nagarajan
Vijayaraj Nagarajan
Kenta Nakai
Isabel A. Nepomuceno
Mohammad Nezami
Anja Nohe
Michael Ochs
Baldomero Oliva
Jose Luis Oliveira
Motonori Ota
David P.
Tun-Wen Pai
Paolo Paradisi
Hyun-Seok Park
Kunsoo Park
Taesung Park
David Pelta
Alexandre Perera
María Del Mar Pérez Gómez
Vinhthuy Phan
Antonio Pinti
Héctor Pomares
Mihail Popescu
Benjarath Pupacdi
Sanguthevar Rajasekaran
Shoba Ranganathan
Jairo Rocha
Fernando Rojas
Jianhua Ruan
Gregorio Rubio
Antonio Rueda
Irena Rusu
Vincent Shin-Mu Tseng

Kunihiko Sadakane
Michael Sadovsky
Belen San Roman
Maria Jose Saez
Hiroto Saigo
José Salavert
Carla Sancho Mestre
Emmanuel Sapin
Kengo Sato
Jean-Marc Schwartz
Russell Schwartz
Jose Antonio Seoane
Xuequn Shang
Piramanayagam Shanmughavel
Xiaoman Shawn
Xinghua Shi
Tetsuo Shibuya
Dong-Guk Shin
Amandeep Sidhu
Istvan Simon
Richard Sinnott
Jiangning Song
Joe Song
Zhengchang Su
Joakim Sundnes
Wing-Kin Sung
Prashanth Suravajhala
Martin Swain
Sing-Hoi Sze
Mehmet Tan
Xing Tan
Li Teng
Tianhai Tian
Pedro Tomas
Carlos Toro
Carolina Torres
Paolo Trunfio
Esko Ukkonen
Lucia Vaira
Paola Velardi
Julio Vera
Konstantinos Votis
Slobodan Vucetic
Ying-Wooi Wan
Chong Wang
Haiying Wang

Jason Wang
Jian Wang
Jianxin Wang
Jiayin Wang
Junbai Wang
Junwen Wang
Lipo Wang
Li-San Wang
Lusheng Wang
Yadong Wang
Yong Wang
Ka-Chun Wong
Ling-Yun Wu
Xintao Wu
Zhonghang Xia
Fang Xiang
Lei Xu
Zhong Xue
Patrick Xuechun
Hui Yang
Zhihao Yang

Jingkai Yu
Hong Yue
Erliang Zeng
Xue-Qiang Zeng
Aidong Zhang
Chi Zhang
Jin Zhang
Jingfen Zhang
Kaizhong Zhang
Shao-Wu Zhang
Xingan Zhang
Zhongming Zhao
Huiru Zheng
Bin Zhou
Shuigeng Zhou
Xuezhong Zhou
Daming Zhu
Dongxiao Zhu
Shanfeng Zhu
Xiaoqin Zou
Xiufen Zou

Additional Reviewers

Alquezar, Rene
Amaya Vazquez, Ivan
Belanche, Lluís
Browne, Fiona
Calabrese, Barbara
Cho, Ken
Coelho, Edgar
Datko, Patrick
Gaiduk, Maksym
Gonzalez-Abril, Luis
Ji, Guomin
Kang, Hong
Langa, Jorge
Liang, Zhewei
Martin-Guerrero, Jose D.

Moulin, Serge
Olier, Iván
Ortega-Martorell, Sandra
Ribas, Vicent
Scherz, Wilhelm Daniel
Sebastiani, Laura
Seong, Giun
Smith, Peter
Treepong, Panisa
Tu, Shikui
Valenzuela, Olga
Vizza, Patrizia
Wu, Yonghui
Yan, Shankai
Zucco, Chiara

Contents – Part II

Computational Proteomics

Computational Systems for Modelling Biological Processes

High-Throughput Bioinformatic Tools for Genomics

Oncological Big Data and New Mathematical Tools

Time Lapse Experiments and Multivariate Biostatistics

Contents – Part I

Biomedical Engineering

Biomedical Image Analysis

Biomedical Signal Analysis

Biomedicine

Challenges Representing Large-Scale Biological Data

Computational Genomics

Investigation of DNA Sequences Utilizing Frequency-Selective Nanopore Structures

Ali Hilal-Alnaqbi[4(✉)], Mahmoud Al Ahmad[1,5], Tahir A. Rizvi[2,5],
and Farah Mustafa[3,5]

[1] Department of Electrical Engineering, College of Engineering,
United Arab Emirates University, Al Ain, UAE
[2] Department of Microbiology and Immunology, College of Medicine
and Health Sciences, United Arab Emirates University, Al Ain, UAE
[3] Department of Biochemistry, College of Medicine and Health Sciences,
United Arab Emirates University, Al Ain, UAE
[4] Department of Mechanical Engineering, College of Engineering,
United Arab Emirates University, P.O. Box 15551, Al Ain, UAE
alihilal@uaeu.ac.ae
[5] Zayed Bin Sultan Center for Health Sciences Division,
United Arab Emirates University, Al Ain, UAE

Abstract. Newer methodologies that are quick, label-free, reliable, and low-cost for DNA sequencing and identification are currently being explored. High frequency based-scattering parameters provide a reliable measurement platform and technique to characterize DNA bases. Using a modeling approach, this work investigates the utilization of high frequency-selective structure coupled with nanopore technology for nucleotide identification and sequencing. The model envisions a coplanar waveguide structure harboring a small hole with an internal diameter of the order of several nanometers to demonstrate the potential use of high frequency to identify and sequence DNA. When DNA molecule enters the pore, it should cause disturbance in the electromagnetic field. This disturbance should result in a shift in the resonance frequency and its corresponding characteristics, thus enabling nucleotide identification. The frequency response of four different single DNA strands composed exclusively of either A, C, G or T were measured and characterized to extract the corresponding dielectric constants and their corresponding base paired strands. These dielectric constant values were then used to model the presence of the corresponding DNA molecules in the nanopore. The conducted simulations revealed distinctions between the single and double-stranded DNA molecules due to their different and distinct electrical properties.

Keywords: DNA · Sequencing · Frequency · Nanopore

1 Introduction

Several electrical approaches are being tested to identify and sequence DNA nucleotides, but a majority of them have limitations of their own, preventing a practical solution to the problem. For example, the conventional nanopore-based DNA sequencing technique

© Springer International Publishing AG 2017
I. Rojas and F. Ortuño (Eds.): IWBBIO 2017, Part II, LNBI 10209, pp. 3–11, 2017.
DOI: 10.1007/978-3-319-56154-7_1

uses transverse current and suffers from high error rates due to the intrinsic noise in the current arising from carrier dispersion along the chain of the molecule; i.e., from the influence of neighboring bases [1]. It has also been proposed that DNA sequencing could be performed by measuring the electron transport properties of individual nucleotides in a DNA molecule by tunneling [2–6]. In fact, this has been used along with statistical identification of the nucleotides based on their electrical conductivity via electron transport, to sequence DNA [7]. However, in direct current measurements, resistance between DNA and the metal electrodes due to contact could potentially arise which may limit the accuracy of identification. Such contacts can be represented by tunneling barriers which require the application of relatively high voltage drops across these barriers [8].

Several other approaches based on micro- and radio waves have been used to study DNA molecules, but each with its own caveats. For example, label-free DNA microarray bioassays using a near-field scanning microwave microscope have been conducted by measuring the change of microwave reflection coefficient at about 4 GHz operating frequency, hybridization between target (free) and capture (immobilized) sequences led to changes in the measured reflection coefficient [9]. On the other hand, Sarkar et al. have investigated the effect of low power microwave in directly analyzing the DNA of the mouse genome [10]. They have demonstrated that exposing DNA to microwave at a power density of 1 mW/cm^2 could change its properties. Cao et al. have reported a notch frequency shift of microwave photonics filter at 100 MHz with higher distinguishing ability [11]. Similarly, Sanggyu Lee et al. have reported that the radiofrequency fields at 2.45 GHz can alter gene expression in cultured human cells through non-thermal mechanisms [12].

Based on the issues outlined above, there appears a need for an advanced electrical approach, which could overcome the problems associated with these techniques. Keeping these caveats in mind, we have investigated a high frequency approach coupled with nanopore technology for its potential for DNA identification and sequencing. The high frequency transmission and reflection scattering parameters are power quantities describing the transmitted and reflected power of a frequency-selective structure [13]. Theoretically, in the absence of DNA, the high frequency selective structure should exhibit its maximum resonance frequency. This frequency could be altered with the present of DNA molecules. On the other side, nanopore technology is suggested to be a reliable technology to enable DNA passage through a very small hole equipped with electrodes setup to enable several electrical measurements [1]. Combining both high frequency and nanopore technologies at the level of integrated device concept would allow for rapid electrical sequencing of long stretches of DNA without the need for labeling.

2 Current Approach

The proposed coupled frequency-selective nanopore structure is depicted in Fig. 1. Briefly, the structure is composed of a coplanar waveguide of two portsin which a nanopore has been pierced between its split signal lines. The coplanar waveguide is used to convey microwave frequency signals; when the signal is transmitted from port one to port two, it passes through the air gap discontinuity within which the nanopore has been punched through the substrate thickness [14]. This discontinuity affects signal transmission and reflection characteristics which can then be measured. The nanopore coplanar waveguide is designed initially (with no DNA) at a predetermined resonance frequency with certain frequency response specifications. When a DNA molecule is passed through the nanopore, the electromagnetic field transmission is disturbed which changes the phase and impedance relationships, altering the resonance frequency. This change in frequency depends on the type of the DNA nucleotide passing through the nanopore with each type of nucleotide having its own characteristic effect on the resonance frequency.

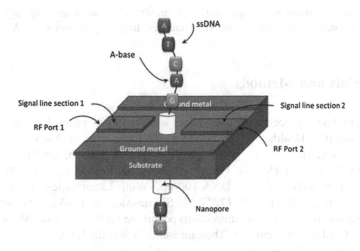

Fig. 1. Schematic representation of the nanopore coplanar waveguide apparatus for DNA strand identification and sequencing.

From the technical point-of-view, the resonance frequency of the structure illustrated in Fig. 1 depends on the length of the signal line and the high frequency equivalent circuit that it represents is shown in Fig. 2(a). The circuit accounts for the nanopore as a series capacitor between the signal line sections, and the nanopore capacitance depends on the distance between the two signal sections and their inner diameter. The presented topology obviates the need to have a direct probing contact as usually is the case in other techniques [2–6]. The corresponding simulated frequency response of this structure is shown in Fig. 2(b), depicting the transmission and reflected scattering parameters. In the absence of DNA, the high frequency selective structure exhibits its maximum resonance frequency around 6.2 GHz as revealed from Fig. 2(b).

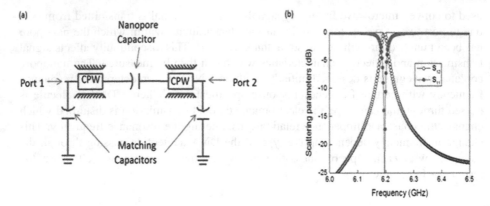

Fig. 2. Nanopore coplanar waveguide equivalent circuit and its simulated frequency response: (a) equivalent circuit representation and (b) simulated frequency response. CPW = coplanar waveguide.

3 Materials and Methods

Reagents. Protran nitrocellulose membrane with 0.45 μm pore size and 300 mm thickness from (GE Healthcare Life Sciences, UK) was used in this work. Tris-base (T6066, Sigma-Aldrich, USA), EDTA (E-5134, Sigma-Aldrich, USA), Tris/Borate/EDTA (10XTBE) buffer (V4251, Promega, USA), 6X loading dye (13526, Norgen, Canada), 100 bp ladder (CSL-MDNA-100 bp, Wolf Laboratories, UK), TEMED (161-0801, BioRad, USA), APS (7727-54, Sigma-Aldrich, USA), 30% acrylamide (A-9099, Sigma-Aldrich, USA), ammonium persulfate (A3678, Sigma-Aldrich, USA) and SYBR®Gold nucleic acid dye (ThermoFischer Scientific, UK).

Oligonucleotide annealing conditions. To observe the electric properties of double-stranded DNA of specific size and composition, four single-stranded oligonucleotides composed of 18 nucleotides (18-mer) of each type of base, A, T, C, and G were obtained commercially (Macrogen Inc., South Korea). The lyophilized oligonucleotide were resuspended in 10 mM Tris, 1 mM EDTA, pH 8 (TE buffer) followed by their dilution in annealing buffer to a working concentration of 100 μM each. This was achieved by mixing equal volumes of single complementary oligonucleotides (A and T/C and G) using an annealing protocol from Metabion, Germany. Briefly, the resuspended complementary oligonucleotides were mixed at equimolar concentrations (4 μM each) in an annealing buffer (10 mM Tris, 1 mM EDTA and 100 mM NaCl, pH 7.5) and heated at 94 °C for 10 min to properly denature the single-stranded oligonucleotides. This was followed by slow cooling at room temperature for one hour to allow the spontaneous generation of double-stranded DNA molecules via hydrogen bonding between complementary bases. The annealed oligonucleotides were aliquoted and stored at −20 °C until further analysis.

4 Results and Discussion

To test our approach as close to reality, we extracted the frequency responses of different types of DNA molecules of known size and chemical composition synthesized commercially. Specifically, each of the DNA strands consisted of 18 building blocks of DNA: deoxyribo nucleotides consisting of deoxyadenosine (dA), deoxythymine (dT), deoxycytosine (dC), or deoxyguanine (dG) phosphates (A, C, G and T). The four single oligonucleotides were then used to create double-stranded DNA molecules of known size and length using the ability of the complementary bases to "anneal" and form hydrogen bonds between the two strands of oligonucleotides. Complementary bases cytosine and guanine form three hydrogen bonds between each other, while adenine and thymine form two hydrogen bonds [15].

To measure the reflection coefficient of the single or double-stranded oligonucleotides, the DNA molecules were immobilized onto a solid support provided by a nitrocellulose membrane, followed by their analysis using the DAKS measurement setup (Fig. 3). The DAKS set up consists of a dielectric assessment kit DAKS-3.5 probe [16] to measure the reflection coefficient of DNA molecules spotted on the nitrocellulose membrane. The DAKS probe is connected to a network analyzer (R&S ZVL) [17] equipped with the DAK software. To ensure consistency among different DNA samples, the same concentration of DNA (4 μM) in the same buffer (10 mM Tris, 1 mM EDTA, 10 mM NaCl, pH 8.0) and volume (20 μl) was used to spot the DNA onto the nitrocellulose membrane that formed a disc of ~5 mm diameter. Each DNA suspension was exposed to a radio-frequency signal with a power of 10 dBm and with a sweep from 200 MHz up to 13.6 GHz (the measurement capability of the equipment). To conduct the RF measurements, the nitrocellulose sheet was supported with a copper-block attached to the DAK probe. The measured frequency responses are shown in Fig. 3(b) and were used to extract the corresponding dielectric constants [13]. As can be seen, the buffer in the absence of DNA molecules as well as each type of single and double-stranded DNA molecule could be distinguished from each other based on their frequency responses except for A which showed a response close to that observed for the G oligonucleotide.

Figure 4 depicts the extracted dielectric constants versus frequency measurements of the single- and double-stranded DNA molecules after de-embedding the control contributions. As can be seen, it was easier to differentiate between the complementary base pairs from of the single oligonucleotides, each exhibiting its own capacitance profile even after subtracting the effects of the buffer medium. The four single oligonucleotides showed lower values when compared to their base paired counterparts. These values were then used to model the presence of the corresponding DNA molecules in the nanopore, which is represented by the nanopore series capacitance of the equivalent circuit shown in Fig. 2(a).

Next, the extracted values of the dielectric constant were used to calculate the corresponding electrical capacitance assuming a gap capacitance model [14]. The basic premise behind this approach is that the dielectric constant is an intrinsic property of the material that measures its ability to store electrical energy when an electric field is

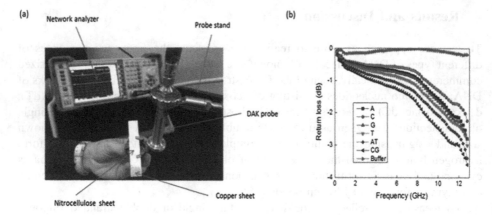

Fig. 3. (a) Depiction of the DAKS measurements setup consisting of the network analyzer, the probe along with its stand holding the nitrocellulose membrane. (b) Graph showing the measured frequency responses of the single- and double-stranded DNA molecules.

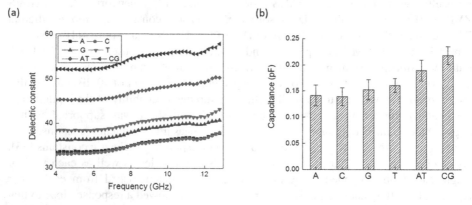

Fig. 4. The extracted electrical parameters of DNA molecules tested using the DAKS setup: (a) dielectric constant versus frequency, and (b) the corresponding relative nanopore capacitance at 5 GHz when DNA is present.

applied. Thus, the capacitance values of the DNA molecules were calculated at 5 GHz and are shown in Fig. 4(b).

The equivalent circuit presented in Fig. 2(a) was then used to generate the corresponding frequency responses when DNA molecules are presented. The simulated reflection and transmission coefficients versus frequency suing sonnet software [18] are plotted in Fig. 5(a) and (b), respectively. As shown, the existence of DNA molecules lowered the resonance frequency. Furthermore, each DNA molecule had a unique RF signature that can be used for detection and identification without the use of labeling.

The current conventional and the suggested "next generation" sequencing methods are either inefficient or very expensive [1–9]. Nanopore technology has a great potential to be used for individual base sequencing, within which a DNA/RNA strands is drawn

(a)

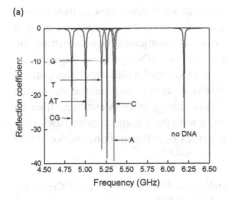

(b)

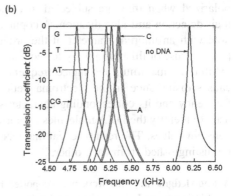

Fig. 5. Simulated frequency response with the presence of DNA molecules: (a) reflection coefficients versus frequency for the tested DNA molecules, and (b) transmission coefficients versus frequency.

through a nanoscopic pore in a conductive material [19]. Electronic properties can be used to identify the individual bases observing the variations in corresponding measurements such as current voltage (IV) [19]. However, the thickness of the nanopore is considered to be very critical issue for the identification of individual nucleobases due to resulting noise and resolution problems [20]. On the other side there were many works report the noise reduction and optimization in scattering parameters based measurements [21]. This usually is done through the proper calibration and the use of deembedding process with proper algorithms [22]. Due to the deembedding process, the high error rates arising from noise as well as carrier dispersion in existing conventional nanopore-based DNA sequencing technique [1–9] are of huge potential to be reduced and optimized, especially in chains of long DNA molecules. The approach of this work would is foreseen to be capable for parallel sequencing of billions of DNA strands with yielding substantially more throughput and minimizing the need for the fragment-cloning methods that are often used in Sanger sequencing of genomes [23].

Now we turn the attention for the use of our proposed approach for DNA sequencing. The process starts with measuring and recording the corresponding scattering parameters for an instant molecule. The resonance frequency is then used to find the corresponding capacitance value of the nanopore capacitance and thus determine which molecule has been measured with the help of Fig. 4(b). The set up does show in Fig. 1 could be also used to measure and identify the double-stranded DNA molecules since the interference from two nucleotides could converge and affect the radio wave in the same manner as single strand does.

5 Conclusion

In summary, this study has investigated the use of an integrated nanopore frequency selective waveguide structure as a potential system for the identification of nucleotide bases and their sequence within a DNA molecule. In principle, DNA bases get polarized when they are subjected to electromagnetic field propagation, thus enable their detection and identification. A coplanar waveguide structure incorporates a small hole with an internal diameter of the order of several nanometers to demonstrate the potential use of this technology to sequence DNA. Without DNA present, the structure exhibits its maximum resonance frequency. When DNA molecules enter the pore, it causes disturbance in the electromagnetic field which results shifting the resonance frequency and its corresponding characteristics. This change in frequency can then be used to identify the DNA molecules by correlating it with their corresponding dielectric constant values. The results reveal that DNA molecules of different composition could be distinguished from each other.

Acknowledgment. This work was supported by funds from *UAEU Sheikh Zayed Center for Health Science to MA and the Start-Up funds from UAEU to FM.*

References

1. Keyser, U.F., et al.: Direct force measurements on DNA in a solid-state nanopore. Nat. Phys. **2**, 473–477 (2006)
2. Albrecht, T.: Electrochemical tunnelling sensors and their potential applications. Nat. Commun. **3**, 829 (2012)
3. D'Orsogna, M.R., Rudnick, J.: Two-level system with a thermally fluctuating transfer matrix element: application to the problem of DNA charge transfer. Phys Rev. E Stat Nonlin Soft Matter. Phys. **66**, 041804 (2002)
4. Lagerqvist, J., Zwolak, M., Di Ventra, M.: Fast DNA sequencing via transverse electronic transport. Nano Lett. **6**, 779–782 (2006)
5. Lagerqvist, J., Zwolak, M., Di Ventra, M.: Influence of the environment and probes on rapid DNA sequencing via transverse electronic transport. Biophys. J. **93**, 2384–2390 (2007)
6. Xu, K.: DNA circuit system and charge transfer mechanism. Engineering **5**, 381–385 (2013)
7. Tsutsui, M., Taniguchi, M., Yokota, K., Kawai, T.: Identifying single nucleotides by tunnelling current. Nat. Nanotechnol. **5**, 86–290 (2010)
8. Porath, D., Bezryadin, A., de Vries, S., Dekker, C.: Direct measurement of electrical transport through DNA molecules. Nature **403**, 635–638 (2000)
9. Lee, K., et al.: Label-free DNA microarray bioassays using a near-field scanning microwave microscope. Biosens. Bioelectron. **42**, 326–331 (2013)
10. Sarkar, S., Ali, S., Behari, J.: Effect of low power microwave on the mouse genome: a direct DNA analysis. Mutat. Res. **320**, 141–147 (1994)
11. Cao, Y., et al.: Resolution-improved in situ DNA hybridization detection based on microwave photonic interrogation. Opt. Express **23**, 27061–27070 (2015)
12. Lee, S., et al.: 2.45 GHz radiofrequency fields alter gene expression in cultured human cells. FEBS Lett. **579**, 4829–4836 (2005)
13. Pozar, D.: Microwave Engineering, 4th edn. Wiley, Hoboken (2011)
14. Simons, R.N.: Coplanar Waveguide Circuits, Components, and Systems. Wiley-IEEE Press, Hoboken (2001)
15. Alberts, B., Johnson, A., Lewis, J., et al.: Molecular Biology of the Cell, 4th edn. Garland Science, New York (2002)
16. DAKS-3.5: DAK System for 200 MHz–14 GHz. http://www.speag.com/products/dak/dak-dielectric-probe-systems/daks-3-5/. Accessed: 28 Feb 2016

17. R&S®ZVL Vector Network Analyzers. https://www.rohde-schwarz.com/vn/product/zvl-productstartpage_63493-9014.html. Accessed 28 Feb 2016
18. Sonnet Software. http://www.sonnetsoftware.com. Accessed: 28 Feb 2016
19. McFarland, H.L.: First-principles investigation of nanopore sequencing using variable voltage bias on graphene-based nanoribbons. J. Phys. Chem. Lett. **6**, 2616–2621 (2015)
20. Maitra, R.D., Kim, J., Dunbar, W.B.: Recent advances in nanopore sequencing. Electrophoresis **33**, 3418–3428 (2012)
21. Mertens, H., et al.: Fast noise prediction for process optimization using only standard DC and S parameter measurements. In: Proceedings of IEEE Bipolar/BiCMOS Circuits and Technology Meeting, BCTM 2012, pp. 1–4 (2012)
22. Chen, C.H., Deen, M.J.: A general noise and S-Parameter deembedding procedure for on-wafer high-frequency noise measurements of MOSFETs. IEEE Trans. Microw. Theor. Techn. **49**, 1004–1005 (2001)
23. Sanger, F., Coulson, A.R., Barrell, B.G., Smith, A.J., Roe, B.A.: Cloning in single-stranded bacteriophage as an aid to rapid DNA sequencing. J. Mol. Biol. **143**, 161–178 (1980)

Uropathogenic Escherichia coli: An Ideal Resource for DNA Microarray Probe Designing

Payam Behzadi$^{(\boxtimes)}$ and Elham Behzadi

Department of Microbiology, College of Basic Sciences, Shahr-e-Qods Branch,
Islamic Azad University, Tehran, Iran
behzadipayam@yahoo.com, elham_behzadi@hotmail.com

Abstract. Uropathogenic Escherichia coli (UPEC) is the most important bacterial agent causing urinary tract infections (UTIs) in patients around the world. The UTIs rank second among different types of infectious diseases. So, there is an urgent need to have a rapid and accurate diagnostic method for detecting UTIs. DNA microarray is an advanced pan-genomic technique which can be used as a rapid and accurate diagnostic method with high specificity and sensitivity. Designing a collection of DNA microarray probes enables us to have a sharp methodology for detection and identification of UPEC pathotypes. For this reason, the authors of the present review literature have tried to represent a vast range of availabilities regarding UPEC recognition by DNA microarray probe designing.

UPEC encompasses a wide range of virulence genes which are contributed to UTIs. This characteristic makes UPEC an ideal resource for DNA microarray probe designing to have a reliable UTIs diagnostic method.

Besides, UPEC possesses different pathogenic strains which are distributed worldwide. And individual strain has its particular conserved sequences within the virulence genes. In other word, DNA microarray probes can be designed for specific UPEC strains within different geographical regions.

In conclusion, UPEC is a living accessible genomic treasure which can be used for designing a diversity of DNA microarray probes with a high specificity and sensitivity.

Keywords: Microarray · Urinary tract infection · Uropathogenic Escherichia coli · Probe · Pangenome

1 Introduction

Urinary tract infections (UTIs) as the second most important infectious diseases are occurred by the well-known pathogenic microorganism of Uropathogenic Escherichia coli (UPEC). As we know, the UTIs are classified into lower and upper parts with asymptomatic or symptomatic, complicated or uncomplicated and acute or chronic characteristics. In addition to human disposing factors for UTIs, the pathogenomic properties of microbial causative agent play an important role for occurring UTIs. Among a wide range of UTIs causative agents, UPEC has a unique situation. Among

© Springer International Publishing AG 2017
I. Rojas and F. Ortuño (Eds.): IWBBIO 2017, Part II, LNBI 10209, pp. 12–19, 2017.
DOI: 10.1007/978-3-319-56154-7_2

patients with UTIs around the world, the incidence of UTIs in women is four-folded in comparison with men. This feature is referred to anatomical characteristics in females. Several global reports confirm the key role of intestinal strains of Escherichia coli (E.coli) as significant resources for recurrent UTIs in women [1–6].

The phylogenetic studies show that, all the pathotypes of E.coli are originated from the commensal and non-pathogenic, strains in their evolutionary pathways. By the time, the intestinal and extra-intestinal pathotypes of E.coli are divided into 5 classes of A, B1, B2, D and E. The A, B1 and E classes are recognized as intestinal strains and the classes of B2 and D involve UPEC strains for the most. However, the B2 group dominates D category. According to previous studies, the UPEC strains are the most well-known etiologic bacterial agents of nosocomial (up to 50%) and community acquired (up to 95%) UTIs [3, 7].

Pangenomics has revealed a huge diversity of genes in non-pathogenic and pathogenic E.coli strains; so the pathogenomics of E.coli may help us to detect and identify different pathotypes of UPEC by advanced molecular and pangenomic technology like DNA Microarray [3, 8].

2 Pangenomics of UPEC

The scientific term of Pangenomics was firstly coined in 2005 by Tettelin and colleagues. Indeed, pangenome denotes common chromosomal sections (known as core or backbone genome) and supplementary genetic elements (adaptive or flexible genome) in a Genus. The core genome encompasses intra-chromosomal gene families including housekeeping and other functional genes which are directly related to normal cellular activities; but the adaptive or supplementary genome comprising mobile genes in the forms of integrons, pathgenicity islands (PAIs), phages, plasmids, retrotransposons and transposons which are in association with vital defense systems. For the most, the flexible genetic repertoire encompasses antimicrobial resistance, heavy metal resistance and virulence genes that pertaining to microorganism environmental adaptation. Today, the microbial Pangenomics is an effective means for detection and identification of microbial causative agents of infectious diseases. Among a wide range of infections, UTIs as common global infectious diseases have an urgent need for an accurate diagnostics and a definite treatment. So, the use of Pangenomics and advanced pangenomic diagnostic techniques is unavoidable. The pangenome (genetic repertoire) of a microorganism like UPEC varies from one strain to another. In accordance with previous studies, the pangenome concept of E.coli varies between 4.5 to 5.5 Mb. The least boarder is recognized in commensal strains of E.coli, while the pathotypes including UPEC strains possess more genomic contents. The precise pangenomic concept for each E.coli strain is accessible through the National Center for Biotechnology Information (NCBI) database (https://www.ncbi.nlm. nih.gov/). NCBI is an important and easy-to-work database which provides us the latest information regarding deciphered microbial pangenomes. Until today, 5363 genome assemblies, 342 sequence reads and 47198 genes are registered for E.coli strains in NCBI [3, 9–16].

Although, there are a large number of genes which are valuable for DNA microarray detection and identification, the unrelated genes (orphan genes) among a mass of gene families within flexible gene pool are appropriate sequences as target goals. The orphan gene sequences prepare an appropriate specificity for influent detection and identification by DNA microarray technology [10, 17, 18].

3 DNA Microarray Technology

Microarray technology is an advanced pangenomic diagnostics which can be used as a reliable, flexible, rapid, accurate, sensitive and specific tool. DNA microarray technique as a miniaturized lab-on-a-chip is appropriate and cost effective when the number of specimens is high. The technique of DNA microarray is principally classified into two separate sections which each part has an operative effect on the other. These sections include dry lab (known as *in silico*) and wet lab (known as *in vitro*). The probe designing section involves *in silico* activities and is completely dependent on databases, servers, tools and software. Probe designing process determines the accuracy, sensitivity and specificity of the method. Employing a skillful bioinformatician is an urgent need to minimize the probable biases. The use of proper analyzers, tools and software affects the final outcomes. Besides, utilization of suitable probe sets is an essential need, too. Noticeably, probe designer software represents several microarray probes which the best ones must be chosen and rechecked for final application. On the other hand, probe spotting, DNA target labeling, hybridization and scanning are the wet lab portion of DNA microarray technique in which the related procedures must be performed in accordance with proper protocols. In brief, all of the procedures and sections must be achieved in correct and precise manner to have an acceptable result [3, 19–24].

In parallel with aforementioned procedures, the more registered recognized microbial genomes in databases such as NCBI, European Molecular Biology Laboratory (EMBL, http://embl.org/) and DNA Data Bank of Japan (DDBJ, http://www.ddbj.nig.ac.jp/), the more accurate and effective designed microarray probes. This is why, we can design very effective and sharp microarray probes for E.coli pathotypes like UPEC [9, 10, 12, 19, 25, 26].

The UPEC pathotypes encompass a wide range of genes (e.g. virulence genes) with a significant diversity. In other word, UPEC is an incredible treasure for designing a huge number of well-designed microarray probes. These properties make DNA microarray an appropriate diagnostic technique for detection and identification of UPEC. The virulence genes may be placed on main chromosome or plasmids. The use of chromosomal genes is suitable for recognition of UPEC pathotypes by phylogenetic properties while the use of plasmid genes is proper for multidrug-resistance UPEC strains [3, 27].

4 Chromosomal Virulence Genes (CVGs)

In accordance with several studies, the presence of virulence genes in UPEC genetic pool makes it easy for the bacterium to adapt itself with peripheral factors. This property enables UPEC to be a successful pathogenic bacterium causing UTIs. Adhesins, invasins, siderophores, iron uptake systems, secretion systems, toxins and etc. are favorable

Table 1. Categorization of different virulence genes relating to adhesins in UPEC strains

Virulence genes	Length (base pair (bp))	Virulence factors	Characteristics	Activity	Location	Type of UTI
afaA	306	Afimbrial adhesion AFA-I	Cluster genes	Adhesin	Exterior	Chronic and recurrent, Lower and Upper UTIs
afaB	798					
afaC (highly conserved)	2583					
afaD (highly conserved)	444					
afaE-1	485					
draA	305	Afimbrial adhesion Dr	The proteins are the same as afa genes products	Adhesin, invasin (by activating signal transduction cascade)	Exterior	Chronic and recurrent, Lower and Upper UTIs
drab	743					
draC	2579					
draD	443					
draE	482					
draP	173					
cgsA	459	Curli fimbriae	Fimbrial shaped	Adhesin	Exterior	All
cgsB	483					
cgsC	333					
cgsD	651					
cgsE	390					
cgsF	417					
cgsG	834					
cfaA	764	CFA/I fimbriae	Pilus	Adhesin	Exterior	Lower and recurrent UTIs
cfaB	547					
cfaC	2914					
cfaD	1217					
cfaE						
fimA	612	Type 1 fimbriae	Bacterial peripheral surrounding fimbriae, cluster genes, conserved genes, chaperone-usher (CU) secretion fimbriae	Adhesin	Exterior	All
fimB	603					
fimC	726					
fimD	2844					
fimE	597					
fimF	544					
fimG	512					
fimH	912					
fimI	692					
focA	543	F1C fimbriae	Chaperone-usher (CU) secretion fimbriae, Cluster genes, in association with cluster genes of S fimbriae	Adhesin	Exterior	All
focB	329					
focC	696					
focD	2679					
focF	528					
focG	504					
focH	1011					
focI	540					

(*continued*)

Table 1. (*continued*)

Virulence genes	Length (base pair (bp))	Virulence factors	Characteristics	Activity	Location	Type of UTI
sfaA	556	S fimbriae	Cluster genes, the S fimbriae compartments are variable in different environmental factors	Adhesin	Exterior	Sever UTIs (mostly in upper UTIs)
sfaB	330					
sfaC	285					
sfaD	549					
sfaE	696					
sfaF	2679					
sfaG	528					
sfaH	969					
sfaS	492					
papA	636	P fimbriae	Gene clusters, molecular clock for recognition of B2 and D strains, Chaperone-usher (CU) secretion fimbriae,	Adhesin	Exterior	Acute lower and upper (especially nephritis) UTIs
papB	315					
papC	2532					
papD	735					
papE	555					
papF	534					
papG	1034					
papH	591					
papI	368					
papJ	582					
papK	540					
papX	555					
elfA	552	Escherichia coli laminin binding fimbriae	Gene cluster, molecular clock for recognition of B1 strains, Recognized in Enterohemorrhagic *E. coli* (EHEC) strains for the most	Adhesin	Exterior	Mostly recurrent UTIs in women
elfC	2601					
elfD	702					
elfG	1072					
hcpA		Hemorrhagic E. coli pilus	Recognized in EHEC strains for the most	Adhesin	Exterior	Mostly recurrent UTIs in women
hcpB						
hcpC						

characteristics for UPEC. The related genes are invaluable treasures for DNA microarray probe designing. Table 1 shows several numbers of chromosomal virulence genes relating to adhesins recognized in a wide range of UPEC pathotypes [3, 27–34]:

As Table 1 shows there are a huge number of adhesins with different genetic characteristics recognized in UPEC strains. The diversity of virulence genes including adhesins enables us to have a vast range of patterns for designing specific, sensitive and flexible DNA microarray probes.

5   Conclusion

E.coli bacteria have divided into two groups of pathogens including Intra-intestinal pathogenic E.coli (InPEC) and Extra-intestinal pathogenic E.coli (ExPEC). The InPEC pathotypes involving Enterohemorrhagic E.coli (EHEC), Enteropathogenic E.coli (EPEC), Enteroinvasive E.coli (EIEC), Enterotoxigenic E.coli (ETEC) etc. encompass different virulence genes and factors. In addition to InPEC, the ExPEC strains including UPEC possess a variety of virulence genes and factors which are contributed to bacterial pathogenicity and adaptation. According to Table 1, there are wide ranges of genes which produce different types of adhesins with different functions and activities. Moreover, the size of genes and their genetic properties are different. However, some of them have huge similarities with each other. So, these genes give us a great opportunity to design an abundance of microarray probes; but, the phylogenetic and structural properties of the genes and their similarities and dissimilarities have a deep influence on the accuracy of the probes. Awareness of the genes characteristics is an essential need for designing sharp, specific, sensitive and flexible microarray probes. Of course, simultaneously the presence of skillful bioinformaticians makes it easier to use proper probe sets and platforms. Best designed microarray probes guarantee an accurate, rapid and reliable diagnosis for a definite treatment.

6   Conflict of Interest

There is no conflict of interest for the authors.

Acknowledgements. We are obliged to **Shahr-e-Qods Branch, Islamic Azad University** as the financial supporter of this paper for presenting in IWBBIO 2017, Granada, Spain.

We also have special thanks to Profs. Ignacio Rojas and Francisco Ortuño for their kind helps and collaborations.

References

1. Behzadi, P., Behzadi, E.: The microbial agents of urinary tract infections at central laboratory of Dr. Shariati Hospital, Tehran, Iran. Turkiye Klinikleri J. Med. Sci. **28**(4), 445–449 (2008)
2. Behzadi, P., Behzadi, E., Ranjbar, R.: Urinary tract infections and Candida albicans. Cent. Eur. J. Urol. **68**(1), 96–101 (2015)
3. Jahandeh, N., et al.: Uropathogenic Escherichia coli virulence genes: invaluable approaches for designing DNA microarray probes. Cent. Eur. J. Urol. **68**, 452–458 (2014)
4. Subashchandrabose, S., Mobley, H.L.T.: Virulence and fitness determinants of Uropathogenic Escherichia coli. Microbiology Spectr. **3**(4) (2015)
5. Matuszewski, M.A., et al.: Uroplakins and their potential applications in urology. Cent. Eur. J. Urol. **69**(3), 252–257 (2016)
6. Behzadi, E., Behzadi, P.: The role of toll-like receptors (TLRs) in urinary tract infections (UTIs). Cent. Eur. J. Urol. **69**, 404–410 (2016)

7. Torres, A.G.: Escherichia coli in the Americas. Springer, Cham (2016)
8. Behzadi, P., Behzadi, E., Ranjbar, R.: IL-12 family cytokines: general characteristics, pathogenic microorganisms, receptors, and signalling pathways. Acta Microbiol. Immunol. Hung. 63(1), 1–25 (2016)
9. Snipen, L.G., Ussery, D.W.: A domain sequence approach to pangenomics: applications to Escherichia coli. F1000Research 1, 19 (2013)
10. Mira, A., et al.: The bacterial pan-genome: a new paradigm in microbiology. Int. Microbiol. 13(2), 45–57 (2010)
11. Sharma, A.: Can "Pan-Genomics" demystify the designation of "Bacterial Species". SM J. Bioinform. Proteomics 1(1), 1001 (2016)
12. Behzadi, P., Behzadi, E., Ranjbar, R.: Basic Modern Molecular Biology, 1st edn. Persian Science & Research Publisher, Tehran (2014)
13. Behzadi, P., Behzadi, E.: Environmental Microbiology, 1st edn. Niktab Publisher, Tehran (2007)
14. Mira, A., Klasson, L., Andersson, S.G.: Microbial genome evolution: sources of variability. Curr. Opin. Microbiol. 5(5), 506–512 (2002)
15. Genome/Escherichia coli (2017). https://www.ncbi.nlm.nih.gov/genome/?term=Escherichia +coli. Accessed 10 Feb 2017
16. Gene/Escherichia coli: National Center for Biotechnology Information, U.S. National Library of Medicine, Bethesda, USA (2017)
17. Hu, P., et al.: Global functional atlas of Escherichia coli encompassing previously uncharacterized proteins. PLoS Biol. 7(4), e1000096 (2009)
18. Cho, B.-K., et al.: The transcription unit architecture of the Escherichia coli genome. Nat. Biotechnol. 27(11), 1043–1049 (2009)
19. Behzadi, P., et al.: Microarray long oligo probe designing for Escherichia coli: an in-silico DNA marker extraction. Cent. Eur. J. Urol. 69, 105–111 (2016)
20. Behzadi, P., Behzadi, E., Ranjbar, R.: Microarray probe set: biology, bioinformatics and biophysics. Alban. Med. J. 2, 78–83 (2015)
21. Behzadi, P., Behzadi, E., Ranjbar, R.: Microarray data analysis. Alban. Med. J. 4, 84–90 (2014)
22. Ranjbar, R., Behzadi, P., Mammina, C.: Respiratory Tularemia: Francisella tularensis and microarray probe designing. Open Microbiol. J. 10, 176–182 (2016)
23. Behzadi, P., Ranjbar, R., Alavian, S.M.: Nucleic acid-based approaches for detection of viral hepatitis. Jundishapur J. Microbiol. 8, e17449 (2015)
24. Behzadi, P., Ranjbar, R.: Microarray long oligo probe designing for Bacteria: an in silico pan-genomic research. Alban. Med. J. 2, 5–11 (2016)
25. Lukjancenko, O., Wassenaar, T.M., Ussery, D.W.: Comparison of 61 sequenced Escherichia coli genomes. Microb. Ecol. 60(4), 708–720 (2010)
26. Mao, B.-H., et al.: Identification of Escherichia coli genes associated with urinary tract infections. J. Clin. Microbiol. JCM 50(2), 449–456 (2011)
27. MOH Key Laboratory of Systems Biology of Pathogens, Institute of Pathogen Biology, CAMS&PUMC, Beijing, China: Virulence factors of pathogenic bacteria, 28 October 2016. http://www.mgc.ac.cn/cgi-bin/VFs/compvfs.cgi?Genus=Escherichia. Accessed 12 Feb 2017
28. Baby, S., Karnaker, V.K., Geetha, R.: Adhesins of Uropathogenic Escherichia coli (UPEC). Int. J. Med. Microbiol. Trop. Dis. 2(1), 10–18 (2016)
29. Zalewska, B., et al.: A surface-exposed DraD protein of Uropathogenic Escherichia coli bearing Dr fimbriae may be expressed and secreted independently from DraC usher and DraE adhesin. Microbiology 151(7), 2477–2486 (2005)
30. Cordeiro, M.A., et al.: Curli fimbria: an Escherichia coli adhesin associated with human cystitis. Braz. J. Microbiol. 47(2), 414–416 (2016)

31. Kwak, M.-J., et al.: Genome sequence of Escherichia coli NCCP15653, a group D strain isolated from a diarrhea patient. Gut Pathog. **8**(1), 7 (2016)
32. Wurpel, D.J., et al.: Comparative proteomics of Uropathogenic Escherichia coli during growth in human urine identify UCA-like (UCL) fimbriae as an adherence factor involved in biofilm formation and binding to uroepithelial cells. J. Proteomics **131**, 177–189 (2016)
33. Archer, C.T., et al.: The genome sequence of E. coli W (ATCC 9637): comparative genome analysis and an improved genome-scale reconstruction of E. coli. BMC Genom. **12**(1), 9 (2011)
34. Roos, V., Klemm, P.: Global gene expression profiling of the asymptomatic bacteriuria Escherichia coli strain 83972 in the human urinary tract. Infect. Immun. **74**(6), 3565–3575 (2006)

Lost Strings in Genomes:
What Sense Do They Make?

Michael Sadovsky[1,3](✉), Jean-Fred Fontaine[2,4], Miguel A. Andrade-Navarro[2,4],
Yury Yakubailik[3], and Natalia Rudenko[3]

[1] Institute of Computational Modelling of SB RAS,
Akademgorodok, 660036 Krasnoyarsk, Russia
msad@icm.krasn.ru
[2] Johannes Gutenberg-Universität Mainz, 55128 Mainz, Germany
fontaine@uni-mainz.de
[3] Institute of Space and Information Technologies, Siberian Federal University,
Kirenskogo Str., 26, 660074 Krasnoyarsk, Russia
yura_yak@mail.ru, nrudnko@gmail.com
[4] Institute of Molecular Biology, 55128 Mainz, Germany
http://icm.krasn.ru

Abstract. We studied the sets of avoided strings to be observed over a
family of genomes. It was found that the length of the minimal avoided
string rarely exceeds 9 nucleotides, with neither respect to a phylogeny
of a genome under consideration. The lists of the avoided strings observed
over the sets of (related) genomes have been analyzed. Very low correlation
between the phylogeny, and the set of those strings has been found.

Keywords: Order · Diversity · Composition · Combinatorics · Evolution · Selection

1 Introduction

A frequency dictionary W_q of nucleotide sequences is claimed to be an entity
bearing a lot of information on that latter [1–6]. A consistent and comprehensive study of frequency dictionaries answers the questions concerning the statistical and information properties of DNA sequences. Let's introduce some basic
definitions. Consider a continuous symbol sequence from four-letter alphabet
$\aleph = \{A, C, G, T\}$ of the length N; the length here is just the total number of
symbols (nucleotides) in a sequence. The sequence is supposed to be relevant to
some genetic entity (genome, chromosome, etc.). No other symbols or gaps in the
sequence take place by supposition. Any coherent string $\omega = \nu_1 \nu_2 \ldots \nu_q$ of the
length q makes a word. A set of all the words occurred within a sequence yields
the support of that latter. Counting the numbers of copies n_ω of the words, one
gets a finite dictionary; changing the numbers for the frequency

$$f_\omega = \frac{n_\omega}{N}$$

© Springer International Publishing AG 2017
I. Rojas and F. Ortuño (Eds.): IWBBIO 2017, Part II, LNBI 10209, pp. 20–29, 2017.
DOI: 10.1007/978-3-319-56154-7_3

one gets the frequency dictionary W_q of the thickness q. This is the main object of our study.

That is a common place that researchers study frequency dictionaries comprising the observed words. Here we make the hypothesis that any string $\omega = \nu_1 \nu_2 \ldots \nu_q$ of the length q to be found in a sequence may have a functional or control value. On the other hand, the total number of words of the length q (in the four-letter alphabet $\aleph$) grows in capacity exponentially:

$$M(q) = 4^q,$$ (1)

where $M(q)$ is the number of all possible words of that length. Obviously, the value determined by (1) becomes to exceed N, as $q > q^*$. This specific figure is determined in very simple way:

$$q^* = \max_q \left\{ 4^q \le N \right\},$$ (2)

and obviously is rather small for any real genetic entity.

Definition. Support Ω of a frequency dictionary W_q is the set of words incorporated into the dictionary, through a search over the given text. If support Ω contains all possible words of the given length q (i.e. $\|\Omega\| = 4^q$), we'll call it ∗-support. $\|\mathcal{A}\|$ means a number of elements in $\mathcal{A}$.

In general for biological sequences, a support Ω is not ∗-support, for sufficiently long q. Suppose, then one to start to develop a series of frequency dictionaries with growing q

$$W_1, \ W_2, \ W_3, \ \ldots, \ W_q.$$ (3)

In commonly studied biological sequences (e.g. genes or genomes), W_1 has usually ∗-support. The same is true for W_2 and W_3 provided a minimal sequence length.

Let now consider some (sufficiently long) genetic sequence. That latter may be a bacterial genome, or a chromosome, if eukaryote is studied, or a genome of organella. Let now develop a series of frequency dictionaries (3) and focus on their supports, solely. Evidently, there exists the shortest length of words $\tilde{q}$ so, that $W_{\tilde{q}-1}$ has ∗-support $\Omega(q-1)$, but $W_{\tilde{q}}$ itself has the support $\Omega(q)$ that is not a ∗-support.

Hence, $\tilde{q}$ is the minimal length of words that yields some lacunae in the support. Consider then this word (that is always easy to determine that latter). Our basic idea is that these words are not occasional, or randomly lost among the other ones; on the contrary, they are lost due to specific (anti)selection. Thus, a researcher can contribute a lot from a (detailed) study of such words.

2 Lost Strings and Evolution

We hypothesize that a set of the lost (or avoided) strings observed over a family of genetic entities is not random, but follows biologically inspired constraints;

indeed, they are a matter of natural selection. This idea, in few various forms, has been formulated earlier [11,12]. Related ideas on the impact of the avoided strings on the structure (and functioning) of cancer genes is discussed in [8]; a comparative study of the avoided stings observed in assembling of a human genome is provided by [7], meanwhile, this study seems to be rather speculative. Finally, more or less theoretically charged paper [9] presents an analysis of evolution on lost strings patterns.

In our study, all genetic sequences have been downloaded from EMBL–bank; any extra symbols falling beyond the alphabet ℵ were omitted, and the parts of a sequence split by those extra symbols were concatenated. The length of a sequence includes the eliminated symbols; an error here does not exceeds 10^{-3}.

2.1   How to Test an Interrelation Between Evolution and Lost Strings

More precisely, if the lost strings were eliminated by natural selection (not randomly) from a sequence, then one should expect that phylogenetically close species must exhibit similarity in the lost strings lists.

Reciprocally, from the analysis of mostly independent genetic entities (whatever one could understand for *mostly* and *independent*) one would be able to observe only elimination of strings.

To test this, one should develop a set of randomly selected entities, and do all the same with it. Surely, the words *randomly selected list of genetic entities* must be defined in some way precisely. For example, one might take a random sampling of the sequences from the list of mitochondrion genomes. These latter are rather short (if plant mitochondria are omitted), have identical function, and the genome consists of a single chromosome.

2.2   Random Test

Another important question is whether an observed list of the losses in the supports of various genetic entities differ from a random one. And one more question here concerns the combinatorics constraints for the "survived" words. These are two different, while strongly related questions.

What kind of a sequence model should be analyzed? Obviously, we shall not study a random non-correlated sequence; on the other hand – why not? What if even a very simple model yields a combinatorial constraints that are pretty close to those observed on some genetic entities?

If a model is not the random non-correlated sequence (Bernoulli process realization), then what type of a model is to be chosen? The very first idea is to compare to some Markov process. So, what parameters of that latter should be applied? And the most important – what is the lowest order of this Markovian process model? Some important results obtained in that direction could be found in [10,12–14].

Table 1. The figures for the length of the shortest avoided strings, for some shorter genomes. N is the length of a genetic entity, L is the least avoided string length, and K is the total number of the lost strings of the length L.

ID	Organism	N	L	K
FR775227	*Salmonella enterica subsp. mitochondrion*	17569	5	2
HQ184045	*Bos taurus isolate Mcg375 mitochondrion*	16340	5	13
JF727176	*Pan troglodytes isolate Flo mitochondrion*	16557	5	13
KC469587	*Sus scrofa domesticus breed pietrain mitochondrion*	16612	5	12
KM061558	*Canis lupus familiaris isolate Cf_stp64 mitochondrion*	16730	5	13
AY217738	*Eimeria tenella*	34750	5	4
AY945289	*Fusarium oxysporum strain F11 mitochondrion*	34477	5	2
DQ508940	*Debaryomyces hansenii mitochondrion*	29462	5	2
DQ642846	*Plasmodium falciparum HB3*	29529	5	92
JQ864234	*Candida albicans strain L296 mitochondrion*	33631	5	8
EU651892	*Hemiselmis andersenii strain CCMP 644 mitochondrion*	60553	5	2
FR775213	*Salmonella enterica subsp. enterica serovar Weltevreden*	64694	6	33
FR775245	*Salmonella enterica subsp. enterica serovar Weltevreden*	63517	5	4
HG004427	*Campylobacter fetus subsp. venerealis*	61142	5	9
KF285530	*Ostreococcus tauri isolate RCC1123 chloroplast*	67681	6	13
AB042240	*Triticum aestivum chloroplast*	134545	7	399
CP00224	*Candidatus Tremblaya princeps PCIT*	138927	6	3
FR775217	*Salmonella enterica subsp. enterica serovar Weltevreden*	131230	6	19
JN861109	*Oryza sativa Indica Group cultivar Hassawi chloroplast*	134448	7	361
X86563	*Zea mays complete chloroplast genome*	140384	7	351
CP000351	*Leptospira borgpetersenii, chromosome 2*	299762	7	85
CP002163	*Candidatus Sulcia muelleri CARI,*	276511	6	99
CP007234	*Candidatus Sulcia muelleri strain TETUND*	270029	6	81
FR775191	*Salmonella enterica*	227697	6	16
FR775236	*Salmonella enterica serovar Weltevreden*	253936	6	9
AY506529	*Zea mays strain NB mitochondrion*	569630	8	1059
CP002243	*Candidatus Moranella endobia PCIT*	538294	7	6
CP003000	*Blattabacterium*	587248	6	10
CP003771	*Mycoplasma genitalium M6282*	579504	6	11
CP006771	*Mycoplasma parvum str. Indiana*	564395	6	8

Table 2. The figures for the length of the shortest avoided strings, for some longer genomic entities. Definitions are the same as in Table 1.

ID	Organism	N	L	K
AL954800	*H. sapiens*, chromosome 14	87 191 216	9	110
CM000265	*H. sapiens*, chromosome 14	87 316 725	9	110
CM000856	*Callithrix jacchus*, chromosome 1	210 400 635	9	3
CM000878	*Callithrix jacchus*, chromosome X	142 054 208	9	62
CM000879	*Callithrix jacchus*, chromosome Y	2 853 901	7	4
CM000001	*Canis lupus familiaris*, chromosome 1	122 678 785	9	4
CM000002	*Canis lupus familiaris*, chromosome 2	85 426 708	9	11
CM000003	*Canis lupus familiaris*, chromosome 3	91 889 043	9	41
CM000004	*Canis lupus familiaris*, chromosome 4	88 276 631	9	63

Table 3. The figures for the length of the shortest avoided strings, for some *E. coli* genomes; K is the total number of the lost shortest strings.

ID	Length	K	ID	Length	K	ID	Length	K
AE005174	5 528 445	86	AP009048	4 646 332	170	CU928163	5 202 090	118
CU928162	5 209 548	95	FM180568	4 965 553	118	U00096	4 641 652	176*
AE014075	5 231 428	92	CU928160	4 700 560	199*	AM946981	4 558 947	213*
CU928161	5 032 268	101	CU928164	5 132 068	134	CP001925	5 386 223	88

3   Some Preliminary Results

Here we provide some preliminary data on the behaviour of the shortest avoided strings in various genomes. Few words should be said on the structure of Table 1. It consists of six blocks, where each block contains five genomes of approximate length; that latter varies from ~15 000 to ~600 000 nucleotides, with approximately equal length step. The idea was to check whether the growth of L is logarithmic, or not. Of course, it might be affected by the genetic material choice; meanwhile, it brings some raw results. Indeed, Table 1 demonstrates that the growth increment is definitely less than $\ln 2$. This pattern is supported, in general, by the data shown in Table 2.

3.1   Lost Strings Sets

Both tables show significant variation in abundance of the lists of lost strings. First of all, let's focus on the dependence of L figure on the length of a sequence under consideration. The growth of L is significantly slower than a typical exponent. A 15 thousand fold growth of the length of a sequence results in a growth of L from $L = 5$ to $L = 9$. This figure seems to be universal: whatever real genetic entity is taken for analysis, one gets $L = 9$; at least for mammalian genomes (probably, some plants genomes, say larch genome, may yield $L = 10$, and one hardly could expect a greater figure).

The number of the avoided strings observed over a sequence is much more sensitive to the length of that former.

3.2   Closely Related Strains and the Avoided Strings

To investigate the impact of phylogeny on the composition of the list of the lost strings observed over a family of genomes, we comprised the lists for a set of *E. coli* genomes, of various strains (see Table 3). Table 3 comprises the genomes of bacteria *E. coli* that is well studied and widely spread object for genetic studies. Actually, EMBL–bank contains 101 genomes of various strains of *E. coli*; we have used twelve of them, randomly chosen.

All the genomes (except three ones) exhibit $L = 8$; Table 3 shows the abundance figures of those losses. There are three genomes (these are AM946981, CU928160, and U00096 entries) that have $L = 7$; meanwhile, it should be said, that unlike the other patterns, here a single absent string has been observed, in each genome. All these entries are marked with asterisk, at the table. This fact seems to be an exclusion itself: indeed, a list of the lost strings is generally several times longer. One might expect that the abundance of lost strings of the given length q should yield the similar figure of the part of the total number of string of the length q. A single string lost among octanucleotides is equal to 16384^{-1} that significantly differs from a typical figure observed for $q = 9$ and greater.

Anyway, a comparison of those lost 7-tipples is of a great interest. These lost strings are GTCTAGG for AM946981, CCTAGGA for CU928160, and GCCTAGG for U00096. First of all, GC-content for these strings is 4/7, 4/7, and 5/7, respectively. Another remarkable feature is that the strings posses a quasi-palindromic structure. All these three septanucleotides could be easily (and obviously aligned: they have the common "kern" $\boxed{\text{CTAGG}}$ of the length $q = 5$.

There are rather few common lost octanucleotides among these twelve bacterial genomes. There are two strings (GAGTCTAG, GGGTCTAG) found in all twelve genomes, two strings (GGCCTAGG, GTCCTAGG) found in eleven genomes, and two strings (ACTAGTCG, ATGCCTAG) found in nine genomes. A pairwise alignment yields very good concordance, for the first and the second couples of the strings: they have common "kerns" $\boxed{\text{GTCTAG}}$ and $\boxed{\text{CCTAGG}}$ of the length $q = 5$, respectively. The last couple exhibits lower concordance level with only tetranucleotide $\boxed{\text{CTAG}}$ common in them. A typical number of the octanucleotides common for several genomes varies from five to seven.

Remarkably, sequence CTAG found in all sequences discussed above is a recognition site of many bacteria or *Archaea* restriction enzymes (e.g. BfaI, CchI, FgoI or Htu [17]). Longer sequence CCTAGG is also a known recognition site of various restriction enzymes (e.g. AvrII and XmaJI).

4   Discussion

To analyze the sets of the lost strings and the properties of those sets one should start from answering the question towards the choice (or definition) of a reference sequence. There might be a number of references, and we start from that one mentioned in Sect. 2.1. Indeed, consider a sequence of the length N from four-letter alphabet $\aleph$, with the following frequencies of symbols: $f_A = \alpha$, $f_C = \beta$, $f_G = \gamma$ and $f_T = \delta$, so that $\alpha + \beta + \gamma + \delta = 1$. Stipulating that the symbols occur independently, and the sequence is not correlated, one has a probability (or a frequency) of a string ω of the length q to be combined from the frequencies of individual symbols through their product. Hence, the least probable string looks like a tract of the same symbol (e.g. A) of the length q with the reciprocal frequency equal to f_A^q.

Obviously, no one has ever observed so far such long lost tracts; this is not a trick, but another evidence of rather non-random structuredness of DNA sequences. Another approach to choose a reference sequence is to implement a Markov model surrogate sequence, with proper frequencies of k-tipples. There is no problem to develop the frequency of l-tipple from k-tipples ($l > k$) (see details in [1–3]):

$$\tilde{f}_{\nu_1\nu_2\nu_3\ldots\nu_{l-1}\nu_l} = \frac{\prod_{j=1}^{l} f_{\nu_j\nu_{j+1}\nu_{j+3}\ldots\nu_{j+k-1}\nu_{j+k}}}{\prod_{j=2}^{l-1} f_{\nu_j\nu_{j+1}\nu_{j+3}\ldots\nu_{j+k-1}\nu_{j+k}}}, \tag{4}$$

a bit more problem is to find out the string ω with the minimal f_ω.

Minimal $f_{\nu_1\nu_2\nu_3\ldots\nu_{l-1}\nu_l}$ value from (4) yet does not provide an answer towards the question on a (minimal) lost string. An order of the reciprocal Markov chain is a point here. Still, there is no natural way to choose some specific order k of Markov chain. Moreover, since a sequence under consideration is finite, then there always exists such order k^*, that provides an absolutely exact matching of a simulated sequences to the given real one.

This point makes a comparative approach rather acute: one should compare the lost strings from two (or more) very closely related sequences, whose bearers are proven to be real close relatives. Such kind of study arises a question on the proximity measure of two (or more) strings. Here alignment looks rather feasible, since the strings under consideration are not longer that 10 nucleotides. Meanwhile, a diversity and specificity of various versions of alignment (both in algorithmic sense, and in software implementations) brings a problem here: probably, one has to revisit an alignment technique to find out the best one, for the short strings. Of course, such choice must be provided with a clear and concise proof of the efficiency of a method.

Another approach looking rather powerful consists in a study of so called trees of the lost strings. Suppose, we have identified the set of the shortest strings within a sequence. At the next step, we should identify the lost strings that are a symbol longer, but they do not inherit found at the first step. Consider two strings ω_1 and ω_2 so that $|\omega_1| = k$ and $|\omega_2| = k + 1$ (here $|\omega_1|$ means the length of a string). A string ω_2 inherits a string ω_1, if $\omega_1 \subset \omega_2$. Thus, suppose the set of the shortest lost strings $\mathcal{A}$ is identified; next, let us identify the set $\overline{\mathcal{A}}$ of the lost strings that are one symbol longer, and no one $\overline{s} \in \overline{\mathcal{A}}$ inherits any $s \in \mathcal{A}$. More detail discussion of all these patterns requires further studies and falls beyond the scope of this paper.

Avoidance in bacterial genomes of palindromes related to restriction enzymes has been mentioned before [15], especially in bacteria using type II restriction-modification systems as defensive mechanism against inappropriate invasion of foreign DNA [16]. Our results show that the shortest avoided strings in 12 E. Coli genomes correspond to such palindromes. With the increasing availability of genome sequences, further work could focus on the identification of additional avoided functional sites in bacterial or animal genomes.

Let's have a more detail look at the formula (4); in particular, the version of the extension of l-tipple into $l + 1$-tipple. Here the formula (4) changes for

$$
\tilde{f}_{\nu_1\nu_2\nu_3...\nu_{l-1}\nu_l\nu_{l+1}} = \frac{f_{\nu_1\nu_2\nu_3...\nu_{l-1}\nu_l} \times f_{\nu_2\nu_3\nu_4...\nu_{l-1}\nu_l\nu_{l+1}}}{f_{\nu_2\nu_3\nu_4...\nu_{l-1}\nu_l}}.
\tag{5}
$$

Formula (5) looks like a Markov process expression, while it is not: it is derived with no hypothesis towards the Markov property of an origin sequence (see [1–6] for details). Thus, another idea to figure out a lost string is to distinguish "inevitably lost" strings from "unexpectedly lost" ones.

Indeed, the formula (5) allows to estimate the expected frequency of q-tipple from the real frequencies of shorter strings, in particular, from the frequency dictionary W_{l-1}. Here two options could be found:

(I) for a given lost string ω of the length q, the expected frequency is

$$
\tilde{f}_{\nu_1\nu_2\nu_3...\nu_{l-1}\nu_l} > 0; \text{ and}
$$

(II) for a given lost string ω of the length q, the expected frequency

$$
\tilde{f}_{\nu_1\nu_2\nu_3...\nu_{l-1}\nu_l} = 0.
$$

Obviously, the option # I holds true always, wherever one seeks for the shortest lost string. Apparently, by definition of the shortest lost string, if you get the shortest lost string at the length q, then the support of the words of the length $q - 1$ is *-support. On the contrary, the option # II may not be met for the *-support bearing the words of the length $q - 1$. Thus, one should distinguish the longer lost strings (say, a symbol or two longer than the shortest one): the biological sense of a string exhibiting the option # II may differ from that one, for a string with non-aero expected frequency.

5 Conclusion

The strings of the minimal length that are not present in a genetic entity may be as informative, as those found in that latter. Preliminary results of the study of such strings show that the sets of the shortest lost strings may not be simulated with any probabilistic, or combinatorial model of a real DNA sequence. The length of the shortest lost strings grows very slowly, as the length of a genetic sequence grows up. Next, phylogenetically close sequences yield rather proximal sets of strings. Finally, the strings comprising a set of the shortest lost strings for a given DNA sequence seem to be rather close each other, in terms of Hamming metrics, or in terms of alignment. Further studies are necessary to figure out the biological charge of such lost strings.

Acknowledgement. This study was supported by a research grant # 14.Y26.31.0004 from the Government of the Russian Federation.

References

1. Bugaenko, N.N., Gorban, A.N., Sadovsky, M.G.: Maximum entropy method in analysis of genetic text and measurement of its information content. Open Syst. Inf. Dyn. **5**, 265–278 (1998)
2. Gorban, A.N., Popova, T.G., Sadovsky, M.G., Wünsch, D.C.: Information content of the frequency dictionaries, reconstruction, transformation, classification of dictionaries, genetic texts. In: Intelligent Engineering Systems through Artificial Neural Networks. Smart Engineering System Design, vol. 11, pp. 657–663. ASME Press, New York (2001)
3. Gorban, A.N., Popova, T.G., Sadovsky, M.G.: Classification of symbol sequences over thier frequency dictionaries: towards the connection between structure and natural taxonomy. Open Syst. Inf. Dyn. **7**, 1–17 (2000)
4. Sadovsky, M.G., Shchepanovsky, A.S., Putintzeva, Y.A.: Genes, information and sense: complexity and knowledge retrieval. Theory Biosci. **127**, 69–78 (2008)
5. Sadovsky, M.G.: Comparison of real frequencies of strings vs. the expected ones reveals the information capacity of macromoleculae. J. Biol. Phys. **29**, 23–38 (2003)
6. Sadovsky, M.G.: Information capacity of nucleotide sequences and its applications. Bull. Math. Biol. **68**, 156–178 (2006)
7. Garcia S.P., Pinho A.J.: Minimal absent words in four human genome assemblies. PLoS One **6**(12), e29344 (2011)
8. Alileche, A., Goswami, J., Bourland, W., Davis, M., Hampikian, G.: Nullomer derived anticancer peptides (NulloPs): differential lethal effects on normal and cancer cells in vitro. Peptides **38**, 302–311 (2012)
9. Acquisti, C., Poste, G., Curtiss, D., Kumar, S.: Nullomers: really a matter of natural selection? PLoS One **10**, e1022 (2007)
10. Aurell, E., Innocenti, N., Zhou, H.-J.: The Bulk and The Tail of Minimal Absent Words in Genome Sequences (2015). arXiv:1509.05188v1
11. Rahman, M.S., Alatabbi, A., Athar, T., Crochemore, M., Rahman, M.S.: Absent words and the (dis)similarity analysis of DNA sequences: an experimental study. BMC Res. Notes **9**, 186 (2016)
12. Garcia, S.P., Pinho, A.P., Rodrigues, J., Bastos, C.A.C., Ferreira, P.: Minimal absent words in prokaryotic, eukaryotic genomes. PLoS One **6**(1), e16065 (2011)
13. Hao, B., Xie, H., Zuguo, Y., Chen, G.: Avoided strings in bacterial complete genomes and a related combinatorial problem. Ann. Comb. **4**, 247–255 (2000)
14. Chairungsee, S., Crochemore, M.: Using minimal absent words to build phylogeny. Theoret. Comput. Sci. **450**, 109–116 (2012)

15. Gelfand, M.S., Koonin, E.V.: Avoidance of palindromic words in bacterial and archaeal genomes: a close connection with restriction enzymes. Nucleic Acids Res. **25**, 2430–2439 (1997)
16. Fuglsang, A.: Distribution of potential type II restriction sites (palindromes) in prokaryotes. Biochem. Biophys. Res. Commun. **310**(2), 280–285 (2003)
17. Roberts, R.J., Vincze, T., Posfai, J., Macelis, D.: REBASE-a database for DNA restriction and modification: enzymes, genes and genomes. Nucleic Acids Res. **43**, D298–D299 (2015)

Comprehensive Study of Instable Regions in *Pseudomonas Aeruginosa*

Dan Wang, Jingyu Li, and Lusheng Wang$^{(\boxtimes)}$

Department of Computer Science, City University of Hong Kong,
Hong Kong, China
cswangl@cityu.edu.hk

Abstract. Pseudomonas aeruginosa is recognized for its intrinsically advanced antibiotic resistance mechanisms. Dispensable genome which includes sequences shared by a subset of strains in a species is important to the study of a species' evolution, antibiotic resistance and infectious potential. In this paper, by using a multiple sequence aligner, we segmented the genomes of 25 Pseudomonas aeruginosa strains into core blocks (shared by all the 25 genomes) and dispensable blocks (shared by a subset of the 25 genomes). In this paper, we use the term instable blocks to refer to dispensable blocks since blocks shared by a subset of the 25 genomes may be vitally important. We then built 25 scaffolds which consisted of core and instable blocks sorted by blocks' starting positions in the chromosomes for each of the 25 strains. In these scaffolds, consecutive instable blocks formed instable regions. We conducted a comprehensive study on these instable regions and found three characteristics of instable regions: instable regions were short, site specific and varied in different strains. We then studied three DNA elements which may contribute to the variation of instable regions: directed repeats (DRs), transposons and integrons. Past studies have shown that sequences flanked by a pair of DRs can be deleted from their host chromosomes or be inserted into new host chromosomes. We developed a pipeline to search for DR pairs on the flank of every instable sequence and found 27 pairs of DRs existing in the instable regions between 21 distinct pairs of core blocks. We also found that in the average, 14% and 12% of instable regions in the 25 scaffolds covered transposase genes and integrase genes, respectively. Our pipeline is at: https://github.com/shever/repeat_finding.

Supplemental Tables available at: https://github.com/shever/repeat_finding/blob/master/Supplemental_Tables.xlsx.

Keywords: *Pseudomonas aeruginosa* · Pan-genome · Dispensable genome · Insertion · Deletion · Homologous recombination · Directed repeats · Transposase · Integrase

1 Introduction

The term pan-genome was first coined by Tettelin in his research of *Streptococcus agalactiae* [1] more than ten years ago. Pan-genome describes the union of

© Springer International Publishing AG 2017
I. Rojas and F. Ortuño (Eds.): IWBBIO 2017, Part II, LNBI 10209, pp. 30–40, 2017.
DOI: 10.1007/978-3-319-56154-7_4

genomes in a clade of interest. The phylogenetic resolution of the clade is from species and serovar to phylum, kingdom and beyond [2]. Pan-genome includes core genome and dispensable genome. Core genome is the intersection of all genomes of interest while dispensable genome includes sequences shared by a subset of the genomes of interest. Core genome typically includes sequences responsible for the major phenotypic traits and basic aspects of the biology of the clade while dispensable genome contributes to the species diversity and persistence in a particular environment [2]. Identification and study of dispensable genome is important to the study of a clade's evolution, niches adaptation, antibiotic resistance, infectious potential and colonization of a new host. For genomes from closely related taxa, their pan-genome can be achieved at the nucleotide sequence level through multiple whole-genome alignment.

Bacterial genomes are dynamic on evolutionary time scale. In bacterial genome, dispensable genome can result from genome rearrangements such as insertions or deletions (indels), or from the activities of mobile DNA elements such as transposons or integrons.

Many past studies reveal that indels can be mediated by directed repeats (DRs) [3,4] and this kind of indels are called homologous recombination. In homologous recombination, DNA strands are exchanged between a pair of similar or identical sequences. Sequences flanked by a pair of DRs can be deleted from the chromosomes or be inserted into new chromosomes, for example, numerous horizontally transferred genes are integrated into their new host chromosomes through homologous recombination [5]. The distances between homologous sequences on the chromosome are varied. The estimated minimal length of homologous sequences which are necessary for the recombination process are between 20 and 100 bp [6].

Activities of mobile DNA elements such as transposons and integrons also contribute to the existence of dispensable genome in bacterial genome. There are two major ways for transposable genetic elements to move from one locus to another within or between genomes. One involves passage through an RNA intermediate prior to synthesis of a DNA copy while the other is limited uniquely to DNA intermediates. For both types of elements, recombination reactions involved in integration are carried out by element-specific enzymes. These are called transposases in the case of DNA elements and integrases in the case of the best-characterized RNA elements, the retroviruses and retrotransposons [7].

In our study, we downloaded the complete genomes of 25 *Pseudomonas aeruginosa* strains which were available in the NCBI GenBank database when we conducted this study. Each of the 25 strains is labeled with a distinct number from 1 to 25 by us and the detailed information of these 25 strains is in Supplemental Table S2. By using multiple sequence aligner Mugsy [8], sequences in the 25 genomes were divided into core blocks (shared by all the 25 genomes) and dispensable blocks (shared by a subset of the 25 genomes). In the following paragraphs, we will use the term instable blocks to refer to dispensable blocks since blocks shared by a subset of the 25 genomes may be vitally important. In our study, we omitted small blocks which is shorter than 1000 bp and strain-specific blocks. To distinguish between different blocks, we labeled each core block with an distinct ID number and labeled each instable block with "D" plus an ID number, for example, "8" is the ID for Core Block 8 and "D9" is the ID for instable Block 9. For each of the 25 strains, we then built 25 scaffolds which consisted of the IDs of core and instable blocks sorted by starting positions in the chromosomes (See Supplemental Table S1). In the scaffolds, consecutive instable blocks flanked by a pair of core blocks formed a *instable region* (InsR) and this pair of core blocks are the *insertion site* for this instable region. DNA sequence inside a instable region is called *instable sequence*. In our study, we found three characteristics of instable regions in *Pseudomonas aeruginosa*: instable regions were short, site specific and varied in different strains. Due to the large variation of instable regions, we then studied three DNA elements which may contribute to the variation of instable regions: DRs, transposons and integrons. We developed a pipeline to find DR pairs which were on the flank of a instable sequence and achieved 27 pairs of DRs existing in the instable regions between 21 distinct pairs of core blocks. We found that averagely, 14% and 12% of instable regions covered transposase genes and integrase genes, respectively.

2   Results

2.1   Characteristics of Instable Regions

Instable Regions Are Short. Among the 25 strains, the number of instable regions varied from 44 to 57 and the total length of instable regions were from 691,941 bp to 1,794,274 bp which accounted for 11% to 24% of their respective whole genome (See Supplemental Table S2). There were a total of 1,231 instable regions among the 25 strains. The median and average length of these 1,231 regions were 6.6 kbp and 20.9 kbp, respectively. The histogram for the length (in bp) of these 1,231 instable regions is in Fig. 1(a). We found that 40.2% of instable regions were shorter than 5 kbp, 77.7% of instable regions were shorter than 20 kbp and only 2.6% of instable regions are longer than 120 kbp (See Fig. 1(a)).

Instable Regions Are Site-Specific. There were 451 core blocks in each of the 25 scaffolds. If the orders of these core blocks are the same in the 25 scaffolds, there will be 450 distinct pairs of adjacent core blocks which can be the potential insertion sites for instable regions. But in fact the orders of core blocks were different in the 25 scaffolds, which may result from genome rearrangements and there were actually a total of 506 distinct pairs of adjacent core blocks which can be the potential insertion sites for instable regions. Among these 506 pairs of adjacent core blocks, we found 134 pairs of core blocks between which instable regions existed (the block IDs of these 134 pairs of core blocks are listed in Supplemental Table S3). For these 134 pairs of core blocks, 90 of them were insertion sites in two or more strains while 44 of them were insertion sites in only one strain (See Fig. 1(b)). There were 9 pairs of core blocks which were insertion sites in all the 25 strains (See Fig. 1(b)).

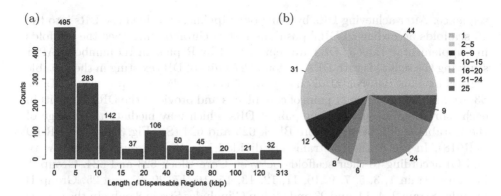

Fig. 1. (a) Histogram for the length of instable regions (kbp). (b) Distribution of the 134 insertion sites. 44 insertion sites exist in one strain, 31, 12, 8, 6, and 24 insertion sites exist in 2 to 5 strains, 6 to 9 strains, 10 to 15 strains, 16 to 20 strains and 21 to 24 strains, respectively, 9 insertion sites exist in 25 strains.

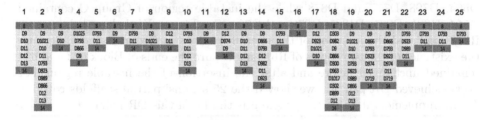

Fig. 2. Instable regions between Core Block 8 and 14 in the 25 strains. The ID of each strain is labeled above its respective column. Column x is the partial scaffold between Core Block 8 and 14 in Strain x, where x = 1, 2,..., 25. Core blocks and instable blocks are respectively represented by green and yellow rectangular boxes with block IDs written on them. (Color figure online)

Instable Regions Vary in Different Strains. Between the same pair of core blocks, the inserted instable region varied a lot in different strains. For example, between Block 8 and 14, the instable regions in the 25 strains were different (See Fig. 2). Among the 90 insertion sites which existed in two or more strains, instable regions in 57 insertion sites were varied in different strains (the block IDs of these 57 insertion sites are in Supplemental Table S3).

2.2 Mechanisms for the Variation of Instable Regions

Due to the large variation of instable regions, we studied three DNA elements which may contribute to the variation of instable regions: directed repeats, transposase genes and integrase genes.

Indels Mediated by Directed Repeats (Homologous Recombination). We developed a pipeline to find DR pairs which were on the flank of a instable sequence. After achieving DRs by using our pipeline, we added the DRs into the 25 scaffolds according to DRs' positions in the chromosomes (See the scaffolds in Supplemental Table 1, DRs are represented by R plus an ID number). After studying the scaffolds with DRs, we found 27 pairs of DRs existing in the instable regions inserted between 21 distinct pairs of core blocks. In Supplemental Table S3, we highlighted these 21 pairs of core blocks and provided the DRs' IDs within each pair. Here, we present one pair of DRs which may mediate the change of the instable region between Core Block 623 and 624 (See Fig. 3 and the DR ID is R618). In Fig. 3, the 25 strains are divided into three groups (Group A, B, and C) according to their scaffolds between Core Block 623 and 624. Group A includes Strain 1, 3, 5, 7, 9, 10, 11, 12, 13, 15, 16, 17, 20, 21 and 25, Group B includes Strain 2, 4, 14, and 23 and Group C includes Strain 8 only. In the strains of Group A, there was only one copy of R618 between Block 623 and 624 while in the strains of Group B and C, there was a instable region flanked by a pair of R618 between Block 623 and 624 (See Fig. 3). The instable regions in Group B and C were different: the instable region of Group B contained instable blocks D564, D798, D777 and D878 while the instable region of Group C contained D564, D881, D971, D1005, D878 and D764 (See Fig. 3). We believed the pair of R618 mediated the insertion of the instable regions in Group B and C because the existence of the two copies of R618 made sure the ends of Block 623 and 624 remained unchanged before and after the insertion of the instable region. For every achieved DR pair Rn, we showed the 25 strains' partial scaffolds covering Rn in Supplemental Table Rn (where n is the ID of this DR pair).

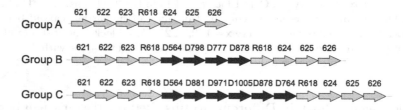

Fig. 3. Insertion of instable regions mediated by R618. The three rows are the partial scaffolds (from Block 621 to 626) of strains in Group A, B and C. Blocks are represented by arrows. Core blocks are in pink, instable blocks are in dark blue and Repeat R618 is in light blue. The number above an arrow is the ID of the corresponding block. (Color figure online)

Transposase and Integrase Genes. We added the genes whose products are transposase or integrase into the scaffolds for all the strains except Strain 15, 23 and 24 which have no annotation information (See the scaffolds in Supplemental Table S1, TSP and IN represent transposase and integrase genes, respectively). We then studied the correlations between these genes and instable regions. We found that more than 60% of transposase and integrase genes were located within instable regions in all the strains except Strain 25 (See Table 1, Column P1 and P2). In Strain 3, 4, 5, 6, 7, 8, 9, 10, 11, 12, 14, 17, 20 and 21, 100% of genes which encode transposase were within instable regions and in Strain 16, 20 and 21, 100% the integrase genes were within instable regions (See Table 1, Column P1 and P2). It was found that 2%–42% of instable regions covered transposase genes and 2%–30% covered integrase genes (See Table 1, Column P3 and P4). On average, 92% of transposase genes were within instable regions, 85% of integrase genes were within instable regions, 14% of instable regions covered transposase genes and 12% of instable regions covered integrase genes (See the last row in Table 1).

Finally, we achieved 25 scaffolds which consisted of core blocks, instable blocks, DRs, transposase and integrase genes sorted by their positions in the

Table 1. Genes of transposase and integrase in instable regions

Strain ID	P1*	P2*	P3*	P4*
1	81% (34/42)	83% (10/12)	27% (15/56)	13% (7/56)
2	96% (48/50)	92% (11/12)	28% (14/50)	14% (7/50)
3	100% (6/6)	88% (7/8)	11% (5/47)	13% (6/47)
4	100% (17/17)	67% (6/9)	17% (9/52)	12% (6/52)
5	100% (5/5)	88% (7/8)	9% (4/45)	11% (5/45)
6	100% (1/1)	71% (5/7)	2% (1/48)	10% (5/48)
7	100% (12/12)	75% (3/4)	16% (7/44)	7% (3/44)
8	100% (2/2)	83% (5/6)	4% (2/52)	8% (4/52)
9	100% (13/13)	90% (9/10)	13% (6/45)	13% (6/45)
10	100% (25/25)	89% (8/9)	20% (11/54)	11% (6/54)
11	100% (8/8)	88% (7/8)	6% (3/51)	14% (7/51)
12	100% (15/15)	92% (12/13)	9% (4/47)	13% (6/47)
13	80% (32/40)	83% (10/12)	27% (15/56)	13% (7/56)
14	100% (5/5)	91% (10/11)	9% (4/47)	11% (5/47)
16	92% (11/12)	100% (10/10)	11% (5/47)	13% (6/47)
17	100% (8/8)	83% (5/6)	11% (5/47)	9% (4/47)
18	95% (21/22)	84% (21/25)	18% (8/45)	24% (11/45)
19	86% (68/79)	90% (26/29)	42% (24/57)	30% (17/57)
20	100% (1/1)	100% (1/1)	2% (1/45)	2% (1/45)
21	100% (1/1)	100% (1/1)	2% (1/45)	2% (1/45)
22	69% (40/58)	82% (9/11)	29% (14/49)	14% (7/49)
25	17% (1/6)	50% (1/2)	2% (1/45)	2% (1/45)
Average	92%	85%	14%	12%

*P1 is the percentage of Tnp genes within InsRs (No. of Tnp in InsRs/Total No. of Tnp). P2 is the percentage of IN genes within InsRs (No. of IN in InsRs/Total No. of IN). P3 is the percentage of InsRs which covers Tnp genes (No. of InsRs which covers Tnp Genes/Total No. of InsRs). P4 is the percentage of InsRs which covers IN genes (No. of InsRs which covers IN Genes/Total No. of InsRs).

strains' chromosomes (See Supplemental Table S1). We plotted Fig. 6 to visualize these 25 scaffolds. From Fig. 6, we can see that DRs, transposase and integrase genes are highly correlated with instable regions.

3 Methods

Building Scaffolds and Finding Instable Regions. By using multiple sequence aligner Mugsy [8], sequences in the 25 genomes were segmented into core blocks, instable blocks and strain-specific blocks. In our study, we omitted small blocks which is shorter than 1000 bp and strain-specific blocks. To distinguish between different blocks, we labeled each core block with an distinct ID number and labeled each instable block with "D" plus an ID number. For each of the 25 strains, we then built 25 scaffolds which consisted of the IDs of core and instable blocks sorted by starting positions in the chromosomes (See Supplemental table S1). In the scaffolds, consecutive instable blocks formed instable regions and we then did a comprehensive study on these instable regions.

Finding Indels Mediated by DRs. With the purpose of finding repeats that are related to indels, we developed a pipeline based on BLAST [9] alignment algorithm that can find the pattern shown in Fig. 4. For all instable blocks, do the following steps to find out a reasonable pair of repeats that may be associated with the insertion or deletion of one instable region.

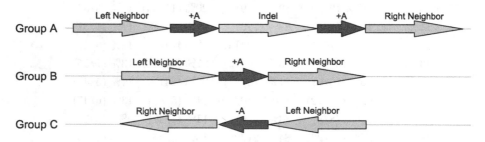

Fig. 4. Layout of the strain groups before and after repeat-mediated insertion or deletion. In this figure, "+" and "−" represent the directions of segments. When deletion of the region in sky blue occurs on group B and group C, sky blue region is removed together with one piece of repeat in dark blue (+A) compared with group A, but the left neighbor and right neighbor in pink keep unchanged. (Color figure online)

First of all, for each instable block (define as *InsB*) on each strain (define as strain *A1*), cut 10,000 base pairs from the left side of *InsB*, and name it as *reA*. Define the start position of *reA* as *reA_s* and the end position as *reA_e*.

Secondly, use BLAST to make an alignment analysis between *reA* and *target segment*. *Target segment* here represents the gene segment starts from *reA_e* and ends on the tail of strain *A1*. The corresponding BLAST alignment result shows as a chart containing several gene segment pairs (name this list as *candA_list* and members in it as *candAs*), each pair of which is located at both sides of *InsB*, concluding that these gene segment pairs may be the cause of *InsB* insertion or deletion. Each line of *candA_list* contains four positions, left *candA*'s start position, left *candA*'s end position, right *candA*'s start position and right *candA*'s end position. We define them as *candAL_s*, *candAL_e*, *candAR_s* and *candAR_e*, respectively. For convenience, the gene segment from *candAL_e* to *candAR_s* is named as *INDEL*, segment from *candAL_s - 10,000* to *candAL_s* is named as *neighborL* and segment from *candAR_e* to *candAR_e + 10,000* is named as *neighborR*.

To figure out which of the members in *candA_list* are related to *InsB* insertion or deletion with the highest possibilities, thirdly, we searched each *candA* together with its corresponding *neighborL* and *neighborR* on strains that lack *InsB*. If *neighborL-candA-neighborR* structure can be found on those strains

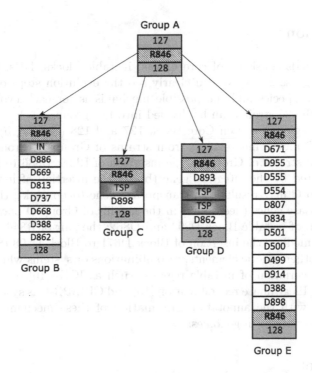

Fig. 5. Evolution of strains through different mechanisms. The five columns are the partial scaffolds (from Block 127 to 128) of strains in Group A, B, C, D and E. Core blocks, instable blocks, DRs, transposase and integrase genes are represented by rectangular boxes in orange, white, gray, deep pink and blue, respectively. Group A includes Strain 1, 2, 3, 5, 6, 8, 9, 10, 11, 12, 13, 14, 15, 16, 19, 22, 24 and 25. Group B includes Strain 4. Group C includes Strain 7 and 18. Group D includes Strain 17, 23. Group E includes Strain 20 and 21 (Color figure online)

and *neighborL-candA-InsB-candA-neighborR* structure can be found on strain *A1* simultaneously, set this *candA* as *RepeatA*.

For further demonstration, fourthly, check the *INDEL* region on all 25 strains. If *INDEL* region always appears together with a pair of *RepeatA* on both sides, *RepeatA* can be regarded as a reasonable repeat that causes *INDEL* region's insertion or deletion.

Our approach checked all the instable blocks in the .maf file and found out all the possible *RepeatA*s that may be the reason of gene indels. While applying BLAST package into our project, we set the E-value threshold as 10^{-10}.

Finding Transposase and Integrase Genes. For each strain, we obtained the positions of genes which encode transposase and integrase from the annotation files in the GFF format. We then add these genes to their strains' scaffolds according to their positions (See supplemental Table S1).

4   Discussion

With the scaffolds consisted of core blocks, instable blocks, DRs, transposase genes and integrase genes, we can clearly see the evolution steps of a instable region within a species and the possible mechanisms for each evolution step. For example, the 25 strains can be divided into five groups (A–E) according to their instable regions between Core Block 127 and 128 (See Fig. 5). Strains of Group B, C, D and E may evolute from strains of Group A through different mechanisms: in strains of Group B, the insertion of Block D886 to Block D862 may be mediated by the integrase gene (blue); the insertion of instable Block D898 in Group C may result from transposase gene (deep pink); the existence of two transposase genes (deep pink) in the strains of Group D may contribute to the insertion of instable Block D893 and D862; the pair of R846 in strains of Group E may mediate the insertion of Block D671 to Block D898 (See Fig. 5).

There are other genetic elements, recombinations or systems which may contribute to the variation of instable regions, such as ICEs [10], resolvases [11], invertases [12], Illegitimate recombination [13] and CRISPR-Cas systems [14, 15]. In the future, with the annotation information of these mechanisms, we can explain more variations in genomes.

5   Data Set

The genome sequences and annotation files were downloaded from the NCBI Genome Database (https://www.ncbi.nlm.nih.gov/genome/genomes/187). The strains names and accession number in the NCBI Genome Database are in Supplemental Table S2.

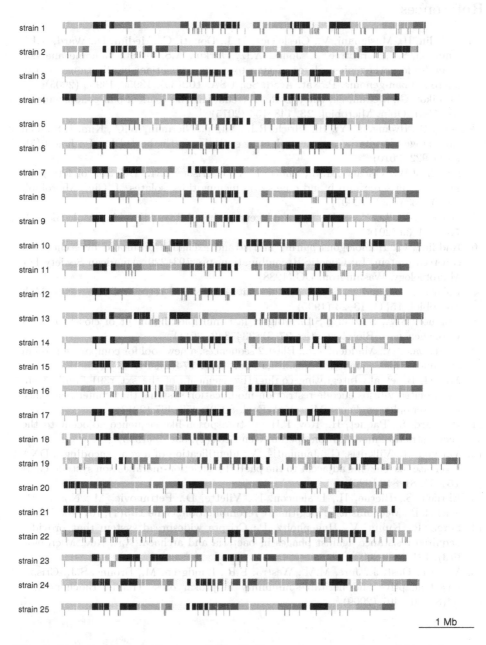

Fig. 6. Visualization of the 25 scaffolds. Each row represents the scaffold of the corresponding strain. Horizontal colored blocks are core blocks while white blocks are instable blocks. Each color represents a distinct set of core blocks. DRs, transposase genes and integrase genes are represented by blue, green and red vertical bars. (Color figure online)

References

1. Tettelin, H., Masignani, V., Cieslewicz, M.J., Donati, C., Medini, D., Ward, N.L., Angiuoli, S.V., Crabtree, J., Jones, A.L., Durkin, A.S., et al.: Genome analysis of multiple pathogenic isolates of streptococcus agalactiae: implications for the microbial pan-genome. P. Nat. Acad. Sci. USA. **102**(39), 13950–13955 (2005)
2. Vernikos, G., Medini, D., Riley, D.R., Tettelin, H.: Ten years of pan-genome analyses. Curr. Opin. Microbiol. **23**, 148–154 (2015)
3. Song, H., Hwang, J., Yi, H., Ulrich, R.L., Yu, Y., Nierman, W.C., Kim, H.S.: The early stage of bacterial genome-reductive evolution in the host. Plos. Pathog. **6**(5), E1000922 (2010)
4. Marsh, J.W., O'leary, M.M., Shutt, K.A., Harrison, L.H.: Deletion of feta gene sequences in serogroup b and c neisseria meningitidis isolates. J. Clin. Microbiol. **45**(4), 1333–1335 (2007)
5. Darmon, E., Leach, D.R.: Bacterial genome instability. Microbiol. Mol. Biol. R. **78**(1), 1–39 (2014)
6. Radding, C.: Homologous pairing and strand exchange promoted by escherichia coli reca protein. In: Genetic Recombination, pp. 193–229. American Society For Microbiology, Washington, DC (1988)
7. Polard, P., Chandler, M.: Bacterial transposases and retroviral integrases. Mol. Microbiol. **15**(1), 13–23 (1995)
8. Angiuoli, S.V., Salzberg, S.L.: Mugsy: fast multiple alignment of closely related whole genomes. Bioinformatics **27**(3), 334–342 (2011)
9. Tatusova, T.A., Madden, T.L.: Blast 2 sequences, a new tool for comparing protein and nucleotide sequences. Fems. Microbiol. Lett. **174**(2), 247–250 (1999)
10. Miguel, B., et al.: Integrating conjugative elements of the sxt/r391 family from fish-isolated vibrios encode restriction-modification systems that confer resistance to bacteriophages. Fems. Microbiol. Ecol. **83**(2), 457–467 (2013)
11. Brassard, S., Paquet, H., Roy, P.H.: A transposon-like sequence adjacent to the acci restriction-modification operon. Gene **157**(1), 69–72 (1995)
12. Vaisvila, R., Vilkaitis, G., Janulaitis, A.: Identification of a gene encoding a DNA invertase-like enzyme adjacent to the paeR7I restriction-modification system. Gene **157**(1), 81–84 (1995)
13. Ehrlich, S., Bierne, H., D'alencon, E., Vilette, D., Petranovic, M., Noirot, P., Michel, B.: Mechanisms of illegitimate recombination. Gene **135**(1), 161–166 (1993)
14. Sorek, R., Kunin, V., Hugenholtz, P.: Crispra widespread system that provides acquired resistance against phages in bacteria and archaea. Nat. Rev. Microbiol. **6**(3), 181–186 (2008)
15. Van Der Oost, J., Jore, M.M., Westra, E.R., Lundgren, M., Brouns, S.J.: Crispr-based adaptive and heritable immunity in prokaryotes. Trends. Biochem. Sci. **34**(8), 401–407 (2009)

Breathogenomics: A Computational Architecture for Screening, Early Diagnosis and Genotyping of Lung Cancer

Emmanuel Adetiba[1,3]([✉]), Marion O. Adebiyi[2,3], and Surendra Thakur[4]

[1] Department of Electrical and Information Engineering, College of Engineering, Covenant University, Ota, Ogun State, Nigeria
emmanuel.adetiba@covenantuniversity.edu.ng
[2] Department of Computer and Information Science, College of Science and Technology, Covenant University, Ota, Ogun State, Nigeria
[3] Covenant University Bioinformatics Research (CUBRe), Ota, Nigeria
[4] KZN e-Skills CoLab, Durban University of Technology, Durban, South Africa

Abstract. The genome sequences of some genes have been implicated to carry various mutations that lead to the initiation and advancement of lung cancer. In addition, it has been scientifically established that anytime we breathe out, chemicals called Volatile Organic Compounds (VOCs) are released from the breath. Hundreds of such VOCs have been uniquely identified from samples of breathe collected from lung cancer patients, which make them viable as chemical biomarkers for lung cancer. Based on the foregoing scientific break-throughs, we developed *breathogenomics*, a computational architecture for screening, early diagnosis and genotyping of lung cancer victims anchored on the analysis of exhaled breath and mutational profiles of genomic biomarkers. The architecture contains two important sub-modules. At the first sub-module, the exhaled breadths of smokers or persons that are at risk of lung cancer are collected and appropriate computational algorithms are employed to determine the presence of any of the VOC biomarkers. Next, a patient with any VOC biomarker in the exhaled breath proceeds to the second sub-module, which contains appropriate computational models for the detection of mutated genes. Once mutations are detected in any of the biomarker genes found in a given patient, such patient is recommended for targeted therapy to promptly curtail the progression of the mutations to advanced stages. The *breathogenomics* archi-tecture serves as a generic template for the development of clinical equipment for breath and genomic based screening, early diagnosis and genotyping of lung cancer. In this paper, we report the preliminary result obtained from the pro-totype that we are currently developing based on the architecture. Constructing a lung cancer early diagnosis/screening system based on the prototype when fully developed will hopefully minimize the current spate of deaths as a result of late diagnosis of the disease.

Keywords: Breathogenomics · Biomarkers · Early diagnosis · Genotyping · Genome · Lung cancer · Screening · VOC

© Springer International Publishing AG 2017
I. Rojas and F. Ortuño (Eds.): IWBBIO 2017, Part II, LNBI 10209, pp. 41–49, 2017.
DOI: 10.1007/978-3-319-56154-7_5

1 Introduction

A wide ranging and collaborative efforts of scientists and researchers from around the world have unraveled the human genome sequence since 2003. The completed nucleotide sequences of the human genome are about three billion characters in length. Also, the sequences of other organisms such as virus, bacteria, archaea, fungi, mouse and several plants have been successfully deciphered. These efforts have generated volume of biological data that keep growing at an astronomical rate almost on a daily basis. Meanwhile, interpreting and making sense of these genomic data is a major challenge that is facing researchers and scientists in fields as varying as molecular biology, agriculture, pharmacy, information engineering, computer science, physics, mathematics, medicine and etc. This is because of the prospects that such interpretation holds for human development, which include innovation of therapeutic procedures, design of novel drugs as well as creation of new diagnostic tools especially for different kinds of cancers such as cervical, breast and lung cancer [1, 2].

Lung cancer is one of the leading causes of mortality globally. The primary reason why the outcome of lung cancer is so poor is because about 70% of patients presented at the first time to pathologists are already at the advanced stage of the disease. Meanwhile, at the moment, therapy at such advanced stage has little positive effect on mortality [3]. In fact, studies have shown that only 14% of lung cancer patients survive after five years of therapy. However, if the cancer can be discovered and treated on time, the rate of survival increases to 48% [4]. These gory statistics have triggered recent research efforts on early detection and screening of lung cancer with a departure from the traditional Low Dose Computer Tomography (LDCT) approach [5, 6]. Some of these recent studies leveraged on exhaled breath analysis [4, 7–11] while other studies adopted genomics [12–16] in an independent manner. In this paper, we hybrided these two recent technological advances to develop *breathogenomics* architecture for screening, early diagnosis and genotyping of lung cancer. This study is motivated based on the fact that breath analysis to detect VOCs can only signal a probability of carcinoma incidence in a subject, whereas further probe into the mutational status of some biomarker genes (in the same subject) will provide certainty as regards the presence or absence of cancer. In the event of mutation detection in such biomarker genes, the subject can promptly be recommended for targeted molecular therapy to forestall progression of the cancer to an advanced stage.

2 Materials and Methods

2.1 The Proposed Architecture

The *breathogenomics* architecture, which is a graphical view of the lung cancer early detection protocol being proposed in this study is shown in Fig. 1. This architecture primarily targets screening of at risk subjects for early detection and prompt arrest of the progression of lung cancer via breath and genomics mutation analysis.

In the first sub-module, breath samples obtained from smokers and persons at risk of lung cancer are analyzed to detect Volatile Organic Compounds (VOC) as

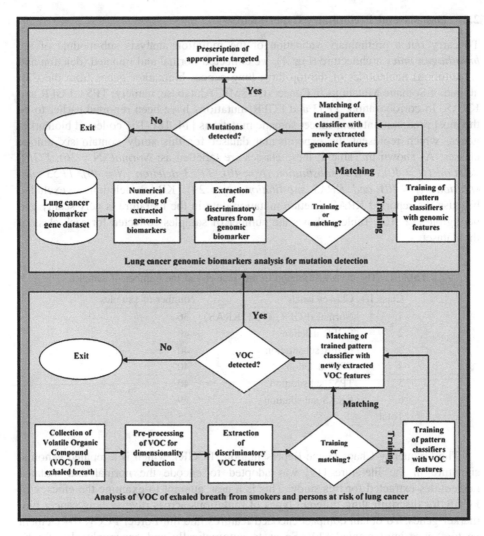

Fig. 1. Breathogenomics architecture

biomarkers. Such VOCs that have been reported in the literature include *styrene, isoprene, benzene, undecane, 1-hexane, hexanal, propyl benzene, 1-2-4 trimethyl benzene, heptanal and methyl cyclopentane* [4, 17]. In the second sub-module, mutational analysis is carried out on genes that have been identified as carriers of different shades of mutations, which lead to the initiation and progression of lung cancer. These genes include *tumor suppressor p53 (TP53), kirsten rat sarcoma viral (KRAS), epidermal growth factor receptor (EGFR), oncogene homolog, cyclin-dependent kinase inhibitor 2A, lysine (K)-specific methyltransferase 2C* and several others [16].

2.2   Dataset and Preliminary Experiments

To carry out a preliminary validation of the mutation analysis sub-module of the **breathogenomics** architecture (Fig. 1), we collected normal and mutated (deletion and substitution) nucleotides of the top three lung cancer biomarker genes from the Catalogue of Somatic Mutations in Cancer (COSMIC) database, namely; TP53, EGFR and KRAS. In corroboration, TP53 and EGFR mutations have been reported earlier to be the most predominant lung cancer somatic mutations [14, 18]. The collected biomarker genes, which represent the experimental dataset for this study contain six unique classes. As shown in Table 1, these classes are labelled as; *Normal* (N = 36), *EGFR deletion (N = 40), EGFR substitution (N = 40), TP53 deletion (N = 40), TP53 substitution (N = 40) and KRAS substitution* (N = 29). KRAS deletion is excluded because the reported KRAS deletion mutation data in the literature is very scanty and negligible [12, 16]. Table 1 shows the number of samples collected for each class of the dataset.

Table 1. The class distribution in the dataset and the number of samples

Class ID	Classes labels	Number of samples
1	Normal (EGFR, TP53, KRAS)	36
2	EGFR deletion	40
3	EGFR substitution	40
4	TP53 deletion	40
5	TP53 substitution	40
6	KRAS substitution	29
Total		225

The Frequency Chaos Game Representation (FCGR), which is a reputed genomic signature in the literature [19] was adopted to encode the normal and mutated nucleotides extracted for this study. This is a new attempt to examine the efficacy of FCGR for discriminating different types of mutations across diverse lung cancer biomarker genes. To obtain compact encoded features, the third order FCGR (64 element vectors) was implemented [20]. So as to automatically and accurately classify the FCGR encoded data, we experimentally compared Multilayer Perceptron Artificial Neural Network (MLP-ANN), Support Vector Machine (SVM) and Naïve Bayes (NB), which are frontline pattern recognition tools in machine learning.

The MLP-ANN contains 64 neurons in the input layer (64 element FCGR), 6 neurons in the output layer (5 mutation classes and 1 normal class) and two hidden layers with the neurons experimentally varied from 10 to 100. The result obtained by varying the neurons in the hidden layer is reported in Sect. 3. All the neurons were configured with **tansig** transfer function. The SVM as well as the NB were configured and trained with five and four different kernel functions respectively. The results obtained for each of the kernel functions are also reported in Sect. 3. The target outputs for each of the classes of the three pattern classifiers are illustrated in Table 2.

Table 2. The class labels and targets for the MLP-ANN, SVM and NB

Classes labels	MLP-ANN target	SVM/NB target
Normal (EGFR/TP53/KRAS)	1 0 0 0 0 0	1
EGFR deletion	0 1 0 0 0 0	2
EGFR substitution	0 0 1 0 0 0	3
TP53 deletion	0 0 0 1 0 0	4
TP53 substitution	0 0 0 0 1 0	5
KRAS substitution	0 0 0 0 0 1	6

The FCGR algorithm as well as the MLP-ANN, SVM and NB pattern classifiers were implemented in MATLAB R2015a. The preliminary experiments carried out in the current study were performed on a PC with Intel Core i5-4210U CPU operating at 2.40 GHz speed, containing 8.00 GB RAM and running 64-bit Windows 8 operating system.

3 Results and Discussion

Table 3 contains the result obtained when the number of hidden layer neurons were varied from 10 to 100. Except for the configuration with 80 neurons in the hidden layer, all the other configurations gave percentage accuracy above 90%. However, the configuration with 90 neurons in the hidden layer produced the best classification result (**Accuracy = 96.89%, MSE = 0.02**) among the different MLP-ANN configurations. As shown in Table 4, which contains the result for the five different SVM kernel functions, the best result among the kernel functions was produced by the quadratic kernel (**Accuracy = 86.36%, MSE = 0.0505**). Table 5 shows the result of the four NB kernel functions. The box kernel function produced the best results (**Accuracy = 84.44%, MSE = 0.16**) among the four NB kernel functions.

Table 3. Multilayer Perceptron Artificial Neural Network result

Number of hidden layer neurons	Accuracy (%)	Mean Square Error
10	92.44	0.02
20	95.56	0.01
30	93.78	0.02
40	95.56	0.01
50	95.11	0.02
60	96.00	0.02
70	93.33	0.02
80	85.78	0.04
90	**96.89**	**0.02**
100	95.11	0.02

Table 4. Support vector machine classifier result

Kernel functions	Accuracy (%)	Mean Square Error
Quadratic	**86.36**	**0.0505**
Polynomial	84.09	0.0493
RBF	50.00	0.0540
Linear	56.82	0.0547
MLP	25.00	0.0445

Table 5. Naïve Bayes classifier result

Kernel functions	Accuracy (%)	Misclassification error
Box	**84.44**	**0.16**
Epanechnikov	56.44	0.44
Normal	49.78	0.50
Triangle	56.44	0.44

By and large, the pattern classifier with the best performance result in this study is the MLP-ANN with 90 neurons in the hidden layer (**Accuracy = 96.89, MSE = 0.02**). The confusion matrix of this MLP-ANN is shown in Fig. 2. The confusion matrix provides details of class-by-class and the overall classification accuracies by the MLP-ANN. For the *normal* class (class 1), all the 36 instances were correctly classified as *normal*. This is also applicable to the *EGFR deletion* (class 2) and *TP53 deletion* (class 4). However, for the *EGFR substitution* (class 3), 38 instances were correctly classified while 1 instance was misclassified as *normal* (class 1) and 1 instance was misclassified as *EGFR deletion* (class 2). 38 instances of the *TP53 substitution* (class 5) were correctly classified, 1 instance was misclassified as *EGFR deletion* (class 1) while 1 instance was misclassified as *TP53 deletion* (class 4). For the *KRAS substitution* (class 6), 26 instances were correctly classified while 3 instances were misclassified as *normal* (class 1).

Furthermore, the Receiver Operating Characteristics (ROC) curves of the different classes classified by the MLP-ANN, which gave the best accuracy in this study is shown in Fig. 3. For perfect classification, the ROC curve is the line connecting (0, 0) to (0, 1) and (0, 1) to (1, 1) [21]. As illustrated in the figure, the curves for *EGFR deletion* and *TP53 deletion* classes are fully aligned with the perfect classification line. However, the curves for *EGFR substitution*, *TP53 substitution* and *KRAS substitution* classes show slight deviations from the perfect classification line. Obviously, the ROC curves agree with the output of the confusion matrix in Fig. 2.

The foregoing results show the efficacy of the FCGR genomic signature with MLP-ANN to discriminate among different mutations and across different lung cancer biomarkers. The perfect classification of all the instances in the deletion mutation classes agrees with the theoretical underpinning of FCGR, in which the frequency of oligonucleotides is used to encode genomic sequences. The fact that deletion mutations substantially alters the oligonucleotide frequency of genomic sequences, thereby enhancing the discriminatory ability of FCGR was shown in the confusion matrix

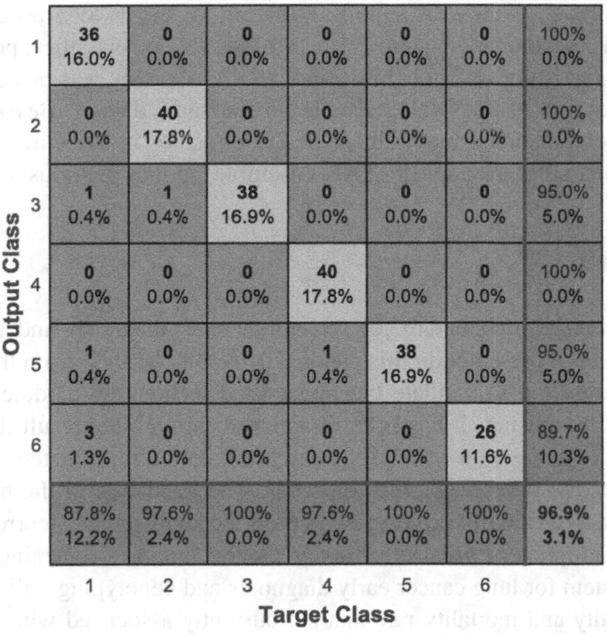

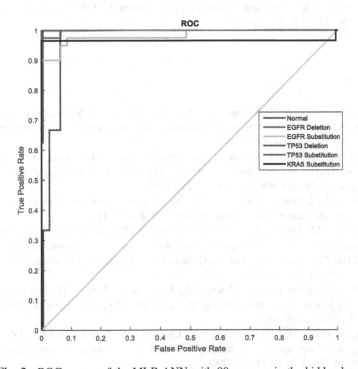

Fig. 2. Confusion matrix of the MLP-ANN with 90 neurons in the hidden layer

Fig. 3. ROC curves of the MLP-ANN with 90 neurons in the hidden layer

output (Fig. 2) as well as the ROC curves. Furthermore, the *96.89%* accuracy obtained using FCGR as genomic signature with MLP-ANN as classifier provides better accuracy than an existing study [12] in which *95.9%* accuracy was posted using HOG genomic features and MLP-ANN ensemble on the same dataset. Beyond the higher accuracy, the computational requirement of the classifier in the current study is far lower than the classifier used (MLP-ANN ensemble) in that previous study.

4 Conclusion

The *breathogenomics* architecture for screening, early diagnosis and genotyping of lung cancer has been presented in this paper. The result obtained from the preliminary experiments carried out to validate the mutation detection sub-module of the architecture is highly promising. It has been established through the result that the FCGR genomic signature along with MLP-ANN can accurately discriminate among different mutations and across different biomarker genes. The validation of the breath analysis sub-module as well as the full prototype based on the architecture is currently ongoing. Ultimately, the adoption of *breathogenomics* architecture as a screening protocol and to develop a system for lung cancer early diagnosis and genotyping will help to curtail the high morbidity and mortality rate that are currently associated with the disease.

Acknowledgement. The publication of this study is supported and funded by the Covenant University Centre for Research, Innovation and Development (CUCRID), Covenant University, Canaanland, Ota, Ogun State, Nigeria.

References

1. Venter, J.C., Adams, M.D., Myers, E.W., Li, P.W., Mural, R.J., Sutton, G.G., Smith, H.O., Yandell, M., Evans, C.A., Holt, R.A., Gocayne, J.D.: The sequence of the human genome. Science **291**(5507), 1304–1351 (2001)
2. Lander, E.S., Linton, L.M., Birren, B., Nusbaum, C., Zody, M.C., Baldwin, J., Devon, K., Dewar, K., Doyle, M., FitzHugh, W., Funke, R.: Initial sequencing and analysis of the human genome. Nature **409**(6822), 860–921 (2001)
3. Baldwin, D.R., Callister, M.E.: What is the optimum screening strategy for the early detection of lung cancer. Clin. Oncol. **28**(11), 672–681 (2016)
4. Chen, X., Cao, M., Hao, Y., Li, Y., Wang, P., Ying, K., Pan, H.: A non-invasive detection of lung cancer combined virtual gas sensors array with imaging recognition technique. In: 2005 IEEE-EMBS 27th Annual International Conference on Engineering in Medicine and Biology Society, pp. 5873–5876 (2006)
5. Marshall, H.M., Bowman, R.V., Vang, I.A., Fong, K.M., Berg, C.D.: Screening for lung cancer with low dose computed tomograph: a review of current status. J. Thorac. Dis. **5** (Suppl. 5), S524–S539 (2013)
6. Cornfield, J., Haenszel, W., Hammond, E.C., Lilienfeld, A.M., Shimkin, M.B., Wynder, E.L.: Smoking and lung cancer: recent evidence and a discussion of some questions. Int. J. Epidemiol. **38**(5), 1175–1191 (2009)

7. Michael, P., Kevin, G.: Volatile organic compounds in breath as markers of lung cancer: a cross-sectional study. Lancet **353**, 1930–1933 (1999)
8. Michael, P., Renee, N.C., Andrew, R.C.C., Anthony, J.G., Kevin, G., Joel, G., Roger, A.M., William, N.R.: Detection of lung cancer with volatile markers in the breath. Chest **123**, 2115–2123 (2003)
9. Szulejko, J.E., McCulloch, M., Jackson, J., McKee, D.L., Walker, J.C., Solouki, T.: Evidence for cancer biomarkers in exhaled breath. IEEE Sens. J. **10**(1), 185–210 (2010)
10. Schmekel, B., Winquist, F., Vikström, A.: Analysis of breath samples for lung cancer survival. Anal. Chim. Acta **20**(840), 82–86 (2014)
11. Dragonieri, S., Annema, J.T., Schot, R., Van der Schee, M.P., Spanevello, A., Carratú, P., Resta, O., Rabe, K.F., Sterk, P.J.: An electronic nose in the discrimination of patients with non-small cell lung cancer and COPD. Lung Cancer **64**(2), 166–170 (2009)
12. Adetiba, E., Olugbara, O.O.: Lung cancer prediction using neural network ensemble with histogram of oriented gradient genomic features. Sci. World J. **2015**(786013), 1–17 (2015)
13. Ramani, R.G., Jacob, S.G.: Improved classification of lung cancer tumors based on structural and physicochemical properties of proteins using data mining models. PLoS ONE **8**(3), e58772 (2013)
14. Chen, Y., Shi, J.-X., Pan, X.-F., Feng, J., Zhao, H.: Identification of candidate genes for lung cancer somatic mutation test kits. Genet. Mol. Biol. **36**(3), 455–464 (2013)
15. Marchevsky, A.M., Tsou, J.A., Laird-Offringa, I.A.: Classification of individual lung cancer cell lines based on DNA methylation markers: use of linear discriminant analysis and artificial neural networks. J. Mol. Diagn. **6**(1), 28–36 (2004)
16. Adetiba, E., Olugbara, O.O.: Improved classification of lung cancer using radial basis function neural network with affine transforms of voss representation. PLoS ONE **10**(12), e0143542 (2015)
17. Adiguzel, Y., Kulah, H.: Breathe sensors for lung cancer diagnosis. Biosens. Bioelectron. **65**, 121–138 (2015)
18. Forbes, S.A., Bindal, N., Bamford, S., Cole, C., Kok, C.Y., Beare, D., et al.: COSMIC: mining complete cancer genomes in the catalogue of somatic mutations in cancer. Nucleic Acids Res. **39**, D945–D950 (2010)
19. Wang, Y., Hill, K., Singh, S., Kari, L.: The spectrum of genomic signatures: from dinucleotides to chaos game representation. Gene **346**, 173–178 (2005)
20. Vijayan, K., Nair, V.V., Gopinath, D.P.: Classification of organisms using frequency-chaos game representation of genomic sequences and ANN. In: 10th National Conference on Technological Trends (NCTT 2009), pp. 6–7 (2009)
21. Zou, K.H., O'Malley, A.J., Mauri, L.: Receiver-operating characteristic analysis for evaluating diagnostic tests and predictive models. Circulation **115**(5), 654–657 (2007)

Mixed-Integer Programming Model for Profiling Disease Biomarkers from Gene Expression Studies

André M. Santiago[1]([✉]), Miguel Rocha[1], António Dourado[2], and Joel P. Arrais[2]

[1] Department of Informatics, University of Minho, Braga, Portugal
ampacsantiago@student.uc.pt, mrocha@di.uminho.pt
[2] CISUC, University of Coimbra, Coimbra, Portugal
{dourado,jpa}@dei.uc.pt
https://www.uminho.pt
http://www.uc.pt

Abstract. Biomedical research has seen great advances in recent years, in great part due to the long-term aid of the ability to identify biological or genetic markers that uniquely match a given disease. Despite several successes stories, the reality is that most diseases still lack an effective way of treatment, and even diagnostic. While the emergence of –omic technologies, enabled the screening of a whole cell at the molecular level, the large quantities of data produced restricted the capability to extract valid outcomes.

In this paper, we propose an optimization model, based of mixed-integer linear programming, capable of identifying a combination of bio-markers for distinguishing between healthy and diseased samples. The model achieves this taking several individuals' gene expression profiles, identifying the most relevant genes for differentiation and discovering the optimal combination of biomarkers that best explains the difference between both states. This model was validated on two different datasets through sampling analysis, achieving an out of sample accuracy up to 93%.

Keywords: Biomarkers · Disease diagnostic · Mixed-integer linear programming

1 Introduction

1.1 Biomarkers and Their Importance

The term biological marker, or biomarker for short, was introduced in 1989 as a MeSH (Medical Subject Heading) term, which can be summarized as a biological parameter, measurable and quantifiable, which is representative of a specific health or disease state. Adding to this, an NIH group went further and standardized the definition of biomarker as a characteristic that can be measured in an objective manner and can be considered an indicator of a series of different

© Springer International Publishing AG 2017
I. Rojas and F. Ortuño (Eds.): IWBBIO 2017, Part II, LNBI 10209, pp. 50–61, 2017.
DOI: 10.1007/978-3-319-56154-7_6

processes, which can be pathogenic as a result of clinical intervention as well as defining several types of biomarkers [1].

Biomarkers may be measured on a biosample, as is the case of blood, urine or saliva. They may also be classified according to their different possible applications as antecedent (evaluates the risk of developing a certain disease condition), screening, diagnostic (recognizing overt disease presence), staging (categorizing the disease severity) and prognostic biomarkers (predicting future disease course) [1]. Considering this, it is important to note that the desirable properties of biomarkers vary according to their intended use [2]. Regardless of its final purpose, all biomarkers need to have high sensitivity and specificity, they should be reproducibly obtainable through standardized methods, acceptable to the patient in question and easily interpretable by clinical staff [3].

1.2 Biomarkers and Disease

As a general rule, early detection of a disease condition plays a crucial role in successful therapy, which, in most cases, the earlier a disease condition is diagnosed, the more likely it can be successfully cured or well maintained. A dramatically reduced severity of the impact of the disease on the patient's life results from this early management of the disease, allowing for the prevention or delay of subsequent complications. However, the majority of systemic disease states are not diagnosed until morbid symptoms emerge at a late stage [4]. Molecular disease biomarkers, be they DNA, RNA or protein molecules, may prove to be the key in overcoming this challenge, being that they act as indicators of particular physiological states and may reveal hidden lethal threats before the disease reaches a state where treatment becomes difficult. To this end, the impact and effectiveness that biomarkers may have for diagnostic use has been demonstrated [5], being able to detect genetic alterations through molecular diagnostics [6], as well as reaching the point of detecting abnormal nucleic acids and proteins in bodily fluids [7]. However, several constraints still limit our capability to be able to effectively recognize the full potential of disease detection, which come down to mainly three [4]: lack of definitive molecular biomarkers for a variety of different diseases, lack of an easy and inexpensive method for sampling and lack of a platform that is portable, accurate and easy-to-use, aiding in early disease detection.

1.3 Biomarkers Prediction

While the field of biomarker discovery has had many interesting developments, there are many different computational technologies available that may yet present themselves as alternative solutions for solving this challenge. Over the last few years, several powerful methods have emerged using knowledge from data mining and statistical bioinformatics allowing the identification of robust and generalizable biomarkers with a high discriminatory ability.

Adding to these, an approach is presented in this paper which reduces the bio-marker discovery problem to a mathematical model, more specifically a mixed-integer linear optimization model, based on the work of Baliban *et al.* [8] and Puthiyedth *et al.* [9], capable of accurately classifying healthy and disease samples by identifying the optimal combination of biomarkers, while also presenting statistical validation for the obtained results obtained. For this reason, the model was tested on two different ArrayExpress datasets that were validated through sampling analysis, by generating 100 complementary training and test sets with the samples of both datasets. Figure 1 provides an illustration for the pipeline developed in this work.

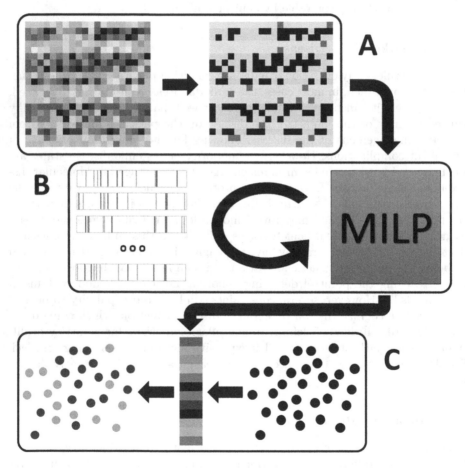

Fig. 1. Overview of the pipeline developed in this work, while illustrating the major steps which include (A) data treatment, (B) training the model and (C) analysis of new data.

2   Materials and Methods

Data Collection. The first dataset used, GSE23400, provided by Su *et al.* [10,11] is the result of a study whose purpose consisted of evaluating gene expression profiles for individuals with the condition esophageal squamous cell carcinoma (ESCC), an aggressive type of tumour with a poor prognosis, and comparing these profiles with the ones of healthy individuals. As such, this dataset consists of a total of 208 samples, of which 104 are considered clinically healthy while the remaining 104 samples are considered clinically diseased.

The second dataset, GSE65194, provided by Dubois [12–14] come from a group of studies whose purpose consisted in building expression profiles for a series of different subtypes of breast cancer (BC), including luminal A and B, triple-negative breast cancer and HER2-positive. This dataset contains a total of 178 samples, of which 11 represent healthy breast tissue.

Data Processing. Data processing was achieved through R and Python. Su *et al.* provide the ESCC dataset described in the form of an ExpressionSet that can be directly loaded, filtered and analysed through the use of a set of R packages, which include "Category" [15], "GOstats" [16], "affy" [17], "genefilter" [18] and "limma" [19], as well as two annotation data packages, "hgu133a.db" [20] and "hgu133b.db" [21], which are all available as part of the Bioconductor project [22]. The existence of two annotation data packages is due to the fact that the dataset is divided into two sets according to one of the two methods used for obtaining the data in the original study, Affymetrix U133A and Affymetrix U133B. The set associated with the first method contained 22283 features and 106 samples while the set associated with the second method contained 22645 features and 102 samples. The features described in both sets consisted of manufacturer identifiers, different between the sets, and, in order to be able to merge both sets into a single one for further analysis, the common features had to be identified. To this end, we cross-linked the manufacturer identifiers in each of the sets with their respective Ensemble gene accession numbers and condensed repeated features within each of the sets by taking the average of their expression values for each sample. Afterwards, we were able to identify the common features between the sets, allowing for their merge, resulting in a single set with 4188 features and 208 samples.

The expression values for all the features within this dataset were then normalized for each sample through the use of Python. Following this normalization step, an additional filtering step was performed by taking a factor multiplied by the median of the standard deviation of the expression values for each feature and rounding the expression values to zero for every feature whose expression values' standard deviation was below said median. The distribution of the expression values is shown on Fig. 2(a). Adding to this, for a threshold of 3, any normalized expression values who were equal or higher than the threshold were rounded to one while all values below said threshold were rounded to zero, as a way to discretize the expression data. The reason behind this step is due to the fact

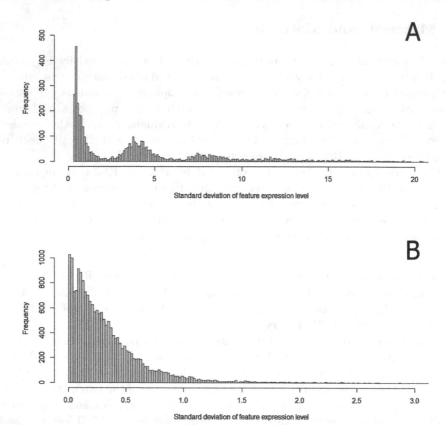

Fig. 2. Histograms illustrating the frequency of the features' expression level standard deviation between all the samples for (A) the ESCC dataset and (B) the BC dataset. Features that are more differentially expressed between the samples will have a higher base value.

that the mathematical model which is going to be presented below only takes into account if a feature is present, or not, in a sample.

The same process was mostly applied to the BC dataset. The difference in this study is that only one method was used for obtaining all of the data, Affymetrix U133 Plus 2.0, which resulted in only one annotation data package being used in R, "hgu133plus2.db" [23]. This simplified the filtering process, as no cross-linking between the samples was needed for identifying the common features. In addition, the dataset was available in the form of a SOFT file, which had to be loaded into the R environment through the use of the package "GEOquery" [24], which is also available through the Bioconductor project [22]. The dataset contained, initially, 54675 features and, after the filtering process, 18732 features. The total number of samples contained in the dataset are 178. Figure 2(b) illustrates the distribution of the standard deviation for the expression values for this dataset.

Mathematical Model for Prediction of Disease Status. The purpose of the objective function for the mathematical model (1) is to discover the combination of biomarkers for which the sum of their weights is the lowest. This means that the selected combination of biomarkers will always try to evaluate all samples as healthy. Adding to this, (1) also favours the minimization of the number of selected features.

$$
\min \left(\sum_{j}^{H} \left(0.1 \cdot \sum_{i} y_i - \sum_{i} w_i \cdot A_{ij} \right) + \sum_{j}^{D} \left(0.1 \cdot \sum_{i} y_i + \sum_{i} w_i \cdot A_{ij} \right) \right) \quad (1)
$$

The model has some constraints to which its variables are subjected to. Before we move on to these, a few parameters and variables need to be defined first. For instance, indexes i and j represent a feature and a sample, respectively. The set H contains all of the healthy samples while the set D contains all of the diseased samples. The variables y_i and z_j are binary variables, with first being 1 if a feature is selected as a biomarker or 0 when not selected, while the second is 1 if a sample is classified as diseased while 0 if otherwise. Another variable that must be taken into consideration is w_i which is continuous and represents the weight associated with a feature i. The only other variables that constitute the model are the slack variables, the score error μ_j^+ and the score margin μ_j^- for each sample j. As a parameter in the model, the matrix A represents the data whose expression values were previously discretized, with A_{ij} being equal to 1 if feature i is present in sample j or equal to 0 if it is not present. Furthermore, N represents the total maximum of potential biomarkers considered by the model, w_i^L and w_i^U represent the lower and upper bounds of the weight variables, specific to the protein i, and U is a constant of value equal to 50.

The first set of constraints set determine the relationship between the score margin and error with the selected potential biomarkers. These are Eqs. (2) and (3) for healthy and diseased samples, respectively.

$$
\sum_{i} w_i \cdot A_{ij} - 1 + \mu_j^+ - \mu_j^- = 0, \quad \forall j \in H \quad (2)
$$

$$
\sum_{i} w_i \cdot A_{ij} + 1 - \mu_j^+ + \mu_j^- = 0, \quad \forall j \in D \quad (3)
$$

Another set of constraints to consider are (5), which help turn the linear problem into a finite one by enforcing upper and lower limits on the score error and score margin, respectively.

$$
\mu_j^+ \geq 0, \qquad \mu_j^- \leq 10 \quad (4)
$$

Before moving on to the remaining constraints, two more sets have to be defined, I_H and I_D. These sets contain features that have been selected as representative of health or disease condition, respectively. This was possible by evaluating, for each of the features, their absolute and relative frequencies in healthy (H_{T_i} and H_{f_i}) and diseased (D_{T_i} and D_{f_i}) samples. All of these have a set cut-off value defined through the data used and the frequencies obtained for each feature, as defined below (5).

$$H_{T_c} = mean\,(H_T) + std\,(H_T), \quad H_{f_c} = mean\,(H_f) + std\,(H_f)$$
$$D_{T_c} = mean\,(D_T) + std\,(D_T), \quad D_{f_c} = mean\,(D_f) + std\,(D_f) \tag{5}$$

The following constraint (6) enforces that, at least, one feature from each set is considered as a potential biomarker.

$$\sum_i^{I_H} y_i \geq 1, \quad \sum_i^{I_D} y_i \geq 1 \tag{6}$$

There is also constraint (7), which turns the problem at hand into a linear problem, by linking y_i to the weight of the features and limiting the value of the latter by a set of bounds. When $y_i = 0$, it forces the weight of feature i to also be 0.

$$y_i \cdot w_i^L \leq w_i \leq y_i \cdot w_i^U, \quad \forall i \tag{7}$$

Another important set of constraints, (8) and (9) bind the sum of the weight parameters to the total number of potential biomarkers, allowing the significance of each feature to translate onto its respective weight as, otherwise, all of the weights would just tend towards the same upper bounds.

$$\sum_i^{I_H} w_i \leq \sum_i^{I_H} y_i \tag{8}$$

$$\sum_i^{I_D} w_i \geq -\sum_i^{I_D} y_i \tag{9}$$

Finally, to ensure that the model predicts all the samples on the training set as either healthy or diseased, constraints (10) and (11) are added to the model.

$$\sum_i w_i \cdot A_{ij} \geq 1 - z_j \cdot U, \quad \forall j \tag{10}$$

$$\sum_i w_i \cdot A_{ij} \leq (1 - z_j) \cdot U - 1, \quad \forall j \tag{11}$$

These constraints work by forcing the sum of the weights of the features present in a sample to either be, at least, -1, when $z_j = 1$, or 1, when $z_j = 0$. Adding to this, they also impose upper and lower bounds, 49, when $z_j = 0$, and -49, when $z_j = 1$.

All of the equations described above, (1)–(11), constitute a complete mixed-integer linear optimization mathematical model which can be solved to global optimality through the use of CPLEX (ILOG 2013) so as to be able to determine the values of the variables y_i and w_i, which are needed for the scoring function presented below.

Scoring Function. A function (12) is also proposed for identifying the status of a sample, be it healthy or diseased, by calculating a score for said sample.

$$S_j = \lambda \cdot \left(\sum_i w_i \cdot A_{ij} \cdot y_i - \theta \right) \tag{12}$$

In this, two new variables are introduced into the function, the offset θ and the multiplier λ. The offset (13) allows us to deviate the mean of the sample scores, potentially centering them on 0, while the multiplier (14) allows us to increase the certainty of a differentiable diagnosis. It is important to note that S represents the scores of the samples in the training set T.

$$\theta = mean\,(S)\,, \quad S \in T \tag{13}$$

$$\lambda = 1000 \cdot std\,(S)\,, \quad S \in T \tag{14}$$

This function's output falls on two main categories, with values of $S_j > 0$ allowing for the characterization of a sample as diseased while values of $S_j < 0$ indicate that a sample is healthy. Considering this, when $S_j = 0$, the condition of the sample is ambiguous, which means that it can not be classified either as healthy or diseased.

3 Results and Discussion

Biomarker prediction for effective disease diagnostic is a topic which has been expanding in recent years, with many different methods and technologies already available which are able to provide groups of biomarkers that can, usually, accurately distinguish between health and disease cases.

The results obtained through the presented model prove to be quite promising, reaching 93% accuracy on the BC dataset. However, the model is quite sensitive to the data used for training, which Fig. 2 clearly illustrates. This is more noticeable with the ESCC dataset, which may be due to the fact that the data was obtained from two different, even if similar, array designs.

Moreover, Table 1 clearly indicates that a maximum combination of 30 potential biomarkers shows a higher accuracy for the test set than with 20, which may be due to over- or·under-fitting, or even 40 biomarkers, in which noise in the expression values may be the culprit. However, for training sets of a smaller size, a maximum combination of 40 potential biomarkers yields the best results, even if only by a very small margin when compared to the results obtained with a maximum of 30 biomarkers. However, Table 2 shows that 20 potential biomarkers represents the optimal maximum combination, for the BC dataset, for any training test size, even if the differences between accuracies are quite small, ranging from less than 1% to, at most, 2%.

When compared to other methods present in the literature, the one showed here demonstrates itself as a viable alternative for identifying potential biomarkers. Taking the example of the one developed by Baliban *et al.*, from which our

Table 1. Sampling accuracy for train and test sets, obtained from the ESCC dataset, and how it varies with changes to the size of the train and test sets and to the maximum number of potential biomarkers (N).

# Samples in training set	# Samples in test set	# Potential biomarkers (N)	Average accuracy (Train set) (%)	Average accuracy (Test set) (%)	Average correct predictions
52	156	20	76.12	68.37	106.66
		30	77.69	69.22	107.99
		40	79.94	71.94	112.23
104	104	20	80.12	75.37	75.37
		30	83.74	80.33	80.33
		40	84.14	81.18	81.18
156	52	20	84.23	81.90	42.59
		30	86.69	85.17	44.29
		40	86.49	83.50	43.42

Table 2. Sampling accuracy for train and test sets, obtained from the BC dataset, and how it varies with changes to the size of the train and test sets and to the maximum number of potential biomarkers (N).

# Samples in training set	# Samples in test set	# Potential biomarkers (N)	Average accuracy (Train set) (%)	Average accuracy (Test set) (%)	Average correct predictions
45	133	20	94.22	86.32	114.81
		30	95.38	85.96	114.33
		40	94.96	85.43	113.62
89	89	20	94.25	90.54	80.58
		30	94.21	90.46	80.51
		40	94.67	90.25	80.32
134	44	20	95.50	93.16	40.99
		30	95.15	91.55	40.28
		40	95.13	92.61	40.75

methodology is inspired, while the predictive accuracy results presented here are lower than the 99% reported in [8], the type of data used in the latter consists of mass spectrometry data while the one used here consists of gene expression data, which are known to be far more noisier by nature. Adding to this, the dataset used in [8] was also unavailable for comparative analysis.

Another method was developed by Sun *et al.* [25], which identified at the time some potential pairs and even triplets of biomarkers for acute lymphoblastic/ myeloid leukemia and colon cancer. While presenting similar predictive accuracy results, the datasets used in [25] considered fewer features and samples in their data, which may have resulted in a larger degree of overfitting. Unfortunately, their datasets were also unavailable to be able to determine how the model developed in this work would perform. Adding to this, the Sun's model is less versatile than ours, considering only a fixed of features in each run while the one showed here is able to consider a varying number of features up to a defined maximum.

Finally, there is also the model developed by Zou *et al.* [26]. Their approach to multi-biomarker panel identification is interesting but suffers from the same issue as Sun's: the number of biomarkers to be identified needs to be fed into the model, while the one presented here takes only into consideration a maximum number, potentially identifying the optimal number and combination of biomarkers. Adding to this, their predictive accuracy results are lower than the ones here demonstrated, reaching only up to 89% accuracy.

Acknowledgments. This work is co-funded by the North Portugal Regional Operational Programme, under the "Portugal 2020", through the European Regional Development Fund (ERDF), within project SISBI- Ref^a NORTE-01-0247-FEDER-003381. This study was also supported by the Portuguese Foundation for Science and Technology (FCT) under the scope of the strategic funding of UID/BIO/04469/2013 unit and COMPETE 2020 (POCI-01-0145-FEDER-006684) and BioTecNorte operation (NORTE-01-0145-FEDER-000004) funded by European Regional Development Fund under the scope of Norte2020 - Programa Operacional Regional do Norte. Joel P. Arrais is funded by CISUC - Center for Informatics and Systems of the University of Coimbra.

References

1. Colburn, W.A., DeGruttola, V.G., DeMets, D.L., Downing, G.J., Hoth, D.F., Oates, J.A., Peck, C.C., Schooley, R.T., Spilker, B.A., Woodcock, J., Zeger, S.L.: Biomarkers definitions working group: biomarkers and surrogate endpoints: preferred definitions and conceptual framework. Clin. Pharmacol. Ther. **69**(3), 89–95 (2001)
2. LaBaer, J.: So, you want to look for biomarkers. J. Proteome Res. **4**(4), 1053–1059 (2005)
3. Manolio, T.: Novel risk markers and clinical practice. New Engl. J. Med. **349**(17), 1587–1589 (2003)
4. Lee, Y.H., Wong, D.T.: Saliva: an emerging biofluid for early detection of diseases. Am. J. Dent. **22**(4), 241–248 (2009)
5. Schrohl, A.S., Würtz, S., Kohn, E., Banks, R.E., Nielsen, H.J., Sweep, F.C.G.J., Brünner, N.: Banking of biological fluids for studies of disease-associated protein biomarkers. Mol. Cell. Proteomics MCP **7**(10), 2061–2066 (2008)
6. Sidransky, D.: Nucleic acid-based methods for the detection of cancer. Science **278**(5340), 1054–1059 (1997). New York
7. Wang, Q., Gao, P., Wang, X., Duan, Y.: Investigation and identification of potential biomarkers in human saliva for the early diagnosis of oral squamous cell carcinoma. Clin. Chim. Acta Int. J. Clin. Chem. **427**, 79–85 (2014)

8. Baliban, R.C., Sakellari, D., Li, Z., Guzman, Y.A., Garcia, B.A., Floudas, C.A.: Discovery of biomarker combinations that predict periodontal health or disease with high accuracy from GCF samples based on high-throughput proteomic analysis and mixed-integer linear optimization. J. Clin. Periodontol. **40**(2), 131–139 (2013)

9. Puthiyedth, N., Riveros, C., Berretta, R., Moscato, P.: A new combinatorial optimization approach for integrated feature selection using different datasets: a prostate cancer transcriptomic study. PloS one **10**(6), e0127702 (2015)

10. Li, W.Q., Hu, N., Burton, V.H., Yang, H.H., Su, H., Conway, C.M., Wang, L., Wang, C., Ding, T., Xu, Y., Giffen, C., Abnet, C.C., Goldstein, A.M., Hewitt, S.M., Taylor, P.R.: PLCE1 mRNA and protein expression and survival of patients with esophageal squamous cell carcinoma and gastric adenocarcinoma. Cancer Epidemiol. Biomarkers Prev. Publ. Am. Assoc. Cancer Res. Cosponsored Am. Soc. Prev. Oncol. **23**(8), 1579–1588 (2014)

11. Su, H., Hu, N., Yang, H.H., Wang, C., Takikita, M., Wang, Q.H., Giffen, C., Clifford, R., Hewitt, S.M., Shou, J.Z., Goldstein, A.M., Lee, M.P., Taylor, P.R.: Global gene expression profiling and validation in esophageal squamous cell carcinoma and its association with clinical phenotypes. Clin. Cancer Res. Official J. Am. Assoc. Cancer Res. **17**(9), 2955–2966 (2011)

12. Maire, V., Némati, F., Richardson, M., Vincent-Salomon, A., Tesson, B., Rigaill, G., Gravier, E., Marty-Prouvost, B., De Koning, L., Lang, G., Gentien, D., Dumont, A., Barillot, E., Marangoni, E., Decaudin, D., Roman-Roman, S., Pierré, A., Cruzalegui, F., Depil, S., Tucker, G.C., Dubois, T.: Polo-like kinase 1: a potential therapeutic option in combination with conventional chemotherapy for the management of patients with triple-negative breast cancer. Cancer Res. **73**(2), 813–823 (2013)

13. Maire, V., Baldeyron, C., Richardson, M., Tesson, B., Vincent-Salomon, A., Gravier, E., Marty-Prouvost, B., De Koning, L., Rigaill, G., Dumont, A., Gentien, D., Barillot, E., Roman-Roman, S., Depil, S., Cruzalegui, F., Pierré, A., Tucker, G.C., Dubois, T.: TTK/hMPS1 is an attractive therapeutic target for triple-negative breast cancer. PloS one **8**(5), e63712 (2013)

14. Maubant, S., Tesson, B., Maire, V., Ye, M., Rigaill, G., Gentien, D., Cruzalegui, F., Tucker, G.C., Roman-Roman, S., Dubois, T.: Transcriptome analysis of Wnt3a-treated triple-negative breast cancer cells. PloS one **10**(4), e0122333 (2015)

15. Falcon, R.G., Sarkar, D.: Category: Category Analysis. R package version 2.34.2

16. Falcon, S., Gentleman, R.: Using GOstats to test gene lists for GO term association. Bioinformatics **23**(2), 257–258 (2007). Oxford, England

17. Gautier, L., Cope, L., Bolstad, B.M., Irizarry, R.A.: affy-analysis of Affymetrix Genechip data at the probe level. Bioinformatics **20**(3), 307–315 (2004). Oxford, England

18. Gentleman, R., Carey, V., Huber, W., Hahne, F.: Genefilter: methods for filtering genes from microarray experiments. R package version 1.50.0

19. Ritchie, M.E., Phipson, B., Wu, D., Hu, Y., Law, C.W., Shi, W., Smyth, G.K.: limma powers differential expression analyses for RNA-sequencing and microarray studies. Nucleic Acids Res. **43**(7), 47 (2015)

20. Carlson, M.: hgu133a.db: Affymetrix Human Genome U133 Set annotation data (chip hgu133a). R package version 3.1.3

21. Carlson, M.: hgu133plus2.db: Affymetrix Human Genome U133 Plus 2.0 Array annotation data (chip hgu133plus2). R package version 3.1.3

22. Gentleman, R.C., Carey, V.J., Bates, D.M., Bolstad, B., Dettling, M., Dudoit, S., Ellis, B., Gautier, L., Ge, Y., Gentry, J., Hornik, K., Hothorn, T., Huber, W., Iacus, S., Irizarry, R., Leisch, F., Li, C., Maechler, M., Rossini, A.J., Sawitzki, G., Smith, C., Smyth, G., Tierney, L., Yang, J.Y.H., Zhang, J.: Bioconductor: open software development for computational biology and bioinformatics. Genome Biol. 5(10), R80 (2004)
23. Carlson, M.: hgu133b.db: Affymetrix Human Genome U133 Set annotation data (chip hgu133b). R package version 3.1.3
24. Davis, S., Meltzer, P.S.: GEOquery: a bridge between the Gene Expression Omnibus (GEO) and BioConductor. Bioinformatics 23(14), 1846–1847 (2007). Oxford, England
25. Sun, M., Xiong, M.: A mathematical programming approach for gene selection and tissue classification. Bioinformatics 19(10), 1243–1251 (2003). Oxford, England
26. Zou, M., Zhang, P.J., Wen, X.Y., Chen, L., Tian, Y.P., Wang, Y.: A novel mixed integer programming for multi-biomarker panel identification by distinguishing malignant from benign colorectal tumors. Methods 83, 3–17 (2015). San Diego, California

Assembly of Gene Expression Networks Based on a Breast Cancer Signature

Dimitrios Apostolos Chalepakis Ntellis[1], Ekaterini S. Bei[1],
Dimitrios Kafetzopoulos[2], and Michalis Zervakis[1(✉)]

[1] School of Electrical and Computer Engineering,
Technical University of Crete, Chania, Hellas
dimchalepakis@gmail.com, abei@isc.tuc.gr,
michalis@display.tuc.gr
[2] Institute of Molecular Biology and Biotechnology, FORTH, Heraklion, Hellas
kafetzo@imbb.forth.gr

Abstract. The aim of this paper is to provide a snapshot of functional networks with central nodes (hubs) relying on the behavior of genes linked to a specific interaction network. Utilizing a 'breast cancer signature', a Bayesian approach is applied for the construction of gene interaction networks from different populations. We demonstrate that the hub genes of the differentiating network between cancer and control states can be regarded as potential markers for breast cancer. Furthermore, the differentiating subnetworks can be informative of the phenotype and provide new knowledge about the functional interactions and molecular pathways involved in breast cancer.

Keywords: Gene expression networks · Bayesian networks · Protein-Protein interactions · Breast cancer markers

1 Introduction

Advanced high-throughput technologies, such as microarrays, that emerged in last decades enabled researchers to go beyond reductionism towards biological complexity. In this context, biological networks proved to be valuable tools for uncovering structural and dynamic complex interactions formed among various constituents of the cell (e.g. DNA, RNA, proteins and small molecules) and determine the biological properties and behavior of a cell over time. Studying a gene interaction network composed of a set of genes (nodes) connected by edges (interactions) can reveal the functional interplay among these genes and its importance for the execution of most biological processes, providing insight into such biological functions in healthy control and disease conditions [1].

In the present study we take advantage of the philosophy of Bayesian networks (BNs) in order to address the biological complexity and interpret the biological information encoded within. They can be efficiently used for modeling complex processes with any elements, such as in medicine and medical research. BNs link statistics with inference in artificial intelligence, with important characteristics as follows. (i) They are graphical models useful for reasoning under uncertainty; (ii) their nodes

I. Rojas and F. Ortuño (Eds.): IWBBIO 2017, Part II, LNBI 10209, pp. 62–73, 2017.
DOI: 10.1007/978-3-319-56154-7_7

represent variables (discrete or continuous) and their arcs represent direct connections (often causal connections) among them, and (iii) they model the strength of connections among variables in a compact and quantitative structure, allowing the inference of probabilistic beliefs about them, which can be updated automatically as new information become available [2]. Bayesian networks have advantages over rule-based systems in that they provide appropriate models to be trained on available datasets while they can still absorb expert knowledge through the tuning of connection weights, which is particularly useful in presence of uncertainty. Bayesian networks overcome inefficiencies of trainable intelligent systems, since they provide reasoning based on the support probabilities while being supported by learning mechanisms for decision support [3].

2 Methodology

2.1 Dataset

We focused on a gene expression microarray dataset of 4,174 genes from 529 tissue samples (425 cancerous breast tissue and 104 non-cancerous, control breast tissue) [4]. Specifically, the original dataset is formed by the integration of five publicly available GEO[1] datasets (GSE22820, GSE19783, GSE31364, GSE9574, GSE18672) and it has been pre-processed with SAM (Significance Analysis of Microarrays) [5]. In addition, we used a gene signature with 77 genes [6] obtained after classification experiments on this dataset, exploring the prediction ability of gene signatures. Subsequently, we augmented the signature with 5 genes (from the initial dataset but not in the original signature) with an important role in breast cancer. These five 'reference' genes were selected for monitoring the algorithmic relevance (*NRG2*) and the positive (*BRCA1*, *ERBB2*) or negative (*NRG2*, *SSBP2*, *PARK7*) association with breast cancer. Overall, we examine a list of 82 (77 + 5) genes and their expression values as the test signature.

2.2 Creating Bayesian Network Structures

We used Bayesian networks because they provide a good and compact representation of gene interactions, expressing joint probability distributions of genes and allowing for inference queries. In particular, they have been found useful for inferring cellular networks, modeling protein signaling pathways, exemplifying system biology, data integration, classification and genetic data analysis [7].

As already mentioned, we used a '77 gene signature' with potential in clinical application [6] along with five 'reference' genes for the construction of Bayesian networks (Table 1). The addition of 'reference' genes supports the verification of algorithmic correctness by tracking if these genes appear as hubs in the constructed networks. In other words, they can be used to assess if the proposed method can detect genes known significant for Breast cancer.

---

[1] GEO: Gene Expression Omnibus; available at https://www.ncbi.nlm.nih.gov/geo/.

Table 1. List of the 82 genes used for the construction of the Bayesian networks.

'77 gene signature' & 5 'reference' genes*
CDKN2A B3GNT3 TUBB3 CLGN HOXB13 OLFM4 SLC34A2 SCGB1D1 PPARGC1A SPINK5 KLK11 CHIT1 WIF1 **PARK7*** COL11A1 COL17A1 COMP CPA3 CSTA SLC26A3 EDN2 EDN3 EGF **ERBB2*** FABP7 FGFR3 FOXJ1 FN1 PRAME **SSBP2*** GREM1 LAMP3 GSTA1 GSTA3 TFCP2L1 NRG1 NR4A1 HSD17B2 AMIGO2 IGHG1 KCNJ16 KRT4 KRT16 LYZ ASCL2 MMP12 CITED1 MSMB NTRK3 OGN ATP6V0A4 VGLL1 PENK PIP ASPN TMEM100 ZNF334 TTYH1 SYT13 ACTA1 S100A7 S100A9 S100P CCL18 CCL19 SIX1 SLC1A1 SOX11 **BRCA1*** TAT TFF1 TNNI2 RERGL TFPI2 C2orf54 CILP DLK1 FGF18 CLDN8 REPS2 CD19 **NRG2***

Given the importance of the molecular interactions in the generation of gene expression networks, we also searched for known interactions of the 82 test genes in the publicly available interaction database BioGRID (Biological General Repository for Interaction Datasets) [8], in order to be used as a priori knowledge for the construction of Bayesian networks. Although BioGRID (version 3.2.108) provided a total of 2,197 interactions for the 82 selected genes only 12 interactions were verified [8]. For the visualization of the two Bayesian networks (control and cancer) and the creation of functional clusters on their structure, we used the clustering algorithms of MCODE, which are available for network analysis through the Cytoscape platform [9]. We constructed Bayesian Networks considering all parameters (expression values) as either discrete or continuous Gaussian variables.

2.3   Discrete Variables

Our data, i.e. the gene expression values, consists of real values and follows a normal distribution. With our discretization procedure, each expression value can take the values 1, 2 or 3. Accordingly, one gene from a particular sample can be defined as underexpressed, normal or overexpressed. To proceed with discretization of the variables, we defined two thresholds based on a statistical t-test for all 4,174 genes. For the effective definition of these thresholds that aim to clearly highlight distinction in expression, we selected the genes with highest differential expression. For this purpose, we calculated the q-values of all genes and selected genes with significant small q-value. Thus, we chose the 900 genes with the lowest q-value (higher differential expression) and computed the means of gene expression values for the cancer and control samples. Overall, two mean values were attributed to each gene, one for controls and one for the cancer group. By ranking the means of all genes, we created a population for low and high means and used them to draw a Gaussian fit for two classes in the same graph (Figs. 1 and 2).

We observed an adequate differentiation between the two classes for the determination of two thresholds. In particular, we determined the 10% of the class of "low means" as the low threshold and the 90% of the "high means" as the high threshold. With this definition of thresholds, if the expression value of a gene is lower that the lower one then it is defined as underexpressed. If its value is between the low and the

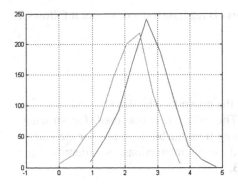

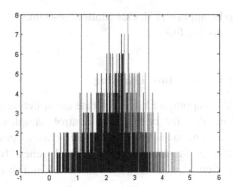

Fig. 1. Gaussian fit: maximum and minimum mean distributions.

Fig. 2. Experimental maximum and minimum mean distributions. The selected thresholds are shown by red lines. (Color figure online)

high threshold then the gene is defined as normal and if its expression value is larger than the high threshold then the gene is defined as overexpressed (Table 2).

According to the above thresholds, we performed discretization of all genes in three layers in order to create the Bayesian networks with discrete variables. These two thresholds were further evaluated using 30 genes known to be increased (*GABRP, FABP4, OLFM4, KRT13, MMP7, MUC1, GLUL, S100A7, S100A9, SOD2, ING4, CDH11, CXCR4, VEGFA, ITM2B*) or decreased (*CAV1, TIMP4, SLC20A1, CD44, HOXB6, CTGF, ECHDC2, SAE1, LCP1, FUCA1, TNFRSF10B, AGR2, ANGPT2, APOL1, EGFR*) in breast cancer, in accordance with the Expression Atlas database [10] and recent studies [11].

Table 2. Thresholds for discretization and expression values after discretization.

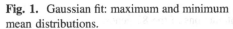

Expression level	Thresholds	Expression levels after discretization
Underexpressed	TE < 1,131	1
Normal	1,131 ≤ TE < 3,48	2
Overexpressed	TE ≥ 3,48	3

2.4 Continuous-Gaussian Variables

Cancer gene expression profiles are not normally-distributed, either on the complete-experiment or on the individual-gene level. Instead, they exhibit complex, heavy-tailed distributions characterized by statistically-significant skewness and kurtosis. The non-Gaussian distribution of this data affects identification of differentially-expressed genes, functional annotation, and prospective molecular classification [12]. For this reason we apply a log base-2 transformation to the expression values. This is beneficial for a number of reasons, including that it stabilizes the variance, compresses the range of data and makes the data more normally distributed, which allows certain statistics to be applied [13]. Our continuous-valued dataset is log transformed, so that it follows a Gaussian fit.

2.5   Structure Learning

At this point we have to create the structure of the networks with discrete and Gaussian variables for cancer and control samples. The initial network used for structure learning, consisted from 12 interactions stemming from the bibliography for the known interactions among our 82 initial genes. In these 12 interactions, we identified the participation of 15 genes presented in Table 3.

Table 3. The twelve known interactions of the 82 genes.

Gene 1	Gene 2
CDKN2A	BRCA1
CDKN2A	TAT
CDKN2A	FN1
CDKN2A	ACTA1
PPARGC1A	PARK7
EGF	ERBB2
FN1	KRT16
FN1	COMP
FN1	IGHG1
NRG1	ERBB2
FGF18	FGFR3
KRT16	IGHG1

For structure learning we used the K2 algorithm combined with a method using the Maximum Weight Spanning Tree (MWST) algorithm, which is beneficial in reducing the complexity without loss of information [14]. To apply the K2 algorithm we have to determine a sorted order of the nodes and the maximum number of parents of each node [15, 16].

The maximum number of parents for each node was set equal to the total number of our genes, since there is not such limitation in nature. The definition of the sorted order is more complex, but forms a crucial issue for the creation of Bayesian networks. We used two different methods to determine this sorted order. The first one results from the application MWST algorithm [14] to the 82 genes. This method applies a weight to each edge, based on the variance of the score when a node becomes the parent of another node. Subsequently, the MWST algorithm produces an undirected tree from the weight array. This undirected tree can be transformed to a directed one, by giving a specific initial root node. Thus, by applying a topological sorting to this tree creates a sorted order of nodes, which we call the MWST order.

The second method we used forms a combination of the previous order and a random sorted order. The 15 (genes that participate in our 12 initially known interactions) were sorted with the MWST method and the rest 67 were randomly sorted. The sorting of the rest 67 genes, even though through a random sorting method, happened to be the best of all the structures we learned and produced the maximum number of edges. We refer to this method as CUSTOM ordering.

2.6   Visualization of the Resulting Networks

After we finished with the structure learning, we had to visualize the structure in order to study network properties and draw our conclusions. From structure learning, we obtained 4 network structures for each type of samples (cancer, healthy). More specifically, for each group of samples we constructed the structures with discrete variables and MWST sorting, discrete variables and CUSTOM sorting and the corresponding ones for Gaussian variables. In total, we constructed 8 network structures. In order to analyze the properties of groups, we merged specific structure. In particular, we created the union of the networks with discrete variables in cancer samples (CD), discrete variables in healthy samples (HD), Gaussian variables in cancer samples (CG) and Gaussian variables in healthy samples (HG). Finally, we created two larger structures with all networks from cancer samples (CU) and all networks from healthy samples (HU). The number of nodes and edges for these summarized networks are presented in Table 4.

Table 4. Established compounds of networks and their key elements (nodes and edges).

Network	Nodes	Edges
CU	82	843
HU	82	605
CD	74	142
HD	77	102
CG	82	765
HG	82	562

3   Results

3.1   Statistical Findings

Small World Phenomenon and Scale Free Networks. The network structures were analyzed for certain characteristics that appear often in biological networks. The first one is the Small World phenomenon [17]. We compared the clustering coefficients of our networks with these of equal-size random graphs (equal number of nodes and edges). In all cases the clustering coefficient of our networks was higher than those of the random graphs, which indicates that the designed networks may reflect the small-world property. Nevertheless, the mean path length was considerably shorter than the logarithm of the number of nodes, so that we cannot verify the small world property with confidence.

The second property was the structure associated with the free-scale networks. We observed that the degree distributions of all our networks followed a power law, which justifies the attribute of free scale for our networks.

Network Comparison. We continued with the comparison of the cancer unity (CU) and the healthy-control unity (HU) networks. At first we found the common edge interactions of the two networks equal to 171. Then we compared the two networks with the method proposed by Gomez and Diaz [18]. This method counts the difference between two networks for a specific distance level. The largest distance between nodes in our networks was found for 4 edges. Thus, we compared the validity of one network with respect to the other as the proportion of correctly inferred edges at each level in the first based on the second network.

In Table 5, we can observe that even for the fourth level the validity of the one networks based on the other is quite low (around 0, 5) indicating that the two networks share little similarity.

Table 5. Comparisons among cancer and healthy-control networks. V is the network validity index [17].

Validity/Network	CU	HU
V1	0,0992	0,1339
V2	0,3215	0,4124
V3	0,4491	0,5088
V4	0,4591	0,5138

Subsequently we created the so called differentiating network composed of the different (non common) edges of the two networks. We applied the MCODE algorithm to this difference-network to obtain clusters and then we analyzed the centralities of these clusters and their pathways (after biological analysis) [19]. Observing the diagrams of means for metrics of degree and betweenness centrality (Figs. 3 and 4, respectively) on the derived clusters of significant pathways, we observe that they range at the same levels, implying that indeed these pathways are significant.

Hub Nodes. We now focus on the derivation of potential Hub nodes, which are the highest degree nodes of a network and usually bear biological significance. The evaluation of each node is based on degree as a local node metric and betweenness centrality as a global metric. Initially we compute the histograms and cumulative distributions the metrics of degree and betweenness centrality for the network. In the cumulative distribution we identify the point where the curve starts flattening. We exploit these values as the minimum attributes of hub nodes in terms of degree and betweenness centrality, respectively. The nodes with the highest degrees and betweennesses are the most significant in our network. Appling this method to both of the CU and HU networks we derive the following significant hub nodes. The common ones are indicated by * (Table 6).

Degree

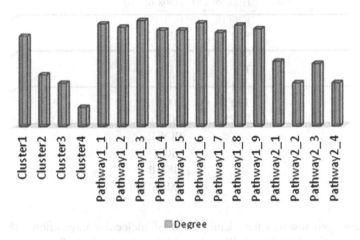

Fig. 3. Degree charts on the differentiating network.

Betweenness Centrality

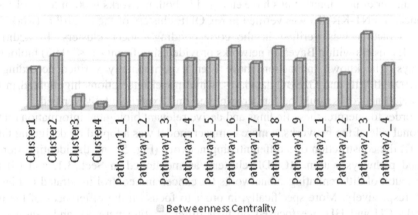

Fig. 4. Betweenness centrality charts on the differentiating network.

3.2 Biological Results

In the present work, the biological evaluation was based on the comparative results between two Bayesian networks, i.e. a 'tumor' network (with expression data from patients with breast cancer) and a 'control' network (with expression data of individuals not suffering from breast cancer) and the functional structures derived from them. Specifically, the publicly available databases BioGRID, HIPPIE (Human Integrated Protein-Protein Interaction rEference) and G2SBC (The Genes to Systems Breast

Table 6. Emergence of final Hubs on "Cancer" and "Healthy" networks.

Hubs of CU	Hubs of HU
KRT16	FN1 (*)
FN1 (*)	ERBB2
OGN (*)	COMP
TFF1	ATP6V0A4
OLFM4	WIF1
TTYH1 (*)	MMP12
REPS2	OGN (*)
	TTYH1 (*)
	NRG2
	MSMB
	TMEM100

Cancer) were used towards the identification of molecular interactions, [8, 20, 21], whereas the database WebGestalt (Web-based Gene Set Analysis Toolkit) was used for the enrichment analysis of biological pathways (KEGG[2] and Wikipathways[3]) [22].

With respect to Bayesian networks, a number of known interactions are found in control (FN1-CDKN2A, FN1-KRT16, FN1-COMP, ERBB2-NRG1, FGFR3-FGF18) and in cancer (ACTA1-CDKN2A, FN1-COMP, KRT16-IGHG1) network, while a large number of new interactions have emerged in both networks (600 in 'control'; 840 in 'cancer'). FN1-KRT16 was verified in MCODE clusters of the control network, but none interaction was verified in the corresponding cancer clusters. In addition, MCODE clusters within Bayesian networks provided significant ($p \leq 0.05$) biological pathways. The known interactions have been experimentally verified according to BioGRID, HIPPIE and G2SBC databases, while novel interactions highlighted in the constructed networks could be further explored and experimentally verified.

In order to interpret these findings and derive relevant biological information on the functional role of the networks, a more compact model was adopted by designing three steps: (i) the construction of a differentiating network (Fig. 5) and the identification of enriched pathways within MCODE clusters, as mentioned in Sect. 3.1, and (ii) the construction of differentiating subnetworks for cancer and control illustrated in Figs. 6 and 7, respectively. More specifically, in order to focus on the differences of the two networks CU and HU, we focused on their differentiating network and examine its structure. The MCODE algorithm was applied to this network and the averages of degree and betweenness centrality were calculated for each derived MCODE cluster, but also for each significant biological pathway identified within these clusters. Figure 5 depicts the differentiating network after processing and organizing clusters with the MCODE algorithm.

In addition, two differentiating subnetworks, one for cancer population (Fig. 6) and one for healthy-control population (Fig. 7), were generated by considering (a) all nodes

---

[2] KEGG: Kyoto Encyclopedia of Genes and Genomes; available at http://www.genome.jp/kegg/.

[3] Wikipathways; available at http://wikipathways.org/index.php/WikiPathways.

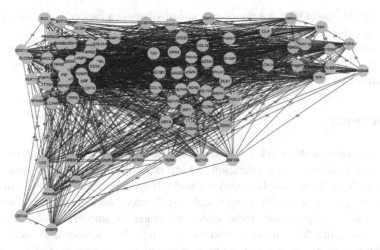

Fig. 5. The differentiating network divided into clusters by the MCODE algorithm.

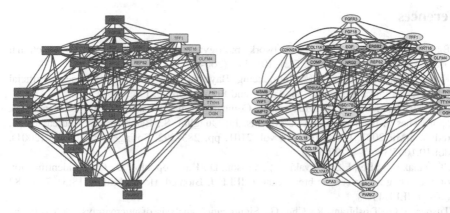

Fig. 6. Cancer differentiating sub-network with central 'cancer' nodes.

Fig. 7. Control differentiating sub-network with central 'healthy' nodes.

that were implicated in significant pathways of the differentiating network, (b) all 'cancer' and 'healthy' hubs, and (c) only the adjacent edges of 'cancer' hubs for the differentiating cancer subnetwork or those of 'healthy' hubs for the differentiating control sub-network, respectively.

According to this compact model, we observed that:

- The differentiating network involves all hub nodes (control and cancer central nodes), so that hub genes can be considered as potential gene markers for breast cancer.
- The differentiating network provides significant ($p \leq 0.05$) pathways that associate with breast cancer, such as ErbB signaling pathway, EGF-EGFR signaling pathway, Pathways in cancer.

- FN1, TTYH1 and OGN are the common hubs between cancer and control differentiating sub-networks (Table 6).
- There are fewer interactions in the cancer differentiating subnetwork compared to the number of interactions in control differentiating subnetwork, despite the fact that the number of genes remains constant.

4 Conclusion

The main outcome of this work is that the constructed networks and subnetworks can give insight into the molecular alterations taking place in different phenotypes (cancer and control). Specifically, the Bayesian networks from cancer and control samples are scale-free comprising nodes with biological significance and substructures formed by functional relationships among these genes governed by important molecular pathways. Moreover, the differentiating subnetworks might be considered as local models to test a biological hypothesis and are more convenient for experimental design.

References

1. Barabási, A.L., Oltvai, Z.N.: Network biology: understanding the cell's functional organization. Nat. Rev. Genet. **5**(2), 101–113 (2004)
2. Korb, K.B., Nicholson, A.E.: Introducing Bayesian networks. In: Bayesian Artificial Intelligence, 2nd edn., pp. 29–54. Taylor and Francis Group, LLC (2011)
3. Oniésko, A., Lucas, P., Druzdzel, M.J.: Comparison of rule-based and Bayesian network approaches in medical diagnostic systems. In: Quaglini, S., Barahona, P., Andreassen, S. (eds.) AIME 2001. LNCS (LNAI), vol. 2101, pp. 283–292. Springer, Heidelberg (2001). doi:10.1007/3-540-48229-6_40
4. Sfakianakis, S., Bei, E.S., Zervakis, M., Vassou, D., Kafetzopoulos, D.: On the identification of circulating tumor cells in breast cancer. IEEE J. Biomed. Health Inform. **18**(3), 773–782 (2014). IEEE Press
5. Tusher, V.G., Tibshirani, R., Chu, G.: Significance analysis of microarrays applied to the ionizing radiation response. Proc. Nat. Acad. Sci. **98**(9), 5116–5121 (2001)
6. Chlis, N.K., Sfakianakis, S., Bei, E.S., Zervakis, M.: A generic framework for the elicitation of stable and reliable gene expression signatures. In: Proceedings of IEEE BIBE, Chania, Greece, pp. 1–4. IEEE Press (2013)
7. Needham, C.J., Bradford, J.R., Bulpitt, A.J., Westhead, D.R.: A primer on learning in Bayesian networks for computational biology. PLoS Comput. Biol. **3**(8), e129 (2007)
8. The biological general repository for interaction datasets. http://thebiogrid.org/
9. Cytoscape: a software environment for integrated models of biomolecular interaction networks. http://www.cytoscape.org/
10. Expression Atlas update - a database of gene and transcript expression from microarray and sequencing-based functional genomics experiments. Nucleic Acids Res. (2014). http://www.ebi.ac.uk/gxa
11. Schummer, M., Green, A., Beatty, J.D., Karlan, B.Y., Karlan, S., Gross, J., Thornton, S., McIntosh, M., Urban, N.: Comparison of breast cancer to healthy control tissue discovers novel markers with potential for prognosis and early detection. PLoS ONE **5**(2), e9122 (2010)

12. Marko, N.F., Weil, R.J.: Non-Gaussian distributions affect identification of expression patterns, functional annotation, and prospective classification in human cancer genomes. PLoS ONE **7**(10), e46935 (2012)
13. Cowley, M., Ying, K.: LogTransform Documentation - a GenePattern module for applying a log transformation on GCT files (not published). Garvan Institute (2011)
14. Bouhamed, H., Masmoudi, A., Lecroq, T., Rebaï, A.: A new approach for Bayesian classifier learning structure via K2 Algorithm. In: Huang, D.-S., Gupta, P., Zhang, X., Premaratne, P. (eds.) ICIC 2012. CCIS, vol. 304, pp. 387–393. Springer, Heidelberg (2012). doi:10.1007/978-3-642-31837-5_56
15. Wei, Z., Xu, H., Li, W., Gui, X., Wu, X.: Improved Bayesian network structure learning with node ordering via K2 Algorithm. In: Huang, D.-S., Jo, K.-H., Wang, L. (eds.) ICIC 2014. LNCS (LNAI), vol. 8589, pp. 44–55. Springer, Heidelberg (2014). doi:10.1007/978-3-319-09339-0_5
16. Al-Akwaa, F.M., Alkhawlani, M.M.: Comparison of the Bayesian network structure learning algorithms. Int. J. Adv. Res. Comput. Sci. Softw. Eng. **2**(3), 404–408 (2012)
17. Watts, D.J., Strogatz, S.H.: Collective dynamics of "small-world" networks. Nature **393**, 440–442 (1998)
18. Gomez-Vela, F., Diaz, N.: Gene network biological validity based on gene-gene interaction relevance. Sci. World J. **2014**(2014), 1–11 (2014)
19. Zhuang, D.Y., Jiang, L., He, Q.Q., Zhou, P., Yue, T.: Identification of hub subnetwork based on topological features of genes in breast cancer. Int. J. Mol. Med. **35**(3), 664–674 (2015)
20. HIPPIE, Human Integrated Protein-Protein Interaction rEference. http://cbdm-01.zdv.uni-mainz.de/~mschaefer/hippie/
21. The Genes-to-Systems Breast Cancer (G2SBC) Database. http://www.itb.cnr.it/breastcancer/
22. Wang, J., Duncan, D., Shi, Z., Zhang, B.: WEB-based GEne SeT AnaLysis Toolkit (WebGestalt): update 2013. Nucleic Acids Res. **41**(Web Server issue), W77–W83 (2013)

Pairwise and Incremental Multi-stage Alignment of Metagenomes: A New Proposal

Esteban Pérez-Wohlfeil, Oscar Torreno, and Oswaldo Trelles[✉]

Department of Computer Architecture, University of Malaga,
Boulevard Louis Pasteur 35, Malaga, Spain
ortrelles@uma.es

Abstract. Traditional comparisons between metagenomes are often performed using reference databases as intermediary templates from which to obtain distance metrics. However, in order to fully exploit the potential of the information contained within metagenomes, it becomes of interest to remove any intermediate agent that is prone to introduce errors or biased results. In this work, we perform an analysis over the state of the art methods and deduce that it is necessary to employ fine-grained methods in order to assess similarity between metagenomes. In addition, we propose our developed method for accurate and fast matching of reads.

Keywords: Next generation sequencing · Metagenome comparison · Read to read comparison · Distance between metagenomes

1 Introduction

New DNA acquisition methods have lowered sequencing costs up to a point where genomic research is producing an exponential growth in the number of sequenced samples, specially in the field of metagenomics. A metagenome is defined as an uncultured sample directly recovered from its original environment. Traditional metagenomic analysis comprises the comparison of one or multiple metagenomic samples against a reference database in order to find out the genomic composition, and to perform further analyses such as functional annotation. However, comparisons performed with a reference database do not fully exploit the data contained in metagenomes (e.g. species that are unknown or highly evolved that do not exactly correspond to those contained) and can lead to errors. For instance, reads belonging to samples that are not present in the reference database will most likely be aligned to their closest representatives. Furthermore, the results of the metagenomic comparison will probably change if the reference database is changed. Thus, the scenario of a direct comparison between metagenomes is gaining interest in order to produce results that are not biased by reference databases.

© Springer International Publishing AG 2017
I. Rojas and F. Ortuño (Eds.): IWBBIO 2017, Part II, LNBI 10209, pp. 74–80, 2017.
DOI: 10.1007/978-3-319-56154-7_8

2   Background

Current methods developed in the field of metagenome-metagenome comparison are mostly based on analyzing the k-mer diversity of the samples (e.g. Compareads [1] or the more recent SIMKA [2]). Alignment-free methods have been reviewed several times (e.g. [3,4]) over the last decades, showing a persistent improvement in the characterization of distributions and handling of random matches. In this line, a variety of statistics have been proposed to compare genomic communities, such as the D2 statistic [5] or the Jaccard index [6]. In particular, most of the metagenomic comparison software use a variant –if not directly– of the Jaccard index (e.g. MASH [7] or SIMKA, along with other ecological distances). This index accounts for the number of shared k-mers between samples divided by the total number of different k-mers.

As shown in [8], analyses based on alignment-free methods can yield fairly accurate estimations of the similarity between samples. While such estimations are mostly used to perform classification of species, they do not serve to ensure relatedness of the reads that compose the metagenomes. In this line, COMMET [9] computes the number of shared reads (instead of k-mers) to produce a read-level similarity. However, the results of COMMET are very sensitive to the used parameters, and still do not show insight on the true correspondence between reads. Figure 1 shows an example where an incorrect classification of reads takes place due to the lack of inter alignments between shared k-mers. On the other hand, methods based on ungapped alignments such as BOWTIE [10] can not model evolutionary events such as insertions or deletions, which comprise a recurrent scenario in metagenomic studies. Typical gapped alignment approaches (e.g. BLAST [11]) often require large execution times and are not specifically designed to align reads to reads.

In this work, we show that the current software approaches for metagenomes comparison are coarse-grained, and that an exhaustive, gapped and fast alignment is required to improve both the assessment of similarity in metagenomes and the execution times.

3   Methods

The proposed method "IMSAME" (Incremental Multi Staged Alignment of MEtagenomes) performs an incremental alignment procedure, illustrated as follows:

1. A hash table of 12-mers is computed for the reference metagenome. The k-mer size is fixed at 12 nucleotides for three reasons: (a) providing sensitivity while retaining robustness [12]; (b) enabling the use of the algorithm without needing to parameterize the k-mer size, which often leads to strongly different results when changed; and (c) avoid the loss of candidate gapped alignments due to seed size to maximize results. In particular, the imposition of a fixed k-mer size avoids including parameters that are not intrinsically related to

the problem (i.e. similarity between samples) that is being solved. Thus, 12 was chosen as a fixed, highly-sensitive but still robust k-mer size.

2. After the hash table for the reference metagenome is computed, the query metagenome is then loaded and distributed to n threads. The algorithm is capable of working with any number of threads, from 1 to as many as the system allows to, thus enabling a massively parallel computation and large reduction of execution times compared to traditional software.

3. Each thread follows a multi-staged alignment step to compute gapped alignments:

   (a) Firstly, the matching 12-mer words between query and reference (hits) are computed.

   (b) A fast, first-approach ungapped alignment is performed for every hit. This is performed by linearly extending the hit in both forward and backward directions and keeping the starting and ending positions that yield the highest score. In this sense, the ungapped alignment continues until a negative score is reached, but the highest one is reported as the final alignment. Every ungapped alignment with an expected value less than the given threshold (typically, near zero, being default 10^{-5}) is considered as candidate for an exhaustive gapped alignment.

   (c) If such ungapped alignment exists, then the gapped alignment is computed using the Needleman-Wunsch global alignment algorithm between the reads that share the particular alignment. To speed up the Needleman-Wunsch algorithm computation time, a heuristic approach is used to enable gap insertion. The heuristic approach uses the stored maximum scores in the given row and column to insert gaps in case the diagonal score decays.

If the alignment produced in the multi-staged alignment step yields a high percentage of identity and coverage[1] (default 80% for both), the pair of reads are considered to be similar (and thus, shared between samples). In order to compute a global similarity measure between metagenomes, two approaches are considered: (1) the number of shared reads is divided by the total number of reads in the two samples, i.e. the Jaccard-index is computed at read level as shown in Eq. 1:

$$J(a,b) = \frac{r_a \cap r_b}{r_a \cup r_b} \tag{1}$$

where a and b are the metagenomes being compared, r_a and r_b are the reads contained in a and b respectively, and $J(a,b)$ is the Jaccard-index at read level. In the second approach (2) the percentage of reads from a contained in b is calculated to indicate the proportion of a that is contained in b. The resulting alignments are optionally written to disk to enable manual verification.

---

[1] Considering coverage as the length of the alignment divided by the length of the query read.

```
> Sample Read 1
CCGATTGCGAAGGCAGCCTGCTA{1}AGCTGCAACTGACATTGAGGCTCGAAAGTG{1}
TGGGTATCAAACAGGATTAGATACCCTGGTAGTCCA{2}CACGGTAAACGATGAATACT
CGCTGTTTGC{2}GATATACAGCAAGCGGCCAAGCGAAAGCGTTAAGTATTCCACCGTG
GGGAGTACGCCGGCAACGGTGAAACTCAAAGGAATTGGACGGGGGCCCGCACAAGCGGA
GGAACATGTGGT{3}TTAATTCGATGATACGCGAGGAACCTTACC{3}CGG

>Sample Read 2
ACCG{2}CACGGTAAACGATGAATACTCGCTGTTTGC{2}TGGCCCAGAACATCGCCTA
CCCACTGCAACTCGATCACTGGCGCAAGGACTATCAGGATCGTCGTGTCAACGAACTCCT
TGAATTCGTGGGCCTCAGTGAGCACGCAAACAAATACCCTTCGCAGCTGTCCGGCGGCCA
GAAG{1}AGCTGCAACTGACATTGAGGCTCGAAAGTG{1}CAGCGCGTCGGCATCGCCC
GCGCACTGGCCACTAATCCGGAGATTCTGCTCGCCGACGAAGCCAC{3}TTAATTCGATG
ATACGCGAGGAACCTTA{3}
```

Fig. 1. Two modified reads (Sample reads 1 and 2) are depicted as an example of problems that could arise from k-mer-based approaches used in software such as COMMET. Notice that the 3 colored k-mers (surrounded by brackets with IDs) are found in both sequences but in different places. In this example, COMMET would classify the sample read 1 as being equal to sample read 2 since it requires T non overlapping k-mers of length K to accept the equalness. The match is accepted even with varying parameters for $K \leq 30$ and $T \leq 3$, which includes default parameters. The sample read 1 has been extracted from the 16S reads contained in the sample run SRR029687 (https://www.ebi.ac.uk/metagenomics/projects/SRP000319/samples/SRS000998/runs/SRR029687/results/versions/1.0). Sample read 2 has been extracted from the full collection of reads contained in the same run.

4 Results and Discussion

In order to show the difficulty of assessing similarities between two metagenomes, a comparison was performed using two samples from the Human Microbiome Project database (HMP). In particular, the runs SRS014475 (in advance M1) and SRS015062 (M2), which correspond to reads extracted from the throat of healthy humans. The details of the sequencing machine used, filtering and trimming protocols can be found at the HMP website[2]. Both metagenomes were compared using COMMET and our proposed approach by computing the number of shared reads and the Jaccard Index. MASH and SIMKA were not included since they do not offer a read-level similarity measure. Additionally, a gapped BLASTn run was executed with a minimum expected value in the alignments of 10^{-5} to contrast results in terms of the number of shared reads and the Jaccard Index. COMMET was run using different parameters (the k-mer size ranging from 20 to 30 and the number of non overlapping k-mers t needed to accept a match ranging from 2 to 3). The proposed method was run using default parameters (5 for open gap penalty and 2 for extension, $+4$ and -4 for match and mismatch, respectively).

---

[2] http://hmpdacc.org/HMASM/#data.

Table 1 shows the number of reads reported to be included from the sample M1 in M2 using our proposed method, COMMET and BLASTn as a reference. Notice that COMMET was run with only one thread since it requires a SGE cluster to run in parallel, which unfortunately is a resource not available at our testing facilities. On the other hand, BLASTn and IMSAME were run using 30 threads since the computation of gapped alignments requires more execution time and their parallelization strategy does not need a specific cluster platform. Although COMMET is reporting a higher number of matched reads for the execution $k = 2$ and $t = 2$ compared to the rest of the applications, it is hard to assess whether these results represent the real similarity between the samples, since a slight change of parameters significantly changes the number of shared reads. On the other hand, our approach showed a higher number of matches than BLASTn with high percentage of identity and coverage (over 80%). Additionally, the gapped alignments were written to the output file to enable careful examination. The proposed method uses the percentage of identity and coverage as indicators of the alignment quality. The parameter values are on the researchers choice depending on the precision they wish to obtain. However, new quality indicators can be easily incorporated. Furthermore, the parameters on which COMMET is based do not allow to set up an experiment with ease, since the recommended size of k and t can strongly change depending on several factors such as the type of reads that are being compared, the type of metagenomes, the machine used to sequence the samples, etc. In the case of BLASTn, the default parameters do not facilitate the insertion of gaps in small sequences, thus producing almost ungapped results. In this sense, the developed method is strictly

Table 1. The number of matched reads is depicted depending on the program used and the set of parameters, along with the time in minutes and the Jaccard similarity index. COMMET is able to compute faster, but the results show high variation with a standard deviation of 188,239 in the number of reads compared to that of IMSAME, 31,326.

Program	Reads matched	Time (minutes)	Jaccard index
BLASTn -evalue 10^{-5}	428,366	469	0.28
BLASTn -evalue 10^{-5} -k 20	391,350	213	0.26
BLASTn -evalue 10^{-5} -k 30	364,889	243	0.24
IMSAME -p 80 -evalue 10^{-5}	546,157	45.5	0.36
IMSAME -p 90 -evalue 10^{-5}	513,079	56.4	0.34
IMSAME -p 95 -evalue 10^{-5}	483,537	158	0.32
COMMET -k 20 -t 2	626,788	45.89	0.41
COMMET -k 20 -t 3	335,513	46.84	0.22
COMMET -k 30 -t 2	311,718	0.58	0.20
COMMET -k 30 -t 3	177,353	1.2	0.11

intended for metagenome-metagenome comparison scenarios, and thus a flexible set of penalty costs should be used to model evolutionary events.

5 Conclusions

In this work we showed the strengths and weaknesses of current metagenome-metagenome comparison software. In addition, we performed an analysis of common approaches for metagenomic studies (BLAST and COMMET) at read level. We showed that COMMET is able to run very fast, but producing highly variable results whose validity is hard to assess in terms of the precise assignation of reads. In addition, we showed that there is no specific software for fine-grained reads to reads alignment. The developed method is able to compute gapped alignments between reads, enabling the modelling of the highly variable microbial communities present in metagenomes. Our approach is able to take advantage of the massively parallel architectures that are available nowadays, which enables our software to compute in reasonable times while maintaining a sensitive and robust detection of matches. We also present an incremental aligning method to reduce running times composed of alignment-free methods, gapped-free methods and finally gapped alignments. Additionally, the developed method is able to report the results at read level, reporting the alignments and thus enabling a careful examination. In terms of future work, we are aiming to produce a distance metric between metagenomes to approximate the number of species present based on clustering methods.

References

1. Maillet, N., et al.: Compareads: comparing huge metagenomic experiments. BMC Bioinform. **13**(19), 1 (2012)
2. Benoit, G., et al.: Multiple Comparative Metagenomics using Multiset k-mer Counting. arXiv preprint arXiv:1604.02412 (2016)
3. Vinga, S., Almeida, J.: Alignment-free sequence comparisona review. Bioinformatics **19**(4), 513–523 (2003)
4. Bonham-Carter, O., Steele, J., Bastola, D.: Alignment-free genetic sequence comparisons: a review of recent approaches by word analysis. Brief. Bioinform. **15**(6), 890–905 (2013). doi:10.1093/bib/bbt052
5. Lippert, R.A., Huang, H., Waterman, M.S.: Distributional regimes for the number of k-word matches between two random sequences. Proc. Natl. Acad. Sci. **99**(22), 13980–13989 (2002)
6. Anne, C., et al.: A new statistical approach for assessing similarity of species composition with incidence and abundance data. Ecology Lett. **8**(2), 148–159 (2005)
7. Ondov, B.D., et al.: Mash: fast genome and metagenome distance estimation using MinHash. BioRxiv (2016). doi:10.1101/029827
8. Yuriy, F., et al.: How independent are the appearances of n-mers in different genomes? Bioinformatics **20**(15), 2421–2428 (2004)
9. Maillet, N., et al.: COMMET: comparing and combining multiple metagenomic datasets. In: 2014 IEEE International Conference on Bioinformatics and Biomedicine (BIBM). IEEE (2014)

10. Langmead, B., Salzberg, S.L.: Fast gapped-read alignment with Bowtie 2. Nature Methods **9**(4), 357–359 (2012)
11. Altschul, S.F., et al.: Gapped BLAST and PSI-BLAST: a new generation of protein database search programs. Nucleic Acids Res. **25**(17), 3389–3402 (1997)
12. Arjona-Medina, J.A., Torreno, O., Chelbat, N., Trelles, O.: Experimental Study of Local Alignment Distributions in the Comparison of Large Genomic Sequences Soibio (2013)

An Expanded Association Approach for Rare Germline Variants with Copy-Number Alternation

Yu Geng[1,4,6], Zhongmeng Zhao[1,4], Daibin Cui[1,4], Tian Zheng[2,3,4],
Xuanping Zhang[1,4], Xiao Xiao[4,5], and Jiayin Wang[2,3,4(✉)]

[1] School of Electronic and Information Engineering,
Xi'an Jiaotong University, Xi'an, China
zmzhao@mail.xjtu.edu.cn
[2] School of Management, Xi'an Jiaotong University, Xi'an, China
wangjiayin@mail.xjtu.edu.cn
[3] Shaanxi Engineering Research Center of Medical and Health Big Data,
Xi'an, China
[4] Institute of Data Science and Information Quality, Xi'an Jiaotong University,
28 West Xianning Road, Xi'an, China
[5] State Key Laboratory of Cancer Biology,
Xijing Hospital of Digestive Diseases, Xi'an, China
[6] Jinzhou Medical University, Jinzhou, China

Abstract. Tumorigenesis is considered as a complex process that is often driven by close interactions between germline variants and accumulated somatic mutational events. Recent studies report that some somatic copy-number alternations show such interactions by harboring germline susceptibility variants under potential selection in clonal expansions. Incorporating these interactions into genetic association approach could be valuable in not only discovering novel susceptibility variants, but providing insight into tumor heterogeneity and clinical implications. To address this need, in this article, we propose *RareProb-G*, an expanded version of a computational method, which is designed for identifying rare germline susceptibility variants located in the somatic allelic amplification or loss of heterozygosity regions. *RareProb-G* is based on a hidden Markov random field model. The interactions among germline variants and somatic events are modeled by a neighborhood system, which is bounded by a *t*-test on variant allelic frequencies. Each variant is assigned four hidden states, which represent the regional status and causal/neutral status, respectively. A hidden Markov model is also introduced to estimate the initial values of the hidden states and unknown model parameters. To verify this approach, we conduct a series of simulation experiments under different configurations, and *RareProb-G* outperforms than *RareProb* on both sensitivity and specificity.

Keywords: Cancer genomics · Association approach · Germline variant · Copy-number alternation · Hidden markov random field model

© Springer International Publishing AG 2017
I. Rojas and F. Ortuño (Eds.): IWBBIO 2017, Part II, LNBI 10209, pp. 81–94, 2017.
DOI: 10.1007/978-3-319-56154-7_9

1 Introduction

It is crucial in cancer genomics to identify deleterious rare germline variants and somatic mutational events. The challenge arises because the minor allele frequencies of rare variants are pretty low, which often leads to a low statistical power and a non-significant odds ratio. Among the existing genetic association approaches, although different methods vary the assumptions on the genetic models, the collapsing strategy is the most common way of handling the rare variants. Such collapsing method merges a collection of given rare variants into a single new "site". And then, the virtual site presents higher minor allele frequency. Unfortunately, the power loss still occurs when "neutral" variants are collapsed along with "causal" ones, even few neutral ones are involved [1–4]. This *Achilles' Heel* greatly limits the usage of rare variant association approaches on the rare germline variants from cancer genomes because few of the germline variants are considered susceptibility [5, 6].

To overcome this weakness, some association approaches adopt an algorithmic selection step to filter the given variants, which could reduce the probability of harboring neutral ones into collapsing step [1–4]. During the filtering, the interactions among the germline variants and somatic mutations are often used [2]. Recent studies report that such interactions may contribute to more than 3% of all cancer cases [5, 7–9], where the importance of one kind of interactions that the rare germline susceptibility variants are harbored by somatic copy-number alternations is emphasized [5, 6]. These copy-number alternations are considered to contribute the potential selective advantages of the susceptibility germline variants. And thus identifying and estimating such patterns may bring the additional information to guide the association approach to select candidate germline susceptibility variants.

Motivated by this, in this article, we propose *RareProb-G*, an expanded method to select susceptibility candidates of rare germline variants with considering the interacting somatic copy-number alterations. *RareProb-G* is implemented by a hidden Markov random field model (HMRF). A neighborhood system of HMRF is established to describe the interactions among germline and somatic variants. For each variant, both the causal/neutral status and regional status are coded as hidden states, while the genotype and variant allelic frequency are observed variables. An EM algorithm is brought to estimate the hidden states and unknown parameters. To demonstrate the performance of the proposed approach, we test *RareProb-G* on a set of simulation datasets generated by different configurations. *RareProb-G* achieves high statistical powers and low type-I error rates, and outperforms the previous version that does not consider the accompanying copy-number alternations.

2 Computational Model

Suppose we are given N samples with half cases and half controls. Each sample is represented by a genotype and let matrix Y represent the given genotypes. Across all samples, let S_N and C_N represent the total numbers of single nucleotide variants (SNVs) and the copy-number alternations (CNAs), respectively. Let Y_s and Y_c denote the sub-matrixes for SNVs and CNAs, respectively. For a single nucleotide site i, let

$Y_{k,s,i} = 1$ denote that sample k carries a single nucleotide variant at site i on at least one haplotype, and $Y_{k,s,i} = 0$ otherwise. For a copy-number alternation region j, let $Y_{k,c,j}$ represent the ploidy status carried by sample k. We define that $Y_{k,c,j} = 1$ if sample k carries an amplified copy-number alternations in region j, $Y_{k,c,j} = -1$ if sample k presents the copy-number loss in region j, and $Y_{k,c,j} = 0$ otherwise.

The proposed hidden Markov random field model consists of three components. First, it establishes a neighborhood system, which models the interactions among germline variants and somatic mutations. Second, the interacted conditional mode is designed, which contains two hidden states, a set of emission probabilities from one hidden state to the observed genotype matrix and a series of transition probabilities among the hidden states. The third component estimates the initial values of the hidden states and unknown model parameters. As it is an expanded version of *RareProb*, the proposed approach is solved by the same EM algorithm described in [1], which trains the unknown parameters and variables.

2.1 Neighborhood System for Interacting Variants

The neighborhood system models the interacting variants. For variant v, let θ_v and ρ_v denote the average variant allelic frequencies (or minor allele frequencies, if the read depths are not available) from the cases and controls. For any variant, we use the same method in [1] to compute a pair-wise t-test (or t-test for minor allele frequency) statistic, denoted by t_v. Let $\omega_{v,v'} = 2t_v t_{v'} / (t_v^2 + t_{v'}^2)$ denote the weight between two variants v and v'. We set a user-defined parameter Ω to filter the interactions, among which two variants are interacted if and only if $\omega_{v,v'} \geq \Omega$.

2.2 Interacted Conditional Mode

The Hidden States. The unknown hidden state is the core of a hidden Markov random field model, which should be estimated from the observed variables. Here we consider two hidden states. The first one, vector X denotes the causal/neutral status of all the variants in Y. To facilitate the computation, we divide X into X_c and X_s, which are corresponding to SNVs and CNAs respectively. For an element in X, we define that $X_{s,i}$ is equal to 1 if SNV i is considered as a susceptibility candidate and 0 otherwise. We also define that $X_{c,j}$ is equal to 1 if CNA j is considered as a susceptibility candidate and 0 otherwise. Suppose that the linkage disequilibrium among somatic CNAs is weak, we have

$$p(Y_c|X_c) = \prod_{i=1}^{C_N} p(Y_{c,i}|X_{c,i}) \text{ and } p(Y_s|X_s) = \prod_{i=1}^{S_N} p(Y_{s,i}|X_{c,i})$$

The other unknown hidden state, vector R denotes the region status of all the variants. Similarly, we divide R into R_c and R_s to simplify the computation, which are corresponding to SNVs and CNAs respectively. Let $R_{\cdot,i} = 1$ indicate that variant i is located in an elevated region and $R_{.,i} = 0$ indicate that variant i is located in a background region. For the elevated and background regions, we follow the same definitions proposed in [1, 2, 10]. Without considering the linkage disequilibrium among somatic CNAs, we have

$$p(X_c|R_c) \propto \prod_{i=1}^{C_N} p(X_{c,i}|R_{c,i}) \text{ and } p(X_s|R_s) \propto \prod_{i=1}^{S_N} p(X_{s,i}|R_{s,i}, X_{c,i}, R_{c,i})$$

By further considering the neighborhood system, an element in a hidden state may be affected by the neighboring elements in the same hidden state. For the element in hidden state R corresponding to CNA i, we have

$$p(R_{c,i}|R_{c,n(i)}; \tau_{c,i}, \upsilon_{c,i}) \propto \exp\left(\tau_{c,i} R_{c,i} + \upsilon_{c,i} \sum_{i' \in n(i)} \omega_{i,i'} R_{c,i'} \right)$$

where i' denotes a neighboring variant, $n(i)$ is the set of neighboring variants that satisfy $\omega_{i,.} \geq \Omega$. $\tau_{c,i}$ and $\upsilon_{c,i}$ are the unknown model parameters, which measure the impacts respectively from $R_{c,i}$ and $R_{c,i'}$ to $R_{c,i}$. For the element in X, we have

$$p(X_{c,i}|X_{c,n(i)}; \gamma_{c,i}, \eta_{c,i}) \propto \exp\left(\gamma_{c,i} p(X_{c,i}|R_{c,i}) X_{c,i} + \eta_{c,i} \sum_{i' \in n(i)} \omega_{i,i'} p(X_{c,i'}|R_{c,i'}) X_{c,i'} \right)$$

where $\gamma_{c,i}$ and $\eta_{c,i}$ are the unknown model parameters, which measure the impacts on $X_{c,i}$ from $X_{c,i}$ and $X_{c,i'}$, respectively.

The hidden states corresponding to the SNVs are also involved the neighborhood system. Moreover, an element in the hidden state R is affected by both the regional status of neighboring elements and the corresponding elements from $X_{c,.}$ and $R_{c,.}$. For the element in hidden state R corresponding to SNV i, we have

$$p(R_{s,i}|R_{s,n(i)}) \propto \exp\left(\tau_{s,i} p(R_{s,i}|X_{c,i}, R_{c,i}) R_{s,i} + \upsilon_{s,i} \sum_{i' \in n(i)} \omega_{i,i'} p(R_{s,i'}|X_{c,i'}, R_{c,i'}) R_{s,i'} \right)$$

where i' denotes a neighboring variant, $n(i)$ is the set of neighboring variants that satisfy $\omega_{i,.} \geq \Omega$. $\tau_{s,i}$ and $\upsilon_{s,i}$ are the unknown model parameters, which measure the impacts on $R_{s,i}$ from $R_{s,i}$ and $R_{s,i'}$, respectively. For the element in X, we have

$$p(X_{s,i}|X_{s,n(i)}) \propto \exp\left(\gamma_{s,i} p(X_{s,i}|R_{s,i}) p(X_{s,i}|X_{c,i}, R_{c,i}) X_{s,i} + \eta_{s,i} \sum_{i' \in n(i)} \omega_{i,i'} p(X_{s,i'}|R_{s,i'}) p(X_{s,i'}|X_{c,i'}, R_{c,i'}) X_{s,i'} \right)$$

where $\gamma_{s,i}$ and $\eta_{s,i}$ are the unknown model parameters, which measure the impacts on $X_{s,i}$ from $X_{s,i}$ and $X_{s,i'}$, respectively. To simplify the model, we assume that each model parameter is identical among all i, respectively.

The Emission Probabilities. The emission probability links the hidden states to the observed variables. The observed variable in this model is the genotype matrix Y and the phenotypes. As we mentioned before, it is considered that the copy-number alternations may contribute to the potential selective advantages of the harbored germline variants. However, according to our knowledge, there is no literature supports that such regional alternations present observed penetrance. So we do not consider the emission probability from the hidden state R to the genotypes and phenotypes. Meanwhile, the causal variants, including both SNVs and CNAs carried by a sample may contribute to the phenotype according to the genetic model and penetrance [6, 11, 12]. So we link the hidden state X to the genotype matrix Y. For any region i, the probability that the heterozygous CNA is supported by a given number of samples follows a Binomial distribution. Let $m_{c,i}^{+}$ and $m_{c,i}^{-}$ represent the numbers of samples carrying CNA i from the case group and control group respectively.

If $X_{c,i} = 1$, then we can draw the following Binomial distributions: $m_{c,i}^{+} \sim \text{Bin}\left(m_{c,i}^{+}; \frac{N}{2}, \theta_{c,i}\right)$ and $m_{c,i}^{-} \sim \text{Bin}\left(m_{c,i}^{-}; \frac{N}{2}, \rho_{c,i}\right)$, where the unknown parameters $\theta_{c,i}$ and $\rho_{c,i}$ indicate the population attribute risks in the case and control groups respectively, which are $\theta_{c,i} = p(X_{c,i} = 1 | R_{c,i} = 1)$ and $\rho_{c,i} = p(X_{c,i} = 1 | R_{c,i} = 0)$. To compute the conditional probability, we first switch them into the conjugate prior distributions, which are $\theta_{c,i} \sim \text{Beta}\left(\alpha_{\theta_{c,i}}, \beta_{\theta_{c,i}}\right)$ and $\rho_{c,i} \sim \text{Beta}\left(\alpha_{\rho_{c,i}}, \beta_{\rho_{c,i}}\right)$. And then, the marginal distribution of $m_{c,i}^{+}$ is

$$
\text{M}\left(m_{c,i}^{+}\right) = C_{\frac{N}{2}}^{m_{c}^{+}} \frac{\Gamma\left(\alpha_{\theta_{c,i}}, \beta_{\theta_{c,i}}\right)}{\Gamma\left(\alpha_{\theta_{c,i}}\right)\Gamma\left(\beta_{\theta_{c,i}}\right)} \frac{\Gamma\left(\alpha_{\theta_{c,i}} + m_{c,i}^{+}\right)\Gamma\left(\frac{N}{2} - m_{c,i}^{+} + \beta_{\theta_{c,i}}\right)}{\Gamma\left(\alpha_{\theta_{c,i}} + \beta_{\theta_{c,i}} + \frac{N}{2}\right)}
$$

The marginal distribution of $m_{c,i}^{+}$ is similar. Thus, we have

$$
p(Y_{c,i} | X_{c,i} = 1) = \frac{\text{M}\left(m_{c,i}^{+}\right)\text{M}\left(m_{c,i}^{-}\right)}{C_{\frac{N}{2}}^{m_{c,i}^{+}} C_{\frac{N}{2}}^{m_{c,i}^{-}}}
$$

On the other hand, if $X_{c,i} = 0$, then the difference on the population attribute risk between $\theta_{c,i}$ and $\rho_{c,i}$ is not considered. Thus, the probability should be equal across the case and control groups, which is

$$
p(Y_{c,i} | X_{c,i} = 0) = \frac{\Gamma\left(\alpha_{\rho_{c,i}}, \beta_{\rho_{c,i}}\right)}{\Gamma\left(\alpha_{\rho_{c,i}}\right)\Gamma\left(\beta_{\rho_{c,i}}\right)} \frac{\Gamma\left(\alpha_{\rho_{c,i}} + m_{c,i}^{+} + m_{c,i}^{-}\right)\Gamma\left(N - m_{c,i}^{+} - m_{c,i}^{-} + \beta_{\rho_{c,i}}\right)}{\Gamma\left(\alpha_{\rho_{c,i}} + \beta_{\rho_{c,i}} + N\right)}
$$

For any single nucleotide site i, the probability that the heterozygous variant is supported by a given number of samples also follows a Binomial distribution.

Let $m_{s,i}^+$ and $m_{s,i}^-$ represent the numbers of samples carrying SNV i from the case group and control group respectively, then we have

$$p(Y_{s,i}|X_{s,i}=1) = \frac{M\left(m_{s,i}^+\right)M\left(m_{s,i}^-\right)}{C_{\frac{N}{2}}^{m_{s,i}^+}\,C_{\frac{N}{2}}^{m_{s,i}^-}}p(X_{s,i}=1|X_{c,i},R_{c,i})p(X_{c,i})p(X_{c,i}|R_{c,i})$$

and

$$p(Y_{s,i}|X_{s,i}=0) = \frac{\Gamma\left(\alpha_{\rho_{s,i}},\beta_{\rho_{s,i}}\right)}{\Gamma\left(\alpha_{\rho_{s,i}}\right)\Gamma\left(\beta_{\rho_{s,i}}\right)}\frac{\Gamma\left(\alpha_{\rho_{s,i}}+m_{s,i}^++m_{s,i}^-\right)\Gamma\left(N-m_{s,i}^+-m_{s,i}^-+\beta_{\rho_{s,i}}\right)}{\Gamma\left(\alpha_{\rho_{s,i}}+\beta_{\rho_{s,i}}+N\right)}p(X_{s,i}=0|X_{c,i},R_{c,i})p(X_{c,i})p(X_{c,i}|R_{c,i})$$

Transition Probabilities. The transition probabilities describe the linkages among the hidden states. We already present part of the transition probabilities $p(X_c|R_c)$ when we define the hidden states. Here we describe the rests. We still first consider region i. Let $C_{X_{c,i}}^+$ and $C_{X_{c,i}}^-$ represent the number of copy-number amplified alternations from the elevated region and background region, respectively. Let $C_{E_{c,i}}$ and $C_{B_{c,i}}$ denote the numbers of CNAs located in the elevated and background regions, respectively. If $X_{c,i}=1$, then we can draw the following Binomial distributions: $C_{X_{c,i}}^+ \sim \text{Bin}(C_{E_{c,i}},\xi_{c,i})$ and $C_{X_{c,i}}^- \sim \text{Bin}(C_{B_{c,i}},\zeta_{c,i})$, where $\xi_{c,i}=p(X_{c,i}=1|R_{c,i}=1)$ and $\zeta_{c,i}=p(X_{c,i}=1|R_{c,i}=0)$. To compute the posterior distributions, we first switch them into the conjugate prior distributions similarly. And then according to the marginal distributions, we have the conditional probability of $X_{c,i}$ given $R_{c,i}$, which is

$$p(X_{c,i}|R_{c,i}) = \frac{M\left(C_{X_{c,i}}^+\right)M\left(C_{X_{c,i}}^-\right)}{C_{\frac{N}{2}}^{C_{X_{c,i}}^+}\,C_{\frac{N}{2}}^{C_{X_{c,i}}^-}}$$

The posterior probability of $\xi_{c,i}$ given $C_{X_{c,i}}^+$ is

$$p\left(\xi_{c,i}|C_{X_{c,i}}^+\right) = \frac{\xi_{c,i}^{\alpha_{\xi_{c,i}}+C_{X_{c,i}}^+-1}\left(1-\xi_{c,i}\right)^{\beta_{\xi_{c,i}}+C_N-C_{X_{c,i}}^+-1}}{\text{Beta}\left(\alpha_{\xi_{c,i}}+C_{X_{c,i}}^+,\beta_{\xi_{c,i}}+C_N-C_{X_{c,i}}^+\right)}$$

Similarly, we have the posterior probability of $\zeta_{c,i}$ given $C_{X_{c,i}}^-$, which is

$$p\left(\zeta_{c,i}|C_{X_{c,i}}^-\right) = \frac{\zeta_{c,i}^{\alpha_{\zeta_{c,i}}+C_{X_{c,i}}^--1}\left(1-\zeta_{c,i}\right)^{\beta_{\zeta_{c,i}}+C_N-C_{X_{c,i}}^--1}}{\text{Beta}\left(\alpha_{\zeta_{c,i}}+C_{X_{c,i}}^-,\beta_{\zeta_{c,i}}+C_N-C_{X_{c,i}}^-\right)}$$

To simplify the model, we assume that each model parameter is identical among all i, respectively.

Then, we consider the SNV i. It is a little more complex because the hidden state $X_{s,i}$ is affected both the hidden state $R_{s,i}$ and the joint hidden states: $X_{c,i}$ and $R_{c,i}$. Let $C^+_{X_{s,i},X_{c,i},R_{c,i}}$ and $C^-_{X_{s,i},X_{c,i},R_{c,i}}$ represent the numbers of the SNVs located in the elevated and background region, respectively. $C_{E_{X_{s,i},R_{s,i}}}$ and $C_{B_{X_{s,i},R_{s,i}}}$ represent the numbers of single nucleotide sites located in the elevated and background regions, respectively. If $X_{s,i} = 1$, then we can draw the following Binomial distributions: $C^+_{X_{s,i},X_{c,i},R_{c,i}} \sim \mathrm{Bin}\left(C_{E_{X_{s,i},R_{s,i}}}, \xi_{X_{s,i},R_{s,i}}\right)$ and $C^-_{X_{s,i},X_{c,i},R_{c,i}} \sim \mathrm{Bin}\left(C_{B_{X_{s,i},R_{s,i}}}, \zeta_{X_{s,i},R_{s,i}}\right)$, where $\xi_{X_{c,i},R_{c,i}} = p\left(X_{s,i} = 1 | R_{s,i} = 1, X_{c,i}, R_{c,i}\right)$ and $\zeta_{X_{c,i},R_{c,i}} = p\left(X_{s,i} = 1 | R_{s,i} = 0, X_{c,i}, R_{c,i}\right)$. We still use the conjugate prior distributions to compute the posterior distributions of $\xi_{X_{s,i},R_{s,i}}$ and $\zeta_{X_{s,i},R_{s,i}}$. Thus, the posterior probability of $\xi_{X_{s,i},R_{s,i}}$ given $C^+_{X_{s,i},X_{c,i},R_{c,i}}$ is

$$p\left(\xi_{X_{s,i},R_{s,i}} \mid C^+_{X_{s,i},X_{c,i},R_{c,i}}\right) = \frac{\xi_{X_{s,i},R_{s,i}}^{\alpha_{\xi_{X_{s,i},R_{s,i}}} + C^+_{X_{s,i},X_{c,i},R_{c,i}}} \left(1 - \xi_{X_{s,i},R_{s,i}}\right)^{\beta_{\xi_{X_{s,i},R_{s,i}}} + S_N - C^+_{X_{s,i},X_{c,i},R_{c,i}} - 1}}{\mathrm{Beta}\left(\alpha_{\xi_{X_{s,i},R_{s,i}}} + C^+_{X_{s,i},X_{c,i},R_{c,i}}, \beta_{\xi_{X_{s,i},R_{s,i}}} + S_N - C^+_{X_{s,i},X_{c,i},R_{c,i}}\right)}$$

And the posterior probability of $\zeta_{X_{s,i},R_{s,i}}$ given $C^-_{X_{s,i},X_{c,i},R_{c,i}}$ is

$$p\left(\zeta_{X_{s,i},R_{s,i}} \mid C^-_{X_{s,i},X_{c,i},R_{c,i}}\right) = \frac{\zeta_{X_{s,i},R_{s,i}}^{\alpha_{\zeta_{X_{s,i},R_{s,i}}} + C^-_{X_{s,i},X_{c,i},R_{c,i}}} \left(1 - \zeta_{X_{s,i},R_{s,i}}\right)^{\beta_{\zeta_{X_{s,i},R_{s,i}}} + S_N - C^-_{X_{s,i},X_{c,i},R_{c,i}} - 1}}{\mathrm{Beta}\left(\alpha_{\zeta_{X_{s,i},R_{s,i}}} + C^-_{X_{s,i},X_{c,i},R_{c,i}}, \beta_{\zeta_{X_{s,i},R_{s,i}}} + S_N - C^-_{X_{s,i},X_{c,i},R_{c,i}}\right)}$$

2.3 Initializing the Hidden States

Initialization the Hidden State X. Better initializing the unknown variables greatly facilitates the convergence of the model training. First, we estimate the initialization values of X. For any variant v, let θ_v represent the minor allele frequency in the case group, and let ρ_v represent the minor allele frequency in the control group. We adopt a z-test on the difference between θ_v and ρ_v, proposed in [4], to estimate the initial value, which is

$$z_v = \frac{\theta_v - \rho_v}{\sqrt{\frac{2}{N}}\sqrt{\frac{\theta_v + \rho_v}{2}\left(1 - \frac{\theta_v + \rho_v}{2}\right)}}$$

where we simply use θ_v as θ_v and use ρ_v as ρ_v. The significant threshold is set to $\alpha = 0.0001$. If the z-test is significant, then we set $X_v = 1$, and $X_v = 0$ otherwise.

Initialization the Hidden State R. Initializing the hidden state R is more complex because R does not has the emission probability. However, as we mentioned before, the allelic imbalance, especially the copy-number amplification may imply the potential selective advantage. Thus, the read depth and variant allelic frequencies may offer the information. For any given SNV s, let $\bar{\varphi}_s$ denote the average variant allelic frequency across all the cases, and let $\bar{v}_s$ denote the average variant allelic frequency across all the controls. Similarly, we adopt a z-test on the difference between $\bar{\varphi}_s$ and $\bar{v}_s$, proposed in [4], to estimate the initial value, which is

$$z_v = \frac{\bar{\varphi}_s - \bar{v}_s}{\sqrt{\frac{2}{N}}\sqrt{\frac{\bar{\varphi}_s + \bar{v}_s}{2}\left(1 - \frac{\bar{\varphi}_s + \bar{v}_s}{2}\right)}}$$

where the significant threshold is set to $\alpha = 0.0001$. If the z-test is significant, then we consider SNV s located in an elevated region and set $R_s = 1$. Otherwise, if the z-test is not significant, then we consider SNV s located in a background region and set $R_s = 1$.

On the other hand, if the variant allelic frequencies are not available, we establish a simple hidden Markov model to achieve this. Let A denote the state transition matrix, which is

$$A = \begin{bmatrix} p(R_v = 0|R_{v-1} = 0) & p(R_v = 0|R_{v-1} = 1) \\ p(R_v = 1|R_{v-1} = 0) & p(R_v = 1|R_{v-1} = 1) \end{bmatrix}$$

where these probabilities are suggested by experts and $v - 1$ is the variant in front of variant v on the genotype sequence.

3   Simulation Experiments

To evaluate the performance of *RareProb-G*, we first use a popular method (fixed number of causal variants) to generate the simulation datasets which is also used in [1, 2, 4], and then we introduce another way (Markov chain-based simulation) of obtaining the artificial datasets, which we may be better modeled the germline variants with copy-number alternations. For each simulation configuration, we repeat 50 times and the average values are reported.

3.1   Generating Simulation Datasets

Fixed Number of Causal Variants. First, we generate the genotypes for the control group. For each variant, the minor allele frequency in the control group is computed by the Wright's distribution:

$$f(\rho_v) = (\rho_v)^{\beta-1}(1 - \rho_v)^{\beta_N-1}e^{\sigma-\rho_v\sigma}$$

where we use the same settings as in [1, 4, 13], which are $\sigma = 12$, $\beta = 0.001$ and $\beta_N = 0.00033$. For the columns in the genotype matrix Y, we randomly select a fixed number of column into Y_c, and the rest columns go to Y_s. And then, we randomly select the fixed numbers of column into Y_c, and the rest columns go to Y_s.

According to the preset number k of causal variants, we randomly select k columns from Y. For causal variant i, the relative risk is computed by $RR_i = \frac{\delta}{(1-\delta)\rho_i} + 1$, where δ is the marginal population attributed risk. And then, the minor allele frequency θ_i is calculated by $\theta_i = \frac{RR_i\rho_i}{(RR_i-1)\rho_i+1}$, and the genotypes are simulated based on θ_i.

Markov Chain-Based Simulation. To better mimic the interacting variants and copy-number alternations, here we propose to use a Markov chain to generate the simulation datasets. This method first generates R_c and X_c by the Markov chain, and then computes R_s and X_s. In the following experiments, we set the transition probability $p(R_c|R_{c-1})$ as:

$$p(R_c = 0|R_{c-1} = 0) = 0.8, \; p(R_c = 1|R_{c-1} = 0) = 0.2$$

$$p(R_c = 0|R_{c-1} = 1) = 0.2, \; p(R_c = 1|R_{c-1} = 1) = 0.8$$

And we set the transition probability $p(X_c|X_{c-1})$ as:

$$p(X_c = 0|X_{c-1} = 0) = 0.8, \; p(X_c = 1|X_{c-1} = 0) = 0.2$$

$$p(X_c = 0|X_{c-1} = 1) = 0.2, \; p(X_c = 1|X_{c-1} = 1) = 0.8$$

We also set the emission probability $p(X_c|R_c)$ as:

$$p(X_c = 0|R_c = 0) = 0.5, \; p(X_c = 0|R_c = 1) = 0.05$$

$$p(X_c = 1|R_c = 0) = 0.5, \; p(X_c = 1|R_c = 1) = 0.95$$

And we set the emission probability $p(X_s|R_c, X_c, R_s)$ as:

$$p(X_s = 0|R_c = 0, X_c = 0, R_s = 0) = 0.5$$

$$p(X_s = 0|R_c = 0, X_c = 0, R_s = 1) = 0.3$$

$$p(X_s = 0|R_c = 0, X_c = 1, R_s = 0) = 0.3$$

$$p(X_s = 0|R_c = 0, X_c = 1, R_s = 1) = 0.2$$

$$p(X_s = 0|R_c = 1, X_c = 0, R_s = 0) = 0.3$$

$$p(X_s = 0|R_c = 1, X_c = 0, R_s = 1) = 0.2$$

$$p(X_s = 0|R_c = 1, X_c = 1, R_s = 0) = 0.2$$

$$p(X_s = 0|R_c = 1, X_c = 1, R_s = 1) = 0.05$$

When X_s is obtained, the causal variants are randomly sampled in the genotypes according to the minor allele frequencies.

3.2   Performance Analysis

We first use the datasets generated by the fixed number of causal variants method. For each dataset, we generate 1000 samples with half cases and half controls. The total number of the copy-number alternations is set to 120. We alter the number of causal CNAs among 50, 70, 90 and 110, respectively. We also vary the population attribute risks among 0.02, 0.03, 0.04 and 0.05. We evaluate the results by two measurements, which are statistical power and type-I error rate. The statistical power of an approach represents the sensitivity and is calculated by the number of significant datasets among all the datasets with the same configuration, using a significance threshold of 2.5 × 10^{-6} based on the Bonferroni correction, assuming 20000 genes genome-wide. The type-I error rate of an approach represents specificity and is equal to the probability of a causal variant not being selected as susceptibility candidate. The performance of *RareProb-G* is shown in Fig. 1.

According to Fig. 1(a), *RareProb-G* always achieves high statistical powers. Along with the increasing of the population attributed risk, the statistical power increases and reaches 100%. We further investigate the cases when the population attributed risk is set to 0.02, the power loss exists because the CNAs in genotypes is relatively rare. According to Fig. 1(b), *RareProb-G* often achieves low type-I error rates.

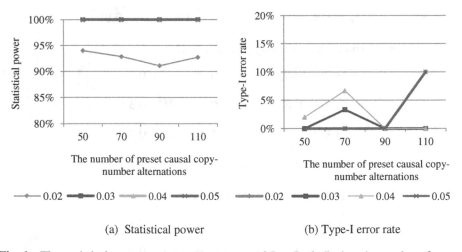

(a) Statistical power   (b) Type-I error rate

Fig. 1. The statistical powers and type-I error rates of *RareProb-G* when the number of preset causal CNAs varies. Lines with different color represent the results under different population attributed risks respectively.

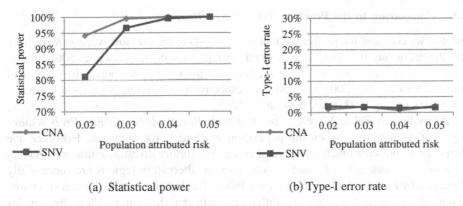

(a) Statistical power   (b) Type-I error rate

Fig. 2. Analysis of the accuracy and the false positive rate (FPR) of pathogens under different PAR

Then we use the Markov chain to generate the simulation datasets. The emission and transition probabilities are mentioned before. The performance of *RareProb-G* is shown in Fig. 2. According to Fig. 2, *RareProb-G* still achieves high statistical powers and low type-I error rates, when the interacting variants are considered.

We mix the SNVs and CNAs, where some SNVs are harbored by CNAs. In this group of experiments, two different methods are used respectively to generate the simulation datasets. We set the total number of CNAs as 120, 50 among which are causal ones. We also set the total number of SNVs as 500, 100 among which are causal ones. Markov chain simulation takes the probability of planting a causal CNA as 0.4167 and the probability of planting a causal CNA as 0.2. We vary the population attributed risk from 0.02 to 0.05 with 0.01 per step. The results are listed in Table 1. From Table 1, we may say that *RareProb-G* presents robust when the pattern of interacting variants changes.

Table 1. The statistical powers and type-I error rates of *RareProb-G* when the mixture of the SNVs and CNAs are preset and the population attributed risk varies. Two methods are used to generate the simulation datasets respectively.

Population attributed risk	Fixed causal		Markov chain	
	Power	Type-I error	Power	Type-I error
0.02	94%	0%	93%	0%
0.03	100%	0%	100%	4%
0.04	100%	2%	100%	0%
0.05	100%	0%	100%	0%
0.02	90%	2%	82%	4%
0.03	99%	1%	96%	4%
0.04	100%	1%	99%	0%
0.05	100%	2%	100%	1%

3.3   Comparing to the Previous Version *RareProb*

Finally, we compare the performance of *RareProb-G* to the previous version *RareProb*, and the results are listed in Table 2. In this group of experiments, we set 500 variants, where 30% of the causal SNVs are amplified by the CNAs. We preset the number of causal SNVs as 100, 200, 300, and 400 respectively, while the population attributed risk varies among 0.02, 0.03, 0.04, and 0.05.

According to Table 2, we can see that both *RareProb-G* and *RareProb* achieve high statistical powers when the population attributed risk increases. However, the *RareProb-G* presents much lower type-I errors. We further investigate this and find that most of the causal SNVs located in copy-number alternation regions are successfully identified by *RareProb-G*. Those variants failed to select by *RareProb* often show low minor allele frequencies or tiny difference between the minor allele frequencies between the cases and controls. It demonstrates the importance of incorporating the allelic imbalance information.

Table 2. Comparisons on statistical power and type-I error rate between *RareProb-G* and *RareProb*, when the number of preset causal variants varies and the population attributed risk alters

Population attributed risk	Causal	Statistical power		Type-I error	
		RareProb-G	*RareProb*	*RareProb-G*	*RareProb*
0.02	100	90%	98%	2%	18%
	200	88%	99%	1%	23%
	300	86%	99%	1%	39%
	400	82%	99%	3%	69%
0.03	100	99%	98%	1%	18%
	200	99%	99%	3%	24%
	300	97%	99%	3%	40%
	400	98%	99%	2%	72%
0.04	100	100%	99%	1%	18%
	200	100%	99%	2%	25%
	300	99%	99%	1%	41%
	400	100%	99%	2%	71%
0.05	100	100%	100%	2%	19%
	200	100%	100%	2%	26%
	300	100%	99%	3%	43%
	400	100%	100%	2%	73%

4   Conclusions

In this article, we present an expanded version of an existing rare variant association approach *RareProb*, to incorporate the information of germline allelic amplification into the selection of rare germline susceptibility variants. This expanded version is named as *RareProb-G*. *RareProb-G* inherits the hidden Markov random field model adopted by *RareProb*, however, it brings several important updates to fit for the copy-number alternations harboring single nucleotide variants.

RareProb-G adopts a two stages of hierarchical iterative mode in the hidden Markov random field. The first stage is designed for copy-number alternations, which consists of the hidden states X_c and R_c, the emission probability $p(Y_c|X_c)$ and the transition probability $p(X_c|R_c)$. For each probability, we use the conjugate prior distribution to compute the marginal distribution, and posterior probability is then calculated from them. The second stage is designed for single nucleotide variants which may locate in a copy-number alternation region. Thus, both the hidden state and transition probability consider X_c and R_c. An EM algorithm adopted by *RareProb* is still used for estimating the hidden states and the unknown model parameters.

For the time-complexity, we conduct the following analysis. This hidden Markov random field model first establishes a neighborhood system, which requires a pair-wise *t*-test for each variant and calculate the weight between each pair of variants. So the time complexity at this step should be $O\left((S_N + C_N)^2\right)$. Then, during each training iteration, the algorithm computes the emission probability and transition probabilities. For each variant, the algorithm calculates one posterior likelihood value for emission probability, and then the time complexity at this step is $O((S_N + C_N)N)$ for N variants. Similarly, the algorithm calculates two posterior likelihood values for transition probabilities, and then the time complexity at this step is $O(2(S_N + C_N)N)$ for N variants. To update the hidden state, the worst case is that none of the interactions are filtered by Ω, and thus the time complexity for this is $O\left((S_N + C_N)^2\right)$. The initialization step is much more straightforward. The algorithm computes twice for X and R is respectively, which is able to achieve in $O(S_N + C_N)$. Thus, the time complexity of *RareProb-G* should be $O\left(((S_N + C_N)N)^k\right)$, where k is the number of iterations, which is often a small number according to our experience.

We conduct a series of performance tests on simulation datasets. According to the results, the statistical power are among 100% and often maintains a low type-I error rates when the numbers of preset causal variants and population attribute risk vary. By introducing the allelic amplification regions, the type-I error rate of *RareProb-G* is greatly reduced comparing to *RareProb*. We may conclude that the proposed approach is useful in the association studies in cancer genomics.

Acknowledgement. This work is supported by the National Science Foundation of China (Grant No: 81400632), Shaanxi Science Plan Project (Grant No: 2014JM8350) and the Fundamental Research Funds for the Central Universities.

References

1. Wang, J., Zhao, Z., Cao, Z., et al.: A probabilistic method for identifying rare variants underlying complex traits. BMC Genomics **14**(1), S11 (2013)
2. Geng, Y., Zhao, Z., Zhang, X., et al.: An improved burden-test pipeline for cancer sequencing data. In: Bourgeois, A., Skums, P., Wan, X., Zelikovsky, A. (eds.) Bioinformatics Research & Applications ISBRA 2016. LNCS (LNBI), vol. 9683, pp. 314–315. Springer, Cham (2016)
3. Bhatia, G., Bansal, V., Harismendy, O., et al.: A covering method for detecting genetic associations between rare variants and common phenotypes. PLoS Comput. Biol. **6**(6), e1000954 (2010)
4. Sul, J., Han, B., He, D., et al.: An optimal weighted aggregated association test for identification of rare variants involved in common diseases. Genetics **188**(1), 181–188 (2011)
5. Lu, C., Xie, M., Wendl, M., Wang, J., Mclellan, M., Leiserson, M., et al.: Patterns and functional implications of rare germline variants across 12 cancer types. Nat. Commun. **6**, 10086 (2015)
6. The Computational Pan-Genomics Consortium: Computational pan-genomics: status, promises and challenges. Brief. Bioinf. (2016, Advance Access published). doi:10.1093/bib/bbw089
7. Kandoth, C., Mclellan, M., Vandin, F., et al.: Mutational landscape and significance across 12 major cancer types. Nature **502**(7471), 333–339 (2013)
8. Ding, L., Raphael, B., Chen, F., et al.: Advances for studying clonal evolution in cancer. Cancer Lett. **340**(2), 212–219 (2013)
9. Rahman, N.: Realizing the promise of cancer predisposition genes. Nature **505**(7483), 302–308 (2014)
10. Dees, N., Zhang, Q., Kandoth, C., et al.: MuSiC: identifying mutational significance in cancer genomes. Genome Res. **22**(8), 1589–1598 (2012)
11. Beckmann, J., Estivill, X., Antonarakis, S.: Copy number variants and genetic traits: closer to the resolution of phenotypic to genotypic variability. Nat. Rev. Genet. **8**(8), 639–646 (2007)
12. Ye, K., Wang, J., Jayasinghe, R., et al.: Systematic discovery of complex indels in human cancers. Nat. Med. **22**(1), 97–104 (2016)
13. Besag, J.: Spatial interaction and the statistical analysis of lattice systems. J. R. Stat. Soc. **36**(2), 192–236 (1974)

Parallelizing Partial Digest Problem
on Multicore System

Hazem M. Bahig[1,2(✉)], Mostafa M. Abbas[3,4],
and M.M. Mohie-Eldin[5]

[1] Computer Science Division, Department of Mathematics, Faculty of Science,
Ain Shams University, Cairo 11566, Egypt
Hazem.m.bahig@gmail.com
[2] College of Computer Science and Engineering,
Hail University, Hail, Kingdom of Saudi Arabia
[3] Qatar Computing Research Institute,
Hamad Bin Khalifa University, Doha, Qatar
mohamza@hbku.edu.qa
[4] KINDI Center for Computing Research, College of Engineering,
Qatar University, Doha, Qatar
[5] Department of Mathematics, Faculty of Science,
Al-Azhar University, Cairo, Egypt

Abstract. The partial digest problem, PDP, is one of the methods used in
restriction mapping to characterize a fragment of DNA. The main challenge of
PDP is the exponential time for the best exact sequential algorithm in the worst
case. In this paper, we reduce the running time for generating the solution of
PDP by designing an efficient parallel algorithm. The algorithm is based on
parallelizing the fastest sequential algorithm for PDP. The experimental study on
a multicore system shows that the running time of the proposed algorithm
decreases with the number of processors increases. Also, the speedup achieved
good scales with increase in the number of processors.

Keywords: Partial digest problem · Parallel algorithm · Scalability · Multicore

1 Introduction

Physical mapping of the genome is one of the fundamental steps in genome studies.
One of the methods that is used in physical mapping is the digestion of DNA with one
restriction enzyme this is the partial digestion process. The enzyme cuts the double
stranded DNA within a specific short sequence of nucleotides called restriction sites.
After that we measure the lengths of obtained fragments and reconstruct the original
ordering of these fragments [1]. For example, the restriction enzyme *TaqI* cuts the
luciferase gene at the *tcga* sequence [2].

Several applications of genomic studies that require the genome mapping are
determining the order of genes or extracting the distinctive short fragments of DNA
sequence, and comparing the genomes of various species [3–7].

© Springer International Publishing AG 2017
I. Rojas and F. Ortuño (Eds.): IWBBIO 2017, Part II, LNBI 10209, pp. 95–104, 2017.
DOI: 10.1007/978-3-319-56154-7_10

The combinatorial problem for partial digestion is called the partial digest problem, PDP. Assume that the set of restriction site locations is represented as the set $X = \{x_0, x_1, \ldots, x_n\}$ and the multiset of lengths of DNA fragments is represented as the multiset $D = \{d_1, d_2, \ldots, d_m\}$. The PDP is defined as follows [8].

Given a multiset $D = \{d_1, d_2, \ldots, d_m\}$. Find the set $X = \{x_0, x_1, \ldots, x_n\}$ such that $\Delta X = \{|x_j - x_i|, 0 \le i < j \le n\} = D$.

For example, the output of the partial digestion process when we use the restriction enzyme *tcga* on the *luciferase* gene is $D = \{9, 30, 100, 170, 293, 302, 393, 402, 462, 562, 632, 732, 855, 864, 945, 954, 975, 984, 1025, 1034, 1247, 1277, 1347, 1377, 1809, 1839, 1979, 2009\}$. The goal of the PDP is to find the set of restriction sites locations which is $X = \{0, 30, 975, 984, 1277, 1377, 1839, 2009\}$ [2].

The complexity analysis of the exact solution for PDP is a still an open problem [9–11]. Many research papers are being introduced to find exact and approximate solutions for PDP [12–20]. The main challenge for finding the exact solution of PDP is the exponential time required for the best known sequential algorithm in case of the worst case. In [21], Zhang gave an example for the worst case instances. Before 2016, the best practical sequential algorithm for PDP is the algorithm designed by Skiena, Smith, and Lemke [18]. Recently, Fomin presented an algorithm for PDP, in [19], which is faster than the Skiena, Smith, and Lemke algorithm in some cases. But still Skiena, Smith, and Lemke algorithm is better than Formin's algorithm in Zhang's instances. In the same year, Abbas and Bahig [20] proposed the fastest exact sequential algorithm for PDP. For Zhang's data, the improvement is greater than 75% over the Skiena, Smith, and Lemke algorithm.

The goal of this research paper is to reduce the running time of the fastest exact sequential algorithm [20] because, with large value of n, the running time of the algorithm that is proposed by Abbas and Bahig is still high. The algorithm takes approximately 19 h for $n = 90$; while the running time of the Skiena, Smith, and Lemke algorithm on the same n is greater than one day. To achieve this goal, we will use high performance computing to speedup the running time of the fastest exact sequential algorithm.

The rest of this paper is as follows. In Sect. 2, we describe briefly the fastest exact sequential for PDP which is the BBb2 algorithm. In Sect. 3, we introduce a new parallel algorithm on multicore system for PDP that is based on the BBb2 algorithm. In Sect. 4, we study the proposed algorithm experimentally according to running time, memory consumed, and scalability in the worst case. Section 5 contains the conclusion of our work.

2   BBb2 Algorithm

The BBb2 algorithm is the proposed algorithm by Abbas and Bahig [20] to find the exact solution of the PDP. The algorithm is based on two main stages. In the first stage, the algorithm applies the breadth-first strategy while using the two bounding conditions that are suggested by Skiena, Smith, and Lemke [18]. In addition, the BBb2 algorithm deletes all repeated subproblems at the same level. For more details about the condition for repeated subproblems, see Theorems 1 and 2 in [20]. The subroutine that is used to traverse the tree level by level is called GenerateNextLevel [20]. Also, in the first stage we will traverse the search tree by using breadth-first strategy for a certain number of levels. This number of levels is determined by using a subroutine called Find_α_M, see [20]. In the second stage of BBb2 algorithm, we solve the subproblems at level α_M, individually, using the breadth first strategy. The values of all elements at the current level are represent by the two lists L_D and L_X. The steps of BBb2 algorithm are as follows.

```
Algorithm BBb2
Input: A multiset of integers, D.
Output: The solution set S.
Begin
    1. Initialize the set S with empty.
    2. Delete the maximum element from the set D and as-
       sign it to the element width.
    3. Assign the set {0, width} to the set X.
    4. Add the set D to the list L_D.
    5. Add the set X to the list L_X.
    6. Find_α_M(N, α_M).
    7. for i = 0 to α_M -1 do
    8.        GenerateNextLevel(L_D, L_X, S)
    9. end for
    10.for each element eD in the list L_D do
    11.    Assign the set eD to eL_D.
    12.    Assign the set eX to eL_X.
    13.    while the list eL_D is not empty do
    14.            GenerateNextLevel(eL_D, eL_X, S).
    15.    end while
    16.end for
End
```

3 Parallel Breadth-Breadth Algorithm

In this section, we propose a parallel algorithm, PBBb2, for PDP based on the algorithm BBb2 under multi-core architecture.

In the PBBb2 algorithm, we parallelize the two main stages of the BBb2 algorithm. In the parallelization of the first stage, we build the solution tree of PDP in the breadth first strategy sequentially till the number of subproblems at a level is greater than or equal to the number of processors, P. After that we assign the subproblems to the processors to work on them till the level α_M. We can summaries the main steps of the parallelization of the first stage as follows.

1. Apply the BBb2 algorithm from line 1–6, where T is a list contains the values of D and X for each subproblem and initially equal to $(D, \{0, maximum(D)\})$.
2. Repeat the following until reaching level α_M.
   (a) If the number of elements of T is less than P, then we apply the procedure GenerateNextLevel many times, at least one, on T until the number of elements of T is greater than or equal to P or until reaching level α_M. If the algorithm reaches to the level α_M, we terminate the process of the first stage and we go to the second stage.
   (b) If the number of elements of T is greater than or equal to P, then we do the following:
      (i) Remove the first $k*P$ elements from the list T and assign it to a new temporary list R, where k is an integer and $k = \lfloor |T|/P \rfloor$.
      (ii) Each processor, p_i, works dynamically on one element, e, from R as follows:

      - Adding the element, e, to a temporary list W_i.
      - Calling the procedure GenerateNextLevel until reaching level α_M and saving the output to the list T_i.

      (iii) The first processor (from the P processors worked on the elements of R) finished with the execution of its work in (ii) will go to Step (a).
3. Add the elements of T_i, $0 \leq i \leq P - 1$, to the list T.
4. Remove the duplication from the list T.

In the parallelization of the second stage, we assign the elements of the list T to the processors and then each processor works on the assigned element until the leaf of the search tree using the breadth first manner or the bounding conditions cut this element. We can summarize the parallelization of the second stage as the following steps.

Repeat the following until the list T is empty.

1. If the number of elements of T is less than P, then we apply the procedure GenerateNextLevel many times, at least one, on T until the number of elements of T is greater than or equal to P or until the list T is empty. In case of the list T is empty, we terminate the second stage.
2. If the number of elements of T is greater than or equal to P, then we do the following:
   (a) Remove the first $k*P$ elements from the list T and assign it to a new temporary list R, where k is an integer and $k = \lfloor |T|/P \rfloor$.
   (b) Each processor, p_i, works dynamically on each element $e \in R$ by executing the steps from line 10 to 16 in BBb2 algorithm. If the processor p_i, found a solution, say s_i, then we add s_i to the set of solutions S if it does not exist in S.
   (c) The first processor (from P processors worked on the elements of R) finished with the execution of its work in (b) will go to Step 1.

4 Performance of PBBb2 Algorithm

In this section, we will evaluate the performance of PBBb2 algorithm by comparing it with the fast sequential algorithm, BBb2, on Zhang data [21] according to running time and memory consumption. We also measure the scalability of PBBb2 algorithm.

To evaluate the performance of PBBb2 algorithm, we use the methodology test that is presented in [20] while we conduct the experiments using multi-core. The machine used in the experimental study is an Intel Xeon E5-2690 and consists of Dual Octa-core processors machine. Each processor has a speed of 2.9 GHz. The memory of the machine is 128 GB RAM, while the cache is 20 MB. Both algorithms were implemented using the C++ language and openMP directives.

In the case of the running time, Table 1 demonstrates the running time of the PBBb2 and BBb2 algorithms for $35 \le n \le 90$, where the symbols 's', 'm' and 'h' are used for second, minute, and hour respectively. The number of processors used to run the PBBb2 algorithm is $P = 2, 4, 6, 8,$ and 10. If $P = 1$, the PBBb2 algorithm is equivalent to the BBb2 algorithm. It is clear that for fixed values of n the running time of the PBBb2 algorithm using $P > 1$ is faster than the BBb2 algorithm. Also, the running time for the PBBb2 algorithm decreases as the number of processors increases as in Fig. 1. In Fig. 1, we use the log scale to represent the running time of PBBb2.

Table 1. Running time for BBb2 and PBBb2 algorithms

n	BBb2	PBBb2				
	P					
	1	2	4	6	8	10
35	0.146 s	0.079 s	0.04 s	0.03 s	0.025 s	0.016 s
40	0.453 s	0.228 s	0.125 s	0.091 s	0.078 s	0.045 s
45	1.558 s	0.793 s	0.441 s	0.315 s	0.28 s	0.192 s
50	3.923 s	1.887 s	0.941 s	0.714 s	0.574 s	0.524 s
55	15.801 s	6.961 s	3.873 s	2.932 s	2.426 s	2.233 s
60	33.134 s	15.736 s	8.685 s	5.956 s	4.884 s	4.239 s
65	1.967 m	0.898 m	0.495 m	0.352 m	0.299 m	0.26 m
70	6.438 m	2.803 m	1.615 m	1.099 m	0.966 m	0.813 m
75	30.027 m	16.109 m	7.999 m	5.507 m	3.388 m	2.969 m
80	44.329 m	22.83 m	10.811 m	6.981 m	5.984 m	5.144 m
85	4.655 h	2.387 h	1.056 h	0.877 h	0.824 h	0.773 h
90	18.525 h	11.52 h	6.104 h	4.546 h	3.74 h	3.526 h

Figure 2 represents the scalability of the PBbb2 algorithm as a function of the number of processors P and the problem size n, where the speedup is defined as the ratio between the fastest sequential algorithm and the parallel algorithm. Figure 2 demonstrates that the speedup achieved good scaling with increasing the number of processors.

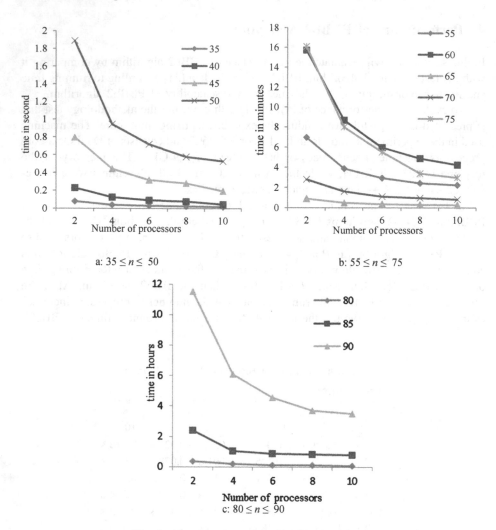

Fig. 1. Running time of the PBBb2 algorithm.

Figure 3 shows the memory consumption of PBBb2 algorithm using different numbers of processors. We used the log scale to represent the memory consumption of the PBBb2 algorithm. From Fig. 3, we can note that in most instances, the memory consumption of the PBBb2 algorithm increases as the number of processors increases. The reason behind increasing the memory consumption in parallelism is that the P processors work on P different subproblems at the same time. In other words, the processors will build different subtrees at the same time in the breadth first manner.

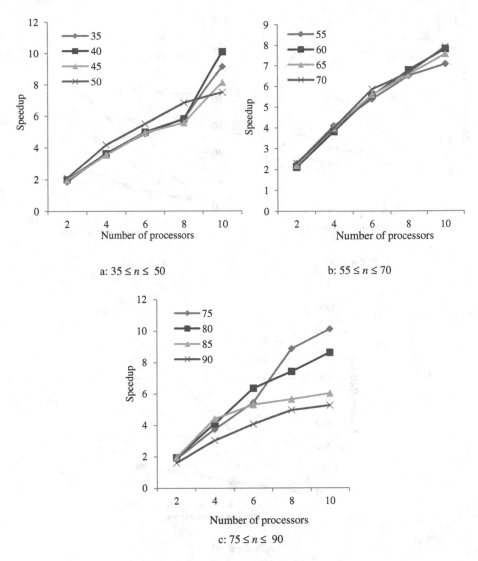

a: $35 \leq n \leq 50$

b: $55 \leq n \leq 70$

c: $75 \leq n \leq 90$

Fig. 2. Scalability of the PBBb2 algorithm.

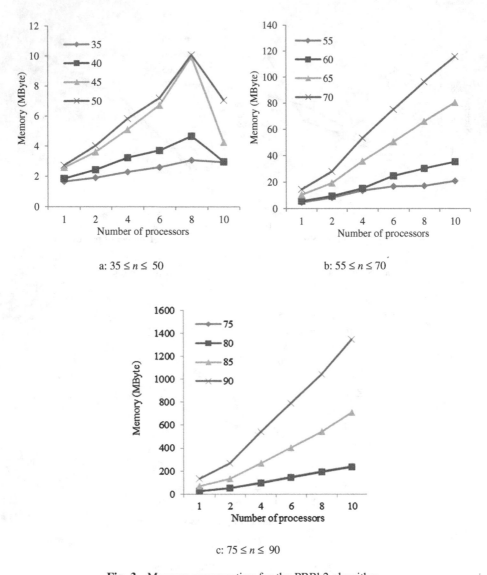

a: $35 \leq n \leq 50$

b: $55 \leq n \leq 70$

c: $75 \leq n \leq 90$

Fig. 3. Memory consumption for the PBBb2 algorithm.

5   Conclusions

In this research paper, we parallelized the fast exact algorithm for the partial digest problem, PDP. The main challenge of PDP is the exponential time for the best exact sequential algorithm in the worst case. The proposed algorithm is based on working on many independent subproblems at the same time and traversing the search tree with the breadth–first strategy. The experimental results on multicore system have shown that the running time of the parallel algorithms decreases as the number of processors increases. The average efficiency of the PBBb2 algorithm is 88.53%. Also, the speedup achieved good scaleing with increasing the number of processors.

References

1. Pevzner, P.: DNA physical mapping and alternating eulerian cycles in colored graphs. Algorithmica **13**(1–2), 77–105 (1995)
2. Devine, J.H., Kutuzova, G.D., Green, V.A., Ugarova, N.N., Baldwin, T.O.: Luciferase from the east European firefly Luciola mingrelica: cloning and nucleotide sequence of the cDNA, overexpression in Escherichia coli and purification of the enzyme. Biochimica et Biophysica Acta (BBA)-Gene Struct. Expr. **1173**(2), 121–132 (1993)
3. Baker, M.: Gene-editing nucleases. Nat. Methods **9**(1), 23–26 (2012)
4. Sambrook, J., Fritsch, E.F., Maniatis, T.: Molecular Cloning. A Laboratory Manual, 2nd edn., pp. 1.63–1.70. Cold Spring Harbor Laboratory Press, Cold Spring Harbor (1989)
5. He, X., Hull, V., Thomas, J.A., Fu, X., Gidwani, S., Gupta, Y.K., Black, L.W., Xu, S.Y.: Expression and purification of a single-chain Type IV restriction enzyme Eco94GmrSD and determination of its substrate preference. Sci. Rep. **5**, 9747 (2015)
6. Narayanan, P.: Bioinformatics: A Primer. New Age International (2005)
7. Dear, P.H.: Genome mapping. eLS (2001)
8. Jones, N.C., Pevzner, P.: An Introduction to Bioinformatics Algorithms. MIT Press, Cambridge (2004)
9. Lemke, P., Werman, M.: On the complexity of inverting the autocorrelation function of a finite integer sequence, and the problem of locating n points on a line, given the (nC2) unlabelled distances between them. Preprint 453 (1988)
10. Daurat, A., Gérard, Y., Nivat, M.: Some necessary clarifications about the chords' problem and the partial digest problem. Theoret. Comput. Sci. **347**(1–2), 432–436 (2005)
11. Cieliebak, M., Eidenbenz, S., Penna, P.: Noisy Data Make the Partial Digest Problem NP-Hard. Springer, Heidelberg (2003)
12. Pandurangan, G., Ramesh, H.: The restriction mapping problem revisited. J. Comput. Syst. Sci. **65**(3), 526–544 (2002)
13. Błażewicz, J., Formanowicz, P., Kasprzak, M., Jaroszewski, M., Markiewicz, W.T.: Construction of DNA restriction maps based on a simplified experiment. Bioinformatics **17**(5), 398–404 (2001)
14. Blazewicz, J., Burke, E.K., Kasprzak, M., Kovalev, A., Kovalyov, M.Y.: Simplified partial digest problem: enumerative and dynamic programming algorithms. IEEE/ACM Trans. Comput. Biol. Bioinf. **4**(4), 668–680 (2007)
15. Karp, R.M., Newberg, L.A.: An algorithm for analysing probed partial digestion experiments. Comput. Appl. Biosci. **11**(3), 229–235 (1995)
16. Nadimi, R., Fathabadi, H.S., Ganjtabesh, M.: A fast algorithm for the partial digest problem. Jpn J. Ind. Appl. Math. **28**(2), 315–325 (2011)
17. Ahrabian, H., Ganjtabesh, M., Nowzari-Dalini, A., Razaghi-Moghadam-Kashani, Z.: Genetic algorithm solution for partial digest problem. Int. J. Bioinform. Res. Appl. **9**(6), 584–594 (2013)
18. Skiena, S.S., Smith, W.D., Lemke, P.: Reconstructing sets from interpoint distances. In: Proceedings of the Sixth Annual Symposium on Computational Geometry, pp. 332–339. ACM (1990)

19. Fomin, E.: A simple approach to the reconstruction of a set of points from the multiset of n2 pairwise distances in n2 steps for the sequencing problem: II algoirthm. J. Comput. Biol. **23**, 1–7 (2016)
20. Abbas, M.M., Bahig, H.M.: A fast exact sequential algorithm for the partial digest problem. BMC Bioinform. **17**, 1365 (2016)
21. Zhang, Z.: An exponential example for a partial digest mapping algorithm. J. Comput. Biol. **1**(3), 235–239 (1994)

Computational Proteomics

Prediction of Calmodulin-Binding Proteins Using Short-Linear Motifs

Yixun Li[1], Mina Maleki[1(✉)], Nicholas J. Carruthers[2], Luis Rueda[1(✉)], Paul M. Stemmer[2], and Alioune Ngom[1]

[1] School of Computer Science, University of Windsor, Windsor, ON, Canada
{li18o,maleki,lrueda,angom}@uwindsor.ca
[2] Institute of Environmental Health Sciences, Wayne State University, Detroit, MI, USA
{aj7682,pmstemmer}@wayne.edu

Abstract. Prediction of Calmodulin-binding (CaM-binding) proteins plays a very important role in the fields of biology and biochemistry, because Calmodulin binds and regulates a multitude of protein targets affecting different cellular processes. Short linear motifs (SLiMs), on the other hand, have been effectively used as features for analyzing protein-protein interactions, though their properties have not been used in the prediction of CaM-binding proteins. In this study, we propose a new method for prediction of CaM-binding proteins based on both the total and average scores of SLiMs in protein sequences using a new scoring method, which we call Sliding Window Scoring (SWS) as features for the prediction. A dataset of 194 manually curated human CaM-binding proteins and 193 Mitochondrial proteins have been obtained and used for testing the proposed model. Multiple EM for Motif Elucidation (MEME) has been used to obtain new motifs from each of the positive and negative datasets individually (the SM approach) and from the combined negative and positive datasets (the CM approach). Moreover, the wrapper criterion with Random Forest for feature selection (FS) has been applied followed by classification using different algorithms such as k-nearest neighbor (k-NN), support vector machine (SVM), and Random Forest (RF), on a 3-fold cross-validation setup. Our proposed method shows promising prediction results and demonstrates how information contained in SLiMs is highly relevant for prediction of CaM-binding proteins.

Keywords: Calmodulin-binding proteins · Short-linear motifs · Sliding window scoring · Classification · Protein interaction

1 Introduction

Calmodulin (CaM) is a calcium-binding protein that is a major transducer of calcium signaling [1]. It has no enzymatic activity of its own but rather acts by binding to and altering the activity on a panel of cellular protein targets. Its targets are structurally and functionally diverse and participate in a wide

© Springer International Publishing AG 2017
I. Rojas and F. Ortuño (Eds.): IWBBIO 2017, Part II, LNBI 10209, pp. 107–117, 2017.
DOI: 10.1007/978-3-319-56154-7_11

range of physiological functions including immune response, muscle contraction and memory formation. Figure 1 is typical of a calcium-dependent interaction where the two halves of CaM bind to opposite sides of the target peptide (the four calcium molecules are green spheres). Identifying CaM target proteins and CaM sites is an important and ongoing research problem because of the great diversity of conformations it uses in its target interactions. This diversity cannot be captured by a single amino acid sequence motif, but instead CaM binding sites are commonly divided into four or more motif classes with different sequence characteristics [2]. Existing algorithms have difficulties in identifying novel CaM-binding proteins.

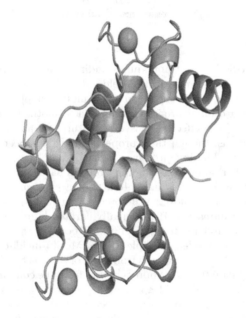

Fig. 1. Structure of CaM (green) interacting with its binding domain from calcineurin (blue). (Color figure online)

On the other hand, Short Linear Motifs (SLiMs), patterns of 3 to 10 amino acids in intrinsically disordered regions of protein sequences, can encode functional aspects of proteins and bind to important domains [3]. They also help regulate many cellular processes, by being interaction sites for other SLiMs in proteins. SLiM-mediated interactions are often transient interactions or utilize additional interaction domains to co-operatively produce stable complexes. Therefore, prediction and analysis of CaM-binding proteins using SLiM profiles has the potential to develop better models for cellular processes such as modulation and regulation of proliferation and apoptosis [4].

Recent studies have focused on the discovery of new SLiMs for the prediction of protein interactions [5]. From different accessible SLiM discovering tools such as SLiMFinder [6], SLiMSearch [7], Minimotif Miner (MnM) [8], and MEME

(Multiple EM for Motif Elucidation) [9], MEME can discover SLiMs through an unsupervised approach and turns out to be a very efficient and successful algorithm for discovering new SLiMs with different number of occurrences in a set of protein sequences. It discovers motifs by optimizing the statistical parameters of the model using the Expectation Maximization (EM) algorithm, and a statistical sequence model to determine the positions and the width of the motif sites in the sequences [10].

In this work, a computational model is being proposed for the prediction and analysis of CaM-binding proteins using SLiM profiles. While in our previous study [11], a list of known CaM-binding motif families taken from [12] were used as features for the prediction, in this study, extracted features are new SLiMs derived from MEME. Two different approaches have been used to discover new motifs using MEME: (a) find SLiMs from each of the positive and negative datasets separately (SM) and (b) find SLiMs from the combined positive and negative datasets (CM). The dataset used in the experiments has been manually curated with 194 human CaM-binding proteins as a positive dataset and 193 Mitochondrial proteins as a negative dataset. For each protein, the extracted features are the counting SLiMs in protein sequences using a new scoring method named Sliding Window Scoring (SWS). Predictions of CaM-binding proteins have been performed in the Waikato Environment for Knowledge Analysis (WEKA) using k-nearest neighbor (k-NN), support vector machine (SVM), and random forest (RF) classifiers. The experimental results indicate that the classification using the SLiMs obtained from CM generally achieve better performance.

2   Materials and Methods

Figure 2 shows a schematic diagram that depicts our method. First of all, we obtain the positive and negative datasets from the protein databases, and download the protein sequences. Then, we obtain the SLiMs in two different ways, and thereafter we use the SWS approach for scoring the sites. Finally, we apply feature selection and classification to the score matrices and analyze the results.

2.1   Dataset

Our manually curated dataset, which contains 194 human CaM-binding proteins collected from the Calmodulin Target Database [2], is used as the positive dataset and 193 Mitochondrial proteins obtained from the Uniprot database as the negative dataset. Mitochondrial proteins were chosen as a negative dataset because no major biochemical function has been demonstrated for CaM in the mitochondria suggesting that the number of CaM-interacting proteins that are localized in the mitochondria will be small relative to other sub-cellular locations. Gene Ontology (GO) cellular component annotations were used to identify mitochondrial proteins so that our negative dataset will include proteins encoded in both the mitochondrial and nuclear genomes. To construct the list all 7,433 proteins

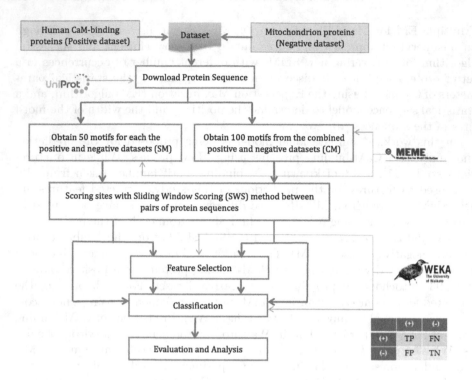

Fig. 2. Diagram of the model being proposed.

that were under the cellular component term Mitochondrion (GO:0005739) and had human taxonomy were downloaded. After filtering out non-reviewed proteins and any proteins with Golgi or Nucleus annotations, 886 proteins were obtained that are strictly mitochondrial as far as GO annotations are concerned. From those remaining Mitochondrial proteins, 193 proteins, which contain few if any CaM binding regions, were selected manually as the negative dataset, yielding a balanced dataset.

2.2   Scoring the Sites

Once the SLiM sets are obtained, MEME outputs files that contain the patterns for the SLiMs, sites found in the protein sequences and their positions, and the probability matrix of the features of each SLiM. Figure 3 shows SLiM No. 57 found in the dataset obtained by CM output by MEME with the sites found in the sequences and the corresponding protein names. Table 1 shows the Position-Specific Probability Matrix (PSPM) of this SLiM. The columns represent the 20 amino acids, while the rows correspond to the scores of the features in the SLiM.

We did not consider the sites in the sequences found by MEME. In contrast, we considered every possible sub-sequence (*l-mer*) in a sequence as a potential site for a motif of the training set. Each sequence is divided into

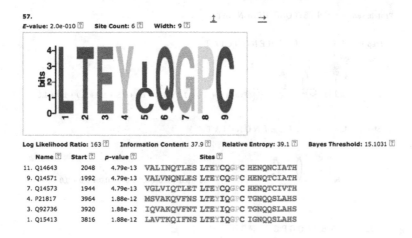

Fig. 3. SLiM No. 57 found by CM

Table 1. Position-specific probability matrix of SLiM No. 57.

Position	A	C	D	E	F	G	H	I	K	L	M	N	P	Q	R	S	T	V	W	Y
1	0	0	0	0	0	0	0	0	0	1.0	0	0	0	0	0	0	0	0	0	0
2	0	0	0	0	0	0	0	0	0	0	0	0	0	0	0	0	1.0	0	0	0
3	0	0	0	1.0	0	0	0	0	0	0	0	0	0	0	0	0	0	0	0	0
4	0	0	0	0	0	0	0	0	0	0	0	0	0	0	0	0	0	0	0	1.0
5	0	0.5	0	0	0	0	0	0.5	0	0	0	0	0	0	0	0	0	0	0	0
6	0	0	0	0	0	0	0	0	0	0	0	0	0	1.0	0	0	0	0	0	0
7	0	0	0	0	0	1.0	0	0	0	0	0	0	0	0	0	0	0	0	0	0
8	0	0	0	0	0	0	0	0	0	0	0	0	1.0	0	0	0	0	0	0	0
9	0	1.0	0	0	0	0	0	0	0	0	0	0	0	0	0	0	0	0	0	0

overlapping l-mers. We designed the SWS method for scoring these sites. Figure 4 shows an example of SWS based on SLiM No. 57 along with its position-specific probability matrix. Let us consider l-mer a in a sequence of length L. We divide the sequence into all possible overlapping l-mers of length W, where l is the length of each SLiM, delivering a total of $\{L - W + 1\}$ l-mers. Then, Eq. (1) is used to calculate the information contained in l-mer a, given a profile X of length L, and a SLiM m of length W:

$$P(a| X) = \sum_{i=1}^{W} P(a_i),\qquad(1)$$

where X is the profile of the motif, $P(a_i)$ is the probability of the amino acid in that profile. Since $P(a| X)$ may be 0 or a very small value if the SLiM and the site have very low similarity, we set a threshold to 60% for $P(a| X)$. Thus, we do

Protein sequence: LTEYCQGPCHENQNCIAT

Step 1:   L T E Y C Q G P C H E N Q N C I A T

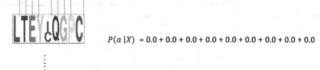

$P(a\,|X)$ = 1.0 + 1.0 + 1.0 + 1.0 + 0.5 + 1.0 + 1.0 + 1.0 + 1.0

Step 2:   L T E Y C Q G P C H E N Q N C I A T

$P(a\,|X)$ = 0.0 + 0.0 + 0.0 + 0.0 + 0.0 + 0.0 + 0.0 + 0.0 + 0.0

⋮

Step 10:   L T E Y C Q G P C H E N Q N C I A T

Fig. 4. Example of the SWS method based on SLiM No. 57 along with its PSPM.

not consider that site and remove the $P(a|\,X)$ score as well. Once the scores for all possible *l-mers* in profile X are obtained, we use Eq. (2) to add up all the scores of the *l-mers* as the score of SLiM m for profile X:

$$P(m|\,X) = \sum_{i=1}^{L-W+1} P(a|\,X).\tag{2}$$

Equation (2) implies that the more frequently that a is a site, the larger the information content is. Thus, in order to erase this effect, we also divide the total information content by the frequency of the site in the sequence, $N \leqslant L - W + 1$, (since we removed some sites with scores lower than 60%):

$$\hat{P}(m|\,X) = \frac{1}{N} \times \sum_{i=1}^{L-W+1} P(a|\,X).\tag{3}$$

Then, we calculate $P(m|\,X)$ and $\hat{P}(m|\,X)$ for all the SLiMs obtained from both SM and CM for each protein sequence. We determine that s_{i1} to s_{in} are the $P(m|\,X)$ scores of each SLiM on every sequence of the protein, while t_{i1} to t_{in} are the $\hat{P}(m|\,X)$ scores of each SLiM on every sequence of the protein, where n is the frequency of SLiMs. Thus, each protein sequence is transformed into two total score vectors of length n as follows:

$$S_{ij} = (s_{i1}, s_{i2}, ..., s_{in}), \text{ and}$$

$$T_{ij} = (t_{i1}, t_{i2}, ..., t_{in}),$$

where S_i and T_i are the $P(m|\ X)$ and $\hat{P}(m|\ X)$ scores of SLiM for n SLiMs on each protein sequence. We call these matrices S score matrix and T score matrix for S_i and T_i respectively. This transformation is applied to each positive pair and each negative pair in the training set with the SLiMs obtained from both the SM and CM approaches.

2.3   Classification

There are a variety of classification methods, of which SVM, RF and k-NN are three of the most well-known ones, and which are used in this study.

Support Vector Machine: SVMs are well known machine learning techniques used for classification, regression and other tasks. The aim of the SVM is to find the hyperplane that ideally separates the feature space into two regions (classes). As this kind of hyperplane is not unique, the SVM chooses the hyperplane that leaves the maximum margin from that hyperplane to the support vectors. The classification by using the SVM is usually inefficient when using a linear classifier, because in general, the data are not linearly separable. Thus, the use of kernels is crucial in mapping the data onto a higher dimensional space in which the classification is more efficient. The effectiveness of the SVM depends on the selection of the kernel, the selection parameters and the soft margin [13]. There are a number of different kernels that can be used in SVMs such as polynomial, radial basis function (RBF) and sigmoid.

Random Forest: RF is a classifier that is based on many decision tree predictors such that each tree depends on the values of a random vector sampled independently and with the same distribution for all trees in the forest. RF achieves excellent performance among current classification algorithms. It also has an effective method for estimating missing data and maintains accuracy when a large proportion of the data is missing [14].

k-Nearest Neighbor: The k-NN rule is among the simplest of all machine learning methods and is a type of instance-based/lazy learning method. To find the class of a test sample, first, the distances between the test sample and each training sample should be calculated and sorted. Then, the most frequent class label in the first "k" training samples (nearest neighbors) is assigned to the test sample. One of the main challenges of this method is to determine the best number of neighbors.

2.4   Feature Selection

The process of choosing the best subset of relevant features that represents the whole dataset efficiently is called feature selection. Applying feature selection before running a classifier is important to reduce the dimensionality of the data by discarding redundant and/or irrelevant features, and, thus, reducing the prediction time while improving the classification performance. In this paper, we applied the wrapper approach with RF for feature selection followed by classification using different algorithms. Wrapper methods embed the model hypothesis search within the feature subset search. In this setup, a search procedure in the space of possible feature subsets is defined, and various subsets of features are generated and evaluated. The evaluation of a specific subset of features is obtained by training and testing a specific classification model, rendering this approach tailored to a specific classification algorithm [15].

3   Results and Discussion

We applied SVM with a Polynomial kernel, RF, and k-NN classifiers on our dataset using WEKA ver. 3.7.11 software [16]. We applied all these classifiers with default parameters: $k = 1$ for k-NN and $Gamma(g) = 0$ and $Cost(c) = 1$ for SVM-Polynomial kernel. 3-fold Cross-Validation is the method we used for training and evaluating all the classifiers. As a performance measure, the accuracy was calculated as follows:

$$Accuracy = \frac{TP + TN}{TP + FP + TN + FN}, \tag{4}$$

where TP is the number of true positives, FP is the number of false positives; TN is the number of true negatives, and FN is the number of false negatives. CaM-binding proteins are positive while Mitochondrial proteins are negative.

3.1   Results

The classification results for the score matrices with SLiMs obtained from SM and CM are shown in Tables 2 and 3, respectively. From the tables, it is noticeable that (a) 1-NN on the S score matrix yields the highest classification accuracy of 80.6% and 78.3% for the SLiMs obtained from SM and CM, respectively; (b) the S score matrix is a better subset of features than the T score matrix for both SM and CM; (c) using the motifs from the combined negative and positive datasets, CM approach, yielded better results than the obtained motifs obtained from each of the positive and negative datasets individually, SM approach, in the most of the experiments.

As another experiment, the wrapper approach with RF was applied to score and rank the features, while SVM, RF and k-NN employed for classification. The performances of the classifiers using different numbers of selected features for S and T score matrices obtained from SM and CM, are shown in Tables 4 and 5, respectively. For the SLiMs obtained from SM, the subset obtained from FS contains seven features for the S score matrix, and nine features for the T score matrix. As for the SLiMs obtained from CM, the subset obtained from FS contains eleven features for the S score matrix, and seven features for the T score matrix. Similarly, from Tables 4 and 5, it is clear that RF on the S

Table 2. Classification results for the score matrices with SLiMs obtained from SM.

Dataset for classification	# Features	Classifier	Accuracy (%)
S score matrix	100	SVM-Polynomial ($c = 1, g - 0$)	72.6
		Random Forest	73.1
		k-NN ($k = 1$)	**80.6**
T score matrix	100	SVM-Polynomial ($c = 1, g = 0$)	55.0
		Random Forest	**68.5**
		k-NN ($k = 1$)	59.7

Table 3. Classification results for the score matrices with SLiMs obtained from CM.

Dataset for classification	# Features	Classifier	Accuracy (%)
S score matrix	100	SVM-Polynomial ($c = 1, g = 0$)	72.6
		Random Forest	74.7
		k-NN ($k = 1$)	**78.3**
T score matrix	100	SVM-Polynomial ($c = 1, g = 0$)	57.6
		Random Forest	**69.3**
		k-NN ($k = 1$)	58.1

Table 4. Classification results for the score matrices with SLiMs obtained from SM using FS.

Dataset for classification	# Features	Classifier	Accuracy (%)
S score matrix	7	SVM-Polynomial ($c = 1, g = 0$)	66.1
		Random Forest	**77.8**
		k-NN ($k = 1$)	77.0
T score matrix	9	SVM-Polynomial ($c = 1, g = 0$)	64.9
		Random Forest	**69.3**
		k-NN ($k = 1$)	66.4

score matrix yield the highest classification accuracy of 77.8% and 80.1% for the SLiMs obtained from SM and CM, respectively. Also, it can be seen that the classification using the SLiMs obtained from CM have better performance than using the SLiMs obtained from SM.

Moreover, comparing the classification results obtained by using the FS method (Tables 4 and 5) with no FS (Tables 2 and 3) demonstrate the strength of the feature selection method in selecting more powerful and discriminating features for classification for the most subsets of features. However, the maximum

Table 5. Classification results for the score matrices with SLiMs obtained from CM using FS.

Dataset for classification	# Features	Classifier	Accuracy (%)
S score matrix	11	SVM-Polynomial ($c = 1, g = 0$)	62.0
		Random Forest	**80.1**
		k-NN ($k = 1$)	78.6
T score matrix	9	SVM-Polynomial ($c = 1, g = 0$)	60.2
		Random Forest	**70.5**
		k-NN ($k = 1$)	68.7

decrease of 6% on the classification performance is still acceptable because the classification performed faster using the smaller number of features.

4   Conclusion

We propose a method for prediction of Calmodulin-Binding using short-linear motifs. Our method shows promising results and demonstrates that information contained in SLiMs is highly relevant for accurate prediction of CaM-binding proteins. The SWS method is useful for scoring the sites and obtaining the datasets for classification. Most of the classifiers perform better on the total scores without dividing by the frequency of the SLiMs. The classification experiments yield good results on the datasets with SLiMs obtained from both of the SM and CM approaches. The 80.6% classification accuracy using 1-NN as the classifier on the total scores obtained from SM is the highest accuracy among all of the experiments. Moreover, the performance of the classifiers improved for most subsets of features by using fewere informative features (motifs) selected by the wrapper approach with RF. Our plan for the future is to investigate the selected motifs obtained by MEME and relate them to existing families of calcium-binding motifs, possibly discovering new motifs or families. Also, possible extension to this works is to investigate the SWS approach on prediction of other types of protein-protein interactions. Another extension to this work is to combine structural and SLiM data in order to achieve a better insight of the location of the motifs on the interface, role on the interaction and other aspects.

References

1. Stevens, F.C.: Calmodulin: an introduction. Can. J. Biochem. Cell Biol. **61**(8), 906–910 (1983)
2. Yap, K.L., Kim, J., Truong, K., Sherman, M., Yuan, T., Ikura, M.: Calmodulin target database. J. Struct. Funct. Genomics **1**(1), 8–14 (2000)

3. Ren, S., Yang, G., He, Y., Wang, Y., Li, Y., Chen, Z.: The conservation pattern of short linear motifs is highly correlated with the function of interacting protein domains. BMC Genomics **9**(1), 452 (2008)
4. Haslam, N.J., Niall, J., Shields, D.C.: Profile-based short linear protein motif discovery. BMC Bioinform. **13**(1), 104 (2012)
5. Rueda, L., Pandit, M.: A model based on minimotifs for classification of stable protein-protein complexes. In: IEEE Symposium on Computational Intelligence in Bioinformatics and Computational Biology (CIBCB), Hawaii, USA (2014)
6. Davey, N.E., Haslam, N.J., Shields, D.C., Edwards, R.J.: SLiMFinder: a web server to find novel, significantly over-represented, short protein motifs. Nucleic Acids Res. **38**, W534–W539 (2010)
7. Davey, N.E., Haslam, N.J., Shields, D.C., Edwards, R.J.: SLiMSearch 2.0: biological context for short linear motifs in proteins. Nucleic Acids Res. **39**(2), W56–W60 (2011)
8. Schiller, M.R., Mi, T., Merlin, J.C., Deverasetty, S., Gryk, M.R., Bill, T.J., Brooks, A.W.: Minimotif Miner 3.0: database expansion and significantly improved reduction of false-positive predictions from consensus sequences. Nucleic Acids Res. **40**, 252–260 (2011)
9. Bailey, T.L., Elkan, C.J.: The value of prior knowledge in discovering motifs with meme. ISMB **3**, 21–29 (1995)
10. Bailey, T.L., Williams, N., Misleh, C., Li, W.: MEME: discovering and analyzing DNA and protein sequence motifs. Nucleic Acids Res. **34**, W369–W373 (2006)
11. Pandit, M., Maleki, M., Carruthers, N.J., Stemmer, P., Rueda, L.: Prediction of calmodulin-binding proteins using canonical motifs. In: Great Lakes Bioinformatics (GLBIO), Toronto, Canada (2016)
12. Mruk, K., Farley, B.M., Ritacco, A.W., Kobertz, W.R.: Calmodulation meta-analysis: Predicting calmodulin binding via canonical motif clustering. J. Gen. Physiol. **144**(1), 105–114 (2014)
13. Duda, R., Hart, P., Stork, D.: Pattern Classification, 2nd edn. Wiley, New York (2000)
14. Sharma, T.C., Jain, M.: WEKA approach for comparative study of classification algorithm. Intl. J. Adv. Res. Comput. Commun. Eng. **2**(4), 1925–1931 (2016)
15. Saeys, Y., Inza, I., Larraaga, P.: A review of feature selection techniques in bioinformatics. Bioinformatics **23**(9), 2507–2517 (2007)
16. Hall, M., Frank, E., Holmes, G., Pfahringer, B., Reutemann, P., Witten, I.H.: The WEKA data mining software: an update. SIGKDD Explor. **11**(1), 10–18 (2009)

Data Mining the Protein Data Bank to Identify and Characterise Chameleon Coil Sequences that Form Symmetric Homodimer β-Sheet Interfaces

Johanna Laibe[1(✉)], Melanie Broutin[2], Aaron Caffrey[1], Barbara Pierscionek[1], and Jean-Christophe Nebel[1]

[1] Faculty of Science, Engineering and Computing, Kingston University London, Kingston-upon-Thames, Surrey KT1 2EE, UK
k1552417@kingston.ac.uk

[2] Department of Bioengineering, Nice Sophia Antipolis University Engineering School, Templiers Campus, 06410 Biot, France

Abstract. A protein's environment may affect its secondary structure. In this study, the focus is on homodimers with symmetric β-sheet interfaces resulting from the conversion of coil sequences into β-strands. All homodimers in the Protein Data Bank relying on those chameleon sequences have been identified. Initial analysis based on sequential and structural features has revealed that many of those dimers display specific properties which could contribute to their detection. Such result is important since it could provide some insight on dimerisation and possibly aggregation mechanisms.

Keywords: Proteins · Homodimerisation · Intermolecular β-strand interfaces · Chameleon sequences · Aggregation

1 Introduction

A protein consists of a chain of amino acids which generally folds spontaneously into a unique three-dimensional conformation corresponding to its global energy minimum [1]. Failure of adopting that structure may lead to loss of function and even harmful effects [6]. As winners of the Paracelsus challenge [18] have shown, a limited number of mutations can dramatically change a protein conformation: a protein which adopts a four helix conformation was designed while retaining 50% identity of a predominantly β-sheet protein [5]. Similarly, it was demonstrated that mutation of a single amino acid could be sufficient to convert a β-strand into an α-helix [23]. In addition to mutations, a protein's environment may also affect its secondary structure. For example, it has been shown that the prion protein, PrP^C, changes its conformation and forms aggregates when interacting with one of its isoforms PrP^{SC} [17]. Those β-sheet aggregates are called amyloid fibrils [4] and have been linked to several human diseases including Alzheimer's, Parkinson's and Creutzfeldt–Jakob's [7].

This study investigates secondary structure alteration resulting from homodimerisation. More specifically, it focuses on coil sequences forming symmetric intermolecular

© Springer International Publishing AG 2017
I. Rojas and F. Ortuño (Eds.): IWBBIO 2017, Part II, LNBI 10209, pp. 118–126, 2017.
DOI: 10.1007/978-3-319-56154-7_12

β-strand interfaces. Following exhaustive search in the Protein Data Bank [3], properties of those 'chameleon' fragments were analysed. This led to the identification of specific features which should contribute to their detection and provide some insight on dimerisation and possibly aggregation mechanisms.

2   Methodology

Since very few proteins displaying that 'chameleon' property have been reported in the literature, with the notable exception of the Met-repressor like family, where all members share a similar ribbon-helix-helix structure that forms a homodimer interface by conversion of their ribbon into a β-strand [9], see Fig. 1, an exhaustive search was conducting using the Protein Data Bank [3]. This was performed according to the following process.

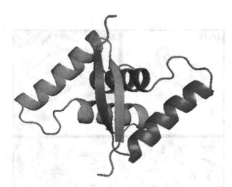

Fig. 1. Met-repressor like family interface (PDB 2P24): this symmetric interface is formed by the interaction of a ribbon-helix-helix pattern (RHH) from each chain. In the process, RHH converts to the β-strand-helix-helix pattern.

Firstly, the whole PDB was filtered to remove entries that don't contain two identical protein chains. Models with sequences with more than 30% identity were also discarded so that the set did not contain homologous proteins.

Secondly, homodimers interacting through at least an interface composed of a β-sheet were identified. This was performed by detecting the presence of amino acids belonging to β-strands from different chains whose C-alphas are within 5Å from each other, i.e. the interaction distance used by the CAPRI community-wide experiment (Critical Assessment of Prediction of Interactions) [10] which corresponds to the distance between two carbons alpha in a hydrogen bond.

Thirdly, for each remaining homodimer interacting through a β-sheet, information available in the 'SHEET' field of the PDB file was extracted to collect the interacting β-strand sequences, their nature, i.e. parallel or anti-parallel, and the number of strands forming the sheet involved in the interface. All anti-parallel interfaces of homodimers were then classified into two categories: the 'chameleon' interfaces, which are formed of exactly two β-strands each of them belonging to a different chain, i.e. the corresponding fragments would have a coil structure in the monomer form, and the 'standard' interfaces, which are formed of a β-sheet composed of at least four β-strands where each chain provides at least two β-strands, i.e. the corresponding fragments would already belong to a β-sheet in the monomer form. Although the existence of 'hybrid' interfaces, i.e. formed by one strand from one chain and two or more strands from the other chain, was also detected, they were not considered further in this study since their mixed environment would not be useful in identifying discriminative properties of chameleon fragments.

Finally, since analysis of the nature of the remaining interaction strands revealed that 90% of 'chameleon' interfaces are anti-parallel, and, among them, 70% are symmetric, it was decided to focus this study on those interfaces. In this work, a β-sheet interface was classified as symmetric, if both strands have the same amino acid sequence. Eventually, this process produced a dataset of 249 anti-parallel symmetric homodimer interfaces from non-homologous proteins: it comprises 80 'chameleon' and 169 'standard' interfaces (Fig. 2).

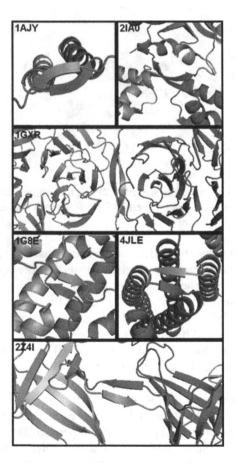

Fig. 2. Example of homodimers displaying symmetric anti-parallel 'chameleon' interfaces

To analyse differences between chameleon and standard fragments, a set of proper-ties was calculated for the two classes of interfaces under consideration. Firstly, since many protein interfaces (~1/3) display a recognizable hydrophobic core [13], hydro-phobicity of those protein interfaces was estimated. This was performed by calculating the grand average of hydropathy (GRAVY) value [11].

For each strand S_i of length n_i, its GRAVY values, G_i, is defined as:

$$G_i = \left(\Sigma_j H_{ij} \right) / n_i \qquad (1)$$

where H_{ij} is the hydropathy value of amino acid j in the strand S_i.

Secondly, given that β-sheets are created by interaction of β-strands through back-bone hydrogen bonds, interface hydrogen bond propensity may be informative about interface type. Using the structural information associated to each homodimer in its PDB file, all hydrogen bonds were retrieved from each β-sheet interface using the RING software with a 3.5Å threshold and the 'Closest' and 'Multiple' parameters, so that all atoms and multiple interactions are considered per residue pair, respectively [15].

Since a backbone residue can form up to 2 hydrogen bonds with an adjacent strand, for each strand S_i of length n_i, its hydrogen bond propensity, HB_i, is defined as:

$$HB_i = (\Sigma_j B_{ij}) / 2n_i \qquad (2)$$

where B_{ij} is the number of backbone hydrogen bonds formed by amino acid j in the strand S_i.

Thirdly, as experiments have shown that stability of antiparallel β-sheets is affected by their length [19], average strand length was calculated for each set. Finally, propen-sities of all amino acids were calculated.

While Table 1 presents average hydrophobicity, hydrogen bond propensity and strand length of chameleon and standard anti-parallel homodimer β-sheet interfaces, Fig. 3 show their amino acid propensities. One observes that neither average hydropho-bicity nor average hydrogen bond propensity is affected by the interface type. On the other hand, chameleon interfaces are much shorter than standard interfaces which are three residues longer in average. Moreover, there are significant differences in their amino acid propensity profiles in particular for aromatic and charged amino acids.

Table 1. Average hydrophobicity, hydrogen bond propensity and strand length of chameleon and standard interfaces

	Chameleon interfaces	Standard interfaces
Average hydrophobicity	0.59	0.52
Average hydrogen bond propensity	0.43	0.42
Average strand length	5.0	7.9

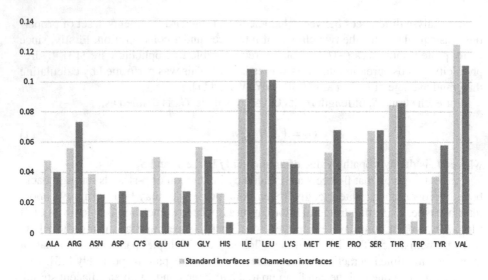

Fig. 3. Amino acid propensities of chameleon and standard interfaces

To explore combinations of features whish may allow discriminating chameleon fragments, unsupervised clustering was performed using different sets of features. More specifically, data were processed using a general purpose clustering tool, CLUTO [16], which has been used in a variety of bioinformatics applications [2, 8, 12, 14]. In order to give each feature equal weight, a normalization process is applied. For each feature F, its values, F_i, are normalised between 1 and −1 [20] as:

$$F_{i\_normalised} = 2\left(F_i - F_{min}\right) / \left(F_{max} - F_{min}\right) - 1 \qquad (3)$$

where F_{max} and F_{min} represent the maximum and minimum values of the feature F.

Using hierarchical partitional clustering, CLUTO produces a binary tree representing similarities between interface profiles and identifies specific clusters within the tree. Note that the quality of each cluster is estimated by its internal similarity (ISim), i.e. the average similarity between the interfaces of the cluster, and its external similarity (ESim), i.e. the average similarity between the interfaces of the cluster and all the other interfaces. The "ideal" cluster would have: ISim = 1.0 and ESim = 0.0.

In addition, CLUTO displays feature values for each interface using a colour palette: shades of green and red indicate feature values between −1 and 1 respectively.

3   Results

Informed by results presented in Table 1 and Fig. 3, all interfaces of interest were clustered using CLUTO and a combination of features including length, hydrophobicity, proline, aromatic (without histidine) and charged amino acid propensities. Figure 4 shows the most discriminative clusters produced using subsets of those properties.

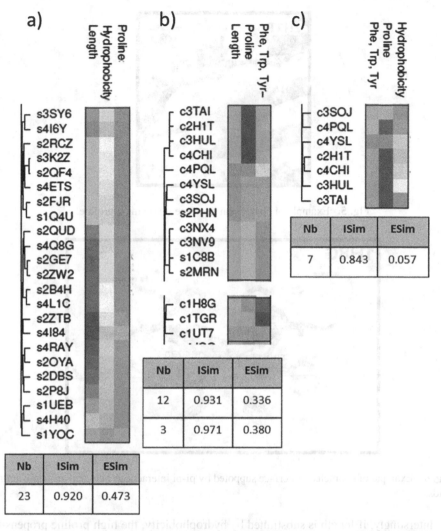

Fig. 4. Good quality interface clusters created by CLUTO according to different sets of properties. The prefix added to PDB ids specifies if an interface is chameleon, "c", or, standard, "s". (Color figure online)

Based on length, hydrophobicity and proline propensity, an homogeneous cluster of relatively good quality allows to discriminate 23 "standard" interfaces, see Fig. 4a. All those interfaces display a high length, low proline propensity and relatively average hydrophobicity. Usage of length, proline and non charged aromatic amino acids (Phe, Trp and Tyr) reveals two good quality clusters, see Fig. 4b, populated mainly of "chameleon" interfaces – 12 "chameleon" and only 3 "standard": both are composed of short interfaces, but one has a high proline propensity, see Fig. 5, while the other one has a high aromatic propensity, see Fig. 6.

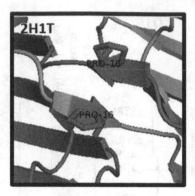

Fig. 5. Example of chameleon interface involving a proline.

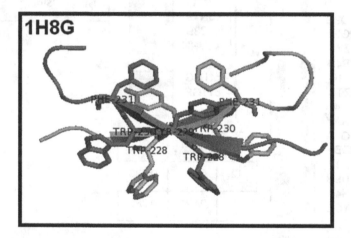

Fig. 6. Example of chameleon interface suppoted by pi-pi interactions between aromatic amino acids.

Interstingly, if length is substituted by hydrophobicity, the high proline propensity group is reduced from 12 to 7 members, but is only composed of "chameleon" proteins, see Fig. 4c. Note that among the only 3 "standard" interfaces with high proline content classified in a largely chameleon cluster, one of them, PDB id 1C8B, displays a sheet structure which is "almost" chamelon, since the non interface strands are much shorter than the interface ones, See Fig. 7.

Since usage of strand length proved useful to produce the clusters shown on Fig. 4a and b, it was also used as sole feature to discriminate between chameleon and standard interfaces: whereas among interfaces based on strands of length 3 amino acids, 91% of them, i.e. 30, are chameleon, all strands of size 10 or more form standard interfaces, i.e. 48.

This initial analysis of chameleon interfaces has revealed that a many of those chameleon dimers (45%) display properties, i.e. short length, high aromatic or proline propensity, allowing to discriminate them from standard ones. Moreover, this study suggests that there are unlikely to form long β-strands since none of them was composed of 10 or more

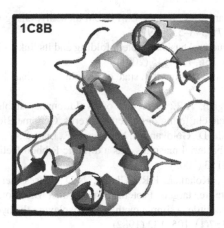

Fig. 7. "Standard" interfaces where non interface strands are much shorter than the interface ones.

residues. There is no doubt that more advanced machine learning approaches, such as support vector machines, neuron networks and decision trees [21, 22], would allow to combine the identified features and others to further characterise chameleon interfaces. Since, many chameleon fragments have been associated to human diseases through aggregation [4, 7, 17], the ability to detect a specific class of chameleon fragments, i.e. those able to form symmetric homodimer β-sheet interfaces, should contribute, not only, to a better insight about homodimerisation, but also in aggregation mechanisms.

4 Conclusion

This study has identified in the Protein Data Bank all symmetric homodimers relying on β-sheet interfaces involving the conversion of coil sequences into β-strands. Initial comparison with standard intermolecular β-strand interfaces has revealed that many of those chameleon dimers display specific properties which should contribute to their detection. When possible, this could provide some insight on homodimerisation and possibly aggregation mechanisms.

References

1. Anfinsen, C.B., et al.: The kinetics of formation of native ribonuclease during oxidation of the reduced polypeptide chain. Proc. Natl. Acad. Sci. U.S.A. **47**, 1309–1314 (1961)
2. Balasubramaniyan, R., Hüllermeier, E., Weskamp, N., Kämper, J.: Clustering of gene expression data using a local shape-based similarity measure. Bioinformatics **21**(7), 1069–1077 (2005)
3. Berman, H.M., et al.: The protein data bank. Nucleic Acids Res. **28**, 235–242 (2000)
4. Chiti, F., Dobson, C.M.: Protein misfolding, functional amyloid, and human disease. Ann. Rev. Biochem. **75**, 333–366 (2006)

5. Dalal, S., Balasubramanian, S., Regan, L.: Protein alchemy: changing β-sheet into α-helix. Nature Struct. Biol. **4**(7), 548–552 (1997)

6. Dobson, C.M.: The structural basis of protein folding and its links with human disease. Phil. Trans. R Soc. Lond. B. **356**, 133–145 (2001)

7. Eisenberg, D., Jucker, M.: The amyloid state of proteins in human diseases. Cell **148**, 1188–1203 (2012)

8. Glazko, G.V., Mushegian, A.R.: Detection of evolutionarily stable fragments of cellular pathways by hierarchical clustering of phyletic patterns. Genome Biol. **5**, R32 (2004)

9. Golovanov, A.P., Barilla, D., Golovanova, M., Hayes, F., Lian, L.-Y.: Parg, a protein required for active partition of bacterial plasmids, has a dimeric Ribbon-Helix-Helix structure. Mol. Microbiol. **50**, 1141–1153 (2003)

10. Jauch, R., Yeo, H.C., Kolatkar, P.R., Clarke, N.D.: Assessment of CASP7 structure predictions for template free targets. Proteins Struct. Funct. Bioinfor. **69**, 57–67 (2007)

11. Kyte, J., Russell, F.: Doolittle. a simple method for displaying the hydropathic character of a protein. J. Mol. Biol. **157**(1), 105–132 (1982)

12. Lanara, Z., Giannopoulou, E., Fullen, M., Kostantinopoulos, E., Nebel, J.-C., Kalofonos, H.P., Patinos, G.P., Pavlidis, C.: Comparative study and meta-analysis of meta-analysis studies for the correlation of genomic markers with early cancer detection. Hum. Genomics **7**, 14 (2013)

13. Larsen, T.A., Olson, A.J., Goodsell, D.S.: Morphology of protein–protein interfaces. Structure **6**, 421–427 (1998)

14. Nebel, J.-C., Herzyk, P., Gilbert, D.R.: Automatic generation of 3D motifs for classification of protein binding sites. BMC Bioinf. **8**, 32 (2007)

15. Piovesan, D., Minervini, G., Tosatto, S.C.E.: The RING 2.0 web server for high quality residue interaction networks. Nucleic Acids Research (2016)

16. Rasmussen, M.D., Deshpande, M.S., Karypis, G., Johnson, J., Crow, J.A., Retzel, E.F.: wCLUTO: a Web-enabled clustering toolkit. Plant Physiol. **133**(2), 510–516 (2003)

17. Roostaee, A., Cote, S., Roucou, X.: Aggregation and amyloid fibril formation induced by chemical dimerization of recombinant prion protein in physiological-like conditions. J. Biol. Chem. **284**(45), 30907–30916 (2009)

18. Rose, G., Creamer, T.: Protein folding: predicting predicting. Proteins Struct. Funct. Genet. **19**(1), 1–3 (1994)

19. Stanger, H.E., Syud, F.A., Espinosa, J.F., Giriat, I., Muir, T., Gellman, S.H.: Length-dependent stability and strand length limits in antiparallel β-sheet secondary structure. Proc. Natl. Acad. Sci. U.S.A. **98**, 12015–12020 (2001)

20. Su, C.-H., Pal, N.R., Lin, K.-L., Chung, I.-F.: Identification of amino acid propensities that are strong determinants of linear b-cell epitope using neural networks. PLoS ONE **7**(2), e30617 (2012). doi:10.1371/journal.pone.0030617

21. Vapnik, V.N., Kotz, S.: Estimation of Dependences Based on Empirical Data. Springer, New York (2006). ISBN 0-387-30865-2

22. Winston, P.H.: Artificial Intelligence, 3rd edn. Addison-Wesley, Boston (1992)

23. Yang, W.Z., et al.: Conversion of a β-strand to an α-Helix induced by a single-site mutation observed in the crystal structure of Fis mutant Pro26Ala. Protein Sci. **7**(9), 1875–1883 (1998)

Associating Gene Ontology Terms with Pfam Protein Domains

Seyed Ziaeddin Alborzi[1,3], Marie-Dominique Devignes[2],
and David W. Ritchie[3(✉)]

[1] Université de Lorraine, LORIA, UMR 7503, 54506 Vandœuvre-lès-Nancy, France
[2] CNRS, LORIA, UMR 7503, 54506 Vandœuvre-lès-Nancy, France
marie-domonique.devignes@loria.fr
[3] Inria Nancy Grand-Est, 54600 Villers-lès-Nancy, France
{seyed-ziaeddin.alborzi,dave.ritchie}@inria.fr

Abstract. With the growing number of three-dimensional protein structures in the protein data bank (PDB), there is a need to annotate these structures at the domain level in order to relate protein structure to protein function. Thanks to the SIFTS database, many PDB chains are now cross-referenced with Pfam domains and Gene ontology (GO) terms. However, these annotations do not include any explicit relationship between individual Pfam domains and GO terms. Therefore, creating a direct mapping between GO terms and Pfam domains will provide a new and more detailed level of protein structure annotation. This article presents a novel content-based filtering method called GODM that can automatically infer associations between GO terms and Pfam domains directly from existing GO-chain/Pfam-chain associations from the SIFTS database and GO-sequence/Pfam-sequence associations from the UniProt databases. Overall, GODM finds a total of 20,318 non-redundant GO-Pfam associations with a F-measure of 0.98 with respect to the InterPro database, which is treated here as a "Gold Standard". These associations could be used to annotate thousands of PDB chains or protein sequences for which their domain composition is known but which currently lack any GO annotation. The GODM database is publicly available at http://godm.loria.fr/.

Keywords: Protein structure · Protein function · Gene Ontology · Content-based filtering

1 Introduction

Proteins carry out many important biological functions. At the molecular level, these functions are often performed by highly conserved regions called "domains". Currently, the Pfam database is one of the most widely used sequence-based classifications of protein domains and domain families [1]. Protein domains may also be considered as building blocks which are combined in different ways in order to endow different proteins with different functions.

© Springer International Publishing AG 2017
I. Rojas and F. Ortuño (Eds.): IWBBIO 2017, Part II, LNBI 10209, pp. 127–138, 2017.
DOI: 10.1007/978-3-319-56154-7_13

A given Pfam domain might exist in several different proteins. It is widely accepted that protein domains often correspond to distinct and stable three-dimensional (3D) structures, and that there is often a close relationship between protein structure and protein function [2]. The Protein Data Bank (PDB) [3,4] contains more than 107,000 3D structures, that have been determined by X-ray crystallography or NMR spectroscopy. As well as sequence-based and structure-based classifications, proteins may also be classified according to their function. For example, the Gene Ontology (GO) [5] organizes a controlled vocabulary describing the biological process (BP), molecular function (MF), and cellular component (CC) aspects of gene annotation. It provides an ontology of defined terms to unify the representation of the gene and protein roles in cells. The GO vocabulary is structured as a rooted Directed Acyclic Graph (rDAG) in which GO terms are nodes connected by different hierarchical relations. Each GO term within the gene ontology has a term name, a distinct alphanumeric identifier, and a namespace indicating to which ontology it belongs.

Although the GO is very useful, it does not generally provide a direct relationship between biological function and a (sequence-based) Pfam domain. Figure 1 illustrates the different kinds of relationships that can occur when considering GO-protein annotations at the domain level. Except for simple single-domain proteins where the mapping is obvious, it is generally not possible to compare and classify structure-function relationships at the domain level. An interesting exception is the dcGO database which provides multiple ontological annotations (Gene Ontology: GO, EC, pathways, phenotype, anatomy and disease ontologies) for protein domains [6]. In dcGO, an association between an ontology term and a domain is inferred from the principle that if a term tends to be attached to proteins in UniProtKB that contain a certain domain, then the term should be associated with that domain. For each Pfam domain, dcGO compares the number of Uniprot sequences containing that domain and annotated with a certain GO term to what could be obtained if association was random. The statistical significance of the association is then assessed using a hypergeometric distribution, followed by multiple hypotheses testing in terms of false discovery rate. Only significant associations are retained in the dcGO database.

Nonetheless, we found that there are several GO-Pfam associations from manually curated data sources (e.g. InterPro) which are not present in dcGO. Moreover, based on our previous ECDomainMiner approach [7,8] to discover associations between EC numbers and protein domains, we found that there are many reliable EC-Pfam associations which are not covered by dcGO. Furthermore, there are thousands of protein structures in the PDB which lack GO annotations. If there is a direct association between protein domains and GO terms, these structures can be annotated through their associated domains. Based on our analysis, we estimated that dcGO associations can only annotate 43% of the unannotated PDB structures. Therefore, we were motivated to develop a more systematic approach, which we call "GODM" ("GO Domain Miner"), with the aim of discovering a much larger set of GO-domain associations than dcGO.

GODM uses a "recommender-based" approach for finding direct associations between GO terms and Pfam domains. We recently developed a similar recommender-based approach called "ECDomainMiner" for assigning enzyme classification (EC) numbers to Pfam domains [8]. Thus, the GODM approach described here represents a natural extension of our previously developed ECDomainMiner approach. Recommender systems are a subclass of information filtering system [9,10] which seek to predict a list of items that might be of interest to an on-line customer, and are divided into two main types. Collaborative filtering approaches make associations by calculating the similarity between activities of users [11,12]. In contrast, content-based filters predict associations between user profiles and description of items by identifying common attributes [10,13]. Here, we use content-based filtering to associate GO terms with Pfam domains from existing GO-chain and Pfam-chain associations from SIFTS [14], and GO-sequence and Pfam-sequence associations from SwissProt and TrEMBL. As well has handling simple one-to-one associations as in dcGO (Fig. 1 part A), GODM can also resolve cases where multiple GO terms are associated with multi-domain chains (Fig. 1 parts B, C, and D).

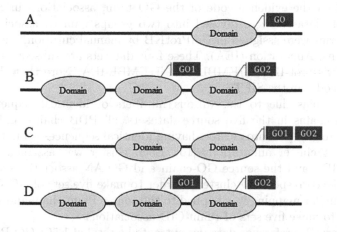

Fig. 1. A graphical representation of different situations of GO-Domain association in a protein sequence or structure.

While SwissProt and TrEMBL were originally developed separately, both databases have since been incorporated in the UniProt resource. SwissProt now represents a non-redundant, high quality, manually curated part of UniProt Knowledge Base (UniProtKB). In contrast, TrEMBL is an automatically annotated and unreviewed part of UniProtKB, and contains around 40 times more entries than SwissProt. In order to parameterise and evaluate our method, we use the InterPro database [15] which contains a large number of manually curated GO-Pfam associations. We assess the performance of our approach against a "Gold Standard" dataset derived from InterPro, and we compare our results with the GO-Pfam associations available from the dcGO database. We also show how our database of more than 20,000 GO-Pfam associations for molecular function ontology can be exploited for automatic annotation purposes.

2   Methods

2.1   Data Preparation

Flat data files of SIFTS (July 2015), Uniprot (July 2015), and InterPro (version 53.0) were downloaded and parsed using in-house Python scripts. From the SIFTS data, associations between PDB chains and GO terms, and associations between PDB chains and Pfam domains were extracted in which each GO term is a leaf in the hierarchy of the Molecular Function ontology (GO-MF) and each Pfam refers either to a Pfam domain or a Pfam family (i.e. Pfam motifs and repeats were excluded). Associations between Uniprot sequence accession numbers (ANs) and GO terms from GO-MF, and AN-Pfam associations were then extracted from the SwissProt and TrEMBL sections of Uniprot to give two datasets of Swissprot associations and TrEMBL associations, respectively. Then, based on the evidence code of the GO term, associations in SwissProt and TrEMBL datasets were divided into two groups namely, associations for which GO terms were assigned in UniProtKB by manual curation, and Inferred from Electronic Annotation (IEA). These four datasets are subsequently called Swissprot, Swissprot-IEA, TrEMBL, and TrEMBL-IEA. Note that there were no evidence codes in the SIFTS.

To reduce bias due to the various numbers of identical sequences and sequences of chains in the five source datasets, all PDB chains and Uniprot sequences were grouped into clusters having identical sequences using the Uniref non-redundant cluster annotations [16]. Each cluster was assigned a unique identifier (CID), and the source GO-chain and GO-AN associations were then mapped to the corresponding cluster in order to make five sets of GO-CID associations. A similar mapping was applied to the source Pfam-chain and Pfam-AN associations to make five sets of Pfam-CID associations.

For the InterPro reference data, we extracted a total of 1,561 GO-Pfam associations in which each GO term is a leaf node of the molecular function ontology and each Pfam refers to either a Pfam domain or a Pfam family. These associations were considered to be "true" associations. However, for training and filtering purposes, we also needed some examples of "false" associations. We therefore selected a set of the lowest-scoring GO-Pfam associations with the same size as InterPro dataset from the other datasets. These associations have to belong to at least two out of five datasets with no intersection with InterPro dataset. Because these associations have very little support in the data, we consider them to be "false" associations. Then, we randomly divided the InterPro dataset and our calculated "false" associations into two "Training" and "Test" subsets of the same size (each having half of the "true" and "false" associations). These two subsets were used for training and evaluation purposes respectively.

In the rest of this article, we will refer to the InterPro dataset as our "Gold Standard" dataset.

2.2 Finding GO-Pfam Associations by Content-Based Filtering

For each of the five datasets, all GO-CID relations are encoded in a binary (GO × CID) matrix, where a 1 represents the presence of a GO annotation and a 0 represents no annotation. This matrix is then row-normalised such that each row has unit magnitude when considered as a vector. Similarly, all CID-Pfam relations are encoded in a second binary (CID × Pfam) matrix which is column-normalised. Consequently, calculating the product of the two normalised matrices corresponds to calculating a matrix of cosine similarity scores between the rows of the first matrix and the columns of the second matrix. Thus, the product matrix represents an array of raw GO-Pfam association scores. Because we wish to draw upon the relations from all five input datasets, we combine the five scores to give a single normalized confidence score (CS):

$$CS_{go,d} = \frac{\sum_i w_i S_i(go, d)}{\sum_i w_i} \tag{1}$$

where $i \in \{SIFTS, Swissprot, Swissprot\text{-}IEA, TrEMBL, TrEMBL\text{-}IEA\}$ enumerates the five datasets, w_i are weight factors, to be determined, and where an individual association score, $S_i(go, d)$ is set to zero whenever there is no data for a given go and d. In order to calculate the weight factors, we calculated Receiver-Operator-Characteristic (ROC) curves [17] using the true associations from the Interpro Training set and all other associations as background associations. The weights were varied from 0.0 to 1.0 in steps of 0.1, and for each combination, associations were scored and ranked, and area under the curve (AUC) was calculated. Finally, we selected the combination of weights that gave the best area under the curve (AUC) of the ROC curve.

2.3 Defining a Confidence Score Threshold

Having determined the best weight for each data source, we next wished to determine a threshold for the confidence score. We scored and ranked the members of the Training set of InterPro, and divided the ranked list into two subsets according to a threshold value that was varied from 0.0 to 1.0 in steps of 0.01. For each threshold value, we counted the number of true associations above the threshold, here called true positives (TPs), false associations above the threshold, false positives (FPs), false associations below the threshold, true negatives (TNs), and true associations below the threshold, false negatives (FNs). We then calculated the "F-measure" which is a harmonic mean of recall and precision using:

$$F = \frac{2 \times TP}{2 \times TP + FP + FN} \tag{2}$$

The score threshold that gave the best F-measure was confirmed by verifying that the F-measure calculated on the Test dataset is also very high. This threshold was thus selected as the best threshold to use for accepting predicted associations.

2.4   Hypergeometric Statistical Analysis

While the above procedure provides a systematic way to infer GO-Pfam associations, we wished to estimate the statistical significance, and thus the degree
of confidence, that might be attached to those predictions. More specifically, we
wished to calculate the probability, or "p-value", that a GO term, go, and a Pfam
domain, d, could be found to be associated simply by chance. For example, it is
natural to suppose such associations can be predicted at random if go or d are
highly represented in the structure/sequence CIDs. In principle, in order to estimate the probability of getting our GO-Pfam associations by chance, one could
generate random datasets by shuffling the relations between GO terms and CIDs
on the one hand, and between Pfam domains and CIDs on the other hand. However, this is quite impractical given the very large numbers of CIDs, GO terms,
and Pfam domains, and the complexity of the filtering procedure that would
have to be repeated for each shuffled version of the dataset. Therefore, following [6], we assume that within each dataset (SIFTS, Swissprot, Swissprot-IEA,
TrEMBL, or TrEMBL-IEA), the random hypothesis for the (go, d) association is
represented by the hypergeometric distribution of the expected number of CIDs
associated with both go and d.

Letting N denote the total number of CIDs, N_d the number of CIDs related
to the Pfam domain d, and N_{go} the number of CIDs related to the GO term go,
the hypergeometric probability distribution is given by

$$p(X_{go,d} \geqslant K_{go,d}) = \frac{\sum_{i=K_{go,d}}^{\min (N_d, N_{go})} \binom{N_{go}}{i}\binom{N-N_{go}}{N_d-i}}{\binom{N}{N_d}}, \tag{3}$$

where $p(X_{go,d} \geqslant K_{go,d})$ represents, in each dataset, the probability of having
a number $X_{go,d}$ equal to or greater than the observed number $K_{go,d}$ of CIDs
associated with both d and go. Traditionally, a p-value of less than 0.05 is taken
to be statistically significant. However, because this test is applied to a large
number of GO-Pfam associations, we apply a Bonferoni correction which takes
into account the so-called family-wise error rate (FWER) [18]. We therefore consider any p-value less than $0.05/T$ as denoting a statistically significant inferred
GO-Pfam association in a dataset, with T the total number of tested GO-Pfam
associations for that dataset.

2.5   Gold, Silver, and Bronze Associations

In order to differentiate associations based on their quality and reliability,
our method categorizes associations into three classes of "Gold", "Silver", and
"Bronze" using their calculated similarity scores and p-values. An association
belongs to the Gold class if all its available p-values are statistically significant.
The Silver class consists of associations for which the number of statistically significant p-values among the five datasets is greater than or equal to the number
of statistically insignificant p-values (e.g. GO-Pfam is a Silver associations if its

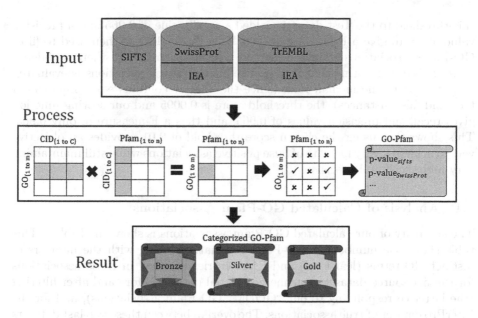

Fig. 2. A schematic overview of the GODM procedure.

p-values are significant in SIFTS, SwissProt, and TrEMBL-IEA). The remaining associations are assigned to the Bronze class. An illustration of the whole procedure is shown in Fig. 2.

3 Results

Our method takes as input five large datasets of MF GO-chain associations from SIFTS, and MF GO-sequence associations from SwissProt, SwissProt-IEA, TrEMBL and TrEMBL-IEA as well as five large datasets of Pfam-Chain and Pfam-sequence associations. These source datasets were merged to give a global dataset of 1,161,372 non-redundant GO-Pfam associations. Using the reference InterPro dataset of 1561 "true" associations against background associations, the best ROC-plot AUC value of 0.99 was obtained with the weights $w_{SIFTS} = 10$, $w_{SwissProt} = 1$, $w_{SwissProt-IEA} = 10$, $w_{TrEMBL} = 1$, and $w_{TrEMBL-IEA} = 8$. These weights clearly give a greater importance to the GO-Pfam associations from SIFTS and the IEA (Inferred from Electronic Annotation) section of SwissProt and TrEMBL compared to those derived from TrEMBL and the manually curated section of SwissProt.

In order to reduce the number of false associations predicted by our approach (and not just to simply optimise the overall AUC performance), various threshold values of the confidence score (using the above weights) were tested on the Training dataset using the F-measure (Sect. 2.3) with respect to the number of true and false associations having scores above or below the threshold. This gave an optimal threshold score of 0.01 for a maximum F-Measure of 0.99. Applying this threshold to the Test dataset yielded a recall value of 0.965 and a precision value of 1.0 to give a F-measure of 0.98. This threshold was then used to filter GO-Pfam associations from the merged dataset according to their confidence score. It is worth noting that if the ranked list of Test associations is evaluated with respect to the median rank (since the dataset contains equal numbers of true and false instances), the threshold score is 0.0095 and our scoring function gives recall and precision values of 0.965, and thus a F-measure of only 0.965. This shows that using the chosen score threshold of 0.01 provides an objective way to achieve a very low rate of false positive associations while still maintaining very high recall and precision.

3.1   Analysis of Calculated GO-Pfam Associations

The summary of our calculated GO-Pfam associations is shown in Table 1. This table shows the numbers of GO-Pfam associations along with the numbers of distinct GO terms (leaf level) and Pfam entries involved in those associations for the five source datasets, our merged global dataset before and after filtering (the latter corresponding to our "GODM" GO-Pfam associations), and for the InterPro dataset of true associations. The overlap between these two last datasets is shown in the last line of the table.

Table 1. Statistics on the given and filtered MF GO-Pfam associations.

Dataset	GO-Pfam associations	GO terms	Pfam entries
SIFTS	10,064	2,763	3,370
SwissProt	22,435	4,220	4,669
SwissProt-IEA	28,982	3,228	4,469
TrEMBL	22,031	2,766	3,613
TrEMBL-IEA	1,136,711	4,254	9,342
Merged	1,161,372	5,510	9,929
Filtered associations (GODM)	20,318	5,047	6,154
Common with InterPro	1,519	586	1,362
InterPro	1,561	591	1,390

Overall, Table 1 shows that our approach yielded a total of 20, 318 GO-Pfam associations that include 1, 519 associations already present in InterPro. While this shows that our method finds 97.3% of the "correct" GO-Pfam associations in InterPro, it also shows that only 2.7% of the correct InterPro associations have confidence scores below our optimal score threshold of 0.01. This relatively high proportion of common associations reflects the fact that our method is designed to give relatively strong support (Confidence Score) to the correct associations in InterPro based on the five input sources. Concerning statistical significance, nearly half of the GO-Pfam associations belong to the Gold class (48%).

3.2 Comparison Between Our GODM and InterPro GO-Pfam Associations

Figure 3 (A) shows the average number of GO-Pfam associations per GO term and Pfam entry both for InterPro (shown in grey) and our calculated GODM dataset (in black). The ratio for our method is higher for GO terms (4.03 versus 2.64) and Pfam entries (3.3 versus 1.12), which reflects: (i) a significant enrichment in the annotation of Pfam domains; and (ii) participation of Pfam domains in different functions as either a single domain or a part of a complex.

Figure 3 (B) shows the distribution of GO terms (in grey) and Pfam entries (in black) according to the number of associations they are involved in. More than 1,800 GO terms and 2,500 Pfam entries are involved in single associations, i.e. associated with a single Pfam domain and a single GO term respectively. Intersection of these single association sets yields a list of 135 one-to-one GO-Pfam associations. Nevertheless, the distribution also shows that our collection of associations rather favours multiple associations, thereby reflecting the complex many-to-many relationships that exist within the original datasets.

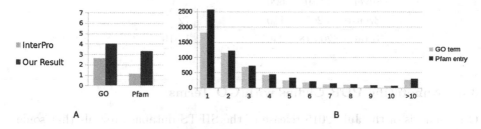

Fig. 3. A: average number of GO-Pfam associations per GO terms and per Pfam entry for the InterPro (grey) and our calculated GODM (black) datasets. B: distribution of GO terms according to their numbers of associations with Pfam entries (grey) and Pfam entries according to their numbers of associations with GO terms (black).

3.3 Comparing GODM and dcGO GO-Pfam Associations

In order to compare our results with dcGO [6], we extracted the Pfam2GO associations from the dcGO website (http://supfam.org/SUPERFAMILY/dcGO) where GO terms are leaves in the MF hierarchy of GO terms. This Pfam2GO dataset includes 3,086 GO-Pfam associations. Figure 4 shows that a total of 2,401 GO-Pfam associations are common to dcGO and our results (overlap B) while only 404 GO-Pfam associations are common between InterPro and dcGO (overlap C). Furthermore, this comparison shows that our GODM dataset contains 17,917 (20,318-2,401) additional GO-Pfam associations that are not available in the dcGO dataset. In a more detailed analysis, the overlap between the GODM and Pfam2GO datasets was studied with respect to our three quality classes. As summarized in the Table 2, the overlap between the two datasets contains 1,621, 600, and 180 Gold, Silver, and Bronze associations, respectively.

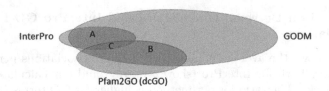

Fig. 4. Venn diagram showing the intersection between Pfam2GO (3,086 associations) from dcGO, our GODM associations (20,318 associations), and manually curated associations (1,561 associations) from InterPro. Region A (1,519 associations) is the overlap between our result and InterPro associations. Region B (2,401 associations) is the common associations between our result and Pfam2GO. Region C (404 associations) is the overlap between Pfam2GO and InterPro associations.

Table 2. Overlap between associations from GODM, Pfam2GO of dcGO, and InterPro.

Dataset	GODM	Overlap	
		With Pfam2GO	With InterPro
Gold	9,771	1,621	922
Silver	4,280	600	455
Bronze	6,267	180	72
Total	20,318	2401	1,519

3.4   Annotating PDB Chains with GO Terms

Our analysis of the July 2015 release of the SIFTS database reveals that some 41% of PDB entries currently lack a leaf GO term annotation. Indeed, we found that a total of 48,409 PDB chains lacking GO annotations in SIFTS include at least one of the 6,154 Pfam domains present in our calculated GODM associations. For those chains, GODM finds 19,371, 7,176 and 12,530 Gold, Silver, and Bronze GO-Pfam associations, respectively, giving a total of 39,077 PDB chains that could benefit from the annotations inferred by GODM. Moreover, 153 PDB chains could benefit from unambiguous one-to-one GO-Pfam associations.

To give an example, GODM finds a Gold association between PF03018 (Dirigent-like protein) and GO term GO:0042349 ("Guiding stereospecific synthesis activity"). Interestingly, the PF03018 domain is present in the PDB chain 4REV A ("Structure of the dirigent protein DRR206") which is not annotated by any GO term from the molecular function ontology. Consequently the GODM recommendation is to annotate the 4REV PDB entry with GO:0042349 term, which explicitly describes the possible function of this protein. Another example is PDB structure 2YRB, which is described only as "the solution structure of the first C2 domain from human KIAA1005 protein", and for which its previously assigned Pfam domain (PF11618) is annotated as a "protein of unknown function (DUF3250)". In this case, GODM finds a Gold association between PF11618 and GO:0031870 (thromboxane A2 receptor binding) thus indicating that this structure could be annotated with that GO term.

4 Conclusion

We have presented a systematic content-based filtering approach for assigning GO terms to protein domains and then categorizing those associations. This was achieved by first collecting existing annotations of protein chains or sequences, namely Pfam domain compositions on one hand and GO-MF leaf term annotations on the other. We then applied the content-based filtering method to find a list of direct associations between GO-MF leaf terms and Pfam domains. Our approach is able to infer a total of 20,318 direct GO-Pfam associations. Thus, compared to the 1,561 manually curated GO-Pfam associations from InterPro database, our approach discovers over 13 times as many associations in a completely automatic way. We have also proposed some possible ways to further analyze the coverage of the our approach. We believe that the large numbers of GO-Pfam associations calculated using our approach can considerably contribute to enriching the annotations of PDB protein chains, and that this will facilitate a better understanding and exploitation of structure-function relationships at the protein domain level.

Acknowledgments. This project is funded by Agence Nationale de la Recherche (grant number ANR-11-MONU-006-02), the Institut National de Recherche en Informatique et Automatique, and Région Lorraine.

References

1. Finn, R.D., Bateman, A., Clements, J., Coggill, P., Eberhardt, R.Y., Eddy, S.R., Heger, A., Hetherington, K., Holm, L., Mistry, J., Sonnhammer, E.L., Tate, J., Punta, M.: Pfam: the protein families database. Nucleic Acids Res. **42**(D1), D222–D230 (2014)
2. Berg, J.M., Tymoczko, J.L., Stryer, L.: Protein Structure and Function. W.H Freeman, New York (2002)
3. Bernstein, F.C., Koetzle, T.F., Williams, G.J., Meyer, E.F., Brice, M.D., Rodgers, J.R., Kennard, O., Shimanouchi, T., Tasumi, M.: The protein data bank. Eur. J. Biochem. **80**(2), 319–324 (1977)
4. Gutmanas, A., Alhroub, Y., Battle, G.M., Berrisford, J.M., Bochet, E., Conroy, M.J., Dana, J.M., Montecelo, M.A.F., van Ginkel, G., Gore, S.P., Haslam, P., Hatherley, R., Hendrickx, P.M.S., Hirshberg, M., Lagerstedt, I., Mir, S., Mukhopadhyay, A., Oldfield, T.J., Patwardhan, A., Rinaldi, L., Sahni, G., Sanz-García, E., Sen, S., Slowley, R.A., Velankar, S., Wainwright, M.E., Kleywegt, G.J.: PDBe: protein data bank in europe. Nucleic Acids Res. **42**(D1), D285–D291 (2014)
5. Ashburner, M., Ball, C.A., Blake, J.A., Botstein, D., Butler, H., Cherry, J.M., Davis, A.P., Dolinski, K., Dwight, S.S., Eppig, J.T., et al.: Gene ontology: tool for the unification of biology. Nature Genet. **25**(1), 25–29 (2000)
6. Fang, H., Gough, J.: dcGO: database of domain-centric ontologies on functions, phenotypes, diseases and more. Nucleic Acids Res. **41**(D1), D536–D544 (2013)
7. Alborzi, S.Z., Devignes, M.D., Ritchie, D.W.: EC-PSI: associating enzyme commission numbers with Pfam domains. bioRxiv, 022343 (2015)

8. Alborzi, S.Z., Devignes, M.D., Ritchie, D.W.: ECDomainminer: discovering hidden associations between enzyme commission numbers and pfam domains. BMC Bioinform. **18**(1), 107 (2017)
9. Hanani, U., Shapira, B., Shoval, P.: Information filtering: overview of issues, research and systems. User Model. User-Adap. Interact. **11**(3), 203–259 (2001)
10. Ricci, F., Rokach, L., Shapira, B.: Introduction to recommender systems handbook. In: Ricci, F., Rokach, L., Shapira, B., Kantor, P.B. (eds.) Recommender Systems Handbook, pp. 1–35. Springer, Heidelberg (2011)
11. Breese, J.S., Heckerman, D., Kadie, C.: Empirical analysis of predictive algorithms for collaborative filtering. In: Proceedings of the Fourteenth Conference on Uncertainty in Artificial Intelligence, pp. 43–52. Morgan Kaufmann Publishers Inc. (1998)
12. Koren, Y., Bell, R.: Advances in collaborative filtering. In: Ricci, F., Rokach, L., Shapira, B. (eds.) Recommender Systems Handbook, pp. 77–118. Springer, Heidelberg (2015)
13. Basu, C., Hirsh, H., Cohen, W., et al.: Recommendation as classification: using social and content-based information in recommendation. In: AAAI/IAAI, pp. 714–720 (1998)
14. Velankar, S., Dana, J.M., Jacobsen, J., van Ginkel, G., Gane, P.J., Luo, J., Oldfield, T.J., ODonovan, C., Martin, M.J., Kleywegt, G.J.: SIFTS: structure integration with function, taxonomy and sequences resource. Nucleic Acids Res. **41**(D1), D483–D489 (2013)
15. Mitchell, A., Chang, H.Y., Daugherty, L., Fraser, M., Hunter, S., Lopez, R., McAnulla, C., McMenamin, C., Nuka, G., Pesseat, S., Sangrador-Vegas, A., Scheremetjew, M., Rato, C., Yong, S.Y., Bateman, A., Punta, M., Attwood, T.K., Sigrist, C.J.A., Redaschi, N., Rivoire, C., Xenarios, I., Kahn, D., Guyot, D., Bork, P., Letunic, I., Gough, J., Oates, M., Haft, D., Huang, H., Natale, D.A., Wu, C.H., Orengo, C., Sillitoe, I., Mi, H., Thomas, P.D., Finn, R.D.: The InterPro protein families database: the classification resource after 15 years. Nucleic Acids Res. **43**(D1), D213–D221 (2015)
16. Suzek, B.E., Huang, H., McGarvey, P., Mazumder, R., Wu, C.H.: UniRef: domprehensive and non-redundant UniProt reference clusters. Bioinformatics **23**(10), 1282–1288 (2007)
17. Fawcett, T.: An introduction to ROC analysis. Pattern Recogn. Lett. **27**(8), 861–874 (2006)
18. Cui, X., Churchill, G.A., et al.: Statistical tests for differential expression in cDNA microarray experiments. Genome Biol. **4**(4), 210 (2003)

3D Protein-Structure-Oriented Discovery of Clinical Relation Across Chronic Lymphocytic Leukemia Patients

Konstantinos Mochament[1(✉)], Andreas Agathangelidis[2], Eleftheria Polychronidou[1],
Christos Palaskas[1], Elias Kalamaras[1(✉)], Panagiotis Moschonas[1],
Kostas Stamatopoulos[2], Anna Chailyan[4], Nanna Overby[3], Paolo Marcatili[3],
Anastasia Hadzidimitriou[2], and Dimitrios Tzovaras[1]

[1] Centre for Research and Technology Hellas, Information Technologies Institute,
Thermi-Thessaloniki, Greece
{k.mochament,kalamar}@iti.gr
[2] Centre for Research and Technology Hellas, Institute of Applied Biosciences,
Thermi-Thessaloniki, Greece
[3] Center for Biological Sequence Analysis, Technical University of Denmark,
Kongens Lyngby, Denmark
[4] Carlsberg Research Laboratory, Copenhagen, Denmark

Abstract. Chronic lymphocytic leukemia (CLL) is the most common adult leukemia with still unclear etiology. Indications of antigenic pressure have been hinted, using sequence and structure-based reasoning. The accuracy of such approaches, and in particular of the ones derived from 3D models obtained from the patients' antibody amino acid sequences, is intimately connected to both the reliability of the models and the quality of the methods used to compare and group them. The proposed work provides a sophisticated method for the classification of CLL patients based on clustering the amino acid sequences of the clonotypic B-cell receptor immunoglobulin, which is the ideal clone-specific marker, critical for clonal behavior and patient outcome. A novel CLL patient clustering method is hereby proposed, combining bioinformatics methods with the extraction of 3D object descriptors, used in machine learning applications. The proposed methodology achieved an efficient and highly informative grouping of CLL patients in accordance to their biological and clinical properties.

1 Introduction

It is well-established that three-dimensional (3D) protein structure plays a pivotal role in functional characterization [1]. On the contrary, prediction of 3D protein structures from amino acid sequences represents one of the most difficult issues in computational structural biology [2]. In the absence of known structure, alternative approaches of comparative modelling can provide a useful 3D model for a protein that can be related to at least one experimentally determined protein structure. The aim of the current study was to categorize chronic lymphocytic leukemia (CLL) patients based on their 3-dimensional protein structures arising from the corresponding clonotypic immunoglobulin amino acid (IG) sequences.

© Springer International Publishing AG 2017
I. Rojas and F. Ortuño (Eds.): IWBBIO 2017, Part II, LNBI 10209, pp. 139–150, 2017.
DOI: 10.1007/978-3-319-56154-7_14

Chronic lymphocytic leukemia (CLL) is the most common adult leukemia, with still unclear etiology. Indications of antigenic pressure have been hinted, using sequence and structure-based reasoning [4]. The categorization of CLL patients has so far been addressed using bioinformatics methods of structural similarity calculation. The novelty of the approach proposed hereby lies on the combination of current state-of-the-art bioinformatics methods with the extraction of machine learning-related features from the protein models. The most up-to-date 3D prediction structure algorithms were implemented to construct patients' models. The proposed combined methodology achieves an efficient grouping of CLL patients in accordance to their biological and clinical features, especially in light of the recently identified stereotyped subsets [5].

The patient clustering was performed following a five-step process, including model construction, structural alignment, 3D feature extraction, similarity measurement, and clustering. The first two steps fall into the pure bioinformatics methods [6], while the rest are considered as part of 3D object segmentation approaches and especially the 3D partial matching approach [7].

2   Related Work

To narrow the gap between the increasing number of proteins with known sequences and the number of proteins with experimentally characterized structure and function, several approaches have been proposed in the literature. These include advanced computational methods developed for modeling structures and functions from sequences, as well as intelligent algorithms for structural alignment, aiming to classify proteins in families and capture evolutionary relationships.

The most comprehensive protein family classifications are included in the CATH [8] and SCOP [9] databases. The classification in these databases is based on the evolutionary relationship between proteins. If 3D protein structures reveal a significantly similar domain, a family relationship is created. Super families are further grouped together if their members share a similar structural fold. The FSSP [10] database classifies structural alignments of proteins in the Protein Data Bank (PDB) based on homology. Recently, the Pfam database [11] demonstrated the use of hidden Markov models (HMM) using the HMMER software in order to create a sustainable system of capturing the diversity in a given set of protein sequences.

In regard to the multiple structural alignment approach (MSA), which is the basic method of similarity identification, there are two significant methodological approaches, namely similarity-based and evolution-based [12]. Similarity-based methods aim to characterize the level of conservation of a particular region or sequence motif compared to other regions/sequences. This class includes methods such as Clustal [13], which is based on progressive pairwise alignment. Muscle [14] added a step in the guide tree that determines the order that sequences are added to a growing MSA. MAFFT [15] has a similar operation, involving fast Fourier transformations to approximate the pairwise alignments and recently was updated with progressive and iterative refinement methods. T-Coffee [16] is currently the most evolved method through the implementation of a more sophisticated scoring function. On the other hand, evolutionary methods mainly involve processes connected to sequence evolution. Insertion, substitution, and deletion are the key features for this type of analysis and they are depicted as an explicit phylogenetic tree [12].

Focusing on the similarity of 3D protein structures, it is defined by the positional deviations of equivalent atoms upon rigid-body superimposition. Numerous scoring functions have been proposed, with TM-score [29] being the most prominent: it measures the global fold similarity and is less sensitive to local structural variations. The aforementioned aligners where applied with the aim to identify similarities between proteins with large conformational changes. Methods for computing the optimal alignment between protein structures implement heuristic algorithms, as in the cases of MAMMOTH [17], LGA/GDT [18] and SAS score & GSAS score [19], or dynamic programming techniques as in the cases of MatAlign [21], LOVOalign [22] or SSM [23].

With reference to 3D protein clustering methodologies, Max Cluster [20] preceded by establishing the ability to process thousands of structures, either against a single reference protein or in an all-all comparison. Max Cluster combines well-known bioinformatics methods and widespread clustering methods in order to conduct structure comparison tasks. More recently, SpacePAC [26] and mutation3D [36] developed mutational oriented methodologies. The former identifies spheres that cover the most number of mutations, and evaluates the results based on p-value and statistically significance of the mutations. mutation3D classifies protein structures based on functional hotspots that hold significant somatic cancer mutations. Although the aforementioned methods point to the formation of protein groups, they, in contrast to the one implemented in the current study, rely on pure bioinformatics methods.

3   Methodology Overview

As a first step, we performed a synthesis of existing bioinformatics methodologies and formed a baseline approach. Subsequently, we created the proposed combined methodology with the addition of extra components regarding the 3D structure of the Ig models.

The baseline approach consists of the following steps. First, the original immunoglobulin sequences of CLL patients were transformed to 3D protein models, using state-of-the-art protein structure prediction tools. This resulted in a set of protein models in PDB format. For each pair of models, a structural similarity score was computed, after the optimal alignment between them was found. The TM-align [24] and TM-score algorithms were used for the alignment and the similarity score computation, respectively. Considering the similarity scores between every pair of models, a similarity matrix was formed, which was ultimately used to organize the proteins into clusters, using various existing clustering methods.

The proposed method modified this baseline approach by replacing the bioinformatics-related TM-score similarity metric between a pair of models with a machine learning-related distance metric between 3D descriptors extracted from these models. Specifically, the Fast Point Feature Histograms (FPFH) [25] descriptor was extracted from each 3D protein model. However, the descriptors for each given pair of models were not extracted from the whole models, but only from their maximally aligned parts.

This was performed in order to improve the discriminative capability of the method. Considering the whole model instead would amplify certain dissimilarities occurring even in models with similar 3D structures, which are due to imperfect modeling, thus reducing the ability of the method to distinguish between truly similar and dissimilar structures. Trials with considering the whole model have also been conducted; however the resulting classification was not as accurate as using only the most aligned parts.

Thus, for each pair of models, the maximally aligned parts were first identified, using again the TM-align algorithm, and then the FPFH descriptors were extracted from only these parts. FPFH descriptors were 'compared using the root mean square deviation (RMSD) distance metric. The distance for every pair of models was computed and all results were incorporated into a distance matrix, which was finally used to model clustering. The multidimensional scaling (MDS) method was used in order to transform the distance matrix into a set of points of relatively low dimensionality, prior to the clustering process. The pipeline diagram of the proposed approach is represented in Fig. 1.

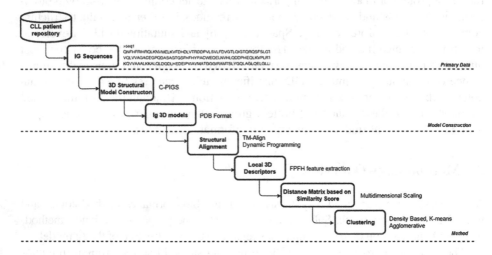

Fig. 1. Block diagram illustrating the proposed methodology.

The motivation behind using a general-purpose 3D descriptor for pairwise distance calculation, instead of relying on bioinformatics-specific similarity scores, was that 3D descriptors such as FPFH have been extensively and successfully used outside the realm of bioinformatics, for solving problems such as 3D object registration and partial 3D matching. Thus, their high capability in describing general 3D structures could be applied to the specific problem of 3D protein structure comparison. Experimental results confirm that the combination of bioinformatics-specific techniques, such as TM-align, and general-purpose 3D descriptors achieve a high discriminative power compared to using only bioinformatics-specific methods [3]. The current paper elaborates on the methodology used in [3] and extends the experimental results, by presenting its performance in an unsupervised setting, for clustering a large number of unlabeled protein models. Furthermore, the promising results obtained by using the FPFH descriptors may motivate further exploration of different types of 3D descriptors, as a future research direction. In the following subsections, the steps of the proposed method are presented in detail.

3.1 Datasets and 3D Models Construction

Two different protein model datasets were formed from 894 Ig sequences, obtained from CLL cases belonging to both known stereotyped subsets and cases with heterogeneous B cell receptor immunoglobulins. The 3D models were built based on the C-PIGS structural prediction method [27]. The number of CLL sequences in each dataset is as follows:

- Dataset D1: 106 patients, 6 predefined subsets (D1.A–D1.F),
- Dataset D2: 894 patients, no predefined subsets.

To efficiently evaluate the clustering methods, the first dataset included sequences from six well-established stereotyped subsets, with the second contains IG sequences from both labeled and unlabeled CLL cases, hence forming a dataset sufficient for unsupervised clustering applications.

Regarding the Immunoglobulin structure prediction, the C-PIGS method was used, which is based on the PIGS approach. In the Prediction of Immunoglobulin Structure (PIGS) [4] approach, antibody models are built using the PIGS web server using the following pipeline: A single template for both VL and VH framework regions was used if displaying greater than 80% sequence identity within both chains; otherwise, the two templates with the highest sequence similarity, measured by the Blosum24 score [35] of both the light and heavy chain were used. The H3 loop was always modeled using the template with the best sequence similarity; the other loops were modeled using a different template only if the corresponding loop in the framework template did not display the same length and canonical structure of the target loop. Finally, the C-PIGS models were built by remodeling the H3 loop of the PIGS models using the template identified by the approach developed Messih et al. [27]. An outstanding study of the customized PIGS methodology is described by Marcatilli et al. at [3].

3.2 Structural Alignment

The TM-align algorithm [28] was used for the Structural Alignment process. As performed by the TM-align algorithm, for the comparison of two IG structures, the second model was rotated and translated appropriately, until the maximum alignment between the two structures was achieved. TM-align identifies the optimal structural alignment between pairs by combining the TM-score [29] rotation matrix and Dynamic Programming (DP) algorithms, and by utilizing internally the Kabsch algorithm [30], which is used to minimize the RMSD distance between structures by optimizing the rotation.

In the TM-align algorithm, the alignment process is performed in three distinct steps. Initially, the alignment is performed over the secondary structures of the IG sequences. Secondly, the alignment is based on the gapless matching of the IG sequences. Finally, the alignment is performed using an equally weighted combination of the previously extracted results from steps one and two. Heuristic iterations are applied, meaning that the steps are repeated until the alignment became stable and the highest TM-score is achieved.

3.3 3D Feature Extraction

Having located the most optimally aligned residues, 3D features were extracted for the formation of robust 3D local descriptors. This was accomplished by comparing each pair of the local descriptors and then forming a combined overall metric of similarity between the source and the target IG sequence. Within this step, features were extracted from aligned parts of the IG sequences using the Fast Point Feature Histogram (FPFH) [25]. FPFH is a robust complexity reduction variation of Point Feature Histogram (PFH) [31] and it was selected as a fast and accurate feature extraction method. Other descriptors have also been tried, such as the Ensemble of shape functions (ESF) descriptor [37]. However, the FPFH descriptor was the one having the most discriminative power for classifying the dataset. Nevertheless, the proposed method is flexible, permitting any type of 3D descriptor to be used instead, thus allowing the investigation of other 3D descriptors in future research.

The PFH descriptor approach is based on the computation of both curvature estimation and surface normals, resulting in a 125-bin histogram for each point. The PFH descriptor uses the relative differences between points, and thus manages to be invariant to scale, orientation and position. FPFH is an improvement of PFH in terms of computational complexity and size. To achieve this improvement, the angles between points were computed only between a point and its k-nearest points. The result was a simplified point feature histogram (SPFH) from where the final FPFH is derived. Thus, FPFH is equal to the sum of the SPFH and the weighted distances of each point p to all its k-neighbor points.

$$FPFH_p = SPFH_p + \frac{1}{k} \sum_{i=1}^{k} SPFH(i)/w_i \qquad (1)$$

The size of the final FPFH descriptor is a 33-bin histogram for each point, while still being invariant to scale, orientation and position, as the original PFH.

3.4 Similarity Metric Extraction

After the computation of the local 3D descriptors, a final indicator was calculated in order to measure the similarity (or dissimilarity) of the compared IG sequences. Using the Root Mean Square Deviation (RMSD) over the compared descriptors, an average distance between the local features of the aligned parts was calculated and used as the quantitative similarity metric between the source and the target IG sequence.

3.5 Clustering

Grouping the set of IG sequences can be achieved in many different ways, using several clustering algorithms with diverse approaches. In this paper, three clustering techniques were considered, namely DBscan, k-means and agglomerative clustering.

Assuming that there is not any previous information regarding the number or the size of the "correct" classes, clustering algorithms produce grouping results, based on internal computations. However, additional information could be provided externally, transforming the procedure from unsupervised to semi-supervised. Among density based spatial clustering over noisy datasets DBscan [32] is one of the most common clustering algorithms and one of the most cited in the scientific literature. The algorithm executes two steps iteratively: it clusters points that are close enough and it labels points that are not in density based areas as outliers. Clustering based on semi-supervised approaches requires a predefined number of clusters. Hence, in contrast to the unsupervised approaches, an additional methodology was introduced for the approximation of the optimal number of clusters. K-means [33] is a widely used algorithm for clustering tasks, when the number of clusters is pre-defined. Finally, the agglomerative clustering approach [34] was used as it is one of the most commonly used semi-supervised approaches.

3.6 Evaluation

Initially the evaluation of the generated clusters was done by comparing the in-stances of the clusters with the externally derived stereotyped subset labeling. Dataset D1 had a predefined annotation and thus the validation was done using this annotated information as ground truth (external evaluation). Additionally, results were also evaluated using an internal evaluation scheme, where the indicator to validate clustering results is made using quantities inherent to the dataset that describes the internal density within clusters and the sparseness with each other (internal evaluation).

3.6.1 External Evaluation

In a first experiment, the performance of the proposed algorithm was compared to the performance of the baseline algorithm, using the TM-score, in terms of clustering accuracy. In order to evaluate the quality of the described methodology, dataset D1, consisting of 106 Ig 3D models originating from 5 CLL stereotyped subsets, was used as ground truth. These subgroups, namely subsets #1, #2, #4, #6, #8, consisted of Ig 3D models sharing highly similar IGs in terms of amino acid sequence. The reasoning behind this choice was, first of all, that homologous Ig primary sequences can be reasonably anticipated to be reflected in overall similar 3D structures and therefore providing a reference for evaluating the developed workflow. Secondly, these subsets were well characterized at both the clinical and biological levels being highly similar at all levels studied [3]. Regarding the size of each subset, the distribution was as follows: Subset #1 (IGHV clan I/IGKV1 (D) 39) n = 37; Subset #2 (IGHV321/IGLV321) n = 15; Subset #4 (IGHV434/IGKV230) n = 19; Subset #6 (IGHV169/IGKV320) n = 12, Subset #7 (IGHV169/IGLV3) n = 12 and Subset #8 (IGHV439/IGKV1 (D) 39) n = 11.

As a measure of clustering accuracy, the true positive and true negative rates were used and the scores are weighed with respect to the size of each subset. The results of the comparison, for various clustering algorithms, are presented in Table 1, where it is verified that the proposed combined TM-3D method achieves a notable improvement compared to simple TM-score method (approximately 10% increase in accuracy).

Table 1. Comparison between simple TM-score and the proposed methodology.

Clustering method	TM-score	TM-3D
k-means	75%	87%
Agglomerative (single-link)	61%	83%
Agglomerative (complete-link)	71%	86%
DBScan	72%	73%

3.6.2 Internal Evaluation

Besides the aforementioned external clustering evaluation, an additional internal metric was used, namely the Davies – Bouldin index (DBindex) [24], in order to measure the compactness and separation of the clusters, in a second experiment. Using the DBindex, the clustering findings are compared to an indicator that does not require any a priori knowledge of existing subsets in the dataset. DBindex describes how closely related the instances within a cluster are and how unrelated the clusters are to each other. The smaller the DBindex value, the more robust the clustering.

Table 2 presents the results of the internal evaluation using different clustering algorithms over the same dataset (D1) and illustrates that both unsupervised and semi-supervised approaches have an optimal DBindex value when the size of cluster is six. Hence besides the evaluation of each method, DBindex was also used for converting the semi-supervised approaches to complete unsupervised, where the selection of the optimal number of clusters was made automatically after iteratively achieving the minimum DBinex value.

Table 2. DBindex values for the various considered clustering methods, using different predefined numbers of clusters.

Clustering method		Number of clusters			
		5	6	7	8
Agglomerative	Single	0.827	**0.709**	0.723	0.71
	Complete	0.698	**0.645**	0.686	0.727
	Average	0.615	**0.546**	0.635	0.668
	Middle	0.827	**0.709**	0.723	0.71
k-means		0.584	**0.532**	0.652	0.707
(Dbscan parameters)		2/0.1 (k = 5)	2/0.15 (**k = 6**)	2/0.2 (k = 7)	2/0.3 (k = 8)
DBscan		0.593	**0.582**	0.591	0.595

3.6.3 Unsupervised Clustering – Biological Interpretation

Finally, the proposed clustering method was evaluated in terms of the biological significance of the resulting clusters, in a larger context. For this experiment, the D2 dataset of 894 pairs of IGH and IGL gene rearrangements from patients with CLL was used. In terms of IG repertoire properties, this dataset was representative of CLL in terms of heavy and light gene repertoire and thus largely informative. In terms of IGH repertoire properties, the current dataset was representative of CLL: IGHV3 cases predominated (414/894, 46.3%) followed by IGHV4 (216/894, 24.2%) and IGHV1 cases (199/894,

22.3%) with only minor exceptions in regard to the frequency of some individual genes: Comparing the cohort used in this study with the general CLL cohort, IGHV4-34 and IGHV4-39 genes were overrepresented, whereas the IGHV1-69 gene was slightly underrepresented. Concerning the light chain usage, around two-thirds of the cases (588/894 cases, 65.8%) expressed kappa chains versus one-third that carried light chains, as previously described in CLL. The kappa and lambda light chain IG gene repertoires were also typical of CLL, characterized by the predominance of the IGHV1-39 and IGLV3-21 genes, respectively.

The clustering method ended up with the formation of 6 subgroups. The distribution of 3D models among these subgroups was as follows: subgroup 0 contained 199/894 models (30.1%), (ii) subgroup 1 comprised 154/894 models (17.2%), (iii) subgroup 2 consisted of 81/894 models (9.1%), (iv) subgroup 3 was formed by 173 models (19.4%), (v) subgroup 4 included 148/894 models (16.6%) and, finally, subgroup 5 involved 69/894 models (7.7%).

In detail, subgroups 0 and 1 contained virtually all IGHV3-expressing 3D models. IGHV3 models with a kappa light chain were assigned to subgroup 0 (257/414, 62.1%), whereas those with a lambda light chain formed subgroup 1 (154/414, 37.2%). In a similar way, subgroups 2 and 3 concerned mostly IGHV1 3D models. The fraction of lambda Ig models was clustered together in subgroup 2 (52/199, 26.1%), whereas the remaining kappa-expressing cases were assigned to subgroup 3 (143/199, 71.9%). Finally, subgroups 4 and 5 included IGHV4 3D models carrying kappa (145/216, 67.1%) and lambda light chains (69/216, 31.9%), respectively. Three-dimensional models from the other, less common IGHV subgroups (namely IGHV2, IGHV5, IGHV6 and IGHV7), even though they did not form distinct subgroups, were also separated based on the isotype of the light chain.

According to results obtained here, the clustering of 3D Ig models clearly reflected the classification of Ig molecules based on the primary sequences of both the heavy and the light chains of the molecule. In detail, we focused our analysis on the Ig models from 3 most common IGHV gene subgroups (namely IGHV1, IGHV3 and IGHV4), which represented 829/894 models (92.8%) in the current cohort. The primary sequence of the IG heavy chain, represented here by the IGHV gene subgroup, was mirrored in the clustering results since 3D models from each IGHV gene subgroup were clearly separated from the others forming individual clusters. Furthermore, the Ig light chain isotype (kappa or lambda) was also found to represent a crucial factor since Ig 3D models from each of the aforementioned IGHV subgroups formed two distinct clusters based on the expression of a kappa or lambda light chain.

4   Discussion

The experimental results of the proposed method are highly relevant for advancing our understanding of CLL development and evolution by efficiently clustering patients with discrete biological and clinical characteristics. Indeed, the possibility to cluster primary IG amino acid sequences from CLL patients into discrete subgroups based on feature extraction from 3D models has implications for refined patient sub-classification but also for the development of novel treatment strategies targeting the BcR IG.

Therefore, even though TM-score and RMSD are used for comparisons between IG sequences, the 3D feature extraction and, subsequently, the partial matching of 3D descriptors added more clarity to the overall similarity process. This could be justified due to the fact that without 3D descriptors, the respective atoms can be compared based only on their position (x, y, z coordinates). On the contrary, with the use of a local 3D descriptor like FPFH, additional information is extracted and hence, each atom (and its local structure) is described with more than three dimensions (33 dimensions in FPFH).

The selection of the appropriate 3D descriptor is an issue worth studying further. Notably, after conducting a small-scale survey regarding the local and global 3D descriptors (PFH, ESF, and VFH), it was evident that FPFH performed better. Nevertheless, it could be argued that FPFH was arbitrarily selected and for that reason a more thorough future survey could investigate the possibility of further improvement through the substitution of the selected 3D descriptor.

Consistent with the previous observation, using only local 3D descriptors and ignoring the information extracted from TM-align will lead to imprecise results. However, by using TM-align, it is ensured that because of the identification of the maximally aligned parts, only the maximum useful information will be taken into account and all other information will be excluded. The above claim was verified by experiments where the input was the entire 3D models (PDB files transformed in OBJ format) and the experimental results were insignificant and thus there was no reason to be further examined.

Finally, regarding the evaluation and results presented based on the internal metric, it should be taken into account that a "good" internal indicator score does not necessarily mean "correct" clusters. The reason for that is because, in contrast to the external evaluation there is no predetermined benchmark classification to set as a reference point and hence the evaluation scores could only be used as indications.

References

1. Webb, B., Sali, A.: Comparative Protein Structure Modeling Using MODELLER. Current Protocols in Bioinformatics. John Wiley & Sons, Inc., (2002)
2. Skwark, M.J., et al.: Improved contact predictions using the recognition of protein like contact patterns. PLoS Comput. Biol. 10(11), e1003889 (2014)
3. Marcatili, P., et al.: Automated clustering analysis of immunoglobulin sequences in chronic lymphocytic leukemia based on 3D structural descriptors. Blood 128(22), 4365 (2016)
4. Marcatili, P., et al.: Antibody structural modeling with prediction of immunoglobulin structure (PIGS). Nat. Protoc. 9(12), 2771–2783 (2014)
5. Agathangelidis, A., et al.: Stereotyped B-cell receptors in one-third of chronic lymphocytic leukemia: a molecular classification with implications for targeted therapies. Blood 119(19), 4467 (2012)
6. Zhang, Y.: I-TASSER server for protein 3D structure prediction. BMC Bioinf. 9, 40 (2008)
7. Liao, S., Jain, A.K., Li, S.Z.: Partial face recognition: alignment-free approach. IEEE Trans. Pattern Anal. Mach. Intell. 35(5), 1193–1205 (2013)

8. Greene, L.H., et al.: The CATH domain structure database: new protocols and classification levels give a more comprehensive resource for exploring evolution. Nucleic Acids Res. **35**(Database issue), D191–D197 (2007)

9. Andreeva, A., et al.: Data growth and its impact on the SCOP database: new developments. Nucleic Acids Res. **36**(Database issue), D419–D425 (2008)

10. Holm, L., Sander, C.: The FSSP database of structurally aligned protein fold families. Nucleic Acids Res. **22**(17), 3600–3609 (1994)

11. Finn, R.D., et al.: The Pfam protein families database: towards a more sustainable future. Nucleic Acids Res. **44**(D1), D179–D185 (2016)

12. Blackburne, B.P., Whelan, S.: Class of multiple sequence alignment algorithm affects genomic analysis. Mol. Biol. Evol. **30**(3), 642–653 (2013)

13. Larkin, M.A., et al.: Clustal W and Clustal X version 2.0. Bioinformatics **23**(21), 2947–2948 (2007)

14. Edgar, R.C.: MUSCLE: a multiple sequence alignment method with reduced time and space complexity. BMC Bioinf. **5**, 113 (2004)

15. Katoh, K., Standley, D.M.: MAFFT multiple sequence alignment software version 7: improvements in performance and usability. Mol. Biol. Evol. **30**(4), 772–780 (2013)

16. Notredame, C., Higgins, D.G., Heringa, J.: T-Coffee: A novel method for fast and accurate multiple sequence alignment. J. Mol. Biol. **302**(1), 205–217 (2000)

17. Ortiz, A.R., Strauss, C.E., Olmea, O.: MAMMOTH (matching molecular models obtained from theory): an automated method for model comparison. Protein Sci. **11**(11), 2606–2621 (2002)

18. Zemla, A.: LGA: a method for finding 3D similarities in protein structures. Nucleic Acids Res. **31**(13), 3370–3374 (2003)

19. Kolodny, R., Koehl, P., Levitt, M.: Comprehensive evaluation of protein structure alignment methods: scoring by geometric measures. J. Mol. Biol. **346**(4), 1173–1188 (2005)

20. Herbert, A., Sternberg, M.J.E.: MaxCluster–a tool for protein structure comparison and clustering (2014). http://www.sbg.bio.ic.ac.uk/~maxcluster

21. Aung, Z., Tan, K.-L.: MatAlign: precise protein structure comparison by matrix alignment. J. Bioinf. Comput. Biol. **04**(06), 1197–1216 (2006)

22. Martínez, L., Andreani, R., Martínez, J.M.: Convergent algorithms for protein structural alignment. BMC Bioinf. **8**(1), 306 (2007)

23. Krissinel, E., Henrick, K.: Secondary-structure matching (SSM), a new tool for fast protein structure alignment in three dimensions. Acta Crystallographica Sect. D **60**(12 Part 1), 2256–2268 (2004)

24. Pandit, S.B., Skolnick, J.: Fr-TM-align: a new protein structural alignment method based on fragment alignments and the TM-score. BMC Bioinf. **9**, 531 (2008)

25. Rusu, R.B., Blodow, N., Beetz, M.: Fast point feature histograms (FPFH) for 3D registration. In: IEEE International Conference on Robotics and Automation, ICRA 2009 (2009)

26. Ryslik, G., Yuwei C., and Hongyu Z. SpacePAC: identifying mutational clusters in 3D protein space using simulation (2013)

27. Messih, M.A., et al.: Improving the accuracy of the structure prediction of the third hypervariable loop of the heavy chains of antibodies. Bioinformatics **30**(19), 2733–2740 (2014)

28. Zhang, Y., Skolnick, J.: TM-align: a protein structure alignment algorithm based on the TM-score. Nucleic acids res. **33**(7), 2302–2309 (2005)

29. Zhang, Y., Skolnick, J.: Scoring function for automated assessment of protein structure template quality. Proteins: Struct., Funct., Bioinf. **57**(4), 702–710 (2004)

30. Louchet, H., Kuzmin, K., Richter, A.: Frequency, phase, and polarization-tracking algorithms for arbitrary four-dimensional signal constellations. In: SPIE OPTO, International Society for Optics and Photonics, pp. 900907–900907 (2013)
31. Rusu, R.B.: Semantic 3D object maps for everyday manipulation in human living environments. KI - Künstliche Intelligenz 24(4), 345–348 (2010)
32. Tran, T.N., Drab, K., Daszykowski, M.: Revised DBSCAN algorithm to cluster data with dense adjacent clusters. Chemometr. Intell. Lab. Syst. 120, 92–96 (2013)
33. Celebi, M.E., Kingravi, H.A., Vela, P.A.: A comparative study of efficient initialization methods for the k-means clustering algorithm. Expert Syst. Appl. 40(1), 200–210 (2013)
34. Meila, M., Heckerman, D.: An experimental comparison of several clustering and initialization methods. arXiv preprint arXiv:1301.7401 (2013)
35. Henikoff, S., Henikoff, J.G.: Amino acid substitution matrices from protein blocks. PNAS 89(22), 10915–10919 (1992). PMC: 50453, PMID: 1438297
36. Meyer, M.J., et al.: mutation3D: cancer gene prediction through atomic clustering of coding variants in the structural proteome. Hum. Mutat. 37(5), 447–456 (2016)
37. Wohlkinger, W., Vincze, M.: Ensemble of shape functions for 3D object classification. In: 2011 IEEE International Conference on Robotics and Biomimetics (ROBIO), pp. 2987–2992. IEEE, December 2011

Application of a Membrane Protein Structure Prediction Web Service GPCRM to a Gastric Inhibitory Polypeptide Receptor Model

Ewelina Rutkowska, Przemyslaw Miszta, Krzysztof Mlynarczyk,
Jakub Jakowiecki, Pawel Pasznik, Slawomir Filipek,
and Dorota Latek[✉]

Faculty of Chemistry, University of Warsaw, Warsaw, Poland
dlatek@chem.uw.edu.pl

Abstract. A novel versatile tool named GPCRM has been developed. It targets structure prediction of a distinct protein family of G protein-coupled receptors (GPCRs). In principle, GPCRM builds a GPCR model using a MODELLER-based homology modeling procedure. In addition, that commonly used procedure was improved by using comparison of sequence profiles, multiple template structures and the extensive loop refinement in Rosetta. We applied our method to predict a three dimensional structure of a gastric inhibitory polypeptide receptor (GIPR) from the secretin-like class B of human GPCRs. The GIPR model was also tested in an ensemble docking study in which we investigated plausible interactions of four potential antagonists with that receptor. Out of those four ligands we suggested ChEMBL_1933363 as the most potent antagonist of GIPR based on the Glide docking results.

Keywords: Gastric inhibitory polypeptide receptor · GIPR · Antagonist · ChEMBL_1933363 · GPCRM · G protein-coupled receptors · Membrane proteins · MODELLER · Rosetta · Structure prediction · Homology modeling

1 Introduction

A distinct, seven transmembrane helices, membrane protein family of G protein-coupled receptors has been extensively studied with a variety of experimental and computational techniques for more than half-century, since G. Wald proved that isomerization of the bound retinal chromophore in rhodopsin is a basis of vision [1]. In 1997 [2] and in 2000 [3], a seven transmembrane helical (7TMH) fold of rhodopsin has been confirmed, but still structures of about 800 other human GPCRs remained unknown. Recent progress in X-ray crystallography of membrane proteins has led to solving structures of another 24 GPCRs. Computational biology [4–7], system biology [8] and bioinformatics [9, 10] have been commonly used to bridge the gap between the number of known GPCR sequences and the number of the solved GPCR structures and in such way to get insights into their cellular function.

The most important computational methods for the GPCR research are: transmembrane regions prediction [11], structure prediction (homology modeling or

© Springer International Publishing AG 2017
I. Rojas and F. Ortuño (Eds.): IWBBIO 2017, Part II, LNBI 10209, pp. 151–162, 2017.
DOI: 10.1007/978-3-319-56154-7_15

threading) [12], virtual screening (structure or ligand-based) [13], molecular dynamics for investigation of, e.g., mechanism of receptor activation [4, 5], phylogenetic analysis for GPCR classification [9] or de-orphanization [14]. Based on the phylogenetic analysis [9] G protein-coupled receptors were grouped into five main classes: the largest and the best known - rhodopsin-like class (A), secretin (B), glutamate (C), adhesion (D) and frizzled/taste2 class (E). High sequence identity can be observed only among members of the same GPCR class [15]. However, the 7TMH bundle is well conserved in all GPCR classes which makes homology modeling of GPCRs possible. Noticeable structural variations are present mainly in loop regions of GPCRs and in orientations, kink and tilt angles of their transmembrane helices. Building a coarse 7TMH model via the homology modeling procedure is possible for nearly every GPCR, unless prediction of transmembrane regions is completely unreliable. However, in some cases, usefulness of such coarse structural models in drug design, which is the most common purpose in the GPCR research, can be highly questionable. This is because low sequence similarity between target and template GPCR sequences is often associated with conformational discrepancies between their active sites. For example, the smoothened receptor (SMO) shares less than 10% identical residues with the class A GPCRs. And indeed, the LY2940680 antagonist binding site of the SMO receptor (PDB id: 4JKV) is long and narrow with much straighter TMH6 comparing the class A binding sites [16]. Another example is the recently solved structure of the human corticotropin-releasing factor receptor 1 (CRFR1) with the binding site of the CP-376395 antagonist located so deeply inside the receptor core that it hardly could be predicted with any homology modeling procedure [17]. Another issue which hampers the homology modeling of GPCRs is the large-scale movement of transmembrane helices during the receptor activation. It was observed for known-to-date GPCR crystal structures from various classes, that there are only few common features of this mechanism, e.g., moving of a part of TMH6 to enable binding of a G protein during activation [18]. Consequently, it is rarely possible to predict an active conformational state of a receptor using an inactive template structure or vice-versa [19]. At present, among less than 30 known crystal GPCR structures which can be used as templates only two are from the B class, one from the C class and one from E, which justify the urgent need for computational methods for GPCR structure prediction.

Many computational methods for GPCR structure prediction were implemented in the form of web services. Such user-friendly tools help to overcome difficulties in setting up the structure modeling procedure on one hand and on the other help to accelerate the research. GPCR structure prediction web services can be broadly divided into two groups [12]: non-MODELLER-based services (e.g. GPCR-I-TASSER [15]), and MODELLER-based services (e.g. GPCR-SFEE [20], GPCR-ModSim [21], GOMoDo [22], GPCRautomodel [23]). MODELLER [24] builds a GPCR model using spatial restraints from template structures. Non-MODELLER methods, such as GPCR-I-TASSER, are based on fold recognition or threading, often combined with experimental data, e.g., from mutagenesis.

Our recently developed tool GPCRM [7, 25] (see Fig. 1) is a noteworthy example of MODELLER-based web services. GPCRM is based on the well-known scaffold of the homology modeling. Sequence similarity is a major factor determining the template structure selection. MUSCLE and CLUSTALW tools are used as engines to generate

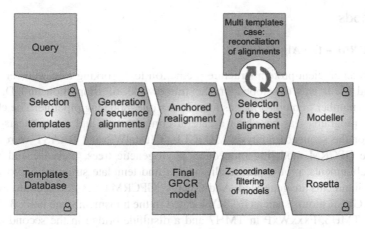

Fig. 1. The flowchart of the GPCR structure modeling procedure implemented in the GPCRM web service.

sequence and profile alignments. MODELLER is used to build a three-dimensional model of GPCR via satisfaction of spatial restraints from a single template or a mixture of template structures by taking into account their local sequence similarity to a target sequence. Rosetta refines loop regions in the obtained GPCR models by the extensive Monte Carlo sampling of the protein backbone fragments library.

Here, we described usage of GPCRM to build a glucose-dependent insulinotropic polypeptide receptor (GIPR). GIPR is a GPCR from the class B and has attracted attention in the field of diabetes [26]. Its endogenous ligand, glucose-dependent insulinotropic polypeptide (or gastric inhibitory polypeptide, GIP), is a forty-two amino acid hormone secreted from the enteroendocrine K cells. GIP activates pancreatic islets to enhance insulin secretion and plays an important role in lipid metabolism and fat deposition by increasing the lipoprotein lipase activity, lipogenesis, fatty acid and glucose uptake in adipocytes [27]. The major obstacle in using GIP in the treatment is the fact, that GIP is easily degraded. Drug discovery of non-peptide, stable ligands is highly desirable in this case. In spite of the recent progress in experimental determination of class B GPCRs structures (two new crystal structures deposited in PDB in 2013) structure prediction of secretin-like GPCRs is still challenging due to their complex topology. They consist of two quite large domains, extracellular and trans-membrane and are activated by hormonal peptides binding to both of domains. For example, GIPR is composed of a large, extracellular N-terminal domain containing six highly conserved cysteine residues forming three disulfide bridges (ECD), a so-called serpentine domain with seven transmembrane helices (TMD) and finally a short intracellular C-terminus. Our efforts described in this manuscript were focused on the homology modeling of GIPR 3D structure and using obtained GIPR models in tracking the interactions of its transmembrane domain with small molecules which could serve in future as drugs.

2   Methods

2.1   GPCRM – the Algorithm

GPCRM is an efficient protein structure prediction tool working as a web service (see Figs. 1 and 2). It was developed on the frameworks of Biopython and Django. It combines a number of bioinformatics tools into a one modeling pipeline. The choice of third-party programs was based on a careful consideration and several tests [7, 25]. Namely, for the sequence analysis MUSCLE, ClustalW2 and BLAST were chosen. During the GPCRM modeling procedure, phylogenetic trees, pairwise and multiple sequence alignments, profile-profile alignments and template structure-based multiple sequence alignments are generated. Additionally, GPCRM takes into account important functional GPCR motifs such as E/DRY located in the transmembrane helix 3 (TMH3), NPxxY in TMH6, FxxxWxP in TMH7 and a disulfide bridge in the second extracellular loop EC2 to anchor automatically generated alignments. Gaps inside the TM helices which are longer than 1 residue are penalized. Such approach to the generation of sequence alignments is necessary to ensure the proper alignment of residues in the active site sequence region. Templates selection is based on the target-template sequence identity. Unless the ClustalW2 identity score exceed 40% only one template is selected as a scaffold to build a three-dimensional structure of GPCR. Otherwise, two templates are used in the MODELLER-based procedure of the model building. If two templates (or even more – the possible option in the Advanced mode of GPCRM) are selected the GPCR model is built as a kind of an average structure. The averaging of template structures depends on the local sequence similarity between the template and the target sequence. There are two separate sets of templates – an active and inactive templates set and the desired activation state of a receptor is taken into account during the modeling procedure. The best MODELLER-generated GPCR models are selected for the extensive loop modeling in Rosetta (the cyclic coordinate descent algorithm - CCD) [28]. Some long loops, e.g., EC2 are divided in two parts separated by a 'cut point' (Cys forming the EC2 disulfide bridge) to ensure the exhaustiveness of the conformational search. In the next step models are assessed by our program for excluding GPCR models with extra and intracellular loops or N and C-terminus penetrating the membrane interior.

Here, the selection criterion is the average Z-coordinate of the loop and termini residues. Z-coordinates are computed with respect to the implicit membrane model derived from the OPM file corresponding to the selected template. The OPM files are downloaded from the OPM database (Orientation of Proteins in Membranes) [29]. Final GPCR models (typically 10 best) are selected based on the Rosetta total score and provided on the GPCRM web-site to download.

All the described modeling steps are joined and processed by Biopython, python, bash, Tcl, xml scripts and procedures from VMD, Pymol, JSmol graphical programs. A user can choose between the 'Automatic' or 'Advanced' task modes. When selecting the second option a user can select templates by himself and change manually the sequence alignment. Recently, an automatic update of the templates database and the 'Fast track' mode for the coarse GPCR modeling were added as new functionalities. The 'Fast track' mode discards the extensive loop modeling in Rosetta providing only

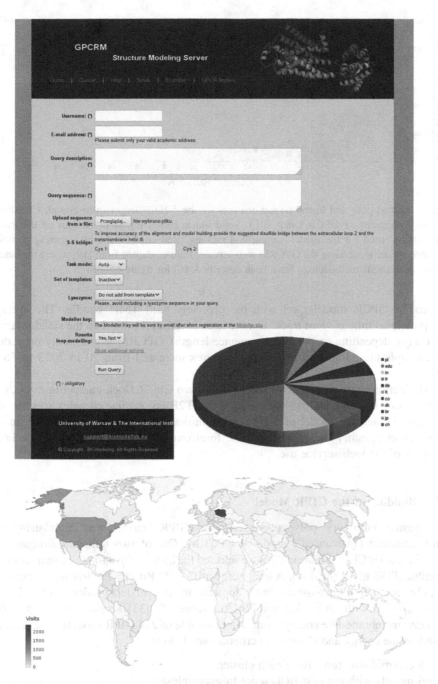

Fig. 2. The web interface of GPCRM with the default option set for a non-expert user (see: http://gpcrm.biomodellab.eu/). Additionally, at the bottom, the statistics of usage is shown. The right panel shows visits per a country domain and the bottom panel shows visits per a country.

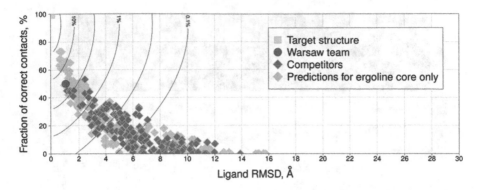

Fig. 3. Ligand pose and ligand pocket contacts for the 5-HT2B/ergotamine complex predicted by the participants of GPCR Dock 2013. Solid black curves represent isolines of the model 'correctness' used for the model scoring and ranking. Red circle is the best scoring complex obtained by our team using the GPCRM web service. Figure adapted from [19] where the details on the assessment methodology have been described. (Color figure online)

the coarse GPCR models, yet with the precisely built TMH bundle. The average computational time of 'Fast track' is more than 20 times shorter than the standard track (3–4 days, depending on the target sequence length). GPCRM is constantly upgraded, for example the number of templates structures increased from 20 in 2013 to 56 in 2016.

The performance of GPCRM was tested in two GPCR Dock competitions in 2010 [30] and in 2013 [19]. In 2013 our model of 5-HT2B-ergotamine complex was the best (see Fig. 3), and the model of 5-HT1B-ergotamine was the third best model. In the SMO target category we were also in the forefront. In Fig. 2 we presented the latest statistics of the web service use.

2.2   Building of the GIPR Model

The gastric inhibitory polypeptide receptor (GIPR) model was built using the semi-automatic ('Advanced') mode of GPCRM. Out of two template structures of secretin-like GPCR available in PDB we selected the human glucagon G-protein-coupled receptor (PDB id: 4L6R) [31]. A total number of 3000 Rosetta models were generated and clustered using the Rosetta cluster application. The models grouped in 66 clusters were assessed with BCL [32] and Rosetta score_jd2 [33] scoring functions which included membrane-like energy terms. An ensemble of 218 GIPR models were selected based on the energy and cluster size criteria - see below:

- 66 centroid structures from each cluster,
- 66 models with the best BCL score in every cluster,
- 66 models with the best Rosetta score in every cluster,
- 10 centroid structures from the cluster with the best average BCL score,
- 10 centroid structures from the cluster with the best average Rosetta score,
- 10 models with the best BCL score from the most populated cluster,

- 10 models with the best Rosetta score from the most populated cluster,
- 10 centroid structures from the most populated and with the best average BCL score clusters.

Models selected according to the above criteria were grouped together after removing duplicates and formed an ensemble of GIPR conformations which was subsequently used in the docking step.

2.3   Small Molecule Docking to the Ensemble of GIPR Conformations

To validate carefully selected GIPR models we challenged the docking of non-peptide ligands reported so far in the literature. GIPR models were prepared for the docking step using the Protein Prep Wizard script from the Schrodinger suite (Schrödinger Release 2015-3). The OPLS_2005 force field was used during the short minimization of the missing sidechains and hydrogen atoms. Four ligands (see Fig. 4) were derived from the BindingDB database [34]. All four ligands were docked to GIPR models and the ligand for which docking poses with the best Glide score were obtained was selected to be presented in details (see Fig. 4B). That selected ligand (name: ChEMBL_1933363) fulfilled the Lipinski's rule of five so it can be used as a drug. Its high potency as a GIPR antagonist was also proved experimentally by its ability to displace a [125I]GIP molecule from the membrane environment similar to this in which hGIPR is expressed [35]. The LigPrep and Epik modules from the Schrodinger suite

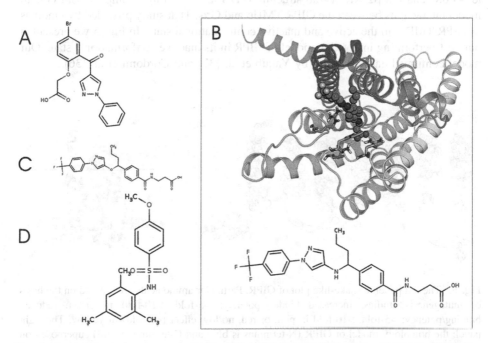

Fig. 4. Potential antagonists of GIPR: (A) ChEMBL_217053, (B) ChEMBL_1933363, (C) ChEMBL_1933360, (D) ChEMBL_3311308.

were used to prepare ligands for docking. The binding site for non-peptide ligands of GIPR is unknown so we had to tackle that problem by an exhaustive search of the whole ligand-accessible receptor space via SiteMap (Schrodinger). As a result, a number of potential ligand binding sites in GIPR was nominated. In the next step, four GIPR non-peptide antagonists, including i.a. ChEMBL_1933363, were docked into each of the proposed binding sites located in every GIPR structure from the ensemble of conformations using the multiple grid arrangement (MGA) approach. That rigid receptor – flexible ligand docking with several possible binding sites was performed using Glide (scripts: xglide_mga.py, xglide.py).

3   Results and Discussion

3.1   The Three Dimensional Protein Model of GIPR

To date two models of the GIPR transmembrane domain has been reported [36, 37]. To identify the putative binding site of GIPR Yaqub et al. [37] performed the protein-protein docking of the extracellular domain (ECD) with co-crystalized GIP and the model of the transmembrane domain (TMD) and conducted the site-directed mutagenesis (summarized in Fig. 5). Yaqub et al. reported that the N-terminal moiety of GIP was interacting with TMH2, TMH3, TMH5 and TMH6 with the crucial Tyr1 residue of GIP interacting with Gln224 (TMH3), Arg300 (TMH5), and Phe357 (TMH6). In the other work, Cordomi et al. [36] used the GCGR crystal structure (PDB id: 4L6R) and the β2-AR crystal structure (PDB id: 3SN6) in a complex with Gαs to model interactions between the GIPR TMH6 and Gαs. That study provided two models of GIPR TMD – in the active and inactive conformational state. In Fig. 5 we presented our best-performing in docking model of GIPR in its inactive conformational state. Our model confirmed earlier studies by Yaqub et al. [37] and Cordomi et al. [36].

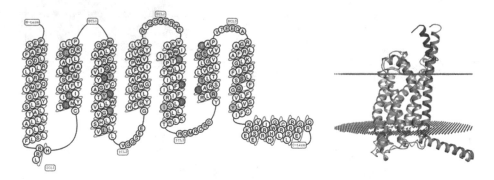

Fig. 5. The left panel: a snake-like plot of GIPR. Particular amino acids are colored on the basis of mutagenesis studies: increased binding/potency: >5-fold, >10-fold in green; reduced binding/potency: >5-fold, >10-fold in pink or red, no/low effect (<5-fold) in yellow. The right panel: the homology model of GIPR (N-terminus is blue and C-terminuc is red) superposed on the GCGR crystal structure (PDB id: 4L6R) (gray) which was used as a template. The membrane location was derived from the OPM database entry for GCGR. (Color figure online)

3.2 Molecular Docking to the GIPR Conformations Ensemble

We docked four potential antagonists to the ensemble of GIPR conformations. The best assessed by Glide score ligand-receptor pairs included ChEMBL_1933363. That ligand interacted with Arg183 (TMH2), Gln220 (TMH3) and Arg289 and Trp287 (EC2) residues (see Fig. 4). Trp287 interacted with the ligand aromatic ring via π-π interactions. Additionally, a carboxylic group of the ligand was engaged in the hydrogen bond network with Gln220 and Arg289. That confirmed earlier studies [36, 37] in which Arg183 (TMH2) and Gln220 (TMH3) were reported as crucial for activity. In particular Arg183 was reported [37] to be a residue participating in the GIP binding. The proposed binding mode of ChEMBL_1933363 provides important insights into its ability to displace [125I]GIP from hGIPR [35].

4 Conclusions

A versatile tool for GPCR structure prediction has been provided to the research community. GPCRM proved to be useful not only in the structure modeling of proteins from the rhodopsin-like class A but also in the research concerning the secretin-like class B of GPCRs. The most important features ensuring the good quality of the generated models are mixing of templates based on their sequence similarity to a target and the exhaustive searching of loop conformations performed by the well-known and multipurpose Rosetta software.

The knowledge of the three-dimensional structures of GPCRs provides invaluable insights into the molecular basis of their functions. The identification of the potential binding sites for a particular druggable receptor is crucial not only to understand the intimate mechanism of the ligand-receptor recognition, but also to speed up the discovery of new, potent and selective GPCR ligands. In the absence of experimental data, generating models using known crystal structures of homologous proteins is the only reliable method to obtain the structural information about GPCRs. However, model quality assessment for GPCRs is still a major problem. In this manuscript we presented one of the ways to improve it. Namely, we used a variety of scoring functions, clustering of similar structures and finally visual inspection to select the most reliable GIPR structures for the docking step. We also showed, that the docking itself was also a good way to track the most probable (of high quality) GPCR models. The GIPR models for which the best docking results (according to the Glide score) were obtained could be subjected to the virtual screening procedure to find novel antagonists in future.

Acknowledgments. The current study was financed by National Science Centre in Poland, the SONATA grant no. DEC-2012/07/D/NZ1/04244.

References

1. Wald, G.: The molecular basis of visual excitation. Nature 219(5156), 800–807 (1968)
2. Unger, V.M., Hargrave, P.A., Baldwin, J.M., Schertler, G.F.: Arrangement of rhodopsin transmembrane alpha-helices. Nature 389(6647), 203–206 (1997). doi:10.1038/38316

3. Palczewski, K., Kumasaka, T., Hori, T., Behnke, C.A., Motoshima, H., Fox, B.A., Le Trong, I., Teller, D.C., Okada, T., Stenkamp, R.E., Yamamoto, M., Miyano, M.: Crystal structure of rhodopsin: a G protein-coupled receptor. Science 289(5480), 739–745 (2000)

4. Yuan, S., Ghoshdastider, U., Trzaskowski, B., Latek, D., Debinski, A., Pulawski, W., Wu, R., Gerke, V., Filipek, S.: The role of water in activation mechanism of human N-formyl peptide receptor 1 (FPR1) based on molecular dynamics simulations. PLoS One 7(11), e47114 (2012). doi:10.1371/journal.pone.0047114

5. Yuan, S., Wu, R., Latek, D., Trzaskowski, B., Filipek, S.: Lipid receptor S1P(1) activation scheme concluded from microsecond all-atom molecular dynamics simulations. PLoS Comput. Biol. 9(10), e1003261 (2013). doi:10.1371/journal.pcbi.1003261

6. Munk, C., Isberg, V., Mordalski, S., Harpsoe, K., Rataj, K., Hauser, A.S., Kolb, P., Bojarski, A.J., Vriend, G., Gloriam, D.E.: GPCRdb: the G protein-coupled receptor database - an introduction. Br. J. Pharmacol. 173(14), 2195–2207 (2016). doi:10.1111/bph.13509

7. Latek, D., Pasznik, P., Carlomagno, T., Filipek, S.: Towards improved quality of GPCR models by usage of multiple templates and profile-profile comparison. PLoS ONE 8(2), e56742 (2013). doi:10.1371/journal.pone.0056742

8. Fridlyand, L.E., Philipson, L.H.: Pancreatic beta cell G-protein coupled receptors and second messenger interactions: a systems biology computational analysis. PLoS ONE 11(5), e0152869 (2016). doi:10.1371/journal.pone.0152869

9. Fredriksson, R., Lagerstrom, M.C., Lundin, L.G., Schioth, H.B.: The G-protein-coupled receptors in the human genome form five main families. Phylogenetic analysis, paralogon groups, and fingerprints. Mol. Pharmacol. 63(6), 1256–1272 (2003). doi:10.1124/mol.63.6.1256

10. Suwa, M.: Bioinformatics tools for predicting GPCR gene functions. Adv. Exp. Med. Biol. 796, 205–224 (2014). doi:10.1007/978-94-007-7423-0_10

11. Wallner, B.: ProQM-resample: improved model quality assessment for membrane proteins by limited conformational sampling. Bioinformatics 30(15), 2221–2223 (2014). doi:10.1093/bioinformatics/btu187

12. Busato, M., Giorgetti, A.: Structural modeling of G-protein coupled receptors: an overview on automatic web-servers. Int. J. Biochem. Cell Biol. 77(Pt B), 264–274 (2016). doi:10.1016/j.biocel.2016.04.004

13. Vass, M., Kooistra, A.J., Ritschel, T., Leurs, R., de Esch, I.J., de Graaf, C.: Molecular interaction fingerprint approaches for GPCR drug discovery. Curr. Opin. Pharmacol. 30, 59–68 (2016). doi:10.1016/j.coph.2016.07.007

14. van der Horst, E., Peironcely, J.E., Ijzerman, A.P., Beukers, M.W., Lane, J.R., van Vlijmen, H.W., Emmerich, M.T., Okuno, Y., Bender, A.: A novel chemogenomics analysis of G protein-coupled receptors (GPCRs) and their ligands: a potential strategy for receptor de-orphanization. BMC Bioinform. 11, 316 (2010). doi:10.1186/1471-2105-11-316

15. Zhang, J., Yang, J., Jang, R., Zhang, Y.: GPCR-I-TASSER: a hybrid approach to G protein-coupled receptor structure modeling and the application to the human genome. Structure 23(8), 1538–1549 (2015). doi:10.1016/j.str.2015.06.007

16. Wang, C., Wu, H., Katritch, V., Han, G.W., Huang, X.P., Liu, W., Siu, F.Y., Roth, B.L., Cherezov, V., Stevens, R.C.: Structure of the human smoothened receptor bound to an antitumour agent. Nature 497(7449), 338–343 (2013). doi:10.1038/nature12167

17. Hollenstein, K., Kean, J., Bortolato, A., Cheng, R.K., Dore, A.S., Jazayeri, A., Cooke, R.M., Weir, M., Marshall, F.H.: Structure of class B GPCR corticotropin-releasing factor receptor 1. Nature 499(7459), 438–443 (2013). doi:10.1038/nature12357

18. Rasmussen, S.G., Choi, H.J., Fung, J.J., Pardon, E., Casarosa, P., Chae, P.S., Devree, B.T., Rosenbaum, D.M., Thian, F.S., Kobilka, T.S., Schnapp, A., Konetzki, I., Sunahara, R.K., Gellman, S.H., Pautsch, A., Steyaert, J., Weis, W.I., Kobilka, B.K.: Structure of a nanobody-stabilized active state of the beta(2) adrenoceptor. Nature 469(7329), 175–180 (2011). doi:10.1038/nature09648

19. Kufareva, I., Katritch, V., Stevens, R.C., Abagyan, R.: Advances in GPCR modeling evaluated by the GPCR Dock 2013 assessment: meeting new challenges. Structure 22(8), 1120–1139 (2014). doi:10.1016/j.str.2014.06.012

20. Worth, C.L., Kleinau, G., Krause, G.: Comparative sequence and structural analyses of G-protein-coupled receptor crystal structures and implications for molecular models. PLoS ONE 4(9), e7011 (2009). doi:10.1371/journal.pone.0007011

21. Gutierrez-de-Teran, H., Bello, X., Rodriguez, D.: Characterization of the dynamic events of GPCRs by automated computational simulations. Biochem. Soc. Trans. 41(1), 205–212 (2013). doi:10.1042/BST20120287

22. Sandal, M., Duy, T.P., Cona, M., Zung, H., Carloni, P., Musiani, F., Giorgetti, A.: GOMoDo: a GPCRs online modeling and docking webserver. PLoS ONE 8(9), e74092 (2013). doi:10.1371/journal.pone.0074092

23. Launay, G., Teletchea, S., Wade, F., Pajot-Augy, E., Gibrat, J.F., Sanz, G.: Automatic modeling of mammalian olfactory receptors and docking of odorants. Protein Eng. Des. Sel. 25(8), 377–386 (2012). doi:10.1093/protein/gzs037

24. Sali, A., Blundell, T.L.: Comparative protein modelling by satisfaction of spatial restraints. J. Mol. Biol. 234(3), 779–815 (1993). doi:10.1006/jmbi.1993.1626

25. Latek, D., Bajda, M., Filipek, S.: A hybrid approach to structure and function modeling of G protein-coupled receptors. J. Chem. Inf. Model. 56(4), 630–641 (2016). doi:10.1021/acs.jcim.5b00451

26. Reimann, F., Gribble, F.M.: G protein-coupled receptors as new therapeutic targets for type 2 diabetes. Diabetologia 59(2), 229–233 (2016). doi:10.1007/s00125-015-3825-z

27. Baggio, L.L., Drucker, D.J.: Biology of incretins: GLP-1 and GIP. Gastroenterology 132(6), 2131–2157 (2007). doi:10.1053/j.gastro.2007.03.054

28. Wang, C., Bradley, P., Baker, D.: Protein-protein docking with backbone flexibility. J. Mol. Biol. 373(2), 503–519 (2007). doi:10.1016/j.jmb.2007.07.050

29. Lomize, M.A., Lomize, A.L., Pogozheva, I.D., Mosberg, H.I.: OPM: orientations of proteins in membranes database. Bioinformatics 22(5), 623–625 (2006). doi:10.1093/bioinformatics/btk023

30. Kufareva, I., Rueda, M., Katritch, V., Stevens, R.C., Abagyan, R.: Status of GPCR modeling and docking as reflected by community-wide GPCR Dock 2010 assessment. Structure 19(8), 1108–1126 (2011). doi:10.1016/j.str.2011.05.012

31. Siu, F.Y., He, M., de Graaf, C., Han, G.W., Yang, D., Zhang, Z., Zhou, C., Xu, Q., Wacker, D., Joseph, J.S., Liu, W., Lau, J., Cherezov, V., Katritch, V., Wang, M.W., Stevens, R.C.: Structure of the human glucagon class B G-protein-coupled receptor. Nature 499(7459), 444–449 (2013). doi:10.1038/nature12393

32. Woetzel, N., Karakas, M., Staritzbichler, R., Muller, R., Weiner, B.E., Meiler, J.: BCL:: score–knowledge based energy potentials for ranking protein models represented by idealized secondary structure elements. PLoS ONE 7(11), e49242 (2012). doi:10.1371/journal.pone.0049242

33. Yarov-Yarovoy, V., Schonbrun, J., Baker, D.: Multipass membrane protein structure prediction using Rosetta. Proteins 62(4), 1010–1025 (2006). doi:10.1002/prot.20817

34. Liu, T., Lin, Y., Wen, X., Jorissen, R.N., Gilson, M.K.: BindingDB: a web-accessible database of experimentally determined protein-ligand binding affinities. Nucleic Acids Res. 35(Database issue), D198–D201 (2007). doi:10.1093/nar/gkl999

35. Filipski, K.J., Bian, J., Ebner, D.C., Lee, E.C., Li, J.C., Sammons, M.F., Wright, S.W., Stevens, B.D., Didiuk, M.T., Tu, M., Perreault, C., Brown, J., Atkinson, K., Tan, B., Salatto, C.T., Litchfield, J., Pfefferkorn, J.A., Guzman-Perez, A.: A novel series of glucagon receptor antagonists with reduced molecular weight and lipophilicity. Bioorg. Med. Chem. Lett. **22**(1), 415–420 (2012). doi:10.1016/j.bmcl.2011.10.113

36. Cordomi, A., Ismail, S., Matsoukas, M.T., Escrieut, C., Gherardi, M.J., Pardo, L., Fourmy, D.: Functional elements of the gastric inhibitory polypeptide receptor: comparison between secretin- and rhodopsin-like G protein-coupled receptors. Biochem. Pharmacol. **96**(3), 237–246 (2015). doi:10.1016/j.bcp.2015.05.015

37. Yaqub, T., Tikhonova, I.G., Lattig, J., Magnan, R., Laval, M., Escrieut, C., Boulegue, C., Hewage, C., Fourmy, D.: Identification of determinants of glucose-dependent insulinotropic polypeptide receptor that interact with N-terminal biologically active region of the natural ligand. Mol. Pharmacol. **77**(4), 547–558 (2010). doi:10.1124/mol.109.060111

In Silico Prediction of 3D Structure of *Anopheles Gambiae* ABCC12 Protein

Marion O. Adebiyi[1](✉), Efejiro Ashano[2], and Emmanuel Adetiba[3]

[1] Department of Computer and Information Sciences and Covenant University
Bioinformatics Research (CUBRe), Covenant University, Ota, Nigeria
marion.adebiyi@covenantuniversity.edu.ng
[2] Department of Biological Sciences, Covenant University, Ota, Nigeria
[3] Department of Electrical and Information Engineering,
Covenant University, Ota, Nigeria

Abstract. In this paper, the *Anopheles gambiae* ABCC12 MRP protein domain sequence which contained 216 residues was obtained from the NCBI database in its fasta format (NCBI entry EAA12438.4). This MRP protein sequence was Gapped Blast using BLOSUM 62 matrix with an E-value cut-off of 0.000001 to identify the closest homologous structure as at the date of the study (Nov 2016). Additionally, the sequence was aligned with three prediction modelers, which are the Modeller v9.15 alignment script, the Swiss Model Server and the Raptor-X server for modelling based on the server's automated choice for a suitable template. The structure predicted by Raptor-X has a higher percentage (90.1%) of residues in the most favored regions as compared to Modeller and Swiss-Model (86.5%). This paper further unveils the quality of structure predicted during homology modeling and the diverse correlation as well as the significance of ABCC12 in drug design for malaria vector.

Keywords: *Anopheles gambiae* · Pyrethroid · Resistance · Homology

1 Introduction

Malaria is a worldwide and deadly parasitic disease that is transmitted by the infectious female Anopheles mosquitoes. The mosquitoes breed in stagnant water found in ditches within living environments and farm lands [1–4]. Among the Millennium Development Goals (MDGs) of the United Nations was to stop and reverse malaria incidence by 2015. This is because the disease is responsible for approximately one million deaths or greater than one death every 30–60 s [5, 6].

However, eradication of mosquito vectors is the most effective strategy to curb malaria transmission. The key methods that are currently in use to eradicate mosquito vectors and consequently control and prevent malaria include Indoor Residual Spraying (IRS), Insecticide Treated Net (ITN) and Long Lasting Insecticide Nets (LLIN). Persistent use of these preventive measures has produced impressive results with respect to drastic reduction in malaria among children and saving of a lot of lives [7]. Nevertheless, *Anopheles gambiae* malaria vector, which is a common African mosquito, has a way of building resistance mechanism against treatment. Some studies that were

© Springer International Publishing AG 2017
I. Rojas and F. Ortuño (Eds.): IWBBIO 2017, Part II, LNBI 10209, pp. 163–172, 2017.
DOI: 10.1007/978-3-319-56154-7_16

conducted in selected African countries have reported that *Anopheles gambiae* clearly showed wide spread resistance to pyrethroid and DDT [8–10]. Resistance of the vector to other insecticides such as organophosphate and carbamate has also been documented in the literature [11–14]. This resistance mechanisms call for monitoring and research studies that will help with effective drug design and sustainable malaria control strategies in Africa [15–22].

Furthermore, it has being proven over and over again that drug design and sustainable Malaria control strategies in Africa can be simplified by using computational approaches. These include whole-genome sequencing and targeted sequencing, which are more cost-effective for investigating areas of interest and comparing result quickly [23]. In the study at hand, our computational investigation focuses on ABCC12 MRP protein domain sequence. The sequence contains 216 residues that are most likely to be involved in effective gene for resistance reconstruction in the vector thereby conserving resources and generating similar comparison models *in silico* [18].

2   Materials and Methods

2.1   Target Sequence

The Anopheles gambiae ABCC12 MRP protein domain sequence used in this study, which contains 216 residues was obtained from the NCBI database in the fasta format (NCBI entry EAA12438.4). The sequence is shown below:

>gb|EAA12438.4|:623-838 AGAP007917-PA, partial [Anopheles gambiae str. PEST]
LAIKDGSFEFETKRSRKELDLVQEDIIDFAFRDLTLQVRQGELVCLEGPVGGG
KSSLLQVIMGYFQCTAGAVAISMDVKEGFGYVAQTPWLQQGTIRDNILWGEI
YDETRYKAVIHACALQYDLDALRGDSTGVGEQGRTLSGGQKARVALARAV
YQNKSIYLLDDILSALDAHVASHIIRHCLFGLLKDKTRIIVTQHSMVLNRATQI
LHVEAGQ

2.2   Template Selection and Criteria

The study was performed in early November. The amino acid sequence of the ABCC-MRP-domain of the ABCC12 protein of the Anopheles gambiae, Accession AGAP007917-PA was submitted to Swiss-model servers and subjected to Gapped Blast using BLOSUM 62 matrix with an E-value cut-off of 0.000001 to identify the closest homologous structure as at that date. The closest template 4c3z.1.A.PDB representing the crystal structure of the nucleotide-binding domain 1 of the human multidrug resistance-associated protein 1 [24–26].

2.3   Sequence Alignment and Modeling

Three models were predicted for the ABCC-MRP domain sequence. First, the template was aligned with the target sequence using Modeler v9.15 alignment script. The 3D homology model of the ABCC-MRP-domain was then created with the Modeler v9.15 auto model class using the alignment. Also, the Swiss Model Server was used to build the 3D structure of the target in automated mode based on the Swiss program self-selected template. Finally, the target sequence was submitted to the Raptor-X servers (http://raptorx.uchicago.edu/StructurePrediction/) allowing the servers to also model based on the servers automated choice for a suitable template [27, 28, 30, 31].

3   Results

3.1   Template

The template which was selected for the query sequence to build all three models was 4C3Z.1.A.PDB which had a p-value of 2.55e−09 (Table 1) [25, 28].

Table 1. Features of the model built from the template 4C3Z1.A using Swiss-Model alignment algorithms

Template	Seq ID	Oligo-state	Method	Coverage
4C3Z.1.A	42.71%	Monomer	X-ray, 2.10 Å	0.92

3.2   Generated Models

Three homology models were predicted and validated, two of which was based on the selected template (Modeller – Fig. 1A; Swiss-Model – Fig. 1B) while the last was built based on a template automatically selected by Raptor-X servers (Raptor-X – Fig. 1C). All models were predicted to be monomers [26].

3.3   Model Evaluation

PROCHECK was used to do an analysis of the Ramachandran plot for the modelled structures (Fig. 2A–C) [29]. From Table 2 and Fig. 2A–C, the structures derived from the three methods were of good quality and were similar to one another. However, the structure predicted by Raptor-X had a higher percentage (90.1%) as compared to Modeler and Swiss-Model (86.5%) of residues in the most favored regions (Table 2) [13]. MolProbity was also used to generate the summary statistics of the best model. Also, the SaliLab Model Evaluation Server (ModEval) was used to estimate the quality of the best model.

3.4   Absolute Quality Estimation of the Protein Structure Models

The essence of determining the quality of any model is to measure the accuracy, stability and reliability of such model, this goes a long way to elucidate its usefulness and suitability for biochemical application [33].

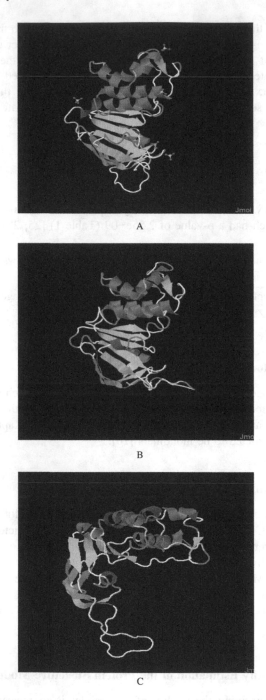

Fig. 1. A–C. Predicted proteins structures modelled by Modeller, Swiss-Model and Raptor-X.

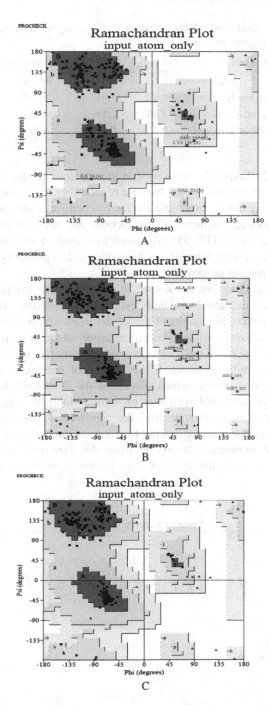

Fig. 2. A–C. Ramachandran plots for predicted structures modelled By Modeler, Swiss-Model and Raptor-X.

Table 2. Summary of results from Ramachandran plot showing amino acids placement with respect to the allowed regions in models

Region	Modeller	Swiss-Model	Raptor-X
Most favored regions	86.5%	86.5%	90.1%
Additionally allowed	9.4%	11.5%	9.9%
Generously allowed	3.6%	2.1%	0.0%
Disallowed regions	0.5%	0.0%	0.0%

QMEANS and QMEANS Z - were used in the absolute quality estimation of this work. QMEANS Z - score was choosen because of its ability to significantly expose the stability of structure modeled homologically for biochemical functions such as; drug design and developments, functional site predictions and analysis, resistance mechanisms prediction and so on [32, 33]. They are tool used in estimating the protein stability of models, its scoring function has been designed to primarily rank alternative models of the same protein sequence.

With or without considering the size of this proteins, the isolated chains of the target protein was given an absolute quality estimation assessment. This was done by directly relating the structural features of the model to its experimental feature of the same size, thereby analyzing various geometrical aspect of the protein. [33–36], Fig. 3 shows the QMEAN Z - score of the local/native quality estimate of best model chosen. The QMEAN Z-score is the estimate of the 'degree of nativeness of the structural features extracted in that model, which elucidates the details of the quality of the model, which can be compared with that of some redundant experimental protein structures in PDB (may not be proteins from the *gambiae*). According to [33], "the Z-scores of the individual terms of the scoring function indicate which structural features of a model exhibit significant deviations from the expected 'native' behaviour, e.g. unexpected solvent accessibility, back-bone geometry, inter-atomic packing, etc".

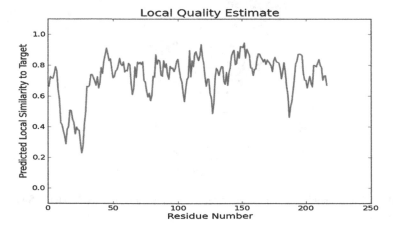

Fig. 3. QMEAN Z - score of the local/native quality estimate of best model chosen.

Proteins with unusually high QMEAN scores (Z-score > 3) marked in green, does not exist in this set. This target model falls in a favourable region of the PDB reference set, that is (1 < |Z-score| < 2). This is suspected to be responsible for why the protein was implicated among the potential insecticital targets identified in [37–39], as highly successful in crystalization, OB & ParCrys was used to estimate protein's propensity to produce diffraction quality crystals with the intention to provide guidance in selecting best drug/insecticide targets [37–39] which further confirms its validation as a potential drug/insecticide target and a useful inhibitable protein to be focussed on during drug design pipeline development. In [33], it is also emphasized that the proteins at the periphery of the QMEAN score spectrum (0.5–1.0) can be assigned to membrane proteins which exist in a fundamentally different environment compared to soluble proteins and extremely stable proteins found in thermophilic organisms (Fig. 4).

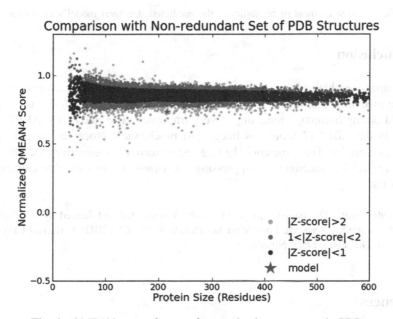

Fig. 4. QMEAN scores for set of non redundant structures in PDB.

Table 3. *Z-score* analysis of the ABCC12 MRP for the isolated chains monomer

Structure	Size	QMEAN	C-β	All-atom	Solvation	Torsion
ABCC12 MRP	600	−2.92	−0.16	−2.07	−2.33	−2.24

In Table 3, the numbers were explicitly extracted to see the detailed Z-score formalism, which was applied for the swiss model isolated protein chain monomer and was found to have below −3.5 in **QMEAN** isolation, which is unfavourable. The modeler isolated protein chain monomer was concluded to have almost meaningless isolation because of the non globular conformation it adopted (Fig. 5).

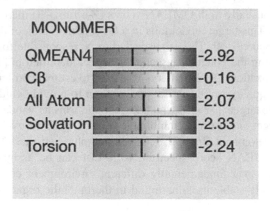

Fig. 5. Measurement of reliability of the predicted structural model's monomer

4   Conclusion

In this paper, we obtained our best model by comparing predicted structures generated from the modeling platforms. We have been able to show that the quality of structure predicted during homology modeling is very important. The role of the ABCC-MRP domain in the ABCC12 protein is integral to biochemical processes carried out by *Anopheles gambiae*. This gene could be targeted to serve as an effective mechanism for malaria control by inhibiting or suppressing the *gambiae's* detoxicative capacity for insecticides.

Acknowledgement. The publication of this study is supported and funded by the Covenant University Centre for Research, Innovation and Development (CUCRID), Covenant University, Canaanland, Ota, Ogun State, Nigeria.

References

1. Sachs, J., Malaney, P.: The economic and social burden of malaria. Nature **415**, 680–685 (2002)
2. Martens, P., Hall, L.: Malaria on the move: human population movement and malaria transmission. Emerg. Infect. Dis. **6**(2), 103–109 (2000)
3. Sinka, M., Bangs, M., Manguin, S.C., Patil, A., Temperley, W., Gething, P., Elyazar, R., Kabaria, C., Harbach, R., Hay, S.: The dominant Anopheles vectors of human malaria in the Asia-Pacific region: occurrence data, distribution maps and bionomic précis. Parasit. Vectors **4**, 89 (2011)
4. Bunnik, E.M., Chung, D.W., Hamilton, M., Ponts, N., Saraf, A., Prudhomme, J., Florens, L., Le Roch, K.G.: Polysome profiling reveals translational control of gene expression in the human malaria parasite Plasmodium Falciparum. Genome Biol. **14**(11), R128 (2013)
5. United Nations. Millennium Development Goals (MDGs): Goal 6: Combat HIV/AIDS, Malaria and other diseases. http://www.who.int/topics/millennium_development_goals/diseases/en/index.html

6. World Health Organization. Malaria. http://www.who.int/malaria/en
7. Ingham, V.A., Jones, C.M., Pignatelli, P., Balabanidou, V., Vontas, J., Wagstaff, S.C., Moore, J.D., Ranson, H.: Dissecting the organ specificity of insecticide resistance candidate genes in Anopheles gambiae: known and novel candidate genes. BMC Genomics **15**(1), 1 (2014)
8. Diabate, A., Brengues, C., Baldet, T., Dabire, R.K., Hougard, M., et al.: The spread of the Leu-Phe kdr mutation through Anopheles gambiae complex in Burkina Faso: genetic introgression and de novo phenomena. Trop. Med. Int. Health **9**, 1267–1273 (2004)
9. Etang, J., Fondjo, E., Chandre, F., Morlais, I., Bregues, C., et al.: First report of the knock down mutations in malaria vectors Anopheles gambiae from cameron. Am. J. Trop. Med. Hyg. **74**, 795–797 (2006)
10. Tripet, F., Wright, J., Cornel, A., Fotana, A., Mcabee, C.M.: Longitudinal survey of knockdown resistance to pyrethroid (kdr) in Mali, West Africa, and evidence of its emergence in the Bamako form of Anopheles gambiae ss. Am. J. Trop. Med. Hyg. **76**(1), 81–87 (2007)
11. N'Guessan, R., Darriet, F., Guillet, P., Traore-Lamizana, M., Corbel, V., et al.: Resistance to carbosulfan in Anopheles gambiae from Ivory Coast, based on reduced sensitivity of acetylcholinesterase. Med. Vet. Entomol. **17**, 19–25 (2003)
12. Weill, M., Lutfalla, K., Mogensen, F., Chandre, F., Berthomieu, A., et al.: Comparative genomics: insecticide resistance in mosquito vectors. Nature **423**, 136–137 (2003)
13. Corbel, V., N'Guessan, R., Brengues, C., Chandre, F., Djogbenou, L., et al.: Multiple insecticide resistance mechanisms in Anopheles gambiae and Culex quinquefasciatus from Benin, West Africa. Acta Trop. **101**, 207–216 (2007)
14. Ranson, H., Abdallah, H., Badolo, A., Guelbeogo, W.M., Kerah-Hinzoumbe, C., et al.: Insecticide resistance in Anopheles gambiae: data from the first year of a multi-country study highlight the extent of the problem. Malar. J. **8**, 299 (2009)
15. Kersey, P.J., Allen, J.E., Christensen, M., Davis, P., Falin, L.J., Grabmueller, C., et al.: Ensembl genomes 2013: scaling up access to genome-wide data. Nucl. Acids Res. **42**, D546–D552 (2014)
16. UniProt-Consortium: Ongoing and future developments at the Universal protein resource. Nucl. Acids Res. **39**, D214–D219 (2011)
17. Müller, P., Donnelly, M.J., Ranson, H.: Transcription profiling of a recently colonised pyrethroid resistant Anopheles gambiae strain from Ghana. BMC Genomics **8**, 36 (2007)
18. Nikou, D., Ranson, H., Hemingway, J.: An adult-specific CYP6 P450 gene is overexpressed in a pyrethroid-resistant strain of the malaria vector, Anopheles gambiae. Gene **318**, 91–102 (2003)
19. Centers for disease control and prevention. http://www.cdc.gov/malaria/about/biology/mosquitoes/
20. Irving, H., Riveron, J., Inbrahim, S., Lobo, N., Wondji, C.: Positional cloning of rp2 QTL associates the P450 genes CYP6Z1, CYP6Z3 and CYP6M7 with pyrethroid resistance in the malaria vector Anopheles funestus. Heredity **109**, 383–392 (2012)
21. Ingham, V.A., Jones, C.M., Pignatelli, P., Balabanidou, V., Vontas, J., Wagstaff, S.C., et al.: Dissecting the organ specificity of insecticide resistance candidate genes in Anopheles gambiae: known and novel candidate genes. BMC Genomics **15**, 1018 (2014)
22. Yang, J., McCart, C., Woods, D.J., Terhzaz, S., Greenwood, K.G., ffrench-Constant, R.H., Dow, J.A.: A Drosophila systems approach to xenobiotic metabolism. Physiol. Genomics **30**, 223–231 (2007)

23. Dharia, N.V., Bright, A.T., Westenberger, S.J., Barnes, S.W., Batalov, S., Kuhen, K., Borboa, R., Federe, G.C., McClean, C.M., Vinetz, J.M., Neyra, V.: Whole-genome sequencing and microarray analysis of ex vivo Plasmodium vivax reveal selective pressure on putative drug resistance genes. Proc. Nat. Acad. Sci. **107**(46), 20045–20050 (2010)

24. Altschul, S.F., Madden, T.L., Schaffer, A.A., Zhang, J., Zhang, Z., Miller, W., Lipman, D.J.: Gapped BLAST and PSI-BLAST: a new generation of protein database search programs. Nucl. Acids Res. **25**, 3389–3402 (1997)

25. Berman, H., Henrick, K., Nakamura, H., Markley, J.L.: The worldwide Protein Data Bank (wwPDB): ensuring a single, uniform archive of PDB data. Nucl. Acids Res. **35**, D301–D303 (2007)

26. Biasini, M., Bienert, S., Waterhouse, A., Arnold, K., Studer, G., Schmidt, T., et al.: SWISS-MODEL: modelling protein tertiary and quaternary structure using evolutionary information. Nucl. Acids Res. **2014**, 1–7 (2014)

27. Sievers, F., Wilm, A., Dineen, D.G., Gibson, T.J., Karplus, K., Li, W., et al.: Fast, scalable generation of high-quality protein multiple sequence alignments using Clustal Omega. Mol. Syst. Biol. **7**, 1–6 (2011). Article number: 539

28. Laskowski, R.A., Chistyakov, V.V., Thornton, J.M.: PDBsum more: new summaries and analyses of the known 3D structures of proteins and nucleic acids. Nucl. Acids Res. **33**, D266–D268 (2005)

29. Laskowski, R.A., MacArthur, M.W., Moss, D.S., Thornton, J.M.: PROCHECK: a program to check the stereochemical quality of protein structures. J. Appl. Cryst. **26**, 283–291 (1993)

30. Larkin, M.A., Blackshields, G., Brown, N.P., Chenna, R., McGettigan, P.A., McWilliam, H., et al.: ClustalW and ClustalX version 2. Bioinformatics **23**, 2947–2948 (2007)

31. Chen, V.B., Arendall III, W.B., Headd, J.J., Keedy, D.A., Immormino, R.M., Kapral, G.J., et al.: MolProbity: all-atom structure validation for macromolecular crystallography. Acta Crystallogr. **D66**, 12–21 (2010)

32. Benkert, P., Tosatto, S., Schwede, T.: Global and local model quality estimation at CASP8 using the scoring functions QMEAN and QMEANclust. Proteins **77**(Suppl 9), 173–180 (2009)

33. Benkert, P., Biasini, M., Schwede, T.: Toward the estimation of the absolute quality of individual protein structure models. Bioinformatics **27**(3), 343–350 (2011). doi:10.1093/bioinformatics/btq662

34. Benkert, P., et al.: QMEAN server for protein model quality estimation. Nucl. Acids Res. **37**, W510–W514 (2009)

35. Benkert, P., et al.: QMEANclust: estimation of protein model quality by combining a composite scoring function with structural density information. BMC Struct. Biol. **9**, 35 (2009)

36. Benkert, P., et al.: QMEAN: a comprehensive scoring function for model quality assessment. Proteins **71**, 261–277 (2008)

37. Adebiyi, M.O.: Computational analysis of Anopheles gambiae metabolism to facilitate insecticidal target and complex resistance mechanism discovery. Diss. Covenant University (2014)

38. http://theses.covenantuniversity.edu.ng/handle/123456789/953

39. Adebiyi, M.O.: Predicting the structure of Anopheles gambiae Cytochrome P450 protein using computational methods (2015)

Exploring Symmetric Substructures in Protein Interaction Networks for Pairwise Alignment

Ahed Elmsallati[1]([✉]), Swarup Roy[2], and Jugal K. Kalita[1]

[1] Department of Computer Science, University of Colorado at Colorado Springs, Colorado Springs, CO 80918, USA
{aelmsall,jkalita}@uccs.edu
[2] Department of Information Technology, North-Eastern Hill University, Shillong 793022, Meghalaya, India
swarup@nehu.ac.in

Abstract. In molecular biology, comparison of multiple Protein Protein Interaction (PPI) networks to extract subnetworks that are conserved during evolution across different species is helpful for studying complex cellular machinery. Most efforts produce promising results in creating alignments that show large regions of biological *or* topological similarity between the PPI networks of various species, but few do both. We present a new pairwise aligner *SSAlign* (Symmetric Substructure Alignment) that extracts maximal substructures from participating PPI networks and uses Gene Ontology Consistency (GOC) as the graph isomorphic function for aligning two subgraphs. We use PPI networks from Isobase data repository for experiments and comparisons. Our results show that in comparison to other contemporary aligners, SSAlign is better at aligning topologically *and* biologically similar subnetworks.

Keywords: Protein protein interaction · Alignment · Symmetric substructure · Network topology · Gene Ontology · Orthology · Homology

1 Introduction

Interplay among proteins macromolecules form an interconnected network called Protein-Protein Interaction (PPI) Network. The current abundance of PPI networks is helping in addressing the challenge of discovering conserved interactions across multiple species. Studies suggest that molecular networks are conserved through evolution [1]. The interactions between protein pairs as well as the overall composition of the network are important for the overall study of cellular functioning of an organism. Comparison of conserved substructures across species helps understand complex biochemical processes. The ultimate goal of network alignment is to predict the protein functions in an unknown species from the known ones. Sequence similarity metrics such as BLAST scores [2] do not provide conclusive evidence for similar protein functions. PPI network alignment can be treated as a supplement to the computation of sequence similarity, providing needed topological information in identifying ortholog proteins.

© Springer International Publishing AG 2017
I. Rojas and F. Ortuño (Eds.): IWBBIO 2017, Part II, LNBI 10209, pp. 173–184, 2017.
DOI: 10.1007/978-3-319-56154-7_17

PPI network alignment is a relatively new research area in computational biology, with the goal of uncovering large shared sub-networks between species and deriving phylogenetic relationships among species by examining the extent of overlap exhibited by different PPI networks [3–5]. Comparing two biological networks is a challenging problem, since conserved sub-networks are expected to possess both topological and biological similarities. A number of aligners has been reported in the literature recently. They either produce promising results in creating biologically or functionally significant alignments, or are effective in producing topological fit, but few achieve both well.

We present a new way of aligning networks using isomorphism of substructures extracted from input PPI networks. We call it the Symmetric Substructure Aligner (SSAlign). GOC score is used as an isomorphic merit function for optimized alignment of symmetric subgraphs. The presence of symmetric substructures contributes to the topological aspect of the alignment and the merit function, GOC, helps achieve biological similarity among the matched proteins.

We organize rest of the paper into four sections. Section 2 presents an introduction and background of the PPI alignment problem. It also discusses a few state-of-the-art aligners. Section 3 introduces our proposed aligner, SSAlign. Experimental results and analysis are reported in Sect. 4. We conclude in Sect. 5.

2   Alignments of PPI Networks

PPI networks are often represented as undirected graphs, the proteins being nodes and molecular interactions between them being edges.

Definition 1 (Protein-Protein-Interaction Network). *A set of proteins* $\mathcal{P}$ *forms a PPI network* $\mathcal{G} = <\mathcal{P}, \mathcal{I}>$ *by virtue of the physical interactions among protein molecules in* $\mathcal{P}$ *due to various biochemical events.* $\mathcal{I}$ *is the symbolic representation of the set of strong interactions between a pair of proteins* P_i *and* $P_j \in \mathcal{P}$.

Comparative analysis of PPI networks across species is an important computational task needed to identify evolutionarily conserved subnetworks. It is often considered a graph alignment problem where proteins from different species are superimposed on one another based on strong evidence accrued from high protein sequence similarity as well as high topological similarity.

Definition 2 (PPI Network Alignment). *Given* k *distinct PPI networks* $\mathcal{G}_1 = <\mathcal{P}_1, \mathcal{I}_1>, \mathcal{G}_2 = <\mathcal{P}_2, \mathcal{I}_2>, \cdots, \mathcal{G}_k = <\mathcal{P}_k, \mathcal{I}_k>$ *from* k *different species, the PPI alignment problem is to find conserved subnetworks within the* k *graphs. This alignment graph* $\mathcal{A}$ *is a subgraph consisting of nodes representing* k *similar proteins (one per species), and edges representing conserved interactions among the species. Each alignment* $\mathcal{A}_i$ *can be represented as:*

$$\mathcal{A}_i = \{<p_{1i}, p_{2i}, \cdots, p_{ki}>|p_{ji} \in \mathcal{P}_j\} \tag{1}$$

Pairwise alignment matches two species, i.e., $k = 2$. Broadly speaking, there are two different classes of alignments. *Global alignment* performs best single superimposition of the whole input networks on top of one another. On the other hand, *local alignment* aligns small subnetworks giving rise to multiple superimposed aligned subgraphs. The overall objective of PPI alignment methods is to maximize the size of the subgraphs, ensuring high protein sequence and topological similarities among the networks. Several pairwise aligners have been proposed in the last few years to achieve alignment with high biological significance.

For instance, PINALOG [6] is a pairwise aligner that identifies dense subgraphs, called communities, within the input networks to find regions of similarity between the two PPI networks. It first finds a mapping from the communities in one graph to the communities in the other and then, for each pair of communities, matches the nodes within them. The similarity between two communities is determined by balancing the sum of the sequence similarities and shared Gene Ontology (GO) terms. NETAL [7] works by recursively defining topological similarity, where nodes are similar if their neighbors are similar. Given this topological similarity matrix, it refines an interaction score matrix by estimating how many interactions will be conserved if the given nodes are aligned. The alignment is constructed greedily by mapping together the nodes with the best interaction scores, and after each pair of nodes is aligned, the interaction score matrix is updated to select the next pair to align. C-GRAAL [8] uses graphlet degree signatures along with another topological metric to heuristically maximize the total number of edges aligned between two networks. C-GRAAL also optionally can take bit scores as input. Topology-based alignment makes C-GRAAL applicable to any type of network. In addition to using graphlet degree signatures, C-GRAAL also uses the presence of common neighbors to inform its seed-and-extend heuristic. GHOST [9] uses spectral graph theory to build a similarity metric that is used as a cost function to align the input networks. The spectral signatures for sub-graphs of different sizes around a specific node are created using the normalized Laplacian and combined with a greedy seed-and-extend algorithm to find the alignment. MAGNA [10] adopts an evolutionary approach using Genetic Algorithms and optimizes randomly selected initial alignments with the help of three different objective scores, namely Induced Conserved Structure (ICS) [9], Edge Correctness (EC) [11], and Symmetric Substructure Score (S^3) [10].

In our study, we observe a few interesting facts about the aligners.

- When proteins interact, they form different network topologies [6]. Hence, homology information seems to be encoded in network topology itself, which leads to the conclusion that sequence information doesn't usually help predict orthology [12].
- Many PPI network aligners that use mainly topological fit to extract the alignment, such as PINALOG [6] and GHOST [9], show high biological scores. However, some other aligners, such as those from the GRAAL family [1,8,12–14] that uses mainly a topological graphlet based approach for alignment, have poor functional enrichment scores for the produced alignments in comparison to PINALOG and GHOST [15].

– State-of-the-art aligners use a cost function based on either sequence or topological similarity or both. They maximize the cost functions to achieve a
functionally enriched alignment. They are able to achieve high biological
(exclusive) or topological similarity scores between two participating networks,
but do not do the both well.

In our work, we develop an aligner that is expected to perform well in terms
of both biological and topological merits.

3   SSAlign: A New Symmetric Substructures Aligner

In this section, we introduce our methodology to obtain a global pairwise network
alignment of PPI networks. Our work is motivated by the fact that the presence
of biologically orthologous proteins during evolution leads to a high structural
similarity, producing better topological fit across multiple species. It is an important ingredient for achieving a good functionally enriched alignment. PINALOG
extracts only clique structures for alignment, while GRAAL and its extensions
align two networks based on graphlet degree signatures, ignoring larger clique-
like structures. We feel that both clique and non-clique structures are equally
important from the point of view of conversation among sub-network structures.
Instead of graphlet degree signatures, we explore all possible inherent network
substructures from both participating networks for alignment.

Definition 3 (k-Substructure). *Given a network $\mathcal{G}$ built with a set of proteins
$\mathcal{P}$ connected by interconnection edges $\mathcal{I}$, a k-substructure is an induced subgraph
extracted from $\mathcal{G}$, consisting of k proteins $\mathcal{P}'$ and a set of edges $\mathcal{I}'$ connecting
these proteins, such that $\mathcal{P}' \subseteq \mathcal{P}$ and $\mathcal{I}' \subseteq \mathcal{I}$*

We extract all substructures or induced subgraphs of size k from a network.
These substructures include complete subgraphs or cliques also. We call them
as k-maximal substructures.

Definition 4 (k-Maximal Substructures). *An induced subgraph or substructure of size k extracted from $\mathcal{G}$ is maximal if no extension of the substructure is
possible. In other words, $\mathcal{G}'_k \subseteq \mathcal{G}$ is maximal if for any two proteins $P_i, P_j \in \mathcal{G}'_k$,
there is an edge $(P_i, P_j) \in \mathcal{I}$ connecting P_i and P_j, then $<P_i, P_j>$ is in $\mathcal{G}'_k$.*

Figure 1 shows two substructures (b) and (c) extracted from (a). The substructure (b) is a k-maximal substructure of $k = 5$ since it contain all the edges
from the graph (a) connecting all selected nodes $\{A_1 \cdots A_5\}$. On the other hand,
(c) is a subset of (b) and so not a maximal subgraph.

Interestingly any clique in a graph is also a form of maximal k-substructure
of size k.

Figure 2 shows the statistics on some of the substructures of size $k = 3$ to 5
nodes extracted from PPI networks of different species.

Once a set of maximal substructures for varying values of k is available
from the participating networks, the task of alignment starts with the finding

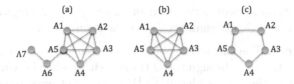

Fig. 1. Example of k-maximal substructure of size 5.

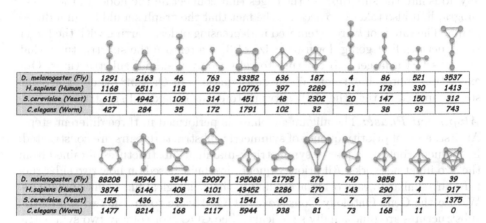

D. melanogaster (Fly)	1291	2163	46	763	33352	636	187	4	86	521	3537
H. sapiens (Human)	1168	6511	118	619	10776	397	2289	11	178	330	1413
S. cerevisiae (Yeast)	615	4942	109	314	451	48	2302	20	147	150	312
C. elegans (Worm)	427	284	35	172	1791	102	32	5	38	93	743

D. melanogaster (Fly)	88208	45946	3544	29097	195088	21795	276	749	3858	73	39
H. sapiens (Human)	3874	6146	408	4101	43452	2286	270	143	290	4	917
S. cerevisiae (Yeast)	155	436	33	231	1541	60	6	7	27	1	1375
C. elegans (Worm)	1477	8214	168	2117	5944	938	81	73	168	11	0

Fig. 2. The total number of different substructures of size up to size 5 extracted from PPI networks of four different species. This statistics are obtained from Isobase dataset used in [16].

of symmetric substructures from the participating networks followed by optimal superimposition of symmetric substructures. We use a simple definition of symmetric substructures for ease of computation.

Definition 5 (Symmetric Substructures). *Two k maximal substructures, $\mathcal{G}'_k = <\mathcal{P}', \mathcal{I}'>$ and $\mathcal{G}''_k = <\mathcal{P}'', \mathcal{I}''>$ are symmetric if $|\mathcal{I}'| = |\mathcal{I}''|$ and $\mathcal{G}'_k$ and $\mathcal{G}''_k$ are isomorphic to each other.*

Superimposition of two symmetric substructures produces graph isomorphism between the two corresponding subgraphs.

Definition 6 (Symmetric Substructures Alignment). *Given two symmetric maximal substructures, $\mathcal{G}'_k \subseteq \mathcal{G}'$ and $\mathcal{G}''_k \subseteq \mathcal{G}''$, where $\mathcal{G}'$ and $\mathcal{G}''$ are two PPI networks of different species, the alignment of symmetric substructures is an appropriate isomorphic mapping function $f : \mathcal{P}' \to \mathcal{P}''$ that maps a pair of corresponding nodes to each other along with the edge between them.*

3.1 Alignment Approach

We divide the overall task of alignment into two major sub-tasks namely, (i) extraction of k-maximal substructures and (ii) optimal alignments of substructures of two different networks using a new gene ontological merit function to achieve a biologically significant alignment outcome.

Extraction Phase: The extraction phase accepts an adjacency matrix of a PPI network as input. It walks through the network to mine all substructures. We use a method similar to the k-walk algorithm [17] to extract maximal substructures of size k. It starts with a set of k target nodes from the original graph and gradually adds into the subgraph all the edges that connect all the nodes of the target subgraph. It also takes into account the fact that the graph should be an induced graph. The values of k are attempted in decreasing order, starting with the larger substructures first going down to a size of 3. We remove the substructures that are subset of a larger substructure to retain only maximal substructures. Our extraction approach extracts all the substructures in $O(|k|.n^3)$ time. The same step is applied to the other PPI network to extract maximal substructures.

Alignment Phase: The alignment phase is performed in three different steps. At first, a set of priority queues of symmetric substructure pairs are constructed. Each queue contains a pair of symmetric k-maximal substructures obtained from the two input networks. All priority queues are built based on how two substructures are similar to each other according to their biological functions similarities. We use the Gene Ontology Consistency (GOC) score to find if two symmetric substructures are functionally significant. The GOC score of any two symmetric substructures, $\mathcal{G}_1$ and $\mathcal{G}_2$ is given as:

$$GOC(\mathcal{G}'_1, \mathcal{G}''_2) = \sum_{(P_i \in \mathcal{G}'_1, P_j \in \mathcal{G}''_2)} \frac{|GO(P_i) \cap GO(P_j)|}{|GO(P_i) \cup GO(P_j)|} \tag{2}$$

where P_i and P_j are two candidate isomorphic proteins to be matched belonging to $\mathcal{G}'_1$ and $\mathcal{G}''_2$ respectively. The set of GO annotations for a protein P_i are denoted by $GO(P_i)$, where all terms are extracted from the GO database with the exclusion of the root terms for each protein. To minimize redundancy, we limit the GO terms used to a maximum distance of five from the root of the ontology.

We optimize the alignments by maximizing the GOC score. Note that the extraction step takes in account all possible orientations to construct any substructure with the same set of nodes and edges. The reason behind this is to maximize the GOC score of the candidates that are being aligned. We perform one-to-one alignment between the proteins from the candidate pair of substructures. We begin the alignment by aligning larger substructures first. A substructure is considered to be larger if the number of nodes (proteins) and edges are greater than the others. During this first phase of alignment, we ignore substructures that are *partially* aligned. The purpose of skipping all partially aligned substructures is to maximize the total number of nodes aligned from both networks and give a chance to the substructures that are fully unaligned to be aligned first. We store all partially aligned substructures in another queue for the second round of alignment. During the second round, we align all the partially aligned substructures left out in the last phase using the same method.

In the final step, we expand the initial alignment obtained from the above steps by aligning the neighborhood of all pairs of nodes that are already aligned. The alignment is based on GOC score. If a pair of nodes in the alignment set touches a set of nodes that are not in the alignment set, this set of nodes is aligned in this phase. This step is motivated by the fact that the star topology is predominant in PPI networks. If the central node is a part of any substructure and other nodes are not, then all these nodes are not aligned. Figure 3 shows an example of a spoke node visualized from one of the PPI networks of a species under study. Another important reason behind performing such a step is to cover all substructures that are excluded from consideration (See Fig. 4). We Should note that SSAlign takes in the worst case $O(n^2)$, where n is the largest number of substructures extracted.

Fig. 3. Example of a spoke node that is part of a substructure in PPI networks of four different species. This sub-network is taken from the visualization of *H. sapiens (Human)* network. The central node, on the left, is part of substructure and all other nodes are not.

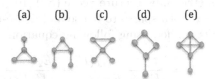

Fig. 4. Excluded substructures extracted from Isobase PPI networks from four different species.

4   Experimental Evaluation

In this section, we assess the performance of our aligner and compare it with the state-of-the-art aligners. Symmetric Substructure Aligner is implemented in Java. We use *JGraphT*[1] library to extract and compute network substructures. All experiments of SSAlign are conducted on a desktop machine, running OS X Yosemite Version 10.10.5, with Quad Core 2.7 GHz Intel Core $i7$ and 16 GB 1600 MHz *DDR3* RAM.

---

[1] http://jgrapht.org/.

4.1   Dataset Used

To evaluate the performance of our proposed method we use real PPI networks. We run our algorithm on four extensively studied species: *Saccharomyces cerevisiae* (Yeast), *Drosophila melanogaster* (Fly), *Caenorhabditis elegans* (Worm), and *Homo sapiens* (Human). All the networks are downloaded from the publicly available IsoBase [18] data repository. Brief details about the networks are shown in Table 1.

Table 1. PPI networks from IsoBase data repository used for experiments

Species	Number of proteins	Number of interactions
Saccharomyces cerevisiae (Yeast)	5499	31898
Drosophila melanogaster (Fly)	7518	25830
Caenorhabditis elegans (Worm)	2805	4572
Homo sapiens (Human)	9633	36386

4.2   Assessment Metrics

We have observed weaknesses in some of the evaluation metrics [11] that are used for assessing the quality of alignments. Aligners score differently when different metrics are used to evaluate the resulting alignment. The Edge Correctness (EC) metric cannot penalize sparse-to-dense alignment. On the other hand Induced Conserved Structure (ICS) is not able to penalize dense-to-sparse alignment. Therefore, we use Symmetric Substructure Score (S^3) [10] for topological assessment, because it penalizes both sparse-to-dense and dense-to-sparse alignments equally. The S^3 metric is defined using following equation:

$$S^3(G_1, G_2, f) = \frac{|f(E_1) \cap E_2|}{|E_1| + |E_{G_2[f(V_1)]}| - |f(E_1) \cap E_2|} \tag{3}$$

where $|E_{G_2[f(V_1)]}|$ denotes the total number of edges of the induced sub-network of G_2 that contains all nodes that have been mapped to by f. The denominator of S^3 can be thought of as the number of unique edges in the composite graph resulting from overlaying the graphs G_1 and G_2 according to the alignment f.

For biological assessment, we evaluate the performance of the aligners using the Gene Ontology Consistency (GOC) [19] score (given in Eq. (2)) which uses Gene Ontology (GO) [20] annotations to enhance the measurement of the alignment quality.

4.3   Results and Analysis

Extraction of cliques and subgraphs are very expensive operations. We extract both in advance as an offline task. All extracted cliques and substructures are stored in a file to speed up the extraction of alignments. For our experiments,

(a)   (b)

Fig. 5. Substructures that do not exist as maximal substructures in PPI networks of four different species under study.

we consider substructures of size up to $k = 5$ only. The reason behind considering such a moderate maximum size is the cost involved in extracting larger substructures in terms of time. We also exclude few of extracted substructures due to the following reasons.

1. There is a large number of possibilities to form such substructures; But, this huge number of such substructures makes the extraction of the final alignment infeasible.
2. Extracting all such substructures does not beget any extra power toward obtaining the final alignment as we can easily expand the final alignment to the neighborhood of aligned node with little cost. Examples of such subgraphs are shown in Fig. 4.
3. A few rare substructures that are usually not present in PPI networks (Isobase). Figure 5 depicts substructures that do not exist in any PPI network as a maximal substructure.

In addition, we extract cliques up to size 11, which is the maximum common clique size among all the networks available in the Isobase [21] repository. We compare performance of our aligner with five different popular aligners namely, C-GRAAL, PINALOG, MAGNA, GHOST and NETAL. C-GRAAL and PINA-LOG that use community-based topological approach similar to our method. MAGNA and GHOST are two recently published aligners that have the best performance among topological aligners [15]. In addition, we choose NETAL as it is one of the fastest topological aligners [15].

The topological scores of the aligners including our proposed aligner (SSAlign) are given in Table 2. The best scores are highlighted in bold. The performance of SSAlign is promising in producing topologically similar alignments for all species pairs. However, NETAL shows marginally superior performance in comparison to SSAlign for three network pairs. Missing data in the Table 2 (and Table 3) indicate the failure of the aligners to generate outcomes.

We measure the biological merit of the alignments using functional enrichment based on the GOC score. Orthologs preserve biological functions during evolution. The more two aligned proteins shared common ontological terms, the more they are significant from an orthology perspective. We report the GOC score from the alignment outcomes produced by aligners in Table 3. In the Table,

Table 2. S^3 scores for different aligners on IsoBase PPI network pairs

Aligner	ce-dm	ce-hs	ce-sc	dm-hs	sc-dm	sc-hs
SSAlign	0.373	**0.464**	**0.357**	0.184	**0.164**	0.134
C-GRAAL	0.142	0.131	0.009	0.119	0.118	0.107
GHOST	0.139	0.148	0.118	0.115	-	-
MAGNA	0.261	0.245	0.246	0.107	0.107	0.116
NETAL	**0.412**	0.396	0.323	**0.187**	0.149	**0.141**
PINALOG	0.006	0.006	0.005	0.005	0.004	0.005

Table 3. GOC scores achieved by the aligners for different PPIN pairs

Aligner	ce-dm	ce-hs	ce-sc	dm-hs	sc-dm	sc-hs
SSAlign	221.17	**125.58**	**136.95**	**310.95**	297.10	**416.69**
C-GRAAL	75.95	41.21	53.48	110.5	100.5	123.2
GHOST	181.5	84.85	95.43	166.9	-	-
MAGNA	12.96	18.64	25.26	37.54	48.84	79.59
NETAL	12.01	18.64	25.26	37.54	48.84	79.59
PINALOG	**229.8**	96.62	136.5	302.3	**363.7**	320.4

we observe that SSAlign is very proficient in producing alignments that are biologically significant as SSAlign takes into consideration GOC scores as an isomorphic merit function during alignment.

The GOC score may be biased towards the number of candidate proteins aligned. The more the number of proteins aligned, the higher the GOC score for the alignment is likely to be. Figure 6 shows statistics for the numbers of nodes aligned by the aligners under review. We note that our method aligns in a judicious way and considers only a moderate number of proteins which are potential candidates for alignments instead of aligning proteins that are biologically and topologically not fit for alignments. In Fig. 6, it is evident that though SSAlign and PINALOG do not align all proteins, GOC scores are still better in comparison to the other aligners where other aligners align all the proteins.

Careful examination of the topological and biological scores achieved by different aligners further reveals that except for SSAlign, no aligner is equally adapt at producing alignments that are concurrently topologically and biologically significant. SSAlign exhibits superior alignment results by obviously outperforming other aligners.

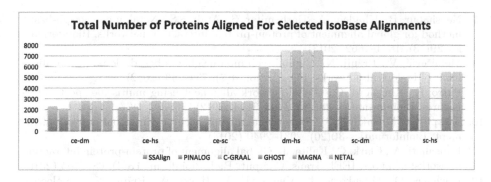

Fig. 6. Total number of protein pairs aligned by the different aligners

5 Conclusion

We present a new and novel approach, SSAlign, for aligning a pair of PPI networks with significant topological and biological similarity. We extract symmetric substructures from two candidate networks and align them using an isomorphic merit function based on gene ontology, GOC. Experimental results exhibit the superiority of SSAlign in producing alignments which are both topologically and biologically fit, where other candidate aligners fail to achieve both simultaneously.

Acknowledgement. AE acknowledges Ministry of Higher Education and Scientific Research, Libya (MOHESR) and University of Tripoli supported this work under grant 972-2007. **SR** thanks Department of Biotechnology, Govt of India for Overseas Research Associateship (vide sanction BT/NE/20/2011) to conduct the research at LINC Lab of UCCS. Special thanks to Dr. Nabeeh Hasan for his helpful comments.

References

1. Milenkovic, T., Ng, W.L., Hayes, W., Przulj, N.: Optimal network alignment with graphlet degree vectors. Cancer Inform. **9**, 121 (2010)
2. Altschul, S.F., Gish, W., Miller, W., Myers, E.W., Lipman, D.J.: Basic local alignment search tool. J. Mol. Biol. **215**(3), 403–410 (1990)
3. Vijayan, V., Saraph, V., Milenković, T.: Magna++: maximizing accuracy in global network alignment via both node and edge conservation. Bioinformatics **31**(14), 2409–2411 (2015). Page btv161
4. Milenković, T., Pržulj, N.: Topological characteristics of molecular networks. In: Koyutürk, M., Subramaniam, S., Grama, A. (eds.) Functional Coherence of Molecular Networks in Bioinformatics, pp. 15–48. Springer, New York (2012)
5. Davis, D., Yaveroglu, Ö.N., Malod-Dognin, N., Stojmirovic, A., Przulj, N.: Topology-function conservation in protein-protein interaction networks. Bioinformatics **31**(10), 1632–1639 (2015)
6. Phan, H.T.T., Sternberg, M.J.E.: Pinalog: a novel approach to align protein interaction networks–implications for complex detection and function prediction. Bioinformatics **28**(9), 1239–1245 (2012)

7. Neyshabur, B., Khadem, A., Hashemifar, S., Arab, S.S.: Netal: a new graph-based method for global alignment of protein-protein interaction networks. Bioinformatics **29**(13), 1654–1662 (2013)
8. Memišević, V., Pržulj, N.: C-graal: common-neighbors-based global graph alignment of biological networks. Integr. Biol. **4**(7), 734–743 (2012)
9. Patro, R., Kingsford, C.: Global network alignment using multiscale spectral signatures. Bioinformatics **28**(23), 3105–3114 (2012)
10. Saraph, V., Milenković, T.: Magna: maximizing accuracy in global network alignment. Bioinformatics **30**(20), 2931–2940 (2014)
11. Elmsallati, A., Clark, C., Kalita, J.: Global alignment of protein-protein interaction networks: a survey. IEEE Trans. Comput. Biol. Bioinform. **13**(4), 689–705 (2016)
12. Kuchaiev, O., Milenković, T., Memišević, V., Hayes, W., Pržulj, N.: Topological network alignment uncovers biological function and phylogeny. J. Royal Soc. Interface **7**(50), 1341–1354 (2010)
13. Malod-Dognin, N., Pržulj, N.: L-graal: Lagrangian graphlet-based network aligner. Bioinformatics **31**(13), 2182–2189 (2015). Page btv130
14. Kuchaiev, O., Pržulj, N.: Integrative network alignment reveals large regions of global network similarity in yeast and human. Bioinformatics **27**(10), 1390–1396 (2011)
15. Clark, C., Kalita, J.: A comparison of algorithms for the pairwise alignment of biological networks. Bioinformatics **30**, 2351–2359 (2014)
16. Clark, C., Kalita, J.: A multiobjective memetic algorithm for PPI network alignment. Bioinformatics **31**, 1988–1998 (2015)
17. Dupont, P., Callut, J., Dooms, G., Monette, J.-N., Deville, Y., Sainte, B.P.: Relevant subgraph extraction from random walks in a graph. Universite catholique de Louvain, UCL/INGI, Number RR, 7 (2006)
18. Park, D., Singh, R., Baym, M., Liao, C.-S., Berger, B.: Isobase: a database of functionally related proteins across PPI networks. Nucleic Acids Res. **39**(suppl 1), D295–D300 (2011)
19. Aladağ, A.E., Erten, C.: Spinal: scalable protein interaction network alignment. Bioinformatics **29**(7), 917–924 (2013)
20. Ashburner, M., Ball, C.A., Blake, J.A., Botstein, D., Butler, H., Cherry, J.M., Davis, A.P., Dolinski, K., Dwight, S.S., Eppig, J.T., et al.: Gene ontology: tool for the unification of biology. Nature Genet. **25**(1), 25–29 (2000)
21. Elmsallati, A., Msalati, A., Kalita, J.: Index-based network aligner of protein-protein interaction networks. IEEE/ACM Trans. Comput. Biol. Bioinform. **PP**, 1 (2016)

Identification and *in silico* Analysis of Glutathione Reductase Transcripts Expressed in Olive (*Olea europaea* L.) Pollen and Pistil

Estefanía García-Quirós[1], Rosario Carmona[1,2], Adoración Zafra[1], M. Gonzalo Claros[2], and Juan de Dios Alché[1(✉)]

[1] Estación Experimental del Zaidín (CSIC), Granada, Spain
{estefania.garcia,dori.zafra,
juandedios.alche}@eez.csic.es
[2] Departamento de Biología Molecular y Bioquímica,
Universidad de Málaga, Málaga, Spain
{rosariocarmona,claros}@uma.es

Abstract. Glutathione (GSH) protects proteins against oxidation of their thiol-containing groups, by alternatively becoming the subject of oxidation, forming glutathione disulfide (GSSG). Appropriate GSH:GSSG levels are maintained by glutathione reductase (GR), a homodimeric flavoprotein which uses NADPH to reduce one GSSG molecule to two of GSH. This enzyme has been characterized in several species, and described as highly conserved, with two isoforms only. Heterogeneity and disctinctiveness of plant reproductive tissues led us to investigate the presence of GR sequences. A *de novo* assembled and annotated olive reproductive transcriptome was subjected to screening, which allowed us to identify at least 11 GR homologues (1 pollen-specific and 10 from pistils). Primers were designed, and full-length sequences were obtained through PCR. *In silico* analysis, including phylogeny, 3-D modeling of N-terminus, and prediction of cellular localization and post-translational modifications was carried out to shed light into the involvement of olive pollen-intrinsic GR in reproductive development.

Keywords: Glutathione reductase · Pollen · Pistil · ROS · Glutathione

1 Introduction

Olive tree is one of the most important crops in the Mediterranean area, representing one 95% of globally cultivated olive trees, and the source of olive oil, accounting the sixth most abundant vegetable oil produced worldwide. It exhibits a high tolerance towards salinity and drought as compared to other salinity susceptible fruit trees as recently reviewed [1]. Resistance to different abiotic stresses is modulated by reactive oxygen species (ROS) metabolism in numerous plant species. Concomitantly, other metabolically related chemicals like total glutathione, reduced glutathione (GSH) and oxidized glutathione (GSSG) contents, GSH/GSSG ratio and glutathione reductase

© Springer International Publishing AG 2017
I. Rojas and F. Ortuño (Eds.): IWBBIO 2017, Part II, LNBI 10209, pp. 185–195, 2017.
DOI: 10.1007/978-3-319-56154-7_18

(GR; EC 1.6.4.2) activity have been described as key actors in the response to salt-stress in plants [2]. Compounds with high ability to oxidize GSH include ROS such as superoxide or hydroxyl radicals. Glutathione reductase (GR) is a flavoprotein enzyme that functions as dimeric disulfide oxidoreductase and utilizes an FAD prosthetic group and NADPH to reduce one molar equivalent of GSSG to two molar equivalents of GSH [3, 4]. Classical fractionation studies described higher GR activities in chloroplasts [5] but it has also been detected in mitochondria, cytosol and peroxisomes [4, 6–9]. Of the four enzymes of glutathione metabolism, GR is by far the best genetically characterized. GR is conserved between all kingdoms. In plant genomes, two GR genes are encoded. Of the two genes encoding GR in Arabidopsis, the second, GR2 (At3g54660), encodes a GR homologous to the GR pea protein with dual localization in plastids and mitochondria [10]. Such dual localization was also demonstrated in Arabidopsis [11]. GR cytosolic activity is encoded by GR 1 (At3g24170), but proteomic analysis also suggested that this gene encoded GR peroxisomal activity [12]. From the results obtained in peas it appears that the peroxisomal enzyme represents a relatively low part of the gene product GR1 [9], responsible for 30 to 60% of total enzyme activity in leaves of Arabidopsis [13]. Available information suggests that GR1 is therefore primarily addressed to the cytosol, although an intriguing question is the physiological significance of its location in peroxisomes, which can be important sources of ROS and related signals [14–17]. Glutathione reductase is a homodimer consisting of 52Kd monomers, each containing 3 domains. GR exhibits single sheet, double layered topology where an anti-parallel beta-sheet is largely exposed to the solvent on one face while being covered by random coils on the other face [18]. This includes a NADPH-binding Domain, FAD-binding domain(s) and a dimerization domain. Each monomer contains 478 residues and one FAD molecule. As mentioned, two GRs are encoded in *Arabidopsis thaliana* [19], both with a isoform in pollen and pistil nonspecific. GR activity is essential for proper pollen development and pollen tube growth, leading to fertilization [20]. Although it is expected that the pollen GR is mainly of type I, and therefore located in the cytosol, a chloroplastidial localization has also been described.

Recently a complete transcriptome of reproductive tissues olive (pollen, pistil and seed) was generated from cDNA using a/Roche 454 Titanium + platform [21]. We have riddled such transcriptome to identify the presence of GR transcripts in the olive reproductive tissues, using the retrieved sequences for further sequence confirmation and bioinformatics analysis.

2 Materials and Methods

2.1 Screening and Identification of GR Transcripts in the Olive Reproductive Transcriptome

Different strategies for selecting these transcripts were defined. The searches were conducted by definition using GO, EC, KEGG and InterPro terms and codes, orthologues and gene names in the annotated transcriptome. BLAST searches were carried on using heterologous sequences available in public databases and library resources well established as TAIR, GenBank and the recent Plant DNA C-values Database and Olive genome and annotation files [22, 23].

2.2   Olive Pollen GR Cloning and Sequencing

Olive pollen transcriptome retrieved sequences corresponding to GR-homologues (partial and complete sequences) were used to design primers in order to amplify the known sequence from pollen cDNA. Plant material was obtained according to [24]. Total RNA from mature pollen was extracted using the RNeasy Plant Total RNA kit (Quiagen, U.S.A.) and first-strand cDNA was synthesized with oligo(dT)19 primer and M-MLV reverse transcriptase (Fermentas). Standard PCR was carried out using Taq polymerase (Promega) and Pfu (Promega) to obtain or confirm nucleotide sequences. Full sequences were obtained by means of both 3'- and 5'-RACE (smarter RACE, Clontech), following manufacturer's specifications. pGEMT-easy (Promega) was used for cloning purposes. Sanger sequencing was achieved in the facilities of the EEZ-CSIC institute in Granada.

2.3   *In silico* Analysis of the Sequences

Nucleotide sequences were aligned using CLUSTAL OMEGA multiple alignment tool with default parameters [25]. Phylogenetic trees were constructed with the aid of the software Seaview [26] using the maximum likelihood (PhyML) method and implementing the most probable nucleotide substitution model (GTR) previously calculated by JmodelTest2 [27]. The branch support was estimated by bootstrap resampling with 100 replications.

For the prediction of protein cell localization, the software Plant-mPloc [28] was used. PhosPhAt [29] and PlantPhos [30] were used to predict serine, threonine and tyrosine phosphorylation. GPS-SNO 1.0 was used for prediction of S-nitrosylation sites [31]. TermiNator was used to predict N-terminal methionine excision, acetylation, myristoylation or palmitoylation [32].

Structure prediction of GR2 N-terminal region was modeled by using the fold recognition-based Phyre2 server [33], with c4dnaA as the template and a 100% confidence. 3D modeling was carried out by PyMol (https://www.pymol.org/).

3   Results

3.1   Recovery, Cloning and Phylogenetic Analysis of GR Sequences from the Olive Reproductive Transcriptome

Eleven entries were identified within the raw data transcriptome, either as partial sequences or as complete sequences corresponding to GR (Table 1) by BLASTing each individual sequence against the GenBank database. On the other hand, these eleven previously selected inputs were individually confronted by BLASTing against the genomic *Olea europaea* assembly and annotation database. Table 1 shows the results obtained in this second BLASTing. Most query sequences have high homology (more than 94%) with the input OE6A000198T2 from genomic *Olea europaea* assembly and annotation database. Noteworthy, GR_Pollen_1 (the only pollen specific sequence obtained from the reproductive transcriptome), and GR_Pistil_8 have full identity with OE6A112067T2, whereas GR_Pistil_1 has also full identity with Oe_s02429. All these olive genome sequences were identified as chloroplastidial GRs (GR2) by blasting [23].

Table 1. Upper panel. Output sequences identified as GR after screening. The lengths of the assembled sequences are indicated. **Lower panel.** Blasting with genome from olive.

Sequence	Sequence Name	Length (pb)
GR_Pistil_1	st11_olive_039691	687
GR_Pistil_2	st11_olive_007100	193
GR_Pistil_3	st11_olive_003323	128
GR_Pistil_4	st11_olive_051636	513
GR_Pistil_5	st11_olive_003115	2020
GR_Pistil_6	st11_olive_003323	1974
GR_Pistil_7	st11_olive_004231	1838
GR_Pistil_8	st11_olive_018143	1049
GR_Pistil_9	st11_olive_004231	1838
GR_Pistil_10	st11_olive_003323	1979
GR_Pollen_1	po11_olive_018706	435

Sequence	Genomic homologous	Homology	Identity
GR_Pistil_1	Oe6_s02429	100%	Glutathione reductase
GR_Pistil_2	Oe6_s07823	95,53%	Uncharacterized
GR_Pistil_3	OE6A000198T2	100%	Glutathione reductase
GR_Pistil_4	OE6A000198T2	94,64%	Glutathione reductase
GR_Pistil_5	OE6A091299T1	98,24%	Glutathione reductase
GR_Pistil_6	OE6A000198T2	94,88%	Glutathione reductase
GR_Pistil_7	OE6A117561T3	99,67%	Glutathione reductase
GR_Pistil_8	OE6A112067T2	99,43%	Glutathione reductase
GR_Pistil_9	OE6A117561T3	99,67%	Glutathione reductase
GR_Pistil_10	OE6A000198T2	94,83%	Glutathione reductase
GR_Pollen_1	OE6A112067T2	100%	Glutathione reductase

Figure 1 shows a phylogenetic analysis including the 11 outputs identified in the annotated transcriptome together with a representation of GRs identified in different taxonomical groups. No clear differentiation between GR1 and GR2 was observed, and the high similarity between both isoforms was particularly evident in one of the homologous species used for phylogenetic study *(T. cacao)*. Most (8 of 11) of the rescued GR sequences from the olive reproductive transcriptome were identified as choroplastidial isoforms of the enzyme, and two of these forms (GR_pistil_9 and 10) clustered together, slightly apart from the remaining sequences. This discovery corroborates the data obtained from recent studies showing that in fact the isoform involved in pollen development is GR2 [13].

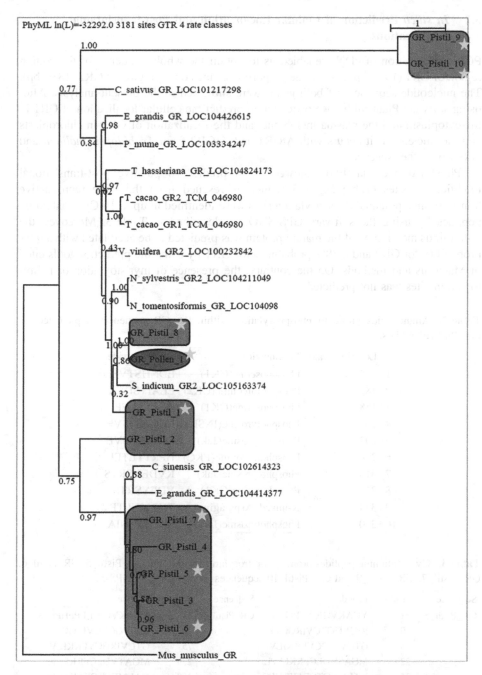

Fig. 1. Phylogenetic relationships between olive pollen GRs and their homologues in vegetative and reproductive tissues. The phylogenetic tree was rooted with a sequence of mouse GR. Yellow stars indicate sequences corresponding to the GR2 (cloroplastidial isoform).

3.2 *In silico* Prediction of Cellular Localization and Post-translational Modifications

PCR amplification of cDNA enabled us to obtain the whole sequence of the coding region of GR1 (1978 bp) and a large proportion of the coding region of GR2 (1839 bp). The nucleotide sequences of both genes were subjected to the present analysis. After using software Plant-mPloc we were able to predict the cellular localization of GR1 in the cytoplasm and the plasma membrane, and the localization of GR2 in chloroplasts and the nucleus, as it occurs with AtGR1 and AtGR2 from *Arabidopsis thaliana* and GRs from other species.

PhosPhAt and PlantPhos allowed us to predict 10 putative post-translational modification sites (Table 2). In 5 sequences rescued from the olive reproductive transcriptome, potential S-nitrosylation sites were identified in up to 10 Cys-containing peptides by using the software GPS-SNO, as described in Table 3. Moreover, the N-terminus methionine of the mature protein was predicted to be acetylated with a 97% probability for GR2 and a 78% probability for GR1 by using the prediction tools cited in Materials and methods. On the contrary, the presence of myristoylation or palmitoylation sites was not predicted.

Table 2. Amino acids prone to phosphorylation within the GR sequence as predicted by PhosPhAt/PlantPhos.

	Locations (aa)	Modification	Substrate side
1	5	Phosphoserine(CK1)	EDNHSTVED
2	182	Phosphothreonine(CK2)	ELAITSDEA
3	188	Phosphoserine(CK1)	DEALSLEEF
4	203	Phosphotyrosine(INSR)	LGGGYISVE
5	203	Phosphotyrosine(Jak)	LGGGYISVE
6	256	Phosphothreonine(PKC)	HPRTTLTEL
7	315	Phosphotyrosine(Jak)	KVDEYSRTS
8	316	Phosphoserine(CK1)	VDEYSRTSV
9	334	N-linked_Asparagine	TNRMNLTPV
10	360	Phosphotyrosine(INSR)	SKPDYSNIA

Table 3. Cys-containing peptides prone to S-nitrosylation within the GR_Pistil_5, GR_Pistil_6, GR_Pistil_7, GR_Pistil_9 and GR_Pistil_10 sequences as predicted by GPS-SNO.

Sequence	Position	Peptide	Sequence	Position	Peptide
GR_Pistil_5	51	YGAKVGICELPFHPI	GR_Pistil_6	51	YGAKVGICELPFHPI
	70	IGGVGGTCVIRGCVP		70	IGGVGGTCVIRGCVP
	75	GTCVIRGCVPKKILV		75	GTCVIRGCVPKKILV
	229	MGASVDLCFRKELPL		229	MGATVDLCFRKELPL
	352	VALMEGTCFAKTVFG		352	VALMEGTCFAKTVFG
	372	PDYSHVPCAVFCIPP		372	PDYSHVPCAVFCIPP
	376	HVPCAVFCIPPLSVV		376	HVPCAVFCIPPLSVV
	438	KVLGASMCGPDAAEI		438	KVLGASMCGPDAAEI
	455	GIAVALKCGATKAQF		455	GIAVALKCGATKAQF

(*continued*)

Table 3. (*continued*)

Sequence	Position	Peptide	Sequence	Position	Peptide
GR_Pistil_7	51	YGAKVGICELPFHPI	GR_Pistil_9	51	YGAKVGICELPFHPI
	70	IGGVGGTCVIRGCVP		70	IGGVGGTCVIRGCVP
	75	GTCVIRGCVPKKILV		75	GTCVIRGCVPKKILV
	277	TENGIKVCTDHGEEL		267	TENGIKVCTDHGEEL
	352	VALMEGSCFAKTVFG		342	VALMEGSCFAKTVFG
	372	PDHTNVPCAVFCIPP		362	PDHTNVPCAVFCIPP
	376	NVPCAVFCIPPLSVV		266	NVPCAVFCIPPLSVV
	438	KVLGASMCGPDAPEI		428	KVLGASMCGPDAPEI
	455	GIAIALKCGATKAQF		445	GIAIALKCGATKAQF
GR_Pistil_10	18	RLAMTLICLLIGAGS			
	44	YGAKVGICELPFHPI			
	63	IGGVGGTCVIRGCVP			
	68	GTCVIRGCVPKKILV			
	222	MGATVDLCFRKELPL			
	345	VALMEGTCFAKTVFG			
	365	PDYSHVPCAVFCIPP			

3.3   2D Structure Prediction and 3D Modeling of the Structure and in Silico Analysis of the Function

After several tests, it was elucidated that both GR1 and GR2 isoforms share the same predicted secondary structure, as expected. The GR_Pistil_10 sequence corresponding to a GR of the cloroplastidial type was chosen to predict the secondary structure (Fig. 2) and 3D modeling shown in Fig. 3. The predicted structure confirmed that the corresponding translated amino acid sequences of GR_Pistil_10 were consistent with available 3D models developed by using folding-recognition software.

4   Discussion

Reproductive tissue transcriptomes differ from their somatic counterpart due to the large number of specific transcripts that are produced during the physiological event that involves reproduction in plants [34]. This can be explained by the requirement of signaling that must take place during pollen tube germination, pollen-pistil interaction... etc., which triggers an increase in transcription during these processes. Thereby, the use of transcriptomes can be an useful way to elucidate enzymatic of signaling networks involved in the development of the micro- and macrosporocytes in Angiosperms, pollen tube germination and pollen-pistil interactions at the time of fertilization [35]. This paper has conducted a bioinformatic screening of exome sequences of olive tree mature pollen and pistil. It has been able to detect the existence of at least two GR genes homologous to other GR genes described above in plant species such as *A. thaliana* or *T. cacao*. The use of PCR and RACE experiments has confirmed the expression of these genes in olive

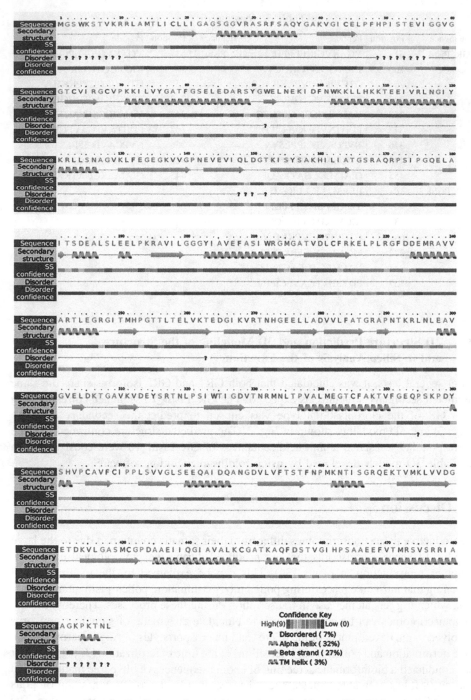

Fig. 2. Prediction of secondary structure in olive pollen GR2.

Fig. 3. (A) Prediction of 3D structure in olive pollen GR2 (by Phyre2). (B) 3D modeling by PyMol of olive pollen GR2.

tree pollen, and led to obtain the complete or a majority of the sequence of GR1 and GR2 genes. Cleaning and filtering of all outputs of the transcriptome to identify those eleven GR sequences obtained proved highly complicated due to the high homology between these two genes. The phylogenetic analysis carried out did not cluster GR1 and GR2 within specific group, thus we cannot suggest distinctive features for these GR homologous in the reproductive tissues. However, by predicting the cellular location of these genes we identified that most of the GR2s are expressed in the chloroplast, thus confirming previous studies which suggested that this is the isoform really involved in adequate pollen development [13].

GSH production by GR is essential for proper pollen tube tip growth [13, 36, 37] controlling both the speed and orientation during pollen tube elongation. Due to the high metabolism rate in olive pollen, grand quantities of ROS are generated and this can be detected by using confocal microscopy [38]. Still many aspects have to be elucidated, such as the importance of GSH in pollen-stigma signaling, and the action of this and GR activity during pollen hydration among others.

Conflict of Interest. The authors confirm that this article content has no conflicts of interest.

Acknowledgments. This work was supported by ERDF-cofunded projects BFU2016-77243-P, P2011-CVI7487, 201540E065, RTC-2015-4181-2 and RTC2016-4824-2. EGQ thanks the MINECO for FPI grant funding.

References

1. Shah, Z.H., Hamooh, B.T., Daur, I., Ha Rehman, H.M., Alghabari, F.: Transcriptomics and biochemical profiling: current dynamics in elucidating the potential attributes of olive. Curr. Issues Mol. Biol. **21**, 73–98 (2017)
2. Kaur, H., Bhatla, S.C.: Melatonin and nitric oxide modulate glutathione content and glutathione reductase activity in sunflower seedling cotyledons accompanying salt stress. Nitric Oxide **59**, 42–53 (2016)

3. Halliwell, B., Foyer, C.H.: Properties and physiological function of a glutathione reductase purified from spinach leaves by affinity chromatography. Planta **139**(1), 9–17 (1978)

4. Edwards, E.A., Rawsthorne, S., Mullineaux, P.M.: Subcellular distribution of multiple forms of glutathione reductase in leaves of pea (Pisum sativum L.). Planta **180**, 278–284 (1990)

5. Foyer, C.H., Halliwell, B.: The presence of glutathione and glutathione reductase in chloroplasts: a proposed role in ascorbic acid metabolism. Planta **133**(1), 21–25 (1976)

6. Rasmusson, A.G., Møller, I.M.: NADP-utilizing enzymes in the matrix of plant mitochondria. Plant Physiol. **94**, 1012–1018 (1990)

7. Jiménez, A., Hernández, J.A., del Río, L.A., Sevilla, F.: Evidence for the presence of the ascorbate-glutathione cycle in mitochondria and peroxisomes of pea leaves. Plant Physiol. **114**, 175–284 (1997)

8. Stevens, R.G., Creissen, G.P., Mullineaux, P.M.: Characterization of pea cytosolic glutathione reductase expressed in transgenic tobacco. Planta **211**, 537–545 (2000)

9. Romero-Puertas, M.C., Corpas, F.J., Sandalio, L.M., Leterrier, M., Rodríguez-Serrano, M., del Río, L.A., Palma, J.M.: Glutathione reductase from pea leaves: response to abiotic stress and characterization of the peroxisomal isozyme. New Phytol. **170**, 43–52 (2006)

10. Creissen, G., Edwards, E.A., Enard, C., Wellburn, A., Mullineaux, P.: Molecular characterization of glutathione reductase cDNAs from pea (*Pisum sativum* L.). Plant J. **2**(1), 129–131 (1992)

11. Chew, O., Whelan, J., Miller, A.H.: Molecular definition of the ascorbate-glutathione cycle in *Arabidopsis* mitochondria reveals dual targeting of antioxidant defences in plants. J. Biol. Chem. **278**(47), 46869–46877 (2003)

12. Kaur, N., Reumann, S., Hu, J.: Peroxisome biogenesis and function. The Arabidopsis Book. American Society of Plant Biologists, Rockville, MD (2009). doi:10.1199/tab.0123

13. Marty, L., Siala, W., Schwarzländer, M., Fricker, M.D., Wirtz, M., Sweetlove, L.J., Meyer, Y., Meyer, A.J., Reichheld, J.P., Hell, R.: The NADPH-dependent thioredoxin system constitutes a functional backup for cytosolic glutathione reductase in Arabidopsis. Proc. Natl. Acad. Sci. U.S.A. **106**, 9109–9114 (2009)

14. Foyer, C.H., Noctor, G.: Redox sensing and signalling associated with reactive oxygen in chloroplasts, peroxisomes and mitochondria. Physiol. Plant. **119**, 355–364 (2003)

15. del Río, L.A., Sandalio, L.M., Corpas, F.J., Palma, J.M., Barroso, J.B.: Reactive oxygen species and reactive nitrogen species in peroxisomes Production scavenging and role in cell signaling. Plant Physiol. **141**, 330–335 (2006)

16. Nyathi, Y., Baker, A.: Plant peroxisomes as a source of signalling molecules. Biochim. Biophys. Acta **1763**, 1478–1495 (2006)

17. Mhamdi, A., Queval, G., Chaouch, S., Vanderauwera, S., Van Breusegem, F., Noctor, G.: Catalase function in plants: a focus on Arabidopsis mutants as stress-mimic models. J. Exp. Bot. **61**, 4197–4220 (2010)

18. Garrett, R.H., Grisham, C.M.: Biochemistry, 3rd edn. Thomson Brooks/Cole, Belmont (2005). ISBN 0534490336

19. Anjum, N.A., Umar, S., Chan, M.T. (eds.): Ascorbate-Glutathione Pathway and Stress Tolerance in Plants. Springer, Netherlands (2010). ISBN 978-90-481-9404-9

20. Cejudo, F.J., Meyer, A.J., Reichheld, J.P., Rouhier, N., Traverso, J.A.: Thiol-Based Redox Homeostasis and Signalling. Frontiers Media SA, Lausanne (2014). ISBN 978-2-88919-284-7

21. Carmona, R., Zafra, A., Seoane, P., Castro, A.J., Guerrero-Fernández, D., Castillo-Castillo, T., Medina-García, A., Cánovas, F.M., Aldana-Montes, J.F., Navas-Delgado, I., Alché, J.D., Claros, M.G.: ReprOlive: a database with linked data for the olive tree (*Olea europaea* L.) reproductive transcriptome. Frontiers Plant Sci. **6**, 625 (2015)

22. Cruz, F., Julca, I., Gómez-Garrido, J., Loska, D., Marcet-Houben, M., Cano, E., Galán, B., Frias, L., Ribeca, P., Derdak, S., Gut, M., Sánchez-Fernández, M., García, J.L., Gut, I.G., Vargas, P., Alioto, T.S., Gabaldón, T.: Genome sequence of the olive tree, *Olea europaea*. GigaScience **5**, 29 (2016)

23. Olive genome and annotation files. http://denovo.cnag.cat/genomes/olive/

24. de Dios Alché, J., M'rani-Alaoui, M., Castro, A.J., Rodríguez-García, M.I.: Ole e 1, the major allergen from olive (*Olea europaea* L.) pollen, increases its expression and is released to the culture medium during *in vitro* germination. Plant Cell Physiol. **45**, 1149–1157 (2004)

25. McWilliam, H., Li, W., Uludag, M., Squizzato, S., Park, Y.M., Buso, N., Cowley, A.P., Lopez, R.: Analysis tool web services from the EMBL-EBI. Nucleic Acids Res. **41**, W597–W600 (2013)

26. Gouy, M., Guindon, S., Gascuel, O.: SeaView version 4: a multiplatform graphical user interface for sequence alignment and phylogenetic tree building. Mol. Biol. Evol. **27**, 221–224 (2010)

27. Darriba, D., Taboada, G.L., Doallo, R., Posada, D.: jModelTest 2: more models, new heuristics and parallel computing. Nat. Meth. **9**, 772 (2012)

28. Chou, K.C., Shen, H.B.: Plant-mPLoc: a top-down strategy to augment the power for predicting plant protein subcellular localization. PLoS ONE **5**, e11335 (2010)

29. Heazlewood, J.L., Durek, P., Hummel, J., Selbig, J., Weckwerth, W., Walther, D., Schulze, W.X.: PhosPhAt: a database of phosphorylation sites in Arabidopsis thaliana and a plant-specific phosphorylation site predictor. Nucleic Acids Res. **36**, D1015–D1021 (2008)

30. Lee, T.-Y., Bretana, N., Lu, C.-T.: PlantPhos: using maximal dependence decomposition to identify plant phosphorylation sites with substrate site specificity. BMC Bioinf. **12**, 261 (2011)

31. Xue, Y., Liu, Z., Gao, X., Jin, C., Wen, L., Yao, X., Ren, J.: GPS-SNO: computational prediction of protein S-nitrosylation sites with a modified GPS algorithm. PLoS ONE **5**, e11290 (2010)

32. Martinez, A., Traverso, J.A., Valot, B., Ferro, M., Espagne, C., Ephritikhine, G., Zivy, M., Giglione, C., Meinnel, T.: Extent of N-terminal modifications in cytosolic proteins from eukaryotes. Proteomics **8**, 2809–2831 (2008)

33. Kelley, L.A., Sternberg, M.J.: Protein structure prediction on the web: a case study using the Phyre server. Nat. Protoc. **4**, 363–371 (2009)

34. Rutley, N., Twell, D.: A decade of pollen transcriptomics. Plant Reprod. **28**, 73–89 (2015)

35. Dukowic-Schulze, S., Chen, C.: The meiotic transcriptome architecture of plants. Frontiers Plant Sci. **5**, 220 (2014)

36. Zechmann, B., Mauch, F., Sticher, L., Müller, M.: Subcellular immunocytochemical analysis detects the highest concentrations of glutathione in mitochondria and not in plastids. J. Exp. Bot. **59**, 4017–4027 (2008)

37. Zechmann, B., Koffler, B.E., Russell, S.D.: Glutathione synthesis is essential for pollen germination in vitro. BMC Plant Biol. **11**, 54 (2011)

38. Zafra, A., Rodriguez-Garcia, M.I., Alche, JdD: Cellular localization of ROS and NO in olive reproductive tissues during flower development. BMC Plant Biol. **10**, 36 (2010)

Computational Systems for Modelling Biological Processes

Cyber Immunity

A Bio-Inspired Cyber Defense System

Peter Wlodarczak[✉]

University of Southern Queensland, West Street, Toowoomba, QLD 4350, Australia
wlodarczak@gmail.com

Abstract. Bio-inspired computing is an active field of research since nature has found solutions for many real-world problems where research so far struggled to develop effective implementations. A new area of research is cyber immunity. Cyber immune systems try to mimic the adaptive immune system of humans and animals because of its capability to detect and fend off new, unseen pathogens. Today's cyber security systems provide an effective defense mechanism against cyber-attacks. However, traditional firewall and intrusion detection systems often struggle to detect and repel so far unknown attacks. A cyber immune system can mitigate this shortcoming by detecting new, unknown cyber-attacks and by providing a powerful defense mechanism. This paper describes the recent advances in cyber immune systems and their underlying, bio-inspired technologies.

Keywords: Bio-inspired computing · Cyber immunity · Adaptive Immune System · Machine Learning

1 Introduction

Bio-inspired computing is an area of research in computer science that aims to build systems modeled after biological phenomenon. Bio-inspired computing uses computers to model nature and simultaneously the study of nature to improve the usage of computers [16]. Bio-inspired computing includes genetic algorithms based on evolutionary processes, Artificial Intelligence (AI) and Artificial Neural Networks (ANN), sensor networks (sensory organs) or artificial immune systems. A recent area of research is cyber immunity based on the human adaptive immune system. Cyber immune systems often use Machine Learning (ML) methods, an area of AI. ML is bio-inspired since it is imitating human learning capabilities on computers. This paper describes an approach for cyber immunity based on AI using ML techniques.

1.1 The Human Immune System

The human body has a remarkably effective defense mechanisms, the immune system (IS), that detects a wide range of harmful agents, called pathogens, such as viruses, parasites, and microbes. The IS has the capability to distinguish pathogens from healthy tissue. The skin fends off external threats to our body similarly to a firewall. It is

© Springer International Publishing AG 2017
I. Rojas and F. Ortuño (Eds.): IWBBIO 2017, Part II, LNBI 10209, pp. 199–208, 2017.
DOI: 10.1007/978-3-319-56154-7_19

constantly renewed and adaptive. Our bodies are police states; the IS constantly monitors the internal environment. In the absence of a working IS, even minor infections can take hold and prove fatal [4].

The IS consists of an innate immune system and a much younger adaptive immune system (AIS). Innate immunity is present in both, vertebrates and invertebrates, whereas the adaptive immune system is found only in invertebrates [4]. If a pathogen breaches physical barriers of the body, the innate immune system provides an immediate, but non-specific response. It is not capable of conferring long-lasting immunity. If a pathogen evades the innate immune system, the AIS is activated. The AIS can generate a pathogen specific, tailored response. It is antigen-specific, and the response is remembered in case the same pathogen enters the body a second time. It can then provide a quick, antigen specific response. The immune system is capable of learning, memory, and pattern recognition [5]. Similarly, the AIS acts as the immune system memory.

The AIS consists of lymphocytes. The major types are B-cells and T-cells. B cells identify pathogens when antibodies on their surface bind to a specific foreign antigen. In the case of an antigen entering the body, the immune system activates signal molecules that attract specialized immune cells, called killer T cells, to the site of infection. The killer T cells destroy cells that are infected by viruses and other pathogens or other dysfunctional cells. An AIS detects and neutralizes malware such as Trojans or ransomware by quarantining or erasing it. This function is analogous to the functions in biological systems of natural killer cells, which kill cells infected with known or unknown viruses, and of macrophages, which phagocytize bacteria [17]. When B cells and T cells are activated, they begin to replicate, and some of their offspring become long-lived memory cells. A cyber IS imitates this behavior. It has adaptive and memory functionality and recognizes cyber threats using pattern recognition.

1.2   Problem Description

In today's world, where more and more devices are connected to the Internet, cyber security has gained unprecedented importance since attacks can be executed instantly, over long distances without the need to transport weapons over distance to deploy them. New forms of cyber-attacks such as Advanced Persistent Threats (APT) and zero-day exploits pose a serious risk for devices such as smart sensors, machines or whole construction sites that communicate through the Internet. The Internet of Things (IoT), or Industry 4.0, where cyber-physical systems communicate and cooperate with each other, expose devices that often use legacy code or firmware that was not originally developed with cyber threats in mind. These facilities were often built without the ability to communicate, and no security measures were implemented. Connectivity was often added after they were constructed exposing them to cyber-threats. The need to open the previously closed production sites offers a whole new attack surface for cyber criminals and the need for effective tools to fence off smart factories, or power grids has never been higher. The amount of data produced on the Internet is ever increasing and analyzing it for malware has become a serious challenge. The Internet of Things (IoT) is about millions of connected, communicating and exchanging objects, scattered all over the world and generating tremendous amounts of data using their sensors every single second [15]. They combine physical devices with the Internet and form cyber-physical systems. Data-injection attacks can degrade the operational reliability and security of any cyber-physical infrastructures [20].

APTs are a set of sophisticated stealth hacking processes used for cyber-espionage or cyber-sabotage. Zero-day exploits are undisclosed vulnerabilities which leave the author with "zero days" to create a patch or find a workaround. They are often traded on the dark web and at the time of attack no security patch exists yet. They are uniquely featured by the stealthy, continuous, sophisticated and well-funded attack process for long-term malicious gain, which render the current defense mechanisms inapplicable [18].

An intrusion occurs when an attacker attempts to gain entry into or disrupt the normal operations of an information system, almost always with the intent to do harm [23]. Today's cyber-attacks fall into three categories:

- attacks by cyber criminals who want to enrich themselves by stealing credit card information, hacking into bank accounts etc.
- hacktivists such as the Anonymous group who have a political agenda
- commercial groups that are engaged in cyber-warfare such as cyber-espionage (The Dukes) [1] or sabotage (Stuxnet) [2]

They are usually highly qualified, highly motivated and are often government or agency-sponsored or supported. For instance, Stuxnet, that destroyed the uranium enrichment facilities in Natanz, Iran, was purportedly developed by the NSA and the Mossad. However, neither country confirmed its involvement. Natanz was not even connected to any external network. Not only was Stuxnet much more complex than any other piece of malware seen before, it also followed a completely new approach that's no longer aligned with conventional confidentiality, integrity, and availability thinking [2].

Conventional cyber security systems such as classic firewalls, intrusion detection systems or end-point protection do not suffice anymore since they cannot detect zero-day attacks or ATP since these exploit new, unknown vulnerabilities. To meet today's cyber security needs, a cyber defense system needs the capability to recognize and repel unknown attacks and adapt to new threats without disrupting the business. Because we have no way of knowing all vulnerabilities, the only resources we can use to learn attack patterns are the sequences of events correlated with the cyber-attacks [3]. Cyber immunity is a bio-inspired approach based on the human AIS that is capable of learning and detecting attacks with unknown signatures (Fig. 1). It mimics the AIS using machine learning techniques by learning normal network behavior and detecting anomalies based on what it has learned. It should be noted that "immune system" is used as a metaphor. There is no analogy between biological and artificial immune systems in the "mechanisms" of innate and adaptive immunity [17].

2 Framework of a Cyber Immune System

Cyber immune systems detect attacks not based on their signature but based on anomalies detected in normal network traffic. They have learning and memory capabilities to detect and remember so far unknown attacks. They belong to the family of cyber security systems. Cyber security is the set of technologies and processes designed to protect computers, networks, programs, and data from attack, unauthorized access, change, or destruction [11]. Cyber immune systems usually adopt ML techniques. ML techniques have the advantage that rules do not need to be programmed, hardcoded, by the programmer, but they can learn the rules themselves. They can also learn new rules during runtime, in case new, so far unknown anomalies occur.

2.1 Machine Learning

ML is a branch of cognitive computing, which is a subarea of Artificial Intelligence (AI). Cognitive science and linguistics emerged as fields at roughly the same time as artificial intelligence, all deeply influenced by the advent of the computer [19]. Cognitive computing has adaptive and learning capabilities, and contextual understanding of elements such as text meaning or domain knowledge. These capabilities make cognitive computing ideal for cyber immunity solutions. They belong to the family of cognitive security solutions. ML is essentially a series of advanced statistics applied to historic data to determine the likely future. In cyber ISs, they predict the probability that a certain network traffic pattern or data transfer is malicious and needs to be discarded. Whereas humans learn from experience, computers learn from data. ML schemes have to be trained using training data. During training, they learn rules inductively; the rules do not need to be programmed. ML techniques are adopted when there are too many rules to be programmed by a developer or when the rules are too complicated.

There are many different ML techniques. Popular ML schemes include naïve Bayes, Artificial Neural Network (ANN), Support Vector Machine (SVM), Logistic Regression (LR) and decision tree induction learners. Usually, several schemes are trained, and the best performing learner is then used for real-time analysis. It is important to notice that many ML schemes such as ANN, SVM and LR are probabilistic models. They return probabilities that a certain behavior is abnormal. Naïve Bayes classifier is a generative, probabilistic model whereas Logistic Regression and SVN are discriminative probabilistic models. They don't allow to generate samples from the joint distribution of the observed variable x and dependent unobserved variable y. Decision trees are non-probabilistic prediction models.

A learning cycle goes through several phases. First, the data is collected and pre-processed. During pre-processing, the data set is purified from irrelevant data. Relevance filtering is an important step since most ML schemes are prone to overfitting. Overfitting happens if the learner gets too complicated and starts to capture noise. The learner is trained by using the purified data set. The training data is also called the truth data or gold data. The basic idea is to find a function that classifies network traffic into normal and deviant, malicious behavior. This is called binary classification. For instance, a ANN consists of perceptrons, the neurons, that are interconnected. The connections represent the axons. The basic idea of a perceptron is to find a linear function f such that:

$$f(x) = w^T x + b$$

where $f(x) > 0$ for one class and $f(x) < 0$ for the other class. The weights $w = (w1, w2, ..., wm)$ and the bias b are adjusted during training until a loss function converges [10].

Perceptrons are organized into a hierarchical structure to form multilayer perceptrons, a type of ANN, to represent nonlinear decision boundaries [10]. The fact that humans can solve many classification problems with astounding ease must lie in the fact that neurons in the brain are massively interconnected, allowing a problem to be decomposed into subproblems that can be solved at the neuron level [9]. This behavior is imitated by an ANN. The data is represented at a higher abstraction level as it passes through the layers of the ANN. In an ANN the connections are weighted. During training, the weights w are adjusted until the classification accuracy is satisfactory.

To evaluate the learners, different measures such as the classification accuracy, f-score, kappa statistics etc. are used. These statistics are compared to select the best performing learner. Training a learner it is a highly iterative process and typically many iterations are needed until satisfactory results are obtained.

2.2 Cyber Immune System

Virus DNA changes, so the immune system has to adapt to recognize the signature of the virus. Similarly, in cyber security, we have to deal with an ever-evolving adversary. Since the attack is unknown, we cannot learn the signature of the attack from previous ones. Instead of learning attack signatures, a cyber IS learns what normal network traffic looks like over an extended period of time. Once trained, it calculates the probability

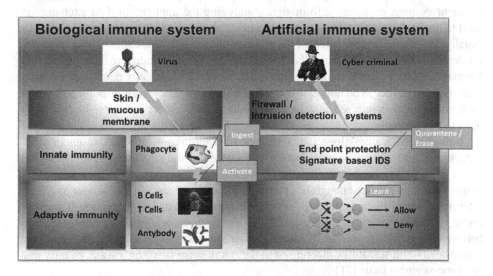

Fig. 1. Comparison of biological and artificial immune system

that a certain deviant pattern is malicious. It constantly updates its results based on new evidence. It can cut off an attacking agent by observing it and detecting what information the agent is after and where it came from.

One of the most important tasks in data mining is to select the observation points. This process is called feature extraction or feature engineering. In many data analysis tasks, it is useful to select and use only the relevant feature [13]. Feature extraction reduces the data volumes that need to be analyzed to data relevant to the problem at hand. It improves both, the training and the classification performance.

Feature Extraction.
Network traffic consists of several layers. The Internet uses the TCP/IP protocol. TCP/IP is a four-layer network protocol. It consists of a link layer, an Internet layer, a Transport and an Application layer. A cyber immune system typically analyzes the Internet and the Application layer. At the Internet layer, IP packets are transmitted. An IP packet is composed of the IP (i.e., transport layer) header and the IP payload. The IP payload might contain data or other, encapsulated higher level protocols such as Network File System (NFS), Server Message Block (SMB), Hypertext Transfer Protocol (HTTP), BitTorrent, Post Office Protocol (POP) Version 3, Network Basic Input/Output System (NetBIOS), telnet, and Trivial File Transfer Protocol (TFTP) [11]. The payload may contain malicious code such as viruses, Trojans or ransomware that can infect a target system and make it unusable. A cyber IS needs to look at the connection data, from where to where packages are sent, and at the payload to detect malicious content. If unusual patterns are detected, it triggers an alarm or interrupts the network connection. Unusual patterns at the Internet layer are for instance connections from countries that usually do not connect to specific servers, or changes in connection frequencies or data transfer volumes that indicate that for instance a cyber espionage attack is under way and data is illegally transmitted to an unknown external system.

At the Application layer, connected end-point programs such as transaction or payment systems exchange information. Analyzing the application data exchange is used to detect for instance fraud, money laundering or online payment scams. Cyber IS usually do not analyze the content of the transactions since there are specialized solutions for detecting cyber fraud. They rather look at the connection patterns to detect abnormal behavior. Contrary to the human adaptive IS, where T killer cells neutralize infected cells, typically cyber IS do not destroy contaminated data. They use end-point protection tools to quarantine infected attachments, or they may trigger an alarm in case they detected an anomaly.

Usually, one feature alone is not enough to determine if an attack is under way. Attacks on web-scale platforms usually have a behavioral signature, made up of the series of steps involved in committing the fraud [21]. For instance, a payment to an untrusted country is not enough to determine cyber fraud since the payment might be legitimate. Many features need to be observed and evaluated to determine if a cyber crime is under way with a high probability. The features are grouped, clustered, into behavioral descriptors such as IP addresses that are associated with users and sessions. These behavioral signatures that represent the features are extracted on a per-entity and per-time-segment basis [21].

Anomaly Detection.
Atypical network on the Internet has a lot of data traffic. To be effectively analyzed, traffic must be grouped, for instance into email, file transfer or streaming traffic. Each group has a different rule set since email attachments and audio and video streaming have a different attack potential. Profiling modules perform clustering algorithms or other data-mining and machine learning methods to group similar network connections and search for dominant behaviors [3]. These clusters are associated with the behavioral signature, the features, to detect anomalies. An anomaly can then be a suspicious attachment, unusual network traffic from an untrusted source or a traffic pattern from a country where a lot of cyber criminality originates.

Clustering is usually done using unsupervised ML techniques. Typical unsupervised methods include k-means clustering and hierarchical clustering. The traffic is grouped into centroids. In k-means clustering, k is the number of centroids. The centroid is the barycenter of the cluster. It partitions observations into k clusters (Fig. 2). The clusters are created using a distance measure such as the squared Euclidean distance. Distance can be the closeness of traffic in terms of time, source address or traffic frequencies. To evaluate the cluster, metrics such as the Dunn index, Davies-Bouldin or Silhouette index can be used. They measure the density of the centroids and the distance between them to assure well separateness. Clustering is a preparation step for the actual anomaly detection.

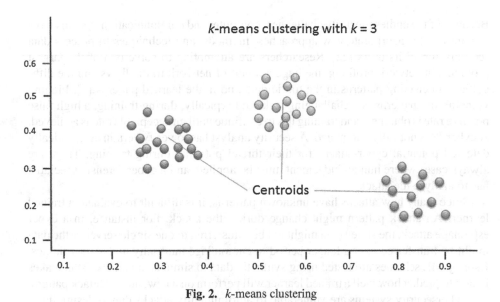

Fig. 2. k-means clustering

Anomaly detection happens in real-time since many attacks are "smash and run". An attacker breaks into a system, collects data and disconnects again. They also often use a third-party system from where they issue the attack to obfuscate the origin. That's why a cyber IS cannot only look at the connection data, i.e. the IP header, but has to scan the payload for malware. An anomaly is a deviation from "normal" network traffic that the learner has learned over a certain period of time. However, an anomaly is not necessarily a threat and often a cyber forensic specialist has to analyze the anomaly to determine if there is really a threat. The analyst feedback is then injected into the training loop of the learner.

The full training lifecycle is depicted in the following Fig. 3:

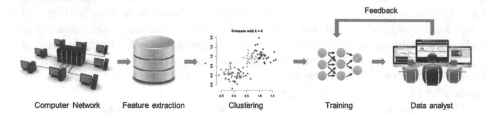

| Computer Network | Feature extraction | Clustering | Training | Data analyst |

Fig. 3. AIS training cycle

The trained and deployed AIS continues to detect new anomalies. As new anomalies are detected, the learner is updated with new thread signatures. The signatures are shared with other AIS to make them "immune" against new threads.

3   Challenges

Because of the endless security challenges, computer and communication systems have to continue to incorporate new approaches, methods, and techniques to process data securely for all its users [22]. Researchers are attempting to solve two of the largest problems in network profiling: the huge amount of network traffic flows and the difficulties in detecting patterns in the traffic data and in the learned patterns [3]. Finding vulnerability patterns is a challenging task, and typically, during training, a high false positive rate is obtained, and training has to continue until an acceptable rate is achieved. A cyber IS is not fully automated. A security analyst has to verify certain automatically detected potential cyber-attacks for their thread potential during training. There are always cases where human judgement must be applied, and a cyber forensic specialist has to analyze the attack.

Since many new attacks have unknown patterns, it is difficult to evaluate a trained learner. An attack pattern might change during the attack. For instance, in a cyber espionage attack, the stolen data might not be transferred to one single server or the data might be transferred over a longer period to camouflage unusually high data volumes. Usually ML schemes are tested using synthetic data to simulate an attack. This makes it hard to predict how well a trained learner will perform on a new, unseen attack pattern.

Cybersecurity systems are vulnerable to autoimmunity attacks due to design flaws or bugs in the source code. In humans, autoimmunity happens when the agent is so similar to components of our body, so the IS cannot distinguish between own and foreign tissue and attacks own organs. This can also happen in a cyber IS. However, autoimmunity in a cyber IS results in false positives, not necessarily in a destruction of the IS system itself.

Lastly, a cyber IS must be Big Data ready to handle the high volumes and speed at which data traffic is generated. The IoT with millions of connected devices will produce unprecedented amounts of data. Interconnected smart things will become the major data producers and consumers instead of humans [15]. The data traffic cannot be indexed such as databases to speed up searches. Instead, they have to apply the pattern matching algorithms directly to the real-time data. This requires enough processing power to handle all the traffic. Also, a pattern might only be suspicious if it repeats over time. The cyber IS has to record the traffic data so it can correlate real-time data with historic data. NoSQL databases that do not have tables and relations such as traditional relational databases have consistently performed better on these tasks. That's why they are the first choice for cyber IS systems.

4 Conclusions

Cyber-intrusions pose a constant threat to today's computer systems, and the number of intrusions increases every year [25]. This creates a demand for cyber security systems that can quickly react to new, unprecedented attack patterns. A cyber immune system can detect and adapt to new, unknown threats. However, it is an addition to classic cyber defense systems, they are not a replacement for traditional firewall, intrusion detection or end-point protection solutions. For instance, whereas it can detect attacks with unknown signatures, it is not capable of detecting unauthorized use of administrator rights since this is considered normal behavior. AIS have to seamlessly integrate into day to day security operations.

Since they learn unseen attack patterns during operations, they often share new, learned anomalies with other cyber IS systems. They exploit the "wisdom of the crowd" by using distributed threat databases and update them with newly detected exploits. Due to the distributed nature and the high availability requirement cloud solutions are a good choice for distributed cyber IS solutions. There are already early implementations of cloud-based cyber immunity solutions with learning capabilities [14]. However, they are still prone to high false positive rates, and human judgement is still required in many cases. However, it is to be expected that cognitive technologies will mature soon enough to significantly slow down cybercriminals and the accuracy, whether or not a current security "offense" can be associated with an attack, will increase in the near future.

Some bio-inspired cyber defense systems such as the one described in this paper are based on the AIS. Other bio-inspired cyber security systems have been based on swarm intelligence [6]. Some are building models mimicking the mechanisms in the biological immune system to better understand its natural processes and simulate its dynamical behavior in the presence of antigens/pathogens [7].

Bio-inspired approaches are highly scalable, use lightweight architectures, and are less resource-constrained compared to traditional security solutions [6]. Whereas there are already cyber immune systems available on the market [12, 21, 24] and there have been patents [8], the area of cyber immunity is still in its infancy and more research is required, specifically to avoid the typically high number of false positives.

References

1. Eronen, P.: Russian hybrid warfare (2016). http://www.defenddemocracy.org/content/uploads/documents/Russian_Hybrid_Warfare.pdf
2. Langner, R.: Stuxnet: dissecting a cyberwarfare weapon. IEEE Secur. Priv. **9**(3), 49–51 (2011)
3. Dua, S.: Data Mining and Machine Learning in Cybersecurity. CRC Press, Boca Raton (2011)
4. Parham, P.: The Immune System. Garland Science, New York (2014)
5. Farmer, J.D., Packard, N.H., Perelson, A.S.: The immune system, adaptation, and machine learning. Physica D **22**(1), 187–204 (1986)
6. Bitam, S., Zeadally, S., Mellouk, A.: Bio-inspired cybersecurity for wireless sensor networks. IEEE Commun. Mag. **54**(6), 68–74 (2016)
7. Dasgupta, D.: Advances in artificial immune systems. IEEE Comput. Intell. Mag. **1**(4), 40–49 (2006)
8. Hill, D.W., Lynn, J.T.: Adaptive system and method for responding to computer network security attacks (2000)
9. Witten, I.H., Frank, E., Hall, M.A.: Data Mining, 3rd edn. Elsevier, Burlington (2011)
10. Wlodarczak, P., Soar, J., Ally, M.: Multimedia data mining using deep learning, pp. 190–196 (2015)
11. Buczak, M.A., Guven, G.: A survey of data mining and machine learning methods for cyber security intrusion detection. IEEE Commun. Surv. Tutor. **18**(2), 1153–1176 (2016)
12. Darktrace: The enterprise immune system (2016). https://www.darktrace.com/
13. Runkler, T.A.: Data Analytics. Springer, Wiesbaden (2012)
14. IBM launches Watson for Cyber security beta program. IBM (2016). http://www-03.ibm.com/press/us/en/pressrelease/51189.wss. Accessed 19 Dec 2016
15. Karkouch, A., et al.: Data quality in internet of things: a state-of-the-art survey. J. Netw. Comput. Appl. **73**, 57–81 (2016)
16. Pintea, C.-M.: Bio-inspired computing. In: Pintea, C.-M. (ed.) Advances in Bio-inspired Computing for Combinatorial Optimization Problems, pp. 3–19. Springer, Heidelberg (2014)
17. Okamoto, T., Tarao, M.: Toward an artificial immune server against cyber attacks. Artif. Life Robot. **21**(3), 351–356 (2016)
18. Hu, P., Li, H., Fu, H., Cansever, D., Mohapatra, P.: Dynamic defense strategy against advanced persistent threat with insiders, pp. 747–755. IEEE Xplore Digital Library (2015)
19. Graves, A., Wayne, G., Danihelka, I.: Neural Turing Machines. Google DeepMind, London (2014)
20. Khalid, H.M., Peng, J.C.H.: A Bayesian algorithm to enhance the resilience of WAMS applications against cyber attacks. IEEE Trans. Smart Grid **7**(4), 2026–2037 (2016)
21. Veeramachaneni, K., Arnaldo, I., Korrapati, V., Bassias, C., Li, K.: AI^2: training a big data machine to defend, pp. 49–54 (2016)
22. Kose, U.: An artificial intelligence perspective on ensuring cyber-assurance for the internet of things. In: Cyber-Assurance for the Internet of Things, p. 249 (2016)
23. Gupta, A., Bhati, B.S., Jain, V.: Artificial intrusion detection techniques: a survey. Int. J. Comput. Netw. Inf. Secur. **6**(9), 51 (2014)
24. Musliner, D.J., Rye, J.M., Thomsen, D., McDonald, D.D., Burstein, M.H., Robertson, P.: FUZZBUSTER: towards adaptive immunity from cyber threats. In: Fifth IEEE Conference on Self-Adaptive and Self-Organizing Systems Workshops, pp. 137–140 (2016)
25. Musliner, D.J., Friedman, S.E., Marble, T., Rye, J.M., Boldt, M.W., Pelican, M.: Self-adaptation metrics for active cybersecurity. In: IEEE 7th International Conference on Self-Adaptation and Self-Organizing Systems Workshops, pp. 53–58 (2013)

An Accurate Database of the Fixation Probabilities for All Undirected Graphs of Order 10 or Less

Fernando Alcalde Cuesta[1], Pablo González Sequeiros[1],
Álvaro Lozano Rojo[2,3(✉)], and Rubén Vigara Benito[2,3]

[1] University of Santiago de Compostela, Santiago de Compostela, Spain
{fernando.alcalde,pablo.gonzalez.sequeiros}@usc.es
[2] Centro Universitario de la Defensa Zaragoza, Zaragoza, Spain
{alozano,rvigara}@unizar.es
[3] IUMA, University of Zaragoza, Zaragoza, Spain

Abstract. We present a extremely precise database of the fixation probabilities of mutant individuals in a non-homogeneous population which are spatially arranged on a small graph. We explore what features of a graph increase the chances of a beneficial allele of a gene to spread over a structured population.

1 Introduction

The fixation probability is a fundamental concept in evolutionary dynamics, representing the probability that a gene spreads over a whole population. An interesting model to study both neutral drift and natural selection on homogeneous population was introduced by P.A.P. Moran in [19]. The Moran process is a Markov chain whose states are the number of individuals with a mutant allele A of a gene starting from an initial population of N residents having the same allele a of the gene at some locus. At each step of time, one individual is selected for reproduction with probability proportional to its *relative fitness* with respect to the resident ones, r or 1 depending on whether it is mutant or resident. Then another individual is randomly chosen (with uniform probability) to be replaced by an identical offspring of the first individual. This Markov chain has two absorbing states corresponding to the *extinction* and *fixation* of the mutant allele A (where all the individuals have the allele a or A respectively). Starting with a single mutant individual in an initial population of N individuals, the fixation probability is

$$\Phi = \frac{r^{N-1}}{r^{N-1} + r^{N-2} + \cdots + 1},\tag{1}$$

for any relative fitness r.

E. Lieberman, C. Hauert and M.A. Nowak generalized this process to structured populations [21]. Let $G = (V, E)$ be an undirected connected graph with

© Springer International Publishing AG 2017
I. Rojas and F. Ortuño (Eds.): IWBBIO 2017, Part II, LNBI 10209, pp. 209–220, 2017.
DOI: 10.1007/978-3-319-56154-7_20

vertex set $V = \{0, 1, \ldots, N - 1\}$ with no loops or multiple edges. Denote by d_i the degree of the vertex i. The Moran process on G with fitness r is the Markov chain obtained as follows. Like for a homogeneous population, we start with a population of N resident individuals occupying the vertex set V. Afterward, one single vertex i_0 is chosen to become occupied by a mutant. At successive steps one vertex i is selected at random with probability $r_i/(r\,m + N - m)$, where r_i is the relative fitness of the individual occupying i (r or 1), and m is the number of vertices occupied by mutants in that moment. Next, a neighbor of i, randomly chosen with uniform probability, is replaced by an identical offspring of i. Now, the fixation probability depends on where the first mutant is placed inside the graph [8], and hence the *average fixation probability* on G, denoted Φ_G^r, is the main concept in the theory of evolutionary graphs. These kind of models have had impact not only in evolutionary genetics but in other areas like invasion dynamics, epidemics of disease, tumor growth or economics and management [2,11,16,20,22]. Actually, if the spreading of favorable innovations can be enhanced by network structures amplifying selection [26], as counterpart, we can find structural properties that increase the robustness of a complex network against invasion [2].

In the case of general graphs there is no closed form for the fixation probability. Some techniques have been proposed to compute this quantity [1,6,9,14,18] and there has been some calculations on small graphs [1,7,12,25]. However, there is no accurate available data for degrees less than or equal to 8, and there is no data available at all for degrees greater than 8.

Here we present an accurate database of the fixation probabilities for all undirected graphs with 10 or less vertices. This dataset could allow to find families of graph structures with interesting evolutionary properties, as already has been done in [3]. Moreover, the database has been enriched with some graph invariants which have been related to the fixation probability, see [2] and the references therein. In this way, exploring these data, it would be possible to shed light on the structural properties of graphs increasing or decreasing their fixation probabilities and the transitions among different evolutionary types. This is a particularly interesting property for biological networks like brain networks or PPI interactomes, as well as for technological ones. Finally, due to its precision, it also could be used as a testing dataset for new methods or computing libraries.

2   The Database

2.1   The Computation

The main steps of the computation process are the following:

Generation of the Graphs. The generation of the edge lists of all undirected graphs (up to isomorphism) with 10 vertices or less was done with Sage [17,23]. Since this is a relatively short computation we made no attempt to parallelize this process. Afterwards we have a binary file with the 11,989,763 connected graphs of order ≤ 10 (we dropped the trivial graph of one vertex).

Description of the Fixation Probability. Consider a connected undirected graph $G = (V, E)$ of order $N \leq 10$, and fix a relative fitness $r > 0$ for the invader mutants. The Markov chain described in the introduction is formalized as follows. The set of states $\mathcal{S}_G$ of the chain is the power set of V, where each set $S \in \mathcal{S}_G$ contains the nodes occupied by mutant individuals. The transition probabilities between $S, S' \in \mathcal{S}_G$ are given by

$$P^r_{S,S'} = \begin{cases} \dfrac{r \sum_{\{i \in S \mid (i,j) \in E\}} \frac{1}{d_i}}{r\#S + N - \#S} & \text{if } S' \setminus S = \{j\}, \\[3mm] \dfrac{\sum_{\{i \in V \setminus S \mid (i,j) \in E\}} \frac{1}{d_i}}{r\#S + N - \#S} & \text{if } S \setminus S' = \{j\}, \\[3mm] \dfrac{r \sum_{\{(i,j) \in E \mid i,j \in S\}} \frac{1}{d_i} + \sum_{\{(i,j) \in E \mid i,j \notin S\}} \frac{1}{d_i}}{r\#S + N - \#S} & \text{if } S = S', \\[3mm] 0 & \text{otherwise.} \end{cases} \quad (2)$$

These values can be arranged in a $2^N \times 2^N$ matrix $\mathbf{P}^r_G = (P^r_{S,S'})$. On the other hand, the fixation probabilities ϕ^r_S associated to all sets $S \in \mathcal{S}_G$ inhabited by mutant individuals are determined by the system of linear equations $\phi^r_S = \sum_{S' \in \mathcal{S}_G} P^r_{S,S'} \phi^r_{S'}$, with the boundary conditions $\phi^r_\emptyset = 0$ and $\phi^r_V = 1$. Equivalently

$$\mathbf{P}^r_G \cdot \begin{pmatrix} 0 \\ \phi^r_S \\ \vdots \\ \phi^r_{S'} \\ 1 \end{pmatrix} = \left(\begin{array}{c|ccc|c} 1 & 0 & \cdots & 0 & 0 \\ \hline P^r_{S,0} & & & & P^r_{S,V} \\ \vdots & & \widehat{\mathbf{P}}^r_G & & \vdots \\ P^r_{S',0} & & & & P^r_{S',V} \\ \hline 0 & 0 & \cdots & 0 & 1 \end{array} \right) \cdot \begin{pmatrix} 0 \\ \phi^r_S \\ \vdots \\ \phi^r_{S'} \\ 1 \end{pmatrix} = \begin{pmatrix} 0 \\ \phi^r_S \\ \vdots \\ \phi^r_{S'} \\ 1 \end{pmatrix}. \quad (3)$$

With this notation, the (average) fixation probability is then

$$\Phi^r_G = \frac{1}{N} \sum_{i \in V} \phi^r_{\{i\}}. \quad (4)$$

To compute this value, we need to solve (3), which can be written as a linear system

$$(\mathbf{I} - \widehat{\mathbf{P}}^r_G) \cdot \phi = \begin{pmatrix} P^r_{S,V} \\ \vdots \\ P^r_{S',V} \end{pmatrix}$$

where $\mathbf{I}$ is the identity matrix of size $2^N - 2$. Looking at (2) it is clear that it is possible to multiply each equation (associated to a state S) by the *reproductive weight* $W^r_S = r\#S + N - \#S$ of S obtaining the final equation $\mathbf{Q}^r_G \cdot \phi = \mathbf{b}^r_G$, where

$$\mathbf{Q}^r_G = \begin{pmatrix} W^r_S & \cdots & 0 \\ \vdots & \ddots & \vdots \\ 0 & \cdots & W^r_{S'} \end{pmatrix} \cdot (\mathbf{I} - \widehat{\mathbf{P}}^r_G), \qquad \mathbf{b}^r_G = \begin{pmatrix} W^r_S \, P^r_{S,V} \\ \vdots \\ W^r_{S'} \, P^r_{S',V} \end{pmatrix}.$$

By construction, the coordinates of $\mathbf{Q}^r_G$ and $\mathbf{b}^r_G$ are degree 1 polynomials on r with rational coefficients.

Computation of the Fixation Probability. The computation of Φ^r_G according to (4) was written in the C programming language. It runs as follows: consider a connected graph G of order $N \leq 10$. Each element in $\mathcal{S}_G$ can be represented by a number of N bits in length, that is an integer between 0 and $2^N - 1$. Given one of those states S, identified with the corresponding number, there are at most $N + 1$ possible transitions: a change in any of the bits of the number, or the number itself if there is no change. Hence, $\mathbf{Q}^r_G$ has at most $2^N \times (N + 1)$ non-zero entries. To compute the row associated to the state S of this matrix, all the possible elections for reproduction are performed and the results are accumulated in the correct positions. Since coordinates of $\mathbf{Q}^r_G$ and $\mathbf{b}^r_G$ are of the form $a\,r + b$ with $a, b \in \mathbb{Q}$, they can be easily represented as a couple of pairs of 64 bits integers. Therefore, we can exactly compute the fundamental matrix $\mathbf{Q}^r_G$ and the vector $\mathbf{b}^r_G$.

Finally, for each value of $r \in \{0.25, 0.5, \ldots, 10\}$, we should find the solution of $\mathbf{Q}^r_G \cdot \boldsymbol{\phi} = \mathbf{b}^r_G$ to compute Φ^r_G, which is the mean of the entrances of $\boldsymbol{\phi}$ corresponding to states with only one mutant (that is, states with a single bit set). To do so, we construct dense matrices evaluating $\mathbf{Q}^r_G$ and $\mathbf{b}^r_G$ for the given fitness r using 64 bit floating point numbers. Note that we construct them dense instead of sparse because of the used linear solver.

To guarantee a high relative precision of the solution that does not depend on the condition number of the matrix $\mathbf{Q}^r_G$, we use a special *LDU* factorization algorithm for *M*-matrices due to Barreras and Peña [5]. This algorithm is slower than iterative algorithms since it asymptotically has the complexity of the Gaussian elimination. However, it is reasonably quick for a single matrix of size 1022×1022, although the amount of graphs and possible values of r forced the parallelization of the computation. This step was trivial since the computations for different graphs are independent. The code is available at [24].

The actual computation was done in the Supercomputer FinisTerrae2 located at CESGA (Spain) using 1024 cores of Haswell 2680v3 CPUs for almost 3 days.

Building the Database. Finally, the database of fixation probabilities already computed was enriched with some graph properties which have been related with its evolutionary behavior. For each undirected graph $G = (V, E)$ we considered the *order*, the number of vertices $N = |V|$, the *size*, the number of edges $|E|$, and some statistics related with the degree distribution. We also computed some global scale measures like the *diameter* $\Delta = \max\{d(i, j)\}_{i,j \in V}$ and the *average path length* $L = \sum_{i,j \in V} \frac{d(i,j)}{N(N-1)}$, where $d(i, j)$ is the length of the shortest path joining the vertices i and j. We also added other measures of 'small-worldness' like the *clustering coefficient* $C = \frac{1}{N} \sum_{i \in V} \frac{2|E_i|}{d_i(d_i-1)}$, with $G_i = (V_i, E_i)$ the subgraph of neighbors of i with the edges between them [29], and the *transitivity* T, which is the ratio of the number of complete subgraphs of order 3 over the number of connected subgraphs of order 3 [28]. Additionally, we added the *heat heterogeneity* of the graph, defined as the variance of the temperature distribution, and the *temperature entropy* which is the entropy of that distribution [26]. Recall that the *temperature* of the vertex i is defined as $T_i = \frac{1}{d_i} \sum_{j \sim i} 1/d_j$ where

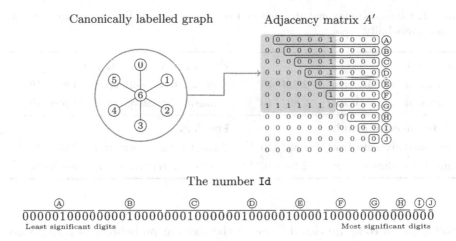

Fig. 1. Generating the Id of a graph.

d_i denotes the degree of i [27]. We completed this property set with three qualitative variables: whether if the graph is a *tree*, *bipartite* and/or *biconnected*. These quantities were computed using the NetworkX software [10].

On the other hand, each graph is identified with a unique 64 bits unsigned integer (the Id) which contains the adjacency matrix of the graph as follows (see Fig. 1): Consider a graph G with $N \leq 11$ vertices. After canonically labeling it (we used the bliss software [15]), consider its adjacency matrix A. Extend this matrix to the right and below as $A' = \left(\begin{smallmatrix} A & 0 \\ 0 & 0 \end{smallmatrix} \right)$ to make it 11×11. In this way we get the adjacency matrix of a graph of order 11 such that the connected component of the first vertex agrees with G, and the rest of connected components are singletons. Since A' is symmetric and the diagonal is always null, all the relevant information belongs to the off-diagonal upper part of A'. Ordering those bits from left to right and from top to bottom we build a number, the Id, with at most 55 bits set. Obviously, the Id constructed in this way is unique for each (isomorphism class of a) graph of order ≤ 11. The step of extending the matrix is required in order to guarantee this condition. In any case, it is trivial to recover the edge list from the Id without previous knowledge of the order or size of the graph (see https://bitbucket.org/snippets/geodynapp/AppyX for C and Python implementations).

2.2 On the Accuracy

Even when the accuracy of the database is granted by the method used to solve the linear problem, we checked the result with the known closed forms for average fixation probabilities. As it has been pointed out the fixation probability for the complete graph K_N is given in (1) for any $r > 0$. Moreover, the Isothermal Theorem [21] implies that constant degree graphs have the same fixation probability as the complete one of the same order. Monk, Green and Paulin devised

Table 1. Estimated maximal absolute and relative errors for the known values of the fixation probability.

For $r = 1$		For $K_{n,m}$ with $n, m \neq 1$	
maximal absolute error	$2.914 \cdot 10^{-16}$	maximal absolute error	$1.665 \cdot 10^{-15}$
maximal relative error	$2.887 \cdot 10^{-15}$	maximal relative error	$1.998 \cdot 10^{-15}$
For Isothermal graphs		**For $K_{1,n}$**	
maximal absolute error	$1.887 \cdot 10^{-15}$	maximal absolute error	$2.220 \cdot 10^{-15}$
maximal relative error	$2.109 \cdot 10^{-15}$	maximal relative error	$2.665 \cdot 10^{-15}$

a technique to compute closed forms of the fixation probability of highly symmetric graphs [18] (see also [27] for star graphs $K_{1,n}$). And finally, in the case of neutral drift ($r = 1$) the fixation probability for a graph of order N is $1/N$ [25]. In those cases it is possible to estimate the absolute and relative errors. Table 1 contains the maximal observed errors.

2.3   The Database File

The database is presented as a single HDF5 file [13] available at [4]. The root group has a dataset called FP which contains the data computed above. The HDF5 format enables us to save column names and data types for each column (composite datatype in the HDF5 jargon). Fractional data is saved as IEEE-754 little-endian 64 bits numbers. Integer are saved with the smallest possible number of bytes, hence, all are one byte little-endian unsigned integers except the Id which uses 64 bits. The dataset FP contains the 58 columns described in Table 2. Notice that the case random drift $r = 1$ has been dropped from the dataset since it is known [25].

3   Results

A careful analysis of this dataset could shed some light to the factors that increase or decrease the fixation probabilities of advantageous mutations in structured population. For the rest of the section we fix a fitness of $r = 2$. By fixing a different fitness we would obtain similar results.

Qualitative Variables. Figure 2 shows the effect in the distribution of the fixation probability of different qualitative properties of graphs. The data is disaggregated by order of the graph. It is remarkable that trees tend to be amplifiers of selection at any order, that is, they have higher fixation probability that the complete graph K_N with the same order N. In fact, in each order, all the trees are above the 75% of the non-tree graphs. Moreover, the star graph $K_{1,n}$ is maximal in its order $N = n + 1$. In the same direction, graphs with cut-points (removing

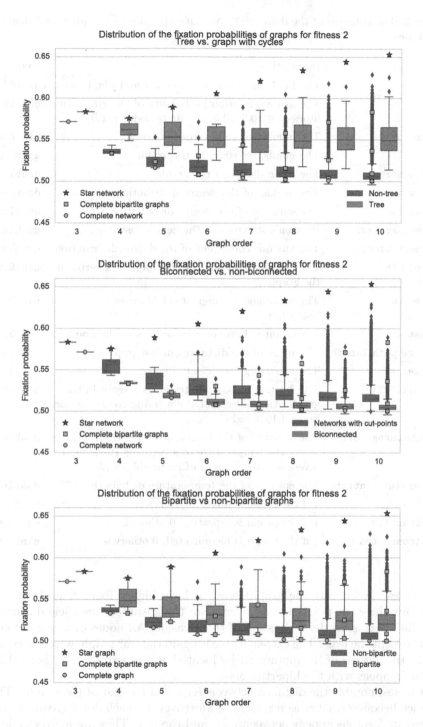

Fig. 2. Qualitative properties of the graphs and the fixation probability.

Table 2. The columns of the dataset `FP`. All datatypes are little-endian and denoted as C types.

Name	Description	Type
`Id`	The id of the graph, see Sect. 2.1 and Fig. 1	`uint64_t`
`FP_r`	The average fixation probability of the graph for fitness $r \in \{0.25, 0.5, \ldots, 10.0\}$ and $r \neq 1.0$	`double`
`order`	The number of vertices of the graph	`uint8_t`
`size`	The number of edges of the graph	`uint8_t`
`degree_mean`	The mean degree of the graph	`double`
`degree_median`	The median of the degree distribution	`double`
`degree_var`	The variance of the degree distribution	`double`
`degree_skewness`	Pearson's skewness of the degree distribution	`double`
`degree_kurtosis`	The kurtosis coefficient of the degree distribution	`double`
`degree_min`	The minimum number of neighbors of a vertex in the graph	`uint8_t`
`degree_max`	The maximum number of neighbors of a vertex in the graph	`uint8_t`
`diametre`	The distance between two vertices in the graph	`uint8_t`
`average_path_length`	The mean of the distances in the graph	`double`
`clustering`	The clustering coefficient [29]	`double`
`transitivity`	The fraction of possible triangles present in the graph. Also called global clustering coefficient [28, Sects. 4.10.3 and 6]	`double`
`heat_heterogeneity`	The variance of the temperature distribution, where the temperature of a vertex is the sum of inverses of the degrees of its neighbors [26]	`double`
`temperature_entropy`	The entropy of the temperature distribution [27]	`double`
`is_tree`	1 if the graph is a tree, 0 otherwise	`uint8_t`
`is_bipartite`	1 if the graph is bipartite, 0 otherwise	`uint8_t`
`is_biconnected`	1 if the graph is biconnected, 0 otherwise	`uint8_t`

this vertex disconnects the graph) tend to have higher fixation probabilities. Probably, these cut-points act as *defenses* of the advantageous allele during its early life, that is, in the early steps of the invasion. All nodes of a tree are cut-points except the leafs. The extremal case is again the star graph, were the center acts as the defense of the mutant alleles located at the periphery [2]. Something similar happens with the bipartite ones.

It is also notable the distance between $K_{1,n}$ and the rest of the graphs. This distance becomes smaller as n increases. Moreover, for each order greater that 6 two isolated graphs appear between $K_{1,n}$ and the rest. They are more visible in the figure about biconnected graphs but can be spotted in the other two plots.

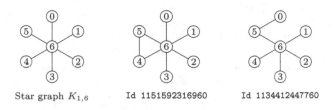

Star graph $K_{1,6}$ Id 1151592316960   Id 1134412447760

Fig. 3. A family of amplifiers of selection ordered from left to right by Φ_G^2.

They corresponds to small modifications to $K_{1,n}$ were an edge is added between two peripheral vertices which could be or not be disconnected from the central one, see Fig. 3.

There are also some suppressors of selection, that is, those with fixation probability (strictly) less than the one of K_N for $r = 2$. For order 6 there are 3 of these graphs (see Fig. 4). This quantity increases rapidly reaching hundreds of thousands for order 10. In [3] a family of graphs obtained from this data has been proved to be a *global suppressor of selection*, i.e., the fixation probability of each of them is below the fixation probability of the complete graph of the same order for any fitness $r > 1$. The discovery of this family is a first step towards finding structural properties increasing the robustness of non homogeneous populations against mutation.

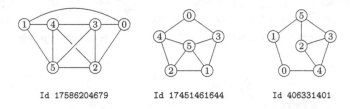

Id 17586204679   Id 17451461644   Id 406331401

Fig. 4. The suppressors of selection of order 6 for $r = 2$.

Quantitative Variables. Among the computed quantitative variables the least interesting one is the temperature entropy [27] which is constant for graphs with the same order. Figure 5 shows the effects of the most interesting quantitative variables on the fixation probability of the graphs of order 10. Similar pictures could be obtained for other orders. It is clear that one by one only the heat heterogeneity [26] explains effect of the population structure in the spread of the mutants, showing a high correlation.

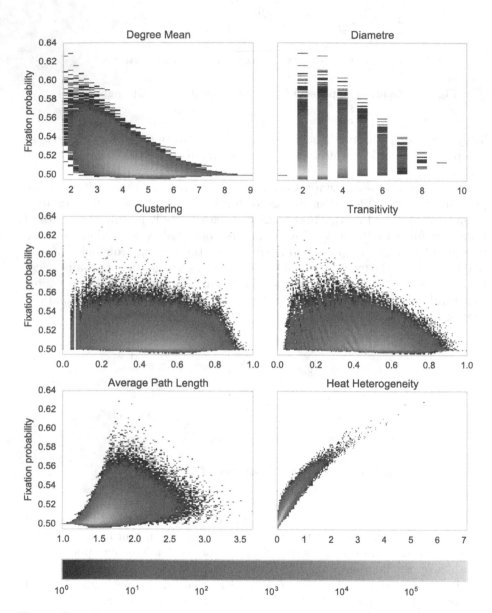

Fig. 5. Quantitative variables versus fixation probability for graphs of order 10 and fitness 2. Color represents the amount of graphs in the corresponding region of the plot. (Color figure online)

4   Conclusions

We present a extremely precise database of the fixation probabilities on small graphs of interest in research areas like invasion dynamics, epidemics of disease, tumor growth or economics and management [2,11,16,20,22]. Other authors have explored the evolutionary dynamics of small graphs [7,12], but this is the first systematic computation for a large family of small graphs, namely all undirected graphs with 10 or less vertices. Exploring this dataset, evolutionary biology researchers could shed some light on the factors and structures affecting the spread of beneficial alleles in larger populations arranged on graphs. Deep statistical techniques are still needed to understand how to design population arrangements with a desired dynamical behavior and this dataset represents the first step in that direction.

Acknowledments. We gratefully thank CESGA for providing access to the FinisTerrae2 supercomputer. FA, PG and ÁL are supported by Spanish Ministry of Economy and Competitiveness and European Social Fund (Grant MTM2013-46337-C2-2-P). ÁL and RV are supported by DGA and European Social Fund (Grant E15 Geometría) and CUDZ (Grant CUD 2015-10). RV is supported by Spanish Ministry of Economy and Competitiveness and European Social Fund (Grant MTM2013-45710-C2).

References

1. Alcalde, F., Sequeiros, P.G., Lozano, Á.: Fast and asymptotic computation of the fixation probability for Moran processes on graphs. BioSystems **129**, 25–35 (2015)
2. Alcalde, F., Sequeiros, P.G., Lozano, Á.: Exploring the topological sources of robustness against invasion in biological and technological networks. Sci. Rep. **6** (2016). Article no. 20666
3. Alcalde, F., Sequeiros, P.G., Lozano, Á.: Suppressors of selection. arXiv:1607.04469
4. Alcalde, F., Sequeiros, P.G., Lozano, Á., Vigara, R.: Data of "an accurate database of the fixation probabilities for all undirected graphs of order 10 or less", Mendeley Data, v2 (2017). http://dx.doi.org/10.17632/587bnf6mt3.2
5. Barreras, A., Peña, J.M.: Accurate and efficient *LDU* decomposition of diagonally dominant *M*-matrices. Electron. J. Linear Algebra **24**, 152–167 (2012/13)
6. Broom, M., Rychtář, J.: An analysis of the fixation probability of a mutant on special classes of non-directed graphs. Proc. Royal Soc. Lond. A **464**(2098), 2609–2627 (2008)
7. Broom, M., Rychtář, J., Stadler, B.T.: Evolutionary dynamics on small-order graphs. J. Interdisciplinary Math. **12**(2), 129–140 (2009)
8. Broom, M., Rychtář, J., Stadler, B.T.: Evolutionary dynamics on graphs - the effect of graph structure and initial placement on mutant spread. J. Stat. Theor. Pract. **5**(3), 369–381 (2011)
9. Díaz, J., Goldberg, L.A., Mertzios, G.B., Richerby, D., Serna, M., Spirakis, P.G.: Approximating fixation probabilities in the generalized Moran process. In: Proceedings of the Twenty-Third Annual ACM-SIAM Symposium on Discrete Algorithms, SODA 2012, pp. 954–960. SIAM (2012)
10. Hagberg, A., Schult, D., Swart, P.: Exploring network structure, dynamics, and function using networkX. In: Varoquaux, G., Vaught, T., Millman, J. (eds.) Proceedings of the 7th Python in Science conference (SciPy 2008), pp. 11–15 (2008)

11. Hindersin, L., Werner, B., Dingli, D., Traulsen, A.: Should tissue structure suppress or amplify selection to minimize cancer risk? Biol. Direct **11**, 41 (2016)
12. Hindersin, L., Traulsen, A.: Most undirected random graphs are amplifiers of selection for birth-death dynamics, but suppressors of selection for death-birth dynamics. PLoS Comput. Biol. **11**, 1–14 (2015)
13. The HDF Group: Hierarchical Data Format, version 5, 1997–2016. http://www.hdfgroup.org/HDF5/
14. Houchmandzadeh, B., Vallade, M.: The fixation probability of a beneficial mutation in a geographically structured population. New J. Phys. **13**, 073020 (2011)
15. Junttila, T., Kaski, P.: Engineering an efficient canonical labeling tool for large and sparse graphs. In: Proceedings of the Ninth Workshop on Algorithm Engineering and Experiments (ALENEX07), pp. 135–149. SIAM (2007)
16. Komarova, N.L., Senguptac, A., Nowaka, M.A.: Mutation-selection networks of cancer initiation: tumor suppressor genes and chromosomal instability. J. Theor. Biol. **223**, 433–450 (2003)
17. McKay, B.D.: Isomorph-free exhaustive generation. J. Algorithms **26**(2), 306–324 (1998)
18. Monk, T., Green, P., Paulin, M.: Martingales and fixation probabilities of evolutionary graphs. Proc. R. Soc. A **470**. Article no. 20130730
19. Moran, P.A.P.: Random processes in genetics. Proc. Cambridge Philos. Soc. **54**, 60–71 (1958)
20. Nowak, M.A.: Evolutionary Dynamics. Harvard University Press, Cambridge (2006)
21. Lieberman, E., Hauert, C., Nowak, M.A.: Evolutionary dynamics on graphs. Nature **433**, 312–316 (2005)
22. Salas-Fumás, V., Sáenz, C., Lozano, Á.: Organisational structure and performance of consensus decisions through mutual influences: a computer simulation approach. Decis. Support Syst. **86**, 61–72 (2016)
23. Stein, W.A., et al.: Sage Mathematics Software (Version 7.0), The Sage Development Team (2016). http://www.sagemath.org
24. MoM software. https://bitbucket.org/geodynapp/mom
25. Shakarian, P., Roos, P., Johnson, A.: A review of evolutionary graph theory with applications to game theory. BioSystems **107**(2), 66–80 (2012)
26. Tan, S., Lu, J.: Characterizing the effect of population heterogeneity on evolutionary dynamics on complex networks. Sci. Rep. **4**, 5034 (2014)
27. Voorhees, B., Murray, A.: Fixation probabilities for simple digraphs. Proc. R. Soc. A **469**, 20120676 (2013)
28. Wasserman, S., Faust, K., Analysis, S.N.: Methods and Applications. Cambridge, New York (1994)
29. Watts, D.J., Strogatz, S.H.: Collective dynamics of "small-world" networks. Nature **393**, 440–442 (1998)

Quantum Computing Based Inference of GRNs

Abhinandan Khan[1]([✉]), Goutam Saha[2], and Rajat Kumar Pal[1]

[1] Department of Computer Science and Engineering, University of Calcutta,
Acharya Prafulla Chandra Roy Siksha Prangan, JD-2, Sector - III,
Saltlake, Kolkata 700 106, India
khan.abhinandan@gmail.com, pal.rajatk@gmail.com
[2] Department of Information Technology, North Eastern Hill University,
Umshing Mawkynroh, Shillong 793 022, Meghalaya, India
dr.goutamsaha@gmail.com

Abstract. The accurate reconstruction of gene regulatory networks from temporal gene expression data is crucial for the identification of genetic inter-regulations at the cellular level. This will help us to comprehend the working of living entities properly. Here, we have proposed a novel quantum computing based technique for the reverse engineering of gene regulatory networks from time-series genetic expression datasets. The dynamics of the temporal expression profiles have been modelled using the recurrent neural network formalism. The corresponding training of model parameters has been realised with the help of the proposed quantum computing methodology based concepts. This is based on entanglement and decoherence concepts. The application of quantum computing technique in this domain of research is comparatively new. The results obtained using this technique is highly satisfactory. We have applied it to a 4-gene artificial genetic network model, which was previously studied by other researchers. Also, a 10-gene and a 20-gene genetic network have been studied using the proposed technique. The obtained results suggest that quantum computing technique significantly reduces the computational time, retaining the accuracy of the inferred gene regulatory networks to a comparatively satisfactory level.

Keywords: Decoherence · Entanglement · Gene regulatory networks · Quantum bit · Quantum computing · Recurrent neural network · Rotating gates

1 Introduction

Genes constitute the functional circuitry or blueprint of living beings. Cellular biochemical reactions occur due to the complex and synergistic behaviour of a group of inter-regulated genes. *Gene regulatory networks* or GRNs characterise these genetic regulatory relationships by directed graphs in the mathematical sense. In GRNs, genes are indicated by nodes and the regulations between the genes, i.e. activation or inhibition are represented by edges in a GRN [1]. With the progress in genetic research, an abundance of quality temporal expression

© Springer International Publishing AG 2017
I. Rojas and F. Ortuño (Eds.): IWBBIO 2017, Part II, LNBI 10209, pp. 221–233, 2017.
DOI: 10.1007/978-3-319-56154-7_21

data has been produced in the form of microarray data [2]. Analysis of these data provides a clear perception of principal cellular activities in the living system. This may also provide a tool for illustrating genetic functions which may be required for disease diagnosis, assessment of drug effects, etc. This provides impetus to the global research community to analyse and develop computational means for the biologically relevant analysis and elucidation of this data. Unfortunately, microarray datasets are generally corrupted with a substantial amount of experimental noise. Furthermore, in real-world large-scale GRNs, the number of genes are far greater than the number of time points. This makes the computational analysis for GRNs very much complex.

The current research work deals with the reverse engineering of GRNs from time-series genetic expression data that contain hidden dynamical information regarding the genetic inter-relationships. The reconfiguration of a GRN denotes the intricate regulatory interactions between the genes that involve DNA, RNA, protein, and other molecules. An interaction may be of two types namely activation and inhibition. Activation enhances the genetic activity, and inhibition does the reverse under the influence of regulatory genes.

Quantum computing, on the other hand, is based on the direct implementation of quantum-mechanical phenomena like *superposition, coherence*, and *entanglement* [3]. General digital computers use data encoded into bits that are in either of the two states 0 or 1. However, quantum computation employs quantum bits or qubits that can be in a superposition of two or more states. Quantum computers have theoretical resemblances with non-deterministic and probabilistic computers. The works of Paul Benioff [4] and Yuri Manin in 1980 [5], Richard Feynman in 1982 [6], and David Deutsch in 1985 [7] introduced the field of quantum computing. Quantum computers have the potential to solve problems efficiently, that no classical computer can solve within a reasonable amount of time.

Quantum computation is the outcome of the implementation of quantum mechanics to the field of algorithms. The ability of parallelism is the fundamental dissimilarity or difference between quantum and classical computation. In calculating the probability, a system is not in an invariable state. On the other hand, the system has a specific probability, and the state probability vector corresponds to all the various possible states. Quantum computing is analogous to this. Probability amplitudes of the quantum states are used, and they are squared normalised. Thus, the calculations based on quantum computations are $\sqrt{N}$ times faster than classical computation. Quantum transformation is realized by quantum rotating gates. There are certain special properties of quantum computation compared to classical computation [3], and a few of these can be implemented to solve traditional optimization problems in an improved manner.

In this paper, we have proposed, for the first time, a novel quantum computation based methodology for reverse engineering GRNs from time-series genetic expression data. Here, we have implemented the recurrent neural network (RNN) [8] formalism for modelling the dynamics of the expression data. We have proposed a novel model parameter training scheme based on the concepts of quantum computing. We have also proposed a new qubit rotation scheme with the help of multiple rotating gates.

We have applied our proposed methodology to three types of networks: (*i*) an artificial 4-gene network previously studied in the context of reconstruction of GRNs [9], (*ii*) a 10-gene network extracted from the GeneNetWeaver (GNW) database [10,11], and (*iii*) a 20-gene network also extracted from the GNW database. The obtained results suggest that the proposed methodology takes less time compared to other methodologies investigated in the contemporary literature, without sacrificing appreciable accuracy.

The rest of this paper has been structured as follows: the scientific background of the relevant topics has been presented in Sect. 2. The proposed methodology has been explained in detail in Sect. 3. Experimental results and related discussions have been presented in Sect. 4. The paper has been concluded with Sect. 5.

2 Scientific Background

2.1 Recurrent Neural Network (RNN)

The dynamics of time-series expression data can be perfectly encapsulated by the RNN formalism [8] as shown in Fig. 1. Nodes represent genes, and edges represent regulatory interactions amongst the genes. Each RNN layer defines the expression level of all genes at a specific time t_i. The expression of a particular gene at the succeeding time point $t_{i+1} = t_i + \Delta t$ can be obtained from the expression levels of all genes at t_i, and w_{ij} denotes the weights of the connecting edges i.e. the nature of control. We have adopted the following RNN formalism [8] for modelling GRNs:

$$x_i(t + \Delta t) = \frac{\Delta t}{\tau_i} * \frac{1}{1 + exp[-\sum_j w_{ij} x_j + \beta_i]} - \left(1 - \frac{\Delta t}{\tau_i}\right) * x_i(t) \qquad (1)$$

where $x_i(t)$ denotes the expression level of gene i at a time point t, and $x_i(t+\Delta t)$ denotes the expression at time point $t + \Delta t$, b_i signifies an external input, and

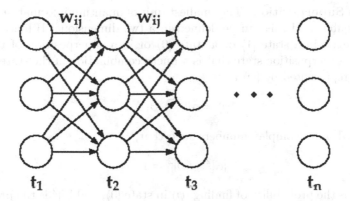

Fig. 1. A simple RNN model.

c_i represents a constant for each gene [8]. In this work, the purpose of RNN is to reproduce the given gene expression profiles truthfully by suitable training of the corresponding model parameters. This, in essence, is an optimisation problem, and we have applied our proposed quantum computing based methodology for this problem. Every optimisation problem necessitates a fitness function which guides the optimisation process. In the present context, *mean square error* (MSE) can be regarded as an appropriate fitness function and is defined as follows:

$$MSE = \frac{1}{NT} \sum_{1}^{N} \sum_{1}^{T} (x_i(t) - \tilde{x}_i(t))^2 \tag{2}$$

where N is the total number of genes in the network, T is the number of time points, $x_i(t)$ is the original expression data, and $\tilde{x}_i(t)$ is the simulated or computed data at a point of time t. Here, parameter optimization of RNN has been secured using a novel optimization tool based on quantum computing. Optimisation algorithms hybridised with quantum computing have been proposed and applied in a few domains of scientific and engineering research. However, in the present context of reverse engineering of GRNs from time-series microarray datasets, this is probably the very first implementation endeavour with quantum computing.

GRN reconstruction is a computationally expensive job. Especially, for large-scale GRNs the raw computational power required is sometimes infeasibly huge. Researchers in this domain investigate different methodologies for reducing the computational burden. The motivation behind the present research endeavour is to reduce the computational complexity of the problem dealt herein. Here, quantum computing seems to be a promising concept that can help in reducing the computational complexity of the GRN reconstruction problem, hopefully, even for large-scale GRNs.

2.2   Quantum Computing

Quantum Superposition: The smallest unit of quantum information processing is the qubit, and this can be defined in a two-dimensional Hilbert space. A qubit can exist in a state $|1\rangle$ or a state $|0\rangle$ or in a superposition of these two states. The superposition state that is a linear combination of the states $|1\rangle$ and $|0\rangle$ can be represented by $|\psi\rangle$ as:

$$|\psi\rangle = \alpha|0\rangle + \beta|1\rangle \tag{3}$$

where α and β are complex numbers such that,

$$|\alpha|^2 + |\beta|^2 = 1 \tag{4}$$

where $|\alpha|^2$ is the probability of finding $|\psi\rangle$ in state $|0\rangle$, and $|\beta|^2$ is the probability of finding $|\psi\rangle$ in state $|1\rangle$. Similarly, an n-bit quantum register can be in a superposition state, $|\phi\rangle$ comprising of 2^n ground states. Any operation on such an n-bit register is equivalent to operating on all the 2^n numbers.

$$|\phi\rangle = \sum_{i=1}^{n} \delta_i |\phi_i\rangle \tag{5}$$

where δ_i denotes the probability amplitude of state $|\phi_i\rangle$, and $|\delta_i|^2$ is the probability of $|\phi\rangle$ to collapse to a ground state $|\phi_i\rangle$ on measurement, such that:

$$\sum_{i=1}^{n} |\delta_i|^2 = 1 \tag{6}$$

Quantum Rotation: The manipulation of any quantum algorithm is achieved through suitable operations on the superposition state or entanglement state. One of the states will be the solution. Our aim is to find that solution state. The quantum rotating gate is an operator that can achieve this. Thus, the construction of suitable quantum rotating gates or operators is the main concern of this methodology. Clearly, selection of quantum gates or operators has a direct effect on the performance of the algorithm. Rotating gates are to be designed according to the problem in question, and we have used two different gates here in this work that can be defined as:

$$U_1(\Delta\theta) = \begin{bmatrix} \cos\Delta\theta & -\sin\Delta\theta \\ \sin\Delta\theta & \cos\Delta\theta \end{bmatrix} \tag{7}$$

$$U_2(\Delta\theta) = \begin{bmatrix} \cos\Delta\theta & \sin\Delta\theta \\ -\sin\Delta\theta & \cos\Delta\theta \end{bmatrix} \tag{8}$$

where U is a unitary operator and thus, can be used as a quantum gate or operator to manipulate the probability amplitudes of any qubit.

Quantum Measurement: An act of measurement perturbs a quantum system in an essential manner, leading to the collapse of the linear superposition state $|\phi\rangle$ onto one of the ground states $|\phi_i\rangle$. This loss of coherence is termed as decoherence [3].

The *No Free Lunch* (NFL) theorem [12] provides further motivation to integrate the concept of quantum computing into the GRN reconstruction problem. According to NFL, there exists no algorithm that is well-suited for all types of problems. Thus, there is always scope for research into updating and improving traditional algorithms using concepts from other domains like quantum mechanics, which we have investigated here.

3   Method

In this section, we have explained our proposed formalism for reverse engineering of GRNs from temporal gene expression data. For N genes in a GRN, there are $(N+2)$ model parameters to be trained for each gene. Thus, the optimisation problem dimension becomes $N\times(N+2)$. This becomes quite computationally expensive for larger GRNs. However, GRNs are known to be biologically sparse [13], i.e. most of the elements in w_{ij} (in Eq. (1)) are zero. However, even this beneficial condition cannot reduce the computational load significantly. Thus, researchers proposed to decompose the $N\times(N+2)$ dimensional problem into N sub-problems of $(N+2)$ dimensions each. Each gene, therefore, has been studied separately, and the corresponding $(N+2)$ parameters of the RNN formalism have been trained for each case.

According to Bolouri and Davidson [14], on an average, a gene is typically regulated by four to eight genes. In this work, we have considered 4-gene, 10-gene, and 20-gene networks and thus, we have assumed the maximum in-degree for a gene to be $m = 4$. This reduces the discrete search space of probable GRNs significantly. Moreover, with a maximum in-degree of 4, there can be a maximum of $\binom{N}{4}$ number of possible candidate GRN structures, where N is the number of genes in a particular GRN. This reduces the overall search space of network structures from 2^N to $\binom{N}{4}$. Furthermore, since all possible combinations of regulators are being considered, the biological plausibility of the candidate GRNs are likely to be maintained to the maximum possible extent.

We have proposed a novel optimisation technique based on the concepts of quantum computing. We have designed a population-based scheme, where a group of multi-qubit systems interact among themselves to find a suitable parameter set for the RNN formalism such that the error defined by Eq. (2) is minimised. We have defined a solution for the RNN model as a 3-tuple $\langle W, B, C \rangle$ where $W = [w_{ij}]_{N\times N}$, $B = [b_i]_{N\times 1}$, and $C = [c_i]_{N\times 1}$. Each of these parameters has been represented by a multi-qubit register.

3.1   Initialization: Multi-qubit Encoding

For the reverse engineering of GRN, we have defined the parameters w_{ij}, b_i, and c_i in the form of multi-qubits q_i as follows,

$$q_i = \begin{bmatrix} \alpha_{i1} \; \alpha_{i2} \; \alpha_{i3} \; \ldots \ldots \; \alpha_{il} \\ \beta_{i1} \; \beta_{i2} \; \beta_{i3} \; \ldots \ldots \; \beta_{il} \end{bmatrix} \tag{9}$$

where l denotes the number of bits used to encode a particular parameter. All α_{ik} and β_{ik} for $k = 1, 2, \ldots, l$ are randomly initialised within the range of $[0, 1]$. In this work, we have assumed $\alpha = \cos\theta$ and $\beta = \sin\theta$, thus, conserving the constraint defined by Eq. (6).

$$q_i = \begin{bmatrix} \cos\theta_{i1} \; \cos\theta_{i2} \; \cos\theta_{i3} \; \ldots \ldots \; \cos\theta_{il} \\ \sin\theta_{i1} \; \sin\theta_{i2} \; \sin\theta_{i3} \; \ldots \ldots \; \sin\theta_{il} \end{bmatrix} \tag{10}$$

3.2 Quantum Measurement

Measurement of the multi-qubit leads to the collapse of quantum encoding onto normal binary encoding. After measurement, we are left with qm_i that is simply an l-bit binary number. For the process of measurement, we have used the following:

$$qm_{ik} = \begin{cases} 0, & \text{if } rand > |\alpha_{ik}|^2 \\ 1, & \text{otherwise} \end{cases} \quad \text{for all } i \text{ and } k \tag{11}$$

3.3 Fitness Calculation

For each of the parameters w_{ij}, b_i, and c_i, of the RNN model, the corresponding qm_i has been converted to decimal value and the fitness has been calculated using Eqs. (1) and (2) for each of the solution tuples.

3.4 Qubit Rotation

We have applied the quantum gates defined by Eqs. (7) and (8) to update the quantum particles q_i according to:

From Eq. (7),

$$q_{ik}^{t+1} = U_1(\Delta\theta_{ik}) \cdot q_{ik}^t$$

$$\begin{bmatrix} \alpha_{ik}^{t+1} \\ \beta_{ik}^{t+1} \end{bmatrix} = \begin{bmatrix} \cos\Delta\theta_{ik} & -\sin\Delta\theta_{ik} \\ \sin\Delta\theta_{ik} & \cos\Delta\theta_{ik} \end{bmatrix} \cdot \begin{bmatrix} \alpha_{ik}^t \\ \beta_{ik}^t \end{bmatrix}$$

$$\begin{bmatrix} \alpha_{ik}^{t+1} \\ \beta_{ik}^{t+1} \end{bmatrix} = \begin{bmatrix} \cos\Delta\theta_{ik} & -\sin\Delta\theta_{ik} \\ \sin\Delta\theta_{ik} & \cos\Delta\theta_{ik} \end{bmatrix} \cdot \begin{bmatrix} \cos\theta_{ik}^t \\ \sin\theta_{ik}^t \end{bmatrix}$$

$$\therefore \begin{bmatrix} \alpha_{ik}^{t+1} \\ \beta_{ik}^{t+1} \end{bmatrix} = \begin{bmatrix} \cos(\Delta\theta_{ik} + \theta_{ik}^t) \\ \sin(\Delta\theta_{ik} + \theta_{ik}^t) \end{bmatrix} \tag{12}$$

From Eq. (8),

$$q_{ik}^{t+1} = U_2(\Delta\theta_{ik}) \cdot q_{ik}^t$$

$$\begin{bmatrix} \alpha_{ik}^{t+1} \\ \beta_{ik}^{t+1} \end{bmatrix} = \begin{bmatrix} \cos\Delta\theta_{ik} & \sin\Delta\theta_{ik} \\ -\sin\Delta\theta_{ik} & \cos\Delta\theta_{ik} \end{bmatrix} \cdot \begin{bmatrix} \alpha_{ik}^t \\ \beta_{ik}^t \end{bmatrix}$$

$$\begin{bmatrix} \alpha_{ik}^{t+1} \\ \beta_{ik}^{t+1} \end{bmatrix} = \begin{bmatrix} \cos\Delta\theta_{ik} & \sin\Delta\theta_{ik} \\ -\sin\Delta\theta_{ik} & \cos\Delta\theta_{ik} \end{bmatrix} \cdot \begin{bmatrix} \cos\theta_{ik}^t \\ \sin\theta_{ik}^t \end{bmatrix}$$

$$\therefore \begin{bmatrix} \alpha_{ik}^{t+1} \\ \beta_{ik}^{t+1} \end{bmatrix} = \begin{bmatrix} \cos(\Delta\theta_{ik} - \theta_{ik}^t) \\ \sin(\Delta\theta_{ik} - \theta_{ik}^t) \end{bmatrix} \tag{13}$$

where the term $\Delta\theta$ controls the rotation of each qubit as illustrated by Fig. 2. This is the crux of the quantum computing based optimization technique, and we have updated $\Delta\theta$ by Eqs. (12) and (13) based on the fitness of a solution tuple.

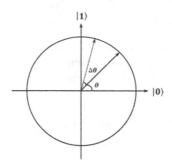

Fig. 2. Qubit rotation.

A pseudo code of the given methodology is as follows:

```
Start
Set number of genes = N, maximum number of regulators
= m, and population size, pop = (N/m)
Set number of bits used for encoding wᵢⱼ, bᵢ, and cᵢ to
lw, lb, and lc, respectively
Initialise Qw = [qw₁, qw₂, ..., qw_pop], Qb = [qb₁, qb₂, ..., qb_pop], and
Qc = [qc₁, qc₂, ..., qc_pop]
Perform quantum measurement on each qwᵢ, qbᵢ, qcᵢ to
produce qwmᵢ, qbmᵢ, qcmᵢ and convert them to decimal
Calculate fitness using Eqs. (1) and (2)
Save the best solution tuple
While termination criterion is not reached
 For i = 1 to pop
  For j = 1 to m × lw
   Using Eq. (12)/(13) generate new values of α and β
   based on the value of the j-th bit of qwmᵢ and
   qwm_best
  End for
  For j = 1 to lb
   Using Eq. (12)/(13) generate new values of α and β
   based on the value of the j-th bit of qbmᵢ and qbm_best
  End for
  For j = 1 to lc
   Using Eq. (12)/(13) generate new values of α and β
   based on the value of the j-th bit of qcmᵢ and qcm_best
  End for
 End for
Perform quantum measurement on each qwᵢ, qbᵢ, qcᵢ to
produce qwmᵢ, qbmᵢ, qcmᵢ and convert them to decimal
Calculate fitness using Eqs. (1) and (2)
```

```
Update the best solution, if a better one is found,
else replace the worst solution of this generation by
the stored best solution
End while
Output the best solution.
End
```

4   Experimental Results and Discussion

We have performed three experiments on the proposed quantum computing based formalism for GRN construction. The evaluation of the proposed algorithm has been done based on certain metrics that have been explained by Eqs. (14)–(17). However, first, we have defined the variables required for the definition of the metrics.

An edge in the final GRN can be categorised as:(i) True Positive (TP), if an existing regulation is correctly identified; (ii) False Positive (FP), if a non-existent regulation is identified as existing in the inferred GRN; (iii) True Negative (TN), if a non-existent regulation is correctly identified; and (iv) False Negative (FN), if an existing regulation is identified as non-existent in the inferred GRN.

The specification of the metrics that have been used for evaluation of the proposed methodology are as follows:

$$\text{Sensitivity}(S_n) = \frac{TP}{TP + FN} \tag{14}$$

$$\text{Specificity}(S_p) = \frac{TN}{TN + FP} \tag{15}$$

$$\text{Accuracy}(ACC) = \frac{TP + TN}{TP + FP + FN + TN} \tag{16}$$

$$\text{F-Score} = \frac{2TP}{2TP + FP + FN} \tag{17}$$

We have applied the proposed quantum computation based reverse engineering methodology on the experimental datasets of an artificial 4-gene network studied previously [9], a 10-gene network extracted from the GNW database, also studied previously [9], and a 20-gene network extracted from the GNW database. The proposed methodology requires 5000 iterations to reach a stable near global solution. Thus, it takes about half the time compared to other traditional optimisation algorithms that have been used in this domain. The simulations have been run on Matlab R2016a, running on an Intel Core i7 processor with 16 GB of DDR4 RAM (2400 MHz). The experimental setups and the parameter values have been given in Tables 1 and 2.

The RNN model parameters for the 4-gene network have been shown in Table 3 and the dynamics in Fig. 3a. The results obtained for the 4 gene network, shown in Fig. 3b, indicate that the proposed quantum computing based

Table 1. Experimental setups (specification of microarray data).

	No. of interactions	No. of time points generated	Maxm. no. of regulators allowed
4-gene artificial network	8	51	2
10-gene network extracted from GNW	12	41	4
20-gene network extracted from GNW	24	51	4

Table 2. Quantum parameter bounds.

	Number of bits used			Range		
	4-gene	10-gene	20-gene	4-gene	10-gene	20-gene
w_{ij}	$(1,1)$	$(1,1)$	$(1,1)$	15	15	15
b_i	$(2,2)$	$(2,2)$	$(2,2)$	6	6	6
c_i	$(1,16)$	$(1,16)$	$(1,32)$	4	4	5

Table 3. RNN model parameters.

w_{ij}				b_i	c_i
20	20	0	0	0	10
15	10	0	0	5	5
0	8	12	0	0	5
0	0	8	12	0	5

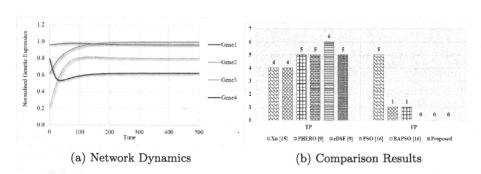

(a) Network Dynamics   (b) Comparison Results

Fig. 3. (a) 4-gene network dynamics, and (b) comparison of results of 4-gene network.

methodology performs better than that of most of the RNN-based techniques available in the contemporary literature.

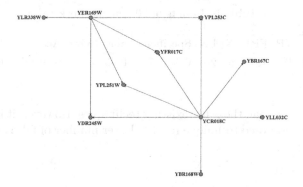

Fig. 4. Original 10-gene network structure.

Table 4. Results of 10-gene network.

Method	TP	TN	FP	FN	TPR	SPC	ACC	F-Score	Graph edges	Time
eDSF [9]	5	77	11	7	0.42	0.88	0.82	0.36	16	-
PSO [16]	5	77	11	7	0.42	0.88	0.82	0.36	16	1.67 h
BAPSO [16]	6	76	12	6	0.50	0.86	0.82	0.40	18	1.69 h
Proposed	5	85	3	7	0.42	0.97	0.90	0.50	8	0.72 h

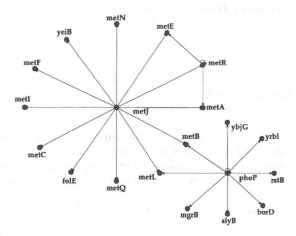

Fig. 5. Original 20-gene network structure.

For the 10-gene network (Fig. 4), the proposed technique gives superior results in terms of reduction of the number of false positives and hence improved accuracy (Table 4). To further gauge the performance of the proposed technique in bigger GRNs, we have implemented it on a 20-gene network (Fig. 5). Although the results (Table 5) are not satisfactory for this larger network, it offers faster

Table 5. Results of 20-gene network.

Method	TP	FP	TN	FN	Specificity	Sensitivity	Accuracy	F-Score
Proposed	1	19	340	20	0.05	0.95	0.90	0.05

computational time than the other existing techniques. However, it is worth mentioning that the proposed technique returns lesser number of false positives here.

5   Conclusion

In this work, we have investigated the reverse engineering of GRNs based on the novel concept of quantum computation from time-series microarray datasets. The obtained results indicate that the inferred GRNs are generally better with respect to all the considered metrics. The proposed formalism requires a lesser number of iterations to train the RNN model parameters, reducing the time complexity of the computation. Thus, this formalism can be suitable for modelling the network dynamics of bigger networks effectively. This novel methodology for reverse engineering of GRN algorithm does not, however, sacrifice much biological plausibility in the inference process. The proposed methodology can be implemented for large-scale GRN construction, as it has the potential to reduce the computational time requirement significantly, hopefully, without sacrificing the accuracy to a considerable extent.

References

1. McLachlan, G., Do, K.-A., Ambroise, C.: Analysing Microarray Gene Expression Data. Wiley, Hoboken (2005)
2. Bar-Joseph, Z.: Analysing time series gene expression data. Bioinformatics **20**(16), 2493–2503 (2004)
3. Nielsen, M.A., Chuang, I.L.: Quantum Computation and Quantum Information. Cambridge University Press, New York (2010)
4. Benioff, P.: The computer as a physical system: a microscopic quantum mechanical hamiltonian model of computers as represented by turing machines. J. Stat. Phys. **22**(5), 563–591 (1980)
5. Manin, Y.: Computable and Uncomputable, p. 128. Sovetskoye Radio, Moscow (1980)
6. Feynman, R.P.: Simulating physics with computers. Int. J. Theor. Phys. **21**(6/7), 467–488 (1982)
7. Deutsch, D.: Quantum theory, the church-turing principle and the universal quantum computer. Proc. Roy. Soc. Lond. A Math. Phys. Eng. Sci. **400**(1818), 97–117 (1985). The Royal Society
8. Vohradsky, J.: Neural model of the genetic network. J. Biol. Chem. **276**(39), 36168–36173 (2001)
9. Kentzoglanakis, K., Poole, M.: A swarm intelligence framework for reconstructing gene networks: searching for biologically plausible architectures. IEEE/ACM Trans. Comput. Biol. Bioinform. **9**(2), 358–371 (2012)

10. Schaffter, T., Marbach, D., Floreano, D.: GeneNetWeaver: in silico benchmark generation and performance profiling of network inference methods. Bioinformatics **27**(16), 2263–2270 (2011)
11. Marbach, D., Schaffter, T., Mattiussi, C., Floreano, D.: Generating realistic in silico gene networks for performance assessment of reverse engineering methods. J. Comput. Biol. **16**(2), 229–239 (2009)
12. Wolpert, D.H., Macready, W.G.: No free lunch theorems for optimization. IEEE Trans. Evol. Comput. **1**(1), 67–82 (1997)
13. D'haeseleer, P.: Reconstructing gene networks from large-scale gene expression data. Ph.D. dissertation, the University of New Mexico (2000)
14. Bolouri, H., Davidson, E.H.: Modelling transcriptional regulatory networks. BioEssays **24**(12), 1118–1129 (2002)
15. Xu, R., Wunsch II, D., Frank, R.: Inference of genetic regulatory networks with recurrent neural network models using particle swarm optimization. IEEE/ACM Trans. Comput. Biol. Bioinform. **4**(4), 681–692 (2007)
16. Khan, A., Datta, P., Pal, R.K., Saha, G.: Gene regulatory networks using bat algorithm inspired particle swarm optimization. In: 2015 IEEE International WIE Conference on Electrical and Computer Engineering (WIECON-ECE), pp. 387–390. IEEE (2015)

Analysis of Soil Data from Eastern of Morocco Based on Data Mining Process

Imane Belabed, Mohammed Talibi Alaoui[(✉)], Abdelmajid Belabed,
and Youssef Talibi Alaoui

Faculty of Sciences, Mohamed I University, Oujda, Morocco
talibialaouim@yahoo.fr

Abstract. Agronomic and agricultural research in East Morocco depends on time and space evolution of the soil, which is an important factor, otherwise, limiting plant cover.

Thus, a chemical and physicochemical soil study was assisted by advanced techniques such as data mining. Today, data exploitation on a global scale is used in a large number of vast agribusiness operating areas. Products of computer operating systems and specific domain data extrapolation are applicable across the disciplines; however they are still relatively new on agriculture.

This work aims to evaluate the feasibility of the conversion of wheat crops into olive crops, based on the evaluation of the soil's ability of the region to withstand the conversion, which is an axis of the Green Morocco Plan for the horizon of 2020.

This paper presents our approach to soil data analysis in order to determine the usefulness of the execution of this strategy, in the eastern of Morocco and its corresponding results.

Keywords: Data mining · Clustering of variables · Wheat-olive reconversion · Association rules

1 Introduction

Currently Data Mining is an area of recent and crucial approach in the world of bioinformatics research and agro-informatics. These techniques are useful to raise a significant and usable knowledge that can be applied by many researchers. Data mining consists of various methodological programs that are mainly produced and used for real decisions of the sectors involved [1]. These techniques are well prepared for their respective areas of knowledge. The use of statistical analysis is both tedious and precise.

In this work, we used our data analysis approach, which includes both the clustering of variables and association rules, to highlight the usefulness of reconversion of wheat crops into olive crops, by answering the following questions: Are the wheat and the olive crops depend on the same elements of soil in order to carry out the conversion without degrading the soil? Also, will we obtain a better yield with the conversion given the history of data that is available?

© Springer International Publishing AG 2017
I. Rojas and F. Ortuño (Eds.): IWBBIO 2017, Part II, LNBI 10209, pp. 234–244, 2017.
DOI: 10.1007/978-3-319-56154-7_22

- Field of study.

We chose to work with data from the ground of eastern Morocco namely 74 plots before data-preprocessing step, with 12 variables being: (SA, TA, C, N, MO, AR, CA, CT, E, K, P, PH) (Table 1).

Table 1. Soil study symbols.

C	Carbon
N	Azote
MO	Magnesium
SA	Salinity
LI	Litium
AR	Argile
CT	Total carbon
CA	Active carbon
PH	PH
P	Phosphore
K	Potassium
E (1/5)	Extract of saturation of soil in water and minerals diluted to 1/5

2 Methodology

In this article, we opted for a data analysis approach that includes both the clustering of variables and association rules, in order to reduce firstly the number of variables by grouping them into clusters, then look for causal rules among our target variables (wheat yields and olive yields) and the previously established groups of variables. This, after the step of discretization of variables, since the association rule requires that the variables of study should be qualitative.

- Our data analysis approach.

   In this section we will detailed our data mining approach step by step.

2.1 Data Pre-processing

We proceeded to the pre-processing of data, which consists of cleaning the noise information, missing values and scaling outliers by normalizing data with the formula (x-x (average)), the details of this manipulation is in the paragraph results.

2.2 Clustering of Variables

The interest of adopting the clustering of variables, despite having a number of variables equal to 12 is that, this work will be the object for another, which aims to make this clustering sequential and dynamic. Thus, we need to have previously the groups of variables formed.

For clustering of variables we used the following concept: to represent a group of variables, the "average" variable in some way, we use a latent variable that is the first factor of the principal component analysis PCA [2]. On this basis, a variable group is represented by the first axis of the PCA.

In this section of clustering we have used three algorithms of clustering adapted for variables to compare their output results. Namely: Hierarchical clustering, clustering with the k-means algorithm and clustering with VARCLUS algorithm.

The Algorithm of Hierarchical Clustering. The algorithm takes a hierarchical approach. The process of building is the hierarchical ascending clustering. At each step, we merge the two groups generating the smallest loss of variability explained and quantified by the difference between the sum of their own first values and the first value of the formed group [8].

Varclus Algorithm. VARCLUS is a top down approach based on a criterion of division of a group of variables into two clusters. The algorithm stops if no cluster can be divided, thus partition the set of variables is divided into a number of disjoint K clusters [8].

K-means Algorithm. K-means for the clustering of variables is a variant of the re-allocation method, adapted to variables. It is still based on the principle of latent components. We set the number of groups that are formed randomly at first. The variables are iteratively assigned to the nearest group, under the square of the correlation with the first factor, until there is convergence. The criterion is to maximize the total variability explained [8].

2.3 Discretization of Variable Clusters

In this section, we first proceeded to the determination of number of ideal classes to be discretized, by comparing different formulas, and then proceeded to discretization using method of equal intervals.

Selection of Number of Clusters. The discretization is to cut clusters in a series of qualitative or quantitative data. In this work, we have discretized quantitative variables such as pretreatment for association rules. The number of clusters was selected by comparing several clues to know the number of ideal clusters for a distribution:

- The method of Brooks-Carruthers [7]:

$$K \leq 5 \log 10 \, n$$

The optimum number of clusters should be less than 9, 35.

- The method of Scott [7]:

Scott's method is supposed to be more accurate, because it involves 'a' minimum and maximum data 'b', and also use other dispersion parameters such as standard deviation.

Scott's method was adopted, since it respects the nature of the distribution of data, also because previous methods gave empty clusters of variables (Table 2).

Table 2. Number of cluster per variable.

Variables	Number of clusters suggested per variable
AR	4,8
Li	4,8
Olive yield	4,36
Wheat yield	3,76
SA	4,67
CT	6,35
CA	5,76
E(1/5)	7,29
PH	4,84
C	4,73
MO	4,52
P	9,30
K	7,97
Arithmetic average	5,6

Discretization Algorithm. Among the discretization methods, we chose the method at equal intervals, i.e. extended clusters are of equal value (constant intervals) [6].

2.4   Association Rules

Association rules are intended to begin specific intelligible relationships between attributes. An association rule is an implication $C1 \rightarrow C2$, where C1 and C2 express conditions on attributes.

In this work we chose to use a supervised method for association rules between target variables and each cluster, taking into account the correlation between class 2 and class 3 (Fig. 2) [9].

3   Results and Interpretation

In this part we will present the corresponding results of each step of our data analysis process (Fig. 1).

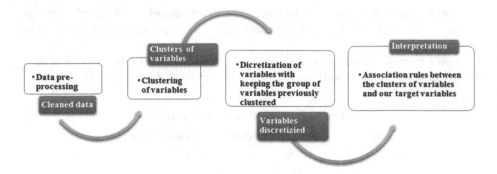

Fig. 1. Our data mining approach.

3.1 Data Pre-processing

In this section, the data was normalized in order to decrease the number of outliers from 29 to 6 after normalization, with the formula (x-x (average)) and then, outliers were removed (Table 3).

Table 3. Outliers before and after normalization.

Number of outliers before normalization	Number of outliers after normalization
29	6

3.2 Clustering of Variables

This paragraph will illustrate the output result of each clustering algorithm, by starting with the algorithm of hierarchical clustering, then the varclus algorithm and finally the k-means algorithm.

The Algorithm of Hierarchical Clustering

Cluster summary

Table 4. List of variable by cluster.

Cluster	Number of members	Variation explained	Proportion explained
1	2	1,8302	0,9151
2	8	5,6867	0,7108
3	2	1,3230	0,6615
Total		8,8399	0,7367

The above table (Table 4) lists the clusters that were built, the number of variables in each cluster, the eigenvalue of the first factor of The PCA, in each group and the proportion of variability explained inside the group. Having, a value of this last near to 1 indicates that, all variability is summarized by the first PCA factor in this group.

Furthermore the sum of the proportion of variability explained indicates the overall quality of the partitioning, which is significant in our case 73, 6%.

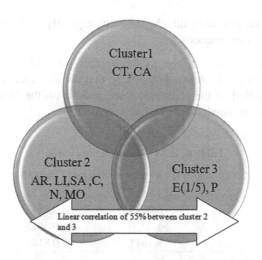

Fig. 2. The variable members of each cluster.

This figure (Fig. 2) illustrates the variables contained in each cluster, and specifies that there is a linear correlation of 55% between cluster 2 and cluster 3. This correlation will be considered in the step of association rules.

Table 5. The correlation of variables with their own cluster.

Attribute	Cluster 1	Cluster 2	Cluster 3
Ar (%)	0,0886	0,7959	0,3574
Li (%)	0,4247	0,8576	0,4548
Sa (%)	-0,2941	-0,9526	-0,4573
CT (%)	0,9566	0,1749	0,2572
CA (%)	0,9566	0,2427	0,1841
E1/5(mg/l)	0,3306	0,4703	0,8133
pH	0,0995	-0,5828	-0,3046
C (%)	0,1811	0,9371	0,6089
N (%)	0,1643	0,9504	0,6417
MO	0,1767	0,9377	0,6113
P	0,0446	0,4253	0,8133
K (ppm)	0,1452	0,6403	0,1403

Cluster correlations – Structure. The above table (Table 5) expresses the significant correlation between each variable and its own cluster, which conclude the good quality of clustering. Nevertheless, this table shows us that, the correlations of PH variable and

K (ppm) variable are low with all clusters; consequently, we removed PH and K variables in the following sections.

Varclus. In this section, at each step, to subdivide a group of variables, we compute the first two factors of the PCA, rotate them, so that the alignment of the variables on the axes is more decided. If the eigenvalue associated with the second axis is greater than 1, the variables are subdivided.

Table 6. List of variable by cluster.

Cluster	Number of members	Variation explained	Proportion explained
1	2	1,8302	0,9151
2	6	5,0079	0,8347
3	2	1,3230	0,6615
Total		8,1611	0,8161

Cluster summary. The above table (Table 6) indicates the repartition of variables into three clusters as the hierarchical clustering algorithm, even with the same variables by cluster with variability explained proportion of 81.6%.

Table 7. Correlations of variables with all clusters.

Attribute	Cluster 1	Cluster 3	Cluster 2
Ar (%)	0,0886	0,3574	0,7807
Li	0,4247	0,4548	0,8664
Sa (%)	-0,2941	-0,4573	-
			0,9470
CT (%)	0,9566	0,2572	0,1994
CA (%)	0,9566	0,1841	0,2650
E1/5 (mg/l)	0,3306	0,8133	0,4979
C (%)	0,1811	0,6089	0,9552
N(%)	0,1643	0,6417	0,9612
MO	0,1767	0,6113	0,9564
P	0,0446	0,8133	0,4405

Cluster correlations – Structure. The above table (Table 7) expresses the significant correlation between each variable and its own cluster, which conclude the good quality of clustering.

K-means Algorithm. In this part, the group of variables is represented by the first axis of the PCA. Each variable is iteratively assigned to the group which is closest to it, in the sense of the square of the correlation. After passing all variables, the main components are updated. The process continues until there is no improvement in the partition, that is, the sum of the eigenvalues of the first factorial axis associated with each group.

Table 8. List of variable by cluster.

Cluster	Number of members	Variation explained	Proportion explained
1	2	1,8302	0,9151
2	6	5,0079	0,8347
3	2	1,323	0,6615
Total		8,1611	0,8161

Cluster Summary. The above table (Table 8) indicates the repartition of variables into three clusters, as the hierarchical clustering algorithm and varclus algorithm with variability explained proportion of 81.1%.

Table 9. Correlation of variables with all clusters.

Attribute	Cluster 2	Cluster 1	Cluster 3
Ar	0,7807	0,0886	0,3574
Li (%)	0,8664	0,4247	0,4548
Sa (%)	-0,9470	- 0,2941	-0,4573
CT	0,1994	0,9566	0,2572
CA (%)	0,2650	0,9566	0,1841
E1/5	0,4979	0,3306	0,8133
C (%)	0,9552	0,1811	0,6089
N(%)	0,9612	0,1643	0,6417
MO	0,9564	0,1767	0,6113
P	0,4405	0,0446	0,8133

Cluster correlations – Structure. The above table (Table 9) expresses the significant correlation between each variable and its own cluster, which conclude the good quality of clustering.

To summarize, in this part we executed the clustering of variables with three algorithms, in order to compare the repartition of variables with each one. As result we obtained the same clusters of variables with the three algorithms. So, we will use this clusters as an input for the next step which is association rules. This, after transforming all variables into qualitative variables, because association rules require qualitative variables as an input.

3.3   Association Rules

In this part, we infer association rules between each cluster and yield of olive crops and wheat crops.

Association rule between cluster 2 and yield of olive trees and wheat. In this part we tried to infer rules between the yield of the olive trees and Cluster 2 (which groups C, N, MO, AR, Li, and SA) and we found the results below:

- The yield of the olive tree is less than 1.32 if MO < 0,944 and C < 0,458 (Table 10).

Table 10. Association rule between cluster 2 and yield of olive trees.

Antecedent	Length	Support	Confidence (%)	Lift
MO < 0,944 and C < 0,458	1	0,392	100	1,321

- The wheat yield is below 11.80 if MO < 0, 944 (Table 11).

Table 11. Association rule between cluster 2 and yield of wheat.

Antecedent	Length	Support	Confidence (%)	Lift
MO < 0,944	1	0,365	79,4	1,199

We deduce from this part that the yield of wheat and olive trees does not depend on the same soil elements, when the yield of the olive crops depends on the carbon (C) and Magnesium (MO), wheat yield depends just on Magnesium (MO).

Association rule between cluster 2 and 3 combined and yield of olive trees and wheat. We chose to combine class 2 and 3 since the correlation between these clusters is significant, and also because there is no association rule taking into account just cluster 3 which contains (E1/5 and P).

- The yield of the olive tree is less than 1.32 if:
  - E1/5 < 0,456 and MO < 0,944 and C < 0,458 or
  - C < 0,458 and E1/5 < 0,456 and P < 49, 51 or
  - E1/5 < 0,456 and P < 49, 51 and MO < 0,944.

We deduce from the table above (Table 12) that the yield of the olive tree also depends on E, P and MO.

Table 12. Association rule between cluster 2 and 3 combined and yield of olive trees.

Antecedent	Length	Support	Confidence (%)	Lift
E1/5 < 0,456 and MO < 0,944 and C < 0,458	3	0,378	100	1,321
C < 0,458 and E1/5 < 0,456 and P < 49,51	3	0,378	100	1,321
E1/5 < 0,456 et P < 49,51 and MO < 0,944	3	0,432	100	1,321

- The wheat yield is below 11.80 if E1/5 < 0,456 and P < 49, 51 and MO < 0,944.

We deduce from the table above (Table 13) that the yield of wheat also depends on E, P and MO.

Table 13. Association rule between cluster 2 and 3 combined and yield of wheat.

Antecedent	Length	Support	Confidence (%)	Lift
E1/5 < 0,456 and P < 49,51 and MO < 0,944	3	0,351	81,3	1,227

We conclude from this part, that the yield of olive crops and wheat crops do not depends on the same soil's element, when the olive crops depends on carbon (C), magnesium (MO), Potassium (P) and E (1/5) (extract of saturation of soil in water and minerals diluted to 1/5), the yield of wheat depends only on E (1/5), magnesium (MO), and Potassium (P).

Therefore, from what has been stated, we deduce that the olive crops needs more soils elements than the wheat crops, so the plots cultivated by wheat cannot support the olive crops. Consequently, the conversion of wheat crops into olive crops is not beneficial neither for soil, nor for crops.

Association rules between the yield of wheat and olives.

- The yield of the olive tree is less than 1.32 if the wheat yield is less than 11, 80.

Taking into account the soil's elements we deduce from the above table that if wheat yield is less than 11, 80 then the yield of the olive tree will be less than 1.32 (Table 14).

Table 14. Association rule between the yield of olive and wheat.

Antecedent	Length	Support	Confidence (%)	Lift
The wheat yield is less than 11, 80	1	0,662	100	1,321

Therefore, we conclude that the conversion of wheat crops into olive crops will not be gainful.

4 Conclusion

In this work, we used a process of data mining that relies on the clustering of variables and association rules, to extract causal rules between soil variables of eastern Morocco and the yield of wheat and olives crops, in order to evaluate the usefulness of the conversion of wheat crops into olive crops, which is part of strategy of the Green Morocco Plan for the horizon of 2020.

Based on components of the ground, it was concluded that the conversion of the wheat crops into olive crops will not be beneficial neither for soil, nor for crops. So, the conversion will not be gainful.

As perspectives, we intend to add several parameters such as, the rainfall and the investment required for each crop in order to evaluate globally this part of the strategy of Green Morocco Plan for the horizon of 2020.

References

1. Kumar A., Kannathasan, N.: A survey on data mining and pattern recognition techniques for soil data mining. IJCSI. Int. J. Comput. Sci. Issues 8(3) (2011)
2. Vigneau, E., Qannari, E.M., Sahmer, K., Ladiray, D.: Classification de variables autour de composantes latentes. Revue de statistique appliquée 54(1), 27–45 (2006)
3. Cunningham, S., Holmes, G.: Developing innovative applications in agriculture using data mining. Department of Computer Science University of Waikato Hamilton, New Zealand, Technical report (1999)
4. Vamanan, R., Ramar, K.: Classification of agricultural land soils a data mining approach. Int. J. Comput. Sci. Eng. 3 (2011). ISSN: 0975-3397
5. Bhargavi, P., Jyothi, S.: Soil classification using data mining techniques: a comparative study. Int. J. Eng. Trends Technol. 2, 55 (2011)
6. Dougherty, J., Kohavi, R., Sahami, M.: Supervised and unsupervised discretization of continuous features (1995)
7. Découpages en classes et discrétisation, Gilles Hunault, Université d'Angers
8. Nakache, J.P., Confais, J.: Approche Pragmatique de la Classification, TECHNIP, chapitre 9, pp. 219–239 (2005)
9. Tanagra Tutoriels: Association rules (2008)

A Multi-objective Optimization Framework for Multiple Sequence Alignment with Metaheuristics

Cristian Zambrano-Vega[1], Antonio J. Nebro[2(✉)], José García-Nieto[2], and José F. Aldana-Montes[2]

[1] Facultad de Ciencias de la Ingeniería, Universidad Técnica Estatal de Quevedo, Quevedo, Los Ríos, Ecuador
czambrano@uteq.edu.ec
[2] Edificio de Investigación Ada Byron, University of Málaga, Málaga, Spain
{antonio,jnieto,jfam}@lcc.uma.es
http://www.uteq.edu.ec/
http://khaos.uma.es/

Abstract. The alignment of more than two biological sequences is a widely used technique in a number of areas of computational biology. However, finding an optimal alignment has been shown to be an NP-complete optimization problem. Furthermore, Multiple Sequence Alignment (MSA) can be formulated according to more than one score function, leading to multi-objective formulations of this problem. Due to these reasons, metaheuristics have been proposed to deal with MSA problems. In this paper, we present jMetalMSA, an Open Source software tool for solving MSA problems with multi-objective metaheuristics. Our motivation here is to offer to the scientific community in computational biology, a platform including state-of-the-art optimization algorithms aimed at solving different formulations of the MSA. We describe the main features of jMetalMSA, including the metaheuristics and scores that are currently available. In addition, we show a working example for illustration purposes.

Keywords: Multiple Sequence Alignment · Computational biology · Multi-objective optimization · Metaheuristics

1 Introduction

A Multiple Sequence Alignment (MSA) is an alignment of three or more biological sequences (ADN, ARN, protein) with the goal of identifying regions of similarity that may be a consequence of functional, structural, or evolution relationships among the sequences. MSA has many applications in field of computational biology, such as phylogenetic tree estimation, secondary structure prediction, and critical residue identification.

Finding an MSA can be defined as an optimization problem where a scoring function is to be maximized (or minimized). MSA is an NP-complete optimization problem [1], where the time complexity of finding an optimal alignment raises

© Springer International Publishing AG 2017
I. Rojas and F. Ortuño (Eds.): IWBBIO 2017, Part II, LNBI 10209, pp. 245–256, 2017.
DOI: 10.1007/978-3-319-56154-7_23

exponentially along with the number of sequences and their lengths. Additionally, there is not a unique way to assess the quality of an MSA, so the problem can have a multi-objective formulation when the goal is to find solutions according to two or more scoring functions. These reasons have led researchers to investigate on the adoption of multi-objective metaheuristics to deal with them [2,3].

Metaheuristics [4] are a family of non-exact, stochastic optimization techniques that have proven to be very effective in finding high quality solutions (optimal or quasi-optimal) in many fields. When they are used to solve a multi-objective problem, instead of searching for a unique solution, the goal is to find the set of optimal solutions, known as the Pareto optimal set (its representation in the objective space is known as the Pareto front), which fulfills the property that there is not any other possible solution improving any of those in the set in all the objectives. The output of a multi-objective algorithm is an approximation to the Pareto optimal set.

Popular multi-objective metaheuristic algorithms include evolutionary algorithms, particle swarm optimization, ant colony optimization, artificial immune systems, and many others [5,6]. Most of proposals of multi-objective metaheuristics applied to MSA are based on evolutionary algorithms [2,3,7–9], but other techniques have been also applied, including artificial bee colony [10] and bacterial foraging optimization [11].

The motivation of our work is that, with a few exceptions (e.g., [3]), the source code of these works is not publicly available, thus hindering interested researchers to apply the proposed algorithms to solve their problems. Our goal is to fill this gap by offering an Open Source software framework, called jMetalMSA, aimed at multi-objective optimization with metaheuristics. jMetalMSA is implemented in Java and includes a number of multi-objective algorithms that are representative of the state-of-the-art, several MSA scores, and specific crossover and mutation operators. The project is publicly available in GitHub[1]. In this paper, we describe the architecture of jMetalMSA and its main features, including a practical use case for validation.

The remaining of this paper is structured as follows. Section 2 describes the architecture of jMetalMSA and its main components. The scores included in the framework are detailed in Sect. 3. Section 4 includes a complete case of use for practical validation. The conclusions and future work are outlined in Sect. 5.

2   Architecture of jMetalMSA

jMetalMSA is based on the jMetal multi-objective optimization framework [12,13], from which it takes most of the core classes. The object-oriented architecture of jMetalMSA is shown in Fig. 1, where we can observe that it is composed of four core classes (Java interfaces). Three of them (*MSAProblem*, *MSAAlgorithm*, and *MSASolution*) inherits from their counterparts in jMetal (the inheritance relationships are omitted in the diagram), and there is a class *Score* to represent a given MSA scoring function.

---

[1] jMetalMSA project in GitHub: http://github.com/jmetal/jmetalmsa.

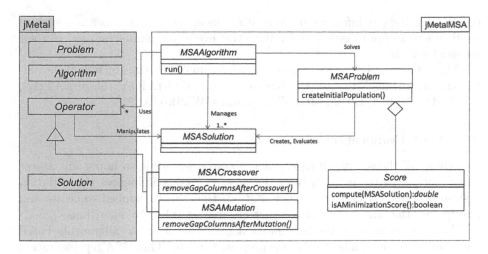

Fig. 1. Architecture of jMetalMSA (some inheritance associations have been not included to simplify the diagram).

Many proposals for solving MSA with metaheuristics include an initialization method based on taking as input a set of pre-computed alignments obtained with other non-metaheuristic methods (such as Clustal-W, MAFFT, MUSCLE, etc.), so the *MSAProblem* class includes a *createInitialPopulationMethod()* that can be used to incorporate these kinds of strategies.

An important issue in current MSA studies is that there is not a consensus about which scoring functions are more adequate to assess the quality of alignments. Multi-objective approaches in jMetalMSA try to overcome this matter to some extent by selecting more than one score, but a review of the existing proposals reveals that most of them use different problem formulations. To provide a high flexibility degree in defining an MSA problem with the particular objective functions, our approach is to have a *Score* class that represents individual MSA evaluation functions, so that a concrete MSA problem formulation is composed of a number of scores, which can be easily set when configuring the algorithms.

When a variation operator is applied to an MSA, there is the chance of getting columns full of gaps. These columns are useless and typically they are removed, but finding them is a time consuming process. Crossover and mutation are typically consecutive operations, and there also exists the possibility of applying more than one mutation operation, so removing gaps at the end of each operator can be very inefficient. The crossover and mutation operators in jMetalMSA are endowed with a method to indicate whether the removing must be carried out after the operators or not.

2.1   Algorithms

As jMetalMSA is based on jMetal, most of algorithms included in the latter can be used in the former. The codification of an MSA (see next section) is based on representing gap information, so continuous optimization algorithms such as particle swarm optimization and differential evolution cannot be used in their classical versions.

The list of metaheuristics currently available in jMetalMSA include the evolutionary algorithms NSGA-II [14], NSGA-III [15], SMS-EMOA [16], SPEA2 [17], PAES [18], MOEA/D [19], MOCell [20], and GWASF-GA [21].

2.2 MSA Codification

The choice of the scheme to represent problem solutions is an important aspect when working with metaheuristics, because it largely influences the variation operators, typically crossover and mutation, that can be applied to create and manipulate the tentative solutions that are managed by the algorithms.

We have included in the framework a codification of the alignments based on groups of gaps, similar to the one proposed in [10]. This MSA representation only stores the positions (begin, end) of the groups of gaps into the sequences, as illustrated in Fig. 2.

Fig. 2. Example of alignment (left) and how it is encoded in jMetalMSA (right).

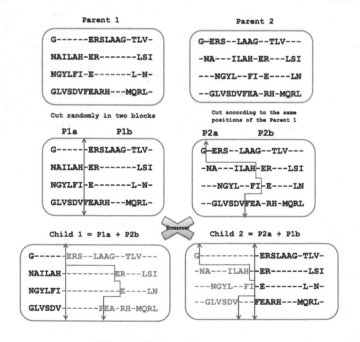

Fig. 3. Single point crossover.

This codification reduces the time of execution of both genetic crossover and mutation operators, since manipulating gap group lists is more efficient than working with large sequences of characters.

2.3   Evolutionary Operators

As commented in the previous section, many metaheuristics (particularly evolutionary algorithms) rely on crossover and mutation operators to search for the problem solutions. Most of presented techniques for MSA use a single-point

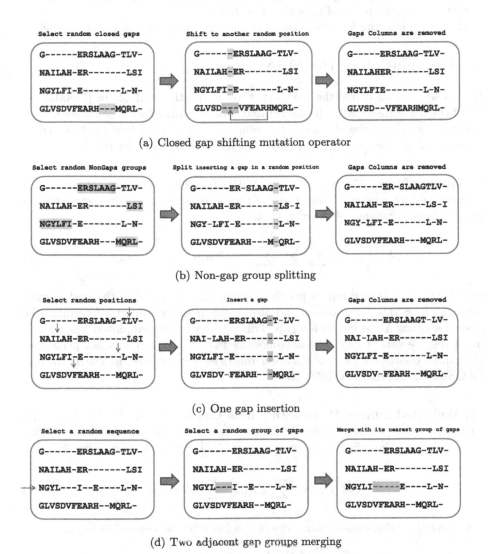

(a) Closed gap shifting mutation operator

(b) Non-gap group splitting

(c) One gap insertion

(d) Two adjacent gap groups merging

Fig. 4. List of mutation operators that are available in jMetalMSA

crossover (e.g., [3]), which is illustrated in Fig. 3. The list of mutation operators included in jMetalMSA are:

- Shift-closed gaps: Closed gaps are randomly chosen and shifted to another position (see Fig. 4a).
- Non-gap group splitting: a non-gap group is selected randomly, and it is split into two groups (see Fig. 4b).
- One gap insertion: Inserts a gap in a random position for each sequence (see Fig. 4c).
- Two adjacent gap groups merging: Selects a random group of gaps and merge with its nearest group of gaps (see Fig. 4d).
- Multiple mutation: Combination of the rest of operators into a single one.

3   Formulation of the MSA Scores

In this section, we define the scores that are currently available in jMetalMSA, all of them are intended to be maximized. For the formulation of theses scores, we have considered the alignment to evaluate S, as a set of k aligned sequences represented as $S = \{s_1, s_2, ..., s_k\}$ all of them with the same length L.

(1) Sum of Pairs. The Sum of pairs (SOP) of an alignment, presented in Eq. 1, is computed by adding all the scores of the pairwise comparisons between each residue in each column of the alignment.

$$SOP(S) = \sum_{i=1}^{k-1} \sum_{j=i+1}^{k} \sum_{c=1}^{L} ScoringMatrix(s_{ic}, s_{jc}) \qquad (1)$$

where $ScoringMatrix$ represents the matrix that determines the cost of substituting a residue for another, and also a gap penalty value to determine the cost of aligning a residue with a gap. This penalty is only employed when aligning a residue with a gap. The alignment of two or more gaps is not penalized. jMetalMSA includes two substitution matrices widely used in the literature: BLOSUM62 [22] and PAM250 [23].

(2) Weighted Sum of Pairs with Affine Gaps. The weighted sum of pairs with affine gaps (wSOP), is the difference between the sum-of-pairs with weights score (SP) and the affine gap penalty score (AGP). It is defined in Eq. 2:

$$wSOP(S) = \sum_{l=1}^{L} SP(l) - \sum_{i=1}^{k} AGP(s_i) \qquad (2)$$

where $SP(l)$ is the sum-of-pair score of the l column as defined in Eq. 3:

$$SP(l) = \sum_{i=1}^{k-1} \sum_{j=i+1}^{k} W_{i,j} \; x \; \delta(s_{i,l}, s_{j,l}) \qquad (3)$$

In Eq. 3, δ represents the substitution matrix used, and $W_{i,j}$ is the sequence weight between two unaligned sequences s'_i and s'_j, as follows:

$$W_{i,j} = 1 - \frac{LD(s'_i, s'_j)}{max(|s'_i|, |s'_j|)} \tag{4}$$

where LD represents the Levenshtein distance between two unaligned sequences s'_i and s'_j (the minimum number of insertions, deletions or substitutions required to change one sequence into the other).

Finally, in the Eq. 2, $AGP(s_i)$ represents the affine gap penalty score of the sequence s_i, which is defined as follows:

$$AGP(s_i) = (g_{open} \; x \; \#gaps) + (g_{extend} \; x \; \#spaces) \tag{5}$$

where g_{open} and g_{extend} are user-defined and represent the weight to open and the weight to extend the gap with one or more spaces, respectively. In jMetalMSA, the default values of theses parameters are: $g_o = 6$ y $g_e = 0.85$.

(3) Single sTRucture Induced Evaluation (STRIKE). Strike [24] is a metric to evaluate the accuracy of an alignment based on structural information of, at least, one sequence of the alignment. This structural information is commonly retrieved from the Protein Data Bank (PDB) [25].

Using the structural information as a source for amino acid frequencies and contacts, a log-odds contact matrix is estimated by measuring the ratio between the frequency of each possible contact and its expectation, given the background frequency of each single amino acid. Given any pair of amino acids i and j, the score for their contacts can be estimated as follows (Eq. 6):

$$M_{ij} = 10 \times \ln(\frac{f_{ij}}{f_i f_j}) \tag{6}$$

where f_{ij} is the frequency of contacts involving i and j across all observed $residue - residue$ contacts, f_i and f_j are the single residue frequencies in the dataset considered.

For the remaining sequences, the pairs of amino acids aligned in the same positions as the previously estimated contacts are retrieved. Such pairs of amino acids are then scored according to a novel scoring matrix. If there are several structures, the final STRIKE score is the average of all the Strike core of each structure.

(4) Percentage of Totally Conserved Columns. The percentage of Totally Conserved Columns (TC) takes into account the number of columns that are fully aligned with exactly the same compound. Maximizing this score ensures more conserved or special regions within the alignment. It can be defined as follows (Eq. 7):

$$TC(S) = 100 \sum_{l=1}^{L} \frac{totalColumn(S_l)}{L} \qquad (7)$$

where S_l is the lth column of S, such that $S_l = s_{il} \ \forall i = 1, ..., k$, and the function $totalColumn(S_l)$ is defined as follows (Eq. 8):

$$totalColumn(S_l) = \begin{cases} 1 & if \ s_{il} = s_{1l} \ \forall i = 2, ..., k \\ 0 & otherwise \end{cases} \qquad (8)$$

(5) Percentage of Non-gaps. The percentage of non-gaps measures the number of residues with respect to the number of gaps into the alignment, as defined in Eq. 9:

$$NonGaps(S) = 100 \sum_{i=1}^{k} \sum_{j=1}^{L} \frac{isNotGap(s_{ij})}{k * L} \qquad (9)$$

where s_{ij} represents the symbol in the j-th position of the i-th sequence in the alignment S. The function $isNotGap$ for a general residue in the alignment is defined as follows (Eq. 10):

$$isNotGap(residue) = \begin{cases} 1 & if \ residue = \text{``}-\text{''}(gap) \\ 0 & otherwise \end{cases} \qquad (10)$$

4   Examples of Use with 2 and 3-Objectives Formulation

To illustrate the use of jMetalMSA, we describe in this section two cases of use in which three multi-objective algorithms are applied to deal with MSA problems of the BAliBASE 3.0 dataset [26].

First, we used MOCell, a multi-objective cellular evolutionary algorithm [20], to solve the BB11001 instance, with the goal of optimizing three objectives: sum of pairs (SOP), TC, and percentage of non-gaps. Once the algorithm has been configured and executed, two files containing the solutions found in terms of approximations to the Pareto optimal set and the Pareto front are generated, so that the first one contains the alignments (in FASTA format) and the second one stores the corresponding score values. Figure 5 shows the Pareto front points that have been obtained when plotting the second file. The biologist then has the choice of selecting her/his preferred alignment. In this regard, Fig. 6 shows the alignments corresponding to the three extreme solutions (the best one per objective). We can observe that the alignment with the best TC score contains the higher number of totally aligned columns, whereas the alignment with the best NonGaps score contains a minor number of gaps into the sequences than the others.

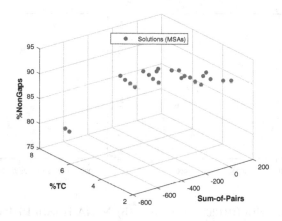

Fig. 5. Pareto front approximation obtained by the algorithm MOCell when solving the BB11001 problem of BAliBASE 3.0. The objectives are: sum of pairs, %TC and percentage of non-gaps.

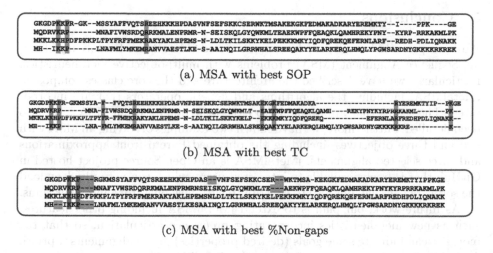

(a) MSA with best SOP

(b) MSA with best TC

(c) MSA with best %Non-gaps

Fig. 6. Extreme solutions of the front obtained (in Fig. 5) when solving the BB11001 problem instance.

The second example consists of applying the algorithms GWASF-GA [21] and NSGA-II [14] to align the BB12001 problem, but using a two-objective formulation: SOP and TC. Figure 7 shows the Pareto front approximations that have been obtained. We can observe that NSGA-II provides a most populated and diverse front of solutions, but GWASF-GA produces a solution with a low SOP value. More detailed algorithmic comparisons with jMetalMSA and the BAliBASE 3.0 dataset can be found in [27,28].

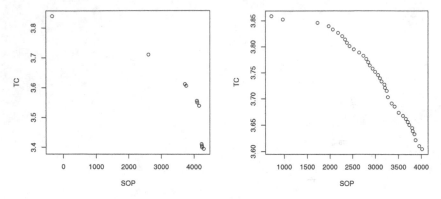

(a) GWASF-GA with BB12001 instance   (b) NSGA-II with BB12001 instance

Fig. 7. 2 objectives Pareto front approximation generated by jMetalMSA using the algorithms NSGAII and GWASFGA resolving two instances of the BAliBASE v3.0.

5   Conclusions

We have presented a software framework called jMetalMSA for solving Multiple Sequence Alignment (MSA) problems with multi-objective metaheuristics. Particularly, we have described the architecture and the core classes composing jMetalMSA, including the algorithms and genetic operators that are incorporated. The scores that are available to assess the quality of an MSA have also been detailed. Furthermore, we have shown two examples of solving an MSA with two and three objectives, including the obtained Pareto front approximations and three selected alignments. jMetalMSA is an Open Source project hosted in GitHub that is freely available to interested users, which can not only download the software package, but also to contribute with new operators and algorithms.

As future work, our plan is to extend jMetalMSA including other features, such as new metaheuristics incorporating preference articulation, so that the biologist can indicate some goals (desired properties) of the alignments a-priori. A study with extensive comparatives and using different combinations of scores is also an ongoing work.

Acknowledgements. This work has been partially supported by the Secretaría Nacional de Educación Superior Ciencia y Tecnología SENESCYT from Ecuador, and Spanish Grants TIN2014-58304-R (Ministerio de Economía y Competitividad), P11-TIC-7529 and P12-TIC-1519 (Plan Andaluz I+D+I - Junta de Andalucía). José García-Nieto is recipient of a Post-Doctoral fellowship of "Captación de Talento para la Investigación" at Universidad de Málaga.

References

1. Wang, L., Jiang, T.: On the complexity of multiple sequence alignment. J. Comput. Biol. **1**, 337–348 (1994)

2. Seeluangsawat, P., Chongstitvatana, P.: A multiple objective evolutionary algorithm for multiple sequence alignment. In: Proceedings of the 7th Conference, GECCO 2005, pp. 477–478. ACM, New York (2005)
3. Ortuño, F., Valenzuela, O., Rojas, F., Pomares, H., Florido, J., Urquiza, J., Rojas, I.: Optimizing multiple sequence alignments using a genetic algorithm based on three objectives: structural information, non-gaps percentage and totally conserved columns. Bioinformatics **29**(17), 2112–2121 (2013). Oxford, New York
4. Blum, C., Roli, A.: Metaheuristics in combinatorial optimization: overview and conceptual comparison. ACM Comput. Surv. **35**(3), 268–308 (2003)
5. Deb, K.: Multi-objective Optimization Using Evolutionary Algorithms. Wiley, New York (2001)
6. Coello, C., Lamont, G., van Veldhuizen, D.: Multi-objective Optimization Using Evolutionary Algorithms, 2nd edn. Wiley, New York (2007)
7. Soto, W., Becerra, D.: A multi-objective evolutionary algorithm for improving multiple sequence alignments. In: Campos, S. (ed.) BSB 2014. LNCS, vol. 8826, pp. 73–82. Springer, Cham (2014). doi:10.1007/978-3-319-12418-6_10
8. Kaya, M., Sarhan, A., Abdullah, R.: Multiple sequence alignment with affine gap by using multi-objective genetic algorithm. Comput. Methods Programs Biomed. **114**(1), 38–49 (2014)
9. Zhu, H., He, Z., Jia, Y.: A novel approach to multiple sequence alignment using multiobjective evolutionary algorithm based on decomposition. IEEE J. Biomed. Health Inform. **20**(2), 717–727 (2016)
10. Rubio-Largo, A., Vega-Rodríguez, M., González-Álvarez, D.: Hybrid multiobjective artificial bee colony for multiple sequence alignment. Appl. Soft Comput. **41**, 157–168 (2016)
11. Rani, R.R., Ramyachitra, D.: Multiple sequence alignment using multi-objective based bacterial foraging optimization algorithm. Biosystems **150**, 177–189 (2016)
12. Durillo, J., Nebro, A.: jMetal: a Java framework for multi-objective optimization. Adv. Eng. Softw. **42**(10), 760–771 (2011)
13. Nebro, A., Durillo, J.J., Vergne, M.: Redesigning the jMetal multi-objective optimization framework. In: Proceedings of the GECCO Companion 2015, pp. 1093–1100. ACM, New York (2015)
14. Deb, K., Pratap, A., Agarwal, S., Meyarivan, T.: A fast and elitist multiobjective genetic algorithm: NSGA-II. IEEE Trans. Evol. Comput. **6**(2), 182–197 (2002)
15. Deb, K., Jain, H.: An evolutionary many-objective optimization algorithm using reference-point-based nondominated sorting approach, Part I: Solving problems with box constraints. IEEE Trans. Evol. Comput. **18**(4), 577–601 (2014)
16. Beume, N., Naujoks, B., Emmerich, M.: SMS-EMOA: multiobjective selection based on dominated hypervolume. Eur. J. Oper. Res. **181**(3), 1653–1669 (2007)
17. Zitzler, E., Thiele, L.: Multiobjective evolutionary algorithms: a comparative case study and the strength Pareto approach. IEEE Trans. Evol. Comput. **3**(4), 257–271 (1999)
18. Knowles, J.D., Corne, D.W.: Approximating the nondominated front using the Pareto archived evolution strategy. Evol. Comput. **8**(2), 149–172 (2000)
19. Zhang, Q., Li, H.: MOEA/D: a multiobjective evolutionary algorithm based on decomposition. IEEE Trans. Evol. Comput. **11**(6), 712–731 (2007)
20. Nebro, A., Durillo, J., Luna, F., Dorronsoro, B., Alba, E.: MOCell: a cellular genetic algorithm for multiobjective optimization. Int. J. Intell. Syst. **24**(7), 723–725 (2009)

21. R. Saborido, A.R., Luque., M.: Global WASF-GA: an evolutionary algorithm in multiobjective optimization to approximate the whole Pareto optimal front. Evol. Comput. (2016, in Press)
22. Henikoff, S., Henikoff, J.: Amino acid substitution matrices from protein blocks. Proc. Natl. Acad. Sci. **89**(22), 10915–10919 (1992)
23. Dayho, M., Schwartz, R., Orcutt, B.C.: A model of evolutionary change in proteins. Atlas Protein Sequences Struct. **5**, 345–352 (1978)
24. Kemena, C., Taly, J., Kleinjung, J., Notredame, C.: Strike: evaluation of protein MSAS using a single 3D structure. Bioinformatics **27**(24), 3385–3391 (2011)
25. Berman, H., Westbrook, J., Feng, Z., Gilliland, G., Bhat, T., Weissig, H., Shindyalov, I., Bourne, P.: The protein data bank. Nucleic Acids Res. **28**(1), 235–242 (2000)
26. Thompson, J., Koehl, P., Poch, O.: Balibase 3.0: latest developments of the multiple sequence alignment benchmark. Proteins **61**, 127–136 (2005)
27. Zambrano-Vega, C., Nebro, A.J., Garca-Nieto, J., Aldana-Montes, J.F.: Comparing multi-objective metaheuristics for solving a three-objective formulation of multiple sequence alignment. Prog. Artif. Intell. 1–16 (2017). Springer
28. Zambrano-Vega, C., Nebro, A.J., Durillo, J.J., Garca-Nieto, J., Aldana-Montes, J.F.: Multiple sequence alignment with multi-objective metaheuristics. A comparative study. Int. J. Intell. Syst. Periodicals (2017, in press). Wiley

Analysis of Gene Expression Discretization Techniques in Microarray Biclustering

J.S. Dussaut, C.A. Gallo, J.A. Carballido, and I. Ponzoni[✉]

Instituto de Ciencias e Ingeniería de la Computación, Universidad Nacional del Sur,
CONICET, Bahía Blanca, Argentina
ip@cs.uns.edu.ar

Abstract. Gene expression biclustering analysis is a commonly used technique to see the interaction between genes under certain experiments or conditions. More specifically in the study of diseases, these methods are used to compare control and affected data in order to identify the involved or relevant genes. In some cases, discretization is needed for these algorithms to work correctly. In this context, the choice of the discretization method is extremely important and has a major impact on the outcome. In this work we analyze several discretization methods for Alzheimer Disease (AD) gene expression data and compare the results of a state-of-art biclustering algorithm after each discretization. The comparison reveals that biclusters obtained from discretized expression values achieve a major coverage and overall enrichment than biclusters generated from real-valued expression data. In a particular experiment, a clustering-based discretization method overcomes all competing techniques for the dataset under study, in statistical terms.

Keywords: Discretization · Data preprocessing · Microarray data analysis · Machine learning

1 Introduction

Gene expression data is broadly used for genomic analysis. In particular, microarray data and RNA sequencing technologies enable the simultaneous measurement of expression levels of thousand genes under certain experiments or conditions. These methods provide a large amount of biological data to be processed. In this context, several machine learning methods [1–7] have been used for the analysis of gene expression data resulting from these technologies. In some cases, input data needs to be discrete [3, 7, 8]. Whenever this is the case, choosing the discretization method for the gene expression data has a major impact on the execution time, parameters, and outcome of the algorithm [9].

In the analysis of microarray data, one of the most common technique used is biclustering of gene expression data [10, 11]. In the case of disease analysis, this method commonly takes a dataset containing the expression levels of genes in at least two samples: usually, there is a sample of control patients, representing those who do not have the disease present, and an affected sample, representing those patients that suffer the disease [12]. In some cases, especially on disease progression analysis, the affected samples are more than one, indicating the stages of the disease. Then, biclustering

I. Rojas and F. Ortuño (Eds.): IWBBIO 2017, Part II, LNBI 10209, pp. 257–266, 2017.
DOI: 10.1007/978-3-319-56154-7_24

methods are used to find a set of genes that have a similar behavior on a set of conditions or experiments. These methods became a popular tool for discovering local patterns on gene expression data since many biological activities are common to a subset of genes and they are co-regulated under certain conditions. The biclustering problem is NP-Hard (non-deterministic polynomial-time hard) [13] and therefore it depends on heuristics to make its performance optimal. In some cases, the discretization of the data improves the execution time of the algorithms and also improves their results [3, 14].

Discretization is a preprocessing step, usually applied to the data in order to reduce the noise [15] that occurs on the technologies used to measure the expression levels of genes. This discretization is performed by mapping the real values of the measurement into a few finite values. The main advantage of data discretization in biological knowledge is that it favors the inference of qualitative models, whereas the real or continuous values allow the inference of quantitative models [3]. Not only the noise is reduced by discretization; moreover, it leads to a better prediction accuracy [16]. The simplification of the data also makes the learning process faster, therefore yielding more compact and shorter results [14]. Discrete values are also easier to understand, use and explain [3, 14].

Choosing a discretization method for the input data has a major impact on the output, also implying some loss of information in exchange for the stated advantages. Moreover, the choice of an appropriate discretization is not a trivial task [9, 15]. The discretization method that is going to be applied, should consider the nature of the biological data and the technology involved in the measurements in order to provide the most accurate representation of the data with the least reduced loss of information. Another issue to be considered are the main features of the computational method involved in the discretization such as the parameters.

1.1   Discretization Algorithms

A complete understanding of the semantics of main discretization approaches is always required to choose the method that best suits a particular case of interest [9]. In this work we analyze several unsupervised algorithms:

1. Median: It is a simple algorithm that has a binary level of discretization, representing the 'activation' and 'inhibition' with two values, and use de median value as a threshold. It has no input parameter, and allows to use the data scope from row, column, or full matrix; this means that the discretization can be determined regarding the gene profile, the experimental conditions or the whole matrix, once the scope is determined, the median is computed to use as a threshold for the activation and inhibition, therefore performing the discretization of the data.

$$A'_{ij} = \left\{ \begin{array}{l} 1 \; if \, A_{ij} > M \\ 0 \; otherwise \end{array} \right. \tag{1}$$

Where A'_{ij} is the element on the row i and the column j of the discrete matrix, and A_{ij} is the element on the row i and the column j of the real-valued matrix; M is the median that is used as the threshold, and it can be M_{IJ}, M_{iJ}, M_{Ij}; this is, the median calculated on the whole matrix, the gene profile or the experimental conditions.

2. Mean: This algorithm is similar to the previous one, in the sense that it has a binary level of discretization, no input parameter, and the same data scope. The main difference is that uses the mean or average as a threshold.
   The equation used by this method is the same as Eq. 1, but here M is the mean, and it can be calculated on the whole matrix, the gene profile or the experimental conditions as well.

3. Mean plus Standard Deviation: This method has the aggregation to the previous one of the standard deviation, that makes the level of discretization ternary, representing the 'activation' or 'up-regulated' as above the mean, the 'inhibition' or 'down-regulated' as below the mean and the 'no-change' as the same value as the mean. The data scope is the same as the previous algorithms. In contrast, this algorithm needs an input parameter called alpha, and it is used to tune the desired deviation. This parameter lets the 'no-change' value to have a range between the mean minus alpha and the mean plus alpha, regarding also the real standard deviation of the data.

$$A'_{ij} = \begin{cases} -1 \ if \ A_{ij} < M - \alpha\mu \\ 1 \ if \ A_{ij} > M + \alpha\mu \\ 0 \ otherwise \end{cases} \tag{2}$$

Where M is the mean, α is the parameter that tunes the desired deviation, and μ is the real deviation of the data, this deviation, as well as the mean, are calculated regarding the data scope; this is on the whole matrix, the gene profile or the experimental conditions.

4. Max - X% Max: This technique has also a binary discretization level, but instead of using a threshold it uses a fixed input parameter X%, this parameter represents a fixed proportion of the maximum value under the desired data scope (row, column or whole matrix); this is, the maximum expression value observed for each gene, for each condition or for the whole matrix. This maximum is then used as a threshold applying the fixed percentage.

$$A'_{ij} = \begin{cases} 1 \ if \ A_{ij} > H(1 - X\%) \\ 0 \ otherwise \end{cases} \tag{3}$$

Where H is the maximum value of the desired data scope, it can be H_{IJ}, H_{iJ}, H_{Ij}; the maximum value for the whole matrix, for each condition or for each gene respectively.

5. Top %X: This algorithm works with an ordered list of every expression value present and performs a binary discretization using as a threshold a fixed input parameter X% as a proportion of the desired data scope (row, column or whole matrix). This is a rank-based algorithm that is applied to an ordered list of the data.

6. Equal Frequency: This method [2] is capable of performing a multilevel discretization. The input parameter is the desired level of discretization and the method tries to split the regarding the data scope (row, column or whole matrix) into the

fixed level of discretization with the constraint that each level maintains an equal frequency. This is, given a fixed amount of symbols (discretization levels), each expression value is arranged in a way that maintains the same number of data points in each symbol. They choose each 'cut-point' p_r in the following way:

$$p_{r+1} = p_r + \frac{(H - U)}{k} \tag{4}$$

Where H is the maximum value, U is the minimum value, k is the parameter that implies the level of discretization and $p_0 = U$ is the starting point. Assigning the corresponding symbol r to those values of A_{ij} that fulfill the condition $p_r < A_{ij} \leq p_{r+1}$.

7. Transitional State: This algorithm performs a binary discretization, and it does not need any input parameter. It performs a discretization using expression variations between time points, and it is generally applicable to time series data.

$$A'_{ij} = \begin{cases} 1 \ if \ A_{ij} - A_{i(j-1)} \geq 0 \\ 0 \ otherwise \end{cases} \tag{5}$$

Where $A_{i(j-1)}$ indicates the value on the previous experimental condition.

8. Soinov's method: This work [17] presents the Soinov's change state that has a ternary level of discretization and uses as an evaluation method the mean and the difference between consecutive samples. Its data scope is only the gene profile (row).

$$S_{ij} = \begin{cases} 1 \ if \ A_{ij} > X_i \\ -1 \ otherwise \end{cases} \tag{6}$$

Where X_i is the average expression level of the gene i among all the samples, and S_{ij} is the state of the gene i over the sample or condition j. With these states they construct a gene regulatory network, discretizing the genes of these interactions via decision trees.

9. Ji and Tan method: This method [18] performs a ternary level of discretization. It is based on clustering that evaluates the ratio between consecutive samples under the gene profile (row). This task is performed by computing variations between successive time instants or experimental conditions and considering these variations significant if they exceed the given threshold. They first assemble a matrix called A'' that has one less column to reflect the variations in the following way:

$$A''_{ij} = \begin{cases} \dfrac{A_{i(j-1)} - A_{ij}}{|A_{ij}|} & if \ A_{ij} \neq 0 \\ 1 \ if \ A_{ij} = 0 \ and \ A_{i(j+1)} > 0 \\ -1 \ if \ A_{ij} = 0 \ and \ A_{i(j+1)} < 0 \\ 0 \ if \ A_{ij} = 0 \ and \ A_{i(j+1)} = 0 \end{cases} \tag{7}$$

Where A'' is the auxiliary matrix, and is calculated according to the variations of the values on the matrix A. Then the matrix A'' is discretized as follows:

$$A_{ij} = \begin{cases} 1 \text{ if } A''_{ij} \geq t \\ -1 \text{ if } A''_{ij} \leq -t \\ 0 \text{ otherwise} \end{cases} \tag{8}$$

Where $t > 0$ is the given threshold.

10. K-means: This algorithm [19] allows multilevel discretization. It is also a cluster-based discretization with an input parameter k as the number of clusters. It uses the Squared Euclidean distance as a similarity measure to assemble the clusters or levels of discretization. The quality metric for the clusters is referred as WCSS (Within-Cluster Sum of Squares) and it is calculated as follows:

$$WCSS(D) = \sum\nolimits_{A_{ij} \in [p_0, p_1]} \left| A_{ij} - \mu_0 \right|^2 + \sum\nolimits_{r=1}^{k-1} \sum\nolimits_{A_{ij} \in (p_r, p_{r+1}]} \left| A_{ij} - \mu_r \right|^2 \tag{9}$$

Where μ_r is the mean of the $A_{ij} \in (p_r, p_{r+1}]$. This represents the Euclidean distance between the elements and the mean of the cluster.

The method then aims to minimize this metric performing an iterative algorithm starting with a set S of points and having k centroids around them. The metric, or distance is calculated and the clusters are formed. The next step is to recalculate the centroids and repeat the process until the stopping criterion is met.

11. Gallo's method: This method [7] performs a binary level discretization, that finds a cut-point in an ordered list that minimizes WCSS in Eq. 9. This is also a cluster-based discretization algorithm that works with a sorted list of the values to be discretized and finds the optimal cut point in order to minimize WCSS. This discretization was applied to discretize gene expression profiles in the inference of gene association rules in the following way: the gene expression profile to be discretized was sorted in a list S and then the optimal cut point that divides S into two groups, S_1 and S_2 was found, trying to minimize the variances from each group, since the discretization of the gene i is calculated as follows:

$$\min\nolimits_{S_1, S_2 \subset S}(var(S_1) + var(S_2)) \tag{10}$$

Where S is the set of sample values for the gene i, S_1 and S_2 are included in S and $S = S_1 + S_2$. Also, the two subsets of S do not share any elements and their cardinality must be greater than one. Basically, the procedure divides the samples of the gene i in the two sets that have the minimum sum of its variances.

2 Methodology

The comparison of above mentioned discretization methods was carried out using the same microarray dataset in order to find the most suitable technique for this particular case. The chosen dataset is a study of Alzheimer's Disease (AD) and contains 4 different datasets indicating the disease progression: incipient, moderate and severe (GSE1297 [12] available at: http://www.ncbi.nlm.nih.gov/geo/). This full matrix has a dimension of 31 samples and more than 22000 probes that translate to about 250 genes.

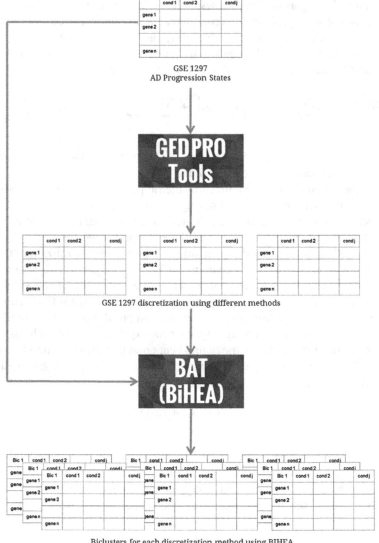

Fig. 1. The methodology workflow using discretized data is presented.

A tool that encapsulates all the discretization techniques was used [9]. The software, called GEDPRO Tools, allows using each discretization method, setting its parameters and saving the discretized data for further use. In order to compare the performance of the discretization method, a biclustering algorithm was used on each discretized data taken from GEDPRO Tools as previously commented. For this, we used the tool BAT [20] that runs the BiHEA algorithm [21]. The stated workflow is shown in Fig. 1. BAT was also executed using the real data without discretization, for comparison purposes.

3   Methodology and Results

After executing BAT [20] we were able to compare the set of biclusters obtained from discretized data and the set of biclusters for the non-discretized data. At a first glance, in order to compare all these results, the coverage for each method was calculated via BAT among the average size of the biclusters, and the amount of biclusters found by the BiHEA [21] algorithm. The results are exposed in Table 1.

Table 1. In the following table, we present the coverage that represents the percentage of the experimental matrix used, in this case, GSE 1297, covered by the biclusters found; the number of biclusters found by the algorithm and their average bicluster size.

Disc. method	Coverage	#Biclusters	Av. Bic. size
Median	55%	194	364,63
Mean	57%	193	343,56
Mean +StdDev	53%	194	410,22
X% Max	65%	194	641,72
Top X%	55%	194	361,54
Equal frequency	51%	195	371,35
K-Means	54%	196	441,10
Trans. state	51%	196	281,84
Soinov's method	34%	191	144
Ji & Tan's method	66%	198	192,49
Gallo's method	74%	198	1008,06
No discretization	16%	198	68,63

These results show a high improvement in the matter of coverage from the use of BAT from non-discretized data to the use of BAT from the different discretization methods, with one method having the best coverage: Gallo's method [7].

Another comparison between the methods was made in the matter of the overall enrichment factor. This factor was calculated using the DAVID [22, 23] tool to retrieve the max enrichment factor for each bicluster. This enrichment factor was taken from the top 10 biclusters, those with the higher score taken from BiHEA, for each method, and then it was normalized by the percentage covered by the respective bicluster on the whole matrix, i.e., it was normalized according to the bicluster size, it is calculated as shown in the following equations.

$$Enrichment(B_i) = MaxEnrichment(B_i) * \frac{|B_i|}{|M|} \tag{11}$$

Where *Enrichment(B_i)* is the enrichment factor *MaxEnrichment(B_i)* taken from DAVID tool and represents the maximum enrichment from the gene set of the bicluster B_i normalized by its size, $|B_i|$, in perspective with the size of the full matrix, $|M|$.

$$EnrichmentFactor(Mth_i) = \frac{\sum_{i=1}^{10} Enrichment(B_i)}{10} \tag{12}$$

Where Mth_i is the method that is being evaluated, *Enrichment(B_i)* is taken from Eq. 1, and the top 10 biclusters are being evaluated. Those enrichment factors were studied in terms of mean (Eq. 12) and standard deviation for each method, and the results are clearly visible in Fig. 2.

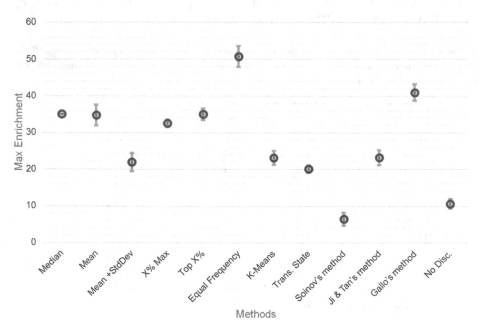

Fig. 2. In the above graphic, we present the maximum enrichment indicators for the top 10 biclusters for each method. These enrichments were taken from the DAVID [3] tool. The graphic shows the median value for these enrichments and its standard deviation.

These results show that applying almost every discretization method yields an overall enrichment factor superior to the one obtained when using non-discretized data. It also shows that the Equal Frequency method throws a better overall enrichment factor, followed by the Gallo's method.

4   Conclusions

This work enlightens the impact of choosing different discretization methods on the prediction accuracy of the biclustering techniques, presenting the coverage and the overall enrichment for each method using a particular dataset. Observing the results, we conclude that for this particular case, using GSE1297 as a microarray expression data, showing the AD progression, the method that has a better outcome in function of the coverage and overall enrichment factor is the Gallo's method [7]. However, the Equal Frequency method [2] has a superior overall enrichment factor, but it also has a bigger standard deviation and a very low coverage in comparison to the previously mentioned method. Even though the number of biclusters found is practically the same, their sizes vary too much. The Equal Frequency method average bicluster size is roughly 1/3 of the average bicluster size for the Gallo's method. For all those reasons we choose the Gallo's method over the Equal Frequency on this particular case.

In a future work, we want to analyze the impact of these methods of discretization on different data scenarios such as time-series data, RNA-sec data, and next-gen data.

Acknowledgements.   This work was supported by CONICET (*Consejo Nacional de Investigaciones Científicas y Técnicas*), grant number: PIP 112-2012-0100471, and UNS (*Universidad Nacional del Sur*), grant number: PGI 24/N042.

References

1. Friedman, N., Goldszmidt, M.: Discretization of continuous attributes while learning Bayesian networks. In: Saitta, L. (ed.) Proceedings of the 13th International Conference on Machine Learning, pp. 157–165. Morgan Kauffman, San Francisco (1996)
2. Dougherty, J., Kohavi, R., Sahami, M.: Supervised and unsupervised discrimination of continuous features. In: Prieditis, A., Russell, S. (eds.) Proceedings of the 12th International Conference on Machine Learning, pp. 194–202. Morgan Kauffman, San Francisco (1995)
3. Karlebach, G., Shamir, R.: Modeling and analysis of gene regulatory networks. Nat. Rev. Mol. Cell Biol. **9**, 770–780 (2008)
4. Alves, R., Rodriguez-Baena, D.S., Aguilar-Ruiz, J.S.: Gene association analysis: a survey of frequent pattern mining from gene expression data. Brief. Bioinform. **11**, 210–224 (2010)
5. Vignes, M., Vandel, J., Allouche, D., Ramadan-Alban, N., Cierco-Ayrolles, C., et al.: Gene regulatory network reconstruction using bayesian networks, the dantzig selector, the lasso and their meta-analysis. PLoS ONE **6**(12), e29165 (2011)
6. Vijesh, N., Chakrabarti, S.K., Sreekumar, J.: Modeling of gene regulatory networks: a review. J. Biomed. Sci. Eng. **6**, 223 (2013)
7. Gallo, C.A., Carballido, J.A., Ponzoni, I.: Discovering time-lagged rules from microarray data using gene profile classifiers. BMC Bioinformatics **12**, 1–21 (2011)
8. Madeira, S.C., Oliveira, A.L.: An evaluation of discretization methods for non-supervised analysis of time-series gene expression data (2005)
9. Gallo, C.A., Cecchini, R.L., Carballido, J.A., et al.: Discretization of gene expression data revised. Brief. Bioinform. **17**, 758–770 (2015)
10. Madeira, S., Oliveira, A.. Biclustering algorithms for biological data analysis: a survey. IEEE/ ACM Trans. Comput. Biol. Bioinform. **1**, 24–45 (2004)

11. Cheng, Y., Church, G.M.: Biclustering of expression data. ISMB **8**, 93–103 (2000)
12. Blalock, E.M., Geddes, J.M., Chen, K.C., et al.: Incipient Alzheimer's disease: microarray correlation analyses reveal major transcriptional and tumor suppressor responses. In: Proceedings of the National Academy of Sciences, pp. 2173–2178 (2004)
13. Aloise, D., Deshpande, A., Hansen, P., Popat, P.: NP-hardness of Euclidean sum-of-squares clustering. Mach. Learn. **75**, 245–248 (2009)
14. Garcia, S., Luengo, J., Sáez, J.A., et al.: A survey of discretization techniques: taxonomy and empirical analysis in supervised learning. IEEE Trans. Knowl. Data Eng. **25**, 734–750 (2013)
15. Dimitrova, E.S., Licona, M.P.V., McGee, J., Laubenbacher, R.: Discretization of time series data. J. Comput. Biol. **17**, 853–868 (2010)
16. Ding, C., Peng, H.: Minimum redundancy feature selection from microarray gene expression data. J. Bioinform. Comput. Biol. **3**, 185–205 (2005)
17. Soinov, L.A., Krestyaninova, M.A., Brazma, A.: Towards reconstruction of gene networks from expression data by supervised learning. Genome Biol. **4**, 1 (2003)
18. Ji, L., Tan, K.L.: Mining gene expression data for positive and negative co-regulated gene clusters. Bioinformatics **20**, 2711–2718 (2004)
19. MacQueen, J.: Some methods for classification and analysis of multivariate observations. In: Proceedings of the Fifth Berkeley Symposium on Mathematical Statistics and Probability, pp. 281–297 (1967)
20. Gallo, C.A., Dussaut, J.S., Carballido, J.A., Ponzoni, I.: BAT: a new biclustering analysis toolbox. In: Ferreira, C.E., Miyano, S., Stadler, P.F. (eds.) BSB 2010. LNCS, vol. 6268, pp. 67–70. Springer, Heidelberg (2010). doi:10.1007/978-3-642-15060-9_8
21. Gallo, C.A., Carballido, J.A., Ponzoni, I.: BiHEA: a hybrid evolutionary approach for microarray biclustering. In: Guimarães, K.S., Panchenko, A., Przytycka, T.M. (eds.) BSB 2009. LNCS, vol. 5676, pp. 36–47. Springer, Heidelberg (2009). doi:10.1007/978-3-642-03223-3_4
22. Huang, D.W., Sherman, B.T., Lempicki, R.A.: Systematic and integrative analysis of large gene lists using DAVID bioinformatics resources. Nat. Protoc. **4**, 44–57 (2009)
23. Huang, D.W., Sherman, B.T., Lempicki, R.A.: Bioinformatics enrichment tools: paths toward the comprehensive functional analysis of large gene lists. Nucleic Acids Res. **37**, 1–13 (2009)

Analysis of Informative Features for Negative Selection in Protein Function Prediction

Marco Frasca, Fabio Lipreri, and Dario Malchiodi[(⊠)]

Dipartimento di Informatica, Università degli Studi di Milano,
Via Comelico 39/41, 20135 Milano, Italy
{marco.frasca,dario.malchiodi}@unimi.it

Abstract. Negative examples in automated protein function prediction (*AFP*), that is proteins known not to possess a given protein function, are usually not directly stored in public proteome and genome databases, such as the Gene Ontology database. Nevertheless, most computational methods need negative examples to infer new predictions. A variety of algorithms has been proposed in *AFP* for negative selection, ranging from network- and feature-based heuristics, to hierarchy-based and hierarchy-less strategies. Moreover, several bio-molecular information sources about proteins, such as gene co-expression, genetic and protein-protein interactions data, are naturally encoded in protein networks, where nodes are proteins and edges connect proteins sharing common characteristics. Although selecting negatives in biological networks is thereby a central and challenging problem in computational biology, detecting the characteristics proteins should have to be considered as negative is still a difficult task. It this work, we show that a few protein features extracted from the network help in detecting reliable negatives. We tested such features in two real world experiments: predicting unreliable negatives with an SVM classifier through temporal holdout on model organisms for *AFP*, and selecting reliable negatives with a clustering-based state-of-the-art negative selection procedure.

Keywords: Negative example selection · Protein function prediction · Biological networks · Fuzzy clustering · Protein features

1 Introduction

Most public databases storing information about gene/gene product functions rarely report the functions a gene does not possess. Examples are the Human Phenotype Ontology [1], the Functional Catalogue [2], the Gene Ontology [3], and the Medical Subject Headings[1]. When a gene is not currently associated with a given function, it is usually considered as negative example for that function; nevertheless, this does not ensure that no studies in future will associate that gene with that function. In particular, this is a central issue in the automated

---

[1] http://www.nlm.nih.gov/mesh.

© Springer International Publishing AG 2017
I. Rojas and F. Ortuño (Eds.): IWBBIO 2017, Part II, LNBI 10209, pp. 267–276, 2017.
DOI: 10.1007/978-3-319-56154-7_25

protein function prediction (*AFP*) problem, where the aim is assigning proteins to biological functions structured as hierarchies with parent-child specialization relationships. In this work we consider the Gene Ontology (GO) hierarchy, which is composed of three branches structured as direct acyclic graphs: biological process (BP), molecular function (MF), and cellular component (CC). GO functions, also referred as terms, often have rare positively annotated proteins, with the consequence that most proteins can be either a negative example or a positive example that has yet to be annotated. On the other hand, computational methodologies require that both positive and negative classes be well represented in order to train accurate predictors. The problem of selecting reliable negative examples is thereby of paramount importance in *AFP* [4,5].

In the literature, attempts for choosing negative examples rely on bootstrap aggregating (bagging) techniques, where the algorithm iteratively trains several binary classifiers to discriminate the known positive examples from random subsamples of the non-positive set [6], on sampling according to positive-negative similarity measures [7,8], or simply randomly sampling non-positive items assuming the probability of getting a false negative to be low [9]. In the specific context of *AFP*, several heuristics for negative selection exploit the GO hierarchy, by considering as negative for a GO term the proteins positive for sibling and/or ancestral GO terms [10], or by computing the empirical conditional probability of the GO term of interest, given the annotation for other GO terms in all three branches, considering solely leaf terms in the hierarchy [11], and the non-leaf ones as well [12].

In this paper we aim at finding significant protein properties which can improve the correct distinction between reliable and non reliable negative instances. Since actually a large bunch of problems in computational biology and medicine exploits network representations, we performed an analysis on protein networks for *AFP*, which can thereby be easily extended to other problems, such as the gene-disease prioritization [13] and the drug repositioning problems [14]. Nodes in the network represent proteins, whereas connections represent precomputed similarity among nodes. Specifically, we extracted six node features from the network, including node degree, connections variance, and positive neighborhood. Subsequently we evaluated in three model organisms the ability of these features in characterizing false negatives through a temporal holdout, that is considering nodes negative in an older release of the Gene Ontology, but positive in a more recent one. We found as result that these six protein/node features provide higher accuracy in detecting false negatives compared with a state-of-the-art algorithm for extracting node features from biological networks. Furthermore, we tested the obtained features in a recently proposed negative selection procedure based on fuzzy clustering techniques [8], to verify whether their capabilities in selecting reliable negatives would improve when representing proteins through the six features found.

The paper is organized as follows: Sect. 2 describes the adopted methodology, while Sect. 3 shows the experiments carried out to validate the proposed features on three model organisms. Some concluding remarks close the paper.

2   Methodology

After having introduced the used notation and the experimental setting, in this section we describe the selected features and we motivate their use.

2.1   Preliminaries

The input of our problem is given by a graph $G\langle V, W \rangle$, where $V = \{1, 2, \ldots, n\}$ is the set of nodes corresponding to proteins, and W is the $n \times n$ connection matrix, where each entry $W_{ij} \in [0, 1]$ represents some notion of functional similarity between proteins i and j ($W_{ij} = 0$ if nodes i and j are not connected). We also consider two matrices $Y, \overline{Y} \in \{0, 1\}^{n \times m}$ of annotations for m given GO functions for nodes V in two different temporal releases (we assume Y is the older one). The k-th column of Y corresponds to the annotation vector for the GO function k, and analogously for $\overline{Y}$. We denote in the rest of the paper by $X_{.r}$ and $X_{i.}$ the r-th column and i-th row of matrix X, respectively, and by $N_i := \{j \in V | W_{ij} \neq 0\}$ the set of neighbors of node i in the graph. For a given GO function $k \in \{1, 2, \ldots, m\}$, assuming for sake of simplicity $y = Y_{.k}$ and $\overline{y} = \overline{Y}_{.k}$, we also denote by $N_i^+ := \{j \in V | W_{ij} \neq 0 \wedge y_j = 1\}$ the set of positive neighbors of node i in the graph. The following subsets of V can be obtained:

$$C_{nn} = \{i \in V | y_i = 0 \wedge \overline{y}_i = 0\},$$
$$C_{np} = \{i \in V | y_i = 0 \wedge \overline{y}_i = 1\},$$
$$C_{pn} = \{i \in V | y_i = 1 \wedge \overline{y}_i = 0\},$$
$$C_{pp} = \{i \in V | y_i = 1 \wedge \overline{y}_i = 1\}.$$

We are interested in extracting protein features from the graph G able to discriminate between proteins in C_{np} (i.e. proteins negative in the older release which became positive in the more recent one) and proteins in C_{nn}.

2.2   Extracting Relevant Proteins Features

For each protein $i \in V$, Table 1 details the following six considered features:

1. *Neighborhood mean* $f1(i)$. The mean of connection weights.
2. *Neighborhood variance* $f2(i)$. The variance of connection weights.
3. *Weighted degree* $f3(i)$. The sum of connection weights.
4. *Number of annotations* $f4(i)$. The number of GO terms the protein i is annotated with (in Y).
5. *Positive neighborhood* $f5(i)$. The sum of connection weights towards positives.
6. *Mean of positive neighborhood* $f6(i)$. The mean of connection weights.

Features $f1$ and $f2$ provide information about the distribution of connections in the node neighborhood, whereas feature $f3$, apart being related to node connectivity, was pointed out as a proxy for gene multifunctionality [15, 16]. Moreover, features $f1$ and $f3$ considered together indirectly carry information about the number of connections, in addition to the weighted degree, which are two main

Table 1. The six selected features.

Symbol	Name	Definition
$f1$	Neighborhood mean	$\dfrac{1}{|N_i|} \sum\limits_{j \in N_i} W_{ij}$
$f2$	Neighborhood variance	$\dfrac{1}{|N_i| - 1} \sum\limits_{j \in N_i} (W_{ij} - f1(i))^2$
$f3$	Weighted degree	$\sum\limits_{j \in N_i} W_{ij}$
$f4$	Number of annotations	$\sum\limits_{k=1}^{m} Y_{ik}$
$f5$	Positive neighborhood	$\sum\limits_{j \in N_i^+} W_{ij}$
$f6$	Mean of positive neighborhood	$\dfrac{1}{|N_i^+|} \sum\limits_{j \in N_i^+} W_{ij}$

measures of node connectivity in graphs [17]. Feature $f4$ allows to understand whether a protein tends to play different roles, and naturally, in order not to introduce bias, we exclude annotations for the current GO term when computing $f4$. Finally, feature $f5$ is a sort of *guilt-by-association* score, exploiting both the number of neighbors positive for the current term and the corresponding connection magnitudes, whereas feature $f6$, alongside feature $f5$, allows to estimate the number of positive ties.

We have considered three model organisms, *Saccaromices cerevisiae* (yeast), *Drosophila melanogaster* (fly) and *Homo sapiens* (human), the first two used to carry out our preliminary protein feature analysis, the latter used as validation dataset in the experiments conducted in Sect. 3. For *yeast* and *fly* we downloaded the input net from [18], where networks covering different types of protein information, including gene co-expression, genetic and physical interactions, protein ontologies, have been already integrated into a consensus network. In order to validate the results obtained on *yeast* and *fly* model organisms, the *human* network has been retrieved from a different database, namely STRING (version 10.0) [19], with the default setting of medium edge confidence scores: the result is a highly informative network differently constructed w.r.t. *yeast* and *fly* nets, where STRING curators already merge several sources of data, including protein homology relationships from different species. Overall, we obtained 5774 yeast, 13064 fly, and 19247 human proteins. For all three organisms, we retrieved two releases of the Gene Ontology, and considered the GO terms in the three GO branches with 5 up to 300 annotated proteins (to have a minimum of information and avoid too generic terms), obtaining 1604, 1862 and 3321 GO terms for *yeast*, *fly*, and *human* data sets respectively. For *yeast* we selected releases 28.03.2013 (forming the matrix Y) and 04.07.2015 ($\overline{Y}$), and we proceeded in a similar way for the remaining two organisms (namely, we selected releases 13.05.2014 and 07.03.2016 for *fly*, and releases 13.05.2014 and

Table 2. GO releases for the selected organisms.

Organism	Older release	Newer release
Yeast	28.03.201	04.07.2015
Fly	3.05.2014	07.03.2016
Human	13.05.2014	29.04.2015

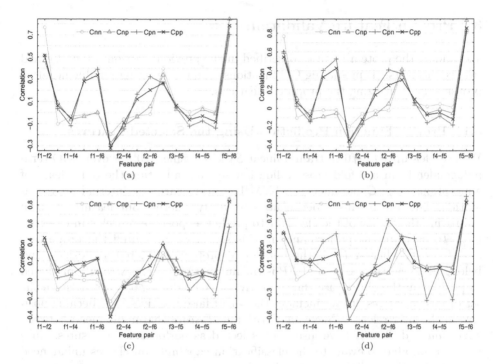

Fig. 1. Correlations between every possible couple of protein features in *yeast* ((a)-mean, (b)-median) and *fly* ((c)-mean, (d)-median) organisms. *Mean* and *median* represent the mean and median of correlations across GO terms.

29.04.2015 for *human*), as detailed in Table 2. Protein annotations with IEA evidence (inferred from electronic annotations) have been excluded.

To determine whether the proposed features had the capability to discriminate between categories C_{nn} and C_{np}, we computed the Pearson's correlation between each couple of protein feature vectors, by averaging across protein categories $C_{nn}, C_{np}, C_{pn}, C_{pp}$ separately. In Fig. 1 we show the results over *yeast* and *fly* organisms. Circles and triangles represent the two categories of interest: negative proteins remained negative, and negative proteins that became positive in the new release. Interestingly, circles show distinguishable behavior from triangles for most feature pairs, and mainly for $f1$–$f2$ (Fig. 1(a,b,d)), $f1$–$f3$ (a,b,c,d), $f1$–$f4$ (c,d), and $f3$–$f4$ (a,c,d). Feature $f5$ and $f6$ show small differences between categories C_{nn} and C_{np}, with some exceptions ($f2$–$f6$ (a, b), $f1$–$f6$ (c), $f1$–$f5$ (b)). Nevertheless, $f5$ and $f6$ are the only term-dependent features, i.e. those sensibly varying with the predicted term.

Overall, this preliminary result thereby suggests the proposed features, and mainly $f1$–$f4$, as relevant for selecting meaningful negatives in AFP; accordingly, in the next section we experimentally test this hypothesis.

3   Protein Feature Validation

To validate the protein features described in the previous section, we performed two experiments: (i) predicting GO functions, and (ii) selecting reliable negative proteins, both utilizing the proposeds features.

3.1   Protein Function Prediction Using the Selected Features

We used features $f1$–$f6$ to train a linear SVM classifier [20] for every GO term independently in a 3-fold cross validation setting, adopting the old release of annotations. The C parameter of SVM has been learnt by internal cross validation. The aim of this experiment is to verify whether our features help in improving the ability of the classifier to predict as positive proteins in category C_{np}. To measure the performance in this sense, we computed the proportion $\nu = FTP/FP$, where FTP is the number of proteins predicted as positive and belonging to category C_{np}, and FP is the number of proteins wrongly predicted as positive in the old release (false positives). Accordingly, $0 \leq \nu \leq 1$, since the set of false positives always includes the set of false positives that became positive in the later release. *Precision*, *Recall* and *F-score* (harmonic mean between *Precision* and *Recall*) have also been selected as performance measures, since they are suitable to evaluate the classification capability on classes unbalanced towards negatives, as those we predicted.

As baseline comparison, we also trained the SVM classifier on the features computed by a state-of-the-art algorithm for extracting feature from biological networks, namely *3Prop* [21], which associates each protein $i \in V$ with three features p_i^j, $j = 1, 2, 3$, where p_i^j is the probability that a random walk of length j starting from positive instances ends at node i. We also combined *3Prop* features and $f1$–$f6$. Finally, features $f1$–$f6$ have been normalized so as to sum up to one across proteins ($\bar{f}k(i) = fk(i)/\sum_{j \in V} fk(j)$, $k = 1, 2, \ldots, 6$), since in principle they have no comparable ranges. The results obtained are in shown Table 3.

In terms of classification performance, SVM classifier trained when proteins are represented using $f1$–$f6$ achieves the best F on *yeast* and *fly*, with statistically significant improvements w.r.t. SVM trained using *3Prop* features (Wilcoxon Signed Rank test, *p-value* < 0.05), which is the worst method. Different is the trend on *human* organism, where SVM with *3Prop* obtain the best performance, and features $f1$–$f6$ are less predictive in this case. Furthermore, the classifier trained using *3Prop* and $f1$–$f6$ features together is always the second

Table 3. SVM results. Best performance for each measure are in boldface.

Features	Prec	Rec	F	ν
Yeast				
3Prop	0.3495	0.1365	0.1833	0.0206
3Prop+f1–f6	0.3983	0.1666	0.2190	0.0256
f1–f6	**0.4062**	**0.2090**	**0.2522**	**0.0343**
Fly				
3Prop	0.1644	0.0638	0.0848	0.0629
3Prop+f1–f6	**0.1796**	0.0704	0.0932	0.0647
f1–f6	0.1550	**0.0949**	**0.1061**	**0.1578**
Human				
3Prop	**0.2284**	**0.0697**	**0.0959**	0.0805
3Prop+f1–f6	0.1737	0.0622	0.0806	0.1258
f1–f6	0.0569	0.0361	0.0391	**0.1349**

method. Interestingly, when evaluating the ability in detecting proteins in category C_{np} (the ν measure), protein features $f1-f6$ are largely the top performing, even on *human* data, whereas SVM using $3Prop$ always achieves the worst result. The improvement in terms of ν w.r.t. the second best result is noticeable and meaningful, and confirms, from a different standpoint, the analysis reported in Sect. 2.2, even on the validation dataset (*human*).

Features $f1-f6$, once stated their contribution in improving SVM ability to better detect proteins in category C_{np}, have been utilized in a state-of-the-art negative selection procedure for *AFP* to discard non-reliable negatives, thus keeping the relevant ones, as shown in next section.

3.2 Selecting Negatives

We investigated the impact of using the proposed features in a procedure devoted to selecting reliable negative examples [8]. This procedure requires as input nodes/proteins represented through feature vectors, computes a fuzzy clusterization of positive examples, and subsequently ranks all negative examples according to their highest membership value σ to the found clusters. Namely, if σ is low the considered point cannot be reasonably associated to the positive instances, thus it is selected as a reliable negative example. Here we operate a reverse selection: since we want to evaluate the ability in detecting proteins in category C_{np}, i.e. the non-reliable negatives, we focus on those negative whose memberships σ is large. Consequently, complementing the results we will obtain the set of reliable negatives.

The experiment consists in feeding, for every organism separately, the mentioned procedure with proteins represented through features $f1-f6$, and getting the corresponding ranking. Such ranking is then used to select a subset of

Table 4. Number of GO terms where one method outperforms the other one in terms of ν_{mean}.

Method	3Prop	3Prop+f1–f6	f1–f6
Yeast			
3Prop	-	657	687
3Prop+f1–f6	846	-	493
f1–f6	826	550	-
Fly			
3Prop	-	503	512
3Prop+f1–f6	636	-	331
f1–f6	627	407	-
Human			
3Prop	-	487	484
3Prop+f1–f6	1254	-	152
f1–f6	1257	183	-

non-reliable negative points: namely, for all values τ in a suitable discretization of $(0, 1)$, we consider only points whose maximum σ exceeds the τ-quantile in the ranking. The resulting set of points is considered as candidate for the class C_{np}. To evaluate the performance, we simply computed the fraction ν of points in this set which belong to C_{np}, that is those that became positive in the novel release $\overline{\mathbf{Y}}$. Again, we repeated such experiment also when representing proteins through $3Prop$ and $3Prop+f1-f6$ features.

We compared for every GO term the mean of ν values (ν_{mean}) obtained with different settings of τ achieved by the negative selection algorithm when using $3Prop$, $3Prop+f1-f6$ and $f1-f6$ respectively. Then, for every pair of methods, we computed the number of wins (GO terms) in terms of ν_{mean} of one method on the other one (Table 4). Even in this experiment features $f1-f6$ win versus both $3Prop$ and $3Prop+f1-f6$ in all the organisms. Particularly meaningful are the results on *human* data, with 1257 wins and 484 loses against $3Prop$, which in turn is the worst method in all the experiments. The proposed features thereby tend to better discriminate negative proteins belonging to categories C_{nn} and C_{np}, by allowing the negative selection procedure to improve the capabilities in detecting proteins negative in the old release and positive in the recent one. Accordingly, such features can be exploited to help learning classifiers based on more reliable negative examples, mainly in the protein function prediction problem, where negative examples largely outnumber the positive ones.

4   Conclusions

In this work we analyzed the informational power retained by a set of six features in AFP problems, whose input is provided as (i) a graph where nodes identify with proteins and edges are labeled with functional similarity between proteins, and (ii) the annotations of the same proteins for the GO functions to be predicted. The use of the proposed features, directly extracted by the above mentioned graph and annotation vectors, was motivated by a temporal hold-out experimental setting involving two different temporal releases of the annotations. In particular, we were interested in discriminating between proteins negatively labeled in the older temporal release which became positively labeled in the more recent one. We assessed the proposed features in two different experimental settings, focusing on their predictive capabilities in the GO function classification problem and on their impact on negative selection procedures. The obtained results show how the proposed features outperform the $3Prop$ state-of-the-art feature extraction algorithm. As a next step, it would be useful to evaluate the contribution of each feature (to identify the most relevant ones), and to apply the same methodology of feature extraction to other real world problems in the realm of medicine and biology characterized by both network-based representation of processed data and not clear definition on negative instances.

References

1. Robinson, P.N., et al.: The human phenotype ontology: a tool for annotating and analyzing human hereditary disease. Am. J. Hum. Genet. **83**(5), 610–615 (2008)
2. Ruepp, A., et al.: The FunCat, a functional annotation scheme for systematic classification of proteins from whole genomes. Nucleic Acids Res. **32**(18), 5539–5545 (2004)
3. Ashburner, M., et al.: Gene ontology: tool for the unification of biology. Nature Genet. **25**(1), 25–29 (2000)
4. Radivojac, P., et al.: A large-scale evaluation of computational protein function prediction. Nat. Methods **10**(3), 221–227 (2013)
5. Jiang, Y., Oron, T.R., et al.: An expanded evaluation of protein function prediction methods shows an improvement in accuracy. Genome Biol. **17**(1), 184 (2016)
6. Mordelet, F., Vert, J.P.: A bagging SVM to learn from positive and unlabeled examples. Pattern Recogn. Lett. **37**, 201–209 (2014)
7. Burghouts, G.J., Schutte, K., Bouma, H., den Hollander, R.J.M.: Selection of negative samples and two-stage combination of multiple features for action detection in thousands of videos. Mach. Vis. Appl. **25**(1), 85–98 (2014)
8. Frasca, M., Malchiodi, D.: Selection of negative examples for node label prediction through fuzzy clustering techniques. In: Bassis, S., Esposito, A., Morabito, F.C., Pasero, E. (eds.) Advances in Neural Networks. SIST, vol. 54, pp. 67–76. Springer, Cham (2016). doi:10.1007/978-3-319-33747-0_7
9. Gomez, S.M., Noble, W.S., Rzhetsky, A.: Learning to predict protein-protein interactions from protein sequences. Bioinformatics **19**(15), 1875–1881 (2003)
10. Mostafavi, S., Morris, Q.: Using the gene ontology hierarchy when predicting gene function. In: Proceedings of the Twenty-Fifth Conference on Uncertainty in Artificial Intelligence, pp. 419–427 (2009)
11. Youngs, N., Penfold-Brown, D., Drew, K., Shasha, D., Bonneau, R.: Parametric bayesian priors and better choice of negative examples improve protein function prediction. Bioinformatics **29**(9), tt10-98 (2013)

12. Youngs, N., Penfold-Brown, D., Bonneau, R., Shasha, D.: Negative example selection for protein function prediction: the NoGO database. PLOS Comput. Biol. **10**(6), 1–12 (2014)
13. Frasca, M., Bassis, S.: Gene-disease prioritization through cost-sensitive graph-based methodologies. In: Ortuño, F., Rojas, I. (eds.) IWBBIO 2016. LNCS, vol. 9656, pp. 739–751. Springer, Heidelberg (2016). doi:10.1007/978-3-319-31744-1_64
14. Ashburn, T.T., Thor, K.B.: Drug repositioning: identifying and developing new uses for existing drugs. Nat. Rev. Drug Discov. **3**(8), 673–683 (2004)
15. Gillis, J., Pavlidis, P.: The impact of multifunctional genes on "Guilt by Association" analysis. PLoS ONE **6**(2), e17258 (2011)
16. Frasca, M.: Automated gene function prediction through gene multifunctionality in biological networks. Neurocomputing **162**, 48–56 (2015)
17. Opsahl, T., Agneessens, F., Skvoretz, J.: Node centrality in weighted networks: generalizing degree and shortest paths. Soc. Netw. **32**(3), 245–251 (2010)
18. Frasca, M., Bertoni, A., et al.: UNIPred: unbalance-aware Network Integration and Prediction of protein functions. J. Comput. Biol. **22**(12), 1057–1074 (2015)
19. Szklarczyk, D., et al.: String v10: proteinprotein interaction networks, integrated over the tree of life. Nucleic Acids Res. **43**(D1), D447–D452 (2015)
20. Cortes, C., Vapnik, V.: Support-vector networks. Mach. Learn. **20**, 273–297 (1995)
21. Mostafavi, S., Goldenberg, A., Morris, Q.: Labeling nodes using three degrees of propagation. PLoS ONE **7**(12), e51947 (2012)

Differential Network Analysis
of Anti-sense Regulation

Marc Legeay[1,2], Béatrice Duval[1(✉)], and Jean-Pierre Renou[2]

[1] LERIA - UNIV Angers - Université Bretagne Loire,
2 bd Lavoisier, 49045 Angers, France
{marc.legeay,beatrice.duval}@univ-angers.fr
[2] IRHS, INRA, AGROCAMPUS-Ouest, Université d'Angers, SFR 4207 QUASAV,
42 rue Georges Morel, 49071 Beaucouzé cedex, France
jean-pierre.renou@inra.fr

Abstract. A challenging task in systems biology is to decipher cell regulation mechanisms. By comparing networks observed in two different situations, the differential network analysis approach enables to highlight interaction differences that reveal specific cellular responses. The aim of our work is to study the role of natural anti-sense transcription on cellular regulation mechanisms. Our proposal is to build and compare networks obtained from two different sets of actors: the "usual" sense actors on one hand and the sense and anti-sense actors on the other hand. Our study only considers the most significant interactions, called an Extended Core Network; therefore our differential analysis identifies important interactions that are impacted by anti-sense transcription. This paper first introduces our inference method of an Extended Core Network; this method is inspired by C3NET, but whereas C3NET only computes one interaction per gene, we propose to consider the most significant interactions for each gene. Secondly, we define the differential network analysis of two extended core networks inferred with and without anti-sense actors. On a local view, this analysis relies on *change motifs* that describe which genes have their most important interactions modified when the anti-sense transcripts are considered; they are called AS-impacted genes. Then from a more global view, we consider how the relationships between these AS-impacted genes are rewired in the network with anti-sense actors. Our analysis is performed by computing Steiner trees that represent minimal subnetworks connecting the AS-impacted genes. We show that the visualisation of these results help the biologists to identify interesting parts of the networks.

1 Introduction

Gene regulation is a key issue in bioinformatics. As micro-array produce large-scale expression datasets, gene network inference is a useful approach to study gene interactions, and a lot of methods have been proposed in the literature for this reverse engineering task [1–4]. Differential network analysis [5–7] proposes to study the cellular response to different situations. In medicine, differential network analysis is used to compare healthy tissues and diseased tissues in

© Springer International Publishing AG 2017
I. Rojas and F. Ortuño (Eds.): IWBBIO 2017, Part II, LNBI 10209, pp. 277–288, 2017.
DOI: 10.1007/978-3-319-56154-7_26

order to reveal network rewiring induced by the disease [8]. In these approaches, the comparison is made by using the same set of genes to construct the gene networks.

Our work has the particularity to study gene networks with sense and anti-sense transcripts. Anti-sense RNAs are endogenous RNA molecules whose partial or entire sequences exhibit complementarity to other transcripts. Their different functional roles are not completely known but several studies suggest that they play an important role in stress response mechanisms [9]. For instance, a significant effect is the post-transcriptional gene silencing: the self-regulatory circuit where the anti-sense transcript hybridizes with the sense transcript to form a double strand RNA (dsRNA) that is degraded in small interfering RNAs (siRNA). Previous studies on *Arabidopsis Thaliana* showed that sense and anti-sense transcripts for a defense gene (RPP5) form dsRNA and generate siRNA which presumably contributes to the sense transcript degradation in the absence of pathogen infection [10].

Our work focuses on apple because a recent study [11] suggests that a large majority of protein coding genes of this organism are actually concerned by anti-sense transcription. The authors have combined microarray analysis with a dedicated chip and high-throughput sequencing of small RNAs to study anti-sense transcription in eight different organs (seed, flower, fruit, ...) of apple (*Malus × domestica*). Their atlas of expression shows several interesting points. Firstly, the percentage of anti-sense expression is higher than that reported in other studies, since they identify anti-sense transcription for 65% of the sense transcripts expressed in at least one organ, while it is about 30% in previous *Arabidopsis Thaliana* studies. Secondly, the anti-sense transcript expression is correlated with the presence of short interfering RNAs. Thirdly, anti-sense expression levels vary depending on both organs and Gene Ontology (GO) categories. It is higher for genes belonging to the "defense" GO category and on fruits and seeds.

The work described in this paper proposes a large-scale analysis of apple transcriptomic data, with measures of anti-sense transcripts in the context of fruit ripening. The fruit ripening is a stress-related condition involving "defense" genes. To highlight the impact of anti-sense transcription, we propose to compare context-specific gene networks that involve two kinds of actors, on one hand the sense transcripts that are usually used in gene networks and on the other hand the sense and anti-sense transcripts. However gene network inference methods generally find many false positive interactions, and some authors have proposed to study the core part of a gene network [2], by only computing for each gene the most significant interaction with another gene. This constraint seems too restrictive and we developed a method where we extend the core network by considering for each gene a small number of significant interactions. We call our gene network inference method the *Extended Core Network*. Then we use this inference method in order to discover which interactions of the core network are modified when we integrate the anti-sense transcripts. We define the notion of *AS-impacted genes* to describe sense genes whose interactions are highly impacted by the integration of anti-sense transcripts in the data. It is also interesting to study the relationships between these AS-impacted genes and we explain how the rewiring around AS-impacted genes can be computed by Steiner trees.

In Sect. 2, we present the Extended Core Network Inference method and we evaluate it on artificial datasets. In Sect. 3, we present our workflow to compare two core gene networks built on different sets of actors and we define the change motifs that highlight significant differences between the interactions. For each step of our workflow, we provide results obtained for the apple data.

2   Extended Core Network Inference Method

2.1   Motivations

Several methods have been proposed to infer gene networks from transcriptomic data. Reviews of these reverse engineering methods can be found in [1, 3, 4]. Some inference methods reconstruct pairwise gene interaction networks by measuring with a statistical criterion whether two genes are co-expressed or co-regulated. This statistical measure can be Spearman or Pearson correlation [12], or mutual information [2, 13], that enables the detection of non-linear relationships. These methods need a step of thresholding to decide which values of the statistical measure are significant. One major drawback of these methods is that many of the predicted interactions are false positives. We can differentiate two types of false positive interactions: an interaction that does not biologically exist, and an indirect interaction. If two genes g_2 and g_3 are regulated by g_1, then mutal information (as well as correlation) between g_2 and g_3 is high and an indirect interaction is put in the inferred network. Indirect interactions lead in large gene networks difficult to interpret by biologists and they must be pruned from the output networks [13, 14]. To avoid false positives, the Conservative Causal Core Network (C3NET) [2] proposes to compute the core of a gene network, by selecting for each gene a unique interaction defined by the maximal mutual information value. Our aim is to compare two inferred networks to identify significant changes in the interactions when we take into account the anti-sense actors. Therefore only considering the maximal interaction for each gene is too restrictive, since several mutual information values may be very close to the maximum, and a strict comparison of the maximal values in two situations is not relevant to compare two networks. So we propose a gene network inference method, based on C3NET and named *Extended Core Network (ECN)*, where for each gene the most significant interactions are put in the inferred network.

2.2   Preprocessing

The microarray data we use are intensity values of each gene in each sample. We first normalize the data using quantile normalization [15, 16]. We also copula-transform the data [17, 18] before estimating the mutual information for each pair of genes.

Extended Core Network uses the mutual information in order to estimate the connections between genes. Given the expression vectors I and J of genes i and j respectively, we estimate the mutual information $M[i,j]$ with the same estimator used in C3NET:

$$M[i,j] = \frac{1}{2} \log(\frac{\sigma_I^2 \sigma_J^2}{|C|})$$

where σ_I^2 and σ_J^2 are the variance of I and J respectively, and $|C|$ is the determinant of the covariance matrix.

As C3NET or ARACNE [13], we test the statistical significance of pairwise mutual information by re-sampling methods. All non-significant values are set to 0 before applying the inference algorithm.

2.3 Algorithm

The Extended Core Network (ECN) algorithm computes an adjacency matrix network for a set of genes G. We first initialize the matrix by considering that there is no connection at all. Then, for each gene, we determine its neighbours. The neighbours of a gene g are the ones with the best mutual information values with g. We use an accepting rate r in order to identify the threshold value that determine if a gene g' has one of the best mutual information with g. The threshold of g is fixed by the maximal mutual information of g with other genes, and the accepting rate r. The accepting rate must be between 0 and 1; 0 means that only the best neighbour will be selected and 1 means that all significant interactions will be selected. When the accepting rate is 0, it is almost the same as in C3NET: if two interactions share the best mutual information, both of them will be selected in ECN whereas only one will be selected in C3NET. When the accepting rate is 1, the interactions of the output network are all the significant mutual information values. Non-significant values are set to 0 in the preprocessing step, so that only non-null values are selected as interactions in the output network. Mutual information is a symmetrical measure, that is why the final step of C3NET is to transform the asymmetrical adjacency matrix into a symmetrical one: the result is thus an undirected graph. Because we want to compare two networks and see which interactions are modified, our algorithm provides a directed network so that we can identify which significant changes occur in the connexions of a gene when we integrate anti-sense actors in the algorithm.

2.4 Evaluation on Artificial Datasets

We now evaluate our Extended Core Network algorithm on simulated data, and compare it with C3NET. We use the same protocol to simulate the artificial data as in [2]. We use biological networks from *E. coli* [19,20] and *S. cerevisiae* [21]. We generate our simulated data with sub-networks of these biological networks thanks to SynTREN [22]. SynTREN allows us to simulate the activity of genes

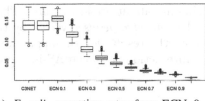

(a) *E. coli*, accepting rates from ECN_0 (0%) to ECN_1 (100%).

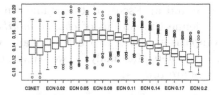

(b) *E. coli*, accepting rates from ECN_0 (0%) to ECN_0.2 (20%).

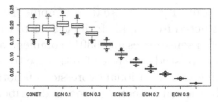

(c) Yeast, accepting rates from ECN_0 (0%) to ECN_1 (100%).

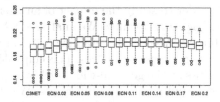

(d) Yeast, accepting rates from ECN_0 (0%) to ECN_0.2 (20%).

Fig. 1. Box plots of F_1 scores for C3NET (first left) and ECN with different accepting rates. The number following ECN indicates the accepting rates. The box plot next to the C3NET one is ECN_0.

from the selected network with noise. We simulate two different expression data: one with *E. coli*, the other with the yeast. SynTREN gives us two expression datasets of 100 samples and 200 genes.

For each simulated expression data, we performed 500 simulations. For each simulation, we selected randomly the samples used to infer the gene networks. To evaluate the error rate of each inference method, we use the F_1 score[1].

Figure 1 shows the box plots of the F_1 scores obtained by the simulations. We use ECN that provides undirected network, as C3NET, to compare them. The ECN_0 method corresponds to the Extended Core Network with a 0% accepting rate. It differs from C3NET if there are several genes g' which share the maximal mutual information with the gene g. ECN has the same complexity as C3NET, that is $\mathcal{O}(n^2)$, where n is the number of genes. We can see that ECN is better than C3NET when the accepting rate is low on *E. coli* (Fig. 1a and b) and yeast (Fig. 1c and d). The score of ECN depends on the accepting rate used. We can see that first, rising the accepting rate improves the inference method, but quickly it degrades it. When we look closer between 0 and 20% of accepting rate (Fig. 1b and d), we can see that an accepting rate of 7% (ECN_0.07) is acceptable on simulated data. We have tested our method on other simulated data and we observed that an accepting rate between 5 and 10% is the best compromise.

3   Differential Network Analysis

We now describe the differential analysis that we perform on two extended core networks to explore the role of anti-sense transcription. We first describe the

---

[1] $F_1 = 2 \cdot \frac{\text{presicion} \cdot \text{recall}}{\text{precision} + \text{recall}}$.

sense and anti-sense data obtained in experiments about apple ripening. Then we define the notions of *change motifs* and *AS-impacted genes* revealed by the comparison of the two networks and we report the results for the apple data. To complete this analysis, we study how the interactions between AS-impacted genes are rewired.

3.1 Biological Material

In order to study the impact of anti-sense transcripts, we use data of apple fruit during fruit ripening. We analyse RNA extracted from apple fruits thanks to the chip AryANE v1.0 containing 63011 predicted sense genes and 63011 complementary anti-sense sequences. This chip allows us to study the role of anti-sense transcripts at the genome-wide level by supplying transcriptional expression on both sense and anti-sense transcripts. We study the fruit ripening process described by two conditions: harvest (H) and 60 days after harvest (60DAH), and for each condition, 22 samples of apple fruit have been analysed. We first identify transcripts displaying significant differences between the two conditions (p-val $< 1\%$). With a further threshold of 1 log change between the two conditions, we found 931 sense and 694 anti-sense transcripts differentially expressed, with among them, 200 transcripts for which both sense and anti-sense fulfil the condition. We use these 1625 transcripts into our analysis. A differential functional analysis has been performed in [23] with those transcripts. The differential functional analysis reveals ontological terms that are significantly represented only when anti-sense transcripts are considered in conjunction with the sense transcripts. The fruit ripening is a stress-related condition involving "defense" genes, and our analysis revealed terms such as hyperosmotic response and response to cold, which are ontological terms related to a stress response. So this functional approach has shown the importance of taking into account the anti-sense information.

3.2 Change Motifs and AS-impacted Genes

Our differential network analysis compares the extended core network inferred from the sense data (S) with the one inferred from the sense and anti-sense data (SAS). We compare these two networks thanks to *AS-impacted genes* and *change motifs*. We define an *AS-impacted gene* as a sense gene s that is linked to one or many other sense actors in the S network, but these interactions are no longer present in the SAS network. In other words, an AS-impacted gene is a sense node s which has no outer link common to the S and the SAS networks. This change of interactions occurs because in the SAS core network, the most significant interactions (greatest values of the mutual information criterion) for this gene s are interactions with anti-sense actors. By merging the S and SAS networks, we define a *change motif* [23,24] as an AS-impacted gene surrounded by all of its neighbours from the S and SAS networks. The change motifs are identified by comparing the two adjacency matrices of S and SAS. Change motifs allow us to identify interactions that are "modified" by adding anti-sense actors into the

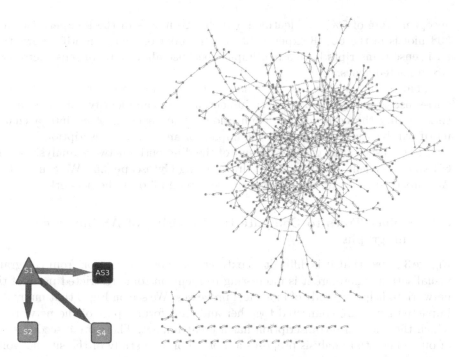

Fig. 2. Illustrations of *change motifs* with Extended Core Network. A sense node is represented in blue, an anti-sense node is represented in purple. The orange triangle-shaped node is an *AS-impacted gene*. A red link is a link only present in the S network. A green link is a link only present in the SAS network. (Color figure online)

Fig. 3. Extended Core Network with a 5% accepting rate for sense-only 60DAH experiment. Orange triangle-shaped nodes represent AS-impacted genes: they are connected to one or several sense nodes in this graph, but in the SAS network, they only have anti-sense neighbours. (Color figure online)

network inference. The information provided by these change motifs would be omitted by a classical gene network inference relying only on sense data. Figure 2 illustrates what we define as a change motif. The S and SAS network are both drawn in the same graph. A sense node is blue and an anti-sense node is purple. A red link is a connection present only in S network, a green link is a connection present only in SAS network and a gray link is a connection present in both networks. Figure 2 represents a change motif because the node $S1$ has no outer link common to the S network ($S1 \rightarrow S2$ and $S1 \rightarrow S4$) and the SAS network ($S1 \rightarrow AS3$). It means that $S1$ shares a greater mutual information with $AS3$ than with $S2$ and $S4$.

The accepting rate of Extended Core Network determines the number of neighbours a node has, and consequently the number of change motifs. With an accepting rate of 5%, we identified 300 change motifs in the H experiment and 308 motifs in the 60DAH experiment. The number of change motifs shows that anti-sense transcripts have a real impact because about 30% of sense actors are AS-impacted genes.

Change motifs allow us to identify areas in the sense core network where anti-sense transcription has a deep impact, and we can identify anti-sense actors that modify the core network connections. The genes involved into a change motif will be studied to analyse the impact of anti-sense transcription.

Figure 3 shows the graphical result of the differential network analysis for the 60DAH experiment. The graph was drawn using Cytoscape [25]. We can see that AS-impacted genes (orange triangles) are spread all over the network.

3.3 Steiner Trees to Compute the Rewiring of AS-impacted Sub-graphs

Figure 3 shows that it is difficult to extract relevant information from the graph visualisation. Therefore it is interesting to focus on some restricted parts of the network to help the study of biological processes. We see on Fig. 3 that many AS-impacted genes are connected together and thus form a part of the network on which the anti-sense transcription has a great impact. Therefore a second stage of our differential analysis proposes a more thorough study of the sub-networks containing many AS-impacted genes. This analysis relies on the Steiner tree problem that we explain below, before presenting our approach.

Given an undirected graph $G = (V, E)$ with a set of vertices V and a set of edges E, the classical Steiner tree problem aims to find, for a set of vertices S called terminal nodes, a sub-graph G' of G containing S such that there exists a path between every pair of vertices of S and with a minimal number of edges. The nodes of $V - S$ necessary in the Steiner tree G' to obtain a connected graph are called Steiner vertices. When dealing with weighted edges, the general Steiner tree problem aims to find a minimal cost sub-graph spanning the terminal nodes. In bioinformatics, the Steiner tree problem can be used to extract information from the large databases of assessed molecular interactions. For example, if a biologist wants to study the implication of a set of proteins in the interactome, it can extract from a dedicated database a Steiner tree with these proteins as terminal nodes. As this combinatorial optimization problem is known to be NP-complete [26], several heuristic methods have been proposed to deal with large graphs. A recent experimental study on a large human protein-protein interaction network can be found in [27].

Our proposal to study the impact of AS-transcription on the gene interactions is the following. Let us recall that we have computed two extended core networks: the network S contains sense actors whereas the network SAS contains sense and anti-sense actors. The comparison of these two networks leads to identify AS-impacted genes, that are nodes of the network S strongly connected to anti-sense transcripts in the network SAS. It is thus interesting to study the impact of anti-sense transcription on groups of AS-impacted genes. We now define an *AS-impacted sub-graph* as a connected sub-graph of the sense network

S restricted to AS-impacted genes and containing at least two such genes. If SG is an AS-impacted sub-graph, that means that its nodes are connected in the sense network S but as these nodes are AS-impacted, these interactions no longer appear in the SAS network. Thus it is interesting to wonder what are the relationships between these nodes in the SAS network. One way to answer this question is to consider the Steiner tree problem. For each AS-impacted sub-graph of S, we look for a Steiner tree in the SAS network connecting its AS-impacted genes. If it can be found, this Steiner tree shows how the interactions present in S between these AS-impacted genes are rewired in SAS. Different ways can be proposed to the biologists to exploit this information. A visualisation of the rewiring of the interactions can help to focus on interesting interactions. A functional analysis of the AS-impacted genes can be realized to decipher the main functions impacted by anti-sense transcription.

To compute the minimum Steiner tree, we use a heuristic approach called the shortest paths based approximation [27]. The methods constructs the Steiner tree ST by successive steps. The first step selects one terminal node as the first node of the Steiner tree ST. Then each following step connects to ST, by a shortest path, one terminal node n that is not yet in the current ST. The node n is chosen as the closest node to all the nodes of the current ST. The algorithm stops when all the terminal nodes are in ST and the resulting tree is pruned by a final step that computes a minimum spanning tree for ST.

As shown experimentally in [27], the shortest paths heuristic is a good compromise between solution quality and computational speed for large graphs such as the ones addressed in bioinformatics networks. SteinerNet is a R package that implements four different approximate and one exact algorithms to solve the minimum Steiner tree problem described in [27]. The SteinerNet package has not been updated since 2013 but its dependencies have changed; so the Steiner-Net package has been deleted from CRAN repository. We updated the package in order to use it with R version 3.2.0. We will be pleased to share this updated package upon request.

We use SteinerNet in order to compute Steiner trees of AS-impacted sub-graphs from the H and 60DAH experiments. The Steiner tree from Fig. 4 is composed by 26 terminal nodes (orange) and 41 Steiner nodes (blue and purple). This tree is the rewiring of the 26 AS-impacted genes from one AS-impacted sub-graph of the 60DAH experiment. The 26 terminal nodes are the 26 AS-impacted genes, meaning that all the genes from the AS-impacted graph were rewired. The red links from Fig. 4 are the connections from the S graph, that is the connections from the AS-impacted sub-graph. This representation allows us to see if the Steiner nodes connecting terminal nodes "replace" an existing connection or "create" a new one between two AS-impacted genes.

Steiner trees allow biologists to identify the rewiring among AS-impacted genes and thus the genes that are highly impacted by anti-sense transcripts. The next question is to find if Steiner nodes that allow this rewiring share biological functions with the AS-impacted genes. We identify GO terms associated with each gene from the Steiner tree and we look for common terms between terms associated with terminal and Steiner nodes. Then we can give to the biologists a list of genes which are associated with the common terms.

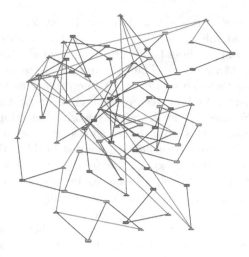

Fig. 4. Steiner tree of an AS-impacted sub-graph from the 60DAH experiment. Orange nodes are AS-impacted genes, blue nodes are Steiner sense nodes, purple nodes are Steiner anti-sense nodes. Gray links are connections from the SAS graph, red links are connections from the S graph.(Color figure online)

From the Steiner tree of Fig. 4, we identified 57 different terms associated to the terminal nodes, and 143 associated to the Steiner nodes, with only 16 terms in common. Because the fruit ripening studied with the 60DAH experiment is a stress for the fruit, the biologists want to study the response of stress. We identify 4 common terms related to a stress response. Our method permits to identify a set of genes which appear in the Steiner trees with a GO term related to our biological function, the stress response. From the 29 AS-impacted sub-graphs of the 60DAH experiment, we computed 35 Steiner trees and created a set of 10 genes related to the response of stress.

4   Conclusion

In order to study on a large scale the impact of anti-sense transcription, we propose to infer gene networks, with the particularity to integrate anti-sense transcripts in the process. We extend the recent ideas introduced in the field of differential network analysis, because in our framework, the main idea is to compare a network inferred from sense data and a network inferred from sense and anti-sense data. To achieve such a comparison, we first propose the Extended Core Network inference method. Extending the idea proposed by C3NET, this method builds a co-expression gene network where only the core of the network is considered by selecting the most important interactions for each gene. Secondly we extend the differential network analysis that we performed in [23] by using the Extended Core Network method instead of C3NET. The comparison of two networks inferred on different sets of actors leads to define change motifs and AS-impacted genes, that highlight the impact of integrating anti-sense transcription in the gene network. We also propose to complete the differential analysis by studying the rewiring of the AS-impacted sub-graphs. The Steiner trees computed by our method provide a more global view to study the impact of anti-sense transcription on biological functions.

Acknowledgement. The authors would like to thank the GRIOTE project, funded by the Pays-de-la-Loire Region. We thank the Institut de Recherche en Horticulture et Semences teams for providing micro-array data on apple with sense and anti-sense probes.

References

1. Bansal, M., Belcastro, V., Ambesi-Impiombato, A., di Bernardo, D.: How to infer gene networks from expression profiles. Mol. Syst. Biol. **3**(1), 78 (2007)
2. Altay, G., Emmert-Streib, F.: Inferring the conservative causal core of gene regulatory networks. BMC Syst. Biol. **4**(1), 132 (2010)
3. Marbach, D., Costello, J.C., Küffner, R., Vega, N.M., Prill, R.J., Camacho, D.M., Allison, K.R., The DREAM5 Consortium. Kellis, M., Collins, J.J., Stolovitzky, G.: Wisdom of crowds for robust gene network inference. Nat. Methods **9**(8), 796–804 (2012)
4. Friedel, S., Usadel, B., von Wiren, N., Sreenivasulu, N.: Reverse engineering: a key component of systems biology to unravel global abiotic stress cross-talk. Front. Plant Sci. **3**, 294 (2012)
5. Sharan, R., Ideker, T.: Modeling cellular machinery through biological network comparison. Nat. Biotechnol. **24**(4), 427–433 (2006)
6. Altay, G., Asim, M., Markowetz, F., Neal, D.E.: Differential C3net reveals disease networks of direct physical interactions. BMC Bioinf. **12**(1), 296 (2011)
7. Ideker, T., Krogan, N.J.: Differential network biology. Mol. Syst. Biol. **8**, 565 (2012)
8. Barabási, A.L., Gulbahce, N., Loscalzo, J.: Network medicine: a network-based approach to human disease. Nat. Rev. Genet. **12**(1), 56–68 (2011)
9. Pelechano, V., Steinmetz, L.M.: Gene regulation by antisense transcription. Nat. Rev. Genet. **14**(12), 880–893 (2013)
10. Yi, H., Richards, E.J.: A cluster of disease resistance genes in arabidopsis is coordinately regulated by transcriptional activation and RNA silencing. Plant Cell **19**(9), 2929–2939 (2007)
11. Celton, J.M., Gaillard, S., Bruneau, M., Pelletier, S., Aubourg, S., Martin-Magniette, M.L., Navarro, L., Laurens, F., Renou, J.P.: Widespread anti-sense transcription in apple is correlated with siRNA production and indicates a large potential for transcriptional and/or post-transcriptional control. New Phytol. **203**(1), 287–299 (2014)
12. Langfelder, P., Horvath, S.: WGCNA: an R package for weighted correlation network analysis. BMC Bioinf. **9**(1), 559 (2008)
13. Margolin, A.A., Nemenman, I., Basso, K., Wiggins, C., Stolovitzky, G., Favera, R.D., Califano, A.: ARACNE: an algorithm for the reconstruction of gene regulatory networks in a mammalian cellular context. BMC Bioinf. **7**(Suppl 1), S7 (2006)
14. Zhang, X., Liu, K., Liu, Z.P., Duval, B., Richer, J.M., Zhao, X.M., Hao, J.K., Chen, L.: NARROMI: a noise and redundancy reduction technique improves accuracy of gene regulatory network inference. Bioinformatics **29**(1), 106–113 (2013)

15. Bullard, J.H., Purdom, E., Hansen, K.D., Dudoit, S.: Evaluation of statistical methods for normalization and differential expression in mRNA-Seq experiments. BMC Bioinf. **11**, 94 (2010)

16. Qiu, X., Wu, H., Hu, R.: The impact of quantile and rank normalization procedures on the testing power of gene differential expression analysis. BMC Bioinf. **14**, 124 (2013)

17. Steuer, R., Kurths, J., Daub, C.O., Weise, J., Selbig, J.: The mutual information: detecting and evaluating dependencies between variables. Bioinformatics **18**(suppl 2), S231–S240 (2002)

18. Kurt, Z., Aydin, N., Altay, G.: A comprehensive comparison of association estimators for gene network inference algorithms. Bioinformatics **30**(15), 2142–2149 (2014)

19. Shen-Orr, S.S., Milo, R., Mangan, S., Alon, U.: Network motifs in the transcriptional regulation network of Escherichia coli. Nat. Genet. **31**(1), 64–68 (2002)

20. Ma, H.W., Kumar, B., Ditges, U., Gunzer, F., Buer, J., Zeng, A.P.: An extended transcriptional regulatory network of Escherichia coli and analysis of its hierarchical structure and network motifs. Nucleic Acids Res. **32**(22), 6643–6649 (2004)

21. Guelzim, N., Bottani, S., Bourgine, P., Kèpés, F.: Topological and causal structure of the yeast transcriptional regulatory network. Nat. Genet. **31**(1), 60–63 (2002)

22. den Bulcke, T.V., Leemput, K.V., Naudts, B., van Remortel, P., Ma, H., Verschoren, A., Moor, B.D., Marchal, K.: SynTReN: a generator of synthetic gene expression data for design and analysis of structure learning algorithms. BMC Bioinf. **7**(1), 43 (2006)

23. Legeay, M., Duval, B., Renou, J.-P.: Differential functional analysis and change motifs in gene networks to explore the role of anti-sense transcription. In: Bourgeois, A., Skums, P., Wan, X., Zelikovsky, A. (eds.) ISBRA 2016. LNCS, vol. 9683, pp. 117–126. Springer, Cham (2016). doi:10.1007/978-3-319-38782-6_10

24. Legeay, M., Duval, B., Renou, J.: Inference and differential analysis of extended core networks: a way to study anti-sense regulation. In: IEEE International Conference on Bioinformatics and Biomedicine, BIBM 2016, Shenzhen, China, 15–18 December 2016, pp. 284–287 (2016)

25. Shannon, P., Markiel, A., Ozier, O., Baliga, N.S., Wang, J.T., Ramage, D., Amin, N., Schwikowski, B., Ideker, T.: Cytoscape: a software environment for integrated models of biomolecular interaction networks. Genome Res. **13**(11), 2498–2504 (2003)

26. Karp, R.M.: Reducibility among combinatorial problems. In: Complexity of Computer Computations. The IBM Research Symposia Series, pp. 85–103. Springer, US, New York (1972)

27. Sadeghi, A., Fröhlich, H.: Steiner tree methods for optimal sub-network identification: an empirical study. BMC Bioinf. **14**, 144 (2013)

Nonlinear Control and Simulation of a Dielectric Elastomer Actuator-Based Compression Bandage on Flexible Human Calf

Shahram Pourazadi[1(✉)], Mehrdad Moallem[2], and Carlo Menon[1(✉)]

[1] MENRVA Research Group, School of Engineering Science,
Simon Fraser University, Burnaby, BC, Canada
{shahramp, c.menon}@sfu.ca
[2] School of Mechatronic Systems Engineering,
Simon Fraser University, Surrey, BC, Canada

Abstract. Compression bandages are widely used for a number of disorders associated with lower human leg including edema, orthostatic hypotension and deep vein thrombosis. In recent years, dielectric elastomer actuators (DEAs) are proposed to be used as active compression bandage to potentially augment or treat the lower leg disorders. DEA bandage applies a variable compression around the leg that varies upon voltage stimulation in the DEA. Prediction and control of the DEA behavior interacting with a soft object like human calf can be a very challenging and complex problem. In this paper a nonlinear analytical model is developed to represent the interaction between a silicon-based DEA compression bandage and soft human calf. An input-output linearization control strategy is utilized to design a controller that applies a desired compression profile to the calf. Lastly, MATLAB Simulink is used to simulate and illustrate the performance of the controller and the model.

Keywords: Active compression bandage · Dielectric elastomer actuator · Soft body · Nonlinear control · Simulation

1 Introduction

Typically, blood tends to accumulate in human lower legs due to capillary filtration and venous pooling [1]. There are certain mechanisms in the lower extremities including muscle contraction and vein valve closure that help the blood return. In some cases these mechanisms do not work properly which results in a decrease in blood flow and an increase in blood volume. Therefore, the amount of blood accumulated in the legs increases and the amount of blood driven to other parts of the body reduces. This venous pooling and reduction in blood flow can cause disorders such as hypotension, blood clots, edema and varicose veins. Each of these disorders has its own stream of significance, method of diagnosis and treatment [2]. In order to compensate for venous pooling, an effective external compression should be applied to the lower extremities [3]. In this way, blood pressure and flow is enhanced and venous pooling is reduced.

While other compression therapy methods such as compression bandages, compression stockings and intermittent pneumatic compression devices are currently

© Springer International Publishing AG 2017
I. Rojas and F. Ortuño (Eds.): IWBBIO 2017, Part II, LNBI 10209, pp. 289–300, 2017.
DOI: 10.1007/978-3-319-56154-7_27

practiced, dielectric elastomer actuators (DEAs) are recently proposed as active compression bandages to overcome some of the shortcomings of the current compression techniques [4, 5]. DEA compression bandage potentially provides a better compliance for the patients and help to alter the amount of compression based on the need of the body and also apply a different range of compression for a wide group of individuals with different leg dimensions and hemodynamic characteristics. Moreover, DEA bandages can actuate very fast and are very light weight and portable. A DEA consists of at least one elastomeric dielectric membrane which is placed in between two compliant electrode layers [6]. Typically the dielectric membrane is made of silicone or acrylic elastomers and the electrodes are made of conductive grease, particles or filaments [7, 8].

The quantity of external compression is an important factor in augmentation and treatment of the mentioned lower leg disorders. For instance, it has been suggested that a graded compression of 18 mmHg at the ankle, 14 mmHg at mid-calf, and 8 mmHg at the knee is required to optimize the venous blood flow in lower leg [9]. Also, it is shown that a compression of 35–40 mmHg at the ankle is required to prevent edema in patients with intense venous disease [10]. Thus, the bandages used for compression therapy should be designed to apply the proper range of compression. Typically, the amount of pressure exerted by elastic bandages is a function of the calf geometry. In practice, the calf is compressible and its volume and geometry change upon compression. Thus, the amount of compression that elastic bandages exert differs depending on the volume and shape of the calf. In this regard, Stenger et al. suggested that volumetric changes of the leg should be considered prior to the design of elastic bandages [3]. Therefore, it is interesting to study and identify how profound the calf compressibility would affect the behavior of the DEA bandage.

In this paper, first, the behavior of the soft human calf is studied and a mathematical model is introduced that relates the external compression to the radius changes (or volume changes) for a human calf. Next, the analytical model for a silicon DEA bandage is provided which calculates the amount DEA applied compression to the calf in terms of DEA properties, applied voltage to the DEA and the calf radius. Then, the two nonlinear models are integrated to describe the interaction between the DEA and the soft human calf. An input-output linearization control method is used to control the nonlinear behavior of the DEA on a human calf. Finally, the analytical model and the controller are implemented in MATLAB Simulink.

2 Analytical Modeling

An illustration of the DEA compression bandage is shown in Fig. 1(a). DEA bandage applies a compression on the calf and the calf deforms based on the amount of this compression. This deformation in the calf results in a variation in the DEA applied compression in turn. In order to explain this coupled behavior, the mathematical model for each component should be defined.

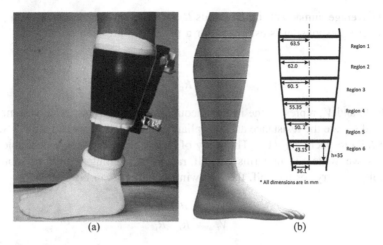

Fig. 1. DEA bandage and the human calf. (a) A prototype of DEA active compression bandage. (b) Dimensions for an average human calf. Reproduced with permission [4, 5].

2.1 Human Calf

The human calf compresses upon application of pressure. The calf volume changes considerably with the application of an external compression and is usually determined by the calf compliance C:

$$C = \frac{\delta \Delta V}{\delta P} \tag{1}$$

where, $\delta \Delta V$ and δP represent volume and external compression changes. Calf compliance can be measured using different methods [11–18]. Common clinical method of calf compliance measurement is by monitoring the calf volume changes using plethysmography while applying an external compression proximal to the knee with congestion cuffs [11–15]. In this method it is assumed that the volume change in the limb is equivalent to volume change of the underlying venous vessel. This assumption however, is argued to be not very accurate [18]. Moreover, this method is believed to not precisely predict the compliance of the calf as the external pressure is applied proximal to the knee and not to the calf itself. A more sophisticated method was however used to accurately measure the calf compliance by monitoring cross sectional changes of the calf under application of an external compression to the calf area [17, 18]. In [5] the compliance data acquired by Thirsk et al. [18] was used and a term was developed for steady state radial strain in the calf:

$$\frac{R_i - R_e}{R_i} = \varepsilon_r = a \cdot e^{b \cdot P} + c \cdot e^{d \cdot P} = f(P) \tag{2}$$

where, ε_r is the radial strain in the calf, R_e is the calf radius in steady state, R_i is the initial calf radius, P is the external compression on the calf and, a, b, c and d are constants that vary for different regions of the calf and are given in [5]. Figure 1(b) shows the average human calf initial radius R_i for different calf regions [5]. In [17] the calf volumetric response was estimated for a given compression on the calf:

$$\frac{d\frac{V}{V_i}}{dt} = \frac{\frac{V_e}{V_i} - \frac{V}{V_i}}{R_c C_c} \tag{3}$$

where, V, V_i and V_e represent the instantaneous, initial and steady state volume of the calf. R_C and C_C are the resistance and compliance in the calf veins and are 3.8 s.kPa/% and 1.1%/kPa respectively. [17]. The study of radius changes is more desirable in this study thus we write (3) in terms of calf radius. Assuming a uniform conical or cylindrical geometry for the calf, the following equations can be written:

$$d\frac{V}{V_i} = 2\frac{R}{R_i}d\frac{R}{R_i} \tag{4}$$

$$\frac{V}{V_i} = (\frac{R}{R_i})^2 \tag{5}$$

$$\frac{\Delta V}{V_i} = 2\frac{\Delta R}{R_i} \Rightarrow \frac{V_i - V_e}{V_i} = 2\frac{R_i - R_e}{R_i} \tag{6}$$

where, R is the instantaneous calf radius. Equation (6) is valid considering a uniform geometrical shape for calf cross section area [19]. Using (2) in (6) results in:

$$\frac{V_e}{V_i} = 1 - 2f(P). \tag{7}$$

Equation (3) can be written in terms of calf radius using Eqs. (4), (5) and (7):

$$\frac{d\frac{R}{R_i}}{dt} = \frac{\frac{1-2f(P)}{\frac{R}{R_i}} - \frac{R}{R_i}}{2R_c C_c}. \tag{8}$$

Equation (8) describes the calf radius changes for a given external compression.

2.2   Silicon DEA Bandage

The total compression applied by the DEA bandage has two components, the mechanical pressure which is the compression resulting from the mechanical stress in the DEA before actuating the DEA, and the actuation pressure, which is the compression variation while actuating the DEA.

Mechanical component. The DEA bandage is assumed to be made out of a flat DEA [4]. This flat DEA can be bent to form a cylindrical bandage with inner and outer radii of A and B. This DEA bandage is stretched radially as it is wrapped around the calf radius R and the following stretch ratios λ_a and λ_b can be obtained:

$$\lambda_a = \frac{R}{A}, \quad \lambda_b = \frac{b}{B} \tag{9}$$

where, b is the DEA outer radii after stretch. Since the geometry of calf is conical, the DEA undergoes different stretch ratios along the height of the calf. As the stretch ratio of the DEA bandage λ changes the amount of its compression on the calf also changes. The amount of this compression is obtained using the following equation [20]:

$$P_m = \int_{\lambda_b}^{\lambda_a} -\frac{\mu_1(\lambda^{\alpha_1} - \lambda^{-\alpha_1}) + \mu_2(\lambda^{\alpha_2} - \lambda^{-\alpha_2})}{\lambda(\lambda^2 - 1)} d\lambda \tag{10}$$

where, $\mu_1, \mu_2, \alpha_1, \alpha_2$ are the ogden parameters for the DEA hyperelastic material which are obtained from tensile tests [4, 20].

Electrical component. The DEA electrical circuit can be simplified as shown in Fig. 2. The DEA can be represented with a simple capacitance and is obtained from [21]:

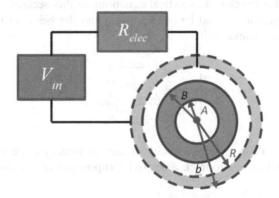

Fig. 2. The electrical circuit and geometry of the DEA bandage.

$$C = \frac{2\pi\varepsilon_0\varepsilon_r L}{\ln(\frac{b}{R})} \tag{11}$$

where, ε_0 is the vacuum permittivity, ε_r is the relative permittivity and L is the height of the DEA cylinder or the calf section. As the DEA compression on the calf changes the calf radius changes and as a result the DEA capacitance changes based on (11). The DEA charging circuit shown in Fig. 2 is a RC circuit with a variable capacitance. Thus, it can be written:

$$V_{in} = V_c + R_{elec}\left(C\frac{dV_c}{dt} + V_c\frac{dC}{dt}\right) = V_c(1 + R_{elec}\frac{dC}{dt}) + R_{elec}C\frac{dV_c}{dt}. \tag{12}$$

R_{elec} represents the DEA equivalent series resistance plus other resistive dissipations in the charging circuit and V_C is voltage across the DEA.

Electromechanical component. The DEA actuation pressure is calculated from the following equation [20]:

$$P_a = \frac{\varepsilon_0 \varepsilon_r V_C^2}{2ln^2(b/R)} \frac{b^2 - R^2}{R^2 b^2}. \tag{13}$$

As discussed earlier the total DEA compression on the calf is composed of P_a and P_m and is given by:

$$P = P_m - P_a \tag{14}$$

As it can be realized for Eq. (14), the DEA compression is reduced upon actuation. Thus, the radius of the calf is increased which in turn results in an increase of the DEA mechanical pressure.

Summarizing the developed analytical equations in this section, a system of coupled, nonlinear equations should be solved to explain the behavior of the DEA performance on the soft human calf:

$$\begin{cases} \frac{d\frac{R}{R_i}}{dt} = \frac{\frac{1-2f(P)}{\frac{R}{R_i}} - \frac{R}{R_i}}{2R_c C_c} \\ \frac{dV_c}{dt} = \frac{V_{in} - V_c(R_{elec}\frac{dC}{dt} + 1)}{R_{elec}C} \\ P = P_m - P_a \end{cases} \tag{15}$$

The following section presents a nonlinear control strategy to design a controller in order for the DEA bandage to track a desired compression on the calf.

3 Controller Design

As mentioned earlier, it is interesting to have a compression bandage that provides a desirable and controlled compression profile on the calf. Equation (14) shows that the DEA compression can be controlled by the actuation pressure and using the input voltage to the DEA. However, the system of equations in (15) is nonlinear and it can be challenging to design a controller that tracks a desires compression profile. While a linear control design can be used to obtain a controller for a small linearized region, more sophisticated, nonlinear controller design methods can be used to obtain a controller for a wider range of the system operation. In order to address the nonlinearities in the system an input-output linearizer controller [22, 23] is utilized in this paper. Using (15) and considering the DEA input voltage V_{in} as the system input and the DEA bandage compression P as the output, a controller can be designed. A nonlinear input control law should be calculated to remove the nonlinearities from the system.

Differentiating the output yields:

$$\dot{y} = \dot{P} = \dot{P}_m - \dot{P}_a = \frac{df(R)}{dt} - \frac{df(V_C, R)}{dt}. \tag{16}$$

For simpler differentiation, P_m and P_a which are nonlinear functions of calf radius R and/or DEA voltage V_C, can be estimated using:

$$P_m = f(R) \approx p_0 R^3 + p_1 R^2 + p_2 R^1 + p_4 R^0 \tag{17}$$

$$P_a = f(V_c, R) \approx \varepsilon_0 \varepsilon_r V_C^2 (k_0 R^{k_1} + k_2) \tag{18}$$

where, p_0, p_1, p_2, p_3, k_0, k_1 and k_2 are constants and are given in Table 1.

Using (12), (17) and (18) in (16) yields:

$$\frac{dy}{dt} = \frac{df(R)}{dt} - 2\varepsilon_0 \varepsilon_r V_c \left(k_0 R^{k_1} + k_2 \right) \frac{V_{in} - V_c \left(R_{elec} \frac{dC}{dt} + 1 \right)}{R_{elec} C} + \varepsilon_0 \varepsilon_r V_c^2 \left(k_0 k_1 \frac{dR}{dt} R^{k_1 - 1} \right). \tag{19}$$

Table 1. Constants obtained for Eq. (17) and (18).

Parameter	Value	Unit	Parameter	Value	Unit
p_0	6.4e3	Pa/m^3	k_0	9.3e-6	1/ m^{k_1+2}
p_1	−2.9e4	Pa/m^2	k_1	−3.9	–
p_2	4.7e4	Pa/m	k_2	4.0e4	1/m^2
p_3	−2.4e4	Pa			

In (19), if the input V_{in} is selected wisely the nonlinearities can be removed. Let us choose:

$$V_{in} = \frac{\left(-v + \frac{df(R)}{dt} + \varepsilon_0 \varepsilon_r V_c^2 (k_0 k_1 \frac{dR}{dt} R^{k_1 - 1}) \right)}{2\varepsilon_0 \varepsilon_r V_c (k_0 R^{k_1} + k_2)} R_{elec} C + V_c \left(R_{elec} \frac{dC}{dt} + 1 \right). \tag{20}$$

By substituting (20) into (19), the output derivative $dP/dt = dy/dt$ is obtained as:

$$\frac{dP}{dt} = \frac{dy}{dt} = v. \tag{21}$$

To make the DEA compression P track a desired profile $P_d(t)$ let us choose:

$$v = \frac{dP_d}{dt} - k(P - P_d). \tag{22}$$

By setting the error as $e = P - P_d$ and using (21) and (22):

$$\frac{de}{dt} + ke = 0 \qquad (23)$$

which ensures the stability of the system by convergence of the error to zero for positive k. However, it should be noted that the selected control input in (20) is singular for $V_C = 0$. Thus, at all times the desired pressure should be in a way to require $V_C \neq 0$.

4   Simulations and Results

The analytical equations and the controller developed in Sects. 2 and 3 were simulated in MATLAB Simulink. Figure 3 shows the DEA bandage and the soft human calf model block diagram.

For this study a silicone DEA with ogden parameters similar to [4] was used. The integration in (10) was solved using a Simpson's numerical integration method constructed into a custom MATLAB *S-function*.

Figure 4 shows the nonlinear controller block diagram. A desired compression profile can be set up and by tuning the controller gain k the DEA bandage can be controlled to exert the desired compression profile on the calf.

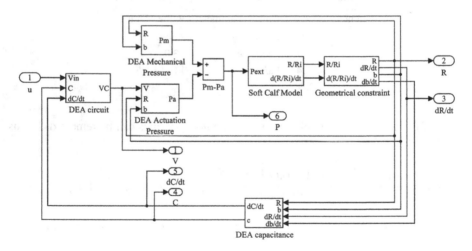

Fig. 3. DEA bandage and soft human calf system block diagram model.

The block diagram in Fig. 4 was simulated in Simulink for a duration of 20 s and using ode23t solver with automatic variable step sizes. The DEA bandage was considered to be wrapped around the calf section region 4 (see Fig. 1(b)) with a height of 35 mm.

Figure 5 shows the performance of the controlled DEA for a compression profile with a step increase and decrease. It is shown that as the gain k is increased the error

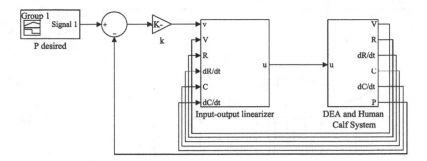

Fig. 4. Nonlinear Controller block diagram.

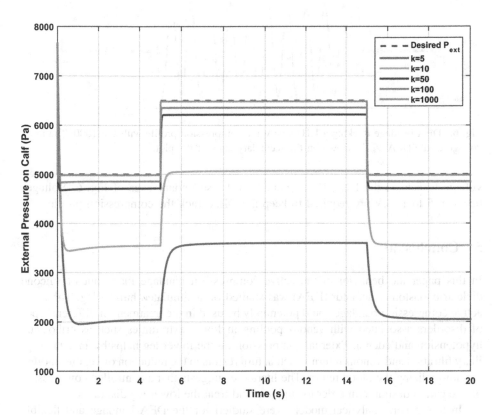

Fig. 5. Step compression profile and the performance of the DEA for different controller gains k.

between the actual and desired compression on the calf is reduced and the DEA bandage tracks the desired profile faster.

Figure 6 shows the simulation of the controlled system for an alternating compression profile. The figure shows that the design is able in controlling the DEA to track an alternating compression profile with errors smaller than 0.05%. The required

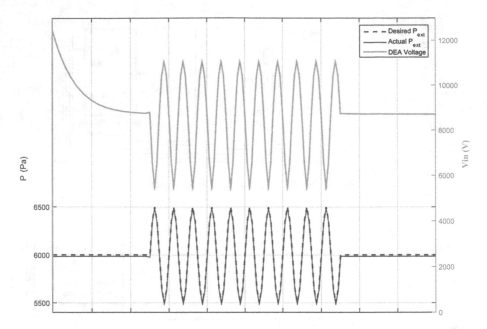

Fig. 6. DEA bandage tracking a 1 Hz sine wave compression profile with $k = 1000$. The input voltage to the DEA V_{in} is shown on the secondary axes of the plot.

voltage to actuate the DEA V_{in} is plotted on the secondary axes of Fig. 6. Voltages between 5 to 11 kV are required to keep the DEA track the compression profile.

5 Conclusion

In this paper the behavior of the active compression bandage made out of silicone dielectric elastomer actuator (DEA) was studied on a simulated human leg. The DEA active compression bandage could potentially be used in the treatment of different types of disorders associated with venous pooling in lower extremities such as orthostatic hypotension and edema. External compression on the lower leg helps balance the capillary filtration and venous return which in turn results in the reduction of venous pooling and normalizing the blood return. The literature suggests that a controlled compression on the calf is desirable in order to augment and treat the lower leg disorders.

In this paper, analytical models were studied for the DEA bandage and flexible human calf. Due to the calf flexibility and compressibility, the size of the calf changes upon application of external compression by the DEA bandage which in turn affect the compression created by the bandage. A system of nonlinear and coupled equations was developed to represent the DEA interaction with the human calf. An input-output linearization control technique was used to design a controller for the DEA behavior on the calf. The designed controller and the system were implemented in MATLAB Simulink for simulations. Simulations were performed for a desired step and alternating compression profiles. It was shown that using the designed controller, the DEA was able to track the desired compression profile with errors less than 0.05%.

Acknowledgment. This research was supported by the Natural Sciences and Engineering Research Council of Canada (NSERC), the Canadian Institutes of Health Research (CIHR) and the Michael Smith Foundation for Health Research (MSFHR).

References

1. Hainsworth, R.: Pathophysiology of syncope. Clin. Auton. Res. **14**(Suppl 1), 18–24 (2004)
2. Lewis, T.: A lecture on vasovagal syncope and the carotid sinus mechanism. Br. Med. J. **1**(932), 873 (1932)
3. Stenger, M.B., Brown, A.K., Lee, S.M.C., Locke, J.P., Platts, S.H.: Gradient compression garments as a countermeasure to post-spaceflight orthostatic intolerance. Aviat. Space Environ. Med. **81**(9), 883–887 (2010)
4. Pourazadi, S., Ahmadi, S., Menon, C.: Towards the development of active compression bandages using dielectric elastomer actuators. Smart Mater. Struct. **23**(6), 65007 (2014)
5. Pourazadi, S., Ahmadi, S., Menon, C.: On the design of a DEA-based device to pot entially assist lower leg disorders: an analytical and FEM investigation accounting for nonlinearities of the leg and device deformations. Biomed. Eng. Online **14**(1), 103 (2015)
6. Bar-Cohen, Y.: Electroactive Polymer (EAP) Actuators as Artificial Muscles: Reality, Potential, and Challenges, 2nd edn. SPIE Optical Engineering Press, Bellingham (2004)
7. Michel, S., Zhang, X.Q., Wissler, M., Löwe, C., Kovacs, G.: A comparison between silicone and acrylic elastomers as dielectric materials in electroactive polymer actuators. Polym. Int. **59**(3), 391–399 (2009)
8. Rosset, S., Shea, H.R.: Flexible and stretchable electrodes for dielectric elastomer actuators. Appl. Phys. A **110**(2), 281–307 (2012)
9. Sigel, B., Edelestein, A., Savitch, L., Hasty, J., Felix, W.: Types of compression for reducing venous stasis. Arch. Surg. **110**, 171–175 (1975)
10. Stemmer, R.: Ambulatory elasto-compressive treatment of the lower extrimities particularly with elastic stokcings. Der Kassenarzt **9**, 1–8 (1969)
11. Halliwill, J.R., Minson, C.T., Joyner, M.J., John, R.: Measurement of limb venous compliance in humans: technical considerations and physiological findings. J. Appl. Physiol. **87**(4), 1555–1563 (1999)
12. Monahan, K.D., Dinenno, F.A., Seals, D.R., Halliwill, J.R.: Smaller age-associated reductions in leg venous compliance in endurance exercise-trained men. Am. J. Physiol. Heart Circ. Physiol. **281**(3), H1267–H1273 (2001)
13. Monahan, K.D., Ray, C.A.: Gender affects calf venous compliance at rest and during baroreceptor unloading in humans. Am. J. Physiol. Heart Circ. Physiol. **286**(3), H895–H901 (2004)
14. Lipp, A., Sandroni, P., Ahlskog, J.E., Maraganore, D.M., Shults, C.W., Low, P.A.: Calf venous compliance in multiple system atrophy. Am. J. Physiol. Heart Circ. Physiol. **293**(1), H260–H265 (2007)
15. Sielatycki, J.A., Shamimi-Noori, S., Pfeiffer, M.P., Monahan, K.D.: Adrenergic mechanisms do not contribute to age-related decreases in calf venous compliance. J. Appl. Physiol. **110**(1), 29–34 (2011)

300   S. Pourazadi et al.

16. Binzoni, T., Quaresima, V., Ferrari, M., Hiltbrand, E., Cerretelli, P.: Human calf microvascular compliance measured by near-infrared spectroscopy. J. Appl. Physiol. **88**(2), 369–372 (2000)
17. Zicot, M., Parker, K.H., Caro, C.G.: Effect of positive external pressure on calf volume and local venous haemodynamics. Phys. Med. Biol. **22**(6), 1146–1159 (1977)
18. Thirsk, R.B., Kamm, R.D., Shapiro, A.H.: Changes in venous blood volume produced by external compression of the lower leg. Med. Biol. Eng. Comput. **18**(5), 650–656 (1980)
19. Whitney, R.J.: The measurement of volume changes in human limbs. J. Physiol. **121**(1), 1–27 (1953)
20. Ahmadi, S., Mattos, A.C., Barbazza, A., Soleimani, M., Boscariol, P., Menon, C.: Fabrication and performance analysis of a DEA cuff designed for dry-suit applications. Smart Mater. Struct. **22**(3), 35002 (2013)
21. Bell, D.A.: Fundamentals of electric circuits, 7th edn. Oxford University Press, Oxford (2009)
22. Zak, S.: Systems and Control. Oxford University Press, Oxford (2003)
23. Slotine, J., Li, W.: Applied Nonlinear Control. Prentice-Hall, Englewood Cliffs (1991)

Dengue Agent-Based Model in South American Temperate Zone

Carlos M. Pais[1(✉)], Maximiliano G. Colazo[1], Maximiliano Fernandez[1],
Silvana Bulatovich[1], and Hugo Fernandez[2]

[1] Cybernetics Laboratory, Department of Bioengineering, School of Engineering, Universidad Nacional de Entre Ríos, Ruta Provincial 11, Km 10, Oro Verde, Entre Ríos, Argentina
cpais@bioingenieria.edu.ar, maxi732@gmail.com,
maxi1991f@hotmail.com, silvana.bulatovich@gmail.com
[2] Administración Nacional de Laboratorios e Institutos de Salud, Instituto de Enfermedades Respiratorias (INER) "Dr. Emilio Coni", Santa Fe, Argentina
fernanhg@outlook.com

Abstract. Dengue is a disease that is increasing yearly in number of cases and severity in temperate zones. Different actions have been taken for controlling this disease in the central region of Argentina, without anticipating the effectiveness of these interventions. Therefore, considering the weather conditions of the zone under study, a mathematical model was implemented that was capable of reproducing the information surveyed about dengue-infected patients in another South American temperate zone. Then, in attempting to reproduce the heterogeneity in population density and in the contact between humans and *Aedes Aegypti* mosquitoes, as well as the impact of randomness on these systems, an Agent-Based Model (ABM) was implemented. Said model is based on data surveyed about the target population and anticipates the possible results of some interventions suggested by epidemiologists.

Keywords: Agent-based modelling · ABM · Dengue · Epidemiology

1 Introduction

Over the last 25 years of the 20th century, the average surface temperature of the Earth increased by 0.4 °C and it is projected to increase 2–3 °C during the 21st century [1]. The consequences of this climate change impact, both directly and indirectly, on human health. Direct impacts are related to floods and extreme heat and cold waves, while some of the indirect consequences are related to the transmission patterns of certain infectious diseases, such as salmonellosis and other vector-borne diseases, including malaria and dengue [2].

Dengue is a viral infection transmitted by Aedes Aegypti mosquitoes within the geographic zone under study. It is also transmitted by the Aedes Albopictus mosquito in other regions. The different forms of dengue, i.e. dengue (with or without warning signs) and severe dengue [3], constitute one of the most relevant re-emerging tropical diseases in this century.

© Springer International Publishing AG 2017
I. Rojas and F. Ortuño (Eds.): IWBBIO 2017, Part II, LNBI 10209, pp. 301–312, 2017.
DOI: 10.1007/978-3-319-56154-7_28

This disease was present only in tropical and subtropical regions; however, over the last 30 years, outbreaks have been evidenced in temperate zones. According to the case report No. 117 issued by the World Health Organization (WHO) in January 2012, nearly half the world's population is at risk [4]. Nowadays, the WHO estimates that there are between 50 and 100 million people infected with dengue worldwide.

It is believed that the current high levels of movement of humans and the effect of population transfers, together with global warming, eased the spread of dengue to areas that were considered fully unreachable by the disease, such as the Oriental Republic of Uruguay [5]. In Europe (Sweden, for example), between 30 and 60 seropositive cases are reported annually, mainly due to overseas traffic [6, 7].

Mosquitoes transmitting dengue typically proliferate in tropical and subtropical zones, and there is fairly refined knowledge regarding the disease and its transmission under these weather conditions. However, evidence shows that this insect has learned to adapt to temperate climate and enters into diapause during the winter season (i.e. overwinters) at egg stage [8].

Given the lack of epidemiological research in temperate zones of South America, it is necessary to predict the number of cases that may emerge each year under these weather conditions. It is also important to anticipate the impact of interventions aimed at controlling this re-emerging disease.

It is now known that the dynamics of dengue is influenced by human behaviour, inasmuch as it has been proved that both spatial and social structures of human populations affect the evolution of the epidemic. It has been demonstrated that the geographical separation of neighborhoods or villages, the separation along lines of social interaction, the socioeconomic level or the social segregation strongly influence the dynamics of the disease [9].

Thus, the accuracy level of modelling an actual dengue epidemic depends on the assumptions of the theoretical framework, as well as on the parameters used to describe the relations between human and mosquito populations, their interaction in the virus transmission process, and the heterogeneity of populations, geographical areas and climates. It is practically impossible to capture these heterogeneity conditions using a global mathematical model based on differential equations.

In order to capture these factors, once the homogeneous mathematical model was capable of reproducing the theoretical premises, a local agent-based model (ABM) was implemented. This model creates a detailed simulation of the features of the actual phenomenon under study. This allows us to predict with sufficient degree of accuracy the dynamics of the system, based on the geographic and demographic characteristics of a limited population in a temperate zone of the central region of Argentina.

2   Materials and Methods

In this study, two models of the same epidemiological phenomenon were developed. First, a global mathematical model was implemented in Matlab®. Then, an agent-based model was written in Repast [10]. In both cases, system input data included the amount of Aedes Aegypti eggs surveyed by the research group in the city of Oro Verde throughout 2015, whereas output data comprised the evolution in the daily number of cases reported by Favier et al. in 2005 in Easter Island, Chile (Fig. 1) [11]. These data were selected to find the value of the parameters and validate the functioning of the models, as there is no serologic information about the number of infected people among the inhabitants of Oro Verde. Furthermore, data reported in [11] correspond to a population akin to that of Oro Verde, given that Easter Island is located in South America, in a temperate zone and has a population number and profile that is similar to the population in the college town of Oro Verde.

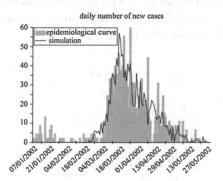

Fig. 1. Cases reported daily in Easter Island, Chile, in 2002. Adapted from Aguilera, X., Olea, A., Mora, J. & Abarca, K. 2002. "Brote de dengue en Isla de Pascua" (Dengue breakout in Easter Island). El Vigía 16, 37–38.

2.1 Deterministic Mathematical Model Based on Kermack–McKendrick Equations [12]

Following the structure of the Kermack–McKendrick theorem, a model was designed as shown in Fig. 2.

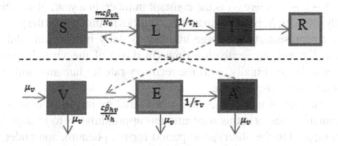

Fig. 2. Diagrammatic dengue model used to formulate the homogeneous mathematical model.

To date, only the Aedes Albopictus mosquito type is known to be capable of vertically transmitting the different chains of dengue virus to its eggs [13]. Consequently, since we only focused on the dynamics of adult Aedes Aegypti mosquitoes and their ability to bear and transmit the disease, the dynamics related to the number of eggs, larvae and pupae was set aside [14].

The diagrammatic model presented in Fig. 2 gives rise to the mathematical model formulated in Eqs. (1) to (6). This homogeneous model is based on the classic hypothesis that there are equal chances for each vector to bite any host.

Each population (humans and vectors) is divided into classes or sub-populations in relation to the disease. In the model, these classes are labelled as follows: Susceptible (S for humans and V for mosquitoes); incubating, Latent or Exposed (L for humans and E for mosquitoes); Infectious (I for humans and A for mosquitoes, as they contain the infectious agent), and immunised, eliminated or Recovered (R, just for humans, as once the mosquito becomes infected, it will carry the virus for life).

$$\frac{dS}{dt} = -\frac{mc\beta_{vh}}{N_v} SA \tag{1}$$

$$\frac{dL}{dt} = \frac{mc\beta_{vh}}{N_v} SA - \frac{1}{\tau_h} L \tag{2}$$

$$\frac{dI}{dt} = \frac{1}{\tau_h} L - \phi_h I \tag{3}$$

$$\frac{dV}{dt} = \mu_v N_v - \frac{c\beta_{hv}}{N_h} VI - \mu_v V \tag{4}$$

$$\frac{dE}{dt} = \frac{c\beta_{hv}}{N_h} VI - \frac{1}{\tau_v} E - \mu_v E \tag{5}$$

$$\frac{dA}{dt} = \frac{1}{\tau_v} E - \mu_v A \tag{6}$$

and $R = N_h - S - L - I$, where N_h is the constant number, in a year, of inhabitants in the city under study. Besides, N_v is the number of mosquitoes resulting from the proportion mN_h.

Parameter c represents the average number of daily bites of an adult mosquito. Constants τ_h and τ_v are the incubation time (in days) of the virus in people and in mosquitoes, respectively. Finally, φ_h is the recovery rate for humans, which is equal to the inverse of average viraemia time.

The vital dynamics of mosquitoes was considered by means of constant μ_v, as the average lifespan of the insect in this zone amounts approximately to 15 days. This period is too short compared to the observation period for the phenomenon under study (365 days), which forces us to take this parameter into account for insects, something that is not necessary for humans.

The Equations System (1) to (6) was integrated using Euler's method, with an step h value of one tenth of a day, and different parameter values were tested in accordance with the literature [15–18] and by minimizing the total squared error between the model output and evolution of reported cases in the epidemic of dengue in Easter Island in 2002 (Fig. 1).

In order to reduce the difference in the dynamics of the deterministic model with regard to the model shown in Fig. 2, randomness was included in a new proposal. In this new model, parameters β_{hv}, β_{vh}, c, τ_h, τ_v, φ_h and μ_v take random values with a normal probability distribution, whose standard deviation and average value match the values surveyed in the aforementioned works.

To study this phenomenon, the average temperature of Oro Verde during a year is included as a system input variable (Fig. 3). This variable affects the system parameters, thus causing changes to the average values and standard derivation throughout a year. As a result thereof, the constants of the model turned into random variables that depended on the daily average temperature (T), evolving throughout the year, according to the following formulas [8]:

$$c = 0.03T + 0.66 \tag{7}$$

$$\beta_{vh} = 0.0729T - 0.9037 \tag{8}$$

$$\beta_{hv} = 0.001044T(T-12.286)(32.461 - T)^{1/2} \tag{9}$$

$$\tau_{lv} = 4 + e^{(5.15-0.123T)} \tag{10}$$

$$\mu_v = 0.8692 - 0.1590T + 0.01116T^2 - 3.408x10 - 4T^3 + 3.809x10 - 6T^4 \tag{11}$$

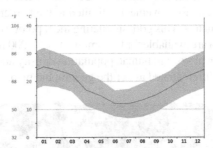

Fig. 3. Climograph for the city of Oro Verde. Data published by the aerodrome of the city of Paraná (10 km away from Oro Verde) at http://es.climate-data.org/location/1897/.

Furthermore, several authors have proved that the diapause capacity of the mosquito during the winter results in a dynamics of the number of dengue-infected people during a year that is completely different from the dynamics observed in tropical or sub-tropical zones. This factor leads to the disappearance of adult mosquitoes during winter in temperate zones, leaving only eggs and, in some cases, larvae [15–17, 19]. This model covers this factor; therefore, mosquitoes are non-existent during winter, and their population growth and decline follow an exponential function. As explained below, this factor determines the duration of the epidemic and results in a chart of the number of reported cases per day for this model that matches, in form and figures, the charts of Easter Island in Chile (Fig. 1).

2.2 Agent-Based Model

This type of model differs from traditional mathematical models mainly because it provides information regarding the global behaviour of the actual system and, at the same time, allows us to examine the consequences of changes in its internal components [20].

ABMs are scientifically relevant, and usually untreatable from a mathematical viewpoint, mainly due to the link between individuals and group behaviour. Although the dynamics of people and insects may be considered relatively simple, its collective behaviour may be complex.

In this study, the ABM was implemented in Eclipse Neon IDE, using the RePast Simphony 2.4.0 plugin. This platform comprises a set of simulation tools for agents and supports a hierarchical modelling approach, which is designed to help those who simulate complex adaptive systems [21].

RePast Simphony, which was developed by the University of Chicago [22], is a free-of-charge and open-source tool that supports multiple languages, such as Java, Groovy, Python and C++.

This model, which leaves the homogeneity premise behind, assumes that the heterogeneity of contacts arises from the structure of vector and host populations in households. Contact between humans and insects from different households is guaranteed due to people movements, as the movement of vectors is limited to the household environment.

This tool helps to simulate the movements of hosts and vectors, under a random walking dynamics [23]. For humans, this movement is limited to the borders of the grid, which has an area of $1,000 \times 1,000$ metres. This grid, simulating the topology of the city of Oro Verde, is presented in Fig. 4. Data are available for this small city of 5,000 inhabitants regarding the inputs to the epidemiological system: human population density and Aedes Aegypti oviposition, both of which are differentiated as per the geographic map.

Fig. 4. Satellite image of the city of Oro Verde depicting the average number of *Aedes Aegypti* mosquito eggs. (Color figure online)

Figure 4 also shows (in yellow) the average numbers of mosquito eggs surveyed in ovitraps throughout the year, which were placed by the research group.

The simulation of the mosquito movement also follows a random walking dynamics, with an average speed that is 32 times lower than that of a human; however, in this case, it is directed towards the places where the insect detects a larger number of humans.

In order to simulate this system based on local modelling, the agents "Human.agent" and "Aedes.agent" were implemented, along with the pseudo-agent "ModelInitial-izer.agent" which is in charge of initializing the model. The computational model code is programmed at a high level using flowcharts (Fig. 5). Based on these flowcharts, Repast Simphony generates the respective Groovy code, which can be translated into C ++ and executed in computer Clusters. This Repast feature allows us to leverage the advantages of parallel code execution [24].

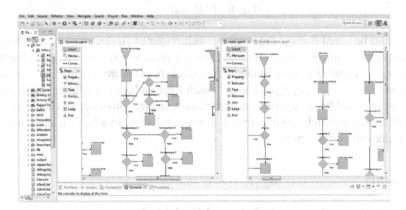

Fig. 5. Flowchart for the ABM model. REPAST Simphony.

The model's GUI records the evolution throughout time of four **data sets**:

- *Mosquito growth*: This set counts the number of mosquitoes existing in the environment per day (every 12 tick counts). This record is used to show the mosquito growth over time.
- *Infected humans*: It records all infected humans per day. This record is used to show the accumulated number of infected humans over time.
- *Recovered humans*: It records all recovered humans per day.
- *Reported cases per day:* It records the number of new infected humans per day (every 12 tick counts). This record is used to show the daily number of reported cases.

Agents develop their tasks in host environments. In Repast Symphony, environments are generated using contexts, which encapsulate a population of agents in a model. The context is a large container of agents that cannot manipulate objects included therein by itself: to this end, tools need to be supplied. These tools are known as projections.

In this model, a single context is used, which contains a grid-type projection. It is worth mentioning that 5-metre width and length dimensions are assigned to each grid cell. There-fore, it is possible for more than one human to occupy the same cell, and the same applies to mosquitoes.

The Aedes and Human Agent shown in Fig. 6 is an abstraction of the Aedes Aegypti mosquito and the people who inhabit the city of Oro Verde.

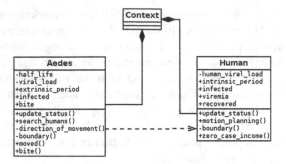

Fig. 6. UML chart.

It should be noted that the "movement" behaviour of Aedes restricts its relocation to a maximum radius of 3 grid positions from the place where they are born, therefore simulating the mobility feature limited to the household environment.

Parameters β_{hv}, β_{vh}, c, τ_h, τ_v, φ_h and μ_v, which were included in different agent behaviours, first matched those reported under the mathematical model case. However, since the results did not match those in Fig. 1, it was necessary to adapt the parameters by median quadratic error minimization.

3   Results

3.1   Stochastic Mathematical Model and Variable Temperature

As a result of the dynamics imprinted on the system by average temperature changes during a year, the stochasticity and the daily variation in the number of mosquitoes, the

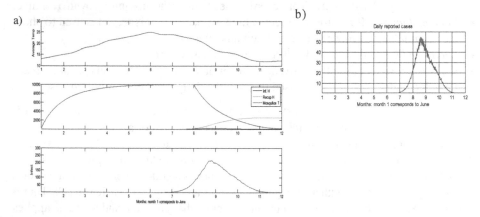

Fig. 7. Outputs of the complete mathematical model, where July is the 1st month. The evolution of the number of mosquitoes throughout the year is presented in (a). (b) shows the short intervals of rapid growth and slow decline in the daily number of reported cases as evidenced in the actual case under Fig. 1.

dynamics levels shown in Fig. 7 were obtained. Figure 7(b) is reasonably similar to Fig. 1. This allows us to assume that any modification to this model will imply some kind of intervention on the actual system.

Consequently, in order to represent the different situations that may occur in the actual system, the following changes were made to the model:

- *Fumigation or junk removal* (by reducing parameter *m*). It was found that, for an epidemic to occur, the minimum value of m should be 1.5 (3 mosquitoes every 2 humans). In other studies, this value was calculated based on *R0*. However, since in this model a large portion of the parameters were replaced with deterministic or random variables, that calculation should not be representative of the phenomenon.
- *Use of mosquito repellents and screens on windows.* The effect of mosquito repellents and other actions aimed at eradicating adult mosquitoes from households influence parameter *c* (number of bites per mosquito and per day). It is evident that this parameter is too sensitive for the model and that any minor reduction in this value, thus leaving it below 0.7, eliminates any possible epidemic occurrence.
- *Effect of medication during viraemia.* There is clinical evidence that, in other pathologies [25], once patients acquire viraemia, certain medications, such as vitamins or dietary supplements, may reduce the length of the disease. This is shown in the model by a reduction of τ_h (inverse to φ_h) in up to a maximum of 2 days in average. It was proved that this action did not lead to significant reductions in the daily number of reported cases.
- *Rise in ambient temperature.* The most interesting result of the changes to the parameters in this model is related to the increase in ambient temperature. This situation is verified year after year in the zone under study. To this end, leaving the remaining system parameters unchanged, and following the forecasts of temperature rise for the first half of this millennium and the zone's temperature profile or climograph, the average temperature was increased by 3°. This change affects all the system's dynamics, taking the number of theoretical reported cases per day from a maximum of 250 cases to an epidemic affecting every individual in the city.

3.2 The ABM

In order to represent, as accurately as possible, the actual situation surveyed in the city of Oro Verde, the ABM was implemented under the initial conditions of number of humans, mosquitoes and their respective locations in the grid (Fig. 8(a)). As shown in said Figure, assuming that each household has an average area of 10 × 10 m, the initial human population was distributed so as to represent the data retrieved from the last survey on households available. Moreover, vectors were initially located in places where the highest density of mosquito eggs was surveyed by the end of fall.

On the basis of the exponential growths and declines proposed in the mathematical model, the number of mosquitoes evolves over the year, checking that mosquito density levels represent those egg density levels found in different grid locations. Figures 8(b) to (d) show the evolution of human and mosquito density levels every 3 months.

a)   b)   c)   d)

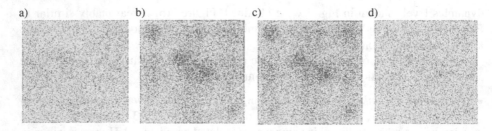

Fig. 8. (a): initial distribution of humans (blue) and mosquitoes (purple) in the grid depicting the city of Oro Verde; (b) number of humans and mosquitoes in the 3rd month of simulation; (c) in the 6th month and (d) in the 9th month. (Color figure online)

Using these initial conditions and the values of the system's constants (considering probability levels), the results shown in Fig. 9 were obtained.

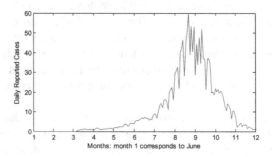

Fig. 9. Daily reported Dengue cases. Mean output of the ABM after 32 runs.

Finally, and similarly to the mathematical model, the ABM was tested by modifying the climograph and the parameters that represent the different actions on the system.

As expected, the results obtained using the ABM (both for the base case and those including intervention) significantly differ from the mathematical model. In all cases, the increase in the number of reported cases per day is much lower than the increase reported by the mathematical model.

4   Conclusion

It was proved that there are clear differences between the results reported by the mathematical model and those generated by the ABM. These differences arise from the influence of the heterogeneity observed in the human and mosquito density levels for the city under study. It may be expected that in cities with a larger population per square metre (i.e. cases which are more similar to a homogeneous density situation) the results from both the mathematical model and the ABM tend to coincide.

Therefore, the ABM presents its most significant results, evidencing that it captures underlying data from the heterogeneity of the human and mosquito density levels for the case under analysis.

Bibliography

1. Intergovernmental Panel on Climate Change, Ed.: Climate Change 2013 - The Physical Science Basis. Cambridge University Press, Cambridge (2014)
2. Patz, J.A., Githeko, A.K., McCarty, J.P., Hussein, S., Confalonieri, U., De Wet, N.: Climate change and infectious diseases. In: Climate Change and Human Health: Risks and Responses, pp. 103–37. World Health Organization (2003)
3. World Health Organization: Dengue: guidelines for diagnosis, treatment, prevention, and control. Spec. Program. Res. Train. Trop. Dis. **147**, x (2009)
4. WHO: WHO|Dengue and severe dengue, Fact Sheet (2016). http://www.who.int/mediacentre/factsheets/fs117/en/. Accessed 16 Nov 2016
5. WHO: Dengue Fever – Uruguay Disease Outbreak News, 10 March 2016 - Uruguay| ReliefWeb, Fact Sheet (2016). http://reliefweb.int/report/uruguay/dengue-fever-uruguay-disease-outbreak-news-10-march-2016
6. Rocklöv, J., Lohr, W., Hjertqvist, M., Wilder-Smith, A.: Attack rates of dengue fever in Swedish travellers. Scand. J. Infect. Dis. **46**(6), 412–417 (2014)
7. Heddini, A., Janzon, R., Linde, A.: Increased number of dengue cases in Swedish travellers to Thailand. Euro. Surveill. **14**(5), 19111 (2009)
8. Helmersson, J.: Mathematical modeling of dengue-temperature effect on vectorial capacity (2012). Phmed.Umu.Se
9. Reiter, P., Lathrop, S., Bunning, M., Biggerstaff, B., Singer, D., Tiwari, T., Baber, L., Amador, M., Thirion, J., Hayes, J., Seca, C., Mendez, J., Ramirez, B., Robinson, J., Rawlings, J., Vorndam, V., Waterman, S., Gubler, D., Clark, G., Hayes, E.: Texas lifestyle limits transmission of dengue virus. Emerg. Infect. Dis. **9**(1), 86–89 (2003)
10. Repast Suite Documentation. https://repast.github.io/. Accessed 06 Feb 2017
11. Favier, C., Schmit, D., Müller-Graf, C.D.M., Cazelles, B., Degallier, N., Mondet, B., Dubois, M.A.: Influence of spatial heterogeneity on an emerging infectious disease: the case of dengue epidemics. Proc. Biol. Sci. **272**(1568), 1171–1177 (2005)
12. Brauer, F., Castillo-Chavez, C.: Mathematical Models in Population Biology and Epidemiology, vol. 40, 2nd edn. Springer, New York (2012)
13. Shroyer, D.A.: Vertical maintenance of dengue-1 virus in sequential generations of Aedes albopictus. J. Am. Mosq. Control Assoc. **6**(2), 312–314 (1990)
14. Santos, L.B.L., Costa, M.C., Pinho, S.T.R., Andrade, R.F.S., Barreto, F.R., Teixeira, M.G., Barreto, M.L.: Periodic forcing in a three-level cellular automata model for a vector-transmitted disease. Phys. Rev. E Stat., Nonlin. Soft Matter Phys. **80**(1), 016102 (2009)
15. Shen, Y.: Mathematical models of dengue fever and measures to control it. Electron. Theses, Treatises Diss., May 2014
16. Coutinhoa, F.A.B., Burattinia, M.N., Lopeza, L.F., Massada, E.: Threshold conditions for a non-autonomous epidemic system describing the population dynamics of dengue. Bull. Math. Biol. **68**(8), 2263–2282 (2006)
17. Massad, E., Coutinho, F.A.B., Lopez, L.F., Da Silva, D.R.: Modeling the impact of global warming on vector-borne infections. Phys. Life Rev. **8**(2), 169–199 (2011)
18. Favier, C., Degallier, N., Boulanger, J.P., Lima, J.R.C.: Early determination of the reproductive number for vector-borne diseases: the case of dengue in Brazil. Trop. Med. Int. Health **11**(3), 332–340 (2006)
19. Lambrechts, L., Paaijmans, K.P., Fansiri, T., Carrington, L.B., Kramer, L.D., Thomas, M.B., Scott, T.W.: Impact of daily temperature fluctuations on dengue virus transmission by Aedes aegypti. Proc. Natl. Acad. Sci. U. S. A. **108**(18), 1–6 (2011)

20. Barnes, D.J., Chu, D.: Introduction to Modeling for Biosciences, vol. 53, no. 9. Springer, London (2010)
21. Macal, C.M., North, M.J.: Tutorial on agent-based modelling and simulation. J. Simul. **4**(3), 151–162 (2010)
22. Collier, N.: Repast: an extensible framework for agent simulation. Nat. Resour. Environ. Issues **8** (2001)
23. Bian, L.: Spatial approaches to modeling dispersion of communicable diseases - a review. Trans. GIS **17**(1), 1–17 (2013)
24. Guyot, P., Drogoul, A.: Designing multi-agent based participatory simulations. In: 5th Workshop on Agent Based Simulations, pp. 1–16 (2004)
25. Cheng-Fa, C., Chin-Teng, C.: Viral load analysis of a biodynamical model of HIV-1 with unknown equilibrium point. In: Proceedings of 2004 IEEE - International Conference on Control Applications, pp. 557–561 (2004)

Data Driven Biology - New Tools, Techniques and Resources

Transcription Control in Human Cell Types by Systematic Analysis of ChIP Sequencing Data from the ENCODE

Guillaume Devailly and Anagha Joshi[✉]

Division of Developmental Biology, the Roslin Institute, University of Edinburgh,
Easter Bush Campus, Midlothian, EH25 9RG, UK
{guillaume.devailly,anagha.joshi}@roslin.ed.ac.uk

Abstract. Transcription control plays a key role during development and disease with trans-acting factors (TFs) regulating expression of genes through DNA interaction. ChIP sequencing is widely used to get the genome wide binding profiles of TFs in a cell type of interest. The reduction in cost of sequencing and the technological improvement has resulted in vast amount of ChIP sequencing data accumulating in the public domain. The ENCODE consortium alone provides 690 publicly available ChIP sequencing data sets across 91 human cell types. We performed a multi-facetted bioinformatics analysis of this data to unravel diverse properties of TFs in the cellular context. Specifically, we characterised genomic location as well as sequence motif preference of the factors. We demonstrated that the distal binding of factors is more cell type specific than the promoter proximal. We identified combinations of factors acting in concert at distinct genomic loci. Finally, we highlighted how this data is of value to associate novel regulators to disease by integrating it with disease-associated gene loci obtained from GWAS studies.

Keywords: ChIP sequencing · Promoter · Transcription · Enhancer · Sequence motif

1   Scientific Background

Chromatin immune-precipitation followed by high throughput sequencing (ChIP-seq) has become the standard method for identifying the binding sites of transcription factors and chromatin modifiers at genome-wide scale. As the data generation is now becoming a routine, the bottleneck has shifted to computational analysis of this data. This explosion of data has therefore led to a new path of discovery moving the field from hypothesis-driven to data-driven analysis. The ENCODE consortium [1] has been one of the leading projects which used same data generation and quality control procedures across multiple labs worldwide to generate diverse genome-wide datasets across human cell types. It has so far produced 690 ChIP sequencing samples for transcription factors and transcription regulators, using 189 different antibodies (163 targeted factors) in 91 cell lines under different cell treatments. The combination of high experimental standards with extensive data release renders this ENCODE dataset invaluable for the scientific community, and serves as model for many other consortia. Notably, the ENCODE released 690 uniformly processed peak files as well as one track combining the peaks

© Springer International Publishing AG 2017
I. Rojas and F. Ortuño (Eds.): IWBBIO 2017, Part II, LNBI 10209, pp. 315–324, 2017.
DOI: 10.1007/978-3-319-56154-7_29

at the UCSC genome browser, frequently used by many labs world-wide. We performed a systematic analysis of this data to understand diverse aspects of transcription control across human cell types.

2   Materials and Methods

Peak lists from ENCODE ChIP sequencing experiments were downloaded from http://hgdownload.cse.ucsc.edu/goldenPath/hg19/encodeDCC/wgEncodeAwgTfb-sUniform/. For each peak, distance to the closest TSS (Transcription start sites) was obtained with bedtools [2] from the list of TSS provided by GENCODE (v21) [3]. Proportion of peaks upstream, overlapping, or downstream a TSS was compute using R, and categorized into peaks localised 10 kb (kilo base pair) or more from the nearest TSS, 5 kb or more, and 1 kb or more. ChIP experiment were clustered according to peak distribution around the nearest TSS using R function kmeans. Overlapping peaks in the 474 experiments done in the six cell lines with over 35 experiments were merged using bedtools, to obtain a binary matrix of 553,211 non overlapping genomic regions × 474 experiments. Two subsets of this matrix were isolated according to distance to the closest TSS: a matrix containing the 300,901 regions at less than 1 kb from the closest TSS and a matrix containing the 144,514 regions at more than 10 kb from the closest TSS. Pearson correlations were then calculated between each pairs of experiments, and the resulting correlation matrices were clustered according to the method described in [4]. From the original matrix containing 553,211 non-overlapping regions, we compute for each cell line Transcription Factor (TF) density in each regions by dividing the number of factors with a peak in that region by the total number of ChIP sequencing experiments done in that cell line, to obtain a matrix of 553.211 regions x TF density in six cell lines. K-means clustering of the regions lead to the identification of 417,597 lowly bound regions in all cell lines, while other regions shown high TF density in one, two or all cell lines. The sequence motif enrichment was performed using HOMER [5] for each transcription factor providing all peak locations as an input. A list of significantly disease associated genomic regions were downloaded from https://www.ebi.ac.uk/gwas/. P values were calculated using hypergeometric test, and were adjusted using Bonferroni correction. The enriched combinatorial patterns were calculated by generating random binding data keeping the same number of peaks for each factor and the significance was estimated compared to 100 randomizations.

3   Results

The ENCODE consortium has generated ChIP sequencing data for transcription factors and transcription regulators providing 690 uniformly processed peak files. These samples were generated using 189 different antibodies (163 targeted factors) in 91 cell lines, with some samples under cell treatments and each sample typically results from merging the analysis of at least two biological replicates. The number of peaks for a factor in a cell type ranges from few hundred to tens of thousands of peaks. ZNF274 in

HeLa-S3 cell type has the least number of peaks called (n = 74) while CFOS in MCF10A cell line has the most number of peaks called (n = 91,953) in this compendium. Here we elaborate diverse aspects of transcription control investigated using this data as well as integrating it together with other genome-wide datasets.

3.1 Genomic Location Binding Preference of Transcription Factors

We annotated all peaks from the 690 ChIP-seq datasets using the closest transcription start site identified by GENCODE (v21) [3]. We then quantified the number of peaks overlapping a TSS as well as the fraction of peaks localised more distally from a TSS. All patterns can be grouped largely into five groups (k-mean clustering) where peaks from group 1 and 2 peaks are preferentially found at distal regions while peaks from group 4 and 5 are preferentially found at promoter proximal regions. Group 4 contains experiments with at least 50% of peaks (Fig. 1A) overlapping a TSS. This includes various forms of RNA polymerase II as well as classical promoter associated transcription factors such as ELK1 or BRCA1 (Fig. 2B). This group defines factors with high affinity for promoters that may play a role in TSS specification. On the other hand, group 1 constitutes of experiments where only about 10% of peaks overlapped a TSS (Fig. 1A).

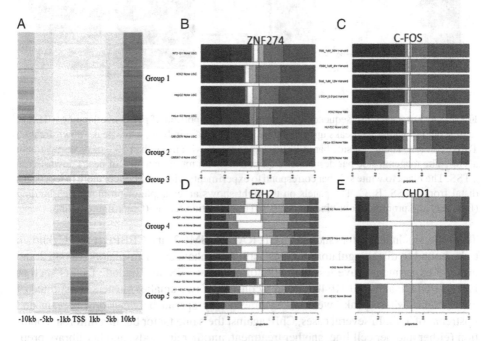

Fig. 1. A. k-means clustering of the full data resulted in five major clusters according to the occupancy of the peaks with respect to transcription start site. Colour code corresponds to fraction of peaks, from low (0 = white) to medium (0.25 = yellow) to high (0.5 = red). B, C, D, E For each ChIP sequencing experiment concerning a given TF (named above each plot), the proportion of peaks overlapping a TSS in represented in white, the peaks upstream the closest TSS are in blue, while peaks downstream a TSS are in red. (Color figure online)

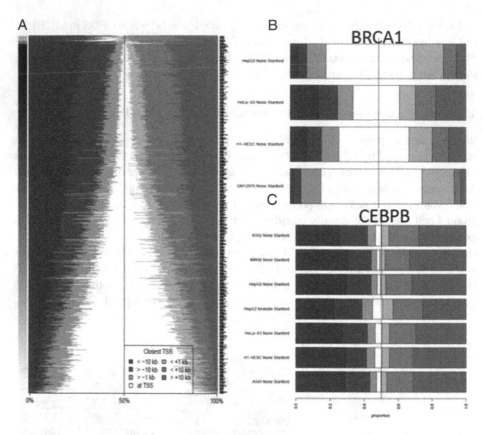

Fig. 2. A. For each ChIP sequencing experiment, the proportion of peaks overlapping a TSS in represented in white, the peaks upstream the closest TSS are in blue, while peaks downstream a TSS are in red. The green side bar indicates experiments with more than 50% of peaks upstream a TSS. The magenta side bar designates experiments with more than 50% of peaks downstream a TSS. The black to white side bar correspond to experiments with respectively more to less peaks distal from a TSS. **B, C.** The panels on the right shows an example of promoter rich binding (BRCA1) and promoter poor binding (CEBPB) transcription factor. (Color figure online)

This group includes factors such as CTCF, RAD21, SMC3, and CEBPB (Fig. 2C) known to be bound to distal regulatory regions acting as enhancers or insulators. Group 2 and group 5 experiments have less than 50% of peaks at TSS, and include factors such as EZH2, POU2F and BHLHE40. Group 3 consists of two smaller clusters constituted of experiments where peaks were not symmetrically spread between downstream and upstream of TSS. In several cases, ChIP against the same factor but in a different condition (either another cell line, another treatment, another antibody, another library preparation protocol or another laboratory) were found in different categories. For example, c-Fos ChIP-seq done by the Yale laboratory in GM12878 cells was found in the first group while those done by Harvard, UCSC or Yale in MCF10-A, HUVEC and HeLa-S3 were in the third group (Fig. 1C). Understanding whether these situations reflect biological differences or experimental artefacts will need further investigations. We also divided experiment in three groups: Green labelled experiments (Fig. 2A) have more peaks located downstream of a TSS than upstream. It includes ChIP sequencing experiments for the elongation specific Pol2 phosphorylated on serine 2, as well as ChIP-seq against ZNF274 (Fig. 1B) which is known to bind preferentially to 3' end of zinc finger genes [6]. In interestingly promoter proximal peaks of ZNF274 are upstream of TSS similar to EZH2 (Fig. 1D) and CHD1 (Fig. 1E). On the other hand, magenta labelled experiments (Fig. 2A) constituted of experiments with higher proportion of peaks located upstream a TSS than downstream. It include RNA polymerase III and its cofactor TFIIIC (and to a lesser extent RPC155 and BRF1). The upstream location of these factors might be due to the presence of unannotated RNA pol III transcribed genes in intragenic regions. The third groups contains the majority of experiments, that contains an about equal number of upstream and downstream peaks (Fig. 2A, sorted according to the proportion of peaks overlapping a TSS).

3.2 Bound Regions Distal to a TSS Are Cell Line Specific

In the ENCODE dataset, six cell lines contain more than 35 ChIP-seq experiments (A549, GM12878, H1-hESC, HeLa-S3, HepG2, and K562), with a total of 474 experiments. From the 553,211 regions bound by at least one factor in the 474 experiments, 300,901 where promoter proximal (<1 kb from the nearest TSS), and 144,514 where far from any TSS (>10 kb). We calculated the Pearson correlation coefficient between each pair of peak lists for the promoter proximal regions (Fig. 3A), and for the distal regions (Fig. 3B). The hierarchical clustering of the resulting correlation matrix for each

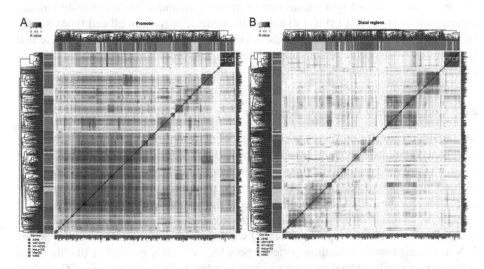

Fig. 3. Correlation matrix across all ChIP-seq experiments in the six main ENCODE cell lines. For each pair of experiments, Pearson correlation coefficient was computed between peak lists. Then the correlation matrix was clustered. Side colours correspond to each cell line. **A.** Correlation matrix for peaks close to a TSS (<1 kb). **B.** Correlation matrix for peaks far from a TSS (>10 kb). (Color figure online)

of the two sets demonstrated that promoter regions share binding sites of many more factors compared to the distal regions. Indeed, correlation value between experiments were on average higher on promoter-proximal sites than on distal sites. Using our colour code, the correlation heatmap of the promoter regions is redder than the correlation heatmaps of the distal regions. Moreover, the factors studied in same cell type clustered together more often at the distal regions than at the promoter regions. The promoter landscape represents three main clusters: a CTCF/SMC3/RAD21 cluster, a large multi-factor cluster including diverse factors, and the third cluster with very low correlation scores, which includes the aforementioned ZNF274 together with repressive complexes such as EZH2/SUZ12 and SETDB1/KAP1. This cluster also includes c-MYC ChIP-seq experiment in untreated H1-hESC by UTA. Other c-MYC samples do not cluster together with this sample putting its experimental quality into doubt. At distal regions, the CTCF/SMC3/RAD21 cluster remains intact but many other experiments clustered according to their cell line of origin, meaning that TF binding at distal regulatory regions tends to be cell line specific.

3.3   Highly Bound Regions in Promoters and Enhancers Show Divergent Properties

From the ENCODE dataset, we again selected six cell lines containing more than 35 ChIP-seq experiments (A549, GM12878, H1-hESC, HeLa-S3, HepG2, and K562). For each non overlapping genomic regions, we computed TF binding density, i.e. the proportion of ChIP-seq experiments done in one cell line with a peak in that region. Clusters were identified by k-means clustering (Fig. 4), and a big low density binding cluster is not shown. It is evident that densely occupied regions in all cell types mostly belong to promoter regions while densely occupied regions in only in any single cell type i.e. cell type specific regions are enriched for non-promoter regions. This is in full agreement with the findings in the previous section.

3.4   Combinatorial Control of Regulators

Mammalian transcription factors are known to work together by binding at the promoter or enhancer regions to activate or repress downstream target genes. To unravel if some regulators are preferentially binding together genome-wide, we built a M*N matrix of all binding events (peaks) in a cell type where M represents the genomic loci bound by at least one factor in that cell type and N being the number of factors studied by ChIP sequencing in that cell type. We build a matrix for each of the six cell types (A549, ES cells, GM12878, HeLa, HepG2 and K562) with N of more than 30. Each cell type has 2 N-1 combinations of binding patterns possible. We then evaluated likelihood of frequency of combinatorial patterns to occur by chance by comparing to 100 random datasets generated such that the total number of binding events for each factor was preserved (Table 1). This analysis re-discovered the transcription factors of the same family (e.g. USF1 and USF2 in ES cells) known to bind to overlapping genomic locations due to highly similar sequence motifs or the components of known complexes or

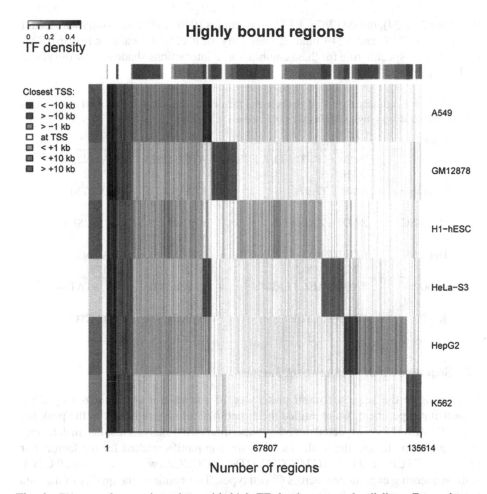

Fig. 4. Clusters of genomic regions with high TF density across 6 cell lines. For each non overlapping regions, we computed the proportion of ChIP-seq experiments done in one cell line with a peak in that region. Clusters were identified by k-means clustering. For visibility, a big cluster of low TF density in every cell type is not shown (417,597 regions). Top colors bars represent the fraction of regions localised at a TSS (white), upstream a TSS (blue) or downstream a TSS (red). (Color figure online)

well studies interactions (e.g. SUZ12, EZH2). CTCF, RAD21 and ZNF143 form a part of cohesin complex and cluster together in the global clustering considering all peaks. Accordingly, they were enriched across multiple cell types. Interestingly, there are a number of cases of combinatorial control where the two factors do not share most of the binding sites therefore do not cluster together in the global clustering tree (Fig. 3) but are significantly co-occupying a relatively small but statistically significant number of gene loci. For example, GTF2F1 and CTCF co-bind 2,233 loci in K562 cell line (P value <1e-256) and these loci are not occupied by any other factor of 150 factor studied. Similarly, in ES cells, 957 genomic loci are co-occupied only by RAD21 and TEAD4

(P value <2.1e-37). In GM12878, ETS1 co-occupies 1090 binding sites only with EGR1 (P value <1e-256) and 1249 binding sites only with P300 (P value <1e-256). This postulates site specific role for these combinatorial interactions shadowed by the global analysis approach.

Table 1. Top 3 significant associations between factors in 6 human cell lines. All associations were predicted at very high significance (all P values < 1e-256).

Rank	1	2	3
A549	HDAC6, P300, ELF1, ETS1, GABP	ATF3, BRF1	RNA pol2, CTCFL
GM12878	SAP30, TAF7	STAT5A, BRG1	P300, ETS1
H1-hESC	RAD21, ZNF143	BACH1, MAFK	USF1, USF2
HeLa-S3	EZH2, RNA pol2 4H8, SIN3A, CJUN, CMYC	GTF2B, NR2F2	RNA pol2, RBBP5
HepG2	TAF1, TAF7, TEAD4	SAP30, ATF1, ATF3	PU1, STAT5A
K562	GTF2F1, CTCF	HDAC1, CJUN	E2F6, CTCF

3.5   Sequence Motif Preference of Factors

To investigate the sequence motif preference of each factor characterized by ChIP sequencing experiment, we identified the enrichment of known motifs in the peak list using HOMER [5]. As expected, for many experiments, the analysis resulted in detecting the sequence motif specific to the factor as the top motif enriched for the factor. For example, a CTCF motif (AYAGTGCCMYCTRGTGGCCA) was top motif of all CTCF ChIP sequencing experiments across 68 cell types. This confirms the quality of the data for downstream characterization. Importantly, CTCF ChIP sequencing in different cell types did not result in enrichment of sequence motifs of cell type specific factors. This demonstrates that CTCF acts mainly as an insulator as well as defining gene regulatory boundaries which are largely independent of cell type. Similarly, the majority of RNA polymerase II experiments across multiple cell types identified ETS as a top enriched motif with only a handful of cases with a cell type motifs enriched such as GATA motif enriched in K562 RNA pol II sample or BZIP motif enriched in HeLa-S3 RNA pol II sample. In ESCs, OCT4-SOX2-TCF-NANOG motif was enriched for the ChIP sequencing of OCT4 and NANOG as expected, moreover this motif was top enriched in P300, BCL11A, HDAC2 and CTBP2 samples as well.

3.6   ChIP Sequencing Binding Overlap with Disease Susceptibility Loci

The ENCODE consortium has demonstrated that a large number of intergenic disease associated gene loci are located in the regulatory regions defined by chromatin modifications and DNase I hypersensitive sites across cell types. To systematically analyse overlap of transcription factors binding loci with Genome Wide Association study (GWAS) high confidence hits, overlap of GWAS hits with ChIP-seq peaks was calculated and the significance of overlap was estimated using bonferroni corrected hypergeometric test. Three of the 690 ChIP sequencing samples: BRCA1 bound loci in GM12878 (2% of peaks), NELFE bound loci in K562 (2.1% of peaks), HDAC8 bound loci in K562 (2%) showed statistically significant overlap with GWAS disease associated loci. All three protein have a well-studied role implied in cancer. As expected, the majority BRCA1 bound loci overlapped with the disease loci identified in breast cancer. Interestingly, 37 BRCA1 target loci overlapped with inflammatory bowel disease. BRCA1 also functions as an important mediator of innate immunity and BRCA1 gene therapy reduces systemic inflammatory response [7]. 12 BRCA1 target loci overlapped with genes linked to childhood obesity. In line with this, it has been shown that without BRCA1, muscle cells store excess fat and start to look diabetic [8]. Taken together, the analysis of transcription targets using ChIP sequencing overlapping with disease associated loci has a power to identify novel factors controlling the disease phenotype.

4   Conclusion

As the next generation sequencing is becoming readily available to the experimental groups world-wide, the major challenge lies in computational analysis of this large resource. This data can be analysed in a plethora of ways integrating together with other datasets, to obtain unexplored biological insights. The ENCODE consortium has taken a big initiative to provide a uniformly processed dataset to the scientific community for computational data integration and analysis. This easy accessibility of pre-processed data facilitates generation of novel biological hypotheses from this resource.

In this paper, we demonstrate six ways of analysing this resource integrating it with other data resources: categorising experiments according to the distribution of distance between peaks and the nearest TSS, correlating experiments on promoter proximal regions only and on distal regions only, identification of common and cell-type specific regions of high TF binding, combinatorial analysis of co-binding, primary and secondary motif analysis for each factor, and identification of factor binding disease related GWAS hits.

These approaches have a potential to develop new hypotheses about transcriptional control mechanisms. We both reproduced and expand some observations previously made in ENCODE companion papers such as [9], and we also developed novel analytic approaches. This shows that the in depth analysis of ENCODE data is still far from complete and must continue. Importantly, all the analysis performed here can be readily transferable to the exploitation of ChIP-Seq datasets in other cellular systems, and thus have the potential to significantly advance our understanding of a wide range of both normal and pathological cellular processes.

Funding. A.J. is a Chancellors Fellow at the University of Edinburgh. This work was supported by the Roslin Institute Strategic Grant funding from the BBSRC.

References

1. Bernstein, B.E., Birney, E., Dunham, I., Green, E.D., Gunter, C., Snyder, M.: An integrated encyclopedia of DNA elements in the human genome. Nature **489**(7414), 57–74 (2012)
2. Quinlan, A.R., Hall, I.M.: BEDTools: a flexible suite of utilities for comparing genomic features. Bioinformatics **26**(6), 841–842 (2010)
3. Harrow, J., Frankish, A., Gonzalez, J.M., Tapanari, E., Diekhans, M., Kokocinski, F., Aken, B.L., Barrell, D., Zadissa, A., Searle, S., Barnes, I., Bignell, A., Boychenko, V., Hunt, T., Kay, M., Mukherjee, G., Rajan, J., Despacio-Reyes, G., Saunders, G., Steward, C., Harte, R., Lin, M., Howald, C., Tanzer, A., Derrien, T., Chrast, J., Walters, N., Balasubramanian, S., Pei, B., Tress, M., Rodriguez, J.M., Ezkurdia, I., van Baren, J., Brent, M., Haussler, D., Kellis, M., Valencia, A., Reymond, A., Gerstein, M., Guigó, R., Hubbard, T.J.: GENCODE: the reference human genome annotation for the ENCODE project. Genome Res. **22**(9), 1760–1774 (2012)
4. Pötzelberger, K., Strasser, H.: Data Compression by Unsupervised Classification. Department of Statistics and Mathematics. WU Vienna University of Economics and Business, Vienna, 11 July 1997
5. Heinz, S., Benner, C., Spann, N., Bertolino, E., Lin, Y.C., Laslo, P., Cheng, J.X., Murre, C., Singh, H., Glass, C.K.: Simple combinations of lineage-determining transcription factors prime cis-regulatory elements required for macrophage and B cell identities. Mol. Cell **38**(4), 576–589 (2010)
6. Frietze, S., O'Geen, H., Blahnik, K.R., Jin, V.X., Farnham, P.J.: ZNF274 recruits the histone methyltransferase SETDB1 to the 3' ends of ZNF genes. PLoS ONE **5**(12), e15082 (2010)
7. Teoh, H., Quan, A., Creighton, A.K., Annie Bang, K.W., Singh, K.K., Shukla, P.C., Gupta, N., Pan, Y., Lovren, F., Leong-Poi, H., Al-Omran, M., Verma, S.: BRCA1 gene therapy reduces systemic inflammatory response and multiple organ failure and improves survival in experimental sepsis. Gene Ther. **20**(1), 51–61 (2013)
8. Jackson, K.C., Gidlund, E.-K., Norrbom, J., Valencia, A.P., Thomson, D.M., Schuh, R.A., Neufer, P.D., Spangenburg, E.E.: BRCA1 is a novel regulator of metabolic function in skeletal muscle. J. Lipid Res. **55**(4), 668–680 (2014)
9. Gerstein, M.B., Kundaje, A., Hariharan, M., Landt, S.G., Yan, K.-K., Cheng, C., Mu, X.J., Khurana, E., Rozowsky, J., Alexander, R., Min, R., Alves, P., Abyzov, A., Addleman, N., Bhardwaj, N., Boyle, A.P., Cayting, P., Charos, A., Chen, D.Z., Cheng, Y., Clarke, D., Eastman, C., Euskirchen, G., Frietze, S., Fu, Y., Gertz, J., Grubert, F., Harmanci, A., Jain, P., Kasowski, M., Lacroute, P., Leng, J., Lian, J., Monahan, H., O'Geen, H., Ouyang, Z., Partridge, E.C., Patacsil, D., Pauli, F., Raha, D., Ramirez, L., Reddy, T.E., Reed, B., Shi, M., Slifer, T., Wang, J., Wu, L., Yang, X., Yip, K.Y., Zilberman-Schapira, G., Batzoglou, S., Sidow, A., Farnham, P.J., Myers, R.M., Weissman, S.M., Snyder, M.: Architecture of the human regulatory network derived from ENCODE data. Nature **489**(7414), 91–100 (2012)

Management of Data Structures Generated During Simulations of the Evolution of Multicellular Systems

Andreea Robu[1(✉)], Mihaela Crisan-Vida[1], Nicolae Robu[1], and Adrian Neagu[2]

[1] Department of Automation and Applied Informatics,
Politehnica University of Timisoara, Timisoara, Romania
{andreea.robu,mihaela.vida,nicolae.robu}@aut.upt.ro
[2] Center for Modeling Biological Systems and Data Analysis,
Victor Babes University of Medicine and Pharmacy, Timisoara, Romania
neagu@umft.ro

Abstract. One major interest in tissue engineering is to model and simulate tissue growth (the auto-organization of cells into tissues) in order to optimize tissue engineering procedures. Therefore, we have developed a computational modeling tool, called SIMMMC, which allows simulating the behaviour of various multicellular systems in the vicinity of biomaterials. SIMMMC relies on the Metropolis Monte Carlo algorithm, taking into account different geometric and energetic input parameters. For a better organization and analysis of the data structures obtained by running SIMMMC and for a secure data storage, we created a cloud app, called SIMMMC Management Tool. Through this app, for each simulation, the model parameters and the output files that contain the new configurations of the biological systems are stored on the cloud. The user can search for specific simulations that are characterized by certain model parameters, visualize the data, download all the associated files on a local computer, or delete unnecessary simulations. Moreover, the SIMMMC Management Tool is convenient for data sharing: it can be accessed by any researcher who is interested in the results of SIMMMC simulations and has received an authentication code.

In this work, we present the architecture of the SIMMMC information system, the functionalities and the architecture of the SIMMMC Management Tool, the implementation of the application, the design of the database, the workflow of the tasks and the user interface.

SIMMMC simulations of multicellular systems generate a wealth of data, making it difficult to retrieve the information of interest. Therefore, the SIMMMC Management Tool is a useful instrument for accessing and for analyzing the big variety of simulation data, helping the optimization of *in vitro* tissue structure creation.

Keywords: Tissue engineering · Multicellular systems · Computational modeling · Manage simulations · Cloud computing · Web app · Database

© Springer International Publishing AG 2017
I. Rojas and F. Ortuño (Eds.): IWBBIO 2017, Part II, LNBI 10209, pp. 325–336, 2017.
DOI: 10.1007/978-3-319-56154-7_30

1 Introduction

Tissue engineering (TE) is an interdisciplinary field whose goal is to create functional tissue constructs in order to regenerate the damaged tissues from the human body [1]. First, cells are isolated from the patient and expanded in culture; then the cells are seeded onto a degradable porous scaffold, which provides mechanical support for anchorage-dependent cells, and fosters their growth. The cell-seeded scaffold is either cultured *in vitro* for a few weeks in order to synthesize extracellular matrix subsequently being implanted into the body, or it is implanted directly into the injured part of the body where tissue regeneration is induced *in vivo* [2].

To be implantable, the new tissue construct has to become functional and sturdy. To reach this goal, tissue engineers may rely on trial-and-error experiments, which are expensive and time consuming. Instead, the evolution of the tissue structure under various conditions can be investigated by predictive computer simulations [3].

Knowledge of the biological mechanisms responsible for cell rearrangements in tissue constructs might reduce the number of laboratory experiments. The modeling of the complex biological systems and the simulation of multicellular self-assembly in the vicinity of biomaterials helps the optimization of TE procedures and products.

To study the auto-organization of cells in tissue engineered constructs, in our previous work, we have developed a modeling and simulation tool, SIMMMC, which describes the evolution of certain artificial tissue structures used in laboratory experiments [4, 5].

The SIMMMC application can model 4 different types of biological systems: (i) a cell aggregate situated next to a biomaterial, (ii) a cell suspension found in the vicinity of a porous scaffold with interconnected spherical pores, (iii) a cell suspension found in the vicinity of a porous scaffold with cubic pores, (iv) a cell aggregate found in the vicinity of a spherical porous scaffold. The simulations are based on Metropolis Monte Carlo algorithms taking into account the works of adhesion between the constituents, as well as chemotaxis [4, 5].

SIMMMC offers essential clues for scaffold design, adhesive interactions, and chemotaxis that assure optimal cell seeding.

Our team ran hundreds of simulations [4, 5], considering multicellular systems made of one or two types of cells, located in the vicinity of different geometrical structures of biocompatible materials. We have taken into consideration different energetic and chemotactic conditions. The data obtained so far occupies about 100 GB and is growing. While we have been trying to analyze the data, we have encountered difficulties in finding specific simulations characterized by certain model and/or simulation outcomes.

In order to solve this problem, we have developed the SIMMMC Management Tool, a web app designed to manage SIMMMC simulation results. It aims to organize resources, to enable search for simulations of interest that meet certain criteria, and to view, download or delete simulation data. Moreover, the application is built to assure a secure data storage on the cloud.

2 Methods

The SIMMMC Software and the SIMMMC Management Tool are parts of a complex system, called SIMMMC Information System. Its architecture is presented in Fig. 1.

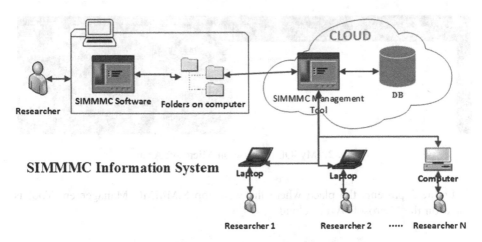

Fig. 1. The architecture of the SIMMMC information system

As shown in Fig. 1, the SIMMMC Software component can be used by researchers to model various biological systems and to simulate their evolution in the vicinity of biomaterials. SIMMMC generates new configurations of the system that can be can be visualized in 3D using the VMD software [6], providing a qualitative analysis of the system's evolution. VMD enables the user to save digital pictures of the system in different views and various styles of representation. For a quantitative analysis of the results, we created dedicated MATLAB scripts that also create output files. All the output files obtained after the simulations are saved on the local computer.

In order to manage the simulation results, we developed the SIMMMC Management Tool, which is a new software component of the SIMMMC Information System. It is designed for managing and saving SIMMMC simulation data in the cloud, resulting in improved data safety and better data processing.

For modeling the SIMMMC Information System, we have used the Business Process Modeling and Notation (BPMN) technique [7] and the Bizagy modeling tool [8], illustrating the entire workflow of the system. BPMN is a graphical notation that describes the logical steps in a business process [7].

To develop the SIMMMC Management Tool, we used PHP language, the WEB Server Apache, and the MySQL database for storing the data [9]. PHP is a server side scripting language designed especially for creating web applications; MySQL is an open source Relational Database Management System (RDBMS) that uses Structured Query Language (SQL) [9].

The SIMMMC Management Tool is a PHP-MySQL web app created and published in the Microsoft Azure cloud. The web app and the MySQL database were created using the Azure Portal and the application was published in Azure using FTP [10].

Figure 2 presents the MySQL database that is available in Microsoft Azure.

Fig. 2. MySQL database on Microsoft Azure

Figure 3 presents the place where the web app SIMMMC Management Tool is stored on the Microsoft Azure cloud.

Fig. 3. SIMMMC Management Tool on Microsoft Azure

Microsoft Azure is a cloud computing platform proposed by Microsoft where the developer can create, deploy or manage applications and services by using a global network of data centers managed by Microsoft [11]. The main advantages of cloud computing are the high processing power, the mobility, the reaction speed, the large storage capacity and high data security. Cloud computing offers great potential for fast information retrieval. The information is available in real time and can be accessed using different applications [12, 13].

3   Results

The BPMN model which describes the SIMMMC Information System workflow is presented in Fig. 4.

The first component, called SIMMMC Software, is used for generating computational models of certain biological systems and for simulating the evolution of different multicellular systems in the vicinity of biomaterials. The SIMMMC Software logical steps are presented in Fig. 4.

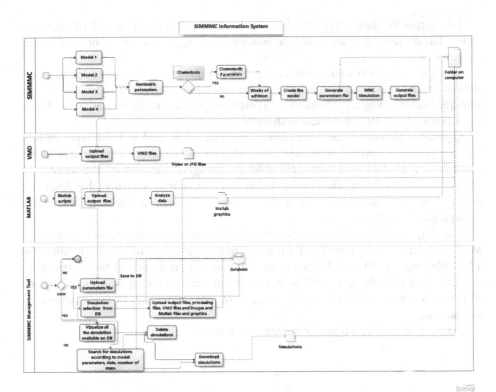

Fig. 4. The workflow of the SIMMMC information system

First, we choose a biological system that we want to model, then we set the geometric parameters, the works of adhesion, and the chemotactic parameters (if chemotactic agents exist in the system). The software builds the computational model and generates a parameter file. During simulation, the SIMMMC application generates output files containing successive configurations of the model system.

The data associated with simulations (the model parameters, the output files containing the biological system's configurations after running the simulation, the VMD files and the images containing 3D models, the MATLAB graphics and scripts for data processing) are stored on the local computer.

For a better management of the enormous amount of information and for an easy retrieval of simulation data, researchers can access the new component of the SIMMMC Information System, called the SIMMMC Management Tool (Fig. 4).

After login, the user can choose from the following operations: upload the parameters file of a certain simulation from the local computer and save the parameters to the database on the cloud; upload all the files according to a selected simulation and save them to the database; visualize all the simulations available on the database, download or delete them; search for certain simulations according to model parameters, date, or number of MCS and download or delete them.

For storing the data, a MySQL database has been used, consisting of 14 relational tables. The database design prevents redundant data storage and assures optimal search tools.

When the user chooses the option of uploading a parameter file associated to a new simulation, the new simulation is saved in the *simulation table*, being characterized by a unique id and a category it belongs to. There are 4 possible categories associated with the 4 types of modeled biological systems.

Geometric parameters are saved in distinct tables dedicated to the 4 types of biological systems SIMMMC deals with. The chemotaxis parameters (if present in the simulation) are saved in a dedicated table; the energetic parameters, the number of Monte Carlo steps (MCS) and the simulation set-up data are saved in a table dedicated to the simulation parameters. All of these tables are related to the simulation table through the simulation id field.

The output files that describe successive configurations of the model system, the MATLAB scripts and other files for processing the data, the MATLAB graphics, the VMD files and the 3D model images, are also saved in the dedicated tables, related to the simulation table through the field that represents the simulation id. In order to store all these files, we used BLOB fields, which allow for saving files of any extension in the database.

Figure 5 shows the MySQL database structure.

Fig. 5. The structure of the MySQL database

Figure 6 presents the activity diagram of the SIMMMC Management Tool, showing the functionalities of the application.

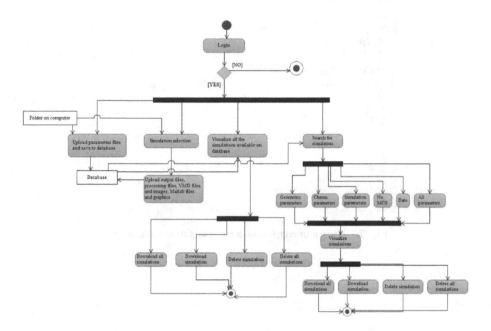

Fig. 6. The activity diagram of the SIMMMC Management Tool

One of the main facilities of the application is the possibility to save to the cloud all data associated to simulations performed using the SIMMMC software. Thus, after the authentication process, one of the user's options is to upload a file with model parameters and to save these parameters in the database.

A parameter file is a text file generated by the SIMMMC software; it contains all the input parameters of a simulation (geometric, energetic and chemotactic parameters), each arranged on a line (Fig. 7). The SIMMMC Management Tool reads the parameter file, line-by-line, and saves all parameters, according to their type, in the tables allotted to them in the database.

The application also offers the possibility to upload the output files, the processing files, the MATLAB files, the VMD files and 3D model images associated to a selected simulation. According to the type of files selected from the local computer, the new application knows where to save them in the database in the appropriate tables.

Another basic function of the application is the visualization of all the simulations saved in the database. The simulations are described by their corresponding parameters. The simulations are structurally divided into 4 categories according to the modeled biological systems.

In order to have access to all the saved files belonging to all the simulations from the database, the application gives the user the possibility to choose the "download all" option. Thus, all the files associated to all the simulations saved in the database will

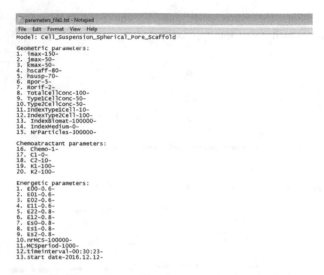

Fig. 7. The parameter file of a representative simulation

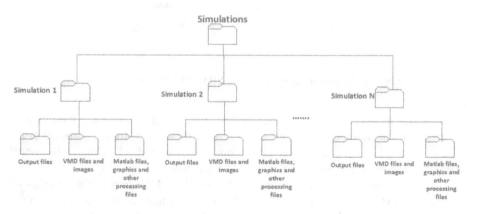

Fig. 8. The folder structure created by downloading simulations

download on the local computer using an archive. The archive (zip, arj, rar etc.) contains a directory and a file structure created dynamically, representative for each simulation (Fig. 8).

First of all, the Simulations Folder is created, and then, within it, a folder for each downloaded simulation is created. According to the type of files that are downloaded for each simulation, a folder with a representative name is created in which they are memorized. Thus, the files from the database are saved on the local computer in a well-organized directory structure. In order to delete all the database simulations, the user has the option "delete all". Moreover, the user has the possibility to download or delete either each simulation separately, or an entire group of simulations.

The most important functionality of the SIMMMC Management Tool consists in searching for the simulations of interest in the database. Searches can be done according to various values of the geometric, chemotaxis, or energetic parameters, according to a specific number of MCS, or according to the time interval when the simulations were performed. The user can also search for the simulations that simultaneously satisfy all the mentioned conditions.

The geometric parameters that characterize the model are different for different biological systems; thus, the geometric search conditions are also different. For instance, if we considered the models of the cell suspensions located in the vicinity of spherical porous scaffolds, the search for simulations can be done according to the radius of the pores and the radius of the orifices that connect them. If the biological system consists of cell aggregates located in the vicinity of spherical porous scaffolds, the geometric parameters through which the app allows the search are the radius of the pores, the radius of the orifices that connect adjacent pores and the radius of the aggregates. If the modeled biological systems are cellular suspensions situated in the vicinity of cubic porous scaffolds, the geometric parameter that determines the search is the side of the cubes. If the models of interest consist in cell aggregates located on biomaterials, the geometric parameters by which the app can make the search is the radius of the aggregate.

The search criteria are dynamically created, so that even if the user does not fill in the values for all the parameters in the search form, the search takes place and it returns the simulations that meet the conditions specified through the completed values. If no value is specified, the program provides all the simulations from the selected category found in the database.

Figure 9 represents two snapshots of the user interface of the SIMMMC Management Tool, which shows the upload of the parameters file option (A) and the upload of all the files belonging to a certain simulation option (B). Any number of files can be selected from the local computer, and, according to their type, they will be saved by the app in the specific tables in the database.

The simulations from the database are organized in four categories according to the four types of biological systems modeled by the SIMMMC software.

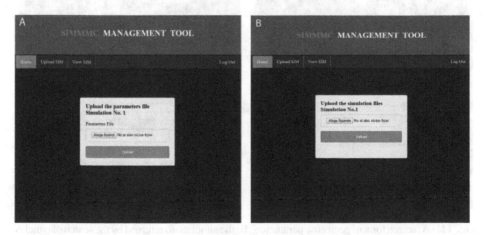

Fig. 9. A: upload the parameters file; B: upload the simulation files

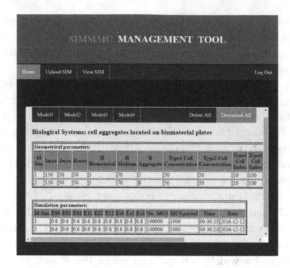

Fig. 10. The geometric and simulation parameters

Figure 10 presents all the parameters that characterize the simulations of the evolution of cell aggregates on the surface of biomaterials. They are displayed in two different tables according to the geometric and simulation parameters. In the case of simulations that also describe chemotaxis, the information is displayed in three tables, the third one containing the chemotaxis parameters. For all the shown simulations, the download and delete options are available, both for simulation groups and individually.

In Fig. 11 we present an example of searching for the simulations that are based on models of cell aggregates located on the surface of biomaterials. We have selected the geometric criteria for searching.

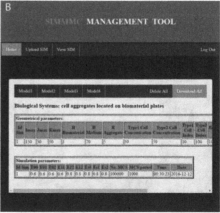

Fig. 11.   A: search form of geometric parameters; B: the simulations returned by the search of panel A

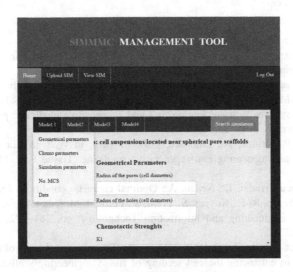

Fig. 12. The search menu for the first category of simulations

Considering that we are interested in cell aggregates with the radius of 5 cell diameters, (Fig. 11A), the app will display all the parameters that characterize the simulations whose model meets the search condition (Fig. 11B).

Shown in Fig. 12 is the search menu for the simulations of the evolution of a cell suspension in the vicinity of a scaffold with spherical pores. The search can be done for different values of the geometric, chemotaxis or energetic parameters, or according to a certain number of MCS or according to a certain interval of time.

4   Conclusions

Computational modeling for tissue engineering is a rapidly developing field that is useful for the optimization of *in vitro* tissue fabrication.

The SIMMMC modeling and simulation software generates a wealth of data, which need to be well organized and easy to find for processing and analysis. They also need to be saved in a secure place. To address the need for secure data storage and effective data retrieval, we developed the SIMMMC Management Tool, which is presented in this article. This tool saves simulation data in a database on the cloud, offers a variety of search options, and gives the possibility to download or delete simulations.

Equipped with an intuitive user interface, the SIMMMC Management Tool is a web app that enables effective data management and analysis. Although it is adapted for the data structure generated by the SIMMMC simulation software, the informatic system developed here might be implemented along the same lines also for other computer simulation programs.

References

1. Lanza, R.P., Langer, R., Vacanti, J.P. (eds.): Principles of Tissue Engineering, 3rd edn. Elsevier Academic Press, Burlington (2007)
2. Mano, J.F., Silva, G.A., Azevedo, H.S., Malafaya, P.B., Sousa, R.A., Silva, S.S., Boesel, L. F., Oliveira, J.M., Santos, T.C., Marques, A.P., Neves, N.M., Reis, R.L.: Natural origin biodegradable systems in tissue engineering and regenerative medicine: present status and some moving trends. J. R. Soc. Interface 4(17), 999–1030 (2007)
3. Semple, J.L., Woolridge, N., Lumsden, C.J.: In vitro, in vivo, in silico: computational systems in tissue engineering and regenerative medicine. Tissue Eng. 11(3–4), 341–356 (2005)
4. Robu, A., Stoicu-Tivadar, L., Neagu, A.: Optimal energetic conditions for cell seeding of scaffolds. In: Precup, R.-E., Kovács, S., Preitl, S., Petriu, E.M. (eds.) Applied Computational Intelligence in Engineering and Information Technology, pp. 261–272. Springer, Berlin Heidelberg (2012)
5. Robu, A., Stoicu-Tivadar, L., Robu, N., Neagu, A.: Computational study of the potential role of chemotaxis in enhancing the cell seeding of tissue engineering scaffolds. In: The 25th European Medical Informatics Conference - MIE2014, Istanbul, Turkey, pp. 735–739 (2014)
6. Humphrey, W., Dalke, A., Schulten, K.: VMD – visual molecular dynamics. J. Mol. Graphics 14(1), 33–38 (1996). www.ks.uiuc.edu/Research/vmd/. Accessed 3 Oct 2014
7. OMG, Business Process Model and Notation (BPMN). version 2.0. http://www.omg.org/spec/BPMN/2.0/. Accessed 10 Oct 2016
8. Bizagy modeling tool. https://www.bizagi.com/. Accessed 10 Dec 2016
9. Filip, I.: Programare WEB, editura Conspress, Bucuresti, p. 65 (2013)
10. Create a PHP-MySQL web app in Azure App Service and deploy using FTP. https://docs.microsoft.com/en-us/azure/app-service-web/web-sites-php-mysql-deploy-use-ftp. Accessed 10 Jan 2016
11. Microsoft Azure. https://en.wikipedia.org/wiki/Microsoft_Azure. Accessed 09 Dec 2016
12. Bollineni, P.K., Neupane, K.: Implications for adopting cloud computing in e-Health. Master's Thesis Computer Science, September 2011
13. Lupşe, O.S., Vida, M.M., Tivadar, L.S.: Cloud computing and interoperability in healthcare information systems. In: INTELLI 2012: The First International Conference on Intelligent Systems and Applications (2012)

Finding Transcripts Associated with Prostate Cancer Gleason Stages Using Next Generation Sequencing and Machine Learning Techniques

Osama Hamzeh[1], Abedalrhman Alkhateeb[1], Iman Rezaeian[1], Aram Karkar[2], and Luis Rueda[1(⊠)]

[1] School of Computer Science, University of Windsor,
401 Sunset Ave, Windsor, ON N9B 3P4, Canada
{hamzeho,alkhate,rezaeia,lrueda}@uwindsor.ca
[2] Schulich School of Medicine and Dentistry, Western University,
1151 Richmond St, London, ON N6A 5C1, Canada
akarkar2019@meds.uwo.ca

Abstract. Prostate cancer is a leading cause of death world-widely and the third leading cause of cancer death in Northen American men. Prostate cancer causes parts of the prostate cells to lose normal control of growth and division. The Gleason classification system is one of the known systems used to grade the aggressiveness of the prostate progression.

In this study, an RNA-Seq dataset of 104 prostate cancer patients with different Gleason stages is analyzed using machine learning techniques to identify transcripts that are linked to prostate progression. The proposed method utilizes information gain as a ranker for feature selection to overcome the curse of dimensionality, because of dealing with a large number of features (41,971 transcripts). Minimum redundancy maximum relevance (MRMR) feature selection was applied on a one-versus-all hierarchical classification model to find the best subset of transcripts that predicts each stage. The Naive Bayes classifier was used at each node of the hierarchical model.

Naive Bayes is compared with support vector machine (SVM) for accuracy as performance measure. The results suggest that Naive Bayes outperforms SVM as a classifier in the hierarchical model. Several transcripts are found to be highly associated with different Gleason stages in prostate cancer patients.

Keywords: Machine learning · Next generation sequencing · Classification · Feature selection · Biomarkers · Transcriptomics · Prostate cancer progression

1 Introduction

Cancer is among the leading causes of death worldwide. In 2013, there were 8.2 million deaths, and 14.9 million incident cancer cases. Prostate cancer was the

© Springer International Publishing AG 2017
I. Rojas and F. Ortuño (Eds.): IWBBIO 2017, Part II, LNBI 10209, pp. 337–348, 2017.
DOI: 10.1007/978-3-319-56154-7_31

leading cause for men cancer incidence 1.4 million [1]. As with all cancers, studying prostate cancer at the molecular level uncovers transcriptional mechanisms of the tumour biology. Traditionally, prostate cancer studies focused primarily on discovering biomarkers for differentiation between benign and malignant tumours. Recently, studies have considered some other aspects of the tumours including progression, metastasis, and recurrence, among others.

The TNM staging system for all solid tumors was proposed by Pierre Denoix between 1943 and 1952, using the size and extension of the primary tumor (T), its lymphatic involvement (N), and the presence of metastasis (M) to classify the progression of cancer [2]. Singireddy et al. proposed a classification method for consecutive TNM stages to identify biomarkers that can identify the progress from each stage to the next one [3]. Gleason score is a grading system for prostate cancer which provides one of the strongest predictors for the prognosis and clinical course of the cancer. This score is assigned to the tumor based on the microscopic patterns of the cells obtained from a sample of the tumor. The score is usually presented as $4 + 3 = 7$, where the first number is the most common pattern of cells in the sample, the second number is the second most common pattern of cells and the third pattern is the overall Gleason score. Generally, a higher Gleason score indicates a more aggressive cancer and worse prognosis [4]. A recent multi-institutional study has proposed grouping Gleason scores into five categories: $6, 3+4 = 7, 4+3 = 7, 8$, and 9–10. This study demonstrated that this grouping corresponded with a 5-year biochemical recurrence free survival of 96%, 88%, 63%, 48%, and 26% [5]. This grouping strategy was shown to effectively indicate the risk of cancer recurrence and provides the basis for the Gleason score categorization used in our study in order to identify transcriptomic biomarkers that could be associated with different stages of prostate cancer progression.

Table 1. Number of samples with different Gleason scores.

Gleason score	Number of samples	Group
$3 + 3$	10	336
$3 + 4$	55	347
$4 + 3$	24	437
$5 + 3$	1	448–538
$4 + 4$	9	448–538
$4 + 5$	2	459–549
$5 + 4$	2	459–549

Table 1 shows the number of samples in each Gleason stage. Due to the low number of samples in some of the cases, we combined samples with Gleason scores of 5 and 6 into one group. Thus, we have five Gleason score groups: $4 + 3, 3 + 4$, below 7 (5 and 6), 8, and above 8 (9 and 10).

Most studies usually use genes to study prostate cancer in different stages; while genes transcribes in many different ways, different transcriptions may lead to a different rate of degradation for the resulting protein. Protein rates of degradation have been found to be correlated with Gleason score in prostate cancer [6]. Therefore, we select the transcripts to identify different Gleason stages within the sample set which provides a higher level of detail than using the gene itself.

RNA-seq produces raw readings that require a significant amount of pre-processing to obtain information. The raw reads need to be aligned against the human genome [7] to remove duplication and assemble the actual genes and transcripts available in the samples taken. Then, each read needs to be counted [7], which will show how many genes or transcripts are expressed in a sample. The measurements are counted in multiple standard measurements. Fragments per kilobase per million (FPKM) [8], reads per kilobase per million of reads (RPKM) [9] or transcripts per million (TPM) are measurements that depend on using paired-end or single end reads. Multiple tools have been created to align reads against the genome. While BLAST [10] was one of the first tools used to align reads, Tophat [11] is one of the mostly used tools. Also RNA-star [12] by Alexander Dobinis has been proven to be the fastest aligner currently available; the latter requires a large amount of computational resources though. On the other hand, machine learning algorithms greatly assist in analyzing the information gathered by the alignment tools, as well as the total number of transcripts found. Gleason score prediction using machine learning classification was studied in [13]. The study showed that the SVM classifier resulted in a slightly higher sensitivity but lower specificity than linear discriminant analysis for final Gleason score prediction with limited prostate cancer patient population. In this paper, we propose a hierarchical model that can be used in classification of prostate cancer tumor's aggressiveness with a very high degree of accuracy. Our approach also detects some transcripts that can be used as biomarkers for progression of prostate cancer.

2   Materials and Methods

2.1   Data Pre-processing

The data set that is used in this paper is publicly available from the National Center for Biotechnology Information (NCBI) with Gene Expression Omnibus (GEO) number GSE54460 [14]. The data was collected using formalin-fixed, paraffin-embedded (FFPE) prostatectomy samples from three independent sites. The Atlanta Veterans Administration Medical Center (AVAMC) provided 61 samples from patients who underwent radical prostatectomy between 1990 and 2000. Another 35 samples were provided by the Sunnybrook Health Sciences Centre at the University of Toronto; these were for patients who were treated between 1998 and 2006. The last ten samples were obtained from the Moffitt Cancer Center (MCC); these samples were obtained from patients who were 21 years or older with pathologically confirmed prostate cancer who underwent radical prostatectomy between 1987 and 2003. The data set contains raw reads from the Illumina HiSeq 2000 NGS; the reads are paired-end sequences with a length of 51 bp each.

First, we pre-processed the data using the workflow illustrated in Fig. 1. The workflow starts from downloading the samples by the SRAtools, aligning the reads against the human genome hg19 using the STAR aligner [12] and then using RSEM [15] to quantify the transcripts based on TPM.

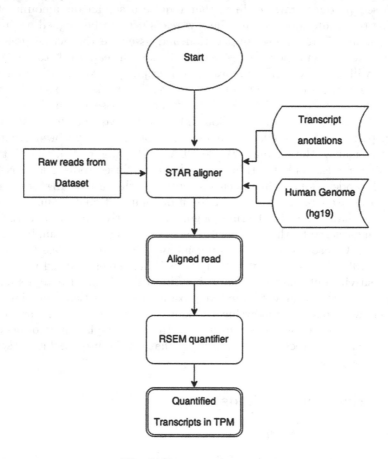

Fig. 1. Preprocessing steps.

2.2   Feature Selection

After the preprocessing step, we obtained 41,971 transcripts with their corresponding TPM values. This is a multi-class classification problem that we modeled following the one-versus-the-rest approach. Since there is a significant amount of features (transcripts in this case), we used feature selection to obtain transcripts that provide the highest information gain first. In this paper, the information gain (IG) attribute evaluator [16] is used to evaluate each attribute.

IG of feature X with respect to class Y is calculated as follows:

$$IG(Y, X) = H(Y) - H(Y|X) \tag{1}$$

where,

$$H(Y) = -\sum_{y \in Y} p(y) \, log_2(p(y)). \tag{2}$$

and

$$H(Y|X) = -\sum_{x \in X} p(x) \sum_{y \in Y} p(y|x) log_2(p(yx)). \tag{3}$$

Here, H(Y) is the entropy of class Y and H(Y|X) is the conditional entropy of Y given X.

Another feature selection method that was used is the minimum redundancy maximum relevance (mRMR) method, which takes features that have minimum redundancy while at the same time have high correlation to the classification variable [17]. The equation for minimizing redundancy (W_i) and maximizing the relevancy (V_i) is as follows:

$$W_i = \frac{1}{|S|^2} \sum_{i,j \in S} I(i,j) \tag{4}$$

and

$$V_i = \frac{1}{|S|} \sum_{i \in S} I(h, i) \tag{5}$$

where S is the set of features, $I(i, j)$ is mutual information between features (i, j), h is the class.

2.3 Identifying Transcripts Within Different Gleason Stages

The Weka [18] open source tool was used to run different classification algorithms on the minimized number of features to identify which transcripts are differentially expressed in the different stages. Naive Bayes is a probabilistic classifier that applies Bayes' theorem with the assumption of independency between the features [19]. SVM classifiers were also used in order to build a classification model based on the features selected in the previous step [20].

3   Results and Discussion

Mapping the reads against the human genome version hg19 was completed for all samples with mapping rates between 88% and 99% uniquely aligned reads. TPM scores for each sample were used through the RSEM tool. The final data set

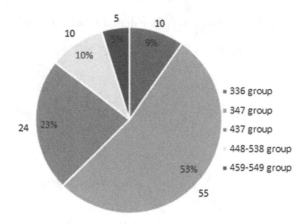

Fig. 2. Gleason stages groups and their distributions.

was combined to include all the transcript TPM values for all the 104 samples. Stated as a classification problem, the classes are five groups of Gleason stages combined. The groups and their distributions are shown in Fig. 2.

We identified 26 transcripts that are differentially expressed among the five different Gleason groups. Tables 2, 3, 4 and 5 show the top discriminative transcripts that can identify the Gleason group stages of prostate cancer by following the hierarchical method shown in Fig. 3. The first node of the hierarchy identifies Gleason stage 347 from the other Gleason groups, with 94% accuracy. The rest of the samples are moved to node 2 where the method identifies samples with Gleason score 437 from the rest, with 98% accuracy. Then the rest are moved to node 3 where samples with Gleason score 336 can be identified from the rest, with 100% accuracy. Finally, the rest of samples are moved to the last node, where the combination of samples that have Gleason scores 448 or 538 can be identified from the combination of samples with Gleason scores 459 and 549, the accuracy is 100% at the last node. All remaining Gleason stages were merged in the last node due to the similarity of the aggression and low number of samples for those stages.

The classifiers used to identify the transcripts and differentiate the classes are shown in Fig. 4. The x-axis represents the different Gleason groups and the y-axis represents the performance of the different classifiers.

As shown in the figure, Naive Bayes outperformed the other classifiers. It distinguished the first Gleason group from the rest by 94% and then distinguished the second Gleason group with a higher accuracy of 98%, while the last two Gleason groups were classified with 100% accuracy.

Many of the genes identified in this work have been previously characterized and described to play some role in prostate cancer as well as other cancers. PIAS3, a gene shown to be associated with group 2 Gleason score, has been shown to enhance the transcriptional activity of androgen receptors in prostate cancer cells which could play an important role in the growth and development

Table 2. Group 1 identified transcripts.

Transcript	Chromosome	Gene	Description
NM_001170880	11	GPR137	G protein-coupled receptor 137 (GPR137), transcript variant 2
NM_001198827	8	C8orf58	Chromosome 8 open reading frame 58 (C8orf58), transcript variant 3
NM_004629	9	9p13.3	Fanconi anemia complementation group G (FANCG)
NM_001098268	13	LIG4S	DNA ligase 4 (LIG4), transcript variant 3
NM_016641	16	GDE1	Glycerophosphodiester phosphodiesterase 1 (GDE1), transcript variant 1
NM_002445	8	MSR1	Macrophage scavenger receptor 1 (MSR1), transcript variant SR-AII
NM_001126337	1	TUFT1	Tuftelin 1 (TUFT1), transcript variant 2
NM_033071	6	SYNE1	Spectrin repeat containing nuclear envelope protein 1(SYNE1), transcript variant 2
NM_052906	22	ELFN2	Extracellular leucine rich repeat and fibronectin typeIII domain containing 2 (ELFN2), transcript variant 1
NM_000714	22	TSPO	Translocator protein (TSPO), transcript variant PBR
NM_004374	8	COX6C	Cytochrome c oxidase subunit 6C (COX6C)
NM_001007544	1	C1orf186	Chromosome 1 open reading frame 186 (C1orf186)
NM_001276438	21	KCNJ15	Potassium voltage-gated channel subfamily J member 15 (KCNJ15), transcript variant 7
NM_001252021	9	TOR2A	Torsin family 2 member A (TOR2A), transcript variant 7
NM_152612	22	CCDC116	Coiled-coil domain containing 116 (CCDC116), transcript variant 1

of prostate cancer [21]. Its reduced expression has been directly indicated in a cellular mechanism that promotes epithelial-to-mesenchymal transition and cell motility, which are important factors in prostate cancer invasion at later stages [22]. Reduced expression of PIAS3 has also been found to play a role in other cancers such as lymphoma [23] and lung cancer [24] cells. UBE2V2, another gene identified in this paper, is a DNA repair gene that has also been previ-

Table 3. Group 2 identified transcripts.

Transcript	Chromosome	Gene	Description
NM_001136224	1	RCOR3	REST corepressor 3 (RCOR3), transcript variant 2
NM_001017967	16	MARVELD3	MARVEL domain containing 3 (MARVELD3), transcript variant 1
NM_006099	1	PIAS3	Protein inhibitor of activated STAT 3 (PIAS3)
NM_152395	3	NUDT16	Nudix hydrolase 16 (NUDT16), transcript variant 2
NM_006473	11	TAF6L	TATA-box binding protein associated factor 6 like (TAF6L)
NM_001145541	11	TCP11L1	T-complex 11 like 1 (TCP11L1), transcript variant 2
NM_182501	2	MTERF4	Mitochondrial transcription termination factor 4 (MTERF4)

Table 4. Group 3 identified transcripts.

Transcript	Chromosome	Gene	Description
NM_003350	8	UBE2V2	Ubiquitin conjugating enzyme E2 V2 (UBE2V2)
NM_153051	22	MTMR3	Myotubularin related protein 3 (MTMR3), transcript variant 2
NM_207445	15	C15orf54	Chromosome 15 open reading frame 54 (C15orf54),

Table 5. Group 4 identified transcripts.

Transcript	Chromosome	Gene	Description
NM_001258330	20	EPB41L1	Erythrocyte membrane protein band 4.1 like 1 (EPB41L1), transcript variant 4

ously identified to play a role prostate cancer, specifically as a variant that is associated with familial prostate cancer [25]. Its differential expression has also been associated with poor prognosis in breast cancer [26]. EPB41L1, a gene that codes for a membrane protein with a role in cytoskeletal organization, is the only gene associated with group 4 Gleason stage. It has previously been shown to be significantly downregulated in prostate cancer [27]. Furthermore, low expression levels of this gene have been associated with an earlier biochemical recurrence. EPB41L1 has also been shown to be differentially expressed in other forms of cancer, such as breast cancer [28] and gastric cancer [29]. Zhenyu

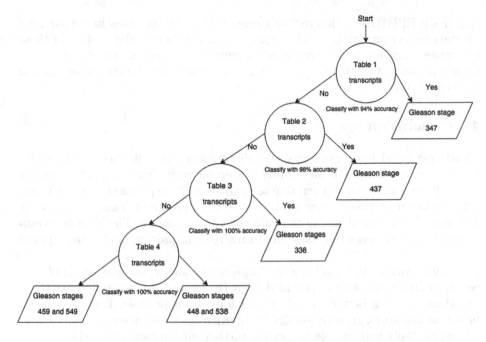

Fig. 3. Classification performance of the four Gleason groups.

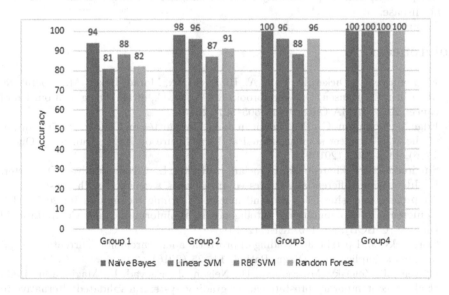

Fig. 4. Classification of the five Gleason group stages.

et al. found that its expression was significantly reduced in highly metastatic lung cancer compared to less metastatic forms of breast cancer [28] They proposed that this was due to the role of EPB41L1 in negative regulation of cell metastasis via inhibition of cell adhesion, migration and invasion. These factors make EPB41L1 an interesting target with potential roles in determining prostate cancer progression and prognosis. The functional relationships of these two genes show important gene activity related to prostate cancer. Therefore, we are planning to investigate more on these genes and the remaining ones in the near future.

4   Conclusion

Identifying novel biomarkers that can distinguish different Gleason stages for prostate cancer samples is an important step for early detection and treatment of the disease. Using next generation sequencing and the power of machine learning, this paper proposes a new method for finding groups of transcripts that are differentially expressed among the different Gleason stages. The identified transcripts belong to genes that play important roles in prostate and other types of cancer.

Future work include utilizing the same methodology for other kinds of cancer. If multiple samples are provided from the same patients in different Gleason stages, then a better analysis would have been possible. Further analysis based on literature, transcriptomics or interactomics databases, as well as wetlab experiments will be able to provide further information about the relevant transcripts that can be potentially used for diagnosis, treatment and prognosis of the disease.

References

1. Fitzmaurice, C., Dicker, D., Pain, A., Hamavid, H., Moradi-Lakeh, M., MacIntyre, M., Allen, C., Hansen, G., Woodbrook, R., Wolfe, C., et al.: The global burden of cancer 2013. JAMA Oncol. 1(4), 505–527 (2015)
2. Edge, S., Compton, C.: The American Joint Committee on Cancer: the 7th edition of the AJCC cancer staging manual and the future of TNM. Ann. Surg. Oncol. 17(6), 1471–1474 (2010)
3. Singireddy, S., Alkhateeb, A., Rezaeian, I., Rueda, L., Cavallo-Medved, D., Porter, L.: Identifying differentially expressed transcripts associated with prostate cancer progression using RNA-Seq and machine learning techniques. In: 2015 IEEE Conference on Computational Intelligence in Bioinformatics and Computational Biology (CIBCB), pp. 1–5. IEEE (2015)
4. Gordetsky, J., Epstein, J.: Grading of prostatic adenocarcinoma: current state and prognostic implications. Diagn. Pathol. 11, 25 (2016)
5. Epstein, J., Zelefsky, M., Sjoberg, D., Nelson, J., Egevad, L., Magi-Galluzzi, C., et al.: A contemporary prostate cancer grading system: a validated alternative to the Gleason score. Eur. Urol. 69(3), 428–435 (2016)
6. Lexander, H., Palmberg, C., Hellman, U., Auer, G., Hellström, M., Franzén, B., Jörnvall, H., Egevad, L.: Correlation of protein expression, gleason score and DNA ploidy in prostate cancer. Proteomics 6(15), 4370–4380 (2006)
7. Trapnell, C., Hendrickson, D., Sauvageau, M., Goff, L., Rinn, J., Pachter, L.: Differential analysis of gene regulation at transcript resolution with RNA-seq. Nat. Biotechnol. 31(1), 46–53 (2013). ISBN 0716776014

8. Trapnell, C., Williams, B., Pertea, G., Mortazavi, A., Kwan, G., van Baren, M.J., et al.: Transcript assembly and quantification by RNA-Seq reveals unannotated transcripts and isoform switching during cell differentiation. Nat. Biotechnol. **28**(5), 5115 (2010). doi:10.1038/nbt.1621

9. Mortazavi, A., Williams, B., McCue, K., Schaeffer, L., Wold, B.: Mapping and quantifying mammalian transcriptomes by RNA-Seq. Nat. Meth. **5**(7), 6218 (2008). doi:10.1038/nmeth.1226

10. Altschul, S., Gish, W., Miller, W., Myers, E., Lipman, D.: Basic local alignment search tool. J. Mol. Biol. **215**(3), 403–410 (1990)

11. Trapnell, C., Pachter, L., Salzberg, S.: TopHat: discovering splice junctions with RNA-Seq. Bioinformatics **25**(9), 1105–1111 (2009)

12. Dobin, A., Davis, C., Schlesinger, F., Drenkow, J., Zaleski, C., Jha, S., Batut, P., Chaisson, M., Gingeras, T.: STAR: ultrafast universal RNA-seq aligner. Bioinformatics **29**(1), 15–21 (2013)

13. Citak-Er, F., Vural, M., Acar, O., Esen, T., Onay, A., Ozturk-Isik, E.: Final Gleason score prediction using discriminant analysis and support vector machine based on preoperative multiparametric MR imaging of prostate cancer at 3T. BioMed Res. Int. **2014**, 690787 (2014)

14. Wei, P., Qiao, B., Li, Q., Han, X., Zhang, H., Huo, Q., Sun, J.: microRNA-340 suppresses tumorigenic potential of prostate cancer cells by targeting high-mobility group nucleosome-binding domain 5. DNA Cell Biol. **35**(1), 33–43 (2016)

15. Li, B., Dewey, C.: RSEM: accurate transcript quantification from RNA-Seq data with or without a reference genome. BMC Bioinform. **12**(1), 1 (2011)

16. Novakovic, J.: Using information gain attribute evaluation to classify sonar targets. In: 17th Telecommunications forum TELFOR, pp. 24–26 (2009)

17. Peng, H., Long, F., Ding, C.: Feature selection based on mutual information criteria of max-dependency, max-relevance, and min-redundancy. IEEE Trans. Pattern Anal. Mach. Intel. **27**(8), 1226–1238 (2005)

18. Frank, E., Hall, M., Witten, I.: The WEKA Workbench. In: Online Appendix for "Data Mining: Practical Machine Learning Tools and Techniques", 4th edn. Morgan Kaufman, Burlington (2016)

19. Domingos, P., Pazzani, M.: On the optimality of the simple Bayesian classifier under zero-one loss. Mach. Learn. **29**(2–3), 103–130 (1997)

20. Cortes, C., Vapnik, V.: Support-vector networks. Mach. Learn. **20**(3), 273–297 (1995)

21. Gross, M., Liu, B., Tan, J., French, F., Carey, M., Shuai, K.: Distinct effects of PIAS proteins on androgen-mediated gene activation in prostate cancer cells. Oncogene **20**(29), 3880 (2001)

22. Izumi, K., Fang, L., Mizokami, A., Namiki, M., Li, L., Lin, W., Chang, C.: Targeting the androgen receptor with siRNA promotes prostate cancer metastasis through enhanced macrophage recruitment via CCL2/CCR2-induced STAT3 activation. EMBO Mol. Med. **5**(9), 1383–1401 (2013)

23. Zhang, Q., Raghunath, P., Xue, L., Majewski, M., Carpentieri, D., Odum, N., Morris, S., Skorski, T., Wasik, M.: Multilevel dysregulation of STAT3 activation in anaplastic lymphoma kinase-positive T/null-cell lymphoma. J. Immunol. **168**(1), 466–474 (2002)

24. Ogata, Y., Osaki, T., Naka, T., Iwahori, K., Furukawa, M., Nagatomo, I., Kijima, T., Kumagai, T., Yoshida, M., Tachibana, I., et al.: Overexpression of PIAS3 suppresses cell growth, restores the drug sensitivity of human lung cancer cells in association with PI3-K/Akt inactivation. Neoplasia **8**(10), 817–825 (2006)

25. Nicolas, E., Arora, S., Zhou, Y., Serebriiskii, I., Andrake, M., Handorf, E., Bodian, D., Vockley, J., Dunbrack, R., Ross, E., et al.: Systematic evaluation of underlying defects in DNA repair as an approach to case-only assessment of familial prostate cancer. Oncotarget **6**(37), 39614 (2015)
26. Santarpia, L., Iwamoto, T., Di Leo, A., Hayashi, N., Bottai, G., Stampfer, M., André, F., Turner, F., Symmans, W., Hortobágyi, G., et al.: DNA repair gene patterns as prognostic and predictive factors in molecular breast cancer subtypes. Oncologist **18**(10), 1063–1073 (2013)
27. Schulz, W., Ingenwerth, M., Djuidje, C., Hader, C., Rahnenführer, J., Engers, R.: Changes in cortical cytoskeletal and extracellular matrix gene expression in prostate cancer are related to oncogenic ERG deregulation. BMC Cancer **10**(1), 505 (2010)
28. Ji, Z., Shi, X., Liu, X., Shi, Y., Zhou, Q., Liu, X., Li, L., Ji, X., Gao, Y., Qi, Y., et al.: The membrane-cytoskeletal protein 4.1N is involved in the process of cell adhesion, migration and invasion of breast cancer cells. Exp. Ther. Med. **4**(4), 736–740 (2012)
29. Seabra, A., Araújo, T., Mello, F., Alcântara, D., De Barros, D., Assumpção, D.E., Montenegro, R., Guimares, A., Demachki, S., Burbano, R.: High-density array comparative genomic hybridization detects novel copy number alterations in gastric adenocarcinoma. Anticancer Res. **34**(11), 6405–6415 (2014)

Bioinformatics from a Big Data Perspective: Meeting the Challenge

Francisco Gomez-Vela[✉], Aurelio López, José A. Lagares, Domingo S. Baena, Carlos D. Barranco, Miguel García-Torres, and Federico Divina

Intelligent Data Analysis (DATAi), Division of Computer Science,
Pablo de Olavide University, 41013 Seville, Spain
{fgomez,dsrodbae,cbarranco,mgarciat,fdiv}@upo.es, alopez@alum.upo.es

Abstract. Recently, the rising of the Big Data paradigm has had a great impact in several fields. Bioformatics is one such field. In fact, Bioinfomatics had to evolve in order to adapt to this phenomenon. The exponential increase of the biological information available, forced the researchers to find new solutions to handle these new challenges.

In this paper we present our point of view on the problems intrinsic to Big Data (volume, velocity, variety and veracity), how they affect the Bioinformatics field, and some solutions that can help Bioinformatics practitioners to deal with the difficulties presented by Big Data.

Keywords: Big Data · Machine learning · Fuzzy sets · Bioinformatics

1 Introduction

The world has changed with the explosion of the WWW. Volume and data transmission grew by generating a total of 1,700 billion bytes of data per minute. Such data come from a wide variety of sources such as servers, communications devices, cameras or sensors [1].

If we consider the studies of the past years with respect to the annual growth for the data generation, it is possible to predict that 44 billion gigabytes will be generated by the year 2020. So the size of the digital universe will be multiplied by 300 [1].

The concept "Big Data" arose to solve the high volume, complexity and diversity problems that characterize the heterogeneous data that is generated, stored and analysed at the moment. Such kind of data require new architectures, methodologies, algorithms and analysis techniques in order to extract useful knowledge from them [2,3].

The complexity of Big Data is represented mainly in four dimensions called "Big Data 4V's" [4]: (a) **Volume** or quantity of data that is generated, (b) **Velocity** with which data is generated, transmitted and/or analysed; (c) **Variety** in the different types of data and (d) **Veracity** or Value of the data that we process.

© Springer International Publishing AG 2017
I. Rojas and F. Ortuño (Eds.): IWBBIO 2017, Part II, LNBI 10209, pp. 349–359, 2017.
DOI: 10.1007/978-3-319-56154-7_32

There are many different fields in which Big Data has become a problem worthy of studying, and one of them is Bioinformatics. Some current Bioinformatics techniques and algorithms are not ready to deal with Big Data's new challenges [3,5]. The application of Big Data to Bioinformatics requires new strategies and algorithms that have high scalability and more efficient processing of large datasets, along with a way to consider the veracity of the data.

In this paper we focus on how Bioinformatics can address the challenge of Big Data. In this sense, we present an overview on the challenges Big Data poses to Bioinformatics practitioners and some ways to tackle these challenges on Bioinformatics field.

1.1 Types of Big Data in Bioinformatics

Due to different new available technologies, the volume of biological data is growing exponentially, since the cost of experiments to obtain the data is much lower in comparison to a few years ago. For example, the European Bioinformatics Institute (EBI), had stored 18 petabytes of biological information in 2013, while in 2015 the EBI stored 75 petabytes [6]. The different types of information that are classically handled in Bioinformatics can be classify into five categories [7]: Gene expression data, DNA and protein sequence, Protein-Protein interaction, Pathway data and Gene Ontology.

Gene Expression Data. This kind of data represents the expression levels of genes over different conditions (including time-series). Over the last few years the Microarray technology has been the most used type of data in the Bioinformatics research [8]. Now in the genome era, there are new types of gene expression data which increase dramatically the number of genes and conditions. As the most representative example we could find the single nucleotide polymorphism (SNP) technology [9]. There are many public resources for gene expression datasets like ArrayExpress [10], Gene Expression Omnibus [11] or Stanford Microarray Database [12].

DNA and Protein Sequence. The sequence of DNA and protein (RNA) are processed using different analytical methods. These methodologies include sequence alignment and database search. This type of data could be used as alternative for the gene expression data or additionally aims like detection of viruses, potential drugs research or identification of Polyadenylation among others. The DNA Data Bank of Japan [13] or RDP databases [14] could be two examples of repositories of this kind of information.

Protein–Protein Interaction. The protein–protein interactions (PPI)s, are crucial information regarding all biological processes. They are rendered in a network topology representing the interactions between the proteins and their biological relationships into a singular process. This kind of data is mainly used for the analysis of different diseases such Alzheimer or cancer. Important PPI repositories are GeneMANIA [15], BioGRID [16] or STRING [17] among others.

Pathway Data. The Pathways represent the molecular reactions into a cell. They are symbolized as graphs composed by proteins, compounds, genes or gene products and their relationships. One of the best pathway database is the Kyoto encyclopaedia of genes and genomes (KEGG) [18], containing information of various organisms. Other important repositories of pathways are Reactome [19] and Pathway Commons [20].

Gene Ontology. The Gene Ontology (GO) [21] database is a widely used biological database because it can provide complete information on attributes across species, and from different databases of genes and gene products. GO uses different vocabularies in order to provide and easy platform to query the gene (or genes') information. Moreover, there are a vast number of tools based on GO to perform analysis to gene sets like AmiGO [22] or DAG-Edit among others.

1.2   Big Data Platforms in Bioinformatics

The aim of this section is to introduce the latest Big Data platforms and how these technologies are focused on the Bioinfomatics.

It is well–known that the evolution of Big Data has undoubtedly marked a before and after resolving high-throughput challenges. Big Data allows to process vast quantity of data using new processing technologies and cloud computing approaches. Those issues related to storage or processing the data that were critical, now are dealt with. More and more researches have been using Big Data infrastructures as their main tool for collection, storage, retrieval, processing and visualization of the data. In the following we give a brief introduction to the most popular Big Data technolgies.

Hadoop. Apache Hadoop [23] was released by Google in 2003. It is an open source approach that allows running distributed applications. Hadoop was the first solution to handle Big Data applications. Hadoop has two specific features: Hadoop Distributed File System (HDFS) and Map Reduce [24] paradigm. Map Reduce divides a huge task into a large quantity of small sub tasks. Map Reduce also handles this sub tasks sending the smaller tasks to a cloud of computer nodes. This is made possible by the use of HDFS.

In [25], Hadoop is used to allow the scientific community to share their computational work-flows. This study focused on the work-flows of Gene Patterns [26]. The use of Hadoop was crucial due to its experiment was planned using cloud computing in different layers: Data Layer, System Layer and Service Layer. These different layers were used to publish the data, organize the work-flows and keep the scientific software up to date.

Another example is the work of Stein [27] where a DNA sequencing study fully oriented using Hadoop as a distributed Cloud Computing approach was presented.

Graphical Processing Units. Graphical Processing Units (GPU)s are used to extend or replace the CPU processing capacity. The biggest advantage of this technology is the fact that this approach has a high-end and high performance with medium cost. One of the most popular standards used in programming heterogeneous environments is the Compute Unified Device Architecture (CUDA) platform created by NVIDIA [28]. An example of this approach can be found in [29,30].

Apache Spark. Apache Spark [31] is a fast engine for large scale data processing. It stores the data in-memory so the running time has been decreased. With this key point, Apache Spark popularity has grown in the last years as it has become one of the main processing tool for massive datasets. Moreover, Apache Spark includes a full data scientist environment with data mining and visualization tools.

We can find several works in literature where Apache Spark was used: in [32] a classification of datasets with unique class distribution is presented. Also, the whole study has been carried out using Hadoop and Apache Spark in order to make a real comparison between the running times of the two frameworks. As a result, Apache Spark outperformed Hadoop. This huge improvement is mainly due to the the data being located in-memory. In the work presented by Boubela et al. [33] a combination of Apache Spark and GPU processing is used for processing functional magnetic resonance imaging (fMRI) Data from the Human Connectome Project. In this particular case, authors needed a high processing Big Data infrastructure to process 4D *Neuroimaging Informatics Technology Initiative* fMRI datasets.

NoSQL Databases. As already stated, the amount of biological data is growing and there are many approaches to process it. Nevertheless, the storage of the data is also an important aspect of Big Data. Typically biological data are characterized by huge size and a heterogeneous structure. The usual solution was to use relational datasets for storing such data. However, these systems became more complicated to adjust and maintain to Big Data approaches. In the meantime, NoSQL schemas were growing, providing a huge scalability for data aggregation. Due to this, the traditionally relational databases are being more and more replaced by NoSQL databases.

One of the mostly used NoSQL open source database is MongoDB [34]. MongoDB is a document oriented database that store the data in collections. Each of these collections contains documents that can be tracked as a pair *<Key, Value>*. The flexibility of storing huge volume of data has become the main characteristic of MongoDB. This database is also a great tool for unstructured, flexible and scalable projects that requires complex queries.

Another example of a NoSQL database is HBase [35]. HBase is a column-oriented data storage which organizes the data in columns and row families in order to obtain the best performance.

Finally, different works use Apache Cassandra as database to store the biological data [36, 37]. As a column-oriented database, Apache Cassandra offers a high availability, partition tolerance and high scalability technology. Moreover, Cassandra allows a distributed storage and parallel processing using largest datasets. So, the processing time of the whole process is reduced.

2 Big Data and Machine Learning in Bioinformatics

In order to deal with the biological issues, Bioinformatic offers some specific techniques and tools. The aim of this section is to present an overview of them. In particular we will focus on: Feature selection, Clustering, Biclustering and Gene Networks.

2.1 Feature Selection

Feature selection (FS) techniques are able to select a subset of most relevant and non-redundant features of an input dataset. Through these techniques, the learning method increase their performance in terms of both computational cost and accuracy. In fact, a FS method can remove irrelevant or redundant features from the input dataset.

With the increase of high dimensional data in Bioinformatics, many researchers have to deal with methods that are not computationally optimized and there is a need to reduce the dimensionality of the input dataset. In this sense, the FS rise as an essential tool to face this problem.

Bioinformatics, FS methods have been used for supervised and unsupervised methods, like the PPI analysis. For example, in [38], the authors proposed a new feature selection strategy that utilizes feature grouping to increase the effectiveness of the search on high dimensional datasets. As feature selection strategy, they proposed a Variable Neighbourhood Search (VNS) metaheuristic. The performance of the method was tested in several gene expression data in order to prove its usefulness. The work presented by Bagyamathi et al. [39] proposeS a new FS algorithm that combines the Improved Harmony Search algorithm with Rough Set theory for Protein sequences in order to tackle the Big Data paradigm. Another example is presented in the work by Zeng et al. [40] where a new Hybrid Distance (HD) in HIS is developed based on the value difference metric, and a novel fuzzy rough set is constructed by information systems. The method has been proved effective compared to others approaches applied on Big Data.

2.2 Clustering

Several clustering methods have been proposed in the fields of Bioinformatics and Machine Learning in the last few years. These methods are usually classified in:

(a) partitional clustering, (b) hierarchical clustering, (c) density-based clustering, (d) graph theoretic clustering among others [7]. The classic approaches, like k-means or CLICK, are not able to deal with the new situation in the Big Data era due to the exponential increase of the size of the gene expression datasets and the multiple sources of data. Therefore, the efforts made by the researchers are focused on the new methods and implementations that can handle with this kind of information.

Parallel clustering methods arise as a solution for the new volume of data while incremental clustering methods handle with velocity problem. For example, two versions of k-means, k-mode and k-prototype, were introduced in the work by Huang [41]. These methods can cope with large scale datasets due to the low complexity of the dissimilarity measure used. Another example is the work presented by Li and Fang [42], where a parallel hierarchical clustering method is proposed. The work by Zhao et al. [43] describes an implementation of k-means based on a Map Reduce architecture. An incremental hierarchical clustering was proposed by Chen et al. in [44]. This approach restructures the region of the object hierarchy by inserting the new object through a sequence of restructuring processes. Finally, in [45] the authors proposed a multi-view version of spectral clustering. This method uses a coregulation-based method to find some coherence from different graphs.

2.3   BiClustering

Biclustering algorithms are essential tools for the study of gene expression data, since they are able to discover subsets of genes with a common biological behaviour under a subset of experimental conditions [46]. However, for the last years, biclustering techniques have been evolving as an answer to new challenges posed by the Big Data era.

The problem of biclustering is known to be NP-hard [47], so most of the biclustering techniques use metaheuristics to obtain an optimal solution from all the possible results. However, in the "Biological Big Data Era" the simple use of metaheuristics is not enough because the huge size of datasets demand great efforts to reduce the computation time. Basically, three methods are the most used methods to accelerate the generation of biclusters from large gene expression datasets: Parallel computing, the use of GPU and the Map Reduce programming paradigm.

Parallel computing implies using a divide and conquer technique while programming an algorithm, so it is divided into parts that can be executed in independent execution threads on multicore processors and/or clusters of computers. So, a crucial point is to locate which part of the algorithm could be parallelized. In [48], a new biclustering algorithm called PBiclustering is shown. In this algorithm, the processing of every bicluster, by adding rows and columns, is executed independently. In other work presented by Jin and Hua [49], Cheng-Chung's algorithm is modified to improve its results and its performance using parallel computing.

There are some important aspects when using GPUs to implement biclustering algorithms [50]: to allocate a predefined amount of memory on GPU, what may be difficult to estimate depending on the input dataset and the undefined number of bicluster results. Besides, due to the size of input matrix, the global memory usage is mandatory, limiting the throughput. Finally, determining the size of the grid of CUDA block and ensuring synchronization between threads of different blocks may be problematic. Despite this, implementations using GPUs show an appreciable improvement in their performance. In [51], a GPU version of the Nonnegative Matrix Factorization algorithm is shown. The FLOC algorithm is another biclustering approach that has been reimplemented using CUDA [52]. The part of FLOC algorithm that computes the mean squared residue of a bicluster is executed in each core of a GPU, producing the calculation for a specific cell of the data matrix and its posterior sum. In [53], a new version of Geometric Biclustering algorithm (GBC) is implemented using GPU-CUDA. Authors compare this GBC version with two new ones: one based on multithreads and the other one implemented using Field-Programmable Gate Array (FPGA). The third method to accelerate the generation of biclusters is based on Map Reduce framework. In [54], the DisCo (Distributed Co-clustering framework) biclustering algorithm is implemented by using Hadoop. It is based on a reorganization of rows and columns of a submatrix, being rows stored in form of (Key, value) pairs. Also, a new version of the Nonnegative Matrix Factorization algorithm is implemented using Map Reduce [55]. Authors have developed an open source version that is able to process sparse matrices up to million-by-million size and can be easily deployed on Amazon Elastic Map Reduce.

2.4 Gene Networks

Another approach to model the biological processes in Bioinformatics are Gene Networks (GN). GNs represent the relationships among genes as a graph, where the nodes represent the genes (or gene products like proteins or compounds) and the edges represent the relationships among them. This approach implies benefits for researchers like an intuitive representation of the information but also some issues that should be considered. There are several approaches for modelling GNs, which can be briefly Boolean, Bayesian, Information theory-based and differential equation models [56].

Due to the difficulties in handling the information (in terms of volume and heterogeneity), the inference of GNs is considered as a Big Data analytic task [57]. The main challenge derives from the huge amount of information that has to be processed. Several GNs inference methods have been proposed in the last few years [57]. However, they are not able to handle datasets with more than a few thousands genes [7]. In addition to this, the performances of these methods decrease exponentially with the increase of the information processed because they do not emphasize computational efficiency [58].

In this sense, only a few of recent work-shave addressed these issues. For example, in [58] the authors present a novel methodology based on a fast computational metaheuristic to generate a correlogram matrix to construct a gene network. Another example of metaheuristic based method to infer GNs is the work of Rau et al. [59]. In this work the author proposed a non-standard adaptation of a simulation-based approach known as Approximate Bayesian Computing based on Markov chain Monte Carlo sampling. The work presented in [8], the authors introduce a new fuzzy approach named FyNE (Fuzzy NEtworks) for modelling of gene association networks. This method is able to handle large datasets, and it uses a fuzzy based function which can incorporate prior biological knowledge into the modelling phase. In [60] used parallel processing to generate a network of 1,000 genes based on artificial data. They achieved a parallel speedup of 20 times with 32 cores.

3   Conclusions and Future Challenges

This paper presented an overview of the impact of the Big Data in Bioinformatics. The problems in the data types and their new volume and variety are diverse. In this sense, new platforms and technologies, like Map–Reduce or the NoSQL databases, offer solutions to some of the problems presented.

This paper shows some inspirational examples of solution proposals to Big Data challenges in Bioinformatics, particularly focusing on Feature Selection, Clustering, BiClustering and Gene Networks inference. Despite the advances presented, there are still many challenges in the field of Bioinformatics regarding Big Data. One of the most relevant refers to the visualization of data, for example, in large genetic networks. Another problem is that most of tools for the biological data analysis, still do not have a high computational performance. Therefore, future efforts should focus on improving the performance of such tools.

Acknowledgement. This work has been funded by the Spanish Ministry of Science and Innovation under grant TIN2015-64776-C3-2-R.

References

1. Zikopoulos, P., Eaton, C.: Understanding Big Data: Analytics for Enterprise Class Hadoop and Streaming Data. McGraw-Hill Osborne Media, IBM, New York (2011)
2. Greene, C., Tan, J., Ung, M., Moore, J., Cheng, C.: Big data bioinformatics. J. Cell. Physiol. **229**(12), 1896–1900 (2014)
3. Marx, V.: Biology: the big challenges of big data. Nature **498**(7453), 255–260 (2013)
4. Bizer, C., Boncz, P., Brodie, M., Erling, O.: The meaningful use of big data: four perspectives-four challenges. ACM SIGMOD Rec. **40**(4), 56–60 (2012)
5. Labrinidis, A., Jagadish, H.: Challenges and opportunities with big data. Proc. VLDB Endowment **5**(12), 2032–2033 (2012)
6. Cook, C., Bergman, M., Finn, R., Cochrane, G., Birney, E., Apweiler, R.: The European Bioinformatics Institute in 2016: data growth and integration. Nucleic Acids Res. **44**(Database Issue), 20–26 (2016)

7. Kashyap, H., Ahmed, H., Hoque, N., Swarup, R., Dhruba Kumar, B.: Big data analytics in bioinformatics: a machine learning perspective. Cornell Univ. Lib. Comput. Eng. Finan. Sci. **13** (2015)
8. Gomez-Vela, F., Barranco, C., Diaz-Diaz, N.: Incorporating biological knowledge for construction of fuzzy networks of gene associations. Appl. Soft Comput. **42**, 144–155 (2016)
9. Liu, Y.: Data Mining Methods for Single Nucleotide Polymorphisms Analysis in Computational Biology. Ph.D. thesis AAI3510948 (2011)
10. Kolesnikov, N., Hastings, E., Keays, M., Melnichuk, O., Tang, Y., Williams, E., Dylag, M., Kurbatova, N., Brandizi, M., Burdett, T., Megy, K., Pilicheva, E., Rustici, G., Tikhonov, A., Parkinson, H., Petryszak, R., Sarkans, U., Brazma, A.: Arrayexpress update-simplifying data submissions. Nucleic Acids Res. **43**(Database Issue), 1113–1116 (2015)
11. Edgar, R., Domrachev, M., Lash, A.E.: Gene expression omnibus: NCBI gene expression and hybridization array data repository. Nucleic Acids Res. **30**(1), 207–210 (2002)
12. Sherlock, G., Boussard, T., Kasarskis, A., Binkley, G., Matese, J., Dwight, S., Kaloper, M., Weng, S., Jin, H., Ball, C., Eisen, M., Spellman, P.: The Stanford Microarray database. Nucleic Acid Res. **29**(1), 152–155 (2001)
13. Tateno, Y., Imanishi, T., Miyazaki, S., Fukami-Kobayashi, K., Saitou, N., Sugawara, H., Gojobori, T.: DNA Data Bank of Japan (DDBJ) for genome scale research in life science. Nucleic Acids Res. **30**(1), 27–30 (2002)
14. Maidak, B., Olsen, G., Larsen, N., Overbeek, R., McCaughey, M., Woese, C.: The RBP (Ribosomal Database Project). Nucleic Acids Res. **25**(1), 109–110 (1997)
15. Warde-Farley, D., Donaldson, S.L., Comes, O., Zuberi, K., Badrawi, R., Chao, P., Franz, M., Grouios, C., Kazi, F., Lopes, C., Maitland, A., Mostafavi, S., Montojo, J., Shao, Q., Wright, G., Bader, G., Morris, Q.: The GeneMANIA prediction server: biological network integration for gene prioritization and predicting gene function. Nucleic Acids Res. **38**(1), 214–220 (2010)
16. Stark, C., Breitkreutz, B., Reguly, T., Boucher, L., Breitkreutz, A., Tyers, M.: BioGRID: a general repository for interaction datasets. Nucleic Acids Res. **34**(Database Issue), 535–539 (2006)
17. Szklarczyk, D., Franceschini, A., Kuhn, M., Simonovic, M., Roth, A., Minguez, P., Doerks, T., Stark, M., Muller, J., Bork, P.: The string database in 2011: functional interaction networks of proteins, globally integrated and scored. Nucleic Acids Res. **39**(Database Issue), 561–568 (2011)
18. Kanehisa, M., Goto, S.: KEGG: kyoto encyclopedia of genes and genomes. Nucleic Acids Res. **28**(1), 27–30 (2000)
19. Fabregat, A., Sidiropoulos, K., Garapati, P., Gillespie, M., Hausmann, K., Haw, R., Jassal, B., Jupe, S., Korninger, F., McKay, S., Matthews, L., May, B., Milacic, M., Rothfels, K., Shamovsky, V., Webber, M., Weiser, J., Williams, M., Wu, G., Stein, L., Hermjakob, H., D'Eustachio, P.: The Reactome pathway knowledgebase. Nucleic Acids Res. **44**(Database Issue), 481–487 (2016)
20. Cerami, E.G., Gross, B.E., Demir, E., Rodchenkov, I., Babur, O., Anwar, N., Schultz, N., Bader, G.D., Sander, C.: Pathway commons, a web resource for biological pathway data. Nucleic Acids Res. **39**(Database Issue), 685–690 (2011)
21. Ashburner, M., Ball, C.A.: Gene ontology: tool for the unification of biology. The Gene Ontology Consortium. Nat. Genet. **25**(1), 25–29 (2000)
22. Carbon, S., Ireland, A., Mungall, C., Shu, S., Marshall, B., Lewis, S.: AmiGO: online access to ontology and annotation data. Bioinformatics **25**(2), 288–289 (2009)

23. Hadoop, A.: Hadoop (2009)
24. Dean, J., Ghemawat, S.: MapReduce: simplified data processing on large clusters. Commun. ACM **51**(1), 107–113 (2008)
25. Dudley, J.T., Butte, A.: Reproducible in silico research in the era of cloud computing. Nature Biotechnol. **28**(11), 1181–1185 (2010)
26. Reich, M., Liefeld, T., Gould, J., Lerner, J., Tamayo, P., Mesirov, J.: Genepattern 2.0. Nat. Genet. **38**(5), 500–501 (2006)
27. Stein, L.: The case for cloud computing in genome informatics. Genome Biol. **11**(5) (2010)
28. NVIDIA: NVIDIA CUDA Programming Guide 2.0 (2008)
29. Sumiyoshi, K., Hirata, K., Hiroi, N., Funahashi, A.: Acceleration of discrete stochastic biochemical simulation using GPGPU. Front. Physiol. **6** (2015)
30. Mane, S.U., Pangu, K.H.: Disease diagnosis using pattern matching algorithm from DNA sequencing: a sequential and GPGPU based approach. In: International Conference on Informatics and Analytics, pp. 1–5 (2016)
31. Spark, A.: Apache spark-lightning-fast cluster computing (2014)
32. Triguero, I., Galar, M., Merino, D., Maillo, J., Bustince, H., Herrera, F.: Evolutionary undersampling for extremely imbalanced big data classification under apache spark. In: IEEE Congress on Evolutionary Computation (CEC), pp. 640–647 (2016)
33. Boubela, R., Kalcher, K., Huf, W., Nasel, C., Moser, E.: Big data approaches for the analysis of large-scale fMRI data using apache spark and GPU processing: a demonstration on resting-state fMRI data from the human connectome project. Front. Neurosci. **9** (2015)
34. Banker, K.: MongoDB in action. Manning Publications Co., Greenwich (2011)
35. Taylor, R.C.: An overview of the Hadoop/MapReduce/HBase framework and its current applications in bioinformatics. BMC Bioinform. **11**(12), S1 (2010)
36. Dudley, J., Butte, A.: A quick guide for developing effective bioinformatics programming skills. PLoS Comput. Biol. **5**(12), e1000589 (2009)
37. Kepner, J., Anderson, C., Arcand, W., Bestor, D., Bergeron, B., Byun, C., Hubbell, M., Michaleas, P., Mullen, J., O'Gwynn, D., Prout, A., Reuther, A., Rosa, A., Yee, C.: D4m 2.0 schema: a general purpose high performance schema for the accumulo database. In: IEEE High Performance Extreme Computing Conference (HPEC), pp. 1–6 (2013)
38. Garcia-Torres, M., Gomez-Vela, F., Melian-Batista, B., Moreno-Vega, J.: High-dimensional feature selection via feature grouping: a variable neighborhood search approach. Inf. Sci. **326**, 102–118 (2016)
39. Bagyamathi, M., Inbarani, H.H.: A novel hybridized rough set and improved harmony search based feature selection for protein sequence classification. In: Hassanien, A.E., Azar, A.T., Snasael, V., Kacprzyk, J., Abawajy, J.H. (eds.) Big Data in Complex Systems. SBD, vol. 9, pp. 173–204. Springer, Cham (2015). doi:10.1007/978-3-319-11056-1_6
40. Zeng, A., Li, T., Liu, D., Zhang, J., Chen, H.: A fuzzy rough set approach for incremental feature selection on hybrid information systems. Fuzzy Sets Syst. **258**, 39–60 (2015)
41. Huang, Z.: Extensions to the k-means algorithm for clustering large data sets with categorical values. Data Min. Knowl. Disc. **2**(3), 283–304 (1998)
42. Li, X., Fang, Z.: Parallel clustering algorithms. Parallel Comput. **11**(3), 275–290 (1989)
43. Zhao, W., Ma, H., He, Q.: Parallel k-means clustering based on MapReduce. In: Proceedings of the 1st International Conference on Cloud Computing, pp. 674–679 (2009)

44. Chen, N., Chen, A., Zhou, L.: An incremental grid density-based clustering algorithm. J. Soft. **13**(1), 1–7 (2002)
45. Kumar, A., Daume, H.: A co-training approach for multi-view spectral clustering. In: Proceedings of the 28th International Conference on Machine Learning (ICML-11), pp. 393–400 (2011)
46. Pontes, B., Giraldez, R., Aguilar-Ruiz, J.: Biclustering on expression data: a review. J. Biomed. Inform. **57**, 163–180 (2015)
47. Madeira, S., Oliveira, A.: Biclustering algorithms for biological data analysis: a survey. IEEE/ACM Trans. Comput. Biol. Bioinformatics **1**(1), 24–45 (2004)
48. Liu, W., Chen, L., Qu, H., Qin, L.: A parallel biclustering algorithm for gene expressing data. In: 2008 Fourth International Conference on Natural Computation, vol. 1, pp. 25–29 (2008)
49. Jin, S., Hua, L.: An improved biclustering algorithm for gene expression data. Open Cybern. Systemics J. **8**, 1141–1144 (2014)
50. Orzechowski, P., Boryczko, K.: Effective biclustering on GPU-capabilities and constraints. Prz Elektrotechniczn **1**, 131–134 (2015)
51. Mejia-Roa, E., Garcia, C., Gomez, J., Prieto, M., Tirado, F., Nogales, R., Pascual-Montano, A.: Biclustering and classification analysis in gene expression using non-negative matrix factorization on multi-GPU systems. In: 11th International Conference on Intelligent Systems Design and Applications, pp. 882–887 (2011)
52. Arnedo-Fdez, J., Zwir, I., Romero-Zaliz, R.: Biclustering of very large datasets with GPU tecnology using cuda. In: Proceedings of V Latin American Symposium on High Performance Computing (2012)
53. Liu, B., Yu, C., Wang, D., Cheung, R., Yan, H.: Design exploration of geometric biclustering for microarray data analysis in data mining. IEEE Trans. Parallel Distrib. Syst. **25**(10), 2540–2550 (2014)
54. Papadimitriou, S., Sun, J.: DisCo: Distributed co-clustering with Map-Reduce: a case study towards petabyte-scale end-to-end mining. In: 2008 Eighth IEEE International Conference on Data Mining, pp. 512–521 (2008)
55. Ruiqi, L., Yifan, Z., Jihong, G., Shuigeng, Z.: CloudNMF: a MapReduce implementation of nonnegative matrix factorization for large-scale biological datasets. Genomics Proteomics Bioinform. **12**(1), 48–51 (2014)
56. Hecker, M., Lambeck, S., Toepfer, S., Van Someren, E., Guthke, R.: Gene regulatory network inference: data integration in dynamic modelsa review. Biosystems **96**(1), 86–103 (2009)
57. Spencer-Angus, T., Yaochu, J.: Reconstructing biological gene regulatory networks: where optimization meets big data. Evol. Intel. **7**(1), 29–47 (2014)
58. Roy, S., Bhattacharyya, D., Kalita, J.: Reconstruction of gene co-expression network from microarray data using local expression patterns. BMC Bioinform. **15**, 1–14 (2014)
59. Rau, A., Jaffrezic, F., Foulley, J., Doerge, R.W.: Reverse engineering gene regulatory networks using approximate Bayesian computation. Stat. Comput. **22**(6), 1257–1271 (2012)
60. Xiao, M., Zhang, L., He, B., Xie, J., Zhang, W.: A parallel algorithm of constructing gene regulatory networks. In: Proceedings of the 3rd International Symposium on Optimization and Systems Biology, pp. 184–188 (2009)

Computational Prediction of Host-Pathogen Interactions Through Omics Data Analysis and Machine Learning

Diogo Manuel Carvalho Leite[1,2(✉)], Xavier Brochet[1,2(✉)],
Grégory Resch[3], Yok-Ai Que[4], Aitana Neves[1,2],
and Carlos Peña-Reyes[1,2(✉)]

[1] School of Business and Engineering Vaud (HEIG-VD),
University of Applied Sciences of Western Switzerland (HES-SO),
Yverdon-Les-Bains, Switzerland
{diogo.leite,xavier.brochet,carlos.pena}@heig-vd.ch
[2] SIB Swiss Institute of Bioinformatics, Lausanne, Switzerland
[3] Department of Fundamental Microbiology,
University of Lausanne, Lausanne, Switzerland
[4] Department of Intensive Care Medicine,
Bern University Hospital (Inselspital), Bern, Switzerland

Abstract. The emergence and rapid dissemination of antibiotic resistance, worldwide, threatens medical progress and calls for innovative approaches for the management of multidrug resistant infections. Phage-therapy, i.e., the use of viruses (phages) that specifically infect and kill bacteria during their life cycle, is a re-emerging and promising alternative to solve this problem. The success of phage therapy mainly relies on the exact matching between the target pathogenic bacteria and the therapeutic phage. Currently, there are only a few tools or methodologies that efficiently predict phage-bacteria interactions suitable for the phage therapy, and the pairs phage-bacterium are thus empirically tested in laboratory. In this paper we present an original methodology, based on an ensemble-learning approach, to predict whether or not a given pair of phage-bacteria would interact. Using publicly available information from Genbank and phagesdb.org, we assembled a dataset containing more than two thousand phage-bacterium interactions with their corresponding genomes. A set of informative features, extracted from these genomes, form the base of the quantitative datasets used to train our predictive models. These features include the distribution of predicted protein-protein interaction scores, as well as the amino acid frequency, the chemical composition, and the molecular weight of such proteins. Using an independent test dataset to evaluate the performance of our methodology, our approach gets encouraging performance with more than 90% of accuracy, specificity, and sensitivity.

Keywords: Ensemble-learning · Genomic · Machine-learning · Phage-therapy · Protein-protein interaction

---

A. Neves and C. Peña-Reyes–Contributed equally to this work.

© Springer International Publishing AG 2017
I. Rojas and F. Ortuño (Eds.): IWBBIO 2017, Part II, LNBI 10209, pp. 360–371, 2017.
DOI: 10.1007/978-3-319-56154-7_33

1 Introduction

The emergence and rapid dissemination of antibiotic resistance worldwide threatens medical progress. As a consequence, in a near future medicine might face a return to the pre-antibiotic era. The paucity of potential new anti-infectives in the pipeline of pharmaceutical industries urges the need for alternatives to fight this public health problem. Phage therapy might represent such an alternative. This re-emerging therapy uses viruses that specifically infect and kill bacteria during their life cycle to reduce/eliminate bacterial load and cure infections. These viruses, called bacterio-phages or phages, have been co-evolving with bacteria for billions of years, controlling bacterial populations and epidemics, and contributing to their genetic exchanges. With the advantage of having low impact on the commensal flora, as they are highly strain specific, some phages might, nevertheless, harbor virulence factors and drive horizontal gene transfer mediating dissemination of pathogenic traits including antibiotic resistance, calling for careful selection before their therapeutic use.

In this context, phage therapy, which was largely shadowed in Occident by the success of antibiotic agents, encounters a renewed interest in the Western world. Phages are ubiquitous environmental, bacterial viruses. Phage therapy success relies on the correct matching of a bacteria and a phage selected among a fully characterized phage library. Currently, rules for phage selection are empirical, but progress has recently been made enabling the use of *in silico* algorithms.

Phages are found in natural and man-made environments, rapidly co-evolving with their bacterial targets. Being highly specific, their infectivity can actually vary drastically across bacterial strains of a same species [1]. Determining phages host ranges is currently achieved using infection tests [2] which, depending on the size of the reference bacterial panel to be tested, may take several days of lab work. Many positive interactions have been uncovered using these tests, revealing that highly phage-sensitive bacteria get infected by phages with both narrow and broad host range, whereas highly phage-resistant bacteria are only infected by broad range phages [3]. Despite these efforts, including ~ 300 phage and bacterial strains, there is still little understanding about for instance phage host ranges [1].

Several studies have investigated how phages actually infect bacteria [4, 5] and how bacteria defend against phage invasions [6]. Phages have receptor-binding proteins (RBPs) that can recognize and bind specifically to receptors on the bacterial cell surface. These bacterial receptors have been experimentally identified in some cases and shown to generally involve both proteins and cell-wall glycopolymers [7]. Once attached to the surface of their bacterial host, phages inject their genomes inside the bacterial cytoplasm. Thus, only phage genomes actually enter target bacteria.

Phages can generally be classified into two categories: virulent phages that enter directly the "lytic cycle", which yields the formation of new phages and their release by cell wall disruption following holins and lysins activity. Temperate phages, in contrast, undergo the "lysogenic cycle", integrating the injected genome into the host chromosome, which then replicates as a prophage at the pace of bacterial divisions. Upon induction (e.g. after cell damage), the prophage eventually excise and enters the lytic cycle. Recently, a machine-learning approach dubbed PHACTS [8] was successfully developed to automatically classify phages lifestyle based on their protein sequences.

Bacteria have evolved sophisticated mechanisms to escape phage infection [4] and phages have likewise developed strategies to defeat bacterial defenses [5]. In order to counteract phage adsorption, bacteria have for example evolved mutations in their phage receptors to render them unrecognizable to the phage. In other cases, phage receptors may be hidden behind physical barriers such as capsules [6]. Some bacteria also developed the ability to block phage DNA injection through superinfection exclusion systems, which block genome injection to the second phage trying to infect a bacterium [7]. Other phages' potential weaknesses include genetically encoded sites that could be targeted by the bacterial restriction-modification system which cuts foreign DNA at specific recognition sites. Another DNA cleavage system evolved by bacteria is the CRISPR-Cas system, which also involves recognition of a specific genomic region on the phage DNA [9]. Other bacteria choose to commit suicide in order to avoid phage dissemination into the clonal bacterial population (abortive infection system [5]). Finally, phages can be defeated by bacteria through phage assembly interference, where bacteria encode phage-inducible chromosomal islands capable of negatively interacting with the assembly of the phage [10]. From the above, it appears that the specificity of a phage for a bacterial host is not only dictated by the phage receptor and lysins, but also depends on the bacterial defense mechanisms.

Being able to predict phage-bacteria infection networks could be particularly beneficial in a clinical context for patients with multidrug resistant infections requiring rapid medical care. Indeed, phage therapy offers a promising alternative to the global crisis resulting from the emergence and dissemination of multi-resistant bacterial strains. Current phage-based antimicrobial strategies include the use of a cocktail of lytic phages to specifically kill a population of bacteria [11], the use of purified lysins to lyse bacteria from the outside [12], and the use of temperate phages to add extra genes to bacteria and render them antibiotic-sensitive [13, 14]. With successes in preclinical and veterinary trials [15], phage therapy still misses access to well-characterized phage libraries as well as methods to rapidly screen for phage candidates given a bacterial strain.

The success of phage therapy mainly relies on the exact matching between both the target pathogenic bacteria and the therapeutic phage. Therefore, having access to a fully characterized phage library is necessary, although no sufficient. An essential and mandatory step to conceive personalized phage-therapy treatments is the capacity to predict the interactions between the target pathogen and its potential phage(s).

To efficiently predict successful phage-bacteria interactions, we will explore *in-silico* methodologies with the goal of enabling optimal selection of phage candidates from an existing phage library that efficiently target a given pathogenic bacteria. To achieve this, we will combine genomic information with state-of-the-art bioinformatics and machine-learning techniques, taking advantage of the growing amount of interaction data already available. In this study we aim at developing an original and relevant approach based on supervised modeling to build predictive classifiers assessing if a given pair of phage-bacterium can interact and/or the likelihood of such an interaction. These predictive models are built on the base of informative features that capture interesting properties of the pair of microorganisms [16]. Thus, one of the most important challenges consists in extracting adequate features based mainly, almost exclusively, on genomic information, and then selecting those features that allows the classifiers to exhibit the best predictive performance. The resulting dimensionality reduction is further extended with the goal of improving the computational cost incurred by the learning algorithms, while maintaining as much of the predictive power as possible.

2 Materials and Methods

2.1 Assembling of the Training Dataset

In phase with our main goal, i.e., predicting phage-bacteria interactions based on genomic data, we retrieved, whenever possible, the corresponding genomes of phages and bacteria from different public databases to create our training dataset, consisting of both positive (i.e., phage-bacteria interactions which have been annotated in public databases) and negative (i.e., pairs of phage bacteria whose interaction is actually unknown and has not been annotated) examples. Complete phage and bacterial genomes were collected from two public databases: PhagesDB [17] and GenBank [18] in February 2016. In total we obtained 1065 phage sequences from which 79 from GenBank [18] and 986 from PhagesDB [17] and 43 host bacteria sequences from GenBank [18], for a final set of 1065 positive phage-bacterium interactions.

Phage Sequences. A first set of complete phage genome sequences was obtained from PhagesDB [17] and gene prediction was performed with GeneMarkS [19], which allowed to retrieve the coding DNA sequence and the protein sequences. A second set of complete genome phages sequences was obtained from the NCBI Genome Resource [20], Nucleotide DataBase [21], whose sequences (genome, coding DNA, and protein) were retrieved using the Entrez service of NCBI [22] with the query "phage [Title] and 'complete genome' ".

Bacterial Sequences. The bacterial host of each phage was obtained from the fields 'Isolation Host' and 'host' respectively in PhagesDB [17] and NCBI Nucleotide [21]. For all bacterial hosts the genome sequences, the coding DNA sequences, and the protein sequences were retrieved by using the Entrez service of the NCBI [22] for 'name of bacteria' [ORGN] AND 'complete genome'. All phages whose host was not known or which did not have a complete sequence genome were removed.

Negative Dataset. The creation of negative interactions, which constitute the negative dataset, was done by randomly selecting phages and bacteria in the positive dataset to constitute pairs of non-interacting phage-bacteria. A given phage-bacterium pair is recorded as negative if it is not present into the positive dataset and if the bacterium belongs to a different species than the phage's known host. Although this method does not fully establish that phage-bacteria pairs do not interact, it includes pairs not known to physically interact that likely don't. Since most phages seem to be specific for one species of bacteria and many are only able to lyse specific strains within a species [23–27] we consider that this method limits possible errors. The negative dataset is of the same size as the positive dataset, 1065 negative interactions.

2.2 Feature Extraction: Protein-Protein Interactions

The efficient infection of a bacterium by a phage is mainly due to the interaction between phage-encoded and bacterial proteins. Indeed, a rapid analysis of our database shows that positive phage-bacterium interactions contain more predicted protein-protein interactions (PPIs) than negative interactions do. This section describes the procedure used to extract two sets of features based on PPIs [28], features that constitute the base of candidate training datasets. In average, the bacteria in our database express 3417 proteins, while phages produce 74 proteins. Thus, the average number of PPIs per phage-bacterium interaction is slightly more than 2.5×10^5, resulting in around 5.4×10^8 PPIs to be characterized. Note that the number of PPIs may be different for each phage-bacterium pair. So, to facilitate the use of predictive machine learning, feature extraction must be done in a way that allows comparing phage-bacterium interactions in spite of the number of PPIs.

Domain-Domain Interaction-Based Scores. Domains are structural or functional subunits of a protein [29, 30] and the interaction between a pair of proteins often involves binding between one or more pairs of their constituting domains. DOMINE [31] is a database of known and predicted protein domain interactions where each domain-domain interaction (DDI) is described by a quality score that measures the reliability of the interaction. Each PPI obtains a score consisting in the sum of all its DDI scores. Given that predicted interactions in DOMINE are based on Pfam domain definitions [32], we used HMMER [33, 44] to detect the domains present in our sequences. For the more than 2.3×10^5 proteins in our database, we obtained around 3.5×10^5 domains. As a result, for each phage-bacterium pair we obtained a vector of PPI scores. Finally, we transformed these vectors into frequencies of PPI-scores (histogram-like) so as to obtain vectors of the same length. For this purpose, two criteria were applied: Use of absolute or normalized frequency, and fixing either the size of the bins or their number. Resulting in four different groups of feature sets:

- 5 sets created using non-normalized frequencies and fixed number of bins: 5, 10, 15, 30, 50 (dubbed NB sets)
- 5 sets with normalized frequencies and fixed number of bins: 5, 10, 15, 30, 50 (NBN sets)
- 5 sets with non-normalized frequencies and fixed-size bins: 1, 5, 10, 15, 20 (TB sets)
- 3 sets with normalized frequencies and fixed-size bins: 1×10^{-6}, 2.5×10^{-6}, 5×10^{-6} (TBN sets)

Primary Protein Sequence Information. Protein-protein interactions may also be characterized directly on the base of the two primary protein sequences and on their associated physicochemical properties [35–38]. We extract and use a set of 27 features from the primary sequence of each protein: 21 features describe the relative abundance of amino-acids in the protein sequence—i.e., the 20 amino-acids and one for unidentified amino-acids—five other features correspond to abundance of some important elements (Carbon, Hydrogen, Nitrogen, Oxygen, and Sulfur) [39] and one more to the molecular weight of the protein. Summarizing, each PPI is described by 54 features and assigned a label. As already mentioned, each phage-bacterium interaction has, in average, more than 250 thousand PPIs, resulting in an extremely high dimensionality. To reduce it, each feature of the interaction is projected—using Principal Component Analysis (PCA)—into its two principal components, which capture already more than 90% of the information of all the PPIs. Thus, the final feature set CHEM contains 108 features for each of the 2130 phage-bacterium interactions.

We have, finally, 19 candidate datasets based on the extracted features, 18 based on DDIs and one based on physicochemical properties of the protein sequences. During the modeling process, described in the next section, the most informative datasets will be selected to constitute a final training dataset.

2.3   Machine Learning Approach and Experimental Setup

We intend to build predictive models able to assess whether or not a given phage-bacterium pair interact (which normally implies that the phage is able to infect the bacterium). To attain such a goal we explore several machine-learning approaches, both individually and as part of an ensemble learning approach, more specifically a Bagging method [39]. The four tested machine-learning methods are: k-Nearest Neighbors (k-NN) [40], Random Forests (RF) [41], Support Vector Machines (SVM) [16], and Artificial Neural Networks (ANN) [16]. The ensemble model applies a simple voting strategy (binary majority) with a bias toward positive predictions in case of vote tie, as illustrated by Fig. 1.

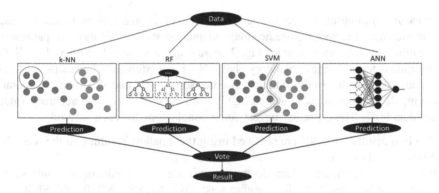

Fig. 1. Bagging ensemble-learning approach, composed of four machine-learning methods and a simple voting strategy.

The modeling process is conducted in three phases: (1) Exploration, where we test several parameter configurations (grid search) and perform a selection of the most informative sets of features; (2) Refinement, where we explore some parameter values of interest on the selected feature sets; and (3) Merge-and-Bagging, where we merge the selected feature sets in a final training dataset and build a final model using the selected parameter values and applying the Bagging voting strategy. Note that all along the modeling process, we assess the classification performance with several metrics—accuracy, f-score, specificity, and sensitivity—and compute it as the average performance on validation obtained with 10-fold cross-validation [16]. Table 1 summarizes the experimental configurations for the three modeling phases.

Table 1. Experimental setup for the three phases of the modeling process.

		Parameter	Modeling phase		
			Exploratory	Refinement	Merge-and-Bagging
Method	k-NN	K	$\{2, 3, 4, 5, 6, 7, 8, 9\}$	$\{1\}$	2
	RF	N-trees	$\{10^2, 10^3, ..., 10^4\}$	–	10^2
		L-size	$\{2, 3, 4\}$	–	3
	SVM	Penalty	$\{10^{-2}, 10^{-1},...,10^4\}$	$\{10^1, 10^3\}$	10^1
		Momentum	$\{10^{-4}, 10^{-3},...,10^4\}$	$\{1\}$	0.1
	ANN	N-neurons	$\{2, 3, 4, 5, 6\}$	$\{7, 8, 9, 10\}$	9
		Epochs	$\{10, 25, 50, 75, 100\}$	$\{10, 25, 50, 75, 100\}$	50
		Momentum	$\{0.1, 0.4, 0.7\}$	$\{0.1, 0.4, 0.7\}$	0.1
Datasets			All 19 sets	TB1, NB50, NB30, CHEM	Merge of NB50, TB1, and CHEM

3   Results: Analysis of Performance

Exploratory modeling. Given the large number of configurations tested for each machine-learning method, we use heatmaps to analyze the combinations of parameters and feature sets. As an example, Fig. 2 shows the accuracy heatmaps for all the configurations tested in the exploratory phase. Note that equivalent heatmaps exist, and were analyzed, for three other metrics: f-score, sensitivity, and specificity. Generally speaking, all methods are able to find highly predictive models, with accuracy values bigger than 90%. These results suggest some conclusions for each method:

- The best results for k-NN are obtained using the smallest k values on the sets TB1, NBN10, and NB10.
- For RF the parameter values do not exhibit significant effects. In contrast, RF obtains the best results on four feature sets: TB1, NBN50, NB50, and NB30.
- In general, SVM performs better for small values of penalty and momentum. The best results are obtained on TB1, NBN50, NB50, and NB30 feature sets.
- The results of ANN suggest that the more neurons in the hidden layer, the better the performance.

On the base of the results of this phase, four feature sets were kept for further modeling: TB1, NB50, NB30, and CHEM. This latter set of features was maintained for the sake of feature diversity in spite of it allowing for relatively low classification performance.

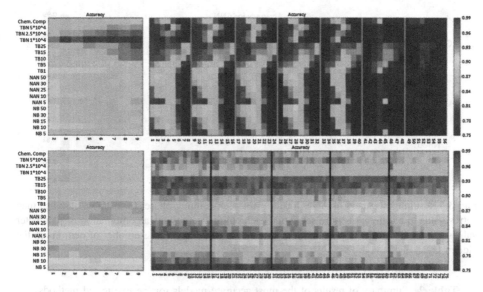

Fig. 2. Resulting accuracy for the exploratory-modeling phase. Each line of the heatmaps corresponds to a different feature set, while each column represents a combination of parameter values, as detailed in Table 1. On the left are represented the results for k-NN (top), and RF (bottom). On the right, the results for SVM (top) and ANN (bottom). The orange (respectively black) lines mark a change on one parameter value, the penalty factor for SVM and the number of neurons for ANN. (Color figure online)

Parameter refining: Among the configurations tested in this phase (see Table 1), the only performance improvement observed occurred for ANN. Figure 3 shows the results obtained by this method on the selected datasets. They confirm the observed trend that the performance increases—reaching accuracy values of around 91%—as the number of neurons in the hidden layer increases. Nevertheless, this former stagnates for 9 and 10 neurons.

Merge-and-Bagging. The ensemble learning applied on the merged dataset (See Table 1 for the experimental setup) attains similar performance values to, although not better than, those obtained in the previous phases with individual methods and datasets.

To illustrate the differences between the five approaches, Table 2 shows the performance figures of the most accurate model for each method (i.e., that exhibiting the highest accuracy value).

One can observe that, although the best results in terms of accuracy and F-score are exhibited by ANN and SVM, the actual differences among four of the methods—all but k-NN—is relatively small and does not allow to make definitive conclusions about them. Looking at sensitivity and specificity, one can notice that K-nn and RF seem to concentrate on the positive interactions while Bagging favors negative cases. Finally, although the ensemble method should obtain better classifiers, the current results ask for further investigation on its real potential.

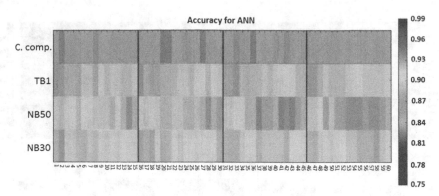

Fig. 3. Resulting accuracy of ANN for the parameter-refining phase. Each line of the heatmap corresponds to a different feature set, while each column represents a combination of parameter values, as detailed in Table 1. The black lines mark a change on the number of neurons in the hidden layer.

Table 2. Summary of results of the most accurate models for the five tested methods.

		Accuracy	F-Score	Sensitivity	Specificity	Dataset
Method	K-nn	89%	91%	96%	83%	TB1
	RF	91%	90%	**97%**	86%	NB30
	SVM	91%	**92%**	93%	91%	NB50
	ANN	**92%**	**92%**	93%	91%	NB50
	Bagging	91%	91%	88%	**95%**	NB50+TB1+CC

4 Discussion

Phage therapy is a promising alternative to fight against antibiotic resistance by using specific bacterial viruses (phages). One important challenge to develop this therapy is to find rapidly and accurately the right phage(s) able to target a given bacterium. In this work we have explored the use of machine learning to build predictive models able to assess whether or not a pair phage-bacterium would interact. The classic approach to this problem is based on experimental techniques ranging from traditional spot assays to recent methods as PhageFISH or microfluidic PCR [42, 43]. These methods are usually time-consuming and require liquid cultures of both bacteria and phages. Recently, some computational approaches have been developed, as presented in a recent paper that reviews and compares several computational approaches to predict phage-bacteria interactions based on genomic information [42]. Finally, a recent bioinformatics tool, available online, has been developed by Volddby Larsen et al. [42]. Dubbed HostPhinder, it predicts the bacterial host of a query phage genome based on its genomic similarity to a database of phage genomes with known host.

In this paper we have proposed a method that extracts two kinds of features, based respectively on domain-domain interactions and proteomic sequences. Then, we have explored the use of several machine-learning methods to build predictive models on the base of these features. The models obtained exhibited more than 90% of accuracy and related metrics such as f-score, sensitivity, and specificity, demonstrating therefore, that the approach is a good road to follow.

Based on the work done, we have been able to identify also some limiting factors of our approach that should be addressed in the near future. Currently, the main weakness concerns the quality of the collected phage-bacterium interactions dataset. First, the diversity of the data is reduced as, from the 1'065 positive interactions, 915 involve the same bacteria (*M. smegmatis*). Second, the dataset does not allow to predict interactions at the level of the bacterial strain but only at the species level. This fact is due to two factors: (1) the annotation of the phages in public databases which generally mentions only the host bacterial species, and (2) the methodology employed to create our negative dataset, as it was built by matching phages with bacteria from a different species than their known host. A third limitation comes from the use of the DOMINE database, whose last update dates back to 2007.

A second group of potential improvements to our method relates with the machine learning approach. On the one hand, the voting strategy for the bagging method could be improved by using weighting factors based on the observed performance of the involved method. On the other hand, the absence of confirmed negative interactions could be addressed by using a one-class learning approach [44].

In spite of these perceived limitations, we have developed an original methodology able to predict phage-bacteria interactions with an encouraging, high level, of accuracy. Given the advances in the understanding of phage biology, the increasing amount of genomic and interaction data available, we consider that machine learning methods associated to bioinformatics tools to predict *in silico* phage-bacteria interactions based on genomic information are timely and will greatly accelerate further developments in the field by enabling the construction of comprehensive and dynamic phage-bacteria infection networks.

References

1. Flores, C.O., Meyer, J.R., Valverde, S., Farr, L., Weitz, J.S.: Statistical structure of host-phage interactions. Proc. Natl. Acad. Sci. **108**, 288–297 (2011)
2. Weitz, J.S., Poisot, T., Meyer, J.R., Flores, C.O., Valverde, S., Sullivan, M.B., Hochberg, M.E.: Phage–bacteria infection networks. Trends Microbiol. **21**, 82–91 (2013)
3. Beckett, S.J., Williams, H.T.P.: Coevolutionary diversification creates nested-modular structure in phage-bacteria interaction networks. Interface Focus 3(6), 20130033 (2013)
4. Labrie, S.J., Samson, J.E., Moineau, S.: Bacteriophage resistance mechanisms. Nat. Rev. Microbiol. **8**, 317–327 (2010)
5. Samson, J.E., Magadán, A.H., Sabri, M., Moineau, S.: Revenge of the phages: defeating bacterial defences. Nat. Rev. Microbiol. **11**, 675–687 (2013)
6. Seed, K.D.: Battling phages: how bacteria defend against viral attack. PLoS Pathog. **11**, e1004847 (2015)
7. Rakhuba, D.V., Kolomiets, E.I., Dey, E.S., Novik, G.I.: Bacteriophage receptors, mechanisms of phage adsorption and penetration into host cell. Polish J. Microbiol. **59**, 145–155 (2010)

8. McNair, K., Bailey, B.A., Edwards, R.A.: PHACTS, a computational approach to classifying the lifestyle of phages. Bioinformatics **28**, 614–618 (2012)

9. Garneau, J.E., Dupuis, M.-È., Villion, M., Romero, D.A., Barrangou, R., Boyaval, P., Fremaux, C., Horvath, P., Magadán, A.H., Moineau, S.: The CRISPR/Cas bacterial immune system cleaves bacteriophage and plasmid DNA. Nature **468**, 67–71 (2010)

10. Ram, G., Chen, J., Kumar, K., Ross, H.F., Ubeda, C., Damle, P.K., Lane, K.D., Penades, J. R., Christie, G.E., Novick, R.P.: Staphylococcal pathogenicity island interference with helper phage reproduction is a paradigm of molecular parasitism. Proc. Natl. Acad. Sci. **109**, 16300–16305 (2012)

11. Matsuzaki, S., Rashel, M., Uchiyama, J., Sakurai, S., Ujihara, T., Kuroda, M., Ikeuchi, M., Tani, T., Fujieda, M., Wakiguchi, H., Imai, S.: Bacteriophage therapy: a revitalized therapy against bacterial infectious diseases. J. Infect. Chemother. **11**, 211–219 (2005)

12. Fischetti, V.A.: Bacteriophage lysins as effective antibacterials. Curr. Opin. Microbiol. **11**, 393–400 (2008)

13. Edgar, R., Friedman, N., Molshanski-Mor, S., Qimron, U.: Reversing bacterial resistance to antibiotics by phage-mediated delivery of dominant sensitive genes. Appl. Environ. Microbiol. **78**, 744–751 (2012)

14. Yosef, I., Kiro, R., Molshanski-Mor, S., Edgar, R., Qimron, U.: Different approaches for using bacteriophages against antibiotic-resistant bacteria. Bacteriophage **4**, e28491 (2014)

15. Lu, T.K., Koeris, M.S.: The next generation of bacteriophage therapy. Curr. Opin. Microbiol. **14**, 524–531 (2011)

16. Han, J., Kamber, M., Pei, J.: Data Mining: Concepts and Techniques, 3rd edn. Morgan Kaufmann Publishers, San Francisco (2011)

17. Hatfull, G., Russell, D., Jacobs-Sera, D., Pop, W.H., Sivanathan, V., Tse, E.: The Actinobacteriophage DataBase at PhagesDB.org. http://phagesdb.org/

18. Benson, D.A., Cavanaugh, M., Clark, K., Karsch-Mizrachi, I., Lipman, D.J., Ostell, J., Sayers, E.W.: GenBank. Nucleic Acids Res. **41**, D36–D42 (2013)

19. Besemer, J., Lomsadze, A., Borodovsky, M.: GeneMarkS: a self-training method for prediction of gene starts in microbial genomes. Implications for finding sequence motifs in regulatory regions. Nucleic Acids Res. **29**, 2607–2618 (2001)

20. NCBI – Genome. https://www.ncbi.nlm.nih.gov/genome/

21. NCBI – Nucleotide. https://www.ncbi.nlm.nih.gov/nucleotide/

22. PubMed Central: Entrez Help (2006)

23. Hyman, P., Abedon, S.T.: Bacteriophage host range and bacterial resistance. Adv. Appl. Microbiol. **70**, 217–48 (2010)

24. Duplessis, M., Moineau, S.: Identification of a genetic determinant responsible for host specificity in streptococcus thermophilus bacteriophages. Mol. Microbiol. **41**, 325–336 (2001)

25. Miklič, A., Rogelj, I.: Characterization of lactococcal bacteriophages isolated from slovenian dairies. Int. J. Food Sci. Technol. **38**, 305–311 (2003)

26. Duckworth, D.H., Gulig, P.A.: Bacteriophages: potential treatment for bacterial infections. BioDrugs **16**, 57–62 (2002)

27. Ben-Hur, A., Noble, W.S.: Choosing negative examples for the prediction of protein-protein interactions. BMC Bioinformatics. **7**(Suppl 1), S2 (2006)

28. Coelho, E.D., Arrais, J.P., Matos, S., Pereira, C., Rosa, N., Correia, M.J., Barros, M., Oliveira, J.L.: Computational prediction of the human-microbial oral interactome. BMC Syst. Biol. **8**, 24 (2014)

29. Parham, P.: Structure des anticorps et origines de la diversité des cellules B. In: De Boeck (ed.) Le système immunitaire, pp. 31–35. De Boeck (2003)

30. Terrapon, N.: Recherche de domaines protéiques divergents à l'aide de modéles de Markov cachées : application à Plasmodium falciparum (2010). https://tel.archives-ouvertes.fr/tel-00811835/document

31. Raghavachari, B., Tasneem, A., Przytycka, T.M., Jothi, R.: DOMINE: a database of protein domain interactions. Nucleic Acids Res. **36**, D656–D661 (2007)

32. Sonnhammer, E., Eddy, S., Birney, E., Bateman, A., Durbin, R.: Pfam: multiple sequence alignments and HMM-profiles of protein domains. Nucleic Acids Res. **26**, 320–322 (1998). Oxford University Press

33. Eddy, S.R., Wheeler, T.J.: HMMER User's Guide. 0–77 (2015)

34. Finn, R.D., Clements, J., Arndt, W., Miller, B.L., Wheeler, T.J., Schreiber, F., Bateman, A., Eddy, S.R.: HMMER web server: 2015 update. Nucleic Acids Res. **43**, W30–W38 (2015)

35. Bock, J.R., Gough, D.A.: Predicting protein–protein interactions from primary structure. Bioinformatics **17**, 455–460 (2001)

36. Shen, J., Zhang, J., Luo, X., Zhu, W., Yu, K., Chen, K., Li, Y., Jiang, H.: Predicting protein-protein interactions based only on sequences information. Proc. Natl. Acad. Sci. USA **104**, 4337–4341 (2007)

37. Xia, J.-F., Han, K., Huang, D.-S.: Sequence-based prediction of protein-protein interactions by means of rotation forest and autocorrelation descriptor. Protein Pept. Lett. **17**, 137–145 (2010)

38. You, Z.-H., Zhu, L., Zheng, C.-H., Yu, H.-J., Deng, S.-P., Ji, Z.: Prediction of protein-protein interactions from amino acid sequences using a novel multi-scale continuous and discontinuous feature set. BMC Bioinformatics. **15**(Suppl 1), S9 (2014)

39. Wade, L.G.: Amino Acids, peptides, and proteins. In: Hall, P. (ed.) Organic Chemistry, pp. 1153–1199 (2003)

40. Cover, T., Hart, P.: Nearest neighbor pattern classification. IEEE Trans. Inf. Theory **13**, 21–27 (1967)

41. Breiman, L.: Random forests. Springer Mach. Learn. **45**, 5–32 (2001)

42. Villarroel, J., Kleinheinz, K.A., Jurtz, V.I., Zschach, H., Lund, O., Nielsen, M., Larsen, M. V.: HostPhinder: a phage host prediction tool. Viruses **8**, 1–22 (2016)

43. Edwards, R.A., McNair, K., Faust, K., Raes, J., Dutilh, B.E.: Computational approaches to predict bacteriophage-host relationships. FEMS Microbiol. Rev. **40**, 258–272 (2016)

44. Khan, S.S., Madden, M.G.: One-class classification: taxonomy of study and review of techniques. Knowl. Eng. Rev. **29**, 1–24 (2004)

Simultaneous Gene Selection and Weighting in Nearest Neighbor Classifier for Gene Expression Data

Antonio Alarcón-Paredes[✉], Gustavo Adolfo Alonso,
Eduardo Cabrera, and René Cuevas-Valencia

School of Engineering, Universidad Autónoma de Guerrero,
México Lázaro Cárdenas Ave., 39070 Chilpancingo, Guerrero, México
{aalarcon, gsilverio, ecabrera, reneecuevas}@uagro.mx

Abstract. Gene expression data, such as microarray data, plays an important role as a biomarker in order to help in the effective cancer diagnosis, tumor classification or drug design at molecular level. However, due to high-dimensionality of microarray datasets, they tend to have irrelevant or redundant features, and may lead to poor classification performance. For this reason, feature selection methods are commonly used to reduce the amount of data and to select relevant genes to improve the accuracy of machine learning methods. In this paper, a simultaneous feature ranking and weighting gene selection method with a nearest neighbor-based classifier is presented. In order to demonstrate the effectiveness of this proposal, a range of experiments over four well-known microarray datasets were carried out. Results showed that our method outperforms previous methods in terms of classification accuracy. Furthermore, evidence for significance of our results by means of non-parametric Friedman test is provided.

Keywords: Gene expression data · Feature selection (FS) · Feature ranking · Feature weighting · Nearest neighbor classifier

1 Introduction

The use of microarray data technology allows monitoring simultaneously thousands of gene expression levels over multiple cases. Gene expression data plays an important role as a biomarker helping in a variety of tasks such as cancer diagnosis, classification of different types of tumors, or effective drug design at a molecular level [1, 2]. Microarray data consists of several thousands of genes (p) and few samples (n), with $p \gg n$. This is a well-known problem referred to as the curse of dimensionality in which most classifiers may lead to overfitting of the data. In other words, when number of features is (much) larger than number of observations, datasets tend to have irrelevant or redundant features; therefore, classifiers may report an extremely poor classification performance on an independent test set [3, 4]. Furthermore, experiments on such high-dimensional domains yield to high computational costs [5]. For this reason, feature selection (FS) methods are commonly used to alleviate these issues. Feature selection is a common pre-processing step in machine learning algorithms in order to

© Springer International Publishing AG 2017
I. Rojas and F. Ortuño (Eds.): IWBBIO 2017, Part II, LNBI 10209, pp. 372–381, 2017.
DOI: 10.1007/978-3-319-56154-7_34

select a small subset from a larger set of variables selecting the most discriminative features, reducing the redundant information to improve the classification accuracy of machine learning methods [6, 7].

Feature selection methods are divided into three main categories: filter, wrapper and embedded methods. Filter methods reduce the dimensionality by evaluating each feature individually, and selecting a percentage of the best ranked features before running the learning algorithm; the challenge is to define a way to measure the feature relevance [8, 9]. On the other hand, wrapper approach employs a learning algorithm and uses its performance to assess the relevance of the selected feature subset. Usually, wrappers lead to better classification performance than filter methods, but also yields to higher computational costs [7, 10]. Embedded methods integrate the feature selection step in the model construction. It can be carried out by maximizing the classification performance whereas minimizing the number of features, or building a classifier on the full training set and iteratively removing the least promising features for the learning algorithm [3, 9, 11].

Some of the previous works use the k-nearest neighbor classifier combining different approaches. For example, the authors of [12] present a ranking and weighting approach based on the chi-squared statistical test for a weighted k-NN classifier. Park and Kim [13] proposed the construction of an ensemble of k-NN classifiers applied to random feature subsets and use an iterative procedure to select the most compact feature set in the same manner as random forests algorithm. In [14], in order to select the optimal size of the neighborhood (k) in the k-NN algorithm and the best feature subset in microarray data, authors use the particle swarm optimization (PSO) as a low computational cost meta-heuristic. Moreover, in [2], authors propose a methodology to accelerate a wrapper-based feature subset selection with an embedded k-NN classifier.

In this paper, a simultaneous feature ranking and weighting gene selection method with a Nearest Neighbor-based classifier is presented. In order to demonstrate the effectiveness of this proposal, a range of experiments over four well-known microarray datasets [15] were carried out. Results showed that our method outperforms most of classic feature selection schemes in terms of classification accuracy. Furthermore, evidence for significance of our results by means of non-parametric test is provided.

The rest of the paper is organized as follows: the proposed algorithm is presented in Sect. 2. Section 3 presents the experimental results and Sect. 4 is devoted to conclusions and future work. Finally, the references are also presented.

2 Proposed Methodology

The proposed method is divided into two stages: the first one performs a feature ranking, and computes their corresponding feature weights. The latter takes these weights and performs a forward feature selection scheme in the ranked data, using a weighted nearest neighbor classifier. In order to explain the aforementioned idea, some notation is needed. Let $X = [\phi_1, \phi_2, \ldots, \phi_p] = [X_1, X_2, \ldots, X_n]^T$ be a matrix of size $n \times p$, with n observations and p features, where the columns are denoted with ϕ_j and the observations by X_i; and let $y = [y_1, y_2, \ldots, y_n]^T$ be a column vector of size

n containing the class of each observation in X. For simplicity and as normalization step, each data column in X is standardized to have zero mean and one standard deviation, as expressed in following equation:

$$\phi_i = \frac{\phi_i - \overline{\phi_i}}{std(\phi_i)} \tag{1}$$

where ϕ_i, $\overline{\phi_i}$ and $std(\phi_i)$ represent the i-th column, its mean, and the i-th standard deviation, respectively.

2.1 Feature Ranking

Classic feature selection filters perform a feature ranking step in order to evaluate the feature relevance. Commonly, the Pearson correlation coefficient, an entropy mutual information scheme, and also some classical statistical tests such as t-test or *chi-squared*, are used to perform the feature ranking. Conversely, here, a simple approach consisting in the Absolute Covariance between the feature vector and its corresponding class is proposed. This procedure ranks the features from best to worse, where high values in the absolute covariance means that there is a high correlation between that feature and the class. The criterion used to rank genes is computed as follows:

$$R(i) = |cov(\phi_i, y)| \tag{2}$$

$$cov(\phi_i, y) = \frac{1}{n}\sum (\phi_i - \overline{\phi_i})(y - \overline{y}) \tag{3}$$

where $|\bullet|$ takes the absolute value, $\overline{\phi_i}$ and $\overline{y}$ are the means of i-th feature vector ϕ_i, and the class vector y, respectively.

2.2 Feature Weighting Stage

A crucial step in this proposal is the feature weighting. It is used to represent the feature relevance in order to enhance the prediction accuracy of the classifier. These weights are used to compute a weighted Euclidean distance in the k-NN procedure. The proposed feature weighting scheme consist of, firstly, obtain the feature ranking $R(i)$ according to what is expressed in Eq. (2); secondly, computes a vector with uniform incremental values in such a manner that the best ranked feature has weight 1, and the worst ranked is closer to 0, described as follows:

$$W(i) = \frac{(p+1) - idx(i)}{p} \tag{4}$$

idx has the sorted ranking index of $R(i)$, and p is the number of features in the data set. For example, let us consider a data set X consisting of $p = 5$ features, and a ranking vector $R = [126.2, 121.1, 32.8, 251.1, 68.4]$, then sorted index is expressed as $idx = [2, 3, 5, 1, 4]$. The result of the Uniform Covariance (UCov) feature weighting for the i-th according to Eq. (4) would be $W = [0.8, 0.6, 0.2, 1, 0.4]$.

2.3 k-Nearest Neighbors

The k-Nearest Neighbor classifier (k-NN) is perhaps the most representative instance-based classifier. It was proposed in 1951 [16] as a non-parametric algorithm for discriminatory analysis. However, it is an effective and robust algorithm for pattern classification. The easiest way to construct a k-Nearest Neighbor classifier is when $k = 1$ (*1*-NN), *i.e.*, the algorithm assigns an unclassified instance to the class of the nearest pre-classified training instance; this closeness is determined by means of a distance function, commonly the Euclidian norm [17].

One of the main assumptions of feature selection is that some features provide more discriminative information to perform a class prediction than others [7]. This must be taken into consideration to construct an improved classifier. For this reason, we used a weighted nearest neighbor classifier implementing a squared feature-weighted Euclidean distance which can be calculated as shown in Eq. (5):

$$d_W\left(X_i, X_j\right)^2 = \sum_{k=1}^{p} W(k) \bullet \left(X_i(k) - X_j(k)\right)^2 \tag{5}$$

where W is a vector with the computed feature weights, X_i and X_j represent two different data observations.

2.4 Forward Selection Method

Wrapper methods use a classification step to evaluate the relevance of feature subsets. In order to guarantee the overall best classification result, an exhaustive analysis of all possible feature subset would be needed, leading to an exponential complexity. Accordingly forward feature selection is used to alleviate this issue. Forward selection is a deterministic wrapper method that works by searching the best feature subset starting from an empty set of features and consecutively adding features one at a time. The feature is added to final subset if it improves the classification accuracy, otherwise it is discarded. This procedure is repeated until there are no more features in the data set [18].

2.5 Feature Selection Algorithm

Let R and W denote the feature ranking and weighting vectors according to Eqs. (2) and (3). The proposed feature selection method, consisting of a forward selection procedure using a weighted *1*-NN classifier is described below and shown in Fig. 1. First, the best classification is initialized as $BC = 0$, and the best subset as an empty set marked as BS. The procedure starts classifying the first feature and sequentially adding ranked features one by one. The classification step is performed by the function f_{1NN}

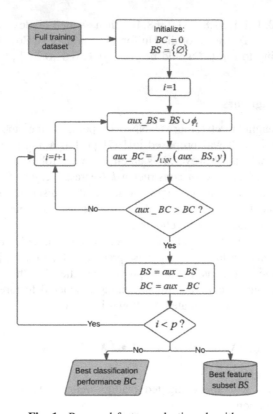

Fig. 1. Proposed feature selection algorithm

using the *1*-NN classifier. Every time the classification performance aux_BC is improved, the algorithm updates the best classification as $BC = aux\_BC$ and the best subset as $BS = aux\_BS$ until all ranked features in the data set are explored.

3   Results

In this paper, an experimental study applied to four well-known gene expression data was carried out. The microarray data sets are summarized in Table 1, and are publicly available in the Kent Ridge Biomedical Data Set Repository [15].

Here we compare our proposed algorithm with other state-of-the-art feature selection algorithms. All experiments were conducted using a personal computer with an Intel Core i3 Processor (3.00 GHz) running on a Windows 7 Ultimate operating system with 4096 MB of RAM.

An experimental study of the proposed simultaneous gene selection and weighting scheme was performed using the ten-fold cross validation method, as used in previous works [2, 19, 20]. The ten-fold cross validation scheme uses each fold as a test set and the remaining nine folds are used for classifier construction; the classification accuracy is obtained by the average of the ten classification results. This validation procedure was repeated 50 times on each data set and the average results were reported, as shown in Table 2.

Table 1. Gene expression data sets description.

Data set	#Features	#Observations	#Classes	Description
Colon	2,000	62	2	Classifies the observations into 40 colon tumors and 22 normal samples
Leukemia	7,129	72	2	Contains observations to distinguish between 25 cases of acute myeloid leukemia (AML) and 47 acute lymphoma leukemia (ALL)
Ovarian	15,154	253	2	Data to discriminate 162 cases of ovarian cancer from the 91 healthy controls
Prostate	12,600	136	2	Distinguishes between 52 prostate tumors and 50 non-tumor prostate samples

Table 2. Classification acuracy results of state-of-the-art feature selection methods in comparison with the proposed method. Best results appear bold-faced.

Data set	Proposed method	FCBF [2]	ReliefF [19]	k-NN Ensemble [19]	BDE-KNN$_{Rank}$ [20]
Colon	**90.3%**	78.0%	74.3%	81.0%	81.3%
Leukemia	**99.2%**	94.7%	93.0%	94.5%	97.1%
Ovarian	**100%**	**100%**	98.8%	**100%**	98.8%
Prostate	**92.8%**	90.0%	87.4%	87.5%	76.5%
Average	**95.57%**	90.7%	88.4%	90.8%	88.4%

In order to show the significance of our results, the non-parametric statistical Friedman test [21] was performed. Statistical significance tests calculate the probability to reject the null hypothesis H_0 which suggest no significant difference between observations.

The proposed feature selection method is compared with existing methods, such as Fast Correlation-Based Filter (FCBF) method, the ReliefF, an ensemble of k-NN classifiers and a hybrid wrapper based on Binary Differential Evolution with a ranking filter method (BDE-kNN$_{rank}$).

The FCBF method was proposed in [22] and evaluates the feature relevance by means of the symmetrical uncertainty measure. If one feature is far enough from other relevant features, the algorithm suggests it has no redundancy and it is selected.

The original Relief [6] method estimates the feature relevance according to how well the algorithm could distinguish between instances that are similar to each other. Besides, the ReliefF [23] algorithm randomly selects an instance and searches for the k nearest observations of the same class (*hits*) and also for the k nearest observations of different classes (*misses*), and updates the feature relevance. The contribution of the *misses* features is then weighted with the prior probability of its corresponding class.

The main difference between ReliefF and the original Relief is the selection of k hits and misses on each random observation.

The k-NN ensemble constructs an ensemble of filter feature selection methods such as CFS, Consistency-based filter, INTERACT, ReliefF and Information Gain. Due to the use of filters, the ensemble is independent of the learning algorithm. However, we take the results using the ensemble with the 1-NN classifier. The authors of this work perform preliminary experiments on a range from 25 to 50 features. Nevertheless they found no classification improvements using 50 features with respect to 25, and decided to use the filter ensemble with only 25 features. A more detailed description can be consulted in [19].

Finally, we also compare our results with a hybrid filter-wrapper approach. First, this method ranks the features according to the information gain measure. Secondly, this algorithm uses a wrapper approach combining the Binary Differential Evolution method with a k-NN classifier (BDE-kNN$_{rank}$) [20].

Table 2 shows the correct classification for this proposal and for the aforementioned methods, where the best classification accuracy for each dataset is bold faced. Results show that the proposed Absolute Covariance Ranking method with the Uniform Feature Weighting for the Nearest Neighbor classifier outperforms the other algorithms in all datasets. Furthermore our proposed method ties with the FCBF and the k-NN ensemble in the Ovarian cancer data set with an accuracy of 100%.

Although the results for some feature selection algorithms report a very similar performance, the Friedman test shows that there are significant differences among them with a value of $p = 0.039$ considering a confidence of 95%. Regarding this, Table 3 depicts the Friedman mean ranks for the feature selection methods, revealing that the best rank was obtained by the proposed method. Note that best method has the lowest mean rank value.

Table 3. Friedman mean ranks for data sets classification accuracy.

Data set	Friedman mean ranks
Proposed method	**1.250**
FCBF [2]	2.750
ReliefF [19]	4.625
k-NN Ensemble [19]	3.000
BDE-KNN$_{Rank}$ [20]	3.375

Table 4 summarizes the number of selected genes by the proposed algorithm and the other feature selection methods. Although the ReliefF and the k-NN ensemble were forced to select 25 features, the BDE-KNN$_{Ranks}$ method selects only two features by own merits. The FCBF and the proposed algorithms show the average of selected genes after repeatedly run the ten-fold cross validation procedure. In Table 5 we present the mean ranks for the selected genes in which our algorithm was the second best ranked. The BDE-KNN$_{Rank}$ algorithm is the best ranked here; however, they get the fourth place regarding the classification accuracy.

Table 4. Average number of selected genes by previous methods in comparison with this proposal. Results for the least selected genes appear bold-faced.

Data set	Proposed method	FCBF [2]	ReliefF [19]	k-NN Ensemble [19]	BDE-KNN$_{Rank}$ [20]
Colon	14.12	15.5	25	25	**2**
Leukemia	7.78	70.6	25	25	**2**
Ovarian	9.80	31.1	25	25	**2**
Prostate	24.32	53.4	25	25	**2**

Figure 2(a, b, c and d) depicts the classification accuracy of the proposed method against the ranked features from best to worse in which the selected features appear with a marker. It is worth mentioning that in the Leukemia and Ovarian cancer data sets, the algorithm selects the most of the best ranked features; in the Colon tumor data set, our algorithm selects two of the medium-ranked features, whereas in the Prostate tumor data set, it selects one of the worst ranked features, and approximately four of the medium-ranked features. However, it can be overcome in future work; since the

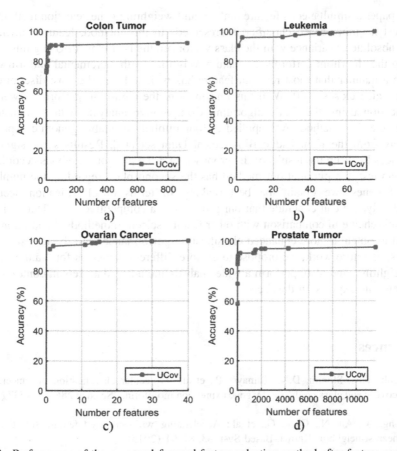

Fig. 2. Performance of the proposed forward feature selection method after feature ranking.

Table 5. Friedman mean ranks for the number of selected genes.

Data set	Friedman mean ranks
Proposed method	2.000
FCBF [2]	4.500
ReliefF [19]	3.750
k-NN Ensemble [19]	3.750
BDE-KNN$_{Rank}$ [20]	**1.000**

medium and the worst ranked features provide a quite small improvement in the classification accuracy, they can be omitted in the selection. Nevertheless, Fig. 2 reveals that the classification performance increases substantially with the best ranked features, which suggest that our Uniform Covariance (UCov) feature ranking approach arranges the most discriminative features at the top of the ranking.

4 Conclusions and Future Work

In this paper a simultaneous feature ranking and weighting gene selection method with a nearest neighbor-based classifier is presented. In this method, features having the higher absolute covariance with the class vector are the best ranked. Using this feature ranking the algorithm performs the feature weighting with incremental uniform values in such a manner that most relevant feature has weight 1 and the least discriminative has the value closest to 0. With this information, the algorithm performs a weighted classification using the *1*-NN algorithm. Comparative analysis with state-of-the-art feature selection methods was applied to four publicly available genomic expression data sets from the Kent Ridge Bio-medical Data set [15]. Results show significant differences between the results of the proposed and the previous methods according to Friedman test. Despite that our method has the second place regarding the number of selected genes, we obtain the best ranking in terms of classification accuracy. Accordingly, we can conclude that our proposal is a competitive and effective feature selection scheme in comparison with other feature selection methods. In addition, with the aim to obtain a more enhanced classification performance with the fewest selected features, as future work, we propose to explore different strategies for feature ranking and weighting, and also, perform a wide analysis including data sets in contexts other than genomic expression data sets.

References

1. Golub, T.R., Slonim, D.K., Tamayo, P., et al.: Molecular classification of cancer: class discovery and class prediction by gene expression monitoring. Science **286**, 531–537 (1999). (80-)
2. Wang, A., An, N., Chen, G., et al.: Accelerating wrapper-based feature selection with K-nearest-neighbor. Knowl.-Based Syst. **83**, 81–91 (2015)

3. Guyon, I., Weston, J., Barnhill, S., Vapnik, V.: Gene selection for cancer classification using support vector machines. Mach. Learn. **46**, 389–422 (2002)
4. James, G., Witten, D., Hastie, T., Tibshirani, R.: An Introduction to Statistical Learning. Springer, New York (2013)
5. Narendra, P.M., Fukunaga, K.: A branch and bound algorithm for feature subset selection. IEEE Trans. Comput. **26**, 917–922 (1977). doi:10.1109/tc.1977.1674939
6. Kira, K., Rendell, L.A.: The feature selection problem: traditional methods and a new algorithm. In: AAAI, pp 129–134 (1992)
7. Guyon, I., Elisseeff, A.: An introduction to variable and feature selection. J. Mach. Learn. Res. **3**, 1157–1182 (2003)
8. Molina, L.C., Belanche, L., Nebot, À.: Feature selection algorithms: a survey and experimental evaluation. In: Proceedings of the 2002 IEEE International Conference on Data Mining, ICDM 2003, pp 306–313. IEEE (2002)
9. Liu, H., Motoda, H.: Computational Methods of Feature Selection. CRC Press, Boca Raton (2007)
10. Kohavi, R., John, G.H.: Wrappers for feature subset selection. Artif. Intell. **97**, 273–324 (1997). doi:10.1016/s0004-3702(97)00043-x
11. Zou, H., Hastie, T.: Regularization and variable selection via the elastic net. J. Royal Stat. Soc. Ser. B (Statistical Methodology) **67**, 301–320 (2005)
12. Vivencio, D.P., Hruschka, E.R., do Carmo Nicoletti, M., et al.: Feature-weighted k-nearest neighbor classifier. In: 2007 IEEE Symposium on Foundations of Computational Intelligence (2007)
13. Park, C.H., Kim, S.B.: Sequential random k-nearest neighbor feature selection for high-dimensional data. Expert Syst. Appl. **42**, 2336–2342 (2015)
14. Kar, S., Das, S.K., Maitra, M.: Gene selection from microarray gene expression data for classification of cancer subgroups employing PSO and adaptive K-nearest neighborhood technique. Expert Syst. Appl. **42**, 612–627 (2015)
15. Kent Ridge Bio-medical Data set. http://datam.i2r.a-star.edu.sg/datasets/krbd/. Accessed 12 Nov 2016
16. Fix, E., Hodges Jr., J.L.: Discriminatory analysis-nonparametric discrimination: consistency properties (1951)
17. Cover, T., Hart, P.: Nearest neighbor pattern classification. IEEE Trans. Inf. Theor. **13**, 21–27 (1967)
18. Hira, Z.M., Gillies, D.F.: A review of feature selection and feature extraction methods applied on microarray data. Adv. Bioinf. **2015**, 13 (2015)
19. Bolón-Canedo, V., Sánchez-Maroño, N., Alonso-Betanzos, A.: An ensemble of filters and classifiers for microarray data classification. Pattern Recogn. **45**, 531–539 (2012)
20. Apolloni, J., Leguizamón, G., Alba, E.: Two hybrid wrapper-filter feature selection algorithms applied to high-dimensional microarray experiments. Appl. Soft Comput. **38**, 922–932 (2016)
21. Friedman, M.: The use of ranks to avoid the assumption of normality implicit in the analysis of variance. J. Am. Stat. Assoc. **32**, 675–701 (1937)
22. Yu, L., Liu, H.: Feature selection for high-dimensional data: a fast correlation-based filter solution. In: ICML, pp 856–863 (2003)
23. Robnik-Šikonja, M., Kononenko, I.: Theoretical and empirical analysis of ReliefF and RReliefF. Mach. Learn. **53**, 23–69 (2003)

Machine Learning Approaches for Predicting High Utilizers in Health Care

Chengliang Yang[1], Chris Delcher[2], Elizabeth Shenkman[2],
and Sanjay Ranka[1(✉)]

[1] Department of Computer and Information Science and Engineering,
University of Florida, Gainesville, FL 32611, USA
ximen14@ufl.edu, ranka@cise.ufl.edu
[2] Department of Health Outcomes and Policy, University of Florida,
Gainesville, FL 32611, USA
{cdelcher,eshenkman}@ufl.edu

Abstract. We applied and developed several machine learning techniques, including linear model, tree-based model, and deep neural networks to forecast expenditures for high utilizers in a very large public health program. The results show promise for predicting health care expenditures for these high utilizers. To improve interpretability, we quantified the contributions of influential input variables to the prediction score. These results help to advance the field toward targeted preventive care to lower overall health care costs.

1 Introduction

The Agency for Healthcare Research and Quality (AHRQ) reported that in 2002, the top 10% of the health care utilizing population accounted for 64% of the overall health care expenditures in the United States [19]. Stakeholders have argued the need for higher efficiency health care for the high utilizers [4]. For example, the deployment of managed care organizations (MCOs) and the capitation payments system [10,16] in the United States public health programs placed incentives for health care providers to deliver services in a more cost-effective way.

Information technology in the Big Data era provides a new, promising way to approach a wide range of health care problems [14]. Hospitals, health care agencies and insurance companies routinely accumulate vast amounts of data from health care-related activities, including electronic medical records, insurance claims, vital signs, patient-reported outcomes, etc. Predictive modeling in this field is new but can reveal important insight for addressing this problem. Specifically, if we could forecast expenditures at the patient-level with acceptable accuracy, we could improve targeted care by anticipating health care needs of high utilizers. Moreover, providers would improve their understanding of the causal pathways leading to expensive events and build strategies to intervene in advance. Preventive care is one of the most effective ways to lower health care expenditures while delivering better quality of care [15].

© Springer International Publishing AG 2017
I. Rojas and F. Ortuño (Eds.): IWBBIO 2017, Part II, LNBI 10209, pp. 382–395, 2017.
DOI: 10.1007/978-3-319-56154-7_35

Researchers have made some effort in the literature to predict health care expenditures. Risk adjustment models based on predictive linear models [16] are a key part of many capitation payment systems. However, these models typically consider only endogenous variables with limited predictive accuracy. Other work has used clinically empirical scores or measures to predict health expenditures [12]. However, these analyses typically focus on a specific group of patients or type of care. [9] discussed the differences of key factors when trying to predict short-, medium- or long-term health care expenditures. The authors in [17] analyzed the temporal utilization pattern of high utilizers in a large public state insurance program. Other efforts of predicting health care expenditures, such as [13], occur at the population-level but not the patient-level. Studies that attempt to predict high utilizing status are lacking. This study aims to address this gap in the literature.

The key contributions of our paper are as follows:

- We applied machine learning models, including linear models, tree-based models and deep neural networks to predict future health expenditures, especially for the high utilizers, defined here as being in the top 10% of expenditures. The methods are scalable to tens of thousands of input variables and millions of patients. The results showed that health care expenditures can be effectively predicted (overall R-squared > 0.7). The prediction error for the high utilizers is lower than the general population, which implies stronger predictability of high utilizers.
- We quantified the contributions of the input variables for each single prediction so that model users can identify potentially modifiable risk factors for possible intervention.
- We systematically compared models on predictive accuracy and model interpretation. We gave concrete suggestions for practical use.

The remaining sections of the paper are organized as follows: In Sect. 2, we introduce the dataset and preprocessing methods used for the study. Section 3 presents detailed explanation of the mechanisms involved with deployed machine learning models. Section 4 provides predictive modeling accuracy, interprets the predictions and discusses multiple strategies to better predict expenditures. In Sect. 5, we provide our overall conclusions and future directions.

2   Data and Preprocessing

2.1   Study Population

In this study, we used the administrative claims dataset from the Medicaid program of the state of Texas, United States. Texas has the third largest Medicaid population (annual enrollment of 4.7 million) in the United States. During the study period (2011–2014), 1,734,896 adults (ages 18–65) were enrolled in the Texas Medicaid program for at least one month. To be included in any analysis, the enrollee must have been enrolled for more than two-thirds of the time of any specific analysis period (either observed period or forecasting period). Length-of-enrollment criteria are applied to avoid modeling errors associated with patients enrolled for very short periods of time, which can skew the results. For this population, total medical expenditure was defined as the sum of professional, institutional, and dental spending from the dataset at the claim level. Pharmacy expenditures were not included in this study. During the study period, the Texas Medicaid program was structured as both fee-for-service (FFS) and MCO-based payment models while FFS was being phased out. For both models, we used the final paid amount by the payor as the expenditures of each claim.

The dataset also contains diagnosis codes (International Classification of Diseases, Ninth Revision, Clinical Modification, ICD-9-CM), procedure codes (ICD-9-CM procedure codes, Current Procedural Terminology [CPT] and Healthcare Common Procedure Coding System [HCPCS]), and medication codes (National Drug Codes [NDC]). During the study period, 3,233 unique ICD-9-CM procedure codes, 21,374 unique ICD-9-CM diagnosis codes, 21,603 unique CPT and HCPCS codes, and 28,366 NDC codes were identified. This study was approved by the IRB of the authors home institution.

2.2 Predicting Objectives

We constructed predictive models to forecast the patients' expenditures of future time periods based on previous time periods. We examined three outcomes:

- per member per month dollar amount (PMPM, total medical expenditure divided by number of months enrolled in Medicaid). This measure is commonly used for expenditure analyses in Medicaid programs [11].
- per member per month dollar amount with log base 10 transformed, logPMPM).
- rank percentiles of the per member per month dollar amount ($pctl$PMPM). This is a continuous measure obtained by dividing the descending ordered rank of PMPM by the number of enrollees in the dataset. Values range from 0 to 1.

We varied the length of time periods in the analysis using base units of 3 months, 6 months and 12 months. This choice of time periods was based on commonly used values for understanding patterns of health care utilization.

2.3 Predictors

We designed multiple features as the input variables of the predictive models. For each previous time period consistent with the desired forecast time period:

- Diagnosis codes (ICD-9-CM) grouped into 283 CCS diagnosis categories [1].
- Procedures codes (CPT and HCPCS) grouped into 231 procedure CCS [1] categories.

- Medication information represented by National Drug Codes (NDC), are grouped by pharmacy classes (893 classes) provided by the U.S. Food and Drug Administration (FDA)'s NDC Directory (Accessed on Oct. 20, 2015).
- Demographic variables such as age, sex, race/ethnicity (White, Black, Hispanic, American Indian or Alaskan, Asian, Unknown/Other), and disabled status.

Taking into account all of these features, each period consists of approximately 1,300 input variables.

3 Methods

3.1 Predictive Models

We applied four predictive models to forecast the patients' expenditures based on the previous time periods, including ordinary least squares linear regression (LR), regularized regression (LASSO), gradient boosting machine (GBM), and recurrent neural networks (RNN). The following section describes the details for these models.

Ordinary Least Squares Linear Regression (LR). LR is the most widely used method in predictive modeling. It serves as the base risk-adjustment model [10,16] for modeling risk-based payment systems in health care. Using the objective to be predicted and input variables described above, we fit a LR model using least squares to predict future expenditures.

Regularized Regression (LASSO). Regularized regression, also known as the least absolute shrinkage and selection operator (LASSO) [20], fits a linear regression model but penalizes solutions with a large number of nonzero coefficients at the same time. It is widely used as the default approach in many supervised machine learning tasks. Given M training instances $\{(\mathbf{x}_i, y_i), \quad i = 1, 2, ..., M\}$, where $\mathbf{x}_i \in \mathbb{R}^N$ is a N-dimensional input variable vector, y_i is the predicting objective, L_1 regularized regression tries to minimize the objective function below:

$$\min_\theta \sum_{i=1}^{M} ||y_i - \theta \mathbf{x}_i||_2^2 + \beta ||\theta||_1 \tag{1}$$

where $\theta \in \mathbb{R}^N$ are the linear coefficients. The first term of the equation above is the objective function that LR tries to minimize. The regularizing term $||\theta||_1$ ensures that a large number of entries of θ are set to zero. This property is advantageous because it makes the model robust to high-dimensional input, as in our case, and selects the most influential input variables. In our study, we use the implementation of LASSO provided by the original authors of LASSO [20]. A 10-fold cross validation is used to select the hyper-parameter β.

Gradient Boosting Machine (GBM). Gradient boosting [8] is another set of successful machine learning techniques to handle high dimensional datasets. It generates an ensemble of decision trees f_t as the predictive model. It learns these trees in an additive manner. In each round, it learns a new tree f_t by optimizing the objective function of:

$$\min_{f_t} \sum_{i=1}^{M} (g_i f_t(\mathbf{x}_i) + \frac{1}{2} h_i f_t^2(\mathbf{x}_i)) + \gamma T + \lambda \sum_{j=1}^{T} w_j^2 \tag{2}$$

where g_i and h_i are the first- and second-order derivatives of some loss function, which, in our case, is squared error between predicted and true values. T is the number of leaves in the decision tree f_t and w_j are the leaf weights. The last two terms are regularizers to control mode complexity.

One advantageous property of GBM is that the information gain of the trees can be aggregated as a measure of input variable importance, which is similar to the coefficients in LASSO. This makes tree methods interpretable in applications. In practice, we use the implementation of GBM provided by [5]. We trained 1,000 decision trees for each GBM. We performed a grid search and five-fold cross validation to decide other hyper-parameters such as learning rate and tree depth.

Recurrent Neural Networks (RNN). Recurrent neural networks are a set of neural network models designed to process sequence data. In health care, RNN models have been used for early detection of heart failure onset from electronic health records [2,7]. The health claims dataset used here to predict medical expenditures is organized as sequential events (e.g. date of diagnosis, date of procedure, and date of medication use). Thus, we applied RNN to model these events as time series rather than including them in the models as unordered events.

For a patient $\{(\mathbf{x}_i, y_i), \quad i = 1, 2, ..., M\}$, where $\mathbf{x}_i \in \mathbb{R}^N$ is the input variable vector, y_i is the predicting objective, we assumed that $\mathbf{x}_i$ consists of T periods. Each period $\mathbf{x}_i^t$ is a K dimensional input variable vector. Also, for non-temporal input variables such as demographics, we denote it as a vector $\mathbf{x}_i^{NT}$ of dimension L. Thus, $\mathbf{x}_i = \{\mathbf{x}_i^1, \mathbf{x}_i^2, ..., \mathbf{x}_i^T, \mathbf{x}_i^{NT}\}$. We used a RNN having similar structure of [6] to perform a regression task to predict y_i. The network structure is depicted in Fig. 1.

Each prediction $\hat{y}_i$ made by RNN can be represented as

$$\hat{y}_i = w(\sum_{t=1}^{T} \alpha^t \beta^t \otimes e^t + e^{NT}) + b \tag{3}$$

w and b are output coefficients and bias respectively.

In our implementation, we used embedding size of 128. Dropout [18] was applied in embedding and context vectors to prevent overfitting. The dropout ratio was set to 0.5. To learn all the parameters, adaptive learning rate method ADADELTA [22] is used when applying back-propagation using stochastic gradient descent.

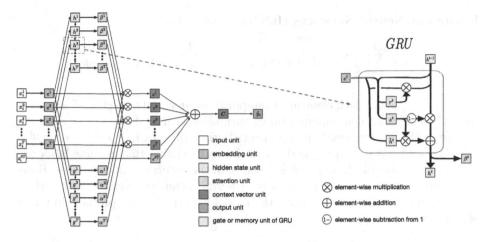

Fig. 1. Schematic diagram of the deployed RNN model. The whole process consists of several steps. **Step 1:** Input variables are embedded by embedding matrices W_T and W_{NT}; **Step 2:** A RNN with single gated recurrent unit (GRU) layer is used to generate attention α^t and β^t from the sequential embeddings, where α^t is scalar that determines that weight of period t and β^t is a vector that determines the importance of elements in each embedding e^t; **Step 3:** Attentions and embeddings are summed to make the context vector. The context vector is later transformed to output.

3.2 Interpreting Predictions

For predictive models in health care, in addition to high accuracy, interpretability is important. Machine learning models are often black-box models focused on pure prediction rather than understanding the degree of variance explained by each component of the model or medical cause-effect relationships. To improve interpretability, we deployed additional strategies to quantify the contribution from each single input variable so that model users can interpret and diagnose the predictions. Below are the approaches that we took for each model.

Ordinary Least Squares Linear Regression (LR) and Regularized Regression (LASSO). For these two linear models, pulling the contributions is straightforward. The product of linear coefficients and variable value are readily converted into the contribution of the corresponding variable.

Gradient Boosting Machine (GBM). As in [21], for each single decision tree learned by the GBM, each prediction is assigned to a leaf following a decision path. The decision path consists of splitting nodes described by input variables. The leaves are assigned weights, which are assigned back to the splitting nodes on the decision path and weighed by the gain in each node. As a result, the input variables in the splitting nodes receive a portion of the weights. The contributions are the sums of these portion weights by input variables across all trees.

Recurrent Neural Networks (RNN). From Eq. 3

$$\hat{y}_i = \sum_{t=1}^{T} \sum_{k=1}^{K} x_{tk} \alpha^t w(\beta^t \otimes W_T[:,k]) + \sum_{l=1}^{L} x_l w W_{NT}[:,l] + b$$

where x_{tk} is the k-th element of temporal input variable vector $\mathbf{x}_i^t$ and x_l is the l-th element of non-temporal input variable vector $\mathbf{x}_i^{NT}$.

From the above equation, the contribution of x_{tk} and x_l are $\alpha^t w(\beta^t \otimes W_T[:,k])$ and $w W_{NT}[:,l]$ respectively. It is worth noting that regular RNN are non-interpretable black-box models because of the recurrent hidden states. However, the recurrences here are used to generate attention weights rather than directly to make predictions. Thus, the model is partially interpretable in terms of input variables.

3.3 Evaluation and Validation

For models with hyperparameters (LASSO, GBM and RNN), we selected best fitting models using cross-validation as described above. To validate these models, we trained the models to predict period t and tested the models to predict period $t + 1$. During training, the information of period $t + 1$ was never accessed. We reported the R-squared and root mean squared error (RMSE) of rank percentiles as the performance measure. Given the policy interests associated with high utilizers [3, 19], we reported the RMSE for this population specifically. Throughout the study, we used the criterion of top 10% of PMPM expenditure to identify high utilizers. In order to get robust results, we used at least three different t values to have multiple times training and testing. We reported the results using the overall average from these multiple experiments. The only exception is for time periods of 12 months. In this case, we didn't have sufficient data, giving us only one training set and one testing set.

4   Experimental Results

We first present the predictive accuracy for each of the predictive methods. We also describe how these models can be used for potential cause-effect analysis for each patient. Finally we discuss the choices between models.

4.1 Prediction Accuracy

In this section, we present the accuracy performance for predicting the expenditures of an individual patient using prior expenditure information. These results use only expenditure data for the immediate preceding period followed by multiple periods (up to 4 periods) and use of other claim level information available for the patient.

Baseline. We present the baseline models below using one prior period and only prior expenditures variables (prior PMPM, *pctlPMPM* or *log*PMPM depending on predicting objective). The options of period length are 3 months, 6 months and 12 months. Due to space considerations, we limit the results to the prediction accuracy on test data and omit them for train data. The results are presented in Fig. 2.

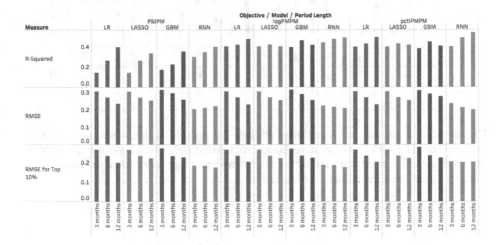

Fig. 2. Baseline: although only using expenditures information from the immediate prior period, the models fit reasonably well from the R-squared values. Generally performance improves when the time period becomes longer. However, GBM and LASSO seem to find their best R-squared when period length = 6 months in predicting *pctl*PMPM and *log*PMPM.

As the R-squared values demonstrate, these baseline models provide a reasonable degree of fit. When the time period increases from 3 months to 12 months, most fit measures improve, with a few exceptions: R-squared for LASSO and GBM in predicting *log*PMPM and *pctl*PMPM; and RMSE for RNN in predicting PMPM. These results generally suggest that predictive models are more effective for longer periods, which is expected, as aggregating over longer periods tend to dampen the short-term variations. That said, reasonable consistency in expenditures is present from one period to the next, which may be useful for predicting high utilizers in practical scenarios.

Using Additional Information. In this subsection, we present results after incorporating additional information to our baseline models. In particular, we added patient-level demographics, diagnoses, medical procedures and medications. Figure 3 shows performance improvement after adding these inputs during a 3 months to 3 months prediction. As expected, almost all the measures improved substantially with this additional information. When we repeated this procedure for periods of 6 months and 12 months, the improvement persisted.

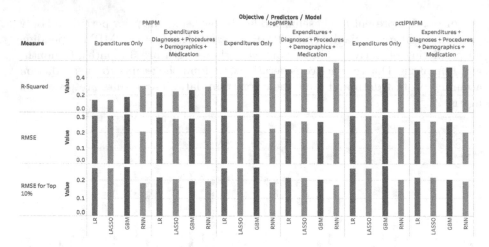

Fig. 3. All four models improved after adding demographics, diagnoses, medical procedures and medications as input variables, suggesting that though prior expenditures already provide a good approximation for future spending, additional information is useful in predictive modeling.

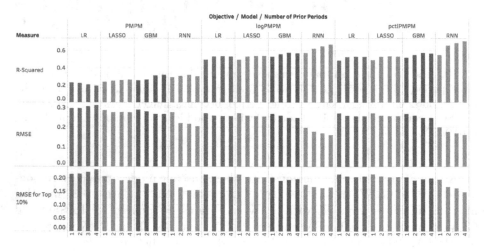

Fig. 4. Performance changes after adding more prior periods. Most measures substantially improved after adding the first three periods. The gain for adding a fourth period to LR, LASSO and GBM is minimal. RNN benefits most, indicating its stronger ability to model temporal relations.

Thus, though prior expenditures provide strong evidence for future costs, additional information, such as demographics, diagnoses, medical procedures and medications, improves prediction accuracy. It is worth noting that the number of input variables used is of the order of 1,000, but the amount of data available to us in terms of the number of patients is approximately 1 million. Thus, as long as the models are well regularized, the possibility of overfitting is limited.

Including Additional Prior Periods. In the previous models, we only used the most recent period prior to the period of prediction. In theory, increasing the number of prior periods added to the model would improve prediction performance. Figure 4 shows the performance changes after adding more periods. We limit these results to quarter-to-quarter prediction because of space considerations. The inputs for each quarter include diagnoses, medical procedures and medications. Demographic variables are input only once as they are assumed to be fixed over the study period. These results show that most measures improve as the number of prior periods increases, but the improvement saturates around three prior periods. RNN benefits the most by the use of additional periods, which is not entirely surprising, as the literature has shown that they are effective in modeling temporal relationships.

The R-squared of linear regression on testing data decreases when the predicting objective is a per member per month expenditure (PMPM). This is because, unlike *log*PMPM and *pctl*PMPM, PMPM is not linearly distributed in the parameter space. It is easy to overfit a linear model in this space, especially when using a large number of parameters and large number of prior periods (which effectively has a multiplicative effect on the number of parameters). We checked the R-squared for the training dataset in the same setting and found it to increase with the number of prior periods. This finding supports our claim of overfitting and implies preference for models robust to it, such as LASSO and GBM. Although the above results apply to quarters, we observed similar trends for 6-month periods.

4.2 Interpreting the Models

For each prediction, we can quantify the contribution from a single input variable in each model used. Figures 5, 6 and 7 present the contributions derived from different models for the same prediction case. We used both expenditures and additional information as well as four prior periods to train the models. The prediction objective is per member per month expenditure percentiles (*pctl*PMPM). To examine the variations within each model, we resampled the training data with replacement 10 times and trained a different model for each of these samples. These models were then used to make 10 different predictions. The average predicted score and its standard deviations (sd) are reported. For the contributions, we plotted each input variable as a circle on the timeline. The center and radius of the circle represent the average contribution and its standard deviation respectively. Results for LR (not presented) are similar to LASSO.

From the figures and predicted values, the single test case demonstrates that all three models make robust predictions. LASSO and GBM have lower standard deviations. LASSO is the most stable model that always gives similar contributions. GBM has a larger standard deviation in contributions but still can derive influential variables. The contribution of each variable given by RNN is very unstable. Clearly this method cannot be effectively used for deriving the importance of input variables. Considering that optimizing the parameters of RNN

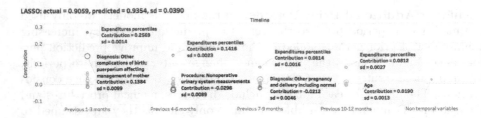

LASSO: actual = 0.9059, predicted = 0.9354, sd = 0.0390

Fig. 5. Contributions derived from a prediction by LASSO. The radius of the circle corresponds to the standard deviation of the contribution.

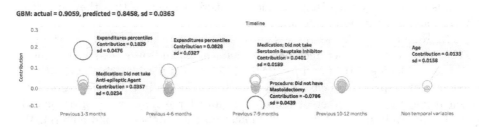

GBM: actual = 0.9059, predicted = 0.8458, sd = 0.0363

Fig. 6. Contributions derived from the same prediction by GBM. We observe a larger variation in contributions. But the variation in predicted value is similar.

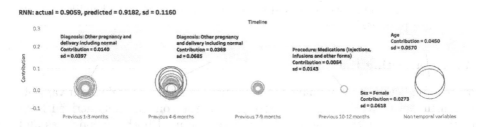

RNN: actual = 0.9059, predicted = 0.9182, sd = 0.1160

Fig. 7. Contributions derived from a prediction by RNN. When comparing LASSO, GBM and RNN, LASSO not only gives stable predicted value, but also generates stable contributions. GBM has consistent predicted values, but is less stable in contributions. RNN is unstable in both, possibly due to its non-convex optimization procedure.

requires a non-convex optimization procedure that may have many local minima, it is not surprising that the stochastic gradient descent algorithm would end up with different solutions (generally corresponding to a different local minimum) each time, leading to a much larger variation in contribution estimates. In conclusion, LASSO and GBM are more useful in generating interpretable contributions than the RNN model.

4.3 Choosing the Best Model

The best model depends on whether the goal is to predict expenditure or understand underlying factors. From Figs. 3 and 4, we observe that RNN is the best model for prediction. GBM is slightly better in R-squared and RMSE for top 10% than LASSO, but GBM has a similar performance for RMSE. However, for interpretation of the underlying prediction, LASSO and GBM are more suitable.

In terms of prediction objective, RNN seems to perform best using *pctl*PMPM. The reason for this could be that *pctl*PMPM is strictly contained in [0,1], which is less likely to cause significant gradient vanishing or exploding issues common in back propagation. For LR, GBM and LASSO, the predicting objective is a task-specific decision. If minimal RMSE for top 10% is the goal, one should use PMPM as predicting objective. If R-squared is the central measure, one should consider *log*PMPM or *pctl*PMPM. All three models are similar in overall RMSE.

5   Conclusions

In this paper, we developed and tested several predictive models to forecast expenditures in the Texas Medicaid program. We started from a baseline case using only expenditures and gradually added input variables to the models. We showed that additional information such as clinical information and demographics are useful to improve prediction performance. In addition, we showed that is beneficial to have historical data from more prior time periods. However, the improvements due to additional prior periods saturates after three to four periods. The prediction accuracy of RNN outperforms LR, LASSO and GBM. In terms of interpretability, LASSO and GBM consistently select similar variables and generate stable contributions independent of the resampling process.

This study is limited in several dimensions. First, we conducted the study within a single public insurance program. The results may vary when the methodologies are applied to different types of health care systems or payment models. Second, we applied only general-purpose machine learning models. Some tailored models may have better performance. Third, the predictive models lack a preventive measure to guide interventions. Finally, health status determined from claims data only is limited. It may be necessary to include additional data sources, such as electronic health records (EHR) or disease severity measures.

Our future work will address some of these limitations. We plan to expand the analysis to different types of health care programs. We will also collect additional data, such as electronic health records, to evaluate predictive performance. Most importantly, we will extend our collaboration with clinicians and policy experts to make the models interactive, such as integrating domain expertise to better direct preventive interventions.

Acknowledgments. This work was supported in part by Texas HHSC and in part through Patient-Centered Outcomes Research Institute (PCORI) (PCO-COORDCTR2013) for development of the National Patient-Centered Clinical Research Network, known as PCORnet. The views, statements and opinions presented in this work are solely the responsibility of the author(s) and do not necessarily represent the views of the Texas HHSC and Patient-Centered Outcomes Research Institute (PCORI), its Board of Governors or Methodology Committee or other participants in PCORnet.

References

1. AHRQ: Agency for healthcare research and quality, clinical classifications software (CCS) (2015)
2. Ashutosh, K., Lee, H., Mohan, C.K., Ranka, S.R., Mehrotra, K., Alexander, C.: Prediction criteria for successful weaning from respiratory support: statistical and connectionist analyses. Crit. Care **20**, 1295–1301 (1992)
3. Berk, M.L., Monheit, A.C.: The concentration of health expenditures: an update. Health Aff. **11**(4), 145–149 (1992)
4. Blumenthal, D., Chernof, B., Fulmer, T., Lumpkin, J., Selberg, J.: Caring for high-need, high-cost patients an urgent priority. N. Engl. J. Med. **375**(10), 909–911 (2016)
5. Chen, T., Guestrin, C.: XGBoost: a scalable tree boosting system. arXiv preprint arXiv:1603.02754 (2016)
6. Choi, E., Bahadori, M.T., Schuetz, A., Stewart, W.F., Sun, J.: Retain: Interpretable predictive model in healthcare using reverse time attention mechanism. In: Advances in Neural Information Processing Systems. NIPS (2016)
7. Choi, E., Schuetz, A., Stewart, W.F., Sun, J.: Using recurrent neural network models for early detection of heart failure onset. J. Am. Med. Inf. Assoc. (2016). doi:10.1093/jamia/ocw112
8. Friedman, J.H.: Greedy function approximation: a gradient boosting machine. Ann. Stat. **29**, 1189–1232 (2001)
9. Getzen, T.: Forecasting health expenditures: short, medium and long (long) term. J. Health Care Financ. **26**(3), 56–72 (2000)
10. Gilmer, T., Kronick, R., Fishman, P., Ganiats, T.G.: The medicaid Rx model: pharmacy-based risk adjustment for public programs. Med. Care **39**(11), 1188–1202 (2001)
11. Harman, J.S., Lemak, C.H., Al-Amin, M., Hall, A.G., Duncan, R.P.: Changes in per member per month expenditures after implementation of Florida's medicaid reform demonstration. Health Serv. Res. **46**(3), 787–804 (2011)
12. Huber, C.A., Schneeweiss, S., Signorell, A., Reich, O.: Improved prediction of medical expenditures and health care utilization using an updated chronic disease score and claims data. J. Clin. Epidemiol. **66**(10), 1118–1127 (2013)
13. Lee, R., Miller, T.: An approach to forecasting health expenditures, with application to the us medicare system. Health Serv. Res. **37**(5), 1365–1386 (2002)
14. Margolis, R., Derr, L., Dunn, M., Huerta, M., Larkin, J., Sheehan, J., Guyer, M., Green, E.D.: The national institutes of health's big data to knowledge (BD2K) initiative: capitalizing on biomedical big data. J. Am. Med. Inform. Assoc. **21**(6), 957–958 (2014)
15. Minnesota Department of Health: An introductory analysis of potentially preventable health care events in Minnesota, p. 1 (2015)
16. Pope, G.C., Kautter, J., Ellis, R.P., Ash, A.S., Ayanian, J.Z., Ingber, M.J., Levy, J.M., Robst, J., et al.: Risk adjustment of medicare capitation payments using the CMS-HCC model. Health Care Financ. Rev. **25**(4), 119–141 (2004)
17. Sen, B., Blackburn, J., Aswani, M.S., Morrisey, M.A., Becker, D.J., Kilgore, M.L., Caldwell, C., Sellers, C., Menachemi, N.: Health expenditure concentration and characteristics of high-cost enrollees in chip. Inquiry J. Health Care Organ. Provis. Financ. **53** (2016). doi:10.1177/0046958016645000
18. Srivastava, N., Hinton, G.E., Krizhevsky, A., Sutskever, I., Salakhutdinov, R.: Dropout: a simple way to prevent neural networks from overfitting. J. Mach. Learn. Res. **15**(1), 1929–1958 (2014)

19. Stanton, M.W., Rutherford, M.: The high concentration of US health care expenditures. Agency for Healthcare Research and Quality, Washington, DC (2006)
20. Tibshirani, R.: Regression shrinkage and selection via the LASSO. J. Roy. Stat. Soc. Ser. B (Methodol.) **58**, 267–288 (1996)
21. Yang, C., Delcher, C., Shenkman, E., Ranka, S.: Predicting 30-day all-cause readmissions from hospital inpatient discharge data. In: 2016 18th International Conference on E-health Networking, Application and Services (HealthCom). IEEE (2016)
22. Zeiler, M.D.: ADADELTA: an adaptive learning rate method. arXiv preprint arXiv:1212.5701 (2012)

A Device Supporting the Self Management of Tinnitus

Pablo Chamoso[1](✉), Fernando De La Prieta[1], Alberto Eibenstein[2],
Daniel Santos-Santos[1], Angelo Tizio[2], and Pierpaolo Vittorini[2]

[1] IBSAL/BISITE Research Group, Edificio I+D+i, University of Salamanca,
Calle Espejo 12, 37007 Salamanca, Spain
{chamoso,fer,daniel_santos}@usal.es
[2] University of L'Aquila, Delta 6, Via G. Petrini, Coppito, 67100 L'Aquila, Italy
pierpaolo.vittorini@univaq.it

Abstract. Tinnitus is an annoying ringing in the ears, in varying shades and intensities. Tinnitus can affect a patient's overall health and social well-being (e.g., sleep problems, trouble concentrating, anxiety, depression and inability to work). Usually, the diagnostic procedure of tinnitus passes through three steps, i.e., audiological examination, psychoacoustic measurement, and disability evaluation. All steps are performed by physicians, by using dedicated hardware/software and administering questionnaires. The paper reports on the results of a one-year running project whose aim is to directly support patients in such a diagnostic procedure by using a specific device and their smartphone.

Keywords: Tinnitus · APP · Device · Audiometry · Acufenometry

1 Introduction

Tinnitus is commonly known as a complex of annoying ringing, buzzing or hissing in the ears, in varying shades and intensities [5]. It can affect a patient's overall health and social well-being at different levels, from sleep problems, trouble concentrating, anxiety, up to ongoing depression and inability to work [3,10]. Usually, the diagnostic procedure consists in three main phases, i.e., diagnosis and accurate audiological examinations, psychoacoustic measurement of tinnitus, and evaluation of disability. All phases are performed by physicians, by using dedicated and complex hardware/software and administering questionnaires.

The paper reports on the first results of a one-year and on going project whose aim is to directly support in such a diagnostic procedure. In particular, the paper describes an hardware device that can be connected to a smartphone/tablet, able to execute (a part of) the procedure to diagnose tinnitus, as controlled by a dedicated APP that automates both the execution of the examinations and the administration of the questionnaires that measure the disability induced by tinnitus.

The novelty of our project is that it directly addresses patients instead of health professionals and provides an integrated tool (hardware/software) able

© Springer International Publishing AG 2017
I. Rojas and F. Ortuño (Eds.): IWBBIO 2017, Part II, LNBI 10209, pp. 399–410, 2017.
DOI: 10.1007/978-3-319-56154-7_36

to perform the audiological and psychoacoustic measurements as well as the evaluation of disability. It is worth remarking that devices controlled by smartphone/tablet APPs are not common and only available to physicians (see, e.g., [7,13] for devices concerning the audiological examination).

The paper is organized as follows. Section 2 introduces the necessary background on tinnitus, its clinical evaluation for measuring the impact of tinnitus in different aspect of quality of life. Section 3 describes the ad-hoc developed hardware device, in terms of its hardware and firmware. Section 4 briefly discusses the APP and how the automated examinations are implemented. Finally, Sect. 6 ends the paper.

2   Background

2.1   Tinnitus

Several definitions of tinnitus are available from the scientific literature. According to Del Bo et al. [5], "[...] tinnitus are sound sensations perceived by the individual, not supported by external sources, acoustic or electric, and acoustic apparatus caused by their activities or by alterations of the mechanisms of sensory processing [...]".

Tinnitus can affects a patient's overall health and social well-being [10]. Even moderate cases can interfere with the ability to work and socialize. The American Tinnitus Association (ATA) conducted a survey of its membership in 2014, to evaluate how ATA members experience tinnitus. Over 1100 people responded to the survey: the majority (about 36%) were annoyed of tinnitus (without a significant impact on life quality) or barely noticed it; others reported on sleep problems, trouble concentrating and anxiety (18%, 16% and 13%, respectively); the remaining – with smaller percentages – reported on social isolation, ongoing depression and inability to work [3].

The subjective nature of the disorder and the limited knowledge of the pathophysiology render extremely problematic the systematization of tinnitus. The classification is therefore an open question, nevertheless in recent years have been proposed "operational" approaches that takes into account the most relevant clinal features as those aspects that have the greatest impact on the quality of life of the subject [4]. Objective tinnitus can be heard by the examiner [9] and can be generated from vascular, muscular or respiratory sources and also from the temporo-mandibular joint. On the basis of the most recent acquisitions, a more detailed "pathogenic" classification for subjective tinnitus was proposed: (i) *Conductive* in case of myocardial twitch of the middle ear or disturbance of tubal ventilation; (ii) *Sensorineural* when we have an anomaly in the inner ear or sensory organ (cochlea and associated structures), vestibulocochlear nerve (cranial nerve VIII) or neural part; and, finally, (iii) *Central* when in presence of primary intracranial tumours, multiple sclerosis, traumatic brain injury and closed secondary "ghost sound".

2.2   Clinical Evaluation

Usually, the diagnostic procedure includes the following three main phases:

1. Diagnosis and accurate audiological examinations performs through an Audiometry (see Subsect. 2.3) and others like Tympanometry and Otoscopy (that are not used in our approach);
2. Psychoacoustic measurement of tinnitus by means of an Acufenometry (see Subsect. 2.4).
3. Evaluation of disability through Questionnaires (see Subsect. 2.5). Together with the previous, this is a further important phase, since it allows to identify patients that are seriously invalidated by the tinnitus [1,16].

2.3   Audiometry

In order to clearly understand how it is performed an audiometry test, it is necessary to introduce some definitions about pure tone, frequency and intensity (loudness). A pure tone is a sound having a single specific "frequency" which is a scientific term to describe a specific aspect of a signal [15]. Intensity of the pure tone is considered the loudness of the sound and is measured in decibels (dB). By far, the most common dB scale used in audiology is the dB HL scale. This scale is referenced to the average hearing of young adults, where "0 dB HL" means the average threshold of audibility of normally-hearing young adults.

Then, the state of the art describes three types of audiometries, as follow:

- Pure Tone Audiometry (PTA) is the procedure that uses pure tones to assess an individual's hearing. Pure tones are generated at different frequencies and intensities by an audiometer and presented to the patient via headphones or, in some cases, through loudspeakers. Depending on the transducer through which the stimuli is presented, the audiometry can be either air-conducted or bone-conducted, detailed below.
- Air Conduction Audiometry (ACA) is defined as the test for assessment the hearing of sound wave travelling through air. This mode of signal presentation assesses the entire auditory system [15]. The general method for air conduction pure tone audiometry goes as follows. The patient is instructed to listen carefully for a beeping sound (pure tone): when heard, even if very softly, he/she is asked to raise the hand. The intensity (loudness) of the tone in each frequency is decreased gradually, until the patient no longer responds. This routine is repeated for all test frequencies in both ears. The Lowest Audible Intensity (LAI) is then defined as the patient's threshold for the particular frequency. Such a procedure then establishes an air conduction pure tone threshold curve for each ear called audiogram. If there is any degree of hearing loss measured at any frequency in either ear, bone conduction pure tone testing must be performed [2].
- Bone Conduction Audiometry (BCA) testing stimulates the cochlea directly, bypassing the outer and middle ear. Bone conduction pure tone audiometry is performed using process as in air conduction audiometry, but the tones

are presented via a bone conduction headset [2]. This type of testing is used to determine whether a hearing lost is reflective of a cochlear/neural deficit (call "sensorineural" [15]) or an outer/middle ear dysfunction ("conductive"). If bone conduction thresholds indicate a hearing loss but one which is less severe than is indicated by air LAI, the loss is termed a "mixed" hearing loss.

2.4 Acufenometry

As mentioned earlier, acufenometry is performed to determine frequency and intensity of the tinnitus. First, the determination of the frequency of the tinnitus is carried out by asking the patient to compare the frequency of a test-sound (i.e., a pure tone) with that of the tinnitus. Two tones are presented alternately to both ears while the frequency is changed (increased or decreased) until the patient finds out the one closest to the tinnitus. Then, the determination of the intensity is instead established by means of the comparison between the test-sound with tinnitus. A pure tone at the previously identified frequency is firstly sent at subliminal levels to the other side ear. Then, the intensity is increased until the patient hears it. In this way the "threshold of perception" of a signal is established and taken as the reference level of 0 dB.

2.5 Questionnaires

Some questionnaires can be distinguished to help in the diagnosis of tinnitus. Most common ones are presented below:

- Pittsburgh Sleep Quality Index (PSQI) is a self-administered questionnaire which assesses sleep quality and disturbances over a 1-month time interval [14]. It is made up nineteen individual items that measure seven domains: subjective sleep quality, sleep latency, sleep duration, habitual sleep efficiency, sleep disturbances, use of sleep medication, and daytime dysfunction over the last month.
- Khalfa Hyperacusis Questionnaire (Khalfa) is a tested and validated tool, suitable to quantify and evaluate various hyperacusis symptoms. It is made up of 14 questions, each with four possible answers ("no", "rarely", "often" and "always"). The scoring procedure yields to a total score and 4 grades of tinnitus [8].
- Tinnitus Handicap Inventory (THI) [11] is a self-administered questionnaire to evaluate the impact of tinnitus on the quality of life. It is made up of 25 questions, each with 3 possible answers ("no", "sometimes" and "yes"). According to the scores THI determines a grade (up 5) of severity.

3 The Device

3.1 The Electronics

A device responsible for generating the pure tone associated to the audiometry and acufenometry processes was designed and developed for this project. The device is composed of a set of components as seen in Fig. 1 and they are described as follow:

- USB Connector -FTDI232- is the component that converts between USB and TTL serial protocols. This component permits to establish the communication between the device and the smartphone.
- Microcontroller Unit (MCU) -ATMEGA328P- is an 8 bits monocomponet of AVR-family. It is where the firmware is execute and it controls the rest of components.
- Signal Generator -AD9850- is a DDS wave generator that generates sinusoidal waves of 1 Vpp and programmable phase by means of serial/parallel protocol.
- Digital Potentiometer -MCP4551- which includes a volatile memory and I2C interface to control the volume. Its function is to attenuate the sinusoidal signal in order to control the level of intensity of the tone.
- Air/Bone Amplifiers -TPA2012- are a D class stereo amplifiers of 2 channels of 1.5 W and 8 Ω. It has 2 digital inputs to adjust output gain from 6 dB to 24 dB.

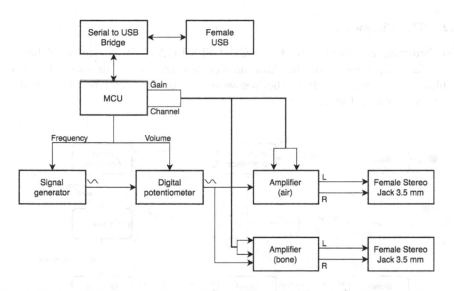

Fig. 1. Device block diagram. This device is able to perform air and bone audimetry.

When the device has to generate a pure tone signal with a given frequency and intensity, the MCU indicates the frequency of the sinusoidal signal to the signal generator module. The signal then passes through the digital potentiometer module, whose resistance is established according to the intensity indicated by the MCU. The output of the digital potentiometer is the input signal modified as to reach the volume.

The resulting signal is then sent to the amplifiers. There are two amplifiers: one associated to air mode and another one associated to bone mode. The amplifier for the selected mode (air or bone) amplifies the signal according to the gain indicated by the MCU; this signal is then sent to the selected channel (left or right), while the remaining channel receives a muted signal. The output is directed to a female stereo jack 3.5 mm module, which is connected to the corresponding headset. The two different and programmable amplifiers are present so to enable the device to be used with different headsets.

This hardware device connects as a peripheral to the patient's smartphone by using the USB On-The-Go (OTG) connection. Thanks to that, the device does not need its own battery, which is itself a cost-reducing benefit, but still maintains its characteristic of mobile device. This is a valuable advantage of the device with respect to common audiometers.

The connection also serves to provide the device with the information required to generate the corresponding signals received from the APP. Between the USB and the MCU there is a Serial to USB bridge that allows two-way communication, similar to a serial port, which facilitates communication between the two devices.

3.2   The Firmware

The firmware is executed by the MCU (ATMEGA328P as explained before) which is compatible with the Arduino bootloader, so the firmware has been implemented using Arduino. The firmware structure is described by the state diagram shown in Fig. 2.

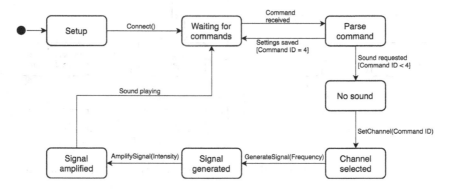

Fig. 2. State diagram of the firmware operation. The device is controlled by the Smartphone

Initially, a preliminary configuration establishes the default settings and mute the device, which continues to wait for a serial communication coming from the smartphone. Once connected, the device enters a state of "Waiting for commands", which finishes when a command sent from the smartphone is received.

Table 1. Command list

ID	Command	Arguments
0	Air-left channel	Frequency: 125–8000 Hz Intensity: 0–125 dB
1	Air-right channel	
2	Bone-left channel	
3	Bone-right channel	
4	Settings	Air mode offset (0–128) Bone mode offset (0–128) Air mode steep (0–128) Bone mode steep (0–128)

Then, the command is parsed and analysed to determine the action to take. All possible commands are described in Table 1. There are 5 possible commands that the firmware can read as shown in Table 1.

3.3   Mechanical Part

We have designed an adapter that integrates both the headphones and a bone conductor, making its use easier and more user-friendly. As you can see in Fig. 3, there are two hinges on each side that adapt the bone conductor to the back of the user's skull.

Fig. 3. Mechanical part design and integration with headphones. 3D rendering (left and centre) and real picture (right).

Both the headphones and the bone conductor have their own male 3.5 mm jack connectors, which are connected to the designed electronic device. The position of each connector is indicated in the device case, as shown in Fig. 4.

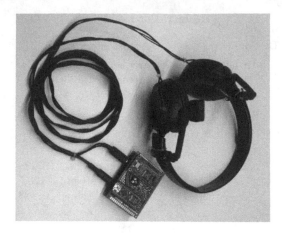

Fig. 4. Electronic device connected to the headphones and bone conductor.

4   The APP

The APP includes the functionalities that are needed for the clinical evaluation described in Subsect. 2.2. Therefore, the APP takes advantage of the device described in Sect. 3 to implement the automated audiometry (Subsect. 4.1) and acufenometry (Subsect. 4.2). Furthermore, the APP proposes and automatically scores the questionnaires for sleep quality, hyperacusis and the impact of tinnitus in life (Subsect. 4.3). The APP is available for smartphones and tablets running Android 5.0 and above. It is written in Java using Android Studio, and is freely available for download[1].

4.1   Automated Audiometry with Reporting

The APP implements the process described in 2.3, for both air and bone conduction audiometry, with the following two exceptions: (i) the patient touches a button placed in the centre of the smartphone when he/she hears the sound; (ii) the decrement/increment of intensity is not performed. It is worth remarking that such a step is present in process so to ensure that the intensity level reported by a patient was actual and that the patient was not "cheating" to the physician. In our case, since the APP is autonomously used by a patient, such a step was considered unnecessary. The process returns a matrix of intensities, i.e., when the patient heard the sound, for both ears, for both ways (i.e., air and bone), for all investigated frequencies, that is given in input to an automated audiometric reporting procedure. Then, the automated audiometric reporting procedure works as follows: initially, for both ears and all investigated frequencies, recalls the intensities; then, deduces the type of problem and its severity; hence, sums up all such information in terms of frequencies ranges, i.e., low frequencies

---

[1] http://vittorini.univaq.it/tinnitus/.

(125 Hz, 250 Hz, 500 Hz), medium frequencies (1 KHz, 2 KHz), and high frequencies (3 KHz, 4 KHz, 6 KHz, 8 KHz); finally decides the location of the problem. Figure 5-(a) shows the results of the automated audiometry and reporting.

4.2 Automated Acufenometry

The APP implements the acufenometry described in Subsect. 2.4, without detecting the "threshold of perception" level. Figure 5-(b) shows the interface used to perform the acufenometry: the switches placed on the top can be used to select which ear has the tinnitus, the horizontal/vertical arrows change the frequency/intensity of the tinnitus, while the central button can be tapped to confirm that the emitted sound actually resembles the tinnitus.

4.3 Questionnaires

The PSQI, Khalfa and THI questionnaires are implemented and automatically scored in the APP. A simple ad-hoc XML document defines and enable the APP to show the questionnaires. The document allows to define:

1. the possible variable types used in the questionnaire, as a specialization of numeric, time and categoric types;
2. the list of questions placed in the questionnaire, their types and if required or optional.

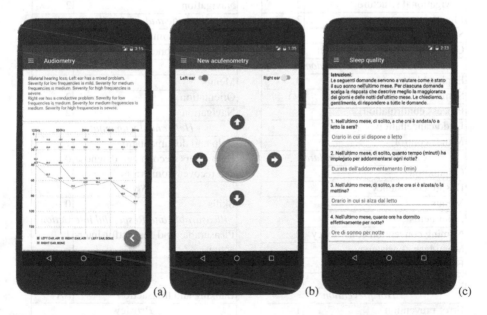

Fig. 5. Snapshots of (a) an audiogram with the automated reporting, (b) an acufenometry in progress, (c) the PSQI

The APP automatically scores the questionnaires and returns in a simple format the results. Figure 5-(c) shows the case of the PSQI questionnaire.

5   Preliminary Results

As preliminary results, we started investigating the APP usability [12]. Since our project was in its first release, coherently with the state of the art, we decided to ask to one usability expert for an heuristic evaluation to generate the initial number of potential usability problems. The expert used a check-list specifically designed to evaluate mobile interfaces, that reuses 69% of literature heuristics, the rest deriving from best-practices and recommendations for mobile

Table 2. Expert-based evaluation result

Item	Score	Item	Score
Visibility of system status		*Recognition rather than recall*	
System status feedback	2	Memory load reduction	2
Location information	NA	General visual cues	1
Response times	2	Input/output data	2
Selection/input of data	1	Menus	2
Presentation adaptation	NA	Navigation	2
Match between system and the real world		*Flexibility and efficiency of use*	
Metaphors/mental models	1	Search	1
Navigational structure	2	Navigation	2
Menus	2	*Aesthetic and minimalist design*	
Simplicity	2	Aesthetic and minimalist design	1
Output of numeric information	NA	Multimedia content	1
User control and freedom		Icons	2
Explorable interfaces	0	Menus	2
Some level of personalization	NA	Orientation	2
Process confirmation	1	Navigation	2
Undo/cancellation	0	*Help and documentation*	
Menus control	2	Help and documentation	1
Consistency and standards		Help users recognize, diagnose and recover from errors	0
Orientation	1	*Skills*	
Design consistency	2	Skills	0
Menus	2	*Pleasurable and Respectful interaction*	
Input fields	2	Pleasurable and respectful interaction	1
Naming convention consistency	2	Input data	1
Menu/task consistency	2	Shopping	NA
Functional goals consistency	1	Banking and transaction	NA
System response consistency	2	*Privacy*	
Error prevention		Privacy	NA
Error prevention	1		
Fat-finger syndrome	1		

interfaces [6]. Such a check-list requires, for each heuristic, to give a score ranging from 0 to 2, where the higher the score, the better the usability (in terms of that heuristic). Table 2 reports the results.

In summary, the median value is 2, i.e., a good usability, with an average of 1.4. We also collected verbose suggestions from the expert, that proposed us to (i) add a clear back/undo button, (ii) more clearly show the goals of each functionality (i.e., audiometry, acufenometry and questionnaires), (iii) implement a font scaling feature[2], (iv) add a search facility, even if it may be not necessary given the shallow navigational structure. These suggestions will be taken into account in the next release of the APP.

6   Conclusions

The proposed usage model, where the device is connected to the users smartphone, has allowed to obtain a low-cost device that includes really powerful functionalities. Although the presented device is only the first prototype, it has been used to evaluate all the parts of the system. It has successfully helped to determine the final configuration to use in an integrated board, with the same electronic components but with smaller size.

As for APP, besides the usability investigation, we also started evaluating also the quality of the automated audiometric reporting procedure, by comparing its results with human reporting. In our preliminary investigation, all the automated audiometric reporting were correct, few of them were even more detailed than the human ones. Beside that, a possible improvement in the automated reporting came out: an audiometry showing a problem only for one frequency is usually reported as an "acoustic hole". Since such a wording is not provided by the APP, it will be added in the next release.

References

1. Aazh, H., McFerran, D., Salvi, R., Prasher, D., Jastrebo, M., Jastrebo, P.: Insights from the first international conference on hyperacusis: causes, evaluation, diagnosis and treatment. Noise Health **16**(69), 123–126 (2014)
2. Association, American Speech-Language-Hearing and others: Guidelines for manual pure-tone threshold audiometry (2005). Bibtex: association_guidelines_2005
3. American Tinnitus Association: Impact of Tinnitus (2016)
4. Axelsson, A., Ringdahl, A.: Tinnitus-a study of its prevalence and characteristics. Br. J. Audiol. **23**(1), 53–62 (1989)
5. Del Bo, M., Giaccai, F., Grisanti, G.: Manuale di audiologia, 3rd edn. Elsevier, Amsterdam (1995). Izione edition
6. Caballero, D.C., Sevillano, J.-L.: Heuristic evaluation on mobile interfaces: a new checklist. Sci. World J. **2014**, e434326 (2014)
7. INVENTIS: Piccolo Portable Audiometer (2016)

---

[2] The font size may be too narrow for some users (especially elderly) if not scalable.

8. Khalfa, S., Dubal, S., Veuillet, E., Perez-Diaz, F., Jouvent, R., Collet, L.: Psychometric normalization of a hyperacusis questionnaire. ORL J. Oto-Rhino-Laryngology Relat. Spec. **64**(6), 436–442 (2002)

9. McCombe, A., Baguley, D., Coles, R., McKenna, L., McKinney, C., Windle-Taylor, P., British Association of Otolaryngologists, Head and Neck Surgeons: Guidelines for the grading of tinnitus severity: the results of a working group commissioned by the British Association of Otolaryngologists, Head and Neck Surgeons, 1999. Clin. Otolaryngol. Allied Sci. **26**(5), 388–393 (2001)

10. Moring, J., Bowen, A., Thomas, J., Bira, L.: The emotional and functional impact of the type of tinnitus sensation. J. Clin. Psychol. Med. Settings **23**, 310–318 (2015)

11. Newman, C.W., Jacobson, G.P., Spitzer, J.B.: Development of the tinnitus handicap inventory. Arch. Otolaryngol. Head Neck Surg. **122**(2), 143–148 (1996)

12. Nielsen, J., Mack, R.L.: Usability Inspection Methods. Wiley, New York (1994)

13. SHOEBOX: SHOEBOX Portable Audiometer and Diagnostic Screening (2016)

14. Smyth, C.A.: Evaluating sleep quality in older adults: the Pittsburgh Sleep Quality Index can be used to detect sleep disturbances or deficits. Am. J. Nurs. **108**(5), 42–50 (2008). Quiz 50–51

15. Vencovsky, V., Rund, F.: Pure tone audiometer. In: 20th Annual Conference Proceeding's Technical Computing, pp. 1–5 (2012). Bibtex: vencovsky_pure_2012

16. Zenner, H.-P.: A systematic classification of tinnitus generator mechanisms. Int. Tinnitus J. **4**(2), 109–113 (1998)

A Case Study on the Integration
of Heterogeneous Data Sources in Public Health

Pierpaolo Vittorini(✉), Anna Maria Angelone, Vincenza Cofini, Leila Fabiani,
Antonella Mattei, and Stefano Necozione

Department of Life, Health and Environmental Sciences,
University of L'Aquila, 67100 Coppito L'aquila, Italy
pierpaolo.vittorini@univaq.it

Abstract. The paper reports on a case study regarding the integration
of heterogeneous data sources, coming from different institutions, needed
to support several studies related to the health status of the population
of L'Aquila (Italy) after the earthquake of April 6th 2009. In detail, the
paper initially describes all the health studies, then the data sources
required to support them, finally proposes a simplified federated architecture and a straightforward technological solution to implement it.

Keywords: Data integration · Public health studies · Earthquake

1 Introduction

On April 6th 2009, the city of L'Aquila (Italy) was hit by a strong earthquake
that caused deaths, mainly as the consequence of the collapse of several buildings.
The effects on public health of the earthquake started to be investigated almost
in the early phases after the natural event, similarly as in other cases (e.g.,
[33]). Nevertheless, the need for further investigations is still actual, as testified
by a recent agreement between the Abruzzo's RHA (Regional Health Agency[1])
and the Dep. of Life, Health and Environmental Sciences of the University of
L'Aquila, with the aim of exploring the effect of the earthquake in terms of public
health. In such a context, the paper reports on the preliminary efforts conducted
by the authors to identify the data sources needed to conduct the needed public
health studies (Sects. 2 and 3) and in particular how to integrate such a data
sources (Sect. 4). Section 5 ends the paper with short conclusions.

2 Studies

1. Respiratory Diseases. In [13], the authors report on the use of Health Discharge Records (HDRs, see also Sect. 3.1 for more details) to investigate the

---

[1] The Abruzzo's RHA is a technical body of the Abruzzo Region that concurs to the
definition of the health policies and strategies.

© Springer International Publishing AG 2017
I. Rojas and F. Ortuño (Eds.): IWBBIO 2017, Part II, LNBI 10209, pp. 411–423, 2017.
DOI: 10.1007/978-3-319-56154-7_37

respiratory health in terms of Pneumonia, Asthma, COPD and respiratory failure. The study regarded the territorial area of the Local Health Unit 1 of the Abruzzo Region – under which L'Aquila is included – in the period 2008–2013. The authors calculated the Standard Hospitalization Rates (SHRs) for different respiratory diseases, and observed different trends in SHRs in different areas, for some of the different respiratory diseases. For calculating the SHRs, the authors relied on demographic data taken from the Italian Institute of Statistics (ISTAT). The authors finally discussed diverse possible causes of the different trends, with a specific focus on a possible relation with the earthquake of the 6th April 2009, since coherent with the related scientific literature. Within the umbrella of the agreement, the proposed study would start from the aforementioned results, and investigate the possible effect of the earthquake considering not only the residence of the patient in different areas, but also including whether and when the patient lived – after the earthquake – in tents, temporary houses, or hotels.

2. *Prevalence of Major Allergic Diseases in Abruzzo in the 2008–2013 Period.*
Allergic diseases are complex identities determined by an interplay of genetic and environmental factors, resulting in the clinical manifestation of the disease [16]. The most common allergic pathologies are Allergic rhinitis (AR), Atopic dermatitis (AD) and asthma. Allergic rhinitis (AR) is a common condition that affects a significant proportion of the global population, across all age groups, and is often a life-long condition. Its impact on direct and indirect healthcare costs is significant and it negatively impacts patients' quality of life, work performance and social functioning. Zuberbier et al. conducted a detailed cost analysis of allergy treatment in the European Union (EU) and found that the cost of insufficiently treated patients ranges between €55 and €151 billion annually from absenteeism and productivity losses. This translates to €2405 per untreated patient annually [43]. Atopic dermatitis is primarily a disease of early childhood. About 20% of all children develop symptoms of atopic dermatitis at some point in their lives [42]. Up to one fourth of subjects with moderate to severe atopic dermatitis in childhood will develop hand eczema to various degrees in adult life [5]. Furthermore, the patients affected by eczema in infancy have higher odds of developing AR and asthma subsequently. The allergic march refers to the natural history of atopic disorders. The allergic march concerns the development of atopic dermatitis and concomitant sensitization to food and aeroallergens in early childhood, progressing to asthma and allergic rhinitis in later childhood or adult life [41]. Many explanations have been put forward for the rapidly increasing prevalence of allergies, asthma and other chronic inflammatory disorders especially in developed countries [2], including the influence of the living environment (rural vs urban, farm vs nonfarm) [32], increased air pollution [28]. The aims of the study are to determine the overall prevalence of AD, AR and asthma in the general population of L'Aquila, the trend of the pathologies during the period of time considered (before and after the earthquake), in the different residential areas such as the New Towns and the original homes and the seasonal trend considering that the population of the city of L'Aquila had to move into newly constructed homes built in areas with different allergenic expositions in

relation to the precedent residential areas. We check the biodiversity hypothesis by analysing the rapport between land use around the homes of study subjects, as a measure of environmental biodiversity, and atopic reaction. Ruokolainen L et at. found that green area within 2–5 km from the home was inversely and significantly associated with atopic sensitization in children 6 years of age and older [36]. Furthermore, the study considered the comparison between the prevalences of the pathologies with other shoreline and outback cities in Abruzzo.

3. *Hospitalization Related to Alcohol and Drug Use Disorders.* Many studies have been focused on psychological distress and unhealthy behaviours on earthquake survivors after the earthquake in L'Aquila, Abruzzo Region in central Italy. Symptoms of psychological distress, as expected after a natural disaster, were highly prevalent in survivors right after the earthquake and the substance abuse was increased among young people [7,14,29]. 3–5 years after the earthquake the frequency of depressive symptoms was decreased but the prevalence of consumption of alcoholic beverages among adults seems to be higher from self-reported data of the Italian Behavioral Risk Factor Surveillance System [24]. However, studies aimed to explain changes in substance use patterns, as a result of disaster experiences, are limited. Our objective is to assess hospitalization related to alcohol and drug use disorders in earthquake survivors comparing data pre/post disaster in Health Unit of L'Aquila and comparing hospitalization data among Health Units of Abruzzo Region. In Italy, the Health System in organized in regional Health Units where the Hospital databases collect Hospital Discharge Records (HDRs) that are sent to the Ministry of Health. The information collected includes the patient's personal characteristics, characteristics of admission according to the International Classification of Diseases and clinical features (main diagnosis, concurrent diagnoses, diagnostic or therapeutic procedures), they represent an important source of epidemiological data about substance use disorders. To investigate about alcohol and drug use disorders in earthquake survivors, we are planning to use data related to survivors' hospitalization using two sources: (i) data collected in Hospital (public and private) in six Health Unit of Abruzzo Region, starting from 2006 (three years before the earthquake) to 2013; (ii) local database of citizens who had government assistance after earthquake and we are going to use it to define the survivors cohort. To define alcohol and drug use disorders, we will use a list of ICD-9-CM diagnostic codes for alcohol and drug use disorders, combining codes from the two Italian studies [6] and [4].

4. *Epidemiological Impact of Vaccine Preventable Diseases.* The prolonged health impact of natural disasters on a community may see the collapse of health facilities and healthcare systems, reduction of surveillance and health programs [25]. Increases in infectious disease transmission and outbreaks following natural disasters are associated with prolonged after-effects of the disaster including displaced populations, high exposure to and proliferation of disease vectors (rodents, mosquitoes), overcrowded shelters, poor water and sanitation conditions, low levels of immunity to vaccine-preventable diseases or insufficient

vaccination coverage, and limited access to healthcare services [25]. All these changes facilitate the occurrence and transmission of infectious diseases, such as acute respiratory infections, diarrhoeal diseases, measles and meningitis, many of which require notification, according to the Information System for Infectious Diseases [19,25]. In particular, measles immunization has been documented as one of the most cost-effective public health interventions in disaster settings and the risk for transmission after a natural disaster are dependent on baseline immunization coverage among the affected population, and in particular among children <15 years old [17]. Gastroenteritis is another vaccine-preventable disease, even if only for rotavirus aetiology that is responsible for 23% of all Italian hospitalizations for gastroenteritis or 63% of all Italian hospitalizations for gastroenteritis of specified aetiology in children aged <6 years old [22,23]. However, the types of infectious diseases may differ between industrialized and developing nations [18]. Understanding the likelihood of an infectious disease outbreak after a natural disaster through risk assessment could aid emergency management teams in the development of appropriate control measures [23]. To achieve this aim, we propose to investigate the spread of vaccine-preventable infectious diseases in Abruzzo, such as measles and acute gastroenteritis, comparing the data of hospitalizations and notifications (for notifiable conditions) for these conditions before the 2009 L'Aquila earthquake with post-earthquake data, to analyse any changes in frequency of these diseases in relation to the earthquake.

5. Analysis of the Determinants of Drugs Consumption in the Sub-regional Areas of Abruzzo. The variability in the prescription of drugs between different geographical areas and over time is an important issue in assessing the quality and efficiency of a health system. Over the years in Italy has been observed a significant regional variability for all prescription indicators, which remains even after taking into account the different distribution of the populations by age and sex. National and regional analyses highlight an increase of about 65% of Defined Daily Dose 1000 persons/ day (DDD_{1000pd}) since 2000 [1]. It was identified a strong geographic variability within each year: the range among regions is about 300 DDD_{1000pd} [38]. The different levels of consumption among regions may be due to differences in health, socio-economic status, organizations and use of health services. In general, the main determinants (age, income) of the drugs consumption explain only part of the observed variability. At different health conditions may correspond different levels of consumption of drugs even though the regional dimension is very broad and not very suitable to detect differences. Therefore it is important to study the variability of drugs consumption at sub-regional level and to identify possible determinants that explain its temporal evolution, including the effects of the earthquake of 2009. The aim is to evaluate the correlations between consumption of specific drugs, determining demographic and socio-economic (age, per capita income, education, employment, relative poverty) and organizational modalities of assistance, related to the supply and induction of consumption (number of primary care physicians and their prescribing attitudes for each type of drugs), also with regard to effects of the earthquake. Statistical analysis can be performed with linear regression models with random effects that describes the possible differences among geographical areas and over time [38]. Longitudinal data (panel data), containing the geographic and temporal dimension and combine the characteristics of cross-sectional data and time series for the evaluation of the association with the consumption of the drug class A-SSN expressed in DDD_{1000pd}.

3 Data Sources

To supply the necessary information for the studies reported in Sect. 2, the authors investigated the needed data sources, and selected the following to be integrated.

3.1 Hospital Discharge Records

Hospital discharge records (HDRs) are a large and useful source of information regarding healthcare. HDRs are maintained by hospitals primarily for management and accounting purposes, but they are currently used nowadays also for epidemiological research [8, 26]. The use of HDRs in epidemiology has strengths, limitations and may introduce biases [40]. The most important strengths are that data already exist, are large and are collected independently from research questions. The limitations are instead that data are pre-collected by non-researchers, with a low or unknown quality and may lack of confounders. Biases may be also introduced, like misclassification as the result of unclear or erroneous clinical documentation, or like the fact that expensive medical procedures are usually documented better than those less costly [26]. Finally, underestimation and misclassification of actual cases are other important limitations, which may affect the estimation of a disease (e.g., [34]). In Italy, HDRs are contained in the so-called "File A" of the health records that are exchanged between the Regions and the State [9], which is in turn divided into "File A1", i.e., the table containing the patients' identity records, and "File A2", i.e., the table containing the patients' health relation information. HDRs will be provided to us by the Abruzzo's RHA. Among all the available metadata contained in these tables, those relevant for the paper are the following:

- File A1
  • general patient information (e.g., tax code, gender, birth date);
- File A2
  • DRG (Diagnosis Related Groups) code, i.e., the treatment assigned to the patient [10];
  • ICD9-CM (International Classification of Diseases) codes for the patient's main and concurrent pathologies [27];
  • dates regarding the hospitalization and discharge.

3.2 Drug Prescription Records

Also Drug Prescription Records (DPRs) are included in the health records that are exchanged between the Regions and the State [9]. Differently from HDRs, they are included in the so-called "File F" which is divided – like "File A" for HDRs – into "File F1" that includes the patients' identity records, and "File F2" that instead contains the patients' health-related information. DPRs will be provided to us by the Abruzzo's RHA. Among all the available metadata, those relevant for the paper are the following:

- File F1
  - general patient information (e.g., tax code, gender, birth date);
- File F2
  - drug code;
  - drug type;
  - supply date.

3.3 Earthquake Residency Records

After the earthquake, the data concerning the residency of the population was initially kept by the Italian Civil Protection, then passed to the Municipality that is keeping these information updated. Among all the available metadata, those relevant for the paper are the following:

- tax code;
- residence
  - type, i.e., tent, hotel, autonomous accommodation[2], C.A.S.E. project[3], MAP[4];
  - start date;
  - end date.

3.4 Notification of Infectious Diseases

In Italy, the surveillance of infectious diseases is supported by the Information System for Infectious Diseases (SIMI) [25]. It is based on notifications made by physicians, then reported to both the local Public Health Department and health authorities (for the adoption of any preventive measures to protect public health), then sent to the Regional Public Health agency, the Central bodies (Ministry of Health, ISTAT, National Institute of Health) and, if applicable, also to International organizations (EU, WHO). Different forms are available for notifications. Nevertheless, the fields relevant for the paper are the following:

- general patient information (e.g., tax code, birth date);
- ICD-9 code for the identified disease;
- date and place of initial symptoms.

---

[2] The case of citizen that provide to an alternative accommodation by themselves.

[3] New houses build for the emergency period by the central government.

[4] Temporary houses built for the emergency period by the municipality.

4   Integrated Database

The problem of integrating heterogeneous databases has been investigated at length, e.g., see [30] for a survey regarding automated schema matching approaches, or [31] for a methodological approach, or [39] for a specific discussion in the context of biomedicine, or [11] for the specific issue of record linkage, or [15, 35] for the use of middleware relying on wrappers, or [37] for a description of federated databases. As known, a federated database system (FDBS) is a collection of cooperating database systems that are autonomous and usually heterogeneous, that can continue its local operations and at the same time participate in the federation, though within a controlled and coordinated manipulation from the so-called federated database management system (FDBMS).

For the aims of our research, we planned to follow the approach of a (simplified) FDBS, i.e., in-between a federated and a wrapper approach. In detail, given the component DBMSs coming from the RHA (HDRs and DPRs), the Municipality (Earthquake Residency Records), and the Public Health Department (Notifications of Infectious Diseases), they will be made accessible – with a specific concern about privacy and security, e.g., through VPN connections – to a central server which will take care of integrating all data, by eventually using the proper record linkage procedures [11]. Figure 1 shows such an architecture.

The next subsections report on the conceptual model of the integrated database (Subsect. 4.1) and on the technological solutions that will be adopted to implement the proposed architecture (Subsect. 4.2).

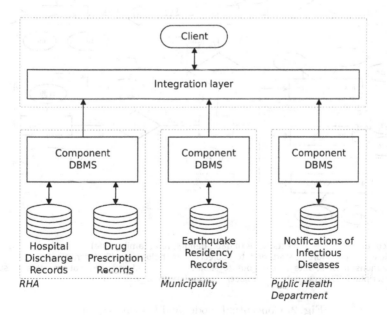

Fig. 1. System architecture

4.1   Conceptual Model

The E/R diagram [12] depicted in Fig. 2 summarizes the conceptual model (with the logical mapping) for the data required to support the studies presented in Sect. 2. In details:

- a subject is identified by his/her tax code (the tc attribute) and we need his/her birth date and gender;
- of a subject, we need to know his/er residencies over the time;
- the residency is identified by an id and contains the address, type, start and end date (in that residency);
- a subject belongs to a certain family, with a given role;
- of a subject, we record a certain number of events, i.e., hospitalizations, notifications of infectious disease, drug prescription
  - hospitalizations are described by the related DRG and the admission and discharge dates;
  - notifications are described by the date and the notifier;
  - drug prescriptions contain the type of administered drug and the date;
- the event is caused by pathologies, of which we record the ICD9 code with description, and its role (e.g., in case of hospitalizations, pathologies are ordered as primary and concomitant).

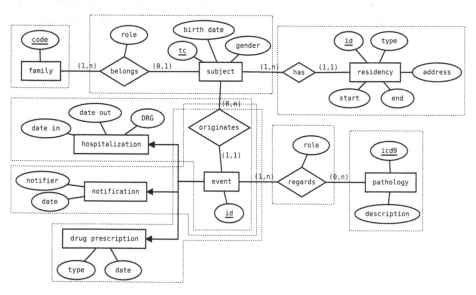

families(code)   subjects(tc, birth date, gender, role, family_code)
residencies(id, type, address, start, end, tc_subject)   pathologies(icd9, description)
hospitalizations(id, DRG, date_in, date_out, tc_subject)   notifications(id, notifier, date, tc_subject)
drug_prescriptions(id, type, date, tc_subject)   regards(icd9, id_event, role)

Fig. 2. Conceptual model and logical mapping

4.2   Technological Solution for the Integration Layer

The integration layer can be implemented through the MariaDB open source DBMS [3]. The reasons for such a choice are manifold:

1. remote tables/views can be directly connected to the central database by using the CONNECT table types [20]. It works either with MariaDB native tables or with ODBC tables, thus making simple the connection between the components DBMSs and the central database;
2. using views over the remote tables/views enables a simple way to automatically populate the central database described in Subsect. 4.1, which in turns supports all the studies discussed in Sect. 2.

Figure 3 shows the current architecture for the integration layer, which closely reflects the rationale summarised above.

4.2.1   Example of Local Connection to Component DBMS

As an instance of a local connection to a component DBMS, and in particular to connect the Municipality of L'Aquila component DBMS to the integration layer, the following SQL statement was used:

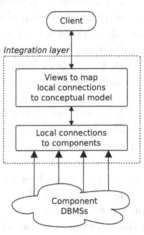

Fig. 3. Integration layer

```
CREATE TABLE residencies (
    ID int(11) NOT NULL,
    TAX_CODE varchar(16),
    TYPE int(11) NOT NULL,
    ADDRESS varchar(100),
    START DATE,
    END DATE,
    FAMILY_CODE int(11)
)
ENGINE=CONNECT
    table_type=ODBC
    tabname='...'
    connection='DSN=...'
OPTION_LIST='User=...,Password=...';
```

4.2.2   Example of Views to Map Local Connections to the Conceptual Model

The following SQL statement:

```
CREATE VIEW families AS
SELECT DISTINCT FAMILY_CODE as code FROM residencies
```

was used – as instance – to populate the `families` table of the central database (see Fig. 2), starting from the `residencies` table, locally connected as above. As a further example, given the locally connected `hdrs` table (from the RHA component DBMS), the following view[5] populates the `hospitalizations` table (see Fig. 2):

---

[5] For sake of clarity, in English, COD_FISCALE is the tax code, DATA_RIC means the date of admission, DATA_DIM means the date of discharge.

```
CREATE VIEW hospitalizations AS
SELECT CONCAT("hdr-", COD_FISCALE, "-", DATA_RIC) as id,
       DRG as DRG,
       DATA_RIC as date_in,
       DATA_DIM as date_out,
       COD_FISCALE as tc_subject
FROM   hdrs
```

5   Conclusions and Future Work

The paper reported on an a case study of integrating different data sources to support different studies in public health, related to the earthquake of April 6th, 2009. In particular, the paper listed the available studies, the needed data sources, and reported on the efforts that lead to set up an architecture that provides the integrated information that supports the studies.

At the time of writing, the system is still under implementation, whereas its feasibility has been tested by implementing all local connections of the component DBMSs with locally available copies of the remote ones. This phase highlighted many issues in the local schemas. For instance, while working on the HDRs, the authors found the data available in different tables, one for each year, with different names for attributes also holding the same semantics. To cope with these issues, the authors took advance of the methodology of Reddy et al. [31] and derived a global and integrated view of HDRs. Such an integrated view is the actual local connection for the component DBMS of the RHA, related to HDRs. Furthermore, as came out from brainstorming meetings with the engineers of the Municipality of L'Aquila, the quality of the residency data source – especially for the data regarding the period close to the earthquake – may be affected by missing data, also of tax codes that are used throughout the logical model as foreign keys. To overcome this problem, probabilistic record linkage procedures [11] are planned to be adopted as user-defined functions [21]. As further work, the authors will define and implement the views that will provide, for each of the aforementioned studies, the needed data.

Acknowledgements. The authors thank Dr. Mascitelli and Dr. Di Candia for their valuable support.

References

1. AIFA. L'uso dei farmaci in Italia - Rapporto OsMed 2015 (2015)
2. Asher, M.I., Montefort, S., Björkstén, B., Lai, C.K.W., Strachan, D.P., Weiland, S.K., Williams, H., ISAAC Phase Three Study Group.: Worldwide time trends in the prevalence of symptoms of asthma, allergic rhinoconjunctivitis, eczema in childhood: ISAAC phases one and three repeat multicountry cross-sectional surveys. Lancet (Lond. Engl.) **368**(9537), 733–743 (2006)
3. Bartholomew, D.: MariaDB Cookbook. Packt Publishing Ltd, Birmingham (2014)

4. Beck, C.A., Southern, D.A., Saitz, R., Knudtson, M.L., Ghali, W.A.: Alcohol and drug use disorders among patients with myocardial infarction: associations with disparities in care and mortality. PloS One **8**(9), e66551 (2013)
5. Bingefors, K., Svensson, A., Isacson, D., Lindberg, M.: Self-reported lifetime prevalence of atopic dermatitis and co-morbidity with asthma and eczema in adulthood: a population-based cross-sectional survey. Acta Dermato-Venereologica **93**(4), 438–441 (2013)
6. Burgio, A., Grippo, F., Pappagallo, M., Crialesi, R.: Hospitalization for drug-related disorders in Italy: trends and comorbidity. Epidemiol. Biostatistics Publ. Health **12**(1), 1–6 (2015)
7. Cofini, V., Carbonelli, A., Cecilia, M.R., Binkin, N., di Orio, F.: Post traumatic stress disorder and coping in a sample of adult survivors of the Italian earthquake. Psychiatry Res. **229**(1–2), 353–358 (2015)
8. Comino, E.J., Hermiz, O., Flack, J., Harris, E., Powell Davies, G., Harris, M.F.: Using population health surveys to provide information on access to and use of quality primary health care. Aust. Health Rev.: Publ. Aust. Hosp. Assoc. **30**(4), 485–495 (2006)
9. Conferenza delle Regioni e delle Provincie Autonome. Accordo Interregionale per la Compensazione della Mobilità Sanitaria, May 2013
10. Davis, K., Anderson, G., Steinberg, E.: Diagnosis related group prospective payment: implications for health care and medical technology. Health Policy (Amst. Neth.) **4**(2), 139–147 (1984)
11. Dusetzina, S.B., Tyree, S., Meyer, A.-M., Meyer, A., Green, L., Carpenter, W.R.: An Overview of Record Linkage Methods. Agency for Healthcare Research and Quality (US), September 2014
12. Elmasri, R., Navathe, S.B.: Fundamentals of Database Systems. Prentice Hall College Div, Boston (2010). 6 har/psc edizione edition
13. Fabiani, L., Mammarella, L., Vittorini, P., Zazzara, F., Di Orio, F.: Preliminary analysis of the respiratory disease, based on hospital discharge records, in the province of L'Aquila, Abruzzo, Italy. Annali Di Igiene: Medicina Preventiva E Di Comunita **28**(6), 392–403 (2016)
14. Gigantesco, A., Mirante, N., Granchelli, C., Diodati, G., Cofini, V., Mancini, C., Carbonelli, A., Tarolla, E., Minardi, V., Salmaso, S., D'Argenio, P.: Psychopathological chronic sequelae of the 2009 earthquake in L'Aquila, Italy. J. Affect. Disord. **148**(2–3), 265–271 (2013)
15. Haas, L., Miller, R., Niswonger, B., Tork Roth, M., Schwarz, P., Wimmers, E.: Transforming heterogeneous data with database middleware: beyond integration. Data Eng. **22**(1), 31–36 (1999)
16. Indinnimeo, L., Porta, D., Forastiere, F., De Vittori, V., De Castro, G., Zicari, A.M., Tancredi, G., Melengu, T., Duse, M.: Prevalence and risk factors for atopic disease in a population of preschool children in Rome: challenges to early intervention. Int. J. Immunopathol. Pharmacol. **29**(2), 308–319 (2016)
17. Watson, J.T., Gayer, M., Connolly, M.A.: Epidemics after natural disasters. Emerg. Infect. Dis. J. **13**(1), 1 (2007)
18. Kawano, T., Hasegawa, K., Watase, H., Morita, H., Yamamura, O.: Infectious disease frequency among evacuees at shelters after the great eastern Japan earthquake and tsunami: a retrospective study. Disaster Med. Publ. Health Preparedness **8**(1), 58–64 (2014)
19. Kouadio, I.K., Aljunid, S., Kamigaki, T., Hammad, K., Oshitani, H.: Infectious diseases following natural disasters: prevention and control measures. Expert Rev. Anti-Infective Therapy **10**(1), 95–104 (2012)

20. MariaDB. CONNECT Table Types, December 2016
21. MariaDB. User-defined Functions, December 2016
22. Mattei, A., Angelone, A.M., Michetti, M., Sbarbati, M., Ceci, R., Murgano, A., di Orio, F.: Epidemiological impact of RV gastroenteritis in the Abruzzo Region: SDO analysis. Annali Di Igiene: Medicina Preventiva E Di Comunita **21**(1), 41–49 (2009)
23. Mattei, A., Sbarbati, M., Fiasca, F., Angelone, A.M., Mazzei, M.C., di Orio, F.: Temporal trends in hospitalization for rotavirus gastroenteritis: a nationwide study in Italy, 2005–2012. Hum. Vaccines Immunotherapeutics **12**(2), 534–539 (2016)
24. Minardi, V., Gigantesco, A., Mancini, C., Quarchioni, E., D'Argenio, P., Cofini, V.: Behavioural risk factors in L'Aquila (Central Italy) 3–5 years after the 2009 earthquake. Epidemiologia E Prevenzione **40**(2 Suppl 1), 34–41 (2016)
25. Ministero della Salute. Sistemi di sorveglianza, May 2016
26. Mohammed, M.A., Stevens, A.: The value of administrative databases. BMJ: Br. Med. J. **334**(7602), 1014–1015 (2007)
27. W. H. Organization. International Statistical Classification of Diseases and Related Health Problems, vol. 1. World Health Organization, Geneva (2004)
28. Parker, J.D., Akinbami, L.J., Woodruff, T.J.: Air pollution and childhood respiratory allergies in the United States. Environ. Health Perspect. **117**(1), 140–147 (2009)
29. Pollice, R., Bianchini, V., Roncone, R., Casacchia, M.: Marked increase in substance use among young people after L'Aquila earthquake. Eur. Child Adolesc. Psychiatry **20**(8), 429–430 (2011)
30. Rahm, E., Bernstein, P.A.: A survey of approaches to automatic schema matching. VLDB J. **10**(4), 334–350 (2001)
31. Reddy, M.P., Prasad, B.E., Reddy, P.G., Gupta, A.: A methodology for integration of heterogeneous databases. IEEE Trans. Knowl. Data Eng. **6**(6), 920–933 (1994)
32. Riedler, J., Braun-Fahrländer, C., Eder, W., Schreuer, M., Waser, M., Maisch, S., Carr, D., Schierl, R., Nowak, R., von Mutius, E., ALEX Study Team.: Exposure to farming in early life, development of asthma, allergy: a cross-sectional survey. Lancet (Lond. Engl.) **358**(9288), 1129–1133 (2001)
33. Ripoll Gallardo, A., Alesina, M., Pacelli, B., Serrone, D., Iacutone, G., Faggiano, F., Della Corte, F., Allara, E.: Medium- and long-term health effects of the L'Aquila earthquake (Central Italy, 2009) and of other earthquakes in high-income Countries: a systematic review. Epidemiologia E Prevenzione **40**(2 Suppl 1), 14–21 (2016)
34. Romanelli, A.M., Raciti, M., Protti, M.A., Prediletto, R., Fornai, E., Faustini, A.: How reliable are current data for assessing the actual prevalence of chronic obstructive pulmonary disease? PloS One **11**(2), e0149302 (2016)
35. Roth, M.T., Schwarz, P.M.: Don't scrap it, wrap it! A wrapper architecture for legacy data sources. In: VLDB, vol. 97, pp. 25–29 (1997)
36. Ruokolainen, L., von Hertzen, L., Fyhrquist, N., Laatikainen, T., Lehtomäki, J., Auvinen, P., Karvonen, A.M., Hyvärinen, A., Tillmann, V., Niemelä, O., Knip, M., Haahtela, T., Pekkanen, J., Hanski, I.: Green areas around homes reduce atopic sensitization in children. Allergy **70**(2), 195–202 (2015)
37. Sheth, A.P., Larson, J.A.: Federated database systems for managing distributed, heterogeneous, and autonomous databases. ACM Comput. Surv. **22**(3), 183–236 (1990)
38. Sorrentino, C.: La dinamica temporale della variabilità regionale, July 2013
39. Sujansky, W.: Heterogeneous database integration in biomedicine. J. Biomed. Inform. **34**(4), 285–298 (2001)

40. Thygesen, L.C., Ersbøll, A.K.: When the entire population is the sample: strengths and limitations in register-based epidemiology. Eur. J. Epidemiol. **29**(8), 551–558 (2014)
41. van der Hulst, A.E., Klip, H., Brand, P.L.P.: Risk of developing asthma in young children with atopic eczema: a systematic review. J. Allergy Clin. Immunol. **120**(3), 565–569 (2007)
42. Williams, H.C.: Clinical practice. Atopic dermatitis. New Engl. J. Med. **352**(22), 2314–2324 (2005)
43. Zuberbier, T., Lötvall, J., Simoens, S., Subramanian, S.V., Church, M.K.: Economic burden of inadequate management of allergic diseases in the European Union: a GA(2) LEN review. Allergy **69**(10), 1275–1279 (2014)

QSAR Classification Models for Predicting Affinity to Blood or Liver of Volatile Organic Compounds in e-Health

Fiorella Cravero[1], María Jimena Martínez[2], Mónica F. Díaz[1], and Ignacio Ponzoni[2(✉)]

[1] Planta Piloto de Ingeniería Química (PLAPIQUI), UNS-CONICET, Complejo CRIBABB, Co. La Carrindanga 7000, CC 717, Bahía Blanca, Argentina
{fcravero,mdiaz}@plapiqui.edu.ar
[2] Instituto de Ciencias e Ingeniería de la Computación (ICIC), UNS-CONICET, Campus Palihue, San Andrés 800, 8000 Bahía Blanca, Argentina
{mjma,ip}@cs.uns.edu.ar

Abstract. In this work, we present Quantitative Structure-Activity Relationship (QSAR) classification models for characterization of molecules affinity to blood or liver for volatile organic compounds (VOCs), using information provided from log P_{liver} measures for VOCs. The models are computed from a dataset of 122 molecules. As a first phase, alternative subsets of relevant molecular descriptors related to the target property are selected by using feature selection methods and visual analytics techniques. From these subsets, several QSAR models are inferred by different machine learning methods. These models allow classifying a new compound as a molecule with affinity to blood, to the liver or equal affinity to both. The model with the highest performance correctly classifies 72.13% of VOCs and has an average receiver operating characteristic area equal to 0.83. As a conclusion, this QSAR model can predict the medium affinity of a VOC, which can help in the development of physiologically based pharmacokinetic computational models required in e-health.

Keywords: VOCs · QSAR modelling · Classification algorithms · Molecular informatics · e-Health

1 Introduction

Indoor air typically contains many volatile organic compounds (VOCs) which are carbon-containing organic chemicals that have high vapour pressure and low water solubility. They come from numerous indoor sources including building materials, furnishings, consumer products, tobacco smoking, etc. Outdoor air is also a source of indoor VOCs. A variety of irritation symptoms is frequently reported by inhabitants related to their periods of occupancy in specific buildings (irritation of eyes, nose, throat, and skin). This increasing evidence results significant, nevertheless, it is insufficient for firm conclusions. It is probably most individual VOCs are not present at

© Springer International Publishing AG 2017
I. Rojas and F. Ortuño (Eds.): IWBBIO 2017, Part II, LNBI 10209, pp. 424–433, 2017.
DOI: 10.1007/978-3-319-56154-7_38

a sufficient concentration in the air of buildings to origin sensory irritation symptoms. However, there is an exception to this for formaldehyde, because it is the best-documented VOC. In addition, some other symptoms such as respiratory symptoms, headaches, and fatigue may be related to concentrations of some specific VOCs presents in buildings, sometimes it is called Sick Building Syndrome. The actual mechanisms for these possible effects are unknown. In spite of all evidence, no federally enforceable standards have been set for VOCs in non-industrial settings, but there is much interest in studying toxicity and human risk estimation. In particular, physiologically based pharmacokinetic (PBPK) modelling is a mathematical modelling method used in health informatics for predicting the absorption, distribution, metabolism and excretion (ADME) of chemical substances in humans and other animal species [1]. The development of PBPK models is a complex task because they require defining submodels based on a high quantity of specific data [2, 3].

For VOCs studies, it is important considering the route of inhalation and, consequently, several respiratory PBPK models have been developed during last decade. In this context, a relevant property under study is the log P_{liver}, which is a blood to liver partition coefficient determined by laborious *in vitro* methods (air:blood and air:liver) [4]. Some computational models for predicting log P_{liver} using Quantitative Structure-Activity Relationship (QSAR) approaches has been proposed in the literature [5, 6]. These regression models have been designed using molecular informatics methods based on machine learning and expert domain knowledge, but none of them is focused on toxicity estimation.

A possible strategy for determining the potential noxiousness of a VOC consists in assessing the compound affinity to a determined medium (oil, water), which can give us a preliminary global view about toxicity coefficients. For this reason, we propose contributing to the PBPK models development for e-health by designing QSAR models for categorising VOCs in terms of their affinity to blood and liver mediums. These prediction models can be obtained by applying classification algorithms provided by machine learning theory. In this work, three well-known classification strategies are explored for inferring the QSAR models: Neural Networks (NNs), Random Forest (RF) and Random Committee (RC), which are freeing available in Weka tool [7]. In next section, the proposed computational methodology, together with VOC data processing, is explained. Then, experimental results obtained by QSAR classifiers are extensively discussed. Finally, main conclusions and future work are summarised.

2 Materials and Methods

The QSAR classification models inferred in this work allow differentiating between compounds related to the liver and those related to blood. Figure 1 describes the methodology carried out. As a first step, a number of molecular descriptors are computed on the VOCs dataset using DRAGON [8]. Once this is done, in order to be able to construct the different models, it is necessary to identify which subsets of the calculated descriptors allow discriminating between the defined classes. In this sense, models with low cardinality (number of descriptors) and straightforward interpretability, in physical-chemical terms, are preferred.

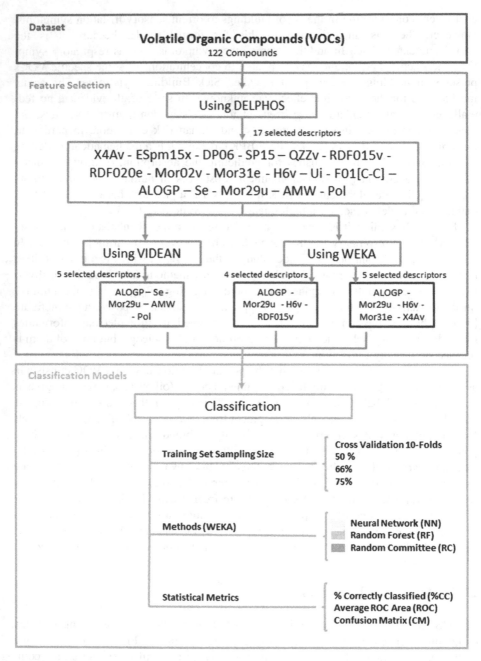

Fig. 1. Methodology and Experimental Design. The colour code used to highlight the subsets of descriptors (Subsets A, B, and C) and the grey scale for the classification methods (NN, RF, RC) correspond to the colours used in Tables 2, 3 and 4 (Color figure online).

For the feature selection stage, DELPHOS tool [9] was used to perform a first ·
selection of descriptors. In this instance, a set of 17 descriptors were obtained. Then,
from this set, a second selection following two different approaches, using the
VIDEAN (Visual and Interactive Descriptor ANalysis) tool [10] and WEKA software
[7], were conducted. In the first case, a subset of 5 descriptors was obtained (Subset A).
In the second case, the selection resulted in two subsets of 4 (Subset B) and 5 (Subset
C) descriptors. For the modelling phase and using this three subsets of descriptors (A,
B and C), different WEKA classification methods were applied: Neural Network (NN),
Random Forest (RF) and Random Committee (RC). Several models were inferred by
varying the size of the training set (50%, 66%, and 75%) and 10-folds cross-validation.
For these models, performance metrics were calculated: the percentage of cases cor-
rectly classified (% CC), average Receiver Operating Characteristic (ROC) area and
confusion matrix (CM).

2.1 Data Processing

The original dataset was taken from Abraham et al. [5]. There are 122 VOCs among
which are hydrocarbons, alkyl halides, alcohols, ethers, esters, ketones, epoxides,
nitriles, halo-benzenes, polycyclic hydrocarbons and benzene derivatives. Each VOC
has associated the *in vitro* blood-to-liver partition coefficient: log P_{liver} (human/rat).
The values of log P_{liver} range from -0.56 to 1.17.

All VOCs structures were drawn using HyperChem [11]. The molecules were
optimised with the same software, in order to find energetically stable conformations.
The structures were pre-optimized with the Force Field Molecular Mechanics (MM+)
procedure. Then, the resulting geometries were further refined by means of the
Semi-Empirical Molecular Orbital Method AM 1 (Austin Model 1) by applying
Polak-Ribiere's algorithm and a gradient norm limit of 0.01 kcal/(Å mol). As a next
step, the HyperChem output files were used by DRAGON [8] to calculate several
classes of descriptors such as: geometrical, constitutional, topological and electrostatic.
Finally, constant descriptors (i.e., variables that take the same value for all samples in
the dataset) and near constants (i.e., variables that take the same value, but allowing
some predetermined small number of samples to take other values) were deleted.

2.2 Thresholds for Target Discretization

We propose in this work to classify the dataset in three groups of VOCs by affinity to
blood (aqueous medium), affinity to the liver (fat medium) and finally they that does
not present a preference between these mediums. A partition coefficient is a ratio
between compound concentrations in each medium after equilibrium has been reached.
In the case of log P_{liver}: high values indicate affinity to fat medium and low values
represent affinity to an aqueous medium.

$$\log Pliver = \log\frac{[VOC\ in\ liver]}{[VOC\ in\ blood]} = \log\frac{affinity\ to\ liver\ (fat\ medium)}{affinity\ to\ blood\ (aqueous\ medium)}$$

A ratio of concentrations equal to 1 indicates the equal medium affinity zone (the same concentration in each medium). Then we considered the range: −0.04 to +0.04 for $\log P_{\text{liver}}$ as "grey zone" class. These values come from the following equations:

$$\log Pliver = \log\frac{[1]}{[1.1]} = -0.04 \qquad \log Pliver = \log\frac{[1.1]}{[1]} = +0.04$$

From this zone, we define the lower values (< −0.04) as "affinity to blood" class and the higher values (> +0.04) as "affinity to liver" class.

In Table 1, it can be observed the distribution of the dataset for this classification. The results are coherent with the nature of compounds (VOCs in the dataset) because high $\log P_{\text{liver}}$ values are presented for non-polar VOCs, and vice versa for polar ones. Please note that the 75% of the dataset has an affinity to the liver; these were the expected result because most of VOCs has low solubility in water.

Table 1. Thresholds for target discretization and percentage of class samples. Three classes are defined in terms of VOCs medium affinity.

Class	Thresholds	Num Molecules	% Molecules
Affinity_to_blood	(-∞ ; -0.04]	15	12.30%
Grey_zone	(-0.04 ; +0.04)	15	12.30%
Affinity_to_liver	[+0.04 ; +∞)	92	75.40%

2.3 Methods and Software Tools

To carry out the different stages of the proposed methodology, four software tools were applied. DRAGON [8] was used to compute the initial set of descriptors. This cheminformatics tool allows calculating thousands of molecular descriptors from a database of compounds. Then, DELPHOS [9] was used for the first instance of descriptors selection. This tool implements a multi-objective optimisation method, based on two phases, to identify alternative relevant subsets of descriptors related to the property under study. In the first phase, a wrapper method is used to perform an exploration within the search space and to find appropriate subsets of descriptors. In the second phase, a final selection is run using the first phase selections and more accurate prediction metrics. Finally, to do the second instance of descriptors selection two tools were used: VIDEAN [10] and WEKA [7].

VIDEAN is a software tool, where domain expertise can be added to the feature selection process by means of an interactive visual exploration of data. Coordinated visual representations are presented for capturing different relationships and interactions among descriptors, target properties and candidate subsets of descriptors. WEKA is a machine learning software that provides a wide variety of algorithms for data analysis and predictive modelling. Different tasks can be done with this tool: data pre-processing, feature selection, regression, and classification, among others.

3 Results and Discussion

In this section, the results obtained for the three different subsets are discussed. These subsets were inferred by feature selection, as explained in the previous section (see Fig. 1). The Subset A is the set of descriptors suggested by means of VIDEAN [10]. On the other hand, from the 17 descriptors selected by DELPHOS, two other subsets were inferred using the WEKA [7]. Wrapper method was used to obtain Subset B, which consists of 4 descriptors, and CfsSubset method was used to obtain Subset C, which consists of 5 descriptors. This is summarised in Table 2.

Wrapper method evaluates feature sets by using a machine learning scheme. Cross validation is used to estimate the accuracy of the learning process for a set of features [12]. CfsSubset is a method that evaluates the worth of a subset of features taking into account the individual predictive ability of each feature along with the degree of redundancy between them. Subsets of features that are highly correlated with the class and that having low intercorrelation are preferred [7].

Table 2. Subsets Inference. The selected descriptors by different feature selection methods are enumerated, according to the colours used in Fig. 1.

Sets		Descriptors
Videan	A (Videan)	ALOGP - Mor29u - AMW - Pol - Se
Weka	B (Wrapper)	ALOGP - Mor29u - H6v - RDF015v
Weka	C (CfsSubset)	ALOGP - Mor29u - H6v - Mor31e - X4Av

Once the subsets were obtained, QSAR classification models were generated with the objective of predict affinity of VOCs to blood or to the liver (see Table 1). The classification models allow interpreting the data differently, like membership to a class. Whereas, a regression model only gives a numerical output, which represents the value of property, in this case, log P_{liver}. However, classification models are not reported in the literature for this property. For this reason, we could not make a performance comparison with other published classification models. The QSAR models were inferred using the following methods: Neural Networks, Random Forest, and Random Committee.

Neural Networks (Multi perceptron): A classifier based on a network that uses backpropagation to classify instances [7].

Random Forest: A classifier based on Random Trees to expand a forest. The Random Trees for inferring a tree considers K randomly chosen features at each node. Performs no pruning. Also, it allows estimation of class probabilities based on a hold-out set (back fitting) [7].

Random Committee: A classifier for generating an ensemble of randomizable base classifiers. Different random number seed is used to build each base classifier. The final prediction is a straight average of all predictions generated [7].

In the experiments, stratified sampling and parameter settings, provided by default for WEKA, were used. Different splitting rates for defining the training and testing set sizes (50/50, 66/34, 75/25) and a 10-folds cross-validation were selected. Several metrics were computed, the percentage of cases correctly classified (%CC), average Receiver Operating Characteristic (ROC) area and confusion matrix (CM). In Table 3, training conditions and these performance metrics are indicated for each dataset.

Correctly Classified (%CC): The percentage of Correctly Classified (%CC) cases is the most simply accuracy measure.

Receiver Operating Characteristic (ROC) area: The concept of a ROC area is based on the notion of a "separator" (or decision) variable. Accuracy is measured by the area under the ROC curve. An area of 1 represents a perfect test; an area of .5 represents a worthless test [13].

Confusion Matrix (CM): In the field of machine learning, more specifically in the problem of statistical classification, a confusion matrix is a specific table layout that allows visualisation of the performance of a classifier. In the matrix, each column represents the instances in an "actual" class, while each row represents the instances in a "predicted" class (or vice versa). The diagonal elements represent the number of samples for which the predicted class is equal to the true class, while off-diagonal elements are those misclassified. When higher the diagonal values of the confusion matrix are, better the classifier performance is [14, 15].

As a first step for defining the best QSAR model, we identify the CMs where for each class most of the samples are correctly classified. Then, we evaluate the %CC totals and finally the value of the average ROC area. Following this criterion, we recommend as best classification model, the one inferred with 50/50 as the size of the training and test set, for Subset A using RC. This model and its statistical metrics are highlighted in Table 3.

In Tables 4 and 5, per each subset and learning algorithm, averages of the metrics computed for the different data sampling are shown. From these tables, it is clear that during the analysis of results, no selected subset of molecular descriptors achieves a predominant performance in terms of %CC and ROC area. Nevertheless, the QSAR classification models obtained by Neural Networks (NN) evidenced, in average, a superior accuracy respect to the other approaches.

Table 3. Performance metrics of the QSAR classification models inferred by Neural Networks (NN), RF and RC from Subsets A, B, and C.

Cross-validation 10-folds

		% CC	ROC	CM			
Subset A	NN	71.31	0.839	2	3	10	blood
				4	0	11	grey
				6	1	85	liver
	RF	68.85	0.820	4	4	7	blood
				5	2	8	grey
				6	8	78	liver
	RC	63.93	0.763	2	4	9	blood
				6	2	7	grey
				12	8	74	liver
Subset B	NN	69.67	0.833	0	5	10	blood
				3	1	11	grey
				4	4	84	liver
	RF	71.31	0.785	5	2	8	blood
				3	4	8	grey
				6	8	78	liver
	RC	68.85	0.734	3	2	10	blood
				4	5	6	grey
				8	8	76	liver
Subset C	NN	71.31	0.790	0	6	9	blood
				2	4	9	grey
				5	4	83	liver
	RF	71.31	0.760	3	2	10	blood
				3	4	8	grey
				5	7	80	liver
	RC	68.03	0.733	3	4	8	blood
				4	5	6	grey
				10	7	75	liver

50 % Training

		% CC	ROC	CM			
Subset A	NN	67.21	0.837	2	0	7	blood
				3	0	5	grey
				4	1	39	liver
	RF	72.13	0.840	4	1	4	blood
				3	2	3	grey
				5	1	38	liver
	RC	**72.13**	**0.826**	5	1	3	blood
				3	3	2	grey
				5	3	36	liver
Subset B	NN	70.49	0.857	5	0	4	blood
				5	0	3	grey
				6	0	38	liver
	RF	70.49	0.809	3	0	6	blood
				1	0	7	grey
				1	3	40	liver
	RC	68.85	0.738	1	4	4	blood
				0	3	5	grey
				2	4	38	liver
Subset C	NN	73.77	0.810	4	0	5	blood
				2	0	6	grey
				2	1	41	liver
	RF	73.77	0.832	3	0	6	blood
				1	1	6	grey
				1	2	41	liver
	RC	70.49	0.693	4	1	4	blood
				1	2	5	grey
				2	5	37	liver

66 % Training

		% CC	ROC	CM			
Subset A	NN	68.29	0.892	0	3	4	blood
				0	1	3	grey
				0	3	27	liver
	RF	65.85	0.794	0	2	5	blood
				1	2	1	grey
				1	4	25	liver
	RC	70.73	0.788	1	2	4	blood
				1	3	0	grey
				1	4	25	liver
Subset B	NN	68.29	0.889	0	3	4	blood
				0	2	2	grey
				0	4	26	liver
	RF	63.41	0.828	0	2	5	blood
				0	1	3	grey
				1	4	25	liver
	RC	65.85	0.758	0	2	5	blood
				1	2	1	grey
				2	3	25	liver
Subset C	NN	70.73	0.901	0	6	1	blood
				0	3	1	grey
				0	4	26	liver
	RF	65.85	0.786	0	3	4	blood
				0	2	2	grey
				2	3	25	liver
	RC	65.85	0.749	0	3	4	blood
				0	2	2	grey
				2	3	25	liver

75 % Training

		% CC	ROC	CM			
Subset A	NN	73.33	0.932	0	1	4	blood
				0	0	3	grey
				0	0	22	liver
	RF	66.67	0.786	1	0	4	blood
				1	1	1	grey
				1	3	18	liver
	RC	70.00	0.721	2	2	1	blood
				1	1	1	grey
				1	3	18	liver
Subset B	NN	73.33	0.899	0	4	1	blood
				0	2	1	grey
				0	2	20	liver
	RF	66.67	0.771	1	0	4	blood
				0	1	2	grey
				1	3	18	liver
	RC	66.67	0.647	1	1	3	blood
				1	1	1	grey
				1	3	18	liver
Subset C	NN	73.33	0.913	0	2	3	blood
				0	2	1	grey
				0	2	20	liver
	RF	70.00	0.800	1	1	3	blood
				0	1	2	grey
				1	2	19	liver
	RC	66.67	0.586	1	1	3	blood
				0	1	2	grey
				1	3	18	liver

Table 4. The average percentage of correctly classified (%CC) samples. In columns two, three and four the average %CC per each subset using the different machine learning methods is reported. Higher averages per subset and machine learning method are highlighted in bold.

	%CC			
	Subset A	Subset B	Subset C	Total Average
Average NN	70.04	70.45	72.29	70.92
Average RF	68.38	67.97	70.23	68.86
Average RC	69.20	67.56	67.76	68.17
Total Average	69.20	68.66	**70.09**	

Table 5. Average of receiver operating characteristic (ROC) areas. In columns two, three and four the average ROC per each subset using the different machine learning methods is reported. Higher averages per subset and machine learning method are highlighted in bold.

	ROC area			
	Subset A	Subset B	Subset C	Total Average
Average NN	0.88	0.87	0.85	**0.87**
Average RF	0.61	0.80	0.79	0.73
Average RC	0.77	0.72	0.69	0.73
Total Average	0.75	**0.80**	0.78	

4 Conclusions

In recent years, several regression models have been reported to predict log P_{liver} in VOCs using QSAR approaches. Of these models, none focused on trying to determine the noxiousness of a VOC. This motivated us to develop a computational model that may provide a preliminary understanding about potential toxicity coefficients. For this task, we design different QSAR classification models that allow predicting if a compound will have an affinity to blood or liver. As a first step, different approaches of feature selection were applied to determine three subsets of molecular descriptors relevant for the property under study. For each of these subsets, several classification methods were used: Neural Networks, Random Forest, and Random Committee, with different validation strategies. After a deep analysis of the results, we decided to report as the best model the one with the highest performance in terms of VOCs correctly classified by the different types of affinities.

In brief, in this paper, we propose a QSAR classification model that can predict the affinity of VOCs to blood or liver mediums, and which may be applied in physiologically based pharmacokinetic (PBPK) modelling for e-health. Another contribution is the definition of the medium-affinity classes. As far as we know, this is the first QSAR classification model derivate from log P_{liver} proposed in the literature. Therefore, our rationale behind the thresholds cut-off criteria can be useful for other QSAR studies related to this property. As future work, we plan to extend the VOCs database and explore other classification methods.

Acknowledgments. This work is kindly supported by CONICET, grant PIP 112-2012-0100471 and UNS, grant PGI 24/N042.

References

1. Buist, H.E., de Wit-Bos, L., Bouwman, T., Vaes, W.H.: Predicting blood: air partition coefficients using basic physicochemical properties. Regul. Toxicol. Pharmacol. **62**(1), 23–28 (2012)
2. Sager, J.E., Yu, J., Raguenau-Majlessi, I., Isoherranen, N.: Physiologically based pharmacokinetic (PBPK) modeling and simulation approaches: a systematic review of published models, applications and model verification. Drug Metab. Dispos. **43**, 1823–1837 (2015)
3. Vork, K., Carlisle, J., Brown, J.P.: Estimating Workplace Air and Worker Blood Lead Concentration using an Updated Physiologically-based Pharmacokinetic (PBPK) Model: Office of Environmental Health Hazard Assessment. California Environmental Protection Agency (2013)
4. Dashtbozorgi, Z., Golmohammadi, H.: Prediction of air to liver partition coefficient for volatile organic compounds using QSAR approaches. Eur. J. Med. Chem. **45**(6), 2182–2190 (2010)
5. Abraham, M.H., Ibrahim, A., Acree, W.E.: Air to liver partition coefficients for volatile organic compounds and blood to liver partition coefficients for volatile organic compounds and drugs. Eur. J. Med. Chem. **42**(6), 743–751 (2007)
6. Palomba, D., Martínez, M.J., Ponzoni, I., Díaz, M.F., Vazquez, G.E., Soto, A.J.: QSPR models for predicting log pliver values for volatile organic compounds combining statistical methods and domain knowledge. Molecules **17**(12), 14937–14953 (2012)
7. Hall, M., Frank, E., Holmes, G., Pfahringer, B., Reutemann, P., Witten, I.H.: The WEKA data mining software: an update. In: ACM SIGKDD Explorations Newsletter, 11(1), pp. 10–18 (2009)
8. DRAGON 5.5 for Windows (Software for Molecular Descriptor Calculations), Version 5.5. Talete srl, Milan, Italy (2007)
9. Soto, A.J., Cecchini, R.L., Vazquez, G.E., Ponzoni, I.: Multi-objective feature selection in QSAR using a machine learning approach. QSAR Comb. Sci. **28**(11–12), 1509–1523 (2009)
10. Martínez, M.J., Ponzoni, I., Díaz, M.F., Vazquez, G.E., Soto, A.J.: Visual analytics in cheminformatics: user-supervised descriptor selection for QSAR methods. J. Cheminform. **7**(1), 1 (2015)
11. HyperChem (TM), Molecular Modeling System, Release 8.0.7 for Windows. Hypercube, Inc., Gainesville, FL, USA (2009)
12. Kohavi, R., John, G.H.: Wrappers for feature subset selection. Artif. Intell. **97**(1), 273–324 (1997)
13. Hajian-Tilaki, K.: Receiver operating characteristic (ROC) curve analysis for medical diagnostic test evaluation. Caspian J. Intern. Med. **4**(2), 627 (2013)
14. Stehman, S.V.: Selecting and interpreting measures of thematic classification accuracy. Remote Sens. Environ. **62**(1), 77–89 (1997)
15. Powers, D.M.: Evaluation: from precision, recall and F-measure to ROC, informedness, markedness and correlation (2011)

Usage of VR Headsets
for Rehabilitation Exergames

Martina Eckert[✉], José Zarco, Juan Meneses,
and José-Fernán Martínez

Centro de Investigación en Tecnologías Software y Sistemas
Multimedia para la Sostenibilidad (CITSEM),
Campus Sur Universidad Politécnica de Madrid (UPM), Madrid 28031, Spain
{Martina.eckert,juan.meneses,jf.martinez}@upm.es,
josezarcotorres@gmail.com

Abstract. The work presented here is part of a large project aimed at finding new ways to tackle exergames used for physical rehabilitation. The preferred user group consists of physically impaired who normally cannot use commercially available games; our approach wants to fill a niche and allow them to get the same playing experience like healthy. Four exercises were implemented with the Blender Game engine and connected to a motion capture device (Kinect) via a modular middleware. The games incorporate special features that enhance weak user movements, such that the avatar reacts in the same way as for persons without physical restrictions. Additionally, virtual reality glasses have been integrated to achieve a more immersive feeling during play. In this work, we compare the results of preliminary user tests, performed with and without VR glasses. Test outcomes are good for motion amplification in some of the games but do not present generally better results when using the VR glasses.

Keywords: Kinect · Motion capture · Virtual reality · Rehabilitation · Exergame · Head Mounted Display · Physical disability · Motor function impairments

1 Introduction

The main goal of our project is to improve the player's sensation of immersion and involvement while playing an exergame, with the special purpose of motivating people who suffer from severe physical impairments and chronical diseases. This is a great challenge, because people notice their physical impairments a lot when using games that are made for the broad mass. Our efforts are focused on implementations that let them forget about their restrictions and have fun, while performing movements that are normally arduous and exhausting but necessary, e.g. for rehabilitation or training. In this way, two objectives can be reached: entertainment and sports for the user.

To achieve this goal, we are currently developing an adventure-like exergaming environment that integrates the Kinect camera with the Blender Game Engine (BGE). The middleware that manages the communication between both is integrated in a modular container that also allows including other devices, such as an Android mobile phone and a Virtual Reality (VR) headset.

© Springer International Publishing AG 2017
I. Rojas and F. Ortuño (Eds.): IWBBIO 2017, Part II, LNBI 10209, pp. 434–442, 2017.
DOI: 10.1007/978-3-319-56154-7_39

The Kinect camera has been proven an excellent medium to promote physical exercises [1]. The problem is, though, that physically impaired people cannot directly play the games offered on the market, which is stated frequently by affected users [2]. Therefore, the exergaming environment underlying the presented work is equipped with special functionalities that cope with physical limitations, e.g. the amplification of weak movements. This way, the player's virtual experience is enhanced and motivation increased. The objective of this work is to find out if the user experience would be even better when playing with a Head Mounted Display (HMD) that immerges the player fully into the game. Our assumption is that more immersion results in longer playing times and more effective treatments. This, for sure, should be tested in long term tests, here we just want to test the suitability for disabled people of wearing a headset and their performance in comparison to using a standard monitor. Four Mini-Games have been implemented to test different movements which are useful for wheelchair users suffering from degenerative muscle diseases. They have been tested for a group of disabled and a control group of healthy people.

Related Work. Compared to EyeToy or Wii, the Kinect seems to be the more natural device and has a great potential to create enjoyable exercises with a low budget [3]. The authors of [4], which is the newest review found, claim that "technologies such as the Microsoft Kinect have the potential to incorporate complex and continually adaptive exercises requiring specific movements and track the extent to which these movements are indeed performed by the players". Some examples of works similar to our approach are mentioned in the following.

REWIRE (Rehabilitative Way out in Responsive home Environments) [5] is a European project that develops a VR-based rehabilitation platform for home exercises. Multiple publications about adaptive games can be found from the consortium members [6, 7]. "Kinect-o-Therapy" was presented by [8]. The authors aimed to present a system that combines entertainment with exercise to motivate patients. The system is also based on four Mini-Games, implemented with the Unity 3D game engine. The authors of [9] propose one of the few systems found for children with disabilities. It is aimed at assisting patients with spastic diplegia and hemiparesis in their rehabilitation process.

Very few literature has been found about works that combine motion capture systems with VR for exergames. The authors of [10] proposed a system for gamifying physical therapy for stroke survivors with an immersive 3D environment. It combines the Kinect, an Oculus Rift goggle and a pair of haptic gloves, which adds the assessment of hands and fingers that the Kinect does not provide. In [11], an exergame environment for cycling is presented that combines the Kinect and an HMD, but the focus is clearly set on the immersive experience and not on motor rehabilitation purposes. Finally, the authors of [12] present "Astrojumper", an immersive VR exergame developed to motivate players to engage in rigorous, full-body exercise. Nevertheless, these exercises are not meant for people with physical impairments.

Concluding, literature reveals that exergames realized with motion capture systems are promising and powerful, but still most works are aimed at restricted groups of patients. Also, to the best of our knowledge, real immersion with VR glasses has not been tested with disabled people, which means that the here presented work is the first of its kind.

2   The Exergaming Environment

The system is based on a modular middleware that joins different sensors, details can be found in [13–15]. The sensors currently included are a motion capture camera (Kinect), a smart phone and an HMD. Others, like e.g. a heart rate detector, could be easily added. The data captured by the sensors is transmitted via their middleware to a 3D video game engine. Here, the movements captured by Kinect are used to control an avatar, and the mobile phone enables wrist rotations not detectable by the camera.

The Kinect middleware has been implemented using version 1.8 of the Kinect for Windows SDK [16]. The spatial positions of 20 joints and the rotations of their bones are captured and transmitted via the OSC (Open Sound Control) protocol [17] to the game engine. The communication module for the HMD, Oculus Rift [18], is based on the OSVR (Open Source Virtual Reality) framework [19]. For game development, the Blender Game Engine (BGE) has been chosen [20]. New functionalities have been integrated as add-ons, as there are three different types of skeletons, auxiliary objects used to distinguish different types of movements and the virtual camera view for the HMD, as shown in Fig. 1 ("New HMD").

Fig. 1.   Add-on (left), camera module attached to the avatar (right).

An internal script is requesting the position and rotation data of the HMD from the middleware 60 times/s and applies it to the virtual camera in real time. In this way, a stereovision image is created by the BGE that corresponds to the avatar's view. It is then sent through the middleware to the glasses, displaying the view to the player. Figure 1 (right side) shows the avatar, with the camera attached to his head, the corresponding view is contained in Fig. 2.

Four mini-games have been implemented for testing the suitability of some selected movements as well as the special functionalities to enhance weak gestures.

Mini-Games Description. The implemented exercises should represent a variety of basic movements, which could be tested by wheelchair users with different diseases, so only upper limb movements have been included. According to our physiotherapist, the aim is to train basic corporal functionalities necessary to perform daily activities and improve quality of life. Therefore, exercises that require lateral body and arm movements with shoulder flexion have been implemented: rowing, climbing, hitting, and

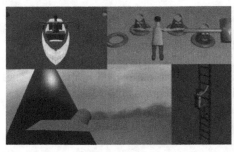

Fig. 2. Four mini-games in mono (left) and stereo view (right) as sent to the VR glasses. From left to right, top down. Left: The Boat, "Whack-a-mole", the Paper-Bird, and the Ladder. Right: the Ladder, the Boat, "Whack-a-mole", and the Paper-Bird.

flying. In the future, further exercises for more precise wrist and forearm movements could be added. To analyze the different experiences, all games have been tested with and without the VR glasses; a visualization of the screens is given in Fig. 2 as normal and VR view.

The Ladder (climbing) challenges the player to reach the end of a long rung ladder by moving the arms (or just hands) up and down alternately. The avatar's arm is copying the user's movements until the moment the user reaches his personal upper limit. At this point, an animation is executed: the avatar grabs the bar and climbs a step, releasing the opposite hand. A counter measures the time needed to get up to the end. In VR view, the player only sees the next rung and the avatar's arms.

The Boat (rowing) shows the avatar inside a rowing boat with the rudders fixed to the hands. Its arms are reacting to the user's simultaneous forward-backwards arm or hand movement. When performed correctly, the boat progresses some meters (simulated by buoys moving past). The aim is to reach the goal as quickly as possible, the time is clocked and appears on the screen. Through the VR glasses, the player sees the frontal part of the boat, the water, and the hands of the avatar when they are moving in front of the body. When the finishing line is reached, a banner appears.

Whack-a-mole (hitting) is an imitation of the widely known homonymous game, where little moles, appearing randomly out of their holes, have to be caught. The player has to move the right arm up and down while pointing in the right direction. If a mole is hit with sufficient speed, it emits a funny suffering sound and disappears. The game gives two minutes of time to strike 20 moles, the number of scored moles is counted. In VR view, the player sees the holes quite near and has not the complete overview of the scene so that it is necessary to turn around more, which implies more exercise.

The Paper-Bird (flying) is the most sophisticated game regarding the movements to perform. The player is conducting a kind of hand-crafted bird that is progressing constantly forward by using arms and trunk. With the arms spread to both sides, the bird is flying straight ahead, lowering one arm produces a sideways turn. The forward and backward motion of the trunk is additionally causing an up and down movement of the bird. The aim is to fly the bird through some yellow rings scattered in the air, without any time limit. In VR view, the player does not see anything of the bird, just the sky, the landscape and the rings.

3 Tests and Results

19 participants (12 male), including 12 individuals with different disabilities, all but two wheelchair users (ages 5–50 years, see Table 1) and 8 children (ages 7–13 years) without any known physical impairments, volunteered to participate in the tests. The participants were recruited in Madrid, Spain, at the Neurological Muscular diseases Association (ASEM) and the Sports Integration Foundation (Fundación TAMBIEN). Approval was obtained from the Review Board at UPM (Universidad Politécnica de Madrid, Spain).

Table 1. Participant information of the target group

Disease	Ages (gender)
SMA (Spinal Muscular Atrophy) type 2	5 (m), 13 (m)
CP (Cerebral Palsy)	11 (m), 12 (m)
Hypertonia	12 (m)
BMD (Becker Muscular Dystrophy)	13 (m)
DMD (Duchenne Muscular Dystrophy)	15 (m), 16 (m)
FSH (Facioscapulohumeral Muscular Dystrophy)	43 (f), 49* (f)
PPS (Post-polio Syndrome)	50* (f)

* No wheelchair

The procedure of the trials has been as follows: Following informed consent, the volunteers (and their parents in case of children) were shortly introduced into our work and the reasons why we need them for the first trials were explained. Before playing, the maximum possible limb movements were recorded to calibrate the system. Then, the users played at first the Mini-Games with a standard monitor of 30 in, from a distance of about 1–2 m, to be in the best detection range for the Kinect. The second session was using the virtual headset. At the end, all participants were asked to fill out a short survey to learn about their experience.

Objective Playing Results. In case of wheelchair users, the Kinect occasionally confused the armrest with the user's arms or the wheels with the user's legs, especially in case of small kids. Therefore, three subjects (both SMA and one CP) had to be excluded from the evaluation. The playing order was always the same: The Ladder, The Boat, Whack-a-mole and The Paper-Bird. Table 2 presents the average results for each game, Fig. 3 visualizes the distribution of individual times and scores.

Particular observations for each of the games were as follows:

The Ladder. The average times are nearly the same for tests performed with a standard monitor. Instead, using the headset, the target group (TG) performed slower than the control group (CG). This means, that the difficulty for both groups seemed to be similar, but the impaired participants struggled more with the VR view. Problems of skeleton detection and game reaction got more evident when the headset was used, this, in turn, led to more confusion, especially in combination with the 3D-experience.

Table 2. Average values per game

	Ladder	Boat	Whack-a-mole	Paper-Bird		
	Time to goal		Success rate	Time	N° rings	
Target group	54.0 s	119.8 s	53%	200.6 s	2.6	Monitor
Control group	55.7 s	75.6 s	81%	124.0 s	5	
Target group	67 s	122 s*	44%	320 s**	4**	HMD
Control group	49 s	71 s	83%	–	7	

*5 users

**3 users

The Boat. While this game was an easy exercise for the CG, some TG users had difficulties with the movement as the configuration turned out not to be sufficiently flexible to capture slightly different movements to the expected ones. Therefore, the TG needed about 50 s longer than the CG when playing with a HMD and 44 s longer when playing with a monitor. Some users were not able to make the game react in the right way in spite of having learned to play it formerly with the monitor view. One user quit the game due to boredom.

Whack-a-mole. As opposed to the other games, this one needs precision and good reaction times. Generally, the monitor view was easier to handle, because the whole scene is visible. In VR view, users found it difficult to find the appearing moles, however, two participants had more success than with the monitor version. With the monitor, the TG achieved an average score of 53%, while the CG got 81%. With the HMD, the TG fell down to 44% and the CG improved up to 83%. All participants finished the game.

The Paper-Bird. This game requires the most complicated body control and is even more difficult to play with VR headset, as the viewing range is restricted and the head control has to be added to the arm and body movements. Only 3 users of the TG played this game with the headset but longer than with the monitor, which can be evaluated as positive outcome as this game has no time limit like the Ladder and the Boat. If players endure longer, it means that they have fun and are motivated to achieve passing through the rings.

Overall, the users of the CG achieved generally better results in the games with HMD than with the standard monitor, while the target users had more difficulties.

Subjective Survey Results. In the survey, the participants awarded 0 to 5 points to general aspects like diversion factor, the ease to play (game response) and aesthetics of each game. Regarding their experience with the VR glasses, the questions were focused on the feelings and if the users found it worth to use the headset.

The average evaluations for those questions are illustrated in Fig. 4 separately for both groups. The Ladder had the best acceptation for the TG, but all users commented the experience as positive. The Boat was the least funny one. Users complained about the VR experience because they saw a detached arm. The view was more monotonous than on the monitor. In Whack-a-mole, all participants stated that they had fun, and liked the action. The Paper-Bird had the worst acceptation for both groups, the

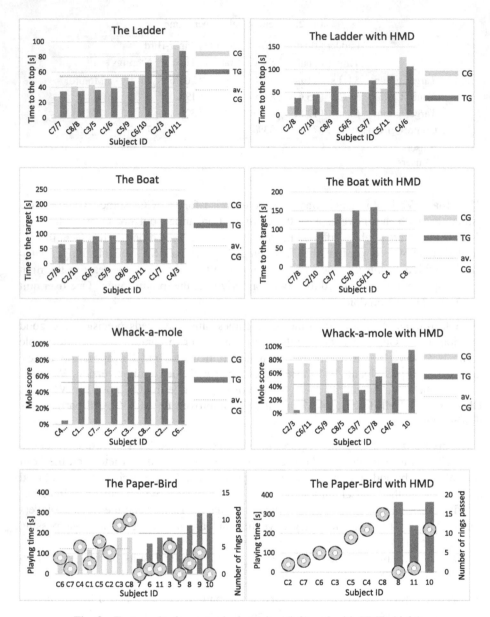

Fig. 3. Test results for a standard monitor (left) and with HMD (right)

movements were generally felt to be little intuitive. However, it wins in aesthetical aspects.

Nobody of the wheelchair users had any sensation of dizziness, whereas the standing users felt some in the Paper-Bird. This could be due to a longer playing time and the change of the vision field provoked by the head movements, which are stronger than seated and quicker than those of the impaired participants.

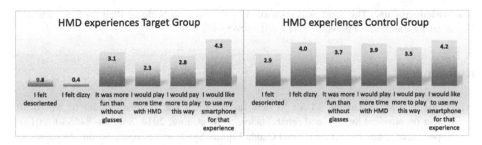

Fig. 4. Evaluation of the VR experiences (scoring from 0 to 5)

The CG had more fun playing with the headset and was more likely to spend time and money for using VR glasses than the TG, although both prefer to use a smartphone if possible (due to the elevated price of VR glasses).

4 Conclusions

We are conscious of the fact that the group of tested persons is too inhomogeneous and the time too short to obtain meaningful results about the effectiveness of the games for the rehabilitation of certain affectations, but this was not the aim at this stage.

First, we wanted to learn about different types of physical restrictions apparent in different diseases, to see how long the disabled endure and if the selected movements are correctly detected depending on the different abilities. The outcome is that most games responded well to the tested users. However, users with muscle weakness in the arms got tired more quickly and reached much lower scores. Overall, the exercises were adequate, but the arm movements required in The Boat seemed to be difficult to perform and the detection rate was low if the movement was not performed sufficiently wide.

Second, we wanted to evaluate the interest of potential users in exergames of the presented characteristics by surveying their experience and observing their reactions. All users had fun and evaluated the games in general positively. Obviously, they liked the action game and the complicated flying game most, as they were highly motivated to pass through the rings. The most "boring" one was The Boat.

The third and most important objective was to test if the VR view has some kind of negative effect like dizziness, if the users are comfortable with the HMD and if they liked the experience more. The outcome was that standing users had more problems with dizziness than wheelchair users. Some complained about the restricted view and implementation errors. Nevertheless, the control group stated that they liked the VR experience more and showed a better game performance with the head set. The target group performed worse than using a standard monitor in spite of the flying game.

Altogether, the experience is positive, but the usefulness of the HMD for a better performance in the physical exercises could not be confirmed. The increased motivation could generally lead to longer playing times, which could lead to an improvement in fitness, but it is not clear if the users would improve the quality of movements. Polishing of the implementation and 3D visualization is necessary, as well as long-term studies, to find this out.

Acknowledgements. This work was sponsored by the Spanish National Plan for Scientific and Technical Research and Innovation: TEC2013-48453-C2-2-R.

References

1. Hondori, H.M., Khademi, M.A.: Review on technical and clinical impact of Microsoft Kinect on physical therapy and rehabilitation. J. Med. Eng. **2014**, 846514–846530 (2014)
2. Apparelyzed. Spinal Cord Injury and Cauda Equina Syndrome Support Forum. http://www. apparelyzed.com/forums/topic/25204-xbox-360-best-kinect-game-for-wheelchair-users-fable-the-journey
3. Webster, D., Celik, O.: Systematic review of Kinect applications in elderly care and stroke rehabilitation. J. Neuroeng. Rehabil. **11**(108), 1–24 (2014)
4. Skjæret, N., et al.: Exercise and rehabilitation delivered through exergames in older adults: an integrative review of technologies, safety and efficacy. Int. J. Med. Inform. **85**, 1–16 (2016)
5. Borghese, P.N.A.: REWIRE. https://sites.google.com/site/projectrewire/
6. Pirovano, M. et al.: Self-adaptive games for rehabilitation at home. In: IEEE Conference on Computational Intelligence and Games, pp. 179–186 (2012)
7. Borghese, N.A., Mainetti, R., Pirovano, M., Lanzi, P.L.: An intelligent game engine for the at-home rehabilitation of stroke patients. In: IEEE 2nd International Conference on Serious Games and Applications for Health, pp. 1–8 (2013)
8. Roy, A.K., Soni, Y., Dubey, S.: Enhancing effectiveness of motor rehabilitation using kinect motion sensing technology. In: IEEE Global Humanitarian Technology Conference: South Asia Satellite (GHTC-SAS), pp. 298–304 (2013)
9. Meleiro, P., Rodrigues, R., Jacob, J., Marques, T.: Natural user interfaces in the motor development of disabled children. Procedia Technol. **13**, 66–75 (2014)
10. Kaminer, C., LeBras, K., McCall, J., Phan, T., Naud, P., Teodorescu, M., Kurniawan, S.: An immersive physical therapy game for stroke survivors. In: Proceedings of the 16th International ACM SIGACCESS Conference on Computers and Accessibility, ASSETS 2014 (2014)
11. Shaw, L.A., et al.: Challenges in virtual reality exergame design. In: 16th Australasian User Interface Conference, Sydney (2015)
12. Finkelstein, S., et al.: Astrojumper: motivating exercise with an immersive virtual reality exergame. Presence Teleoperators Virtual Environ. **20**, 78–92 (2011). MIT Press
13. Eckert, M., Gómez-Martinho, I., Meneses, J., Martinez, J.F.: A multi functional plug-in for exergames. In: International Symposium on Consumer Electronics, Madrid, pp. 4–6 (2015)
14. Eckert, M., Gómez-Martinho, I., Meneses, J., Martínez, J.F.: A modular middleware approach for exergaming. In: IEEE International Conference on Consumer Electronics, Berlin, pp. 172–176 (2016)
15. Eckert, M., Gómez-Martinho, I., Meneses, J., Martínez, J.F.: New approaches to exciting exergame-experiences for people with motor function impairments. Sensors **17**(15), 1–22 (2017)
16. Kinect for Windows SDK v1.8. https://www.microsoft.com/en-us/download/
17. Introduction to OSC. http://opensoundcontrol.org/introduction-osc
18. Oculus Rift V.R. https://www.oculus.com/
19. OSVR Developer Portal. http://osvr.github.io
20. Blender Foundation Blender.org. http://www.blender.org/

High-Throughput Bioinformatic Tools for Genomics

Search of Regions with Periodicity Using Random Position Weight Matrices in the Genome of *C. elegans*

E.V. Korotkov[1,2(✉)] and M.A. Korotkova[2]

[1] Institute of Bioengineering, Research Center of Biotechnology
of the Russian Academy of Sciences,
Leninsky Ave. 33, bld. 2, 119071 Moscow, Russia
genekorotkov@gmail.com
[2] National Research Nuclear University "MEPhI",
Kashirskoe shosse, 31, Moscow 115409, Russia

Abstract. A mathematical method was developed in this study to determine tandem repeats in a DNA sequence. A multiple alignment of periods was calculated by direct optimization of the position-weight matrix (PWM) without using pairwise alignments or searching for similarity between periods. Random PWMs were used to develop a new mathematical algorithm for periodicity search. The developed algorithm was applied to analyze the DNA sequences of C. elegans genome. 25360 regions having a periodicity with length of 2 to 50 bases were found. On the average, a periodicity of ~ 4000 nucleotides was found to be associated with each region. A significant portion of the revealed regions have periods consisting of 10 and 11 nucleotides, multiple to 10 nucleotides and periods in the vicinity of 35 nucleotides. Only $\sim 30\%$ of the periods found were discovered early. This study discussed the origin of periodicity with insertions and deletions.

Keywords: Period · Sequence · Random matrix · Alignment

1 Introduction

Periodicity is one of the structural regularities of sequences and is widely represented in amino and DNA sequences [1]. A periodicity is considered as latent, if the similarity between any two periods is not statistically significant or if it belongs to the twilight zone [2]. Perfect periodicity can become latent periodicity, if it accumulates over 1.0 mutation per nucleotide in the studied DNA sequence [3]. The distinctive property of latent periodicity is that it cannot be detected by pairwise comparisons of nucleotide sequences. However, latent periodicity can be found if a mathematical method is applied to directly detect the alignment of nucleotide sequences without constructing pairwise alignments. The periods of a sequence with latent periodicity are sequences for multiple alignment that is statistically significant. The aim of this study was to

---

This work was supported by Russian Science Foundation.

I. Rojas and F. Ortuño (Eds.): IWBBIO 2017, Part II, LNBI 10209, pp. 445–456, 2017.
DOI: 10.1007/978-3-319-56154-7_40

develop a mathematical method which allows finding the periodicity of DNA sequences as well as latent periodicity.

At present, there is a significant gap in the mathematical approaches developed in search for periodicities in symbolic and numeric sequences (sequence-based methods). Spectral approaches enable the finding of adequate "fuzzy" periodicity in nucleotide sequences without the insertion(s) or deletion(s) of nucleotides. Fourier transform, Wavelet transform, information decomposition and some other methods can be attributed to the number of spectral methods [4, 5]. However, these approaches have a significant limitation – they do not allow the detection of a periodicity with insertions and deletions.

On the other hand, methods based on pairwise alignment and HMM can accurately find insertions and deletions [6–8]. However, these methods cannot detect a latent periodicity, in a situation where the statistical significance of similarity between any two periodic sequences is small [1, 9]. This is due to the fact that the periodicity of DNA sequences (with the number of periods greater than or equal to 4) is detected by pairwise similarity between periods. In the absence of statistically significant pairwise similarity, these approaches are incapable of finding latent periodicity. First, it involves algorithms and programs, such as TRF [6], Mreps [10], TRStalker [11], ATRHunter [12], T-REKS [13], IMEX [14, 15], CRISPRs [16], SWAN [17] and some others [18, 19], because the similarity between different periods is very low in the case of latent periodicity. This leads to lack of seeds and identical short strings. Therefore, this study proposes a mathematical method that considers this gap and finds the latent periodicity of any symbolic sequence in the presence of insertions and deletions (in unknown positions of the analyzed sequence) and in the absence of a known position-weight matrix (PWM).

Any periodicity of the sequence S with length N can be characterized by either the frequency matrix [20] or created on its base, the PWM M [21]. The signs of the rows of this matrix are nucleotides, and the signs of the columns are the positions of the period. The element of this matrix $m(i, j)$ indicates the weight $m(i, j)$ which has the nucleotide i in position j of the period. The positions of the period vary from 1 to n. The sequence S_1 of length N, which is an artificial periodic sequence 1, 2, ..., n, 1, 2, ..., n, ... is introduced. Here, the numbers are treated as symbols and columns in the matrix M are consistent with them. For period equal to n, the sequence S corresponds to a certain frequency matrix and PWM $M(4, n)$. The problem is formulated as follows: This study has a sequence S with length N. It is necessary to find such optimal PWM M_0, where the local alignment [2] of sequences S_1 and S have the greatest statistical significance. Under the statistical significance, the probability P is that $F_r > mF_{max}$, where mF_{max} is the maximum weight of a local alignment of sequences S and S_1, using the optimal matrix M_0. Here, F_r represents the maximum weight of a local alignment randomly mixed sequence S and sequence S_1, using the optimal matrix M_r. The search is for matrix M_0, which has the lowest probability P. It is always possible to set the threshold level of the probability P_0 and if the probability $P(F_r > mF_{max})$ will be less than P_0, then the local alignment found of sequences S and S_1, using the optimum matrix M_0 can be considered as statistically significant.

It is possible to use a local alignment algorithm for alignment of the nucleotide sequence S and an artificial periodic sequence S_1, relative to the known PWM [22]. It is necessary to find the optimal PWM M_0 by any means. Therefore, the aim of this study was to develop a mathematical approach for finding the matrix M_0, as well as a method for assessing the probability P. To determine the optimal PWM, an optimization procedure was used, as well as a local alignment algorithm. To estimate the probability P, the Monte Carlo method was used.

A mathematical method was developed in this study to find more than 4 tandem repeats in the DNA sequence. The multiple alignment of periods was calculated by direct optimization of the PWM without using pairwise alignments or a search for similarity between periods. This means that for each n, a matrix M_0 was found, the probability P was estimated and the alignment of the sequences S and S_1 was built using the M_0 matrix. It is not the goal of this study to analyze all the known DNA sequences, since the developed method requires large computer resources. The developed algorithm was applied to search for periodicity with insertions and deletions in the *C. elegans* genome. This study showed the presence of periodicity with insertions and deletions in the *C. elegans* genome regions for which the presence of periodicity was not previously known.

2 Methods and Algorithms

In this study, a window which equals 600 base pairs was used to search for periods in the chromosomes of *C. elegans* genome. This window moved with step equal to 10 base pairs from the beginning to the end of each chromosome of *C. elegans*. chromosome. The DNA sequences in the window were denoted as S. To search for periodicity with insertions and deletions in sequence S, the algorithm shown in Fig. 1 was used. As seen from the algorithm, firstly, a set of random matrices Q_n (Fig. 1, step 2) of size $4 \times n$ was generated, where n is the length of the period, and 4 is the alphabet size of the studied sequence. Then, the matrices were converted since the distribution of the similarity function F_{max} for each of the matrices in the set of all random sequences (set Sr, paragraph 2.4) ought to be similar. Then, a local alignment of the studied sequence S was built relative to each converted random matrix (Fig. 1, step 4). Local alignment was used to determine the similarity function F_{max} for each converted matrix. The converted matrix having the highest value of the similarity function F_{max}, with the studied sequence S, was chosen as it was done in [23].

Thereafter, this matrix was optimized to achieve the highest value of the similarity function F_{max} (mF_{max}) with the studied sequence S (Fig. 1, step 5) and the optimized matrix was called M_0. If $mF_{max}(n)$ is more than the cutoff level F_0 then the sequence S contains the region with periodicity equal to n. In this study, periodicity in the interval from 2 to 50 base pairs was evaluated. If several periods have $mF_{max}(n) > F_0$, n which has the maximum value of $mF_{max}(n)$ was selected (Fig. 1, step 6). Selection of the level of F_0 is considered in paragraph 2.4. Subsequently, the window was moved for 10 base pairs along the *C. elegans* chromosome and the calculations were repeated (Fig. 1, step 7). As a result of the algorithm, the dependence of mF_{max} on n was obtained for sequence S with help of a local alignment. This means that the boundaries of the regions with $mF_{max}(n)$ may differ from the beginning and end of the sequence S. It also means that the values of $mF_{max}(n)$ for different n can be obtained for different

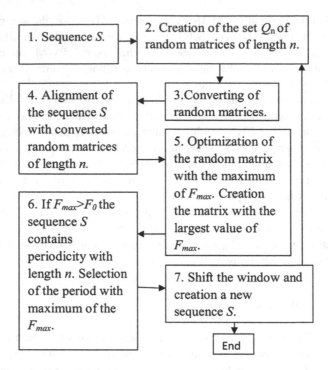

Fig. 1. The main stages of the algorithm used for calculation $mF_{max}(n)$ for analyzed sequence S.

fragments of the studied sequence S. The boundaries of the fragments, obtained for relevant values of $mF_{max}(n)$ are shown. Some steps of the algorithm shown in Fig. 1 were considered in [23] and some new details are examined below (paragraph 2.1–2.4).

2.1   Creation of a Set Q_n of Random Matrices with Length N

Random matrices Q_n with dimension $4 \times n$ were used, where n is the length of the period (Fig. 1, step 2). Each matrix can be viewed as a point in space $4 \times n$ and elements of a matrix are real random numbers. A set of random matrices Q_n was created when the distance between them in the space $4 \times n$ was not less than a certain value. To calculate the differences between the two matrices $m_1(i, j)$ and $m_2(i, j)$, the information measure was used [24]:

$$
\begin{aligned}
I_j(M_1, M_2) = &\sum_{i=1}^{20} m_1(i,j) \ln(m_1(i,j)) + \sum_{i=1}^{20} m_2(i,j) \ln(m_2(i,j)) \\
&- \sum_{i=1}^{20} (m_1(i,j) + m_2(i,j)) \ln(m_1(i,j) + m_2(i,j)) + \\
&(s_1(j) + s_2(j)) \ln(s_1(j) + s_2(j)) - s_1(j) \ln(s_1(j)) - s_2(j) \ln(s_2(j))
\end{aligned}
\tag{1}
$$

where $s_k(j) = \sum_{i=1}^{20} m_k(i,j) \cdot 2I_j$ has an asymptotic chi-square distribution with 3-th degrees of freedom [24]. Then we calculated:

$$I(M_1, M_2) = \sum_{j=1}^{n} I_j(M_1, M_2) \tag{2}$$

Hence, $2I(M_1, M_2)$ has an approximate $\chi^2(df)$, and df equal to $3n$ since $I_1(M_1, M_2)$, $I_2(M_1, M_2), \ldots, I_{n-1}(M_1, M_2)$ are independent and $I_n(M_1, M_2)$ is completely determined by $I_1(M_1, M_2)$, $I_2(M_1, M_2), \ldots, I_{n-1}(M_1, M_2)$ [24]. Then the chi-square distribution was approximated by means of the normal distribution:

$$x(M_1, M_2) = \sqrt{4I(M_1, M_2)} - \sqrt{2df - 1} \tag{3}$$

The value $x(M_1, M_2) \sim N(0, 1)$, где $N(0, 1)$ is the standard normal distribution. $N(0, 1)$ is very useful as a measure of the differences between matrices $m_1(i, j)$ and $m_2(i, j)$. The probability $p = P(x > x(M_1, M_2))$ shows that differences between the matrices $m_1(i, j)$ and $m_2(i, j)$ are determined by random factors. If the difference between the matrices $m_1(i, j)$ and $m_2(i, j)$ increases, then $x(M_1, M_2)$ becomes larger. The difference between matrices $L = x(M_1, M_2)$ not less than 1.0 was chosen.

Here, an algorithm was used to generate the matrices. Each element of the matrix $m(i, j)$, $i = 1, \ldots, 4$, $j = 1, \ldots, n$ was randomly filled with equal probability of either 0 or 1. The matrix was then compared with all matrices that were already included in the set Q_n. If at least one matrix has a difference less than $L = 1.0$, than the generated matrix was not included in the set Q_n. If the difference was greater than $L = 1.0$ for all matrices from the set Q_n, then the matrix is included in the set Q_n. The 10^6 of such matrices were created for each period length n.

2.2   Conversion of Random Matrixes

The calculation of the converted matrix M' and alignment of nucleotide sequence with converted random matrices was done as it was for amino acid sequences (paragraph 2.7 and 2.6 in [23]).

2.3   Optimization of a Random Matrix with the Largest Value of Similarity Function

For all matrices from the set Q_n the modified matrix max(m'), which had the highest value of the similarity function F_{max} was determined. Let call this value as mF_{max}. Thus, the alignment was calculated and the coordinates of the alignment were determined (Fig. 1, step 5). However, despite the use of a very large number of matrices, the matrix max(m') may have the value mF_{max}, which is not the largest for a sequence S and for length of period n. This indicates that the largest value can be achieved for matrix M_0, which lies at some distance from the matrix max(m'), that is less than the chosen threshold $L = 1.0$ (paragraph 2.1). Therefore, approximately 10^6 matrices were created, having distance L from the matrix $max(m')$ from $1.0-0.1*i$ to $0.9-0.1*i$ (for $i = 0$). These matrices were also used as indicated in paragraph 2 and a new matrix max (m') was chosen which had the highest value mF_{max}. This procedure was repeated for i from 1 to 9 and $max(m')$ for $i = 9$ was chosen as M_0 matrix.

2.4   Generation of Random Sequences and Selection of F_0

A set Sr of random sequences was created by random shuffling of the sequence S and the set Sr containing 200 sequences. To generate one random symbolic sequence, a random number sequence of length $N = 600$ was generated by the random number generator. Then, a random number sequence was arranged in ascending order, storing the generated permutations. The produced permutations were used to mix the sequence S, and as a result of this mixing, the random symbolic sequence from the set Sr was created.

In this study, threshold F_0 was determined as follows: Firstly, the sequences of C. elegans chromosomes were obtained and mixed randomly as carried out during the creation of set Sr. Thereafter, using the algorithm illustrated in Fig. 1, we determined the number of sequences $Hr(F)$, which have $mF_{max}(n) > F$ for every n in the range of 2 to 50 bases. The length of the window, as in the case of the analysis of C. elegans chromosomes, was equal to 600 nucleotides. Simultaneously, the number of sequences $H(F)$, which have $mF_{max}(n) > F$ for sequences of the C. elegans chromosomes was determined. After that, F_0, which has the ratio $Hr(F_0)/H(F_0) \leq 0.05$, was chosen. This choice of F_0 gives the number of false positives (errors of the first kind) less than 5%. In this study, $F_0 = 390.0$ and it provides $Hr(F_0)/H(F_0) \leq 0.05$, for analysis of the C. elegans genome.

This study did not analyze the period which had 3 nucleotides. This means that each window was checked for the presence of a period which equals 3 nucleotides. To do this, the mutual information between the sequence S and artificial periodic sequence $S_2 = \{123\}_{200}$ was calculated. Thereafter, the matrix of the triplet periodicity was calculated and with the help of this matrix, the correlation between S and S_2 sequences was determined as shown previously (Frenkel and Korotkov, 2008). For the measurement of correlation, the argument of normal distribution X was selected. The higher value of X corresponds to higher correlation between sequences S и S_2. It was identified that if $X < 3.0$, it indicates the absence of a period equal to 3 bases in the sequence S and the search for periods was carried out using this study's algorithm (Fig. 1). However, $X \geq 3.0$ indicated that the sequence S was not analyzed and the window was shifted by 10 nucleotides.

3   Results and Discussion

In general, 5 autosomes and X chromosome with a total length of about 100 million bases were analyzed in this study. Sequences were obtained from the website ftp://ftp.ncbi. nlm.nih.gov/genomes/archive/old_genbank/Eukaryotes/invertebrates/Caenorhabditis_ elegans/. The calculations were performed at the supercomputer cluster of the Russian

Academy of Sciences (http://www.jscc.ru/eng/index.shtml). In C. elegans genome, 25360 regions having a periodicity with length of 2 to 50 bases were found. On the average, a periodicity of ~4000 nucleotides was found to be associated with each region. The sequences found were collected in a data bank from the website: http://victoria. biengi.ac.ru/cgi-bin/indelper/index.cgi. It is interesting to consider the distribution of the lengths of periods found in C. elegans. This distribution is shown in Fig. 2.

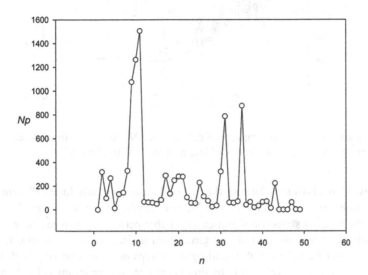

Fig. 2. Length distribution of the periods found in genome C. elegans. Np is a number of periods, n is a period length.

From this figure, it is obvious that the distribution is very nonuniform and a significant portion of the revealed regions have lengths of periods equal to 10 and 11 nucleotides. Also, there are many regions with period multiple of 10 nucleotides (20 and 40 nucleotides) and Fig. 2 shows the peak of the periods in the vicinity of 35 nucleotides. The small peak represents a period equal to 2 bases. Figure 2 also shows the absence of a significant number of regions with period equal to 3 bases. This is due to the fact that DNA with period equal to 3 bases was not analyzed. In this study, some number of regions with triplet periodicity were determined in a situation in which the original X was less than 3.0, and the period equal or multiple to 3 bases arose after the creation of alignment with insertions or deletions.

In this study, one region with periods was considered as examples. The region had a period length of 4 nucleotides, and this period can be detected only in the presence of deletions or insertions. The spectrum of $mFmax(n)$ is shown in Fig. 3.

This region was found in the first chromosome of the C. elegans genome, in sequence NC_003279.8. $mF_{max}(4) = 1348.52$. This period was not detected by TRF [6] and T-REKs [13] programs. These programs revealed an insignificant periodicity equal to 10 bases. TRF found 2.6 periods while T-REKs found 3 periods equal to 10

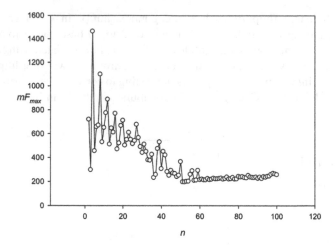

Fig. 3. mFmax(n) spectrum for fragment of the sequence NC_003279.8 from chromosome 1 of the C. elegans genome. The coordinates of fragment are: 887101-887688.

nucleotides. Mreps [10] found a period equal to 4 bases, but only for 3 different parts of the region with error rate of more than 0.37. The lengths of segments are 54, 61 and 54 bases and the level of statistical significance of this periodicity was not determined by the program. However, the period in region equal to 588 bases was found. It is about four times greater than the length found by the Mreps program and very high levels of statistical significance were found. In this sequence, the program ATR hunter [12] found 2 periods with length of 10 bases and 2 periods with length of 12 bases and completely did not see a period equal to 4 bases. Program TRStalker [11] found 4 repeats with length of 80 bases and 3 repeats with length 12 bases but did not find 4 base repeats. The program Repfind [25] found dispersed perfect repeats with repeat length equal to 4 bases and these repeats had a lower level of statistical significance (about 10–13). According to this study's estimates, mFmax(4) = 1348.52, it corresponds to P(mFmax > 1348.50) < 10–50, because the average value of mFmax for random sequences Sr is about 152.8 and $\sigma \sim 59.3$. The BWT program [26] found no repeats in the sequence. The resulting alignment and the resulting matrix M0 can be received from http://victoria.biengi.ac.ru/cgi-bin/indelper/index.cgi. A consensus period with length equal to 4 nucleotides is T(A/G)C(A/C). This period was repeated more than 140 times in the region found and the period equal to 4 bases had the highest statistical significance.

In this study, the influence of random base substitutions on the mF_{max} level was evaluated. To do this, sequences with lengths 600 and 400 nucleotides long and period equal to 20 nucleotides were used. Random positions were selected in these sequences and random replacements of the nucleotides were made on any of a, t, c, and g with equal probability. Thereafter, $mF_{max}(20)$ was calculated. The resulting function is shown in Fig. 4. It can be seen that $F_0 = 390$ is equal to approximately 1.6 and 1.0 random substitutions per nucleotide, for sequences with lengths equal to 600 and 400

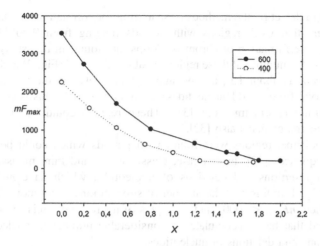

Fig. 4. Influence of base changes on $mF_{max}(20)$ for sequences 400 and 600 base pairs. X is the number of base changes per 1 nucleotide. The period length equals to 10 b.p.

nucleotides, respectively. This result shows the upper boundary of the accumulation of random substitutions in the discovered regions and this bound is 1.6 substitutions per nucleotide.

The results of this study were compared with that of the T-REKs program. To this end, intervals were introduced: 500–600, 900–1000, 1400–1500, 1900–2000, 2400–2500, 2900–3000. For these intervals, all the sequences with periods found in this study were chosen. For each sequence, the period length n was found. Thereafter, the periods in these sequences were searched by the program T-REKs. T-REKs is one of the best tools for finding tandem repeats in DNA sequences. It is believed that the T-REKs program reveals the same period, if it detects a period length which has a difference of no more than ±1 base from our period. This interval was chosen, due to the fact that we have developed a method which may make insertions, deletions and closed periods to have statistically important mF_{max}. It was also felt that the program T-REKs, finds the same period, if the number of detectable periods is not less than $L/2n$, where L is the length of the sequence with period equal to n. As a result, the proportion of regions detected by the program T-REKs for different intervals was calculated. This function is shown in Fig. 5. From this graph, it is clear that before $mF_{max} = 1500$, the program T-REKs can find less than 30% regions and only for $mF_{max} > 2200$ did the program reveal more than 50% of the regions.

There is a natural question about the biological significance of the periods found. It applies primarily to periods of 10 and 11 nucleotides long, as well as to the nucleotides of multiple periods. There are earlier suggestions that the periodicity length of 10 and 11 nucleotides has a relationship with the α-helices in proteins, as well as with the processes of DNA compaction [27, 28]. In this study, sequences without period equal to 3 bases were analyzed which is specific for the protein-coding regions. This means that most parts of the detected regions could be linked with DNA compaction [29, 30]. Also, this study identified regions with periods (with insertions and deletions) which are impossible to detect by the methods of searching for correlations in DNA [27, 28]. It is very likely that work regions with periods ranging from 9 to 11 bases and associated with the formation of chromatin loops, are found in this study. If we take into account that the number of these regions is about 2×10^3 (Fig. 2) and the genome size of *C. elegans* is about 10^8, the average distance between these regions (having periods in interval from 9 to 11 nucleotides) is about 5×10^4. This is consistent with the size of 30 nm chromatin loops [31]. These regions could be "hot spots" for chromosomal rearrangements also [32].

At the same time, regions were found with periods which could be micro- and minisatellite sequences [33]. In this case, classic micro and mini minisatellites were identified with insertions and deletions of nucleotides which have $mF_{max} > 2000$. According to Fig. 4, in this case the number of substitutions is not more than 50% per nucleotide. When $mF_{max} < 1000$, ancient copies of micro- and minisatellite sequences were discovered that have accumulated a considerable number of nucleotide substitutions, insertions and deletions of nucleotides.

It is also interesting to estimate the part of the *C. elegans* genome which has period regions. The average length of the region which was found with the periods is 400 bases and the number of regions found is 25360. This corresponds to a total length equal to about 10^7 nucleotides, which is $\sim 10\%$ of the total length of the *C. elegans* genome.

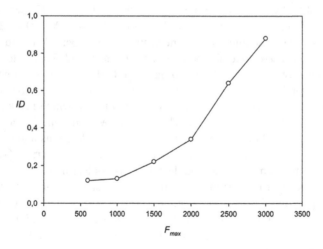

Fig. 5. Comparison of developed algorithm with the program T-REKs [13]. *ID* shows the part of periodicities regions which can find the T-REKs. We can assume that the results are the same if the T-REKs detects at least 50% of the number of periods and the period length differs not by more than one base.

Acknowledgements. This work was supported by Competitiveness Growth Program of the Federal Autonomous Educational Institution of Higher Professional Education National Research Nuclear University MEPhI (Moscow Engineering Physics Institute).

References

1. Korotkov, E.V., Korotkova, M.A., Kudryashov, N.A.: Information decomposition method to analyze symbolical sequences. Phys. Lett. Sect. A Gen. At. Solid State Phys. **312**, 198–210 (2003)
2. Durbin, R., Eddy, S., Krogh, A., Mitchison, G.: Biological Sequence Analysis: Probabilistic Models of Proteins and Nucleic Acids. Cambridge University Press, Cambridge (1998). doi:10.1017/CBO9780511790492
3. Suvorova, Y.M., Korotkova, M.A., Korotkov, E.V.: Comparative analysis of periodicity search methods in DNA sequences. Comput. Biol. Chem. **53**(PA), 43–48 (2014). doi:10.1016/j.compbiolchem.2014.08.008
4. Tiwari, S., Ramachandran, S., Bhattacharya, A., Bhattacharya, S., Ramaswamy, R.: Prediction of probable genes by Fourier analysis of genomic sequences. Comput. Appl. Biosci. CABIOS **13**, 263–270 (1997)
5. Lobzin, V.V., Chechetkin, V.R.: Order and correlations in genomic DNA sequences. The spectral approach. Uspekhi Fiz Nauk **170**, 57 (2000)
6. Benson, G.: Tandem repeats finder: a program to analyze DNA sequences. Nucleic Acids Res. **27**, 573–580 (1999)
7. Parisi, V., De Fonzo, V., Aluffi-Pentini, F.: STRING: finding tandem repeats in DNA sequences. Bioinformatics **19**, 1733–1738 (2003)
8. Anisimova, M., Pečerska, J., Schaper, E.: Statistical approaches to detecting and analyzing tandem repeats in genomic sequences. Front. Bioeng. Biotechnol. **3**, 31 (2015). doi:10.3389/fbioe.2015.00031
9. Turutina, V.P., Laskin, A.A., Kudryashov, N.A., Skryabin, K.G., Korotkov, E.V.: Identification of amino acid latent periodicity within 94 protein families. J. Comput. Biol. **13**, 946–964 (2006). doi:10.1089/cmb.2006.13.946
10. Kolpakov, R., Bana, G., Kucherov, G.: Mreps: efficient and flexible detection of tandem repeats in DNA. Nucleic Acids Res. **31**, 3672–3678 (2003)
11. Pellegrini, M., Renda, M.E., Vecchio, A.: TRStalker: an efficient heuristic for finding fuzzy tandem repeats. Bioinformatics **26**, i358–i366 (2010). doi:10.1093/bioinformatics/btq209
12. Wexler, Y., Yakhini, Z., Kashi, Y., Geiger, D.: Finding approximate tandem repeats in genomic sequences. J. Comput. Biol. **12**, 928–942 (2005). doi:10.1089/cmb.2005.12.928
13. Jorda, J., Kajava, A.V.: T-REKS: identification of Tandem REpeats in sequences with a K-meanS based algorithm. Bioinformatics **25**, 2632–2638 (2009)
14. Mudunuri, S.B., Kumar, P., Rao, A.A., Pallamsetty, S., Nagarajaram, H.A.: G-IMEx: a comprehensive software tool for detection of microsatellites from genome sequences. Bioinformation **5**, 221–223 (2010)
15. Mudunuri, S.B., Nagarajaram, H.A.: IMEx: imperfect microsatellite extractor. Bioinformatics **23**, 1181–1187 (2007). doi:10.1093/bioinformatics/btm097
16. Grissa, I., Vergnaud, G., Pourcel, C.: CRISPRFinder: a web tool to identify clustered regularly interspaced short palindromic repeats. Nucleic Acids Res. **35**, W52–W57 (2007). doi:10.1093/nar/gkm360
17. Boeva, V., Regnier, M., Papatsenko, D., Makeev, V.: Short fuzzy tandem repeats in genomic sequences, identification, and possible role in regulation of gene expression. Bioinformatics **22**, 676–684 (2006). doi:10.1093/bioinformatics/btk032
18. Lim, K.G., Kwoh, C.K., Hsu, L.Y., Wirawan, A.: Review of tandem repeat search tools: a systematic approach to evaluating algorithmic performance. Brief Bioinform. **14**, 67–81 (2013). doi:10.1093/bib/bbs023

19. Moniruzzaman, M., Khatun, R., Yaakob, Z., Khan, M.S., Mintoo, A.A.: Development of microsatellites: a powerful genetic marker. Agriculturists **13**, 152 (2016). doi:10.3329/agric. v13i1.26559

20. Korotkov, E.V., Korotkova, M.A., Kudryashov, N.A.: The informational concept of searching for periodicity in symbol sequences. Mol. Biol. (Mosk) **37**, 436–451 (2003)

21. Shelenkov, A., Skryabin, K., Korotkov, E.: Search and classification of potential minisatellite sequences from bacterial genomes. DNA Res. **13**, 89–102 (2006). doi:10. 1093/dnares/dsl004

22. Smith, T.F., Waterman, M.S.: Identification of common molecular subsequences. J. Mol. Biol. **147**, 195–197 (1981)

23. Pugacheva, V.M., Korotkov, A.E., Korotkov, E.V.: Search of latent periodicity in amino acid sequences by means of genetic algorithm and dynamic programming. Stat. Appl. Genet. Mol. Biol. **15**, 381–400 (2016)

24. Kullback, S.: Information Theory and Statistics. Dover publications, New York (1997)

25. Betley, J.N., Frith, M.C., Graber, J.H., Choo, S., Deshler, J.O.: A ubiquitous and conserved signal for RNA localization in chordates. Curr. Biol. **12**, 1756–1761 (2002)

26. Pokrzywa, R., Polanski, A.: BWtrs: a tool for searching for tandem repeats in DNA sequences based on the Burrows-Wheeler transform. Genomics **96**, 316–321 (2010). doi:10. 1016/j.ygeno.2010.08.001

27. Herzel, H., Weiss, O., Trifonov, E.N.: 10–11 bp periodicities in complete genomes reflect protein structure and DNA folding. Bioinformatics **15**, 187–193 (1999)

28. Larsabal, E., Danchin, A.: Genomes are covered with ubiquitous 11 bp periodic patterns, the "class A flexible patterns". BMC Bioinform. **6**, 206 (2005). doi:10.1186/1471-2105-6-206

29. Schieg, P., Herzel, H.: Periodicities of 10–11 bp as indicators of the supercoiled state of genomic DNA. J. Mol. Biol. **343**, 891–901 (2004). doi:10.1016/j.jmb.2004.08.068

30. Kumar, L., Futschik, M., Herzel, H.: DNA motifs and sequence periodicities. Silico. Biol. **6**, 71–78 (2006)

31. Kadauke, S., Blobel, G.A.: Chromatin loops in gene regulation. Biochim. Biophys. Acta **1789**, 17–25 (2009). doi:10.1016/j.bbagrm.2008.07.002

32. Kantidze, O.L., Razin, S.V.: Chromatin loops, illegitimate recombination, and genome evolution. BioEssays **31**, 278–286 (2009). doi:10.1002/bies.200800165

33. Richard, G.-F., Kerrest, A., Dujon, B.: Comparative genomics and molecular dynamics of DNA repeats in eukaryotes. Microbiol. Mol. Biol. Rev. **72**, 686–727 (2008). doi:10.1128/ MMBR.00011-08

RNA Sequencing Analysis of Neural Cell Lines: Impact of Normalization and Technical Replication

V. Bleu Knight[(⊠)] and Elba E. Serrano[(⊠)]

Department of Biology, New Mexico State University, Las Cruces, NM, USA
{bleu, serrano}@nmsu.edu

Abstract. RNA sequencing offers a versatile platform for profiling biological samples. The novelty and flexibility of this technology has the potential to introduce technical variation at distinct points in the experimental workflow. We evaluated variation in RNA sequencing data acquired from commercially available cell lines cultured in our laboratory: human neural stem cells and normal human astrocytes. After normalizing data with three different methods, we used principal variance component analysis to estimate the contribution to technical variance from replicate cell lots, library preparations, and flow cells. Differentially expressed genes were evaluated using ANOVA analysis. Results indicate that the largest component of technical variance was library preparation. Moreover, comparative analysis of RNA sequencing data from the two cell types showed that the identification of differentially expressed genes and the contributions to variance are strongly influenced by the normalization method. Our results underscore the necessity for technical replication in RNA-seq experiments.

Keywords: RNA-seq · Human neural stem cells · Normalization · Technical replicate · Next generation sequencing · Principal variance component analysis

1 Introduction

The flexibility of RNA sequencing technology (RNA-seq) gives rise to the potential for complex technical variation to obscure the experimental findings. While pioneering advancements have been made in data generation and statistical analyses, other technical processes in the RNA-seq workflow have received relatively less attention (reviewed in [1]). Experiments have shown that the results of RNA-seq experiments can be affected by the quality and amount of RNA [2, 3], library preparation [4–6], sequencing platform [7], normalization method [8, 9], and choice of software for data analysis [10]. The versatility and novelty of RNA-seq pose challenges for the researcher because standardized best practices have yet to be universally adopted [11].

Advances in RNA-seq technology have substantially reduced the cost of sequencing and facilitated replication in the design of RNA-seq experiments. Studies have emphasized the importance of biological replication for minimizing the variation from technical replication [8, 11, 12]. However, biological replication is not always possible. In our lab, for instance, we are interested in using RNA-seq in conjunction

© Springer International Publishing AG 2017
I. Rojas and F. Ortuño (Eds.): IWBBIO 2017, Part II, LNBI 10209, pp. 457–468, 2017.
DOI: 10.1007/978-3-319-56154-7_41

with the federally approved H9-derived human neural stem cell line (hNSC) to investigate mechanisms underlying the specification of neural cell fate. Because this cell line is derived from a single biological donor, we evaluated the variation of technical replicates at distinct steps in the RNA-seq workflow in order to facilitate the design of a robust RNA-seq experiment. In particular, we compared the transcriptomes of two distinct neural cell types, hNSC and Normal Human Astrocytes (NHA), using replicate cell lots, library preparations, and flow cells. Principal variance component analysis revealed that that the greatest contributor to technical variance was library preparation. Moreover, we found that technical variance can outweigh the biological variance between these two cell types, depending on the choice of normalization method. Our results underscore the necessity for technical replication in RNA-seq experiments.

2 Methods

2.1 Cell Culture

Human Neural Stem Cells. Gibco® H9 hESC-Derived Human Neural Stem Cells (hNSC; ThermoFisher Scientific, N7800100) from two different lots (lots 1402001 and 1408001; referred to hereafter as lot A and lot B, respectively) were cultured according to the manufacturer's specifications. Complete NSC serum free media (100 ml) comprised 97 mL Knockout DMEM/F-12 (Gibco®, 12660-012), 1 mL Glutamax (Gibco®, 35050-061), 2 µg bFGF (Gibco®, PH60024), 2 µg EGF (Gibco®, PHG0314), and 2 mL StemPro neural supplement (Gibco®, A10508). Media were sterilized by filtration through a 0.2 µm porous membrane and stored in 10 ml aliquots to avoid reheating. Frozen ampules of cells were thawed and transferred to sterile 15 ml centrifuge tubes along with 4 ml of media, then centrifuged at 210 G for 5 min. Supernatant was removed to rid cells of cryoprotectant, then cells were resuspended in media and plated into T-25 flasks (one flask per ampule) coated with CellStart (Gibco®, A10142). The flasks of hNSCs (passage 0) were incubated at 37°C, 5% CO2. Media were replenished every 48 h that followed. Cells were subcultured by partial digestion when cultures reached 80% confluence according to the manufacturer's specifications. Briefly, media were removed and cells were rinsed in DPBS without calcium and magnesium before adding 2 mL of pre-warmed StemPRO Accutase (Gibco®, A11105-01) to each flask. Cells were observed for detachment under the microscope, then transferred to centrifuge tubes along with 9 ml of media. Cells were centrifuged at 210G for 5 min. Supernatant was removed, then cells were resuspended in 2 ml of prewarmed media and triturated. Cells were plated on flasks coated with CellStart.

Normal Human Astrocytes. Normal Human Astrocytes (NHA; Lonza, CC-2565) from two different lots (lots 0000412568 and 0000402839; referred to hereafter as lot A and lot B, respectively) were cultured as described previously [13]. Frozen ampules of cells were thawed and plated into T-25 flasks (passage 0) and incubated at 37°C, 5% CO_2. When cultures reached 80% confluence, 5 days after plating (passage 1), cells were subcultured by partial digestion and plated in T-25 flasks coated with CellStart as recommended for hNSCs. On the first media change that followed the first passage, media were changed to Complete NSC serum free media and NHA were cultured in conditions that paralleled those of hNSCs thereafter.

Spontaneous Differentiation. The second passage of cells from both NHA (2) and hNSC (2) lots were passaged as described above using StemPro Accutase, tittered using a hemocytometer, and plated at a density of 2500 cells/cm^2 on T-25 flasks coated with poly-L-ornithine (Sigma P3655) and laminin (ThermoFisher, 23017015) for differentiation, as recommended by the manufacturer. Spontaneous differentiation media were prepared with 97% Knockout DMEM/F-12, 1% Glutamax, and 2% StemPro neural supplement. 24 h after plating, complete NSC serum free media were replaced with spontaneous differentiation media for NHA and hNSC cell lines. During the ten day differentiation protocol recommended by the manufacturer, media was replenished every 48 h with 75% media changes, taking care not to expose cells to air.

2.2   Live Cell Imaging with Phase Contrast Microscopy

Before isolating RNA, phase contrast images of live cells were captured with Metavue image capture software (Molecular Devices), which controlled a Coolsnap HQ CCD camera (Photometrics) attached to an inverted Nikon TE-2000 microscope.

2.3   Sample Preparation and Sequencing

RNA was isolated using the instructions for the PureLink® RNA Mini Kit (Ambion). DNA was removed using the DNA-free™ kit (Ambion) according to the manufacturer's specifications. RNA quality was assessed with an Agilent 2100 Bioanalyzer. Each RNA sample was divided in half, so that two different libraries were prepared per RNA sample. RNA samples with RIN Values greater than 8.9 were submitted to the BioMicro Center at the Massachusetts Institute of Technology for library preparation and sequencing. The SMART-Seq v3 Ultra Low Input RNA Kit for Sequencing (Clonetech) was used to produce cDNA from 10 ng of total RNA in accordance with the manufacturer's specifications. Samples were fragmented and transferred to the SPRI-works for ligation to BioMicro Center adapters, size selection, multiplex barcoding, and enrichment using BioMicro Center PCR primers. Libraries were assessed for fragment size and distribution on an AATI Fragment Analyzer™ (Advanced Analytical). Samples were multiplexed and sequenced twice on an Illumina NextSeq (Illumina) sequencer in accordance with the protocol for 150 base pair (bp) paired end (PE) reads. The library preparation was not strand-specific; therefore, reads mapping to the forward and reverse strands are pooled.

2.4   Quality Control, Filtering, and Alignment

Phred scores were assessed with FastQC to evaluate base call accuracy. All average reads across every nucleotide position had minimum quality scores corresponding to 99% base call accuracy (Phred score > 20). Fastq files were aligned to the human genome (hg 19) using the Burrows-Wheeler Aligner (BWA-MEM, v0.7.10). SAM files were imported into JMP Genomics (v 8.0, SAS Institute, Inc.) which summarized counts by gene using the UCSC human genome annotation (hg 19). A detection threshold for genes was applied to ensure that all samples had a minimum raw read count > 10 per gene; however, this filter did not result in the removal of any genes. We focused on overall gene expression, rather than isoform-specific expression, to facilitate the evaluation of data using standard techniques.

2.5   Normalization

In order to compare differential expression between (groups of) lanes, we normalized read counts to adjust for varying lane sequencing depths and other potential technical effects. Sequence data was normalized by reads per kilobase per million mapped reads (RPKM), Upper-quartile scaling (UQS), and trimmed means of M component (TMM) using JMP Genomics (v 8.0, SAS Institute Inc.). The raw count data was log_2 transformed to facilitate comparison with the normalized expression data.

2.6   Statistical Analyses

The raw count data and UQS normalized data were fit to a Poisson model as part of the ANOVA analysis in JMP Genomics. Statistical analysis was performed with JMP Genomics, and p-values were adjusted for multiple comparisons with the step-up false discovery rate (FDR) method of Benjamini and Hochberg [14]. Principal variance component analysis with JMP Genomics calculated the variance contribution from cell type (fixed variable), as well as the random variables that included library preparation, flow cell, and cell lot (Table 1). Because it was our expectation that a low number of genes would be differentially expressed, we used a false discovery rate of 10% and a $|log_2\,fold\,change|$ value of 0.5 to evaluate potential differences. For a detailed analysis of the statistical analyses used in this study, refer to [15, 16] and the literature for JMP Genomics v 8.0.

Table 1. Different experimental parameters contributing to variation in the study

Component	Number of replicates	Type of variation	Effect
Cell type	2 (NHA, hNSC)	Biological	Fixed
Lot	2 per cell type	Technical	Random
Library preparation	2 per lot	Technical	Random
Flow cell	2 per library	Technical	Random

2.7   Figure Preparation

Figures were prepared with JMP Genomics (v 8.0, SAS Institute Inc.), JMP Pro (v 11.0, SAS Institute Inc.), and Excel (Microsoft, 2010). Additional image assembly, labelling, and alignment were completed with Photoshop (Adobe, CS6).

2.8   Responsible Conduct and Reproducibility

Guidelines for preclinical research set forth by the NIH were used to guide our experimental design. The hNSC line was derived from the WA09 (H9) embryonic stem cell line (NIH approval number NIHhESC-10-0062). The NIH registry for human embryonic stem cells retains the information about the WA09 (H9) stem cell line from which the hNSC lines was derived. NHA were produced and de-identified by Lonza, who retains a signed record of consent from the donor. Both cell lines were used within 10 population doublings (3 passages) as recommended by the respective manufacturers. RNA-seq results (see below) verified the human species origins of both cell lines. The use of commercially available, de-identified human cell lines produced before 2015 is exempt from review by the Institutional Review Board. Nevertheless, the protocol used in these experiments was approved by the NMSU Institutional Biosafety Committee (approval # 1401SE2F0103). Official gene symbols were used in accordance with the standards developed by the HUGO Gene Nomenclature Committee. Sequence data were collected with a single blind protocol at the MIT BioMicro center, without prior knowledge of the nature of the biological samples. Researchers at NMSU assigned groups and assessed the outcomes.

3   Results

3.1   Live Cell Imaging with Phase Contrast Microscopy

Phase contrast images of cells demonstrate morphological differences between hNSCs and NHA cultured in a spontaneous differentiation environment (Fig. 1). hNSCs (Fig. 2A, C) appear smaller and more populated in comparison to NHA (Fig. 2B, D). NHA appear as stellate cells, which is more typical of astrocyte cultures. Long processes, typical of neurons, are seen in cultures of both cell types but are more prevalent in hNSC cultures.

3.2   Evaluation of Sequence Quality and Read Distribution

Assessment of RNA-seq read quality revealed that the average Phred scores for all read positions met the criteria for 99% base call accuracy (Phred score > 20; data not shown). The read output from the Illumina NextSeq varied depending on the sample, library preparation and flow cell (Fig. 2). The number of reads from the fourth flow cell was considerably greater than the other flow cells (more than twice the amount of other flow cells, in some cases). Evaluation of the data revealed that the increase in reads from the fourth cell was distributed throughout the transcriptome and did not preferentially distribute reads among larger or smaller genes.

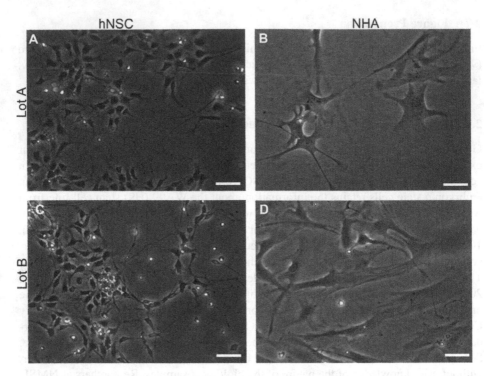

Fig. 1. Phase contrast images depict neural stem cells (hNSC; A, C) and normal human astrocytes (NHA; B,D). Cells were cultured under spontaneous differentiation conditions for 10 days and were imaged live, prior to RNA extraction for transcriptome analysis. Representative images are shown from two lots for both cell types: Lot A (A, B); Lot B (C, D). Scale bar = 50 μm

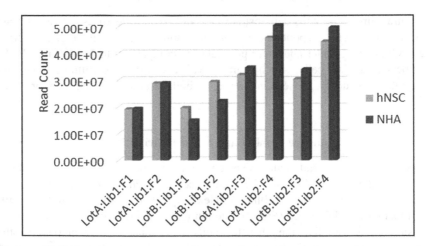

Fig. 2. Bar plots show the total read counts per lane for each combination of cell type (hNSC, NHA), lot (LotA, LotB), library number (Lib1, Lib2) and flow-cell (F1, F2, F3, F4).

3.3   Normalization

Normalization methods were evaluated based on the reduction of variation between technical replicates and the retention of differentially expressed genes between the two cell types. Heat maps of sample-to-sample correlation coefficients were clustered using the Ward method. The hierarchical cluster analysis illustrates the effects of normalization on sample data (Fig. 3). Clustering associated samples based on cell type for $\log_2$ transformed data, RPKM normalized data, and UQS normalized data. For TMM normalized data, samples did not cluster based on any of the factors used to group the data. The most coherent heat maps are observed in the UQS normalized data (Fig. 3D).

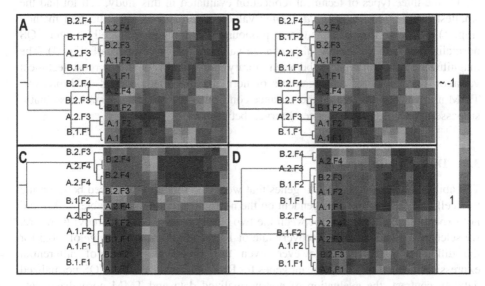

Fig. 3. Heat maps depict Pearson correlation coefficients and hierarchical cluster analysis with the Ward method (Green, hNSC; Violet, NHA). Pearson correlation coefficients (r values; Blue $\sim$ -1, Red = 1) and clusters were calculated for $\log_2$ transformed data (A), RPKM-normalized data (B), TMM normalized data (C) and UQS normalized data (D). Sample annotation reflects cell lot (A, B), library (1, 2), and flow cell (F1, F2, F3, F4) and is depicted in the order: lot.library.flow cell (Color figure online).

3.4   Lot, Library Preparation, and Flow Cell Effects

Principal variance component analysis with JMP Genomics was used to determine the proportions of variance from different parameters (Table 1). The proportions of variance contributed by the experimental components differed depending on the choice of normalization method (Fig. 4). RPKM normalization did not substantially impact the percentage of variation contributed by the experimental components as compared to the log2 transformed data (< 2% difference). In contrast, TMM normalization increased the proportion of variance contributed by library preparation to over 50%, as compared to under 20% with other normalization methods. To compensate for this increase, the percentage of variance based on cell type decreased from over 50% with other methods to 16% with TMM normalization. UQS normalization also substantially impacted the components of experimental variance by increasing the contribution from cell type to 75%, greater than the cell type contribution found with other methods (16%, TMM normalized; 57%, log_2 transformed; 58% RPKM normalized). Moreover, the contribution from the cell lot to the experimental variance was 5% following UQS normalization, almost double the contribution from lot in non-normalized data (3%) and in data normalized with other methods (3%, RPKM; 3%, TMM). The increase in the contributions from cell type and lot after UQS normalization were accompanied by a decrease in the contributions from the other experimental components relative to other normalization methods (library preparation, 11%; flow cell, 6%; residual, 5%).

Of the three types of technical replication evaluated in this study, cell lot had the smallest distribution of *log_2 fold change* values, regardless of normalization method (Fig. 4). Replicate flow cells had more pronounced differences than cell lot, and UQS normalization reduced the *log_2 fold change* values for flow cell by half (Fig. 4). The magnitude of the differences between library preparations is the largest technical variation that we observed, regardless of normalization method (Fig. 4). Moreover, TMM normalization increased the variance contributed by library preparation so that it surpassed the biological variance observed between cell types.

3.5 Differential Expression

The ability to uncover significant genes that were differentially expressed between the two cell types was strongly dependent on the normalization method (Table 2). We did not expect drastic differences between the two cell types used in this study; therefore, we selected a generous false discovery rate of 10% and a *log_2 fold change* of ±0.5 for the difference calculation. However, even this broad assessment of differential expression resulted in no significant genes for RPKM normalized and UQS normalized data. In contrast, the evaluation of non-normalized data and TMM normalized data resulted in 276 and 232 differentially expressed genes, respectively. The ten genes with the largest |*log_2 fold change*| values for each normalization method are listed in Table 2. The ten largest |*log_2 fold change*| values ranged from 0.59 – 0.62 for non-normalized, 0.46 – 0.49 for RPKM normalized data, 0.59 – 0.61 for TMM normalized data, 0.37 – 0.41 for UQS normalized data.

Of the top ten genes with the largest |*log_2 fold change*| values between cell type, four genes were common among all normalization methods (*italicized*), three genes were on three of the four lists, while six genes were on two lists, and one gene was only on one list (Table 2). Nine out of the top ten genes with the largest |*log_2 fold change*| values that were calculated with log_2 transformed and TMM normalized data were identical. RPKM normalized data shared seven out of the top ten genes with the largest |*log_2 fold change*| values with both TMM normalized and UQS normalized data. The genes with the largest |*log_2 fold change*| values ranged in size from 413 bp to ~ 28 kb and the sizes were well distributed (> 10 kb, 7.5%; 5 – 10 kb, 30%; 1 – 5 kb, 67.5%; < 1 kb, 7.5%).

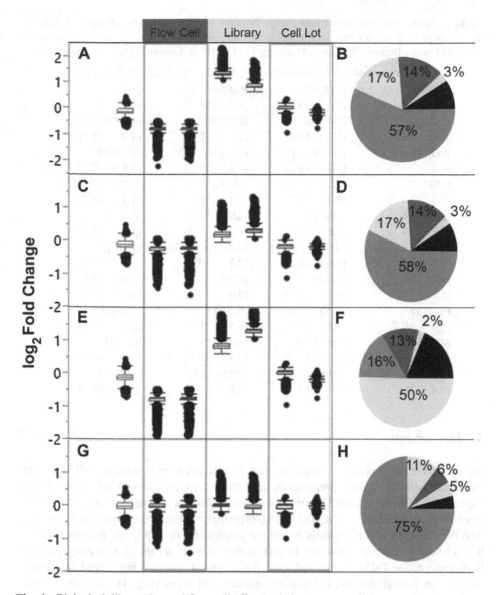

Fig. 4. Biological, library, lot, and flow cell affect variation among replicate samples. Box plots (A,C,E,G) portray $\log_2$ fold-changes between technical replicates for both cell types (hNSC, green; NHA, violet) and between cell types (blue). Pie charts illustrate components contributing to sample variance (B,D,F,H). Principal variance component analysis revealed the contribution from cell type (blue), flow cell (red), library preparation (yellow), cell lot (light green), and residual variance (black). Estimates are presented for $\log_2$ transformed data (A,B), RPKM normalized data (C,D), TMM normalized data (E,F) and UQS normalized data (G, H) (Color figure online).

Table 2. Genes with the largest fold change between hNSC and NHA cell types (p-values). Significant** differentially expressed genes were only uncovered following $\log_2$ transformation and TMM normalization (Benjamini-Hochberg FDR = 10%). Common genes shown in italics.

**Log$_2$ transformed	RPKM normalized	**TMM normalized	UQS normalized
PARD6G-AS1 (4.7×10^{-3})	*HNRNPU* (2.8×10^{-10})	*PARD6G-AS1* (4.7×10^{-3})	ZNF496 (6.1×10^{-7})
HNRNPU (1.6×10^{-7})	ZNF496 (2.2×10^{-8})	*HNRNPU* (1.5×10^{-7})	*PARD6G-AS1* (1.9×10^{-2})
RNF126 (4.8×10^{-4})	*PARD6G-AS1* (5.0×10^{-3})	*RNF126* (4.9×10^{-4})	*HNRNPU* (1.8×10^{-10})
OR4F17 (2.8×10^{-3})	*RNF126* (2.0×10^{-4})	*OR4F17* (2.8×10^{-3})	ACTB (1.1×10^{-3})
LINC01002 (2.3×10^{-3})	*OR4F17* (2.6×10^{-3})	LINC01002 (2.3×10^{-3})	*RNF126* (1.6×10^{-3})
LOC105376854 (2.7×10^{-4})	LOC105376854 (9.6×10^{-5})	LOC105376854 (2.7×10^{-4})	*OR4F17* (1.2×10^{-2})
HCN2 (5.8×10^{-4})	MBD3 (6.1×10^{-6})	HCN2 (5.9×10^{-4})	LINC01002 (1.0×10^{-2})
PRSS57 (5.3×10^{-4})	ARHGAP45 (1.2×10^{-5})	PRSS57 (5.5×10^{-4})	ARHGAP45 (2.1×10^{-4})
MBD3 (3.9×10^{-5})	CIRBP-AS1 (8.0×10^{-6})	MBD3 (4.1×10^{-5})	ADAMTSL5 (6.5×10^{-5})
ARID3A (2.4×10^{-4})	ADAMTSL5 (8.3×10^{-6})	CIRBP-AS1 (4.8×10^{-5})	ARID3A (7.2×10^{-4})

4 Discussion

We examined the contribution to experimental variation from biological (cell type) and technical (cell lot, library preparation, and flow cell) components of an RNA-seq experiment (Table 1). JMP Genomics was employed for normalization of data with three methods (RPKM, TMM, and UQS), principal variance component analysis, and ANOVA analysis of differentially expressed genes between cell types. Results indicate that library preparation was the largest contributor to technical variance (Fig. 4). Furthermore, for TMM normalized data, the technical variance introduced by library preparation outweighed the differences between cell types (Fig. 4). Although TMM normalized data had the greatest percentage of technical variance, it was the only normalization method that permitted the detection of significant differentially expressed genes (FDR = 10%, |*log2 fold change*| > 0.5; n = 232; Table 2).

This study was prompted in part by concerns about concerns about replication and reproducibility in next generation sequencing analyses [17]. The nature of RNA-seq permits the replication of samples at many points during the experiment. Our findings underscore the need for technical replication in RNA-seq experiments, especially with regard to library preparation. This result is congruent with the recommendation by the SEQC Consortium for replicate library preparation [18]. Moreover, we have added to the evidence that emphasizes the impact of normalization method on experimental findings in gene expression studies [9, 19]. The results and corresponding data set from this experiment will be used to build undergraduate skills that incorporate replication and reproducibility in experimental design. In particular, we will produce freely available online training modules that teach bioinformatics and statistics using JMP Genomics.

Acknowledgements. Research was supported by NIH R25NS080685 and the NMSU Manasse endowment. We are grateful to Dr. Stuart Levine and Dr. Shmulik Motola of the Massachusetts Institute of Technology BioMicro Center for technical support. The content of this manuscript is solely the responsibility of the authors and does not necessarily represent the official views of the National Institutes of Health or NMSU.

References

1. Kukurba, K.R., Montgomery, S.B.: RNA sequencing and analysis. Cold Spring Harb. Protoc. **2015**, 951–969 (2015)
2. Chen, E.A., Souaiaia, T., Herstein, J.S., Evgrafov, O.V, Spitsyna, V.N., Rebolini, D.F., Knowles, J.A.: Effect of RNA integrity on uniquely mapped reads in RNA-Seq, pp. 5–7 (2014)
3. Adiconis, X., Berlin, A.M., Borges-Rivera, D., Busby, M.A., DeLuca, D.S., Fennell, T., Gnirke, A., Levin, J.Z., Pochet, N., Regev, A., Satija, R., Sivachenko, A., Thompson, D.A., Wysoker, A.: Comprehensive comparative analysis of RNA sequencing methods for degraded or low input samples. Nat. Methods **10**, 1–20 (2013)
4. Lahens, N.F., Kavakli, I.H., Zhang, R., Hayer, K., Black, M.B., Dueck, H., Pizarro, A., Kim, J., Irizarry, R., Thomas, R.S., Grant, G.R., Hogenesch, J.B.: IVT-seq reveals extreme bias in RNA sequencing. Genome Biol. **15**, 1–15 (2014)
5. Tsompana, M., Valiyaparambil, S., Bard, J., Marzullo, B., Nowak, N., Buck, M.J.: An automated method for efficient, accurate and reproducible construction of RNA-seq libraries. BMC Res. Notes **8**, 1–5 (2015)
6. Bhargava, V., Head, S.R., Ordoukhanian, P., Mercola, M., Subramaniam, S.: Technical variations in low-input RNA-seq methodologies. Sci Rep. **4**, 3678 (2014)
7. Ross, M.G., Russ, C., Costello, M., Hollinger, A., Lennon, N.J., Hegarty, R., Nusbaum, C., Jaffe, D.B.: Characterizing and measuring bias in sequence data. Genome Biol. **14**, R51 (2013)
8. Bullard, J.H., Purdom, E., Hansen, K.D., Dudoit, S.: Evaluation of statistical methods for normalization and differential expression in mRNA-Seq experiments. BMC Bioinform. **11**, 1471–2105 (2010)
9. Risso, D., Ngai, J., Speed, T., Dudoit, S.: Normalization of RNA-seq data using factor analysis of control genes or samples. Nat. Biotechnol. **32**, 896–902 (2014)
10. Robles, A., Qureshi, S.E., Stephen, S.J., Wilson, S.R., Burden, C.J., Taylor, J.M.: Efficient experimental design and analysis strategies for the detection of differential expression using RNA-sequencing. BMC Genom. **13**, 1471–2164 (2012)
11. Conesa, A., Madrigal, P., Tarazona, S., Gomez-Cabrero, D., Cervera, A., McPherson, A., Szcześniak, M.W., Gaffney, D.J., Elo, L.L., Zhang, X., Mortazavi, A.: A survey of best practices for RNA-seq data analysis. Genome Biol. **17**, 13 (2016)
12. Marioni, J.C., Mason, C.E., Mane, S.M., Stephens, M., Gilad, Y.: RNA-seq: an assessment of technical reproducibility and comparison with gene expression arrays. Genome Res. **18**, 1509–1517 (2008)

13. Knight, V.B., Serrano, E.E.: Hydrogel scaffolds promote neural gene expression and structural reorganization in human astrocyte cultures. PeerJ **5**, e2829 (2017)
14. Benjamini, Y., Hochberg, Y.: Controlling the false discovery rate: a practical and powerful approach to multiple testing. J. R. Stat. Soc. Ser. B **57**, 289–300 (1995)
15. Scherer, A.: Batch Effect and Experimental Noise in Microarray Studies: Sources and Solutions. Wiley, Chichester (2009)
16. Boedigheimer, M.J., Wolfinger, R.D., Bass, M.B., Bushel, P.R., Chou, J.W., Cooper, M., Corton, J.C., Fostel, J., Hester, S., Lee, J.S., Liu, F., Liu, J., Qian, R., Quackenbush, J., Pettit, S., Thompson, K.L.: Sources of variation in baseline gene expression levels from toxicogenomics study control animals across multiple laboratories. BMC Genom. **16**, 1–16 (2008)
17. Endrullat, C., Glökler, J., Franke, P., Frohme, M.: Standardization and quality management in next-generation sequencing. Appl. Transl. Genomics **10**, 2–9 (2016)
18. SEQC-Consortium: A comprehensive assessment of RNA-seq accuracy, reproducibility and information content by the Sequencing Quality Control consortium. Nat. Biotechnol. 32, 903–914 (2014)
19. Qin, S., Kim, J., Arafat, D., Gibson, G.: Effect of normalization on statistical and biological interpretation of gene expression profiles. Front Genet. **3**, 1–11 (2013)

Towards a Universal Genomic Positioning System: Phylogenetics and Species IDentification

Max H. Garzon[✉] and Sambriddhi Mainali

Department of Computer Science, The University of Memphis, 38152, Tennessee, USA
{mgarzon,smainali}@memphis.edu

Abstract. Technology to gather biomic data now far exceeds the capabilities of tools to extract useful information and knowledge from it, a challenging predicament facing demands in our time, such as personalized medicine. We propose a new family of data structures to represent and process omics data in a way that is more anchored in biological reality and processed by algorithms that are more consistent with it, so that DNA itself can be used to process it to extract useful knowledge, organize and store it as needed. These structures enable much more efficient crunching of genomic and proteomics data and can be used as a foundation of a truly universal Genomic Positioning System (GenIS). The power of this approach is illustrated by applications to two important problems in biology, a new universal set of biomarkers and methods to do phylogenetic analysis and species identification and classification. We show that certain metrics on these representations can be used to obtain *ab initio*, from genomic data alone (possibly including full genomes), in a matter of minutes or hours, well established and accepted phylogenies crafted in biology (such as the 16S rRNA-based plylogenies) in the course of the last 50 years. We also show how the same representation can also be used to solve recognition problems associated with genomic data, which includes in particular the problem of species identification and a solution to the problem of storing large genomes into compact representations while preserving the ability to query them efficiently. We also sketch other applications to be explored in the future, including objective criteria to produce biological taxonomies to produce a truly universal and comprehensive "Atlas of Life", as it is or as it could be on earth.

Keywords: Genomic/proteomic analysis · Noncrosshybridizing oligonucleotide bases · Microarray analytics · Phylogenetic analysis · nxh chips · Species identification

1 Introduction

A fundamental problem in our time, and particularly in bioinformatics, has been brought forth by our ability to generate enormous amounts of biological data through, e.g., genomics (e.g., the human genome project, Next-Generation Sequencing (NGS)), proteomics and metabolomics. On the other hand, our ability to process the available data to produce useful information and actionable knowledge has lagged behind. Although basic biological mechanisms are well understood, the enormous adaptivity, complexity and diversity of their implementation by specific organisms in specific

© Springer International Publishing AG 2017
I. Rojas and F. Ortuño (Eds.): IWBBIO 2017, Part II, LNBI 10209, pp. 469–479, 2017.
DOI: 10.1007/978-3-319-56154-7_42

environmental conditions present an enormous challenge to do so, particularly in the face of disease and desirable outcomes, such as personalized medicine. Common methods that have been crafted to tackle such problems are usually focused on peculiar methods that are not scalable to other problems, suffer from the curse of dimensionality of DNA spaces and/or combinatorial explosion in computational analyses.

By contrast, computational sciences have developed the basic knowledge and appropriate tools to tackle big data sets, but they lack the biological knowledge and specificity to handle bioinformatic problems. The Geographic Positioning System (GPS) is an excellent example. Methods developed for computer networks (such as the internet, the web, and wireless communication) have enabled billions of people on the planet to use a cell phone to communicate. This requires, in particular, the ability of the systems to determine the location of the phone anywhere on planet earth so as to establish routing paths to send messages through. Another example is relational database systems. They have enabled our ability to store terabytes of digital data in very small spaces. The storage, however, is not specific to the content of the data, so that semantically related data can be stored in arbitrarily different and remote locations. That is in sharp contrast to what biological organisms do (e.g., living cells and brains), where location, physical proximity and obstruction represent hard anchoring constraints that are exploited for biological function, such as cell membranes, organs and organisms. Without them, biological reality, in particular organs and living organisms as we know them, would be impossible.

In order to address this gap, we propose in this paper a new data structure to represent and process omics data in a way that is more anchored in biological reality and processed by algorithms that are more consistent with it. These ideas were implicit in ongoing attempts to build computers out of DNA molecules, as in Garzon and Yan [10]; Adleman [1]. We extend this idea to create appropriate representations of biomic information in DNA, so that DNA itself can be used to process it to extract useful knowledge, organize, store and further process it as needed. The framework and data structures are briefly described in Sect. 2, but more thorough descriptions can be found in Garzon and Bobba [8] and Garzon et al. [9]. We have furthered the technique in order to achieve a truly universal Genomic Positioning System (GenIS) by introducing a novel set of noncrosshybridizing (nxh) bases (technical definitions are given below in Sect. 2) that define universal data structures for arbitrary genomic data. We show how every genomic sequence x gets associated a so-called digital signature on every given nxh basis that contains a wealth of information about the organism. To illustrate the power of this approach, we present applications to two important problems in biology. In Sect. 3, we show, for example, that certain metrics on these signatures can be used to obtain *ab initio*, from genomic data alone (possibly including a full genome), in a matter of minutes or hours, well established and accepted phylogenies crafted in biology (such as the 16S rRNA-based phylogenies), in the course of the last 50 years. In Sect. 4, we show how the same representations can also be used to solve a version of the difficult general recognition problem in computer science for biomic data, which includes in particular the problem of species identification and a solution to the problem of storing large genomes compactly while preserving the ability to query them efficiently. In Sect. 5, we discuss other applications to be explored in the future, including objective criteria to produce biological taxonomies to produce a truly universal atlas of life.

2   GenIS: A Framework for DNA Indexing on DNA Chips

This section sketches the theoretical foundations of the underlying method, DNA indexing (DNI). We focus on the relevant novel methods to be used later in applications in the remainder of the paper. The framework was motivated by the field of DNA computing, inspired by the ideas of using DNA itself as a computational medium in Adleman [1] and as smart glue for self-assembly applications in Seeman [18] and Winfree et al. [22]. These developments have stimulated the development and analysis of various techniques aimed at finding large sets of short oligonucleotides with noncrosshybridizing (nxh) properties by several groups, particularly the nxh bases described in Garzon and Bobba [8]; Garzon et al. [9]; Deaton et al. [5]. In this paper, we continue the exploration of these new tools and show how recent solutions to the nxh chip design problem in DNA Computing [8, 9] can yield the results below.

Currently, the most powerful way to capture genomic data is DNA microarrays (Schena [17]; Stekel [19]) and DNA chips. The fundamental problem with standard DNA microarrays is that useful information is often deeply buried under mountains of "noisy", redundant and incomplete data. An analysis of microarray reliability has been presented in Garzon and Mainali [6] that makes evident why this problem occurs and provides an effective way to address it using a next generation of microarrays, namely nxh chips. They have resulted in demonstrable increases in the confidence, accuracy and reliability (false positives/negatives) of DNA chips, which become even more prominent in applications such as species identification and diagnosis, as shown below. To make this paper self-contained, we summarize the basic facts next.

The key property behind the operation of microarrays is hybridization between Watson-Crick complements, the fundamental characteristic property of DNA. If DNA chips are to produce highly informative "pixels" (spots), knowledge of the structural properties of what we term "DNA spaces" would help. We conducted preliminary studies in prior work (Garzon et al. [11–13]). DNA chips usually require uniform length short oligos of up to $n = 60$-mers or so. Let B be a set of DNA n-mer species affixed to the spots of a chip. These oligo(nucleotide)s will be called probes (or in the case of a very special type of set, an encoding basis, a term to be precisely defined below.) In practice, this set is a judicious target-free selection of oligos that provide full but tight covering of the DNA space of all oligos of said size n, unlike a full selection in a target organism, as is the case for ordinary microarrays. The key property is named *noncrosshybridization* (nxh), which comes in degrees of quality determined by a numeric parameter τ that controls for hybridization stringency of reaction conditions. The best choice would be the Gibbs energy, or approximations thereof, using Huget et al. [21], typically set at $\tau = -6$ kcal/mole (the standard minimum free energy required for hybridization of two strands to occur.) Such a selection requires deep knowledge of the structural properties of the Gibbs energy landscapes of the full set of all n-mers (DNA_n). The key tool to understand this structure is a refinement of the notion of Gibbs energy to the so-called hybridization distance (h-distance.) Four key properties will nearly optimize noise removal from DNA chips:

(a) The h-distance is defined in such a way that low values of $h(x, y)$ are indicative of the degree of WC-complementarity of two oligos x, y or their WC-complements (e.g. $h(x, y) = 0$ if an only if x, y are identical or - *perfect* complements, i.e. h does not distinguish perfect WC-complements);

(b) the h-distance is a *true distance*, i.e. it satisfies the characteristic properties of Euclidean distance (*reflexivity:* $h(x, x) = 0$; *symmetry* $h(x, y) = h(y, x)$; and the familiar geometric *triangle inequality:* $h(x, z) \leq h(x, y) + h(y, z)$, for arbitrary n-mers x, y, z);

(c) A basis B is selected so that its sequences satisfy the property $h(b_i, b_j) \geq \tau$, for distinct probes b_i, b_j in B. A noncrosshybrizing set is an "orthogonal" set of oligos for hybridization, i.e., have no redundancies in terms of hybridization. Moreover, a hybridization decision for a pair of oligos x, y based on h agrees with one made on the Nearest Neighbor (NN) model of Gibbs energy about 80% of the time, as shown in Garzon and Bobba [8];

(d) The fourth key property of the chip is that it should be *complete*, i.e., provide full coverage of the space of all n-mers, so that an arbitrary random n-mer fragment x_i in a target x will hybridize to at least one (and hence, by property (c), exactly one) probe b_j if and only if $h(b_j, x_i) < \tau$ for a given level of hybridization stringency τ of chip B.

Given a nxh basis B, an nxh chip can be built on solid surface using standard biotechnology developed for DNA/RNA microarrays, as shown in Fig. 1.

nxhChipDesign

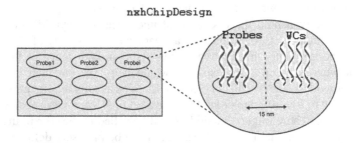

Fig. 1. Design of virtually noise-free nxh chips as proposed in Garzon and Mainali [6] based on a judicious selection of probes according to properties (a)–(d). It consists of a number of spots in 1-1 correspondence with the oligos in an nxh basis B. Each spot consists of two subsets of probes, one containing a certain number of copies (say 1,000,000) of the oligos in B to some degree ofresolution and the other containing the same number of copies of their WC complements (WCs) at sufficient separation to avoid cross-hybridization to their neighbor complements. Digital *signatures* can be obtained for arbitrary genomes x when shredded into fragments (also called *shreds* below) of comparable size (as would be obtained as output of a NGS sequencing of x) to that of the probes. The signature of x serves as a **data structure** to store x because a number of analyses of the original genome (perhaps bypassing the process in NGS), and even nongenomic information as in Bobba et al. [3, 15, 16], can be extracted directly from the signatures.

Properties (a)–(d) guarantee that on an nxh chip, a random oligo x_i of comparable probe size is much more likely to hybridize to fewer probes, and under appropriate stringency τ (directly related to the minimum separating distance between oligos in B and other conditions) to ensure hybridization to at most one probe. This so-called nxh property immediately translates into the desirable properties to the problems mentioned above. The noise is notably reduced (in fact, it is completely eliminated under ideal conditions), results will be more predictable, and the corresponding analyses will be much more reliable, as demonstrated in Garzon and Mainali [6].

On any given DNA chip, a unique hybridization pattern (the digital signature) can be produced as described in Fig. 1. A given (possibly unknown) target is digested, usually tagged, and poured over the chip under appropriate reaction conditions enforcing stringency τ. We emphasize once again that this reduction in the number of probes is typically considered a loss of redundancy, and hence a loss of signal-to-noise ratio in standard analyses [18, 19] that would diminish the payoff of the readout. The results below provide additional evidence to the counterintuitive fact that, to the contrary, this judicious selection of probes will improve the sensitivity of a chip for target genome discrimination, this time in the field of phylogenetic analysis. Therefore, signatures will exhibit minimum variability and noise, assuming saturation conditions in target concentrations and long enough relaxation time to obtain a full signal (order of hours.) This result has been already demonstrated in simulation in Bobba et al. [3] with a selection of genes important in seven human diseases as well as *in vitro* while extracting an nxh basis by a PCR-base selection protocol fully described, e.g., in Bi et al. [2].

The application of this indexing technique to phylogenetics thus requires addressing a fundamental general question: how much of the biomic information contained in an original genome x is captured by its digital signature? This question will be partially resolved in the following sections. As discussed in Sect. 5, a full answer appears possible, but will require deeper investigation of the structure of DNA spaces.

3   Phylogenetic Analyses by DNA Indexing (DNI)

In this Section we describe the new method for taxonomical classification based on DNI. The theoretical and experimental foundations of the method have been established in prior work in Garzon et al. [9]; Garzon et al. [13]; Bi et al. [2], but the main features are again summarized here to make this paper self-contained. A previous study in Garzon and Wong [7] developed and validated bacterial phylogenies for a family of important bacterial genomes using nxh sets specifically designed for these targets and therefore cannot be extended to arbitrary genomes. The families of universal nxh sets described above in Sect. 2 can now be used to address this shortcoming. Here we described the results of further analogous studies to explore the feasibility and biological validity of the resulting phylogenies.

The fundamental problem of phylogenetics is to elucidate the genealogical (ancestor/descendant) relationships among biological organisms, as they appear to be related among all life forms on earth and as they change through time. As such, phylogenetics has traditionally remained a related but essentially different field from taxonomics, the science of naming and classifying the diversity of organisms, and from other subfields of systematics such as comparative phylogenetics and biodiversity. The basic principle of systematics is that members of higher-level groups share fewer characteristics than those in the lower-lever groups. Conventional methods classify different organisms by grouping them into a different "species", "genus", "family", "orders", "class", "phylum", "kingdom" or "domain", as in Henning [14] or Woese and Fox [23], to illustrate the level of difference between living organisms. Two organisms belonging to the same species level would share more similarities than organisms of different species. Phylogenetics, on the other hand, compares the similarity of DNA, the blue print of life. With the advances in DNA sequencing technologies, the potential in providing a nonbiased method to understand the natural history of all living things on earth has now emerged as much more feasible. Phylogenetics studies the relationships of living organisms based on their evolutionary histories. The most commonly used methods for phylogenetics to infer DNA relatedness between species include parsimony, maximum likelihood, and MCMC-based Bayesian inference. The general technique is to align DNA or amino acid sequences in some metric to estimate the similarity between organisms. The general assumption is that evolution is a branching process, whereby individuals may undergo genetic changes (mutations) that may diverge into separate clades. The rate of mutation, and thus speciation, is dictated by many environmental factors. Survival of a species is subject to natural selection. Thus evolution may be visualized as a multi-dimensional feature-space that any population (species or genus, for example) moves through over time. In order to come up with robust models, biologists carefully but subjectively select a set of so-called most conserved genes (e.g. those on ribosomal 16S rRNA, or biomarkers such as the mitochondrial cytochrome c.oxidase subunit (1-COI)) to analyze the phylogenetic relationship between organisms. These genes serve certain vital functions for the cell. The rate of successful mutation in these genes must be rare because mutations of these genes would likely be harmful to the cell. Species having a defective vital function would be eliminated by natural selection. Among the many conserved genes, ribosomal RNA (rRNA) genes are the most commonly recognized [24]. However, comparing the DNA metrics of the rRNA sequences alone is problematic because this multidimensional network cannot pin point the time of occurrence of each event. Biologists solved this problem by attaching some known proteins of which their natural histories are better understood to correlate the rRNA-based metric and build the tree of life. Using a different set of biomarkers or proteins as guide could produce a different tree of life. Additionally, the resolution of 16S rRNA at the lower lower taxonomic groups is very poor. What other genes should be used as a guide to construct the rRNA-based tree of life? How to select suitable biomarkers to locate organisms in a tree of life? These problems remain unresolved in Biology.

The set of oligos in a nxh DNA basis provide a solution to this problem that is objective (experimenter independent) and can be processed in vitro at speeds much faster than ordinary HCP simulations on an actual DNA chip, built as described above. To test their biological soundness, a study of a group of 17 bacteria, used in a previous study by Garzon & Wong [7] and shown in Fig. 2, was reproduced by substituting the nxh set originally designed for the set by the universal set in a DNA chip design obtained by previous work in Garzon and Bobba [8], here referred to as bases 3mE3-2-0.45 and 4mersPolar3-3. Their genomic signatures described below were obtained in simulation on a High Performance Parallel Computer System (HPC). The metrics used to compare signatures to obtain the dendrograms were *the simple Euclidean distances between them*, regarded as high dimensional vector in Euclidean space after normalizing them (dividing by the total numbers of hits of the target fragments to each probe). Alternatively, we might use the angle, or the statistical correlation between them. As can be observed from the resulting phylogenies in Fig. 2, both the Euclidean distance and angle as similarity criteria both produce the most commonly accepted phylogeny among biologists, as described in Garzon and Wong [7], even on very low resolutions nxh bases (3-mers and 4-mers), while Pearson statistical correlation (not shown, but see Fig. 3) produces phylogenies inconsistent with biological knowledge. These results substantially improve results obtained by the frequency methods using signatures on a full set of 3-mers in Chen et al. [4], as well as in Garzon and Wong [7], and can clearly be scaled to produce phylogenies of arbitrary biological families, phyla and even the full DNA or RNA biome. It is remarkable that this result has been obtained purely on genomic analysis, independently of the many other considerations that led to the original tree of life 16S sRNA.

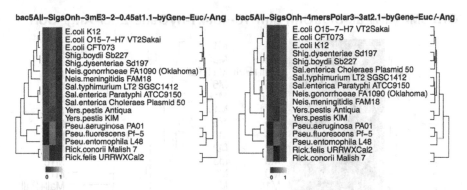

Fig. 2. Genomic signatures of 17 bacteria clades on two nxh DNA chips of three 3-mers at stringency $\tau = 2$ and three 4-mers at $\tau = 3$. The signatures of their full genomes reproduce the phylogenetic trees using simple metrics such as the Euclidean distance (**-Euc**) and angle (**-Ang**) between the corresponding 3D vectors (normalized to a common full range of their values in the interval [0, 1].) Both produce the most commonly accepted phylogeny among biologists [7, 20], whereas the metric of Pearson statistical correlations (see Fig. 3 too) among the vectors does not.

An additional consequence is worth pointing out. These results also provide evidence to counter the superficial observation that genomic signatures cannot provide an adequate tool for genetic comparison since they ignore the transcription process altogether (and so do not involve protein or protein expression directly.) 16S rRNA is essentially protein-based and yet, the phylogenetic tree obtained by the contrast metric from Fig. 2 shows that it is well approximated by analyses of genomic signatures alone on a 16-mer nxh DNA chip built *a priori* and designed for the collection of 6 subfamilies of plasmids and bacteria used in this validation. In fact, these analysis are essentially high dimensional and allow for far more complex phylogenetic trees to be expressed, a possibility pointed out in Garzon and Wong [7].

4 Species Identification

A related but essentially different problem is the *recognition problem* for genome sequences. This is a well-known and difficult problem in computer science, usually unsolvable or NP-complete in full generality. For finite sets, such as the genes belonging to a specific organism or chromosome, it becomes more tractable, but the problem remains scruffy and not amenable to efficient approaches, especially if the genomes are to be represented by compact data structures. The problem is important because, in general, it is known in computer science that it may allow a great deal of compression for storage of a set representing genomic information for a TOL. To illustrate the point, even *infinite* sets of low complexity (such as sets describable by regular expressions or context-free grammars) can be represented by *finite* data structures. Here we show another application of digital genomic signatures that show promise in helping the problem, at least to a good degree of

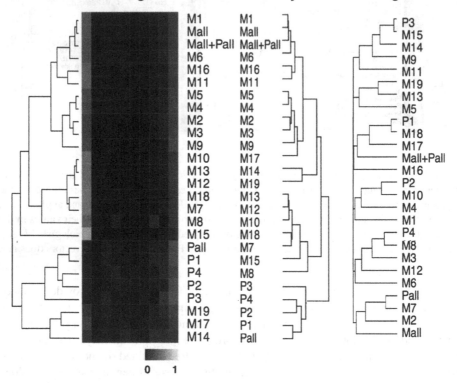

Fig. 3. Digital genomic signatures can also produce a solution to the species ID and classification problem of genomic sequences, again using the Euclidean distance and angle metrics, but not the correlation metric. These examples show classifications of 23 genes in *s. cerevisiae* into two categories (mitochondrial M and plasmid P) for which GenIS has not been trained at all. Similar results are obtained with the full set of genes in other chromosomes, even if considered together in a "single-pot" run.

approximation, for the more complex data genomic sets in bioinformatics, where a particular case of the problem is called *species identification.*

A scalable solution the problem of species identification problem is illustrated as follows. Given a full genome as a set of genomic sequences (for example, the 6000+ genes in the 12 chromosomes I-XII of *s. cerevisiae* (*s.c.* below), or its mitochondrial DNA and the plasmids therein), the problem is to design a compact representation of the sequences that allows for biologically significant discrimination of the sequences from one another and their membership to a specific set of sequences (e.g., gene, a chromosome, a genome, a genus, and so forth.)

A GenIS solution is shown below in Fig. 3. The signatures of the Open Reading Frames (ORFs) of the entire mitochondrium (19 genes, Mito M) of s.c. and their four plasmids P were calculated separately (first three rows in Fig. 3) on a higher resolution nxh basis 8mP10-4 with 10 probes at minimum distance $\tau = 4$. Additionally, each of the $23 = 19 + 4$ genes were calculated individually and shown in rows 4–26. The signal in the first and last probes show clearly the difference in membership to Mito (rows 4–22) and to Plasmids (rows 23–26), although signals from the other probes do contribute to the classification. The dendrograms again classify the genes appropriately into corresponding categories, one for Ps and several for Ms, by either -**Euc** or -**Ang** metrics. At this time, the biological significance of these subgroups is not clear. Once again, the statistical correlation is not nearly as good a metric for a dendrogram to tell the difference in membership of the various genes into the two categories M and P.

5   Conclusions and Future Work

This paper extends the hypothesis-free, whole-genome type of phylogenetic analysis proposed in Garzon and Wong [7] for a specific subset in the phyllum of bacteria to general methods that are universally applicable to the full biome, in principle. The solution includes a comprehensive framework and a Genomic Information System GenIS for bioinformatic analysis. The major goal of this research is to develop a methodology for enabling arbitrary species classification based on so-called universal nxh chips that cover the entire spectrum of organisms, known or unknown. These DNA chips can be regarded as generalized barcodes. Although barcodes are, at best, expected to discriminate among species, they do not provide an insight into the more general and complex relationships among genomes captured by phylogentic relationships, or their complex intra- or inter-genomic relationships, let alone a basis for comparative genome wide analysis. On the other hand, GenISs have the potential to provide a *universal coordinate system* to characterize very large groups (genera, families, and even phyla) of organisms on a common reference system, a veritable comprehensive "Atlas of Life", as it is or as it could be on earth.

These results raise a number of questions of interest to the bioinformatics analysis of the biome. The most important question is perhaps the full biological validation and significance of the resulting phylogenies and TOL. For example, they can be obtained on arbitrary genomic sequences, such as COIs, 16S rRNA, or the full genome. Traditionally, 16S rRNA or COIs are the standard sequences, but the limitations that led to their wide-spread use are no longer present in this framework; it might be reasonable to assume that full genomes may produce more reliable phylogenies, but a close comparison of the phylogenies may give cause for pause and thus an answer requires a more principled comparison. Second, the application to species identification and the recognition problem for genomic sequences and the properties exhibited by this methodology naturally suggest that it can be used advantageously in other problems. To mention a few, first the Cartesian coordinatization provided by GenIS enables the calculation of evolutionary rates in absolute time and, more importantly allow for an extrapolation of where evolution is taking a given family of genomes (e.g. E. Coli), by extrapolation of their trajectory (as described by the phylogenetic tree) in the universal reference frame provided by GenIS. This is akin to determining the trajectory traced in the sky by observations of the planets at various times and enabling a determination of their "next move": Where is evolution going with each species? A third problem is an analogous application to metabolomics using time series of gene expression on mRNA for example. These and other problems are worthy of further exploration.

Acknowledgements. Many thanks to the High Performance Computing Center (HPC) at the U of Memphis for the time to compute the digital signatures.

References

1. Adleman, L.: Molecular computation of solutions of combinatorial problems. Science **266**, 1021–1024 (1994)
2. Bi, H., Chen, J., Deaton, R., Garzon, M., Rubin, H., Wood, D.H.: A PCR protocol for *in Vitro* selection of non-crosshybridizing oligonucleotides. J. Nat. Comput. **2**(3), 417–426 (2003)
3. Bobba, K.C., Neel, A.J., Phan, V., Garzon, M.H.: "Reasoning" and "Talking" DNA: can DNA understand English? In: Mao, C., Yokomori, T. (eds.) DNA 2006. LNCS, vol. 4287, pp. 337–349. Springer, Heidelberg (2006). doi:10.1007/11925903_26
4. Chen, J., Chen, S., Deng, L.-Y., Bowman, D., Shiau, J.-J., Wong, T.-Y., Madahian, B., Henry, L.: Phylogenetic tree construction using Trinucleotide Usage Profile (TUP). BMC **17**(13), 381 (2016)
5. Deaton, J., Chen, J., Garzon, M., Wood, D.H.: Test Tube Selection of Large Independent Sets of DNA Oligonucleotides R, pp. 152–166. World Publishing Co. Singapore (Volume dedicated to Ned Seeman on occasion of his 60th birthday)
6. Garzon, M.H., Mainali, S.: Towards reliable microarray analysis and design. In: The 9th International Conference on Bioinformatics and Computational Biology (2017)
7. Garzon, M.H., Wong, T.-Y., Garzon, M.H., Wong, T.Y.: DNA chips for species identification and biological phylogenies. Nat. Comput. **10**, 375–389 (2011)
8. Garzon, M.H., Bobba, K.C.: A geometric approach to Gibbs energy landscapes and optimal DNA codeword design. In: Stefanovic, D., Turberfield, A. (eds.) DNA 2012. LNCS, vol. 7433, pp. 73–85. Springer, Heidelberg (2012). doi:10.1007/978-3-642-32208-2_6
9. Garzon, M.H., Phan, V., Neel, A.: Optimal codes for computing and self-assembly. Int. J. Nanotechnol. Mol. Comput. **1**, 1–17 (2009)
10. Garzon, M.H., Yan, H. (eds.): DNA 2007. LNCS, vol. 4848. Springer, Heidelberg (2008). doi:10.1007/978-3-540-77962-9

11. Garzon, M.H., Phan, V., Roy, S., Neel, A.J.: In search of optimal codes for DNA computing. In: Mao, C., Yokomori, T. (eds.) DNA 2006. LNCS, vol. 4287, pp. 143–156. Springer, Heidelberg (2006). doi:10.1007/11925903_11

12. Garzon, M.H., Phan, V., Bobba, K.C., Kontham, R.: Sensitivity and capacity of microarray encodings. In: Carbone, A., Pierce, N.A. (eds.) DNA 2005. LNCS, vol. 3892, pp. 81–95. Springer, Heidelberg (2006). doi:10.1007/11753681_7

13. Garzon, M.H., Blain, D., Neel, A.J.: Virtual test tubes for biomolecular computing. J. Nat. Comput. 3(4), 461–477 (2004)

14. Hennig, W.: Grundzüge einer Theorie der Phylogenetischen Systematik (1950). English revision, *Phylogenetic Systematics*. (tr. D. Davis and R. Zangerl), Univ. of Illinois Press, Urbana 1966, reprinted 1979

15. Neel, A.J., Garzon, M.H.: DNA-based memories: a survey. In: Bel-Enguix, G., Jiménez-López, M.D., Martín-Vide, C. (eds.) New Developments in Formal Languages and Applications. SCI, vol. 113, pp. 259–275. Springer, Heidelberg (2008)

16. Reif,H., LaBean, T.H., Pirrung, M., Rana, V.S., Guo, B., Kingsford, C., Wickham, G.S.: Experimental construction of very large scale DNA databases with associative search capability. In: Jonoska, N., Seeman, N.C. (eds.) DNA 2001. LNCS, vol. 2340, pp. 231–247. Springer, Heidelberg (2002). doi:10.1007/3-540-48017-X_22

17. Schena, M.: Microarray Analysis. Wiley, Hoboken (2003)

18. Seeman, N.: DNA in a material world. Nature 421, 427–431 (2003)

19. Stekel, D.: Microarray Bioinformatics. Cambridge University Press, Cambridge (2003)

20. Volff, J.N., Altenbuchner, J.: A new beginning with new ends: linearisation of circular chromosomes during bacterial evolution. FEMS Microbiol. Lett. 186(2), 143–150 (2000)

21. Huget, J.M., Bizarro, C.V., Forns, N., Smith, S.B., Bustamante, C.A., Ritort, F.: Single-molecule derivation of salt-dependent base-pair free energies in DNA. PNAS 107(35), 15431–15436 (2010)

22. Winfree, E., Liu, F., Wenzler, L.A., Seeman, N.C.: Design and self-assembly of two-dimensional DNA crystals. Nature 394, 539–544 (1998)

23. Woese, C., Fox, G.: Phylogenetic structure of the prokaryotic domain: the primary kingdoms. Proc. Natl. Acad. Sci. U.S.A. 74, 5088–5090 (1977)

24. http://en.wikipedia.org/wiki/Phylogenetics (2008). Accessed Feb 2017

A High Performance Storage Appliance
for Genomic Data

Gaurav Kaul[1], Zeeshan Ali Shah[2,3], and Mohamed Abouelhoda[2,3]([✉])

[1] Corp. (UK) Ltd., London, UK
[2] King Faisal Specialist Hospital and Research Center (KFSHRC),
Riyadh, Saudi Arabia
mabouelhoda@yahoo.com
[3] Saudi Human Genome Program,
King Abdulaziz City for Science and Technology (KACST), Riyadh, Saudi Arabia

Abstract. Rapid advancements in the area of next generation sequencing is revolutionizing the way in which biologists and now increasingly, clinicians analyze genomic data. These advances have substantially decreased the time and the cost it takes to sequence the genomes of new patients, thereby making genomic techniques more mainstream and giving rise to the new era of precision medicine. National scale genome programs have been launched in various parts of the world such as USA, the United Kingdom, and Saudi Arabia to name a few. One of the key insights out of this mainstream adoption is that even though the time and cost of generating sequence data has decreased dramatically, the cost of analyzing the data to yield clinically relevant information has not proportionally decreased. On the contrary, downstream analysis of the genomic data now dominates the cost in terms of time, effort and monetary value. This could be attributed to a number of factors: the sheer volume of data, limited knowledge of phenotypic, regulatory and epigenetic artifacts within the genome, and limited computational capabilities of existing data analysis tools and infrastructure. Overcoming these challenges is central to realize a more accurate, sophisticated and cost-effective genomic medicine. Another challenge, related to the limited analytic capabilities of existing computational and storage infrastructure is what we address in this paper. We discuss how novel trends in hardware, including the emergence of cheap, high performance and endurance solid-state storage associated with low latency interconnect and software defined orchestration, can help creating a high performance storage tier which improves data acquisition, storage, transmission and analysis over the current commercial alternatives.

Keywords: Bioinformatics · High performance storage · Next generation sequencing · Sequence analysis

1 Introduction

The development of new high-throughput and massively parallel DNA sequencing technologies has reduced both the cost and the time required to sequence

© Springer International Publishing AG 2017
I. Rojas and F. Ortuño (Eds.): IWBBIO 2017, Part II, LNBI 10209, pp. 480–488, 2017.
DOI: 10.1007/978-3-319-56154-7_43

Table 1. Big data of the genomic sciences compared to astronomy.

Data phase	Astronomy	YouTube	Genomics
Acquisition	25 zetta-bytes/year	500–900 million hours/year	1 zetta bases/year
Storage	1 EB/year	1–2 EB/year	2–40 EB/year
Analysis	Realtime/in situ reduction	Limited	Genome alignments/variant calling
Distribution	Dedicated lines from antennae (600 TB/s)	10 MB/s	Many small (10 MB/s) and some (10 TB/s) data transfer

an entire human genome. The Human Genome Project took around 13 years to sequence the first human genome and cost up to a one billion US dollars. Today, the same process can be completed within weeks for a few hundred dollars. Specifically, the price for one million base-pairs or 1 Mbp has dropped to a few cents in 2013. This is thanks to the advent of Next Generation Technology (NGS) which came in to wider use since 2008. This progress exceeds Moore's law, stating the empirical observation that the unit cost for the same performance of processor halves every 18 months or so. This makes the sequencing much more prevalent and within the reach of regional labs, universities, as well as clinics and hospitals [1,2]. The comparison of genomic sequencing with Moore's law has, however, an important implication. The amount of data generated and the cost of generating that data far outstrips the cost and performance of computing required to analyze the data. This exacerbates issues in all the key four phases identified in a data-intensive field – acquisition, storage, analysis and distribution. Table 1 (adapted from from [3]) compares genomic data along these four phases to other domains. In [3], it has been concluded that genomics will shortly overshadow other domains with respect to these phases.

In this paper, we address some of these challenges related to genomic data acquisition, storage, analysis, and distribution by building a high performance storage tier using commodity hardware. Our solution leverages the capabilities of emerging solid-state devices, high performance interconnect, the Lustre parallel file system and novel orchestration software. Using this high performance tier and interfacing it between the genomic sequencers at one side and the compute infrastructure, has enabled us to overcome the above challenges in a cost efficient manner.

This paper is organized as follows: In the following section, we review the basic challenges associated with managing huge genomic datasets. In Sect. 3, we introduce the features of our system and its implementation details. Sections 4 and 5 include experiments and conclusions, respectively.

2   Background

Before presenting the approach we took to architect and evaluate the high performance storage tier for genomic data analysis, we explore in this section the existing hardware and software bottlenecks, pinpointing the limitations of the existing storage architecture when it comes to handle large scale genomic data. We also look at the current hardware trends regarding storage and how these are leading to a more data centric deployment model compared to a compute centric model.

2.1 Saudi Human Genome Project – Key Challenges

The basic motivation behind this work is the challenges we faced in the Saudi Human Genome Project (SHGP), which aims to sequence the genomes of over 100,000 individuals from across the Kingdom. The project is one of the largest national scale sequencing projects in the region. It aims at identifying genetic basis (genes and mutations) of inherited diseases in the Saudi population utilizing state of the art genome sequencing and bioinformatics. Sample collection, preparation and sequencing is done in 15 regional centers across the Kingdom. All the data are then sent for analysis at a central site in King Abdulaziz City for Science and Technology (KACST) in Riyadh. The project is an excellent domain where the big data challenges can be seen firsthand.

2.2 Limitations of Current Infrastructure

Traditionally, large scale genomic projects have used high performance compute and storage model for providing required computational services. Storage solutions can be classified as direct attached Storage (DAS), network attached storage (NAS), or storage area network (SAN) storage as shown in Fig. 1. DAS are connected directly to the servers without extra switches in between, and can be considered as an extension of local server disks. NAS and SAN are connected through switch; making them scalable and the IO speed depends on the network speed and can reach a throughput of 100 GB using commercially available switches.

These storage models work well in most settings; yet they give rise to bottlenecks when it comes to big data. These bottlenecks are observed during IO

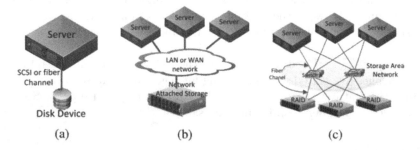

Fig. 1. Storage systems: (a) direct attached (DAS), (b) network attached storage (NAS), and (c) storage area network (SAN).

intensive phases in bioinformatics pipelines, especially when multiple processes write to a common shared storage which leads to waiting and delay in processing. The root cause for the delay at the shared storage is the reliance on traditional rotating hard disk drives (HDD) that can only handle a few hundreds of I/O operations per second (IOPS). (The network speed is no longer considered as a major bottleneck due to the recent advancement in networking devices.) To increase the performance and also to provide certain levels of fault tolerance, storage vendors have utilized exceedingly clever ways to cluster individual disks in storage arrays and use them in parallel.

2.3 Novel Trends in Storage and Interconnect

NAND-Flash-based SSD devices are gaining attraction as a faster alternative to disks, and close the performance gap between DRAM and persistent storage. SSDs are an order of magnitude cheaper compared to DRAM, and an order of magnitude faster compared to magnetic disk. Many existing database and analytics software has shown improved performance with SSDs. Many commercial SSD devices have adopted high-performance PCIe interface in order to overcome the slower SATA bus interface designed for disk. Attempts to use flash as a persistent DRAM alternative by plugging it into a RAM slot are also being explored. SSD storage devices have been largely developed to be a faster drop-in replacement for disk drives. This backward compatibility has helped their widespread adoption. However, additional software and hardware is required to hide the difference in device characteristics. Even with the high performance of SSDs, any inefficiency in the storage management software significantly affects the performance. Hence, optimizing such software is an active research topic. Due to their high performance, SSDs also affect fabric requirements. The latency to access disk over Ethernet is dominated by the disk seek latency (2). However, in a SSD-based cluster the storage access latency could even be lower than network access. These concerns are being addressed by faster network fabrics such as Infiniband and new Intel Omni-Path fabric. Low-overhead software protocols such as RDMA or user-level TCP stacks that bypass the operating system are also widely used.

2.4 Emergence of Near Data Processing

The emergence of cheap and fast SSD storage has given a new term – Storage Class Memory or SCM. One such storage class memory announced by Intel is the 3DXpoint memory. Bringing the access time of storage closer to that of the memory allows us to create a large pool of storage class memory and hence do processing "near" to the storage. This concept gives rise to the potential of near data processing or in this case, near storage processing. In this paper, we take a first step of building large capacity storage using high performance NVMe NAND SSDs and interconnecting it with remote high performance Intel Omni-Path fabric. This binds the compute and storage closely together allowing execution of certain data intensive workloads on the storage.

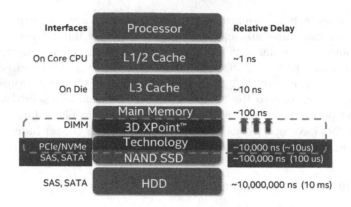

Fig. 2. Different levels of latency between processor and storage.

3   Building a High Performance Storage Tier

In this section, we cover the building blocks of our high performance storage tier that has been built from existing commodity hardware and software. To build all flash array storage, we used a high density NVMe drive 2U chassis manufactured by Supermicro [4]. This chassis can host up to 46 NVMe's. (The current capacity of Intel NVMe is 2 TB.) All the NVMe's are attached to the the motherboard through PCI-Express Gen 3 interface. The motherboard can have up to 2x Intel Xeon processor E5-2600, and up to 3 TB RAM. The flash array can be directly connected to the servers via a high speed 100G switch. (We used Intel Omni-Path interconnect as the fabric switch). To increase the throughput, multiple flash arrays can be grouped together and the management software will handle the parallel read and write operations. This was achieved through the use of the Intel Enterprise Edition of Lustre 3.0 file system management software.

One of the key challenges was the interoperability with the existing storage with minimal disruption to the existing pipelines. To solve this issue, we used

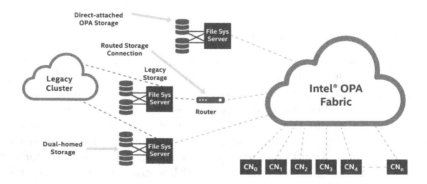

Fig. 3. Design of our high performance storage. CN_0 ,.., CN_N refer to compute nodes.

the Intel switch along with proper setting of the Luster 3.0 software. Figure 3 shows the structure of our storage system, where Intel 100G OPA is the central networking switch(s) connecting flash array, compute nodes and legacy storage together.

4 Workloads and Evaluation

To evaluate the performance of the storage solution, we conducted two experiments: In the first experiment, we measured the IO performance using standard benchmarking tools. In the second experiment, we measured the performance of a basic task that is critical for large scale sequencing projects; namely (reference-based) compression of FastQ and BAM files.

4.1 Experiment 1: Standard Benchmarking

In this experiment, we used the IOzone program [5] to measure the IO performance of the flash storage. The IOzone program was installed directly on the flash storage node (the CPU on the chassis motherboard) and was invoked from there. The usual SAS direct attached storage was connected to a server with similar processor speed. Figure 4 shows the IO performance for the two types of storage solutions. For write operations, the flash array is 1.5 to 2.5 times faster that the HDD storage. For write operations, the flash array is 1.3 to 1.6 times faster.

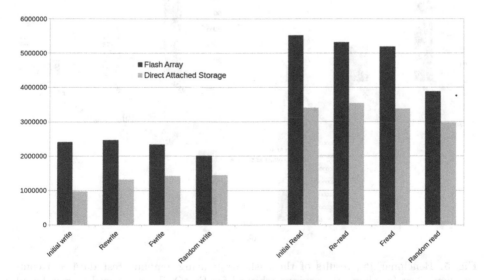

Fig. 4. Benchmarking results of the flash array compared to hard disk based storage. The y axis is the number of read/write operations per second. Fread and Fwrite is the performance of the fread and fwrite functions.

4.2 Experiment 2: Genome Compression

As the key bottleneck for large scale genomic project is storage space, it makes economic sense to do lossless compression on the raw genomic data in the high performance tier where it is first received and processed from the sequencer. The performance of the compression algorithm becomes critical to finish as quickly as possible and to save as much storage space as possible. In our evaluation we used the FastQ and BAM compression algorithm from PetaGene package [6,7].

PetaGene provides the PetaSuite framework for genomic data compression. If a reference genome is available, then one can achieve higher compression rates by using relative compression mode, where the NGS sequences (reads) can be compressed based on their alignment to the reference genome. In this case, one actually compresses their alignment. On decompression, the original sequences are reconstructed from decompressed alignment information.

For this experiment, we used the human NGS dataset (NA12878) from the Coriell Cell Repository [2]. The sequencing was based on Illumina system using TruSeq Nano V2.5 kit. The dataset is composed of 4 parts with total size of about 130 GB. As a reference genome, we used the human genome version 19 (hg19).

In Parts (a) and (b) of Fig. 5, we show the compression results. The figures show the importance of using reference based compression compared to usual

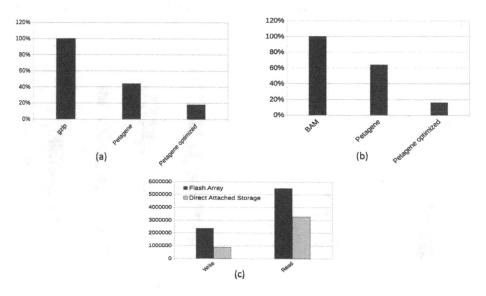

Fig. 5. Benchmarking results of the flash array using genomic test data and compression. Part (a) shows the savings achieved for FastQ files compared to gzip-based compression, Part (b) shows the savings achieved compared to the built-in compression in the BAM file, and Part (c) shows the speed of reading and writing the file (FastQ only) using flash array and direct attached storage. The y-axis is the number of read/written operations per second.

gzip and BAM compression (BAM format is for storing aligned reads and it is also based on gzip format). PetaGene's Lossless FasterQ compression gave a 66% space saving over the FastQ.gz files, while its optimized Quality-Lossless FasterQ compression gave an 82% space saving. Equivalent BAM files were also compressed and gave 36% space saving, while its Quality-Lossless compression gave 84% space saving.

In addition to comparing the compression ratios, we also compared the compression time using our flash array solution against HDD direct attached storage. In Part (c) of Fig. 5, we show the performance of the flash array compared to the usual hard disk based storage. The write speed of the flash array is 2.6 times faster than the hard disk based and the read speed is 1.7 times faster. This shows how efficient to use flash array for genomic dataset management and processing.

5 Conclusions

In this paper, we presented a rack-size appliance for genomic data processing that uses flash storage, near-store processing and integrated networks for cost-effective analytics. We have demonstrated the performance benefits by running standard benchmarking and by using an example of genomic data compression on large amounts of data using the PetaSuite compression technology from PetaGene UK. The solution we suggest is very low-cost compared to other commercially packaged solutions. Our solution is also easy to operate and manage using Luster software or any similar open source product.

In future, we expect much reduction in the cost of NVMe's and larger capacities beyond 2 TB. This will dramatically affect the cost and improve the performance.

It is also worth mentioning that the use of flash array is very useful for setting up private clouds and it can boost the performance of variant analysis pipelines using either cloud middle ware like OpenStack or Docker containers. Genomic analysis workflows in these environments was presented in [8,9].

While the current version of our work focus on a highly modular software-defined architecture for the next generation genomic analytics, the future version will specify, design and prototype a novel hardware architecture where SoC (System on Chip) based microservers, memory modules and accelerators, will be placed in separated modular server trays interconnected via a high-speed, low-latency opto-electronic system fabric. These resouces will be allocated in arbitrary sets, as driven by t-for-purpose resource/power management software. These blocks will employ state-of-the-art low-power components and be amenable to deployment in various integration form factors and target scenarios. Using this architecture, we aim at delivering a full-edged, vertically integrated datacentre-in-a-box prototype to showcase the superiority of disaggregation in terms of scalability, effciency, reliability, performance and energy reduction which will be demonstrated in future pilot use-cases.

Acknowledgments. This publication was supported by the Saudi Human Genome Project, King Abdulaziz City for Science and Technology (KACST). Our thanks to Majed Alelaiwi, Gabriele Paciucci, Adam Roe and Ahmad Al-jeshi of Intel for their collaboration throughout the project. Our thanks to Majed Alelaiwi, Gabriele Paciucci, Craig Rhodes, Adam Roe and Ahmad Al-jeshi of Intel for their collaboration throughout the project. We would also like to thank Faheem Karim and Martin Galle from Supermicro on their advice on chassis and configuration. We would like to thank Vaughn Wittorff and Dan Greenfield of PetaGene (Cambridge, UK) for allowing us to use their test compression runs.

References

1. Gonzalez-Garay, M.: The road from next-generation sequencing to personalized medicine. Pers. Med. **11**(5), 523–544 (2014)
2. DePristo, M., Banks, E., et al.: A framework for variation discovery and genotyping using next-generation DNA sequencing data. Nat. Genet. **43**(5), 491–498 (2011)
3. Stephens, Z., Lee, S., Faghri, F., Campbell, R., Zhai, C., Efron, M., et al.: Big data: Astronomical or genomical? PLoS Biol. **13**(7) (2015)
4. Supermicro (2016). www.supermicro.com
5. IOzone: File system benchmarking (2016). www.iozone.org
6. PetaGene (2016). www.petagene.com/
7. Greenfield, D., Stegle, O., Rrustemi, A.: GeneCodeq: quality score compression and improved genotyping using a bayesian framework. Bioinformatics **32**(20), 3124–3132 (2016)
8. Ali, A.A., El-Kalioby, M., Abouelhoda, M.: Supporting bioinformatics applications with hybrid multi-cloud services. In: Ortuño, F., Rojas, I. (eds.) IWBBIO 2015. LNCS, vol. 9043, pp. 415–425. Springer, Heidelberg (2015). doi:10.1007/978-3-319-16483-0_41
9. Ali, A.A., El-Kalioby, M., Abouelhoda, M.: The case for docker in multicloud enabled bioinformatics applications. In: Ortuño, F., Rojas, I. (eds.) IWBBIO 2016. LNCS, vol. 9656, pp. 587–601. Springer, Heidelberg (2016). doi:10.1007/978-3-319-31744-1_52

Obtaining the Most Accurate *de novo* Transcriptomes for Non-model Organisms: The Case of *Castanea sativa*

Marina Espigares[1], Pedro Seoane[1], Rocío Bautista[2], Julia Quintana[3], Luis Gómez[3,4], and M. Gonzalo Claros[1(✉)]

[1] Departamento de Biología Molecular y Bioquímica,
Universidad de Málaga, Malaga, Spain
claros@uma.es
[2] Plataforma Andaluza de Bioinformática,
Universidad de Málaga, Malaga, Spain
[3] Departamento de Sistemas y Recursos Naturales,
ETSI Forestal, de Montes y del Medio Natural,
Universidad Politécnica de Madrid, 28040 Madrid, Spain
[4] CBGP UPM-INIA, Universidad Politécnica de Madrid,
Campus de Montegancedo, 28223 Pozuelo de Alarcón, Madrid, Spain

Abstract. Gene expression analyses of non-model organisms must start with the construction of a high accurate *de novo* transcriptome as a reference. The best way to determine the suitability of any *de novo* transcriptome assembling is its comparison with other well-known "reference" transcriptomes. In this study, we took six complete plant transcriptomes (*Arabidopsis thaliana, Vitis vinifera, Zea mays, Populus trichocarpa, Triticum aestivum* and *Oryza sativa*) and compared all of them using a series of metrics system for a principal component analysis, resulting that *A. thaliana* and *P. trichocarpa* were the best references. This has been automated using AutoFlow. A primary assembly of short reads from Illumina Platform (50 nt, single reads) and long reads from Roche-454 technology from *Castanea sativa* was performed individually using *k*-mers from 25 to 35 and different assemblers (Oases v2, SOAPdenovoTrans, RAY, MIRA4 and MINIMUS). The resulting contigs were then reconciled with the aim of obtaining the best transcriptome. Oases and SOAP were used for the assembling of short reads, MIRA and MINIMUS for the assembling of long reads or the reconciliations, and RAY, that can compute *de novo* transcript assembling from heterogeneous (long and short reads) next-generation sequencing data, was included to avoid the reconciliation step. A total of 90 different assemblies were generated in a single run of the pipeline. A hierarchical clustering on the PCA components (HCPC) was implemented to automatically identify the best assembling strategies based on the shortest distance in HCPC to the two plant reference transcriptomes is selected. In this approach, reconciliation of Roche/454 long reads with Illumina contigs produce more complete and accurate gene reconstructions than other combinations. Surprisingly, reconstructions based only on Illumina and the ones creates with RAY seem to be less accurate. For this specific study, the most complete and accurate transcriptome corresponds to the Illumina contigs obtained with SOAPdenovoTrans and reassembled with 454 long reads using MIRA4. This is only a one example of a

© Springer International Publishing AG 2017
I. Rojas and F. Ortuño (Eds.): IWBBIO 2017, Part II, LNBI 10209, pp. 489–499, 2017.
DOI: 10.1007/978-3-319-56154-7_44

transcriptome building. Many other assembling can be performed just changing parameters, *k*-mers, sequencing technology, assemblers, reference organisms, etc. The pipeline in AutoFlow is easily customizable for those purposes.

Keywords: Transcriptome · Assembling · Workflow · PCA · Non-model organism

1 Background

RNA-seq is becoming a standard analysis for gene expression. In species where no genome sequence is available, the first step is the procurement of a *de novo* reference transcriptome, preferably from reads that will not be used in the future analysis or, less suitably, from the same reads if no other possibility is available. A problem is how to evaluate the suitability of this *de novo* transcriptome. The hypothesis that will be tested in this work is that reliable *de novo* transcriptomes should resemble well-characterised transcriptomes of model species.

Non-model species are very important from the economical or ecological point of view. For example, the European chestnut (*Castanea sativa*) is a forest tree having an important impact on local economy [1]. The main problems affecting its sustainable exploitation are maintaining genetic diversity, growing a superior nut and timber as well as attaining resistance to pests and disease such as canker blight. El Bierzo (Spain) is especially sensitive to chestnut yields because of its abundance in this region and because about 80% of their grafted trees are affected by canker blight disease, with important consequences in nut and timber production.

As an example, we have already dedicated some efforts in identifying most prominent genes involved in the defense response of chestnut against the blight disease [1]. It was carried out using RNA-seq analyses with samples from a virulent and a non-virulent strain of *C. parasitica* on natural chestnuts, revealing that the jasmonate and ethylene pathways are activated on *C. parasitica* infection, including the ROS production [1].

The objective of the present study is to obtain an automated, reproducible and flexible workflow that served to obtain the most complete and reliable *de novo* transcriptome of a non-model species. It is illustrated with the European chestnut reference transcriptome and will be extended to another plant species, such as olive tree or grapevine.

2 Methods

2.1 Reference Transcriptomes

The evaluation of external reference transcriptomes was based on the comparison of six plant transcriptomes downloaded from Phytozome. The model plant selected were *Arabidopsis thaliana, Vitis vinifera, Zea mays, Populus trichocarpa, Triticum aestivum and Oryza sativa.*

2.2 High Throughput Sequences

Four libraries from the RNA-seq experiment described by Quintana [1] were used, that account for a total of 90,549,382 single-end reads of 50 bp, generated by Illumina High-Seq 2000 system. Additionally, a total of 263,165 long reads from Roche-454 technology available at NCBI with the accession SRR954861 were also incorporated.

2.3 Assembling Workflow

The assembling workflow was constructed using AutoFlow [2], a workflow manager developed in our laboratory that has been successfully used in other studies to ensure both reproducibility and reliability [3, 4]. It is Ruby-based and accepts integration and automatic distribution of tasks to a queue system. It also has the capability of splitting big jobs into small parallel tasks, decision taking on-the-fly, dependence handling and quality control checks. The general outline of the assembling workflow is illustrated in Fig. 1, where the three main blocks are clearly defined, as explained in next sections.

2.3.1 Raw Data Pre-processing

Since raw reads came from two different technologies, they are pre-processed separately using SeqTrim-Next [1]. Pre-processing includes filtering low-quality sequences, identification of specific features (such as poly-A or poly-T tails, terminal transferase tails, and adaptors), removal of contaminant sequences (from vector to any other artefacts) and trimming (not masking) the undesired segments. As a result, the single-end and paired-end sequences were obtained as FastQ files that will be subsequently used in the next step.

2.3.2 Assembling Approaches

This module is intended to combine different assemblers based on different algorithms adapted for only one type of reads, or that can accept different types of reads: Oases [5], SOAPdenovoTrans [6] and Ray [7] for Illumina reads, Euler-SR [8] and MIRA4 [9] for Roche-454 reads, and MIRA4 and RAY for both reads, although only RAY can handle them simultaneously. Ray, CAP3 [10] and Miminus [11] were used for reconciliation of contigs derived from assembled reads.

The Illumina reads were assembled using Bruijn algorithms. We chose Oases [5] and SOAPdenovo-Trans [6] and RAY [7], both designed to assemble RNA-seq reads in the absence of a reference genome. They were executed using k-mers 25 and 35. Mainly due to the read length, the assemblers used for Illumina will not suit for the Roche-454 reads (except RAY). We therefore run a combination of two completely different algorithms with the idea that better results can be produced compared to each algorithm individually. One is MIRA4 (Mimicking Intelligent Read Assembly, based on overlap-layout-consensus algorithm) using the RNASeq parameters [9], and the other is Euler-SR (based on de Bruijn graphs using k-mer 29) [8]. The Euler-SR and MIRA4 resulting contigs are then reconciled with CAP3 [10] to improve the contig result.

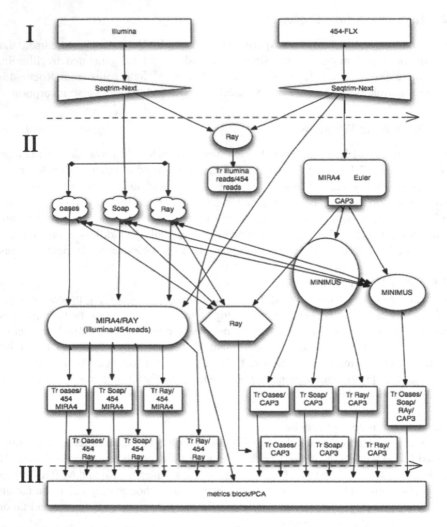

Fig. 1. Workflow designed for the construction of the best *C. sativa* transcriptome by combining different assemblers and different assembling strategies. The workflow can be divided in three parts: (I) Pre-processing of the raw data, (II) assembling approaches, and (III) validation using a set of metrics for further principal component analyses.

With the rationale of using long read for extending contigs derived from short reads, Roche-454 reads were assembled with the basic Illumina assemblies of Oases, SOAPdenovoTrans and RAY. We selected MIRA4 and RAY again to do so because of their ability to assemble sequences from different technologies and different lengths.

In another strategy, Illumina contigs were reconciled with Roche-454 contigs using Minimus. This assembler addresses the challenges of large whole-genome sequencing projects and produces significantly fewer assembly errors, at the cost of generating a more fragmented assembly [11]. Going further, we also used RAY with the same aim.

This assembler improves assemblies on Illumina in comparison with others short-read assemblers and is the first assembler that can simultaneously assemble reads or contigs from a mix of sequencing technologies [7].

The last approach is based on the idea that assemblers produce better results when combining raw reads than when pre-assembled reads are used. However, due to the huge number of reads that are obtained with the current technologies and the very different properties of Roche-454 and Illumina outputs, it is not an easy task. Fortunately, RAY Meta can compute *de novo* genome or transcript assemblies with huge, heterogeneous amounts of next-generation sequencing data [12]. We therefore use it for such a purpose without any further reconciliation.

2.3.3 Validation

The final part of the workflow is dedicated to produce metrics for the evaluation of the accuracy and precision of the assembled transcriptomes. The metrics are calculated for all assemblies as well as the two reference transcriptomes (*A. thaliana* and *P. thichocarpa*) to have a targeting metric. The statistical analysis is based on principal component analysis (PCA) to convert all the factors and variables gained into a set of values of linearly uncorrelated variables.

To compile metrics and statistical information, the following software is executed for each assembling (or reference transcriptome): (i) BUSCO (Benchmarking Universal Single-Copy Orthologs) to measure for quantitative assessment of genome assembly and annotation completeness [13]; (ii) Full-Lenghter-Next to classify transcript by their different full-length status, which transcripts are misassembled or chimeras, and to provide orthology annotation; (iii) Bowtie mapper to provide the percentage mapping reads; and (iv) a Ruby script to calculate N50, N90, number of contigs, transcript mean length, total number of Ns, and the number of gaps. As a result, the complete set of data for each assembling consist of factors (program used, task, *k*-mers and sequencing technology) and variables (number of contigs, Contigs > 500 nt, transcript mean length, N50, total Ns, number of gaps, number of transcripts coding a complete protein, number of partial transcripts, number of misassembled transcripts, number of chimerical transcripts, number of different orthologs, number of proteins annotated, mapping rates by sequencing technology, number of transcripts having reads from different sequencing technologies). All these are PCA analysed with FactoMineR [14] to find which variables are most strongly correlated and which assembling is closer to the *P. trichocarpa* and *A. thaliana* complete transcriptomes. Also, a hierarchical clustering on the PCA components (HCPC) [15] were implemented to identify the best assembling strategies. It was selected as the one having the least distance in HCPC to the two plant reference transcriptomes.

3   Results and Discussion

3.1   Plant Reference Transcriptomes

Whereas the construction of *de novo* transcriptomes has already been widely tackled in many studies before [2–4], it has never been taken into account the fact that every sequencing library could potentially have different optimal methods for its *de novo* assembling. We have then conceived a very complex strategy, comparing different assemblers based on different algorithms or using different parameters with the purpose of not being bound with one algorithm in order to finding the most accurate transcriptome for a non-model organism. The process must be automated due to its inherent complexity and the indispensable repeatability in future uses. Hence, the assembling combinations described in Fig. 1 enable the generation of 90 different transcriptomes to test against a target, reference transcriptome.

Having a good reference is crucial to evaluate the *de novo* transcriptomes. Six complete plant transcriptomes (*Arabidopsis thaliana, Vitis vinifera, Zea mays, Populus trichocarpa, Triticum aestivum and Oryza sativa*) were evaluated using the metric system, PCA and HCPC analyses described in part III of Fig. 1. The results (Fig. 2) indicate that only the transcriptomes of *P. trichocarpa* and *A. thaliana* are clustering in a region where the optimal parameters reached a maximum (Fig. 3). This suggests that only these transcriptomes must be used as reference for further analyses.

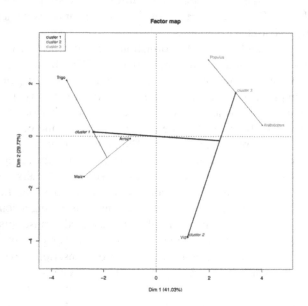

Fig. 2. Plot of the two main dimensions of the PCA comparing the six plant reference genomes. Colours represent clusters, the green being in the positive area for both dimension, suggesting that these two transcriptomes are the most reliable ones. (Color figure online)

3.2   Towards the Best Chestnut Transcriptome

Using the two reference transcriptomes, the best *Castanea sativa de novo* assembling was investigated. A total of 88.4 million reads from the Illumina platform were recovered as useful, and 147.700 from the Roche-454 technology. This means a 97 and 56% of recovery, respectively. Reads were assembled as indicated in Fig. 1-(ii) and then validated (Fig. 1-(iii)). The results of the validation by PCA were first analyzed as

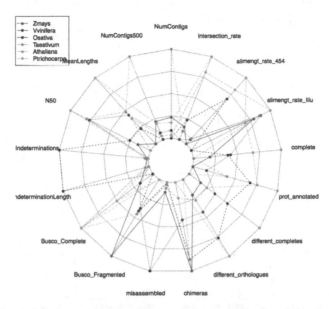

Fig. 3. Scheme comparing the six references according to metrics described in this study. *A. thaliana* and *P. trichocarpa* present metrics such as complete contigs or protein annotated outside of the circle, while indeterminations go inside, confirming that they seem better than others.

a hierarchical clustering (Fig. 4). Control transcriptomes were clustered together (blue box in Fig. 4), and the two closer assemblies correspond to the Roche-454 primary assemblies reconciled with Illumina primary assemblies (irrespective of the assembler; Fig. 4, green box).

When plotting the first two dimensions of the PCA, first dimension accounts for the 52.4% of variability, and the second for the 25.0% (Fig. 5). In all cases, the reconciliation of Illumina assemblies with the two *k*-mers with SOAPdenovoTrans and the Roche-454 reads is the transcriptome better targeting the control transcriptomes. However, in Fig. 5, the red block of assemblies seems also close to the reference transcriptomes, but this is a visual artefact since both axes are of equal magnitude even though they represent very different variability. This effect becomes minimized in Fig. 6, where a 3-dimension graph combines Figs. 4 and 5.

Since the HCPC method is able to rank the average distances from the new transcriptomes to the references (not shown), showing that the assembling with the shortest distance to the references corresponds to the Roche-454 long reads and Illumina contigs assembled with SOAPdenovoTrans. This reconciliation using MIRA4, produces more complete and accurate reconstruction of genes than the assemblies reconstructed using Minimus or even RAY from 454 and Illumina primary assemblies (Fig. 6). Also, the use of SOAPdenovoTrans as Illumina assembler when *k*-mers 25 and 35 are merged by Minimus, is the best option from all the tested approaches.

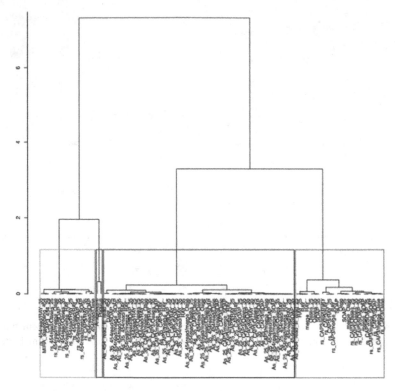

Fig. 4. Hierarchical clustering of assemblies after the PCA. In red and black, Illumina primary assemblies and Illumina assemblies reconciled with Roche-454 assembly, also Illumina and Roche-454 raw reads assembly; in blue, references (*P. trichocarpa* and *A. thaliana* complete transcriptomes); in green, Roche-454 primary assembly; in blue, reconciliation of Roche-454 reads and Illumina assemblies. (Color figure online)

3.3   Important Variables for Assembling Chestnut Transcriptome

In order to discover which is the contribution of each factor and variable in the PCA for this particular analysis, we represented them as a circular diagram (Fig. 7) divided into four areas, two for each dimension with either positive or negative values. For instance, the annotated proteins and the number of contigs are the variables with more significance in the study, according to the first dimension, and they are negatively correlated. In fact, the first dimension is highly correlated to the protein annotated in the transcriptomes. Notice that less number of contigs implies more protein annotated and that the most important contributions for the first principal component are the variables obtained from Full-LenghterNext and the script of metrics. The most important contribution for the second dimension are the number of complete transcripts predicted with BUSCO [13]. On the other hand, other variables seem to contribute equally, such as mean length and the number of different complete proteins (longer transcripts are more probably producing complete proteins), and the N50 and the total number of complete proteins (the same as above). No extrapolation can be done for future assembling since

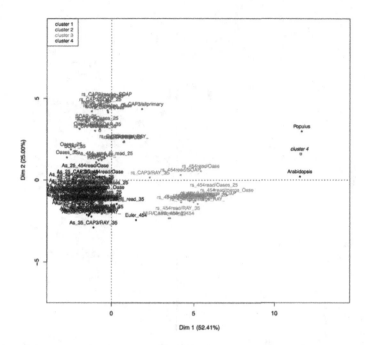

Fig. 5. Plot of the two main dimensions of the PCA. Colours are as in Fig. 4. (Color figure online)

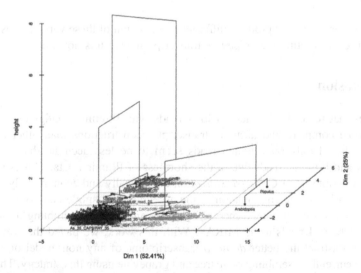

Fig. 6. 3D hierarchical clustering where colours are as in Fig. 5. Clearly, the blue collection of assemblies produce metrics more resembling the control transcriptomes. (Color figure online)

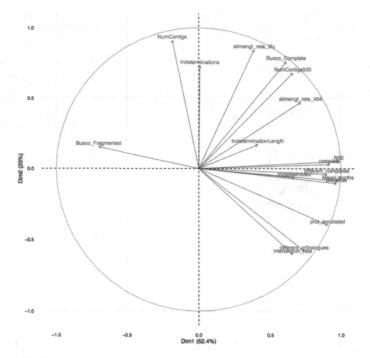

Fig. 7. Variable factor map to explain the contribution of the different factors and variables used in the PCA.

new kind of raw reads will produce different contribution of these variables, as we have seen when reconstructing the grapevine transcriptome (results not shown).

4   Conclusion

Assemblies that reconcile Roche-454 long reads with Illumina contigs using MIRA4 produce more complete and accurate transcript reconstructions than other strategies. Assemblies based only on Illumina reads seem to be less accurate than using both Roche-454 and Illumina reads due to the shortness of Illumina data. This is illustrated with the transcriptome of *Castanea sativa*, an ecologically and economically important species. It consists in the combination of the assembling of single-end 50 nt reads with SOAPdenovoTrans combining *k*-mers 25 and 35, and then reassembling these contigs with the Roche-454 reads using MIRA4. With this work, we paved the way to automatically reconstruct the better *de novo* transcriptome of any non-model organism. In fact, we are currently assembling olive tree and grapevine using this strategy. That is why we state that this workflow can be tailored to the user's convenience in order to include more assemblers and parameters for any kind of sequencing data and organism, extending the application of *de novo* assemblint to any species lacking of genomic information.

Acknowledgments. This work has been supported by co-funding from the ERDF (European Regional Development Fund) 2014-2020 "Programa Operativo de Crecimiento Inteligente" to the grant RTA2013-00068-C03 of the Spanish INIA and MINECO. The authors also thankfully acknowledge the computer resources and the technical support provided by the Plataforma Andaluza de Bioinformática of the University of Málaga.

References

1. Quintana, J.: Molecular tools to improve chestnut management: El Bierzo as case of study, Ph.D. Thesis (2015)
2. Seoane, P., Ocaña, S., Carmona, R., Bautista, R., Madrid, E., Torres, A.M., Claros, M.G.: AutoFlow, a versatile workflow engine illustrated by assembling an optimised de novo transcriptome for a non-model species, such as Faba Bean (*Vicia faba*). Curr. Bioinform. **11** (4), 440–450 (2016)
3. Ocana, S., Seoane, P., Bautista, R., Palomino, C., Claros, G.M., Torres, A.M., Madrid, E.: Large-scale transcriptome analysis in Faba Bean (Vicia Faba L.) under ascochyta fabae infection. PLoS ONE **10**(8), 1–17 (2015)
4. Carmona, R., Zafra, A., Seoane, P., Castro, A., Guerrero-Fernández, D., Castillo-Castillo, T., Medina-García, A., Cánovas, F.M., Aldana-Montes, J.F., Navas-Delgado, I., Alché, J.D., Claros, M.G.: ReprOlive: a database with linked data for the olive tree (*Olea europaea* L.) reproductive transcriptome. Front. Plant Sci. **6**, 625 (2015)
5. Schulz, M.H., Zerbino, D.R., Vingron, M., Birney, E.: Oases: robust de novo RNA-seq assembly across the dynamic range of expression levels. Bioinformatics **28**, 1086–1092 (2012)
6. Luo, R., Liu, B., Xie, Y., Li, Z., Huang, W., Yuan, J., He, G., Chen, Y., Pan, Q., Liu, Y.Y.Y.Y., Tang, J., Wu, G., Zhang, H., Shi, Y., Liu, Y.Y.Y.Y., Yu, C., Wang, B., Lu, Y., Han, C., Cheung, D.W., Yiu, S.-M., Peng, S., Xiaoqian, Z., Liu, G., Liao, X., Li, Y., Yang, H., Wang, J.J., Lam, T.-W., Wang, J.J.: SOAPdenovo2: an empirically improved memory-efficient short-read de novo assembler. Gigascience **1**(1), 18 (2012)
7. Boisvert, S., Laviolette, F., Corbeil, J.: Ray: simultaneous assembly of reads from a mix of high-throughput sequencing technologies. J. Comput. Biol. **17**(11), 1519–1533 (2010)
8. Pevzner, P.A., Tang, H., Waterman, M.S.: An Eulerian path approach to DNA fragment assembly. Proc. Natl. Acad. Sci. U.S.A. **98**, 9748–9753 (2001)
9. Chevreux, B., Pfisterer, T., Drescher, B., Driesel, A.J., Müller, W.E.G., Wetter, T., Suhai, S.: Using the miraEST assembler for reliable and automated mRNA transcript assembly and SNP detection in sequenced ESTs. Genome Res. **14**(6), 1147–1159 (2004)
10. Huang, X., Madan, A.: CAP3: a DNA sequence assembly program. Genome Res. **9**(9), 868–877 (1999)
11. Sommer, D.D., Delcher, A.L., Salzberg, S.L., Pop, M.: Minimus: a fast, lightweight genome assembler. BMC Bioinform. **8**, 64 (2007)
12. Boisvert, S., Raymond, F., Godzaridis, E., Laviolette, F., Corbeil, J.: Ray Meta: scalable de novo metagenome assembly and profiling. Genome Biol. **13**(12), R122 (2012)
13. Simão, F.A., Waterhouse, R.M., Ioannidis, P., Kriventseva, E.V., Zdobnov, E.M.: BUSCO: Assessing genome assembly and annotation completeness with single-copy orthologs. Bioinformatics **31**(19), 3210–3212 (2015)
14. Lê, S., Josse, J., Husson, F.: FactoMineR: an R package for multivariate analysis. J. Stat. Softw. **25**(1), 1–18 (2008)
15. Husson, F., Josse, J., Pagès, J.: Principal component methods - hierarchical clustering - partitional clustering: why would we need to choose for visualizing data? Technical report, pp. 1–17 (2010)

Accelerating Smith-Waterman Alignment of Long DNA Sequences with OpenCL on FPGA

Enzo Rucci[1], Carlos Garcia[2($\boxtimes$)], Guillermo Botella[2], Armando De Giusti[1], Marcelo Naiouf[3], and Manuel Prieto-Matias[2]

[1] III-LIDI, CONICET, Facultad de Informática,
Universidad Nacional de La Plata, 1900 La Plata, Buenos Aires, Argentina
{erucci,degiusti}@lidi.info.unlp.edu.ar
[2] Depto. Arquitectura de Computadores y Automática,
Universidad Complutense de Madrid, 28040 Madrid, Spain
{garsanca,gbotella,mpmatias}@ucm.es
[3] III-LIDI, Facultad de Informática, Universidad Nacional de La Plata,
1900 La Plata, Buenos Aires, Argentina
mnaiouf@lidi.info.unlp.edu.ar

Abstract. With the greater importance of parallel architectures such as GPUs or Xeon Phi accelerators, the scientific community has developed efficient solutions in the bioinformatics field. In this context, FPGAs begin to stand out as high performance devices with moderate power consumption. This paper presents and evaluates a parallel strategy of the well-known Smith-Waterman algorithm using OpenCL on Intel/Altera's FPGA for long DNA sequences. We efficiently exploit data and pipeline parallelism on a Intel/Altera Stratix V FPGA reaching upto 114 GCUPS in less than 25 watt power requirements.

1 Introduction

In recent years, genome research projects have produced a vast amount of biological data. In fact, biologists are working in conjunction with computer scientists to extract relevant biological information from these experiments. The comparison of millions of sequences [13] is one of the most useful mechanisms known in Bioinformatics, commonly solved by heuristic methods.

Smith-Waterman (SW) algorithm compares two sequence in an exact way and corresponds to the so-called local methods because it focuses on small similar regions only. Besides, this method has been used as the basis for many subsequent algorithms and is often used as basic pattern to compare different alignment techniques. However, one of the main drawbacks is the cost of this approach in computing time and memory requirements which makes it unsuitable in some cases.

Regarding on the performance aspect, many approaches, such as BLAST [1] and FASTA [14] considerably reduce execution time at the expense of not guaranteeing the optimal result. Nevertheless, accelerating the SW algorithm is still a great challenge for the scientific community. SW is usually used to align two DNA

© Springer International Publishing AG 2017
I. Rojas and F. Ortuño (Eds.): IWBBIO 2017, Part II, LNBI 10209, pp. 500–511, 2017.
DOI: 10.1007/978-3-319-56154-7_45

sequences or a protein sequence to a genomic database. A single matrix must be built for each alignment and the matrix size depends on sequence lengths. In DNA alignment, the matrix can be huge since these sequences can have upto hundreds of million nucleotides. As protein sequences are shorter, multiple small matrices are usually computed simultaneously since they are independent among them [16].

In the last years, we have witnessed attempts to parallelize for both schemes. These efforts reduce SW execution time through the exploitation of High-Performance Computing (HPC) architectures. However, most implementations focus on short sequences, particularly protein sequences [15]. For very long sequences, as the DNA case, few implementations are nowadays available. In the hardware accelerators scenario, we highlight *SW#* [9] and *CUDAlign* [18] (and its newer versions [4,5]) that focus on the alignment of huge DNA sequences with multi CUDA-enabled Graphics Processor Units (GPUs). Meanwhile, Liu et al. have presented *SWAPHI-LS* [12] a highly optimized hand-tuned implementation for Intel Xeon Phi accelerators. In addition, several proposals based on Field Programmable Gate Array (FPGA) speedup sequence alignments [2,22] in the context of DNA comparison. Moreover, Wienbrandt presents a study of multi-FPGAs version in [21].

Nowadays, HPC capabilities are changing in data-centers scenario. FPGAs are being integrated with CPUs due to accelerators consolidation in HPC community as a way of improving performance while keeping power efficiency. Recently, Microsoft announced that their data-centers equipped with FPGAs increase dramatically Bing's engine capacities, and they have also incorporated them to Azure ecosystem [6]. In the same way, Amazon has included FPGAs in its Amazon Web Services [10]. With the acquisition of Altera in 2015, Intel plans to incorporate FPGA capabilities in the next Xeon's server processors because they expect to be used by at least 30% of data-center in next years [11].

Our paper proposes and evaluates a SW implementation, capable of aligning DNA sequences of unrestricted size, for Altera's FPGA. Unlike previous works on literature, we employed the novel OpenCL paradigm on FPGAs. Although Altera's developers promote a similar implementation in [19], no-real sequence data of fixed, limited size ($n = 256$ residues) are used. As this issue can radically differ from real bioinformatic contexts, the observed behavior becomes unpredictable for these scenarios, especially in alignment of long DNA sequences. We would like to point out that, unlike GPU or Xeon Phi accelerators which need to be purchased separately, the hardware used in this study will be available soon in the next processor generation at null cost. In that sense, this study can be taken as starting point for future hybrid CPU-FPGA solutions.

The rest of the paper is organized as follows. Section 2 describes SW algorithm. Section 3 introduces Altera's OpenCL programming extension while Sect. 4 addresses our parallelization of the SW algorithm using OpenCL on FPGAs. Section 5 presents experimental results and finally Sect. 6 outlines conclusions and future lines of work for this novel viability study.

2 Smith-Waterman Algorithm

The SW algorithm is used to obtain the optimal local alignment between two sequences and was proposed by Smith and Waterman [20]. This method employs a dynamic programming approach and its high sensitivity comes from exploring all the possible alignments between two sequences.

Given two sequences S_1 and S_2, with sizes $|S_1| = m$ and $|S_2| = n$, the recurrence relations for the SW algorithm with affine gap penalties [7] are defined below.

$$H_{i,j} = max\{0, H_{i-1,j-1} + SM(S_1[i], S_2[j]), E_{i,j}, F_{i,j}\} \tag{1}$$

$$E_{i,j} = max\{H_{i,j-1} - G_{oe}, E_{i,j-1} - G_e\} \tag{2}$$

$$F_{i,j} = max\{H_{i-1,j} - G_{oe}, F_{i-1,j} - G_e\} \tag{3}$$

$H_{i,j}$ contains the score for aligning the prefixes $S_1[1..i]$ and $S_2[1..j]$. $E_{i,j}$ and $F_{i,j}$ are the scores of prefix $S_1[1..i]$ aligned to a gap and prefix $S_2[1..j]$ aligned to a gap, respectively. SM is the *scoring matrix* which defines match/mismatch scores between nucleotides. G_{oe} is the sum of gap open and gap extension penalties while G_e is the gap extension penalty. The recurrences should be calculated with $1 \leq i \leq m$ and $1 \leq j \leq n$, after initialising H, E and F with 0 when $i = 0$ or $j = 0$. The maximum value in the alignment matrix H is the optimal local alignment score.

It is important to note that any cell of the matrix H can be computed only after the values of the upper, left and upper-left cells are known, as shown in Fig. 1. These dependences restrict the ways in that H can be computed.

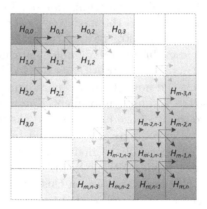

Fig. 1. Data dependences in the alignment matrix H.

3 OpenCL Extension on Altera's FPGA

OpenCL is a framework for developing parallel programs across heterogeneous platforms. It is currently supported by several hardware devices such as CPUs,

GPUs, DSPs and FPGAs. The OpenCL is based on host-device model, where host is in charge of managing device memory, transferring data from/to device and launching the kernel code.

Kernel corresponds to a piece of code which expresses the parallelism of a program. OpenCL programming model divides a program workload into *work-groups* and *work-items*. Usually, in the denoted *data parallel programming* model, *work-items* are grouped into *work-groups*, which are executed independently on a processing element. Data-level parallelism is ordinarily exploited in SIMD way, where each *work-item* is mapped to a lane width of the target device.

OpenCL memory model uses a particular memory hierarchy which is also characterized by the access type, performance and scope. Global memory is read-write accessible by all *work-items* which implies a high latency memory access. Local memory is a shared read-write memory that can be accessed from all *work-items* of a single *work-group*. Besides, it usually involves a low latency memory access. Constant memory is a read-only memory visible by all *work-items* across all *work-groups*, and private memory as the name suggests it is only accessible by a single *work-item*.

OpenCL allows a programmer to express parallelism abstracting the target platform details. We can highlight portability and reduction in development time as the main advantages. FPGAs permit programming networks composed of logic elements, memory blocks and specific DSP blocks. In order to verify and create digital designs, Hardware Description Languages (HDLs) are generally used, which are complex, error prone and affected by an extra abstraction layer as they contain the additional concept of time.

Regarding the execution model, Altera's OpenCL SDK [3] recommends the use of *task parallel programming* model, where the kernel consists of a single *work-group* with a unique *work-item*. The Altera OpenCL Compiler (AOC) is capable of extracting parallelism from each loop iteration in a loop-pipelined way allowing to process the loop in a high-throughput fashion.

Altera's OpenCL extension also take advantage of I/O channels and kernel channels as in OpenCL 2.0 by means of pipes [8]. Altera's channel extension allows to transfer data between work-item's in the same kernel or between different kernels without host interaction.

4   SW Implementation

In this section we will address the programming aspects and optimizations applied to our implementations on FPGA accelerated platforms. Algorithm 1 shows the pseudo-code for the host implementation. Memory management is performed in OpenCL by means of *clCreateBuffer* (memory allocation and initialization) and *clEnqueueReadBuffer* (memory transfer to host). Kernels are invoked through the *clEnqueueTask* function.

The kernel is implemented following the *task parallel programming* model mentioned in Sect. 3. Algorithm 2 shows the pseudo-code for our kernel. The alignment matrix is divided into vertical blocks (*BW* means *Block Width*) and

Algorithm 1. Host pseudo-code

1: clCreateBuffer's(...)	▷ Create buffers + transfer sequences
2: $NB = n/BW$	▷ NB is the number of vertical blocks
3: **for** $b \leq NB$ **do**	
4: clEnqueueTask(...)	▷ Compute b-th block
5: swap($prevLastColH, curLastColH$)	
6: swap($prevLastColE, curLastColE$)	
7: **end for**	
8: clEnqueueReadBuffer($maxScore$)	

Algorithm 2. Pseudo-code for Smith-Waterman kernel

1: _kernel void SW_kernel (S_1, S_2, m, b, $match$, $mismatch$, G_{oe}, G_e, $prevLastColH$, $curLastColH$, $prevLastColE$, $curLastColE$, $maxScore$) {
2: Load the BW residues of S_2 corresponding to b-th block from global memory to private memory
3: **for** $i \leq m$ **do**   ▷ each row
4:     Load the i-th residue of S_1 from global memory to private memory
5:     Read previous block data from global memory ($prevLastColH$ and $prevLastColE$)
6:     #pragma unroll
7:     **for** $j \leq BW$ **do**
8:         Calculate $H_{i,j}$ in private memory
9:     **end for**
10:    Write data for next block to global memory ($curLastColH$ and $curLastColE$)
11: **end for**
12: Update $maxScore$ in global memory (if appropiate)
13: }

each block is computed in row-by-row manner: from top to bottom, left to right direction (see Fig. 2). Besides improving the data locality, this blocking technique reduces the memory requirements for block execution, which favors the exploitation of the private low-latency memory. In that sense, we employed two buffers to store one row for matrices H and F. Additionally, both sequences are partially copied to private memory.

Fully unrolling of the inner loop represents an essential aspect of this kernel from performance point of view. This technique allows the AOC to exploit loop instruction pipelining and, in consequence, more operations per clock cycle are performed. As the compiler needs to know the number of iterations at compile phase, S_2 sequence must be extended with dummy symbols to make its length a multiple of fixed BW value.

Due to the data dependencies mentioned in Sect. 2, each block needs the last column H and E values of the previous block. Global memory buffers are employed to communicate these data. To avoid read-write dependences in global memory, separate buffers are used: one for reading the values from the previous block and one for writing the values for the next block, so after each kernel invocation buffers are swapped in the host. It is important to mention that,

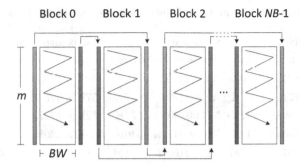

Fig. 2. Schematic representation of our OpenCL kernel implementation.

although in *OSWALD* implementation [17] Altera OpenCL channels are used to exchange these information, the use of this technique is not affordable in DNA context with million of nucleotide bases involved.

Moreover, host-side buffers are allocated to be 64-byte aligned. This fact improves data transfer efficiency by means of Direct Memory Access. Both sequence are transferred when creating the device buffers and optimal score is retrieved after all kernels finished.

5 Experimental Results

5.1 Experimental Platforms and Tests Carried Out

Tests have been performed on different platforms running CentOS (release 6.6):

- A server with two Intel Xeon CPU E5-2670 8-core 2.60 GHz, 32 GB main memory and an Altera Stratix V GSD5 Half-Length PCIe Board with Dual DDR3 (two banks of 4 GByte DDR3).
- A server with two Intel Xeon CPU E5-2695 v3 16-core 2.30 GHz, 64 GB main memory and two NVIDIA GPU cards: one Tesla K20c (Kepler architecture, 2496 CUDA cores, 5 GB dedicated memory and Compute Capability 3.5) and one GTX 980 (Maxwell architecture, 2048 CUDA cores, 4 GB dedicated memory and Compute Capability 5.0).
- A server with two Intel Xeon CPU E5-2695 v3 16-core 2.30 GHz, 128 GB main memory and a single Xeon Phi 3120P coprocessor card (57 cores with 4 hw thread per core, 6 GB dedicated memory).

We have used the Intel's ICC compiler (version 16.0.3) with the *-O3* optimization level by default. The synthesis tool used is Quartus II DKE V12.0 2 with OpenCL SDK v14.0 and the CUDA SDK version is 7.5.

To provide the most relevant study, tests were made with the real DNA sequences retrieved from the National Center for Biotechnology Information (NCBI)[1], ranging from thousands to millions of nucleotide bases. The accession numbers and sizes of the sequences are presented in Table 1. Also, for the

---

[1] Sequences are available in http://www.ncbi.nlm.nih.gov.

Table 1. Information of the sequences used in the tests.

Sequence 1		Sequence 2		Matrix size	Score
Accesion	Size	Accesion	Size		
AF133821.1	10K	AY352275.1	10K	10K × 10K	5027
NC_001715.1	57K	AF494279.1	57K	57K × 57K	51
NC_000898	162K	NC_007605	172K	162K × 172K	18
NC_003064.2	543K	NC_000914.1	536K	543K × 536K	48
CP000051.1	1M	AE002160.2	1M	1M × 1M	82091
BA000035.2	3M	BX927147.1	3M	3M × 3M	3888
AE016879.1	5M	AE017225.1	5M	5M × 5M	5220775
NC_005027.1	7M	NC_003997.3	5M	7M × 5M	157
NC_017186.1	10M	NC_014318.1	10M	10M × 10M	10235056
NT_033779.4	23M	NT_037436.3	25M	23M × 25M	9059

sake of validation, optimal alignment scores are included in Table 1. The scoring scheme used was: +1 for match; −3 for mismatch; −5 for gap open; and −2 for gap extension. Each particular test was run ten times and performance was calculated with the average of ten execution times to avoid variability.

5.2 Performance and Resource Usage Evaluation

The metric GCUPS is used to performance assessment in the Smith-Waterman scenario [18]. In order to evaluate FPGA performance rates, we have considered different kernel implementations according to integer data type and BW value. We detail below the main differences:

- The name prefix denotes the integer data type used; i.e. *int*, *short* and *char* represent 32, 16 and 8 bit integer data types, respectively.
- The name suffix denotes the BW value used; e.g. *bw32* means that BW value was set to 32.

Table 2 presents FPGA resource utilization and the performance achieved for our OpenCL kernels implementations. Larger BW means better performance but also higher resource consumption. Adaptive logic modules (ALMs) are the most influenced resources; registers (Regs) and RAM blocks (RAMs) are slightly increased while DSP blocks (DSPs) stay intact. In spite of this fact, there are still available resources that allow increasing BW parameter. Nevertheless, larger values could not be used because AOC reports non-real read-write dependences in private memory associated to H and F matrices. In fact, these *false* dependences reduce the number of operations per clock cycle decreasing considerably performance rates achieved.

Regarding integer data type, as can be observed, smaller data type not only improves performance but also decreases resource consumption. This behavior

Table 2. Performance and resource usage comparison for different OpenCL kernel implementations.

Kernel	int_bw32	int_bw64	int_bw128	int_bw256	short_bw256	short_bw512	char_bw512	char_bw768
Integer type	int (32 bits)				short (16 bits)		char (8 bits)	
Maximum value	2147483647				32767		127	
BW	32	64	128	256	256	512	512	768
ALMs	30%	36%	48%	69%	50%	76%	49%	68%
Regs	11%	12%	12%	12%	10%	13%	10%	15%
RAM	22%	22%	22%	25%	21%	24%	20%	33%
DSPs	0%	0%	0%	0%	0%	0%	0%	0%
10K × 10K	2.22	4.60	7.45	14.47	18.43	23.23	-	-
57K × 57K	4.27	8.42	16.32	30.35	38.01	58.21	73.07	91.89
162K × 172K	4.75	9.33	18.04	33.45	41.85	66.34	84.67	109.29
543K × 536K	4.97	9.66	18.86	35.51	43.36	70.21	86.01	113.78
1M × 1M	5.12	9.93	19.44	36.52	-	-	-	-
3M × 3M	5.24	10.14	19.87	37.32	45.58	73.39	-	-
5M × 5M	5.27	10.18	19.93	37.49	-	-	-	-
7M × 7M	5.28	10.20	20	37.56	45.86	73.73	-	-
10M × 10M	5.28	10.21	20.03	37.61	-	-	-	-
23M × 25M	5.29	10.23	20.07	37.68	46.01	74	-	-
Matrix size	GCUPS							

is clearly exposed when comparing *int_bw256* and *short_bw256* kernels: using the same *BW* configuration, *short_bw256* reports an increase from up to 1.22× in performance with a reduction from up to 0−0.28× in resource usage against to *int_bw256* counterpart. A similar behavior is observed with *short_bw512* and *char_bw512* kernels: *char_bw512* presents an increase from up to 1.28× in performance with a reduction from up to 0−0.36× in resource usage respect to *short_bw512*. However, it is also important to take into account that the use of narrower integer data types also involves an important reduction in representation range. Due to this fact, there are three alignment scores out of ten that can not be represented when using 16 bit integer data. It is also observed for the experiments with 8-bits data type where only three experiments can be carried out[2].

From sequence length point of view, all kernels benefit from larger workloads regardless of sequences similarity. The best performances achieved are 37.68, 74 and 113.78 GCUPS for *int*, *short* and *char* kernels, respectively.

5.3 Performance and Power Efficiency Comparison with Other SW Implementations

In this subsection our developed version is compared with other optimized SW implementations in the HPC scenario: the Xeon Phi-based *SWAPHI-LS* program (v1.0.12) [12] and the GPU-based *SW#* [9] and *CUDAlign* (v3.9.1.1024) [4] programs. We would like to note that both GPU implementations were configured to perform only their score version.

---

[2] The symbol '-' indicates an alignment that can not be computed because the optimal score exceeds the corresponding maximum value.

Table 3. Performance of other SW implementations.

Implementation	SWAPHI-LS	SW#	CUDAlign	SW#	CUDAlign
Device	Intel Xeon Phi 3120P	NVIDIA Tesla K20c		NVIDIA GTX 980	
10K × 10K	0.42	0.16	0.06	0.3	0.03
57K × 57K	7.69	5.86	1.57	7.62	1.08
162K × 172K	21.24	32.78	10.23	33.33	8.18
543K × 536K	30.67	42.36	29.71	64.53	45.89
1M × 1M	32.84	51.01	39.64	75.24	79.21
3M × 3M	33.9	43.48	39.73	69.54	84.05
5M × 5M	34.16	90.15	79.53	120.92	160.79
7M × 5M	34.38	43.64	39.55	68.84	84.43
10M × 10M	33.19	90.22	79.96	118.81	163.77
23M × 25M	30.36	44.19	40.69	67.55	84.84
Matrix size	GCUPS				

Table 3 presents performances for the *SWAPHI-LS*, *SW#* and *CUDAlign* implementations. Comparing our FPGA implementation against Xeon-Phi accelerator, *SWAPHI-LS* yields an average performance of 25.89 GCUPS and a peak of 34.38 GCUPS being outperformed by *int_bw256* version in all cases. In particular, the most impressive performance difference occurs for smaller matrix sizes where *int_bw256* runs on average 21.8× faster. For the rest of the tests, the performance gain decreases but still improves on average 1.45×.

Unlike our FPGA kernels, both GPU implementations are sensitive to sequences similarity; better results are obtained on alignments with higher scores. On Tesla K20c, *SW#* achieves an average performance of 44.38 GCUPS and a maximum performance of 90.15 GCUPS, outperforming *CUDAlign* by a factor of 1.78× on average. *int_bw256* improves *SW#* on the three shortest alignments while the latter runs 1.62× faster on average than the former on the largest ones. In contrast, *SW#* only beats *CUDAlign* on the four shortest alignments on GTX 980. *CUDAlign* reports 71.23 GCUPS on average and a peak of 163.77 GCUPS, achieving an average speedup of 1.24× on the six largest matrix sizes. Just as the previous case, *int_bw256* improves both GPU implementations on the three shortest alignments while is outperformed in the rest. In that sense, *CUDAlign* runs 2.76× faster on average than *int_bw256*.

In FPGA context, a theoretical comparison with other OpenCL implementations is the only possibility since the absence of available source codes. We would like to note that the Altera staff implementation [19] uses 32-bit integer data type and imposes a fixed, limited size ($n = 256$ residues) to S_2 sequence, which substantially differs from the common ones used in DNA analysis. In spite of this fact, Altera's staff implementation reported a peak of 24.7 GCUPS on Stratix V meanwhile our approach achieves an average of 33.8 GCUPs.

Table 4. Power efficiency comparison.

Implementation	Device	GCUPS	Watts	GCUPS/Watts
int_bw256	Altera Stratix V	37.68	25	1.51
short_bw512		74		2.96
char_bw768		113.78		4.55
SWAPHI-LS	Intel Xeon Phi 3120P	34.38	270	0.13
SW#	NVIDIA Tesla K20c	90.22	225	0.4
CUDAlign		79.96		0.36
SW#	NVIDIA GTX 980	120.92	165	0.73
CUDAlign		160.79		0.97

In order to complete this study, we have also analyzed the power efficiency. Table 4 presents power efficiency ratios considering the GCPUS peak performance and Thermal Design Power (TDP) in each accelerator. It can be seen that the worst ratio observed is for *SWAPHI-LS* due to low performance rates and high TDP on the Xeon Phi. GPU implementations place at an intermediate position because they obtain the highest performance peak but at the expense also of higher power requirements. As expected, both GPU implementations obtain better GCUPS/Watts ratios on GTX 980 in comparison with Tesla K20c since its better performance and power capacities. We would like to remark that FPGA implementation reaches the best GCUPS/Watts ratios. Despite not achieving the highest performance peak, its low power consumption leads to the best choice taking this aspect into account. Further, the lowest performance FPGA kernel considering integer data type (*int_bw256*) outperforms *SWAPHI-LS* implementation by a factor of 11.62× and to the GPU implementations by a range of 1.56–3.78×. It is important to mention that if this analysis is carried out considering average GCUPS instead of GCUPS peak, larger differences in favor of FPGA implementations will be found.

6 Conclusions

In this paper, we addressed the benefits of a parallel SW implementation using OpenCL on Intel/Altera's FPGA, not only from performance perspective but also from power efficiency point of view. To the best of the author's knowledge, this is the first time that a paper examines an implementation of this kind with real, long DNA sequences. In addition, as Intel will incorporate FPGA capabilities in its next Xeon processors in a free manner, this study can be useful as an starting point for future hybrid CPU-FPGA implementations.

Among main contributions of this research we can summarize:

- Data type exploitation is crucial to achieve successful performance rates. Narrower data-types reports better performance rates with less resource impact, but at expense of decreasing representation width. In fact, a peak of 114 GCUPS is reached when using 8 bit integers.

– From performance perspective, our most successful 32-bit implementation reaches 37.7 GCUPS peak, running 1.53× faster than Altera's staff implementation [19]. Despite being competitive in performance terms respect to other solutions on accelerators (such as GPUs or Xeon Phi), our implementation is significantly better from power efficiency perspective. In particular, the fastest 32-bit FPGA kernel outperforms *SWAPHI-LS* by a factor of 11.62× and the GPU implementations by a range of 1.56–3.78× in efficiency.

Taking into account these encouraging results, as future lines we will consider three aspects:

– As all kernels developed still have free available resources, we will try to exploit them to improve performance rates. On one hand, we expect to solve performance limitations imposed by AOC with larger BW values. On the other hand, we plan to combine distinct integer data width kernels to explore different configurations in order to find the best performance-representation width trade-off.
– As OpenCL allows multiple devices exploitation, we would like to extend this work to a multi-FPGA implementation and explore most successful workload distribution.
– As real power consumption on accelerators can differ from TDP values due to a variety of reasons, we plan to measure the instant power consumption in all devices in order to make a more fairly performance vs. power analysis.

Acknowledgments. This work has been partially supported by Spanish government through research contract TIN2015-65277-R and CAPAP-H5 network (TIN2014-53522).

References

1. Altschul, S.F., Madden, T.L., Schffer, A.A., Zhang, J., Zhang, Z., Miller, W., Lipman, D.J.: Gapped BLAST and PSIBLAST: a new generation of protein database search programs. Nucleic Acid Res. **25**(17), 3389–3402 (1997)
2. Caffarena, G., Pedreira, C.E., Carreras, C., Bojanic, S., Nieto-Taladriz, O.: FPGA acceleration for DNA sequence alignment. J. Circuits Syst. Comput. **16**(2), 245–266 (2007)
3. Altera Corporation: Altera SDK for OpenCL Programming Guide, v14.0 (2014)
4. de Oliveira Sandes, E.F., Miranda, G., Alves de Melo, A.C.M., Martorell, X., Ayguadé, E.: CUDAlign 3.0: parallel biological sequence comparison in large GPU clusters. In: CCGRID, pp. 160–169. IEEE Computer Society (2014)
5. de Oliveira Sandes, E.F., Miranda, G., Martorell, X., Ayguad, E., Teodoro, G., Alves de Melo, A.C.M.: CUDAlign 4.0: incremental speculative traceback for exact chromosome-wide alignment in GPU clusters. IEEE Trans. Parallel Distrib. Syst. **27**(10), 2838–2850 (2016)
6. Feldman, M.: Microsoft goes all in for FPGAs to build out AI cloud (2016). https://www.top500.org/news/microsoft-goes-all-in-for-fpgas-to-build-out-cloud-based-ai/

7. Gotoh, O.: An improved algorithm for matching biological sequences. J. Mol. Biol. **162**, 705–708 (1981)
8. Khronos Group: The OpenCL Specification, version 2 (2014)
9. Korpar, M., Sikic, M.: SW# - GPU-enabled exact alignments on genome scale. Bioinformatics **29**(19), 2494–2495 (2013)
10. Leopold, G.: AWS Embraces FPGAs, Elastic GPUs (2016). https://www.hpcwire.com/2016/12/02/aws-embraces-fpgas-elastic-gpus/
11. Leopold, G.: Intels FPGAs target datacenters, networking (2016). https://www.hpcwire.com/2016/10/06/intels-fpgas-target-datacenters-networking/
12. Liu, Y., Tran, T.T., Lauenroth, F., Schmidt, B.: SWAPHI-LS: Smith-Waterman Algorithm on Xeon Phi coprocessors for long DNA sequences. In: IEEE International Conference on Cluster Computing (CLUSTER), pp. 257–265 (2014)
13. Mount, D.W.: Bioinformatics: Sequence and Genome Analysis. Cold Spring Harbor Laboratory Press, Mount (2004)
14. Pearson, W.R., Lipman, D.J.: Improved tools for biological sequence comparison. Proc. Nat. Acad. Sci. U.S.A. **85**(8), 2444–2448 (1988)
15. Rucci, E., García, C., Botella, G., Giusti, A., Naiouf, M., Prieto-Matías, M.: State-of-the-Art in Smith–Waterman protein database search on HPC platforms. In: Wong, K.-C. (ed.) Big Data Analytics in Genomics, pp. 197–223. Springer, Cham (2016). doi:10.1007/978-3-319-41279-5_6
16. Rucci, E., Garca, C., Botella, G., De Giusti, A., Naiouf, M., Prieto-Matas, M.: An energy-aware performance analysis of SWIMM: SmithWaterman implementation on Intel's multicore and manycore architectures. Concurrency Comput. Pract. Exp. **27**(18), 5517–5537 (2015)
17. Rucci, E., Garca, C., Botella, G., De Giusti, A., Naiouf, M., Prieto-Matas, M.: OSWALD: OpenCL Smith-Waterman Algorithm on Altera FPGA for large protein databases. Int. J. High Perform. Comput. Appl. (2016). doi:10.1177/1094342016654215
18. de Oliveira Sandes, E.F., Alves de Melo, A.C.M: CUDAlign: using GPU to accelerate the comparison of megabase genomic sequences. In: Proceedings of the 15th ACM SIGPLAN Symposium on Principles and Practice of Parallel Computing, PPoPP 2010, pp. 137–146. ACM, New York (2010)
19. Settle, S.O.: High-performance dynamic programming on FPGAs with OpenCL. In: IEEE High Performance Extreme Computing Conference (HPEC 2013), pp. 1–6 (2013)
20. Smith, T.F., Waterman, M.S.: Identification of common molecular subsequences. J. Mol. Biol. **147**(1), 195–197 (1981)
21. Wienbrandt, L.: Bioinformatics applications on the FPGA-based high-performance computer RIVYERA. In: Vanderbauwhede, W., Benkrid, K. (eds.) High-Performance Computing Using FPGAs, pp. 81–103. Springer, New York (2013). doi:10.1007/978-1-4614-1791-0_3
22. Yamaguchi, Y., Tsoi, H.K., Luk, W.: FPGA-based Smith-Waterman Algorithm: analysis and novel design. In: Koch, A., Krishnamurthy, R., McAllister, J., Woods, R., El-Ghazawi, T. (eds.) ARC 2011. LNCS, vol. 6578, pp. 181–192. Springer, Heidelberg (2011). doi:10.1007/978-3-642-19475-7_20

Smith-Waterman Acceleration in Multi-GPUs: A Performance per Watt Analysis

Jesús Pérez Serrano[1], Edans Flavius De Oliveira Sandes[2],
Alba Cristina Magalhaes Alves de Melo[2], and Manuel Ujaldón[1(✉)]

[1] Computer Architecture Department, University of Malaga, Málaga, Spain
{jperezserrano,ujaldon}@uma.es
[2] Computer Science Department, University of Brasilia, Brasília, Brazil
{edans,alves}@unb.br

Abstract. We present a performance per watt analysis of CUDAlign 4.0, a parallel strategy to obtain the optimal alignment of huge DNA sequences in multi-GPU platforms using the exact Smith-Waterman method. Speed-up factors and energy consumption are monitored on different stages of the algorithm with the goal of identifying advantageous scenarios to maximize acceleration and minimize power consumption. Experimental results using CUDA on a set of GeForce GTX 980 GPUs illustrate their capabilities as high-performance and low-power devices, with a energy cost to be more attractive when increasing the number of GPUs. Overall, our results demonstrate a good correlation between the performance attained and the extra energy required, even in scenarios where multi-GPUs do not show great scalability.

Keywords: GPGPU · CUDA · DNA sequences alignment · Energy costs

1 Introduction

The advent of the Human Genome Project has brought to the foreground of parallel computing a broad spectrum of data intensive biomedical applications where biology and computer science join as a happy alliance between demanding software and powerful hardware. Since then, the bioinformatics community generates computational solutions to support genomic research in many subfields such as gene structure prediction [5], phylogenetic trees [32], protein docking [23], and sequence alignment [12], just to mention a few of an extensive list.

Huge volumes of data produced by genotyping technology pose challenges in our capacity to process and understand data. Ultra high density microarrays now contain more than 5 million genetic markers, and next generation sequencing is enabling the search for causal relationship of variation close to the single nucleotide level. Furthermore, current clinical studies include hundreds of thousands of patients instead of thousands genetically fingerprinted few years ago, transforming bioinformatics into one of the flagships of the big data era.

© Springer International Publishing AG 2017
I. Rojas and F. Ortuño (Eds.): IWBBIO 2017, Part II, LNBI 10209, pp. 512–523, 2017.
DOI: 10.1007/978-3-319-56154-7_46

In modern times of computing, when data volume pose a computational challenge, the GPU immediately comes to our minds. CUDA (Compute Unified Device Architecture) [21] and OpenCL [31] have established the mechanisms for data intensive general purpose applications to exploit GPUs extraordinary power in terms of TFLOPS (Tera Floating-Point Operations Per Second) and data bandwidth. Being GPUs the natural platform for large-scale bioinformatics, researchers have already analyzed raw performance and suggest optimizations for the most popular applications. This work extends the study to energy consumption, an issue of growing interest in the HPC community once GPUs have recently conquered the green500.org supercomputers list.

Our work focuses on biological sequences alignment in order to find the degree of similarity between them. Within this context, we may distinguish two basic approaches: Global alignment, in an attempt to align the entire length of the sequence when a pair of sequences are very similar in content and size, and local alignment, where regions of similarity between the two sequences are identified. Needleman-Wunsch (NW) [20] proposed a method for global comparison based on dynamic programming (DP), and Smith-Waterman (SW) [30] modified the NW algorithm to deal with local alignments. Computational requirements for SW are overwhelming, so researchers either relax them using heuristics as in the well-know BLAST tool [16], or rely on high performance computing to shorten the execution time. We have chosen commodity GPUs to explore the latter.

The rest of this paper is organized as follows. Section 2 completes this section with some related work. Section 3 describes the problem of comparing two DNA sequences. Section 4 summarizes our previous studies. Sections 5 and 6 introduce our infrastructure for measuring the experimental numbers, which are later analyzed in Sect. 7. Finally, Sect. 8 draws conclusions of this work.

2 Related Work

SW has become very popular over the last decade to compute (1) the exact pairwise comparison of DNA/RNA sequences or (2) a protein sequence (query) to a genomic database involving a bunch of them. Both scenarios have been parallelized in the literature [8], but fine-grained parallelism applies better to the first scenario, and therefore fits better into many-core platforms like Intel Xeon Phis [13], Nvidia GPUs using CUDA [26], and even multi-GPU using CUDAlign 4.0 [27], which is our departure point to analyze performance, power, energy and cost along this work.

On the other hand, energy consumption is gaining relevance within sequence alignment, which promotes methodologies to measure energy in genomic sequence comparison tools.

In [4], it is minimized the power consumption of a sequence alignment accelerator using application specific integrated circuit (ASIC) design flow. To decrease the energy budget, authors reduce clock cycle and scale frequency.

Hasan and Zafar [9] present performance versus power consumption for bioinformatics sequence alignment using different field programmable gate arrays (FPGAs) platforms implementing the SW algorithm as a linear systolic array.

Zou et al. [33] analyze performance and power for SW on FPGA, CPU and GPU, declaring the FPGA as the overall winner. However, they do not measure real-time power dynamically, but simplify with a static value for the whole run. Moreover, they use models from the first and second Nvidia GPU generations (GTX 280 and 470), which are, by far, the most inefficient CUDA families as far as energy consumption is concerned. Our analysis measures watts on physical wires and using Maxwell GPUs, the fourth generation where the energy budget has been optimized up to 40 GFLOPS/W, down from 15–17 GFLOPS/W in the third generation and just 4–6 GFLOPS/W in the previous ones.

3   DNA Sequence Comparison

A DNA sequence is represented by an ordered list of nucleotide bases. DNA sequences are treated as strings composed of characters of the alphabet $\sigma = A, T, G, C$. To compare two sequences, we place one sequence above the other, possibly introducing spaces, making clear the correspondence between similar characters [14]. The result of this placement is an alignment.

Given an alignment between sequences S0 and S1, a score is assigned to it as follows. For each pair of characters, we associate (a) a punctuation ma, if both characters are identical (match); or (b) a penalty mi, if the characters are different (mismatch); or (c) a penalty g, if one of the characters is a space (gap). The score is the addition of all these values. Figure 1 presents one possible alignment between two DNA sequences, where ma $= +1$, mi $= 1$ and g $= 2$.

$$A \quad T \quad A \quad C \quad T \quad C \quad C \quad A$$
$$A \quad T \quad A \quad - \quad T \quad C \quad C \quad A$$
$$\underbrace{+1 + 1 + 1 - 2 + 1 + 1 + 1 + 1}$$

$$score = 5$$

Fig. 1. Example of alignment and score.

3.1   Smith-Waterman

The SW algorithm [30] is based on dynamic programming (DP), obtaining the optimal pairwise local alignment in quadratic time and space. It is divided in two phases: calculate the DP matrix and obtain the alignment (traceback).

Phase 1. This phase receives as input sequences S0 and S1, with sizes $|S_0| = m$ and $|S_1| = n$. The DP matrix is denoted $H_{m+1,n+1}$, where $H_{i,j}$ contains the score between prefixes S0[1..i] and S1[1..j]. At the beginning, the first row and column are filled with zeroes. The remaining elements of H are obtained from Eq. 1.

$$H_{i,j} = \max \begin{cases} H_{i-1,j-1} + (\text{if } S_0[i] = S_1[j] \text{ then } ma \text{ else } mi) \\ H_{i,j-1} + g \\ H_{i-1,j} + g \\ 0 \end{cases} \tag{1}$$

In addition, each cell $H_{i,j}$ contains information about the cell that was used to produce the value. The highest value in $H_{i,j}$ is the optimal score.

Phase 2 (Traceback). The second phase of SW obtains the optimal local alignment, using the outputs of the first phase. The computation starts from the cell that has the highest value in H, following the path that produced the optimal score until the value zero is reached.

Figure 2 presents a DP matrix with $score = 5$. The arrows indicate the alignment path when two DNA sequences with sizes $m = 12$ and $n = 8$ are compared, resulting in a 913 DP matrix. In order to compare Megabase sequences of, say, 60 Million Base Pairs (MBP), a matrix of size 60,000,001 × 60,000,001 (3.6 Peta cells) is calculated.

	*	C	T	C	G	A	T	A	C	T	C	C	A
*	0	0	0	0	0	0	0	0	0	0	0	0	0
A	0	0	0	0	0	1	0	1	0	0	0	0	1
T	0	0	1	0	0	0	2	0	0	1	0	0	0
A	0	0	0	0	0	1	0	3	1	0	0	0	1
T	0	0	1	0	0	0	2	1	2	2	0	0	0
C	0	1	0	2	0	0	0	1	2	1	3	1	0
C	0	1	0	1	1	0	0	0	2	1	2	4	2
A	0	0	0	0	0	2	0	1	0	1	0	2	5
A	0	0	0	0	0	1	1	1	0	0	0	0	3

Fig. 2. DP matrix for sequences S0 and S1, with optimal $score = 5$. The arrows represent the optimal alignment.

The original SW algorithm assigns a constant cost g to each gap. However, gaps tend to occur together rather than individually. For this reason, a higher penalty is usually associated to the first gap and a lower penalty is given to the remaining ones (affine-gap model). Gotoh [7] proposed an algorithm based on SW that implements the affine-gap model by calculating three values for each cell in the DP matrix: H, E and F, where values E and F keep track of gaps in each sequence. As in the original SW algorithm, time and space complexities of the Gotoh algorithm are quadratic.

3.2 Parallel Smith-Waterman

In SW, most of the time is spent calculating the DP matrices and, therefore, is a candidate process to be parallelized. From Eq. 1, we can see that cell $H_{i,j}$ depends on three other cells: $H_{i-1,j}$, $H_{i-1,j-1}$ and $H_{i,j-1}$. This kind of dependency is well suited to be parallelized using the wavefront method [22], where the DP matrix is calculated by diagonals and all cells on each diagonal can be computed in parallel.

Figure 3 illustrates the wavefront method. In step 1, only one cell is calculated in diagonal d_1. In step 2, diagonal d_2 has two cells, that can be calculated in parallel. In the further steps, the number of cells that can be calculated in parallel increases until it reaches the maximum parallelism in diagonals d_5 to d_9, where five cells are calculated in parallel. In diagonals d_{10} to d_{12}, the parallelism decreases until only one cell is calculated in diagonal d_{13}. The wavefront strategy limits the amount of parallelism during the beginning of the calculation (filling the wavefront) and the end of the computation (emptying the wavefront).

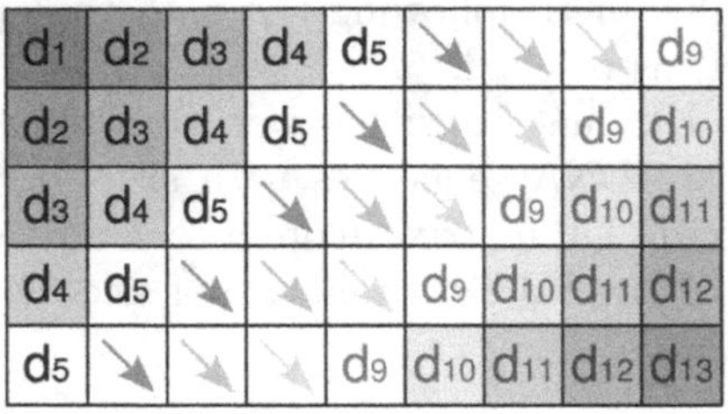

Fig. 3. The wavefront method.

4 CUDAlign Implementation on GPUs

GPUs calculate a single SW matrix using all many-cores, but data dependencies force neighbour cores to communicate in order to exchange border elements. For Megabase DNA sequences, the SW matrix is several Petabytes long, and so, very few GPU strategies [11,26] allow the comparison of Megabase sequences longer than 10 Million Base Pairs (MBP). SW# [11] is able to use 2 GPUs in a single Megabase comparison to calculate the Myers-Miller [15] linear space variant of SW. CUDAlign [26] obtains the alignment of Megabase sequences with a combined SW and Myers-Miller strategy. When compared to SW#, CUDAlign presents shorter execution times for huge sequences on a single GPU [11].

Comparing Megabase DNA sequences in multiple GPUs is more challenging. GPUs are arranged logically in a linear way so that each GPU calculates a subset of columns of the SW matrix, sending the border column elements to the next GPU. Asynchronous CPU threads will send/receive data to/from neighbor GPUs while GPUs keep computing, that way overlapping the required communications with effective computations whenever feasible.

4.1 CUDAlign Versions

CUDAlign was implemented using CUDA, C++ and pthreads. Experimental results collected in a large GPU cluster using real DNA sequences demonstrate good scalability for up to 16 GPUs [27]. For example, using the input data set described in Sect. 5, execution time was reduced from 33 h and 20 min on a single GPU to 2 h and 13 min on 16 GPUs (14.8 × speedup).

Table 1 summarizes the set of improvements and optimizations performed on CUDAlign since its inception, and Table 2 describes all stages and phases for the 4.0 version, the one used along this work.

Table 1. Summary of CUDAlign versions.

Version	Major contributions	Ref
1.0	Compares on GPUs sequences of unrestricted size using the affine gap model of SW. It provides the optimal score and the end coordinates of the optimal alignment, but not the full alignment	[24]
2.0	Incorporates the Myers-Miller (MM) algorithm to retrieve the full alignment of two sequences in linear space	[25]
2.1	Improvements on six stages: 1–3 run on GPUs, 4–6 on CPUs	[26]
3.0	Multi-GPU for SW phase 1 to distribute the DP matrix, and overlap computations with communications to the CPU	[29]
4.0	Multi-GPU for SW phase 2, including Pipeline Traceback (PT) and Incremental Speculative Traceback (IST) to estimate the point where optimal alignment will cross border columns	[27]
MASA	Multi-platform Architecture for Sequence Aligner, enabling versions to run on (1) a serial CPU, (2) multicore CPU using OmpsSs, (3) manycore GPU using CUDA, and (4) Xeon Phi using OpenMP	[28]

Table 2. Summary of CUDAlign 4.0 stages, including the SW phase it belongs to and the processor where it is executed.

Stage	Description	Phase	Who
1	Obtains the optimal score	1	GPU
2	Partial traceback	2	GPU
3	Splitting partitions	2	GPU
4	Myers-Miller with balanced splitting and orthogonal exec	2	CPU
5	Obtaining the full alignment	2	CPU
6	External visualization (optional)	2	CPU

5   Experimental Setup

We have conducted an experimental survey on a computer endowed with an Intel Xeon server and an Nvidia GeForce GTX 980 GPU from Maxwell generation. See Table 3 for a summary of major features.

For the input data set, we have used real DNA sequences coming from the National Center for Biotechnology (NCBI) [19] database. Results shown in Table 4 use sequences from assorted chromosomes, whereas Tables 5 and 6 use as input a pair of sequences from the chromosome 22 comparison between the human (50.82 MBP - see [18], accession number NC_000022.11) and the chimpanzee (37.82 MBP - see [17], accession number NC_006489.4).

To execute the required stages on a multi-GPU environment, we performed two modifications in CUDAlign 4.0: (1) a subpartitioning strategy to fit each partition in texture memory, and (2) writing extra rows in the file system as

Table 3. Characterization of the infrastructure used along our experimental analysis.

	CPU	GPU
Processor	Xeon E5-2620 v4 8 cores @ 2100 MHz	GeForce GTX 980 2048 cores @ 1126 MHz
Memory	64 GB DDR4 @ 2400 MHz 256 bits, 76.8 GB/s	4 GB GDDR5 @ 7000 MHz 384 bits, 336 GB/s
Software	O.S. Ubuntu 14.04.4 LTS 64 bits	CUDA 8.0

marks to be used in later stages to find the crosses with the optimal alignment. Moreover, we focus our experimental analysis on the first three stages of the SW algorithm, which are the ones extensively executed on GPUs as Table 2 reflects.

6   Monitoring Energy

We have built a system to measure current, voltage and wattage based on a Beaglebone Black, an open-source hardware [3] combined with the Accelpower module [6], which has eight INA219 sensors [1]. Inspired by [10], wires taken into account are two power pins on the PCI-express slot (12 and 3.3 volts) plus six external 12 volts pins coming from the power supply unit (PSU) in the form of two supplementary 6-pin connectors (half of the pins used for grounding).

Accelpower uses a modified version of `pmlib` library [2], a software package specifically created for monitoring energy. It consists of a server daemon that collects power data from devices and sends them to the clients, together with a client library for communication and synchronization with the server (Fig. 4).

The methodology for measuring energy begins with a start-up of the server daemon. Then, the source code of the application where the energy wants to be measured has to be modified to (1) declare `pmlib` variables, (2) clear and set the wires which are connected to the server, (3) create a counter and (4) start it.

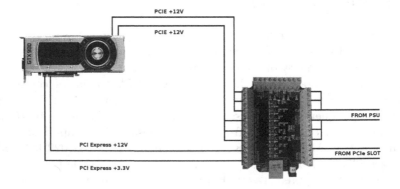

Fig. 4. Wires, slots, cables and connectors for measuring energy on GPUs.

Table 4. Power, execution times and energy consumption on four GPUs for different alignment sequences.

Sequence	Stage 1	Stage 2	Stage 3		
Average power (watts per GPU)					
chr22	101.11 W	116.26 W	77.27 W		
chr21	102.11 W	116.47 W	78.89 W		
47M	104.37 W	117.12 W	76.33 W		
chrY	103.25 W	119.63 W	0.00 W		
Execution time (seconds)			**Total time**		
chr22	11161.92 s	185.20 s	14.25 s	11361.38 s	
chr21	9687.36 s	61.49 s	11.03 s	9759.89 s	
47M	6694.95 s	88.25 s	9.05 s	6792.26 s	
chrY	6798.12 s	3.99 s	0.00 s	6802.11 s	
Energy consumption (kilojules per GPU)			**Total energy**	**Total cost**[*]	
chr22	1128.63 kJ	21.53 kJ	1.10 kJ	4x 1151.27 kJ	0.1660 €
chr21	989.26 kJ	7.16 kJ	0.87 kJ	4x 997.29 kJ	0.1440 €
47M	698.82 kJ	10.34 kJ	0.69 kJ	4x 709.85 kJ	0.1024 €
chrY	701.94 kJ	0.48 kJ	0.00 kJ	4x 702.42 kJ	0.1012 €

[*] Energy costs are shown for all four GPUs and on an average fare of 0.13 €/kWh.

Once the code is over, we (5) stop the counter, (6) get the data, (7) save them to a .csv file, and (8) finalize the counter.

7   Experimental Results

We start showing execution times and energy spent by four different sequences on a multi-GPU environment composed of four GeForce GTX 980 GPUs. Those sequences require around 5–6 h on a single GPU, and the time is reduced to less than a half using 4 GPUs. It is not a great scalability, but we already anticipated the existence of dependencies among GPUs, thus hurting parallelism.

Table 4 includes the numbers coming from this initial experiment. We can see that stage 1 predominates for the execution time, and that wattage keeps stable around 100 W for all sequences. Power goes down to less than 80 W in the third stage, but its weight is low (negligible for the case of the chrY sequence, where stage 2 also takes little time).

Once we have seen the behaviour of all these sequences, we have selected just chr22 as the more stable to characterize SW from now on.

Table 5 shows the results for chr22 when SW is executed on a multi-GPU environment. As expected, power consumed by each GPU remains stable regardless of the number of GPUs active during the parallelization process. Execution times keep showing the already announced scalability on stage 1. Those times are somehow unstable for stage 2, and finally reach good scalability on stage 3. Because GPUs keep computing on stage 1 most of the time, the overall energy cost is heavily influenced by this stage. Basically, entering multi-GPU from a

Table 5. Power, execution times and energy consumption on different number of GPUs for the chr22 alignment sequence.

No. GPUs	Stage 1	Stage 2	Stage 3		
Average power (watts per GPU)					
4	101.11 W	116.26 W	77.27 W		
3	101.53 W	108.16 W	78.79 W		
2	100.30 W	114.68 W	76.74 W		
1	102.95 W	114.44 W	81.27 W		
Execution time (seconds)			**Total time**		
4	11161.92 s	185.20 s	14.25 s	11361.38 s	
3	14719.32 s	253.72 s	17.70 s	14990.76 s	
2	22080.04 s	159.77 s	23.17 s	22262.99 s	
1	22302.24 s	291.50 s	46.65 s	22640.40 s	
Energy consumption (kilojules per GPU)			**Total energy**	**Total cost**[*]	
4	1128.63 kJ	21.53 kJ	1.10 kJ	4x 1151.27 kJ	0.1660 €
3	1494.60 kJ	27.45 kJ	1.40 kJ	3x 1523.44 kJ	0.1650 €
2	2214.77 kJ	18.32 kJ	1.78 kJ	2x 2234.88 kJ	0.1614 €
1	2296.22 kJ	33.36 kJ	3.79 kJ	2333.37 kJ	0.0842 €

[*] Energy costs are shown for all GPUs involved and on an average fare of 0.13 €/kWh.

Table 6. Savings (in execution time) and penalties (in energy cost) when accelerating SW chr22 sequence comparison on 4, 3 and 2 GPUs versus a baseline on a single GPU.

	Stage 1		Stage 2		Stage 3		Total	
No. GPUs	Savings (time)	Penalty (energy)	Savings (time)	Penalty (energy)	Savings (time)	Penalty (energy)	Savings (time)	Penalty (energy)
4	49.96%	96.60%	36.47%	158.15%	69.46%	6.09%	49.82%	97.35%
3	34.01%	95.26%	12.97%	146.85%	62.06%	0.81%	33.79%	95.86%
2	1.00%	92.90%	45.20%	9.83%	50.34%	−6.07%	1.67%	91.55%

single GPU execution doubles the energy cost, and then remains stable for 3 and 4 GPUs, where execution times are greatly reduces. That way, the performance per watt ratio is disappointing when moving from single to twin GPUs, but then evolves nicely for 3 and 4 GPUs.

Table 6 summarizes gains (in time reduction) and losses (as extra energy costs) on all scenarios of our multi-GPU execution for the chr22 sequence comparison. Stage 3 is the more rewarding one with the highest time savings and the lowest energy penalties, but unfortunately, SW keeps computing there just a marginal period of time. Stage 2 sets records in energy costs, and stage 1 keeps on an intermediate position, which is what finally characterizes the whole execution given its heavy workload. The sweetest scenario is stage 2 using 2 GPUs, where we are able to cut time in half and spend less energy overall. In the opposite side, the worst case goes to stage 1 using 2 GPUs, where time is reduced just one percent to almost double the energy spent. Finally, we have a solid conclusion on four GPUs, with

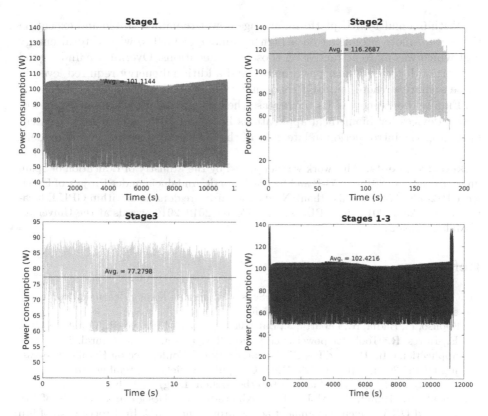

Fig. 5. Power consumption in four GPUs for stages 1, 2 and 3 of the chr22 sequence comparison. The last chart shows the global results involving all of them.

time being reduced 50% at the expense of doubling the energy budget. Figure 5 provides details about the dynamic behaviour over time for each of the stages when running the chr22 sequence comparison on four GPUs.

8   Conclusions

Along this paper, we have studied GPU acceleration and power consumption on a multi-GPU environment for the Smith-Waterman method to compute, via CUDAlign 4.0, the biological sequence alignment for a set of real DNA sequences coming from human and chimpanzee homologous chromosomes retrieved from the National Center for Biotechnology Information (NCBI).

CUDAlign 4.0 comprises six stages, with the first three accelerated using GPUs. On a stage by stage analysis, the first one is more demanding and takes the bulk of the computational time, with data dependencies sometimes disabling parallelism and affecting performance. On the other hand, power consumption was kept more stable across executions of different alignment sequences, though it suffered deviations of up to 30% across different stages.

Within a multi-GPU platform, average power remained stable and execution times were more promising on a higher number of GPUs, with a total energy cost which was more attractive on those last executions. Overall, we find a good correlation between higher performance and additional energy required, even in those scenarios where multi-GPUs do not exhibit good scalability.

Finally, we expect GPUs to increase their role as high performance and low power devices for biomedical applications in future GPU generations, particularly after the introduction in late 2016 of the 3D memory within Pascal models.

Acknowledgments. This work was supported by the Ministry of Education of Spain under ProjectTIN2013-42253-P and by the Junta de Andalucia under Project of Excellence P12-TIC-1741. We also thank Nvidia for hardware donations within GPU Education Center 2011–2016 and GPU Research Center 2012–2016 awards at the University of Malaga (Spain).

References

1. Ada, L.: Adafruit INA219 Current Sensor Breakout. https://learn.adafruit.com/adafruit-ina219-current-sensor-breakout
2. Alonso, P., Badía, R., Labarta, J., Barreda, M., Dolz, M., Mayo, R., Quintana-Ortí, E., Reyes, R.: Tools for power-energy modelling and analysis of parallel scientific applications. In: Proceedings of 41st International Conference on Parallel Processing (ICPP 2012), pp. 420–429. IEEE Computer Society, September 2012
3. BeagleBone: Beaglebone black. http://beagleboard.org/BLACK
4. Cheah, R., Halim, A., Al-Junid, S., Khairudin, N.: Design and analysis of low powered DNA sequence alignment accelerator using ASIC. In: Proceedings of 9th WSEAS International Conference on Microelectronics, Nanoelectronics and Optoelectronics (MINO 2010), pp. 107–113, March 2010
5. Deng, X., Li, J., Cheng, J.: Predicting protein model quality from sequence alignments by support vector machines. J. Proteomics Bioinform. **9**(2) (2013)
6. González-Rincón, J.: Sistema basado en open source hardware para la monitorización del consumo de un computador. Master thesis Project. Universidad Complutense de Madrid (2015)
7. Gotoh, O.: An improved algorithm for matching biological sequences. J. Mol. Biol. **162**(3), 705–708 (1982)
8. Hamidouche, K., Machado, F., Falcou, J., Melo, A., Etiemble, D.: Parallel Smith-Waterman comparison on multicore and manycore computing platforms with BSP++. Int. J. Parallel Program. **41**(1), 111–136 (2013)
9. Hasan, L., Zafar, H.: Performance versus power analysis for bioinformatics sequence alignment. J. Appl. Res. Technol. **10**(6), 920–928 (2012)
10. Igual, F., Jara, L., Gómez, J., Piñuel, L., Prieto, M.: A power measurement environment for PCIe accelerators. Comput. Sci. - Res. Develop. **30**(2), 115–124 (2015)
11. Korpar, M., Sikic, M.: SW# GPU-enabled exact alignmens on genome scale. J. Bioinform. **29**(19), 2494–2495 (2013)
12. Li, H., Homer, N.: A survey of sequence alignment algorithms for next-generation sequencing. Briefings Bioinform. **11**(5), 473–483 (2010)
13. Liu, Y., Tam, T., Lauenroth, F., Schmidt, B.: SWAPHI-LS: Smith-Waterman algorithm on xeon phi coprocessors for long DNA sequences. In: IEEE Cluster, pp. 257–265 (2014)

14. Mount, D.: Bioinformatics: Sequence and Genome Analysis. CSHL Press, New York (2004)
15. Myers, E., Miller, W.: Optimal alignment in linear space. Comput. Appl. Biosci. (CABIOS) **4**(1), 11–17 (1988)
16. NCBI: Blast: Basic local alignment search tool (2017). https://blast.ncbi.nlm.nih.gov/Blast.cgi
17. NCBI: NCBI Chimpanze Website (2017). https://www.ncbi.nlm.nih.gov/genome/gdv/?context=genome&acc=GCF_000001515.7&chr=22
18. NCBI: NCBI Human Website (2017). https://www.ncbi.nlm.nih.gov/genome/gdv/?context=genome&acc=GCF_000001405.36&chr=22
19. NCBI: NCBI Web Site (2017). http://www.ncbi.nlm.nih.gov
20. Needleman, S., Wunsch, C.: A general method applicable to the search for similarities in the aminoacid sequence of two proteins. J. Mol. Biol. **48**(3), 443–453 (1970)
21. Nvidia: CUDA Home Page, October 2010. http://developer.nvidia.com/object/cuda.html
22. Pfister, G.: In Search of Clusters: The Coming Battle in Lowly Parallel Computing. Prentice Hall, Upper Saddle River (1995)
23. Pierce, B., Wiehe, K., Hwang, H., Kim, B., Vreven, T., Weng, Z.: ZDOCK server: interactive docking prediction of protein-protein complexes and symmetric multimers. J. Bioinform. **30**(12), 1771–1773 (2014)
24. Sandes, E., Melo, A.: CUDAlign: using GPU to accelerate the comparison of megabase genomic sequences. In: Proceedings of 15th ACM SIGPLAN Symposium on Principles and Practice of Parallel Programming (PPoPP 2010), pp. 137–146 (2010)
25. Sandes, E., Melo, A.: Smith-Waterman alignment of huge sequences with GPU in linear space. In: Proceedings of IEEE International Parallel and Distributed Processing Symposium (IPDPS 2011), pp. 1199–1211 (2011)
26. Sandes, E., Melo, A.: Retrieving Smith-Waterman alignments with optimizations for megabase biological sequences using GPU. IEEE Trans. Parallel Distrib. Syst. **24**(5), 1009–1021 (2013)
27. Sandes, E., Miranda, G., Martorell, X., Ayguadé, E., Teodoro, G., Melo, A.: CUDAlign 4.0: incremental speculative traceback for exact chromosome-wide alignment in GPU clusters. IEEE Trans. Parallel Distrib. Syst. **27**(10), 2838–2850 (2016)
28. Sandes, E., Miranda, G., Martorell, X., Ayguadé, E., Teodoro, G., Melo, A.: MASA: a multi-platform architecture for sequence aligners with block pruning. ACM Trans. Parallel Comput. **2**(4), 28:1–28:31 (2016)
29. Sandes, E., Miranda, G., Melo, A., Martorell, X., Ayguadé, E.: CUDAlign 3.0: parallel biological sequence comparison in large GPU clusters. In: Proceedings of IEEE/ACM CCGrid 2014, pp. 160–169 (2014)
30. Smith, T., Waterman, M.: Identification of common molecular sequences. J. Mol. Biol. **127**(1), 195–197 (1981)
31. The Khronos Group: The OpenCL Core API Specification, Headers and Documentation (2009). http://www.khronos.org/registry/cl
32. Wan, P., Che, D.: Constructing phylogenetic trees using interacting pathways. Bioinformation **9**(7), 363–367 (2013)
33. Zou, D., Dou, Y., Xia, F.: Optimization schemes and performance evaluation of Smith-Waterman algorithm on CPU, GPU and FPGA. Concurrency Comput. Practice Experience **24**, 1625–1644 (2012)

A Deep Learning Network for Exploiting Positional Information in Nucleosome Related Sequences

Mattia Antonino Di Gangi[1,2], Salvatore Gaglio[3], Claudio La Bua[3],
Giosué Lo Bosco[4,5(✉)], and Riccardo Rizzo[6]

[1] Fondazione Bruno Kessler, Trento, Italy
[2] ICT International Doctoral School, University of Trento, Trento, Italy
[3] Dipartimento dell'Innovazione Industriale e Digitale,
Universitá degli studi di Palermo, Palermo, Italy
[4] Dipartimento di Matematica e Informatica,
Universitá degli studi di Palermo, Palermo, Italy
giosue.lobosco@unipa.it
[5] Dipartimento di Scienze per l'Innovazione e le Tecnologie Abilitanti,
Istituto Euro Mediterraneo di Scienza e Tecnologia, Palermo, Italy
[6] ICAR-CNR - National Research Council of Italy, Palermo, Italy

Abstract. A nucleosome is a DNA-histone complex, wrapping about 150 pairs of double-stranded DNA. The role of nucleosomes is to pack the DNA into the nucleus of the Eukaryote cells to form the Chromatin. Nucleosome positioning genome wide play an important role in the regulation of cell type-specific gene activities. Several biological studies have shown sequence specificity of nucleosome presence, clearly underlined by the organization of precise nucleotides substrings. Taking into consideration such advances, the identification of nucleosomes on a genomic scale has been successfully performed by DNA sequence features representation and classical supervised classification methods such as Support Vector Machines and Logistic regression. The goal of this work is to propose a classification method for nucleosome positioning that, differently from the proposed method so far, does not make any use of a sequence feature extraction step. Deep neural networks (DNN) or deep learning models, were proved to be able to extract automatically useful features from input patterns. Under this framework, Long Short-Term Memory (LSTM) is a recurrent unit that reads a sequence one step at a time and can exploit long range relations. In this work, we propose a DNN model for nucleosome identification on sequences from three different species. Our experiments show that it outperforms classical methods in two of the three data sets and give promising results also for the other.

Keywords: Nucleosome classification · Epigenetic · Deep learning networks · Recurrent Neural Networks

© Springer International Publishing AG 2017
I. Rojas and F. Ortuño (Eds.): IWBBIO 2017, Part II, LNBI 10209, pp. 524–533, 2017.
DOI: 10.1007/978-3-319-56154-7_47

1 Introduction

Nucleosomes in eukaryotes wrap 150 bp DNA or about 1.7 turns, and this mainly allows to organize a large and complex genome in the nucleus. From the regulation point of view, nucleosome positions can also block the access of transcription factors and other proteins to DNA. As an example, we recall the behavior of Pho5 promoter in yeast, that under normal conditions is occupied by well-positioned nucleosomes, preventing the transcription factor Pho4 from binding to its target binding site. When induced by phosphate starvation, the nucleosomes are depleted from the promoter region so that Pho4 can bind to its target DNA binding sequence thus activating the Pho5 gene transcription [1]. However, nucleosome binding can sometimes enhance transcription by bringing distant DNA regulatory elements. Genome-wide studies have found that in general, the transcription activity in promoter regions is inversely proportional to nucleosome depletion. The interesting aspect from the computational point of view, is that distinct DNA sequence features have been identified to be associated with nucleosome presence [2]. Interestingly, the most informative sequence features are those that are traditionally viewed as "degenerative", such as CpG density and poly-A tract [3]. For these reasons, machine learning methods which process sequence features, have been developed to predict genome-wide nucleosome patterns, sometimes with great accuracy [4]. What is clear in the machine learning field, is that the performance of the related methods depend on feature representations, which are typically designed by experts. As a subsequent phase, it is necessary to identify which features are more appropriate to face with a given task, and this remains actually a very crucial and difficult step. For the special case of nucleosome identification, the most promising machine learning methodologies that have been proposed involve a feature extraction phase such as spectral representation of DNA sequences [5–10,13]. Informally, such representation associates a sequence with a feature vector of fixed length, whose components count the frequency of each substrings belonging to a finite set of words. The main advantage is that the sequence is represented into a numerical space making possible the application of several effective machine learning methods.

Recently deep neural networks or deep learning models [14], were proved to be able to automatically extract useful features from input patterns with no *a priori* information. As a result, deep learning models has given significant contributions in several basic but arduous artificial intelligence tasks. Very important advancements have been made in computer vision [15,16], natural language processing (NLP) [17,18] and machine translation [19–21], where deep learning techniques represent now the state of the art.

The two main categories of deep neural models are *Convolutional Neural Networks (CNN)* and *Recurrent Neural Networks (RNN)*. CNNs are characterized by an initial layer of convolutional filters, followed by a non Linearity, a sub-sampling, and a fully connected layer which realized the final classification [22]. Collobert et al. [23] firstly shown that CNNs can be used effectively also for sequence analysis, in the case of a generic text. RNNs are explicitly designed for processing sequential data. With respect to other kind of networks, they store an hidden state, which works as an internal memory. There are several architectures for RNNs, but they all have cyclic connections directed through the hidden state, which is therefore updated after each time step for remembering what happened in previous time steps. Differently from other architectures, RNNs are capable of handling sequence information over time, and this is an outstanding property in the case of sequence classification.

The use of deep neural network for DNA sequence classification is not new, both for generic sequences [24,25] and for the case of nucleosome related ones [26]. Anyway, the proposed methods make use of the k-mers representation as a preprocessing step, and a CNN as deep learning network.

Conversely, in this work we want to avoid the feature extraction step in order to fully exploit the capabilities of DNNs, making use of a convolutional layer for extracting features from local sequences of nucleotides, and an LSTM to take into account longer-range positional information.

Results computed on three datasets of nucleosome forming and inibiting sequences show the effectiveness of the proposed model, which improve significantly the state of the art in two of the three datasets.

In the next Section RNNs and LSTM are introduced and described, in Sect. 3 are described the data-sets and the experiments, while the results are reported in Sect. 4. Conclusions and remarks are discussed in Sect. 5.

2   Methods

2.1   Convolutional Layer

The convolutional layers are the core components of a CNN. They calculate L one-dimensional convolutions between the kernel vectors w^l, of size $2n + 1$, and the input signal x:

$$q^l(i) = \sum_{u=-n}^{n} w^l(u)x(i - u) \tag{1}$$

In Eq. 1 $q^l(i)$ is the component i of the l-th output vector and $w^l(u)$ is the component u of the l-th kernel vector. After a bias term b^l is added and the nonlinear function g is applied:

$$h^l(i) = g(q^l(i) + b^l). \tag{2}$$

2.2   Long Short-Term Memory Layer

Recurrent Neural Networks are generally used for processing sequences of data which evolves along the time axis. The simpler version of a RNN owns an internal state h_t which is a summary of the sequence seen until the previous time step $(t - 1)$ and it is used together with the new input x_t:

$$h_t = \sigma(W_h x_t + U_h h_{t-1} + b_h)$$

$$y_t = \sigma(W_y h_t + b_y)$$

where W_h and U_h are respectively the weight matrices for the input and the internal state, W_y is the weight matrix for producing the output from the internal state, and the two b are bias vectors.

The learning capability of this kind of RNNs is limited by the *vanishing gradient problem* [12], which prevent the learning of long term dependencies. *Long Short-Term Memory (LSTM)* [27] is a variant of RNN designed with the explicit intent of preventing the vanishing gradient, and it is defined by the following equations.

$$f_t = \sigma(W^f x_t + U^f h_{t-1} + b^f)$$

$$i_t = \sigma(W^i x_t + U^i h_{t-1} + b^i)$$

$$o_t = \sigma(W^o x_t + U^o h_{t-1} + b^o)$$

$$c_t = \tanh(W^c x_t + U^c h_{t-1} + b^c)$$

$$s_t = f_t \cdot s_{t-1} + i_t \cdot c_t$$

$$h_t = \tanh(s_t) \cdot o_t$$

where i_t, f_t, o_t are, respectively, the input, forget and output gates, with values between 0 and 1, which decide what part of the input, of the previous hidden state and of the candidate output should flow through the network. The vanishing gradient is due to the derivative of the tanh function that is always strictly less than 1. In LSTM, the derivatives depend also on the gates, so that they are not anymore limited.

2.3 Softmax Layer

In general, in order to compute a classification task, a Deep Learning Neural Network needs the so called *Softmax layer* [28] It is mainly composed by K units, where K is the number of different classes. Each unit is densely connected with the previous layer and computes the probability that an element is of class k by means of the formula:

$$softmax_k(x) = \frac{e^{W_k x + b_k}}{\sum_{l=1}^{k} e^{W_l x + b_l}}$$

where W_l is the weight matrix connecting the l-th unit to the previous layer, x is the output of the previous layer and b_l is the bias for the l-th unit. Softmax is widely used by deep learning practitioners as a classification layer because of the normalized probability distribution it outputs, which proves particularly useful during back-propagation.

2.4   Sub-sampling and Regularization Layers

Another important component of a deep neural network is the max-pooling layer, that usually follows the recurrent or convolutional layers in the computation flow. It is a non-linear down-sampling layer that partitions the input vector into a set of non-overlapping regions and, for each sub-region, the maximum value is considered as output. This processing layer reduces the complexity for the following layers and operates a sort of translational invariance. The dropout layer [11] randomly sets to zero the output from the preceding layer during training, with a probability p given as a fixed parameter. When $p = 0.5$, it is equivalent to training $2^{|W|}$ networks with shared parameters, where $|W|$ is the number of neurons subject to dropout. This results in a strong regularization effect, which helps in preventing overfitting.

3   Experiments

3.1   Dataset Description

The dataset used in this work is composed by three sets of DNA sequences underlying nucleosomes from the following three species: (i) *Homo sapiens (HM)*; (ii) *Caenorhabditis elegans (CE)* and (iii) *Drosophila melanogaster (DM)*. Details about all the step of data extraction and filtering of the three datasets can be found in the work by Guo et al. [29] and in the references therein. Each of the three datasets is composed by two classes of samples: the nucleosome-forming sequence samples (positive data) and the linkers or nucleosome-inhibiting sequence samples (negative data). The HM dataset contains $2,273$ positives and $2,300$ negatives, the CE $2,567$ positives and $2,608$ negatives and the DM $2,900$ positives and $2,850$ negatives. The length of a generic sequence is 147 bp. Starting from a dataset S of $|S|$ sequences, the training and test sets of the proposed classifier have been selected using a 10 fold cross validation schema.

In this work we evaluate models for classifying genomic sequences without providing a-priori information by means of feature engineering. One method for achieving this is represented by the use of *character-level one-hot encoding*. This representation takes into account each character i of the alphabet by a vector of length equal to the alphabet size, having all zero entries except for a single one in position i. This method leads to a sparse representation of the input.

Finally, the sequence of the dataset are represented by using a matrix of dimensions 4×147. Figure 1 shows an example of the representation for a short sequence.

3.2   The Classifier Architectures

We propose three kind of architectures, obtained by the composition of six kinds of neural layers: a convolutional layer, a max pooling layer, a dropout layer, a long short-term memory (LSTM) layer, a fully connected layer and a softmax layer. The convolutional layer is used to extract features from the input matrix

$$A = \begin{bmatrix} 1 \\ 0 \\ 0 \\ 0 \end{bmatrix} \quad C = \begin{bmatrix} 0 \\ 1 \\ 0 \\ 0 \end{bmatrix} \quad G = \begin{bmatrix} 0 \\ 0 \\ 1 \\ 0 \end{bmatrix} \quad T = \begin{bmatrix} 0 \\ 0 \\ 0 \\ 1 \end{bmatrix}$$

$$ACGCTGCACAA = \begin{bmatrix} 1 & 0 & 0 & 0 & 0 & 0 & 0 & 1 & 0 & 1 & 1 \\ 0 & 1 & 0 & 1 & 1 & 0 & 0 & 0 & 1 & 0 & 0 \\ 0 & 0 & 1 & 0 & 0 & 0 & 1 & 0 & 0 & 0 & 0 \\ 0 & 0 & 0 & 0 & 0 & 1 & 0 & 0 & 0 & 0 & 0 \end{bmatrix}$$

Fig. 1. The representation used for the sequences. In the upper row the vectors used to represent each base, in the lower row the representation of the small sequence fragment.

x of 4×147 binary values. This layer calculates $L = 50$ $1D$ convolutions [22] between the kernel vectors w^l $l = 1, 2, \dots L$ and the input matrix x. This layer extracts features from local fixed-length sequences of nucleotides, basically producing a sequence of k-mers in a representation useful for the following layers. The Max Pooling operation with width and stride 2 helps to capture the most salient features extracted by the previous layer and reduces the output size from 145 to 72 vectors. The Dropout operation with probability $p = 0.5$ is used to prevent overfitting during the training phase. The LSTM layer scans the sequential features output of the previous layer and outputs its hidden state at each time step. The purpose is to find relations between the time steps along all the sequence. A fully-connected layer with softmax activation produce the actual classification, based on the position of the features.

The above mentioned layers are combined to define three deep neural architectures of increasing level of completeness. The simplest one is the LSTM-FCX1 (see Fig. 2(a)) that uses LSTM, droup-out and a softmax layer. The medium one is the LSTM-FCX2 (see Fig. 2(b)) that increases the complexity of LSTM-FCX2 (see Fig. 2(c)) by adding a fully connected layer and a drop out before the softmax. Finally, the CONV-LSTM-FCX2 uses a convolutional layer plus a max pooling that precedes a LSTM-FCX2 like network.

4 Results

We have computed a total of 3 indexes to measure the performance of the classifiers: *Accuracy (A), Precision (Pr) and Recall (Re)*:

$$A = \frac{tp + tn}{tp + tn + fp + fn} \tag{3}$$

$$Pr = \frac{tp}{tp + fp} \tag{4}$$

$$Re = \frac{tp}{tp + fn} \tag{5}$$

were tp are the true positive, tn are the true negative, fp are the false positives and fn are the false negatives.

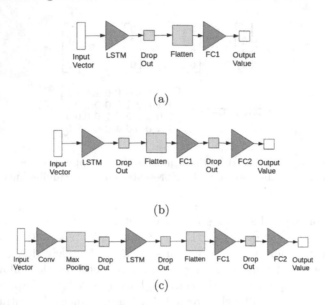

Fig. 2. The three architectures used in this work, LSTM-FCX1 (a), LSTM-FCX2 (b) and CONV-LSTM-FCX2 (c). The computing blocks are the dark triangles: Convolutional Network (Conv), LSTM network and Fully Connected Multi Layer Perceptrons (FC); the light rectangles are Max Pooling, Dropout and Flattening operations.

In Table 1 we report means (μ) and variance (σ) of the accuracy, precision and recall values (in percentage) reached by the used deep learning classifier.

We notice that the best architecture is the CONV-LSTM-FCX2. In this case, differently from the other architectures, LSTM becomes robust to small translations due to the fact that it discovers the position of important local sequences in a coarse-grained fashion by put before convolution and max pooling.

Note that the same authors that provide the three datasets have also developed a successful method for nucleosome prediction called *iNuc-PseKNC* [29]. In particular, it uses a Support vector machine with a radial basis function kernel, and a novel feature-vector that incorporates six DNA local structural properties. They have observed that by a cross-validation tests on the three datasets that the overall accuracy rates achieved by their method were 86.90%, 86.27% and 79.97% for C. Elegans, H. Sapiens and D. Melanogaster respectively. Moreover, they report also Sensitivity (recall) rates of 90.30%, 87.86% and 78.31%. Surprisingly, without using any feature representation process, our architectures seem to outperform their model in terms of accuracy on C. Elegans and D. Melanogaster datasets with a relative increment of 3% and 7% respectively, and in terms of recall on every dataset. Anyway, the comparison needs further investigation because authors of iNuc-PseKNC method have computed the performance metrics using jacknife cross-validation rather than 10 fold. Anyway, it is clear that the nucleosome identification problem can benefit from the proposed LSTM architecture. Note also that the same dataset has been recently processed by a Convolutional Neural network [26] but the results are really far from those shown here.

Table 1. The results obtained using the proposed classifier. All the values are in percentage

	Accuracy		Precision		Recall	
	μ	σ	μ	σ	μ	σ
C.Elegans-CONV-LSTM-FCX2	90.03	0.91	86.44	2.38	93.28	1.68
H.Sapiens-CONV-LSTM-FCX2	84.10	1.07	78.85	4.69	88.54	4.04
D.Melanogaster-CONV-LSTM-FCX2	85.57	1.65	83.83	2.97	86.66	1.96
C.Elegans-LSTM-FCX2	88.27	0.69	87.07	2.34	89.61	1.35
H.Sapiens-LSTM-FCX2	83.57	2.13	82.49	2.70	85.10	4.89
D.Melanogaster-LSTM-FCX2	83.63	1.64	83.20	2.27	84.86	1.89
C.Elegans-LSTM-FCX1	89.27	1.60	88.16	2.65	90.63	2.72
H.Sapiens-LSTM-FCX1	82.39	2.34	79.68	3.21	86.94	6.53
D.Melanogaster-LSTM-FCX1	84.17	1.24	84.45	2.87	84.28	2.75

5   Conclusion

In this study, deep neural networks for the automatic classification of nucleosome forming sequences have been proposed. These architectures exploit the information about nucleotides position in a sequence, by using convolutional and recurrent layers. The effectiveness of the models are proved in terms of both accuracy and recall, for which they push forward the state of the art in two out of three considered data sets. Further work would go into the direction of exploiting more contextual information, for example by using bidirectional LSTMs, which consider both the left and the right context of each position, instead of only the left context. Moreover, as the previous work obtained good results by using only fixed-size feature vectors, we need to explore ways of combining position-based and global information. Unfortunately, analogously to many other methodologies, the used deep neural architecture have limitation on biological interpretation of the captured patterns contributing to the positive predictions. In order to investigate on this issue, we plan to adopt the so called autoencoders, whose main advantage is the automatic extraction of relevant features of the items to classify.

References

1. Svaren, J., Horz, W.: Transcription factors vs. nucleosomes: regulation of the PHO5 promoter in yeast. Trends Biochem. Sci. **22**, 93–97 (1997)
2. Struhl, K., Segal, E.: Determinants of nucleosome positioning. Nat. Struct. Mol. Biol. **20**(3), 267–273 (2013)
3. Yuan, G.C.: Linking genome to epigenome. Wiley Interdisc. Rev.: Syst. Biol. Med. **4**(3), 297–309 (2012)
4. Pinello, L., Lo Bosco, G., Yuan, G.-C.: Applications of alignment-free methods in epigenomics. Briefings Bioinform. **15**(3), 419–430 (2014)

5. Kuksa, P., Pavlovic, V.: Efficient alignment-free DNA barcode analytics. BMC Bioinform. **10**(Suppl. 14), S9 (2009)
6. Pinello, L., Lo Bosco, G., Hanlon, B., Yuan, G-C.: A motif-independent metric for DNA sequence specificity. BMC Bioinform. **12**, Article No. 408 (2011)
7. Giosué, L.B., Luca, P.: A new feature selection methodology for k-mers representation of DNA sequences. In: Serio, C., Liò, P., Nonis, A., Tagliaferri, R. (eds.) CIBB 2014. LNCS, vol. 8623, pp. 99–108. Springer, Heidelberg (2015). doi:10.1007/978-3-319-24462-4_9
8. Rizzo, R., Fiannaca, A., Rosa, M., Urso, A.: The general regression neural network to classify barcode and mini-barcode DNA. In: Serio, C., Liò, P., Nonis, A., Tagliaferri, R. (eds.) CIBB 2014. LNCS, vol. 8623, pp. 142–155. Springer, Heidelberg (2015). doi:10.1007/978-3-319-24462-4_13
9. Lo Bosco, G.: Alignment free dissimilarities for nucleosome classification. In: Angelini, C., Rancoita, P.M.V., Rovetta, S. (eds.) CIBB 2015. LNCS, vol. 9874, pp. 114–128. Springer, Heidelberg (2016). doi:10.1007/978-3-319-44332-4_9
10. Fiannaca, A., La Rosa, M., Rizzo, R., Urso, A.: Analysis of DNA barcode sequences using neural gas and spectral representation. In: Iliadis, L., Papadopoulos, H., Jayne, C. (eds.) EANN 2013. CCIS, vol. 384, pp. 212–221. Springer, Heidelberg (2013). doi:10.1007/978-3-642-41016-1_23
11. Srivastava, N., Hinton, G.E., Krizhevsky, A., Sutskever, I., Salakhutdinov, R.: Dropout: a simple way to prevent neural networks from overfitting. J. Mach. Learn. Res. **15**(1), 1929–1958 (2014)
12. Hochreiter, S., Bengio, Y., Frasconi, P., Schmidhuber, J.: Gradient flow in recurrent nets: the difficulty of learning long-term dependencies. In: Kremer, S.C., Kolen, J.F. (eds.) A Field Guide to Dynamical Recurrent Neural Networks. IEEE Press, New York (2001)
13. Fiannaca, A., Rosa, M., Rizzo, R., Urso, A.: A k-mer-based barcode DNA classification methodology based on spectral representation and a neural gas network. Artif. Intell. Med. **64**(3), 173–184 (2015)
14. Bengio, Y.: Learning deep architectures for AI. Found. Trends Mach. Learn. **2**(1), 1–127 (2009)
15. Farabet, C., Couprie, C., Najman, L., et al.: Learning hierarchical features for scene labeling. IEEE Trans. Pattern Anal. Mach. Intell. **35**(8), 1915–1929 (2013)
16. Tompson, J.J., Jain, A., LeCun, Y., et al.: Joint training of a convolutional network and a graphical model for human pose estimation. In: Advances in Neural Information Processing Systems, pp. 1799–1807 (2014)
17. Kiros, R., Zhu, Y., Salakhutdinov, R.R., et al.: Skip-thought vectors. In: Advances in Neural Information Processing Systems, pp. 3276–3284 (2015)
18. Li, J., Luong, M-T., Jurafsky, D.: A hierarchical neural autoencoder for paragraphs and documents. In: Proceedings of the 53rd Annual Meeting of the Association for Computational Linguistics and the 7th International Joint Conference on Natural Language Processing, pp. 1106–1115 (2015)
19. Luong, M-T., Pham, H., Manning, C.D.: Effective approaches attention-based neural machine translation. In: Proceedings of the Conference on Empirical Methods in Natural Language Processing, pp. 1412–1421 (2015)
20. Cho, K., Van Merriënboer, B., Gulcehre, C., et al.: Learning phrase representations using RNN encoder-decoder for statistical machine translation. In: Proceedings of the Conference on Empirical Methods in Natural Language Processing (EMNLP), pp. 1724–1734 (2014)

21. Chatterjee, R., Farajian, M.A., Conforti, C., Jalalvand, S., Balaraman, V., Di Gangi, M.A., Ataman, D., Turchi, M., Negri, M., Federico, M.: FBK's neural machine translation systems for IWSLT. In: Proceedings of 13th International Workshop on Spoken Language Translation (IWSLT 2016) (2016)

22. LeCun, Y., Bottou, L., Bengio, Y., Haffner, P.: Gradient-based learning applied to document recognition. Proc. IEEE **86**(11), 2278–2324 (1998)

23. Collobert, R., Weston, J., Bottou, L., Karlen, M., Kavukcuoglu, K., Kuksa, P.: Natural language processing (almost) from scratch. J. Mach. Learn. Res. **12**, 2493–2537 (2011)

24. Rizzo, R., Fiannaca, A., La Rosa, M., Urso, A.: A deep learning approach to DNA sequence classification. In: Angelini, C., Rancoita, P.M.V., Rovetta, S. (eds.) CIBB 2015. LNCS, vol. 9874, pp. 129–140. Springer, Heidelberg (2016). doi:10.1007/978-3-319-44332-4_10

25. Lo Bosco, G., Di Gangi, M.A.: Deep learning architectures for DNA sequence classification. In: Petrosino, A., Loia, V., Pedrycz, W. (eds.) WILF 2016. LNCS (LNAI), vol. 10147, pp. 162–171. Springer, Cham (2017). doi:10.1007/978-3-319-52962-2_14

26. Lo Bosco, G., Rizzo, R., Fiannaca, A., La Rosa, M., Urso, A.: A deep learning model for epigenomic studies. In: SITIS The 12th International Conference on Signal Image Technology & Internet Systems, pp. 688–692 (2016, to appear)

27. Hochreiter, S., Schmidhuber, J.: Long short-term memory. Neural Comput. **9**(8), 1735–1780 (1997)

28. Bridle, J.S.: Probabilistic interpretation of feedforward classification network outputs, with relationships to statistical pattern recognition. In: Soulié, F.F., Hérault, J. (eds.) Neurocomputing, pp. 227–236. Springer, Heidelberg (1990)

29. Guo, S.-H., Deng, E.-Z., Xu, L.-Q., Ding, H., Lin, H., Chen, W., Chou, K.-C.: iNuc-PseKNC: a sequence-based predictor for predicting nucleosome positioning in genomes with pseudo k-tuple nucleotide composition. Bioinformatics **30**(11), 1522–1529 (2014)

Oncological Big Data and New Mathematical Tools

A Clinical Tool for Automated Flow Cytometry Based on Machine Learning Methods

Claude Takenga[1]([⊠]), Michael Dworzak[2,3], Markus Diem[1,2],
Rolf-Dietrich Berndt[4], Erling Si[4], Michael Brandstoetter[5],
Leonid Karawajew[6], Melanie Gau[1], and Martin Kampel[1]

[1] Computer Vision Lab, TU Wien,
Favoritenstraße 9/183-2, 1040 Vienna, Austria
{takenga,mdiem,mgau,kampel}@caa.tuwien.ac.at
[2] Labdia Labordiagnostik GmbH, Zimmermannplatz 8, 1090 Vienna, Austria
michael.dworzak@ccri.at
[3] Children's Cancer Research Institute, Medical University of Vienna,
Zimmermannplatz10, 1090 Vienna, Austria
[4] Infokom GmbH, Johannesstrasse 8, 17034 Neubrandenburg, Germany
{rberndt,esi}@infokom.de
[5] CogVis Software and Consulting GmbH,
Wiedner Hauptstrasse 17, Vienna, Austria
brandstoetter@cogvis.at
[6] Charité – Universitaetmedizin Berlin, Charitéplatz 1, 10117 Berlin, Germany
Leonid.karawajew@charite.de

Abstract. Clinical researchers working in flow cytometry (FCM) nowadays experience increasing demands to perform experiments that involve high throughput, rare event analysis and detailed immunophenotyping. Beckman Coulter and Becton Dickinson offer multi-use flow cytometry sorters that can analyze up to 70K EPS (events per seconds) with more than nine parameters enabled. While this multi-parametric feature provides a great power for hypothesis testing, it also generates a vast amount of data, which is analyzed manually through a processing called gating. For large experiments, this manual gating turns out to be time consuming and requires intensive operator training and experience. The lack of required expertise leads to wrong interpretation of data, thus a wrong therapy course for the case of patients with acute lymphoblastic leukemia (ALL) is followed. This paper aims to present a pipeline-software, as a ready-to-use machine learning based automated FCM assessment tool for the daily clinical practice for patients with ALL. The new system increases accuracy in assessment of FCM based on minimal residual disease (MRD) method in samples analyzed by conventional operator-based gating since computer-aided analysis potentially has a higher power due to the use of the whole multi-parametric FCM-data space at once instead of using methods restricted to two-dimensional decision rules. The tool is implemented as a telemedical network for analysis, clinical follow-up, treatment monitoring of leukemia and allows dissemination of automated FCM-MRD analysis to medical centres in the world.

Keywords: Clinical tool · Flow cytometry · Leukemia · Machine learning · Telematics platform · Telemedicine

© Springer International Publishing AG 2017
I. Rojas and F. Ortuño (Eds.): IWBBIO 2017, Part II, LNBI 10209, pp. 537–548, 2017.
DOI: 10.1007/978-3-319-56154-7_48

1 Introduction

Traditional methods for FCM data processing rely on subjective manual gating. While FCM data acquisition can be harmonized straightforwardly, data analysis and interpretation rely largely on operator skills and experience. Hence, these requirements represent the current bottleneck of safely applying the FCM-Minimal Residual Disease (MRD) methodology in a growing community of diagnostic laboratories. Thus, machine learning methods and support vector machines represent an objective methodology for automated analysis of FCM data to cluster cell populations and determine MRD [1–3].

In the last years, several groups of researchers have focused their works on developing computational methods for identifying cell populations in multi-dimensional FCM data [4–6]. In [7] we focused on the evaluation of different strategies for automatic FCM analysis on a newly collected database: the annotated data set derives from samples of ALL patients that have been treated uniformly according the AIEOP-BFM 2009 treatment protocol [8]. These data base were used in our work [9], with the objectives to develop a ready-to-use automated FCM-based MRD-assessment tool, to reduce subjectivity caused by manual operator gating and to increase result comparability through automation and standardization.

This paper applies results produced from previous works [1, 2, 7, 9, 10] and presents a ready-to-use software implemented as a telemedical network system for FCM automated data analysis in clinical practice. The developed system can connect worldwide laboratories to the expert system and increase the quality of FCM data analysis even for countries with less expertise in the field. The system solves the issue that current multi-parametric FCM data analysis approaches still remain subjective due to the need of experienced operator for both data analysis and interpretation of the results and presented the new automated system based on machine learning algorithms to analyze multi-parametric FCM data samples.

2 State-of-the-Art in FCM Data Analysis: Case of ALL

2.1 Acute Lymphoblastic Leukemia (ALL)

ALL is a disseminated malignancy of B- or T-lymphoblasts which imposes a rapid and accurate diagnostic process to support an optimal risk-oriented therapy and thus increase the curability rate. In this type of cancer, the bone marrow makes too many immature lymphocytes (a type of white blood cell, Fig. 1). A blood stem cell may become a myeloid stem cell or a lymphoid stem cell [4]. A myeloid stem cell becomes one of three types of mature blood cells: Red blood cells that carry oxygen and other substances to all tissues of the body, platelets that form blood clots to stop bleeding and white blood cells that fight infection and disease. A lymphoid stem cell becomes a lymphoblast cell and then one of three types of lymphocytes (white blood cells): B lymphocytes that make antibodies to help fight infection, T lymphocytes that help B lymphocytes make the antibodies that help fight infection and natural killer cells that attack cancer cells and viruses, Fig. 1. In a child with ALL, through genetic mutation, precursor cells become malignant lymphoblasts of either B-cells or T-cells lineage. The cells do not work like normal lymphocytes and are not able to fight infection. These cells are cancer - leukemia cells. Also, as the number of leukemia cells increases in the blood and bone marrow, there is less room for healthy white blood cells, red blood cells, and platelets. This may lead to infection, anemia, and easy bleeding [4].

There are different types of treatment for ALL. Children with ALL should have

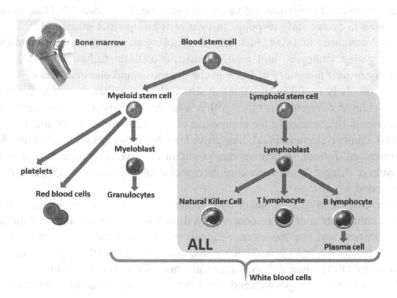

Fig. 1. Blood cell development. Steps from a blood stem cell to one of the three cells: red blood cell, platelet, white blood cell. ALL cells of interest.

their treatment planned by a team of doctors who are experts in treating childhood leukemia. Four types of standard treatment are used: chemotherapy, radiation therapy, chemotherapy with stem cell transplant and targeted therapy. Treatment is given to kill leukemia cells that have spread or may spread to the brain, spinal cord, or testicles. The treatment of childhood ALL is done in phases: Remission induction, Consolidation/intensification and Maintenance [4].

2.2 Flow Cytometry (FCM) and Minimal Residual Disease (MRD) for ALL

A variety of methods to detect submicroscopic levels of leukemia in patients with ALL have been developed in the last decades. Flow cytometry (FCM) and polymerase chain reaction (PCR) have emerged as the most promising methods for detecting submicroscopic levels of leukemia [4, 11]. FCM has evolved to an indispensable tool in biology and medicine, with a significant impact on hematology. Recently, FCM has also become an attractive approach for evaluation of therapy response and especially detection of MRD [3]. Flow cytometric method affords the detection of one leukemic cell among 10.000 normal bone marrow cells, and can be currently applied to at least two thirds of all patients with acute leukemia. Detection of MRD with these methods during clinical remission appears to be independently associated with treatment outcome. The measurement of MRD at critical intervals during the disease course is a new tool to gauge the effectiveness of therapy in children with ALL [13]. Prospective studies in large series of patients have demonstrated a strong correlation between MRD levels during clinical remission and treatment outcome. Therefore, MRD assays can be reliably used to assess early response to treatment and predict relapse [12].

Several advances in FCM, including availability of new monoclonal antibodies, improved gating strategies, and multi-parametric analytic techniques, have all dramatically improved the utility of FCM in the diagnosis and classification of leukemia. FCM is particularly useful in situations in which the researcher wishes to discriminate between multiple different cell types within a heterogeneous population and measure their individual frequencies or the expression of a specific molecule of interest, Fig. 2. In several clinical studies, FCM has proved to be indispensable, allowing for the measurement of distinct, sometimes rare populations of cells that are indicative of the progression of disease or the success of therapeutic intervention [14, 15]. Details about FCM is found in [14].

For the case of an ALL, blood or bone marrow cells of a patient are incubated with a specific combination of fluorescence-labeled antibodies. These bind to their respective cell antigens and the fluorescence signals are then detected by a sensor in a flow cytometer for each single cell. That means, we will accurately identify these cells, while discriminating other cell types that are not of interest. Leukemic cells can be distinguished by their antigen expression patterns which are different to those of normal

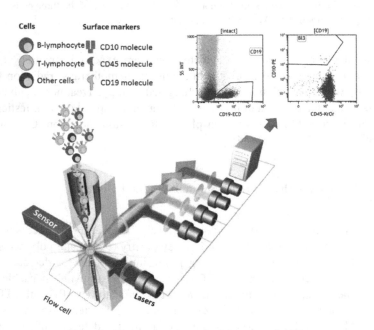

Fig. 2. Flow cytometry principle, case of ALL

bone marrow or blood cells. This sample is then transferred to the flow cytometer's flow cell which allows them to pass through the laser beam one at a time. The fluorescent dyes are then excited by the laser, and their emitted spectrum is detected by sensors, which digitize the information and visualizes it as two-dimensional dot plots, or one-dimensional histograms.

Flow cytometry provides a quantitative cell description by a number of variables, including cell size, granularity and expression of cell-surface and intracellular proteins. Due to the continuous development of flow cytometric techniques, their readouts have become increasingly complex and require adequate analysis methods [3].

2.3   Manual Gating for FCM Data Analysis

One of the most basic principles of FCM analysis is "gating," which is the sequential identification and refinement of a cellular population of interest using a panel of molecules (also known as markers) that are visualized by fluorescence in a unique emission spectrum [14]. Current diagnostic evaluation of flow cytometry readouts relies on simplistic two-dimensional analysis techniques. The basis is a labor-intensive gating procedure. In a series of two-dimensional dot plots, leukemic cells are manually flagged by drawing polygons around regions, which are known to contain mostly leukemic cells. A large number of two-dimensional plots need to be inspected and several regions need to be defined manually. Finally, candidates for leukemic cells are those inside a Boolean combination of drawn regions, called gate [3].

The challenge of MRD detection is to identify a small population of residual leukemic cells in a background of thousands of other cells of different types and developmental stages. With 2D graphical plot of otherwise multi-dimensional FCM data the operator aims at labeling relevant cell populations in the FCM data set. This procedure is performed in a sequential and hierarchical way and is called gating. Figure 3 shows an example of the hierarchical gating procedure as performed in a diagnostic laboratory. Several hundred thousand observations (also called events) of a sample are classified manually by this gating procedure. Gates are polygons that are drawn by the operator around cell clusters in different scatter plots showing the 2D projection of the higher dimensional observation space onto different dimensions. Events within the gates drawn in different plots are combined by Boolean operations to obtain specific biologically meaningful cell populations. For example leukemic blast cells lie in the intact gate (they are intact cells) and the CD19 gate (a high level of CD19 antigen is found only on the surface of B-lymphocyte cells) and the Bl3 gate (a high level of CD10 on B-lymphocytes indicates leukemic blast cells), Fig. 3. In ALL samples the goal is to identify the latter population of leukemic blast cells. The four plots are an example of hierarchical gating performed manually by a medical expert. The gating hierarchy follows the order (a)–(d) of Fig. 3, only events that are gated in the previous step are visible in the following one [2, 17].

The entire gating hierarchy is illustrated in Fig. 4. This analysis is done manually and therefore time-consuming, highly subjective, and dependent on the operator's experience. The same hierarchy of gates was used for creating our data base. Colored clusters (blue) are cells that are inside the polygon region of the gate drawn in the

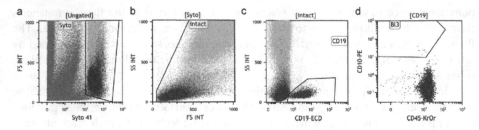

Fig. 3. Example of hierarchical gating procedure as performed in a diagnostic laboratory following (a)-(b)-(c)-(d) steps, [7]

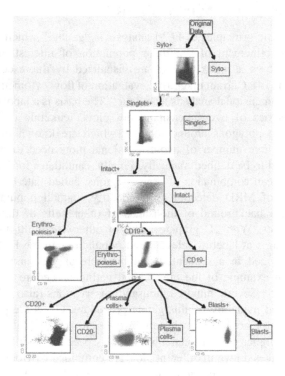

Fig. 4. Entire gating hierarchy (Color figure online)

respective plot. All cells outside the gate are discarded and not shown in the next scatter plot of the subsequent gating steps.

The first scatter plot (left) with parameters FSC-A vs. Syto41 shows all events (intact cells, clumped cells, debris, etc.). In the above scatter plot 'Syto +' all colored events are the selection of the 'Syto +' Gate. 'Syto −' denote not selected events (not intact cells, debris). The Syto gate is followed by Singlets, Intact, CD19 and Erythropoiesis, CD20, Plasma. Finally, the leukemic blast cells are colored in the scatter plot 'Blasts +', the non-colored events are all other 'CD19 +' cells, that are not leukemic cells (leukemic cells are a subset of 'CD19 +' cells). The above gates have the following biological meaning:

'Syto +' refers to cells with DNA (used to discard bubbles and debris), 'Singlets +' discards clumped cells (too big), 'Intact +' rejects dead cells (too small), and 'CD19 +' refers to B cells. Leukemic cells are inside 'Blasts +', CD20 positive cells are inside 'CD20 +' and Erythropoiesis and Plasma cells are 'Erythropoiesis +' and 'Plasma +' respectively, [17].

3 Contribution Beyond the State-of-the-Art

3.1 Machine Learning in Autogating and MRD Assessment for FCM Analysis

The gating procedure requires high level of expertise in the field and is time consuming as we have shown this in the previous sub-point. Those are the primary reasons that researchers should consider implementing automated gating and computation assistance for the FCM analysis. The time needed to manually analyze an FCM experiment is dependent on the number of markers or parameters needed to gate and on the number of experimental units to process.

There is an increasing number of approaches to automate FCM analysis. In the AutoFLOW project, we aimed to develop a ready-to-use software package which assists and automates the analysis and gating process in order to reduce FCM analysis time effort [2, 9, 10, 17, 18]. Research interests are moving towards this new direction and address this issue [6]. Existing learning approaches can be categorized in four groups: supervised/unsupervised, parametric/non-parametric, generative/discriminative and probabilistic/non-probabilistic, [10]. A parametric approach based on mixture of Gaussian functions was tested in our works [9, 17]. Preliminary results were good, blasts were modelled separately. No blast event is missed in the modelling phase. By merging more than 150 samples, result of modeling using Gaussian Mixture Model (GMM) are shown on Fig. 5 [9].

A new database consisting of 200 BCP-ALL (B-cell precursor ALL, excluding pro-B ALL) cases treated according to AIEOP-BFM ALL 2009 protocol has been established [8, 10, 21] and used in this work. Flow cytometric MRD measurement was performed at predefined time-points during treatment (day 15, day 33, day 78), and up to three different antibody combinations were used per time-point. FCM data were processed using a uniform gating strategy, defining leukemic blasts, and normal (re-generating) cell populations. Gated in FACSDiva™: Hierarchical polygon filters (Gates) follows the following steps: Syto - Intact – Singlets - CD19 – Erythropoiesis – Blasts.

To compare automated algorithms with the current practice of manual gating, experienced flow cytometrists from two of the most known cancer research centers in Europa: Children's Cancer Research Institute - Medical University of Vienna in Austria and Charité – Universitaetsmedizin Berlin in Germany, followed the established set of guidelines [8] to construct manual gates for all discernible populations. Results in our work [2] show reasonable results in detecting the number of cells affected with ALL.

The supervised GMM approach used to produce the above results was presented in [17]. It enables the implementation of an automatic clustering of FCM samples with

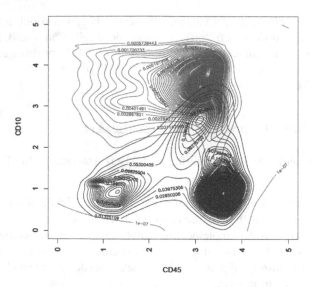

Fig. 5. GMM result for merging more than 150 samples, [9] class conditional density in 9-dimensional observation space plots of 2-dimensional marginal densities (prior = 50:50) more than 150 samples merged: blasts (red) vs. non-blasts (blue) (Color figure online)

complex distributed cell populations with focus on MRD assessment. This approach applies clinical expert knowledge about the structure of samples of different phenotypes inherent in the training samples. The model produces useful results both on new samples acquired with the same machine as used for training data acquisition as well as samples coming from different cytometer types.

3.2 Implementation of the System as a Telemedical Network

It has been demonstrated that clinical decision support system can help to improve the quality of medical care and analysis [18]. This can conveniently and effectively assist doctors, clinicians in complex tasks such as disease diagnosis, treatment, prevention and in our particularly case, decision making in the analysis of FCM data, task which requires a high level of expertise. In order to enable the use of this system in a clinical practice and by several laboratories, we conceived in this work a tool and developed it as a telemedical network for an automatic analysis of FCM data. It allows the dissemination of automated FCM-MRD analysis to medical centres without long-term experience worldwide.

The telematics platform plays a central role in the implementation of the system, Fig. 6. It is a secure and scalable framework which enables the implementation of health-related applications and their provision based on SaaS (software as a service) technology [19]. The platform is composed of a set of modules. Each module is for a particular service or group of services responsible. These services can be consumed at the application layer by the end-users as SaaS-model (Software as a Service) or as modular business solution. During the conception phase of the platform, special consideration has been paid on the security and the privacy issues. Therefore, session models, high-level data encryption methods, authentication, authorization control and methods ensuring confidentiality have been designed and implemented. Furthermore, medical data and personal data are encrypted and stored separately. The platform includes an authorization mechanism that controls the access. This function sets the authorization to applications, services and some areas of the data according to the user category and privacy constraints. Several databases are stores in a secure environment and access to them is controlled by the main central platform, Fig. 6. FCM data base (FCM DB), which contains data needed for FCM-result representation on the clinician application, is stored separately from the data base with administrative data (Admin DB) which focuses more on user management tasks. Data used for logging procedure are stored in a separate data base. This way, logging data, medical data and administrative data are stored separately.

The AutoFLOW system developed in this project is illustrated in Fig. 6. It is composed by three client applications: FlowADMIN, FlowONLINE and FlowVIEW.

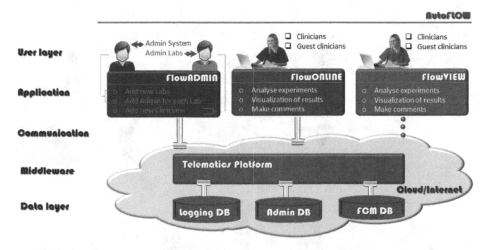

Fig. 6. Telemedical concept of the automated FCM Software-AutoFLOW

FlowADMIN is a web tool used by the Administrator of the system to perform two main tasks: register new labs to the system and create their corresponding account for the administrator of the lab. The administrator of each lab accesses the same application FlowADMIN and has the authority to: register new clinicians and guest clinicians to the lab and perform all the management tasks over the users of the lab he manages, Fig. 6. Guest clinicians can be given temporary access to some chosen experiments in order to share their opinions on an experiment or group of experiments.

The system provides in addition two versions of client software (web and desktop) to be accessed by clinicians in order to remain flexible and take benefits of both technologies. New web technologies were applied in the implementation phase of the web clients of the system. Methods of the responsive web design using media queries,

HTML5, CSS3 and JavaScript in order to have an application adapted for all screen sizes of user's device (Smartphone or personal computer) have been applied.

FlowVIEW is an interactive visualization software for AutoGating, presented in our work [1]. Users can load Flow Cytometry (fcs) data and visualize it similar to conventional software such as Kaluza Analysis. While the number of Blasts and therefore the MRD is nowadays determined through a manual gating process, FlowVIEW allows the AutoGating. Hence, users need to push a button which triggers an automated labelling process. The classification of cells into Blasts and Non-Blasts is based on aligning GMMs sampled from the new population with multiple GMMs from the training set [9]. Figure 7 shows a screenshot of FlowVIEW where blast cells are colored in red. The left AutoGating panel shows the statistics of all three populations if the model's priors are set to default (0.33). Functionalities of the developed client FlowVIEW can be watched at the link [20].

FlowONLINE, Fig. 6, is an online web client to be accessed by clinicians for analyzing FCM data without installation effort. The backend of FlowONLINE is hosted

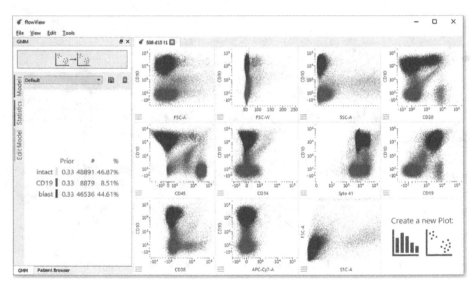

Fig. 7. FlowVIEW-desktop version of the visualization software for AutoGating: blasts (red) and non-blasts (other colors). (Color figure online)

by Python Django connected with a SQLite database on the same server. The computation and visualization of FCM data is conducted by a module called FlowVIEWEx, which is a Python wrapper for FlowVIEW C++ libraries, developed for this project. Sharing the same library between the desktop client FlowVIEW and the Web Client FlowONLINE, reduces development time and effort needed to produce identical results in both clients.

4   Conclusion

This paper has presented a machine learning based system developed for clinical practice in assisting the analysis and the therapy process of children with ALL. The system is implemented as ready-to-use telemedical software for FCM automated data analysis and can connect worldwide laboratories to the expert system. Thus enhancing the quality of FCM data analysis, improving quality-assured MRD-assessment even for countries with less expertise in the field. This reduces work-load and lab-costs for operator training, reduces the error in assessment of FCM-MRD in samples which are difficult to analyze by conventional operator-based gating and also reduces subjectivity caused by manual operator gating. The system applies machine learning techniques, a supervised approach for automatic FCM data analysis with focus on MRD assessment. The approach uses clinical expert knowledge about the features of the training samples of different phenotypes. It is based on the representation of sample by a GMM which parameters are represented as a linear combination of multiple reference GMM. The tool developed can be used for the daily clinical practice and will allow dissemination of automated FCM-MRD analysis worldwide even to medical centres without long-term experience in the field.

Acknowledgments. The AutoFLOW project is funded by Marie Curie Industry Academia Partnerships & Pathways (FP7-MarieCurie-PEOPLE-2013- IAPP) under the grant no.610872.

References

1. Reiter, M., Kleber, F., Rota, P., Groeneveld-Krentz, S., Schumich, A., Dworzak, M., Gau, M.: An automated flow cytometry data analysis support system. In: CYTO 2015, Glasgow (2015)
2. Rota, P., Kleber, F., Reiter, M., Groeneveld-Krentz, S., Kampel, M.: The role of machine learning in medical data analysis. A case study: flow cytometry. In: Proceedings of the 11th Joint Conference on Computer Vision, Imaging and Computer Graphics Theory and Applications, VISAPP, vol. 3, pp. 303–310. ISBN 978-989-758-175-5. doi:10.5220/0005675903030310
3. Toedling, J., Rhein, P., Ratei, R., Karawajew, L., Spang, R.: Automated in-silico detection of cell populations in flow cytometry readouts and its application to leukemia disease monitoring. BMC Bioinform. **7**, 282 (2006). doi:10.1186/1471-2105-7-282
4. PDQ® Pediatric Treatment Editorial Board: PDQ Childhood Acute Lymphoblastic Leukemia Treatment. National Cancer Institute, Bethesda. http://www.cancer.gov/types/leukemia/patient/child-all-treatment-pdq. [PMID: 26389385]. Accessed 28 Oct 2016
5. Costa, E.S., Arroyo, M.E., Pedreira, C.E., García-Marcos, M.A., Tabernero, M.D., Almeida, J., Orfao, A.: A new automated flow cytometry data analysis approach for the diagnostic screening of neoplastic B-cell disorders in peripheral blood samples with absolute lymphocytosis. J. Leukemia **20**, 1221–1230 (2006). doi:10.1038/sj.leu.2404241. Published online 25 May 2006
6. Aghaeepour, N., Finak, G., Hoos, H., Mosmann, T.R., Brinkman, R., Gottardo, R., Scheuermann, R.H.: Critical assessment of automated flow cytometry data analysis techniques. Nat. Methods **10**(3), 228–238 (2013). doi:10.1038/nmeth.2365

7. Rota, P., Groeneveld-Krentz, S., Reiter, M.: On automated flow cytometric analysis for MRD estimation of acute lymphoblastic leukaemia: a comparison among different approaches. In: The Proceedings of Bioinformatics and Biomedicine (BIBM 2015), pp. 438–441 (2015)
8. AIEOP-BFM ALL 2009 treatment protocol. http://www.bfm-international.org/aieop/aieop_index.html
9. Reiter, M., Kleber, F., Hoffmann, J., Groeneveld-Krentz, S., Groeneveld, T., Dworzak, M., Schuumich, A., Kampel, M., Gau, M.: Status report of the automated software-based MRD-assessment procedure. In: 28th Annual Meeting of the International BFM (Berlin-Frankfurt-Muenster) Study Group, Budapest, Hungary (2015)
10. Reiter, M., Ho-mann, J., Kleber, F., Schumich, A., Peter, G., Kromp, F., Kampel, M., Dworzak, M.: Towards automation of flow cytometric analysis for quality assured follow-up assessment to guide curative therapy for acute lymphoblastic leukaemia in children. Memo Mag. Eur. Med. Oncol. **7**(4), 219–226 (2014). Springer
11. Bonner, W.A., Hulett, H.R., Sweet, R.G., Herzenberg, L.A.: Fluorescence activated cell sorting. Rev. Sci. Instrum. **43**(3), 404–409 (1972). doi:10.1063/1.1685647
12. Campana, D., Coustan-Smith, E.: Detection of minimal residual disease in acute leukemia by flow cytometry. J. Cytometry Part A **38**(4), 139–152 (1999). doi:10.1002/(SICI)1097-0320 (19990815)38:4<139:AID-CYTO1>3.0.CO;2-H
13. Karawajew, L., Dworzak, M., et al.: Minimal residual disease analysis by eight-color flow cytometry in relapsed childhood acute lymphoblastic leukemia. J. Haematol. **100**, 935–944 (2015). doi:10.3324/haematol.2014.116707
14. Verschoor, C.P., Lelic, A., Bramson, J.L., Bowdish, D.M.E.: An introduction to automated flow cytometry gating tools and their implementation. J. Front. Immunol. **8**, 1–8 (2015). doi:10.3389/fimmu.2015.00380. http://www.frontiersin.org
15. Nunes, C., Wong, R., Mason, M., Fegan, C., Man, S., Pepper, C.: Expansion of a CD8(+) PD-1(+) replicative senescence phenotype in early stage CLL patients is associated with inverted CD4: CD8 ratios and disease progression. Clin. Cancer Res. **18**(3), 678–687 (2012). doi:10.1158/1078-0432.CCR-11-2630
16. Dalmazzo, L.F.F., Jácomo, R.H., Marinato, A.F., Figueiredo-Pontes, L.L., Cunha, R.L.G., Garcia, A.B., et al.: The presence of CD56/CD16 in T-cell acute lymphoblastic leukaemia correlates with the expression of cytotoxic molecules and is associated with worse response to treatment. Br. J. Haematol. **144**(2), 223–229 (2009). doi:10.1111/j.1365-2141.2008.07457
17. Reiter, M., Rota, P., Kleber, F., Diem, M., Groeneveld-Krentz, S., Dworzak, M.: Clustering of cell populations in flow cytometry data using a combination of Gaussian mixtures. J. Pattern Recogn. **60**, 1029–1040 (2016). Elsevier
18. Chu, H., Yang, Y., Li, Q., Xu, Y., Wei, H.: A scalable clinical intelligent decision support system. In: Chang, C.K., Chiari, L., Cao, Yu., Jin, H., Mokhtari, M., Aloulou, H. (eds.) ICOST 2016. LNCS, vol. 9677, pp. 159–165. Springer, Cham (2016). doi:10.1007/978-3-319-39601-9_14
19. Berndt, R.D., Takenga, C., et. al.: SaaS-platform for mobile health applications. In: Proceedings of the IEEE-International Multi-conference on Systems, Signals and Devices, Chemnitz, Germany, pp. 1–4, March 2012, doi:10.1109/SSD.2012.6198120
20. FlowVIEW Functionalities. https://www.youtube.com/watch?v=fu0V76Cppa4
21. Cario, G., Rhein, P.: High CD45 surface expression determines relapse risk in children with precursor B-cell and T-cell acute lymphoblastic leukemia treated according to the ALL-BFM 2000 protocol. Haematologica **99**(1), 103–110 (2014). doi:10.3324/haematol.2013.090225

Prognostic Modeling and Analysis of Tumor Response to Fractionated Radiotherapy for Patients with Squamous Cell Lung Cancer

Hualiang Zhong[✉], Hoda Sharifi, Haisen Li, Weihua Mao, and Indrin J. Chetty

Department of Radiation Oncology,
Henry Ford Health System, Detroit, MI 48202, USA
Hzhongl@hfhs.org

Abstract. The efficacy of radiotherapy depends on tumor response to radiation. The purpose of this study is to derive three basic tumor-response parameters from a sequence of measured tumor volumes, and then use these parameters to predict treatment outcome for patients with squamous cell lung cancer. 43 patients were investigated in this study. A tumor kinetic model was iterated daily, following the fractionation scheme of radiation treatment for each patient. By minimizing the difference between the model-predicted tumor regression and the measured changes in tumor volumes, three parameters were calibrated and then used to calculate tumor control probability (TCP). The results showed that the major portion of the patients have tumor doubling time (DT) less than 183 days and half time (HT) for dead cell resolving less than 50 days. The modeling differences from measured tumor volumes were summated over the 43 patients, and the sum reaches its minimum with DT at 90 days and HT at 30 days. The behavior of the kinetic model is stable to DT and HT as the modeling difference has only one minimum within the clinically meaningful ranges for the two parameters. One third of these patients have the cell survival fraction at 12 Gy (SF_{12}) less than 0.1, and 80% of them with SF_{12} less than 0.6. In summary, kinetic modeling of tumor response may help quantify radiobiological parameters which can be used to predict treatment outcomes for individual patients.

Keywords: Kinetic modeling · Radiobiological parameters · Tumor control probability · Tumor regression

1 Introduction

Tumor control probability (TCP) has been used in clinic as a guideline for treatment of cancer patients (Allen et al. 2012). Most of TCP models were developed based on the Poisson distribution of cell survival fractions which characterizes the surviving ratios of clonogenic tumor cells. If the initial number of clonogens is N_0, and the cell surviving fraction is SF_D under the fractional dose D, then the Poissonian TCP can be computed by $TCP = \exp\{-N_0 \times (SF_D)^m\}$, where m is the number of treatment fractions (Zaider and Minerbo 2000). This model has been improved by incorporating clonogen proliferation kinetics (Hanin 2001), and generalized by incorporating cell cycle effects

© Springer International Publishing AG 2017
I. Rojas and F. Ortuño (Eds.): IWBBIO 2017, Part II, LNBI 10209, pp. 549–559, 2017.
DOI: 10.1007/978-3-319-56154-7_49

(Dawson and Hillen 2006). To cope with the diversity of tumor response to radiation, Webb and Nahum incorporated the heterogeneity of tumor cell density into the Poissonian TCP model (Webb and Nahum 1993), and optimized the parameter of cell density for different groups of patients (Webb, 1994). However, most of these complex models either lack simple analytical solutions, or their parameters are often unavailable for patients in clinic. Therefore, only empirical models that have manageable parameters were recommended as a biological modeling tool for clinical practice (Allen et al. 2012). For example, a typical TCP model used in clinic has two parameters D_{50}, the dose at which 50% of tumors are controlled, and normalized dose-response gradient γ evaluated at $D = D_{50}$, derived directly from a group of clinical outcomes (Brahme 1984). The empirical TCP models have served as an important guideline for radiation treatment planning and outcome assessment (Allen et al. 2012).

Radiation treatment for cancers is a dynamic cell birth-and-death process coupled with multi-fractional radiation dose inputs. The efficacy of radiation therapy (RT) is subject to many factors such as cell repairing, repopulation or re-oxygenation. For a given dose size, the exact surviving fraction of cells may depend on tumor radiosensitivity which is commonly described by the linear-quadratic (LQ) model (Dale 1985; Sachs and Brenner 1998). With this model, the dose required to achieve a desired tumor control probability (TCP) can be calculated, and the biologically equivalent dose (BED) for different fractionation/protraction schemes can be derived (Brenner 2008; Guerrero and Carlone 2010). For patients treated with stereotactic body radiation therapy (SBRT), the LQ model was extended to a linear-quadratic-linear (LQL) form which is appropriate for modeling of tumor response to large dose sizes (Carlone et al. 2005; Guerrero and Li 2004). With the improved model, the survival fraction of clonogens can be estimated more accurately for different dose rates (Brenner et al. 1998) and for different dose sizes (Guerrero and Li 2004). However, radiobiological parameters used in these models were derived from outcomes of a group of patients. Without accounting for heterogeneities in cell proliferation and radio-sensitivity for individual patients, the accuracy of these models could be limited (Hall and Giaccia 2012).

Bhaumik and Jain divided cells into the proliferating and non-proliferating compartments, and investigated the tumor growth kinetics on the basis of interactions between the two compartments (Bhaumik and Jain 1987). Gupta et al. verified that there are interstate conversions between stem and non-stem cells (Gupta et al. 2011). Sachs et al. developed differential equations to capture the essential features of tumor/vasculature dynamics, and modeled radiation-induced changes in the number of DNA double strand breaks per cell and the number of cells per tumor (Sachs et al. 2001). Since radiation-induced dead cells may remain at the site for a period of time after treatment, Okumura et al. developed an analytic, two-compartment cell population model to simulate the tumor regression process (Okumura et al. 1977). In this regard, Lim et al. and Chvetsov developed discrete kinetic models for modeling of complex clinical regimens (Lim et al. 2008; Chvetsov et al. 2009), and Zhong et al. showed that these discrete models are approximate to the solution of the Okumura simulation system (Zhong and Chetty 2014).

In this study, we incorporated the discrete kinetic model (Zhong and Chetty 2014) into the analysis of clinical outcomes for a cohort of lung cancer patients treated with

SBRT. Three explicit parameters, cell survival fraction, tumor doubling time (DT), and half time (HT) for dead cell resolving, were estimated from follow-up CT images. With the derived parameters, tumor control probability was computed and analyzed for different groups of these patients.

2 Materials and Methods

Radiation-induced DNA damage may be repaired or lead to apoptosis, and the resultant changes in the numbers of living and dead cells are correlated to the variation of tumor volumes during the course of radiation treatment.

(a) Modeling of tumor response to fractionated radiotherapy

Let S(t) and K(t) represent the numbers of the surviving and dead cancer cells after irradiation, respectively. The mathematical relationship between the two components can be described by (Okumura et al. 1977; Zhong and Chetty 2014):

$$\dot{S}(t) = (a_1 - a_2)S(t), \quad \dot{K}(t) = cK(t) + a_2 S(t) \tag{1}$$

where a_1, a_2 and c represents cell proliferation, killing and resolving rates. The analytical solution of these equations was obtained and used to construct a discrete kinetic model represented by

$$S_i = S_{i-1}e^{\ln 2/T_d + \ln(SF_D)}$$

$$K_i = \left(K_{i-1} + \frac{\ln(SF_D)}{\ln 2/T_d + \ln(SF_D) + \ln 2/T_r}S_{i-1}\right)e^{-\ln 2/T_r} - \frac{\ln(SF_D)}{\ln 2/T_d + \ln(SF_D) + \ln 2/T_r}S_{i-1}e^{\ln 2/T_d + \ln(SF_D)} \tag{2}$$

where S_i and K_i represent the numbers of the surviving and dead cancer cells after treatment fraction i, and T_d and T_r denote cell doubling time and half time for dead-cell resolving, respectively. SF_D is cell survival fraction after exposure of radiation dose D. In our hospital, all the SBRT patients were treated 12 Gy per fractions for 4 fractions within 10 days. Let D equal 12 Gy for treatment days, and 0 Gy for non-treatment days, then SF_D can be uniformly represented by $SF_D = \exp\{-bD\}$ with D assigned by either 12 or 0. The variable replacement makes it convenient to implement a day-to-day manageable treatment scheme in the discrete model for individual patients.

(b) Computation of tumor control probability

While cancer cells can be killed or inactivated by ionizing radiation, damaged cells may get repaired and continued to repopulate. By incorporating the proliferation effect, the TCP model can be generalized as (Li et al. 2012)

$$TCP_T = \exp\{-N_0(SF_D)^m \exp(T\ln(2)/T_d)\}. \tag{3}$$

where N_0 is the number of cancer stem cells which drives tumor growth and are responsible for recurrence (Visvader and Lindeman 2008). As the relative proportions of stem and non-stem cells are in equilibrium (Gupta et al. 2011), N_0

is assumed to be proportional to the total number of the tumor cells during the course of treatment. The radio-sensitivity (SF$_D$) and tumor doubling time (T$_d$) can be derived by fitting the kinetic model (2) with a sequence of measured tumor volumes.

(c) Derivation of radiobiological parameters

Since tumor volume is proportional to the number of its clonogens (Bentzen and Thames 1996), the regression of the tumor volume may represent changes in the total number of the tumor cells characterized by Eq. (2). This allows the parameters T$_d$, T$_r$, and SF$_D$ to be calibrated from the measured tumor volumes (Lim et al. 2008; Huang et al. 2010). It was assumed that the tumor volumes are unchanged between the first image scan and the first treatment fraction. All the tumor volumes measured from the subsequent images were normalized to the initial volume to obtain the tumor regression ratios. Since the tumor volume is proportional to the number of its clonogens with a linear constant coefficient (Brenner 1993), an objective function that describes the average deviation of the modeling can be defined by

$$Q(SF_D, T_r, T_d) = 100 \times \sqrt{\sum_{i=1,...,n} \frac{1}{n} |(S_i + K_i)/S_0 - M_i/M_0|^2} \qquad (4)$$

where S_0 and M_0 are the number of the tumor cells and the tumor volume measured before treatment. M_i is the tumor volume measured at the i-th assessment, S_i and K_i are the numbers of the living and dead cells on day i, predicted by the kinetic model (2), respectively. For the measurement M_i performed on day i, dose D on day j, $0 < j < i$, can be assumed either 12 or 0 Gy, depending on whether or not treatment was delivered on that day. S_i and K_i can be calculated by iterating Eq. (2) i times. By moving the parameters T$_d$, T$_r$ and SF$_D$ through their clinically potential ranges, the objective function Q was minimized and the optimal values for these parameters were found. This search program was implemented with C++ on a linux computer.

Since the minimum of the function Q could be sensitive to uncertainties in tumor volume measurements for individual patients, we also calculate the sum of the objective function Q over all the patients with each of the three parameters incremented alternatively:

$$R = \frac{1}{L} \min \left\{ \sum_{j=1,...,L} Q_j | SF_D, T_r, T_d \right\}, \qquad (5)$$

where L is the number of the patients. The optimal parameters were then analyzed based on the fitness of the model to the measured tumor volumes.

(d) Specification of patient data and treatment regimens

43 patients with squamous-cell lung cancer were treated at Henry Ford Health System with 12 Gy × 4 fractions of SBRT using 6 MV photon beams. Treatment outcomes for each of these patients were investigated using the developed model. Specifically, the initial tumor volumes of these patients were measured from planning images, as shown in Fig. 1a, and their average is 4.6 cm^3. Follow-up

evaluations were performed about 48, 142 and 240 days after their first treatment fraction, and exact dates may vary due to patient schedule changes. From the three follow-up measurements, the average tumor volumes were reduced to 2.8, 2.3, and 2.5 cm^3, respectively. For each of these patients, the relative tumor volumes in their follow-up images to their initial volumes were shown in Fig. 1b.

From Fig. 1b, it can be found that there are about 15% of patients with the tumor growth observed in their last assessment which was conducted about 240 days after treatment. From the measured tumor volumes, the radiobiological parameters were calibrated and used to estimate tumor control probabilities for each patients.

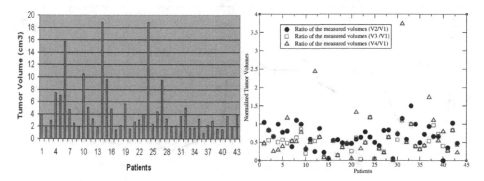

Fig. 1. (a) Tumor volumes measured from the initial CT scans of the 43 lung cancer patients; (b) the ratio of the tumor volumes after treatment and the initial tumor volumes for these patients.

3   Experiments and Results

(a)   Variation of radiobiological parameters in the kinetic model

With T_r fixed on day 5n, n = 1, ..., 60, the parameters b and T_d were increased from 0 to 1.0 and 1000, with the step sizes of 0.002 and 10, respectively. Since $SF_{12} = e^{-12b}$, the interval (0, 1.0) for the parameter b corresponds to the range of 6×10^{-6} to 1.0 for SF_{12}. Therefore, the searched ranges include all potential values for the parameters SF_{12} and T_d. With the parameters selected in these ranges, the function R in Eq. (5) was computed with the minimum residues shown in Fig. 2a. The computed function R has only one local minimum for T_r in the range of 300 days (Fig. 2a). Similarly, T_d was fixed on day 10n, n = 1, ..., 100, and the parameters b and T_r were increased with the steps of 0.002 and 5. The resultant R values were shown in Fig. 2b. It can be found that the R function assumes the minimum at T_r equal to 30 and T_d equal to 90 days, respectively. Unlike T_r or T_d, the parameter SF_D(b) showed complex behaviors with multiple local minimums found for the function R. Figure 2 (c,d) shows the residues of R with respect to b and SF_D, respectively, with the coefficient b incremented 0.002 each time.

(b)   Derivation of optimal parameters for individual patients

Based on the measured tumor volumes, the three parameters T_d, T_r and SF_d were optimized for each of the 43 patients. The residues of the function Q optimized for

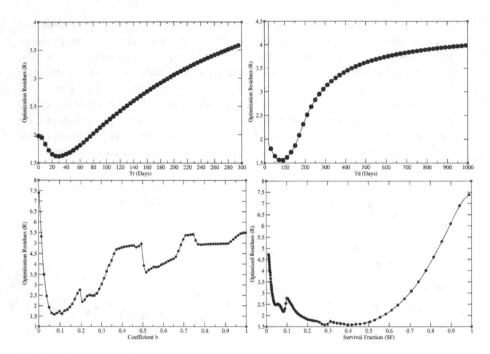

Fig. 2. The residues of the function R averaged over the forty-three lung cancer patients: the parameters T_r and T_d were increased stepwise in (a) and (b), and the coefficient b was incremented in (c) and (d).

each patient were shown in Fig. 3. The model was fitted with the measured tumor volumes with the fitting residues equal to 1.5% on average.

Among the 43 patients, patients 10 and 18 have small fitting residues. Their fitted curves were displayed in Fig. 4a, and the parameters used in the fitted model were shown in Table 1. With these parameters, the Poissonian TCP were computed with Eq. (3) and the results were shown in Table 1. Compared with P_{18}, patient 10 is more sensitive to radiation with the survival fraction equal to 0.21. The TCP computed without accounting for proliferation is 0.62, better than that for P_{18}. But due to the short tumor double time (61 days), the TCP value was reduced dramatically one year later for patient 10. This is consistent to the trend of the tumor regression curve shown in Fig. 4a. For patients with small T_d, the prediction of tumor control is strongly influenced by the time variable T.

Table 1. The TCPs and tumor regression rates computed for the two patients with the response model best fitted to their measured tumor volumes.

Patients	T_d (days)	T_r (days)	SF_{12}	M_3/M_0	TCP_0	TCP_{10}	TCP_{365}
P_{10}	61	31	0.21	0.34	0.62	0.58	8.9×10^{-14}
P_{18}	431	31	0.53	0.36	0.54	0.53	0.32

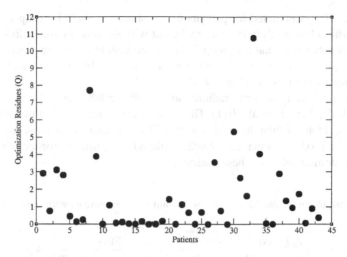

Fig. 3. The residues of the optimized function Q for 43 lung cancer patients.

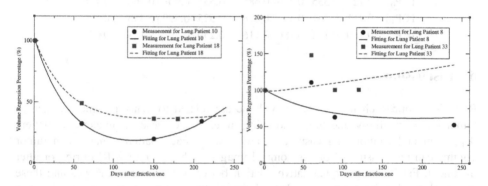

Fig. 4. (a) patients with the response model best fitted to their measured tumor volumes; (b) patients with the response model worst fitted to their measured tumor volumes.

From Fig. 3, it can be found that patients 8 and 33 have large fitting errors. This is because the tumor volumes measured in their second CT images are greater than their initial tumor volumes (Fig. 1b). The temporary increase of tumor volume during or after treatment could be induced by side effects of ionizing radiation (Sofia et al. 2010). This behavior, however, was not modeled in the current system. In addition, uncertainties in the tumor volume measurements may also contribute to the large fitting errors. Due to these uncertainties, 13 patients have their optimal parameters found outside the potential parameter ranges. The statistics for the other 30 patients were shown in Table 2.

It was reported that tumor doubling time for lung cancer is in the range of 30–1077 days (Usuda et al. 1994), and the major portion (60%) of patients with squamous cell cancers have the doubling time less than 183 days (Wilson et al. 2012). These results are consistent to what we obtained in this study. Without accounting for cell proliferation, TCP_0 calculated for patients with a shorter T_d (<183, group 1) are higher than those with a longer T_d (>183, group 2), but with the time factor included, TCP_{365} for group 2 is better than that for group 1. This is consistent to the previous observation that the survival rate is lower for patients with a shorter doubling time than those with a longer doubling time (Usuda et al. 1994).

The squamous cell cancer is radioresistant, and the lab measured results showed that $SF_5 = 0.65$ (Hedman et al. 2011). This study showed that most of the patients have $SF_{12} < 0.6$, and part of them have $SF_{12} < 0.1$. TCPs calculated for these radiosensitive patients are up to 0.9 as shown in Table 2, while with the time factor included, TCP_{365} was reduced dramatically for these patients.

Table 2. Distributions of the 30 saquamous cell lung cancer patients with different parameters calibrated.

	T_d (days)		T_r (days)			SF_{12}		
Groups	<183	>183	<50	50–150	>150	<0.1	0.1–0.6	>0.6
n	18	12	16	11	3	10	14	6
TCP_0	0.723	0.335	0.436	0.685	0.534	0.90	0.467	0.198
TCP_{10}	0.711	0.331	0.427	0.678	0.525	0.899	0.453	0.187
TCP_{365}	0.160	0.184	0.186	0.116	0.250	0.194	0.151	0.041

4   Discussion

Radiation-induced changes in tumor volume are related to many parameters such as tumor radio-sensitivity and tumor doubling time. While these parameters are patient dependent and cannot be measured in advance for each patient, changes in tumor volume can be measured from anatomical images such as CT or MRI during or after treatment. Establishing a quantitative relation between the volume changes and these radiobiological parameters may help understand the underlying mechanism of tumor response and is important to development of adaptive treatment plans for individual patients. In this study, we derived three radiobiological parameters from measured tumor volumes, and based on the calibrated parameters we calculated the tumor control probability for 43 patients who had squamous cell lung cancer.

Tumor doubling time and radio-sensitivity are two crucial factors in determining the efficacy of radiation treatment. Compared to other types of lung cancer, squamous cells have shorter doubling time and are more radio-resistant. It was reported that squamous cell cancer comprised 60% of a group with rapid tumor growth (DT < 183 days) and 3.3% of a group with slow growth (DT > 365 days) (Wilson et al. 2012). The maximum doubling time is 449 days for squamous carcinoma vs 4263 days for other tumor cells (Wilson et al. 2012). In our study, it was found that the model calibrated T_d is less than 183 days and T_r is less than 50 days for the major portion of the 43 squamous cell cancer patients. The doubling time is consistent to the results measured directly from anatomical images (Arai et al. 1994; Wilson et al. 2012). The SF_{12} calibrated in this study is less than 0.6 for most of these squamous cell lung cancer patients. This estimation is comparable to the surviving rate of 0.56 measured by

Hedman et al. (2011) for cells irradiated by 2×5 Gy. However, as demonstrated in Fig. 2d, the fitting function R for the 43 patients have several local minimums in the range of 0.01–0.6 for SF_{12}. This is different from the case of T_r or T_d, where the function R has only one local minimum. Therefore, SF_{12} may have more calibration uncertainties than the other two parameters. In general, the kinetic modeling method allows these radiobiological parameters to be investigated systematically, and based on these parameters TCP can be calculated for individual patients.

It should be mentioned that some tumor responses were not modeled in this study. For example, vascularization or reoxygenation may have impact on treatment outcomes (Chaplain 1996; Toma-Dasu and Dasu 2013); the interconversion of cancer cells between different states may also influence the proliferation rate of these cells (Hillen et al. 2013; Gupta et al. 2011). Without modeling these components, the robustness and accuracy of the model may be compromised. Figure 4b showed that tumor regression cannot be modeled accurately when there are radiotherapy-stimulated tumor growth or uncertainties in the measured tumor volumes. To address these issues, 13 patients with large fitting errors were excluded from the statistical analysis performed in this study. Development of a more robust model to simulate the tumor regression and growing process may help estimate basic radiobiological parameters which in turn could help predict treatment outcomes for individual patients.

5 Conclusion

With the kinetic tumor response model, three radiobiological parameters have been calibrated from measured tumor volumes for patients with squamous cell lung cancer. The derived parameters may help predict treatment outcomes and are useful for development of patient-specific treatment regimens for lung cancer patients.

Acknowledgement. The authors gratefully acknowledge the financial support from the National Institutes of Health Grant No. R01 CA140341.

References

Allen, L., Alber, M.X., Deasy, J.O., Jackson, A., Jee, K.W.K., Marks, L.B., Martel, M.K., Mayo, C., Moiseenko, V., Nahum, A.E., Niemierko, A., Semenenko, V.A., Yorke, E.D.: The use and QA of biologically related models for treatment planning: short report of the TG-166 of the therapy physics committee of the AAPM. Med. Phys. **39**, 1386–1409 (2012)
Arai, T., Kuroishi, T., Saito, Y., Kurita, Y., Naruke, T., Kaneko, M.: Tumor doubling time and prognosis in lung cancer patients: evaluation from chest films and clinical follow-up study. Japanese lung cancer screening research group. Japan. J. Clin. Oncol. **24**, 199–204 (1994)
Bentzen, S.M., Thames, H.D.: Tumor volume and local control probability: clinical data and radiobiological interpretations. Int. J. Radiat. Oncol. Biol. Phys. **36**, 247–251 (1996)
Bhaumik, K., Jain, V.K.: Mathematical models for optimizing tumour radiotherapy I: a simple two compartment cell-kinetic model for the unperturbed growth of transplantable tumours. J. Biosci. **12**, 153–164 (1987)

Brahme, A.: Dosimetric precision requirements in radiation therapy. Acta Radiol. Oncol. **23**, 379–391 (1984)

Brenner, D.J.: Dose, volume, and tumor-control predictions in radiotherapy. Int. J. Radiat. Oncol. Biol. Phys. **26**, 171–179 (1993)

Brenner, D.J.: The linear-quadratic model is an appropriate methodology for determining isoeffective doses at large doses per fraction. Semin. Radiat. Oncol. **18**, 234–239 (2008)

Brenner, D.J., Hlatky, L.R., Hahnfeldt, P.J., Huang, Y., Sachs, R.K.: The linear-quadratic model and most other common radiobiological models result in similar predictions of time-dose relationships. Radiat. Res. **150**, 83–91 (1998)

Carlone, M., Wilkins, D., Raaphorst, P.: The modified linear-quadratic model of Guerrero and Li can be derived from a mechanistic basis and exhibits linear-quadratic-linear behaviour. Phys. Med. Biol. **50**, L9–L13 (2005)

Chaplain, M.A.J.: Avascular growth, angiogenesis and vascular growth in solid tumours: the mathematical modeling of the stages of tumour development. Math. Comp. Mod. **23**, 47–87 (1996)

Chvetsov, A.V., Dong, L., Palta, J.R., Amdur, R.J.: Tumor-volume simulation during radiotherapy for head-and-neck cancer using a four-level cell population model. Int. J. Radiat. Oncol. Biol. Phys. **75**, 595–602 (2009)

Dale, R.G.: The application of the linear-quadratic dose-effect equation to fractionated and protracted radiotherapy. Br. J. Radiol. **58**, 515–528 (1985)

Dawson, A., Hillen, T.: Deriation of the tumor control probability (TCP) from a cell cycle model. Comput. Math. Methods Med. **7**, 121–141 (2006)

Guerrero, M., Carlone, M.: Mechanistic formulation of a lineal-quadratic-linear (LQL) model: split-dose experiments and exponentially decaying sources. Med. Phys. **37**, 4173–4181 (2010)

Guerrero, M., Li, X.A.: Extending the linear-quadratic model for large fraction doses pertinent to stereotactic radiotherapy. Phys. Med. Biol. **49**, 4825–4835 (2004)

Gupta, P.B., Fillmore, C.M., Jiang, G., Shapira, S.D., Tao, K., Kuperwasser, C., Lander, E.S.: Stochastic state transitions give rise to phenotypic equilibrium in populations of cancer cells. Cell **146**, 633–644 (2011)

Hall, E.J., Giaccia, A.J.: Radiobiology for the Radiologist. Lippincott Williams & Wilkins, Philadelphia (2012)

Hanin, L.G.: Iterated birth and death process as a model of radiation cell survival. Math. Biosci. **169**, 89–107 (2001)

Hedman, M., Bergqvist, M., Brattstrom, D., Brodin, O.: Fractionated irradiation of five human lung cancer cell lines and prediction of survival according to a radiobiology model. Anticancer Res. **31**, 1125–1130 (2011)

Hillen, T., Enderling, H., Hahnfeldt, P.: The tumor growth paradox and immune system-mediated selection for cancer stem cells. Bull. Math. Biol. **75**, 161–184 (2013)

Huang, Z., Mayr, N.A., Yuh, W.T., Lo, S.S., Montebello, J.F., Grecula, J.C., Lu, L., Li, K., Zhang, H., Gupta, N., Wang, J.Z.: Predicting outcomes in cervical cancer: a kinetic model of tumor regression during radiation therapy. Cancer Res. **70**, 463–470 (2010)

Lim, K., Chan, P., Dinniwell, R., Fyles, A., Haider, M., Cho, Y.B., Jaffray, D., Manchul, L., Levin, W., Hill, R.P., Milosevic, M.: Cervical cancer regression measured using weekly magnetic resonance imaging during fractionated radiotherapy: radiobiologic modeling and correlation with tumor hypoxia. Int. J. Radiat. Oncol. Biol. Phys. **70**, 126–133 (2008)

Okumura, Y., Ueda, T., Mori, T., Kitabatake, T.: Kinetic analysis of tumor regression during the course of radiotherapy. Strahlentherapie **153**, 35–39 (1977)

Sachs, R.K., Brenner, D.J.: The mechanistic basis of the linear-quadratic formalism. Med. Phys. **25**, 2071–2073 (1998)

Sachs, R.K., Hlatky, L.R., Hahnfeldt, P.: Simple ODE models of tumor growth and anti-angiogenic or radiation treatment. Math. Comput. Model. **33**, 1297–1305 (2001)

Sofia, V.I., Martins, L.R., Imaizumi, N., Nunes, R.J., Rino, J., Kuonen, F., Carvalho, L.M., Ruegg, C., Grillo, I.M., Barata, J.T., Mareel, M., Santos, S.C.: Low doses of ionizing radiation promote tumor growth and metastasis by enhancing angiogenesis. PLoS One **5**, e11222 (2010)

Toma-Dasu, I., Dasu, A.: Modelling tumour oxygenation, reoxygenation and implications on treatment outcome. Comput. Math. Methods Med. **2013**, 141087 (2013)

Usuda, K., Saito, Y., Sagawa, M., Sato, M., Kanma, K., Takahashi, S., Endo, C., Chen, Y., Sakurada, A., Fujimura, S.: Tumor doubling time and prognostic assessment of patients with primary lung cancer. Cancer **74**, 2239–2244 (1994)

Visvader, J.E., Lindeman, G.J.: Cancer stem cells in solid tumours: accumulating evidence and unresolved questions. Nat. Rev. Cancer **8**, 755–768 (2008)

Webb, S., Nahum, A.E.: A model for calculating tumour control probability in radiotherapy including the effects of inhomogeneous distributions of dose and clonogenic cell density. Phys. Med. Biol. **38**, 653–666 (1993)

Webb, S.: Optimum parameters in a model for tumour control probability including interpatient heterogeneity. Phys. Med. Biol. **39**, 1895–1914 (1994)

Wilson, D.O., Ryan, A., Fuhrman, C., Schuchert, M., Shapiro, S., Siegfried, J.M., Weissfeld, J.: Doubling times and CT screen-detected lung cancers in the Pittsburgh lung screening study. Am. J. Respir. Crit. Care Med. **185**, 85–89 (2012)

Zaider, M., Minerbo, G.N.: Tumour control probability: a formulation applicable to any temporal protocol of dose delivery. Phys. Med. Biol. **45**, 279–293 (2000)

Zhong, H., Chetty, I.: A note on modeling of tumor regression for estimation of radiobiological parameters. Med. Phys. **41**, 081702 (2014)

Uncertainty Quantification for Meningococcus B Carriers Prediction

Luis Acedo, Clara Burgos, Juan-Carlos Cortés, and Rafael J. Villanueva[✉]

Instituto Universitario de Matemática Multidisciplinar,
Universitat Politecnica de Valencia, Valencia, Spain
rjvillan@imm.upv.es

Abstract. Competition among the different classes of genogroups of meningococcal and further bacteria takes place in the ecosystem they constitute within the human body environment. To the best of our knowledge, most of the epidemiological models neglect competition that could highly influence the dynamics of the meningococcal as well as its outbreak. In this paper, we propose a susceptible-carrier-susceptilbe epidemiological model in order to describe the evolution of the percentage of carriers in the population over the time. Moreover, we introduce a competition model to study the dynamics of the meningococcal genogroups in Spain among the carriers. As we are just interested in the meningococcus B, the latter model considers two subpopulations, meningococcal genogroup B and the rest. Both models depend on a number of parameters that are adjusted using real data from the distribution of the genogroups in Spain in years 2011 and 2012. Since these data survey contain sample errors, we further quantify the uncertainties associated to the model parameters taking advantage of a probabilistic fitting method that has been recently proposed by some of the authors. This allows us to give a probabilistic prediction for the evolution of the carriers of the two genogroup over the next few years by constructing a 95% confidence interval for the percentage of carriers of meningococcus B.

Keywords: Uncertainty quantification · Epidemiological model · Competition model · Meningococcal B

1 Introduction

Meningitis is a severe infectious disease with very bad prognosis is some cases. Mortality rates can be up to 10% and many other survivors have sequelae [1,2]. Nowadays, the main cause of Meningitis is the bacterium Neisseria meningitidis which can infect the brain and the spinal cord and, even, lead to acute septicemia. The disease is not a zoonosis but it affects only human populations and it transmits mainly by contact with carrier individuals and, consequently, it is a major concern in adolescents where the risk behaviour is maximized by sharing glasses, cigarette butts, kisses etc.

© Springer International Publishing AG 2017
I. Rojas and F. Ortuño (Eds.): IWBBIO 2017, Part II, LNBI 10209, pp. 560–569, 2017.
DOI: 10.1007/978-3-319-56154-7_50

In Spain, the major cause of meninginitis in the last decades has been genogroup C (MenC) with most of the other cases corresponding to meningococci genogroup B. Recently, there have been important Public Health actions towards the control of Meningitis C by improving the vaccination campaign by booster doses and the vaccination of children at age 12. After that, it is predicted that the pandemic of MenC would be controlled in the future and no new cases have appeared since the vaccination plan was implemented [3]. Recurrent epidemics are also originated by serogroup A, in the so-called sub-saharan african meningitis belt [4], whereas serogroups B, C and Y are more common in Europe and North America. Recently, there have been also a great concern due to the emergence of serogroup W135 in countries such as Chile [5], UK and other European countries [6].

In Spain, genogroups B and C account for most infections but in the case of B there have been major difficulties in order to develop an effective vaccine because the group B capsular polysaccharide does not stimulate an immune response due to its similarity with the human polysialic acid, a modification of neural cell adhesion molecules. However, recent research has overcome this incovenience [7] and a vaccination campaign started in UK in 2015 and it is now also programmed in Spain [8].

For these reasons, it is timely to assess the risk of a meningitis B outbreak in the next few years in the case no vaccination program is implemented. To achieve this objective we have proposed a combined model involving transmission of the meningococci in the human hosts and competition among the different genogroups. The propagation of the disease is described in terms of a Susceptible-Carrier-Susceptible compartmental model satisfying a Ricatti difference equation.

The competition among genogroup B and other genogroups is studied by a Lotka-Volterra model [9] in which the time derivative is replaced by a fractional operator [10] to take into account the memory effects arising from DNA recombination with other genogroups that coexist and compete with this one [11]. The parameters of the model are obtained by a probabilistic fitting technique described elsewhere [12].

The paper is organized as follows. In Sect. 2, we introduce two mathematical epidemiological models that will be considered to describe, firstly, the dynamics of the susceptibles and carriers individuals of the population and, secondly, the dynamics of meningococci genogroup B (Men B) and the rest of meningococcus. Sample data are also introduced in this section. Section 3 is devoted to construct probabilistic predictions of the Men B (target variable) for Spain over the next few years. This predictions are given by means of the mean and 95% confidence intervals of the target variable. The proposed approach also allows us to construct the probability density function for each of the model parameters. An estimate of the probability of that the percentage of carriers of Men B be greater that a fixed threshold is also given in Sect. 3. Conclusions are drawn in Sect. 4.

2   Mathematical Models

This section is addressed to introduce the two mathematical models that we will consider to describe, firstly, the evolution of susceptible-carrier-susceptible (SCS) individuals and, secondly, the dynamics of meningococcal genogroups over the time. Although both models are simplifications of the complex ecosystem makes up by all the meningococci genogroups and the human hosts, as we shall see later, these models are based on well-founded paradigms that have been extensively used mathematical epidemiology. Both models will be applied to real data obtained in the seroepidemiological study for all meningococcal genogroups and, in particular, the Men B strain.

In this study, we use real data supplied by the Reference Laboratory for Meningococci of the Spanish Institute of Health Carlos III, collected in Table 1, corresponding to December, 2011 and December, 2012. We point out that these are currently the only available data.

Table 1. Sample sizes and percentages of susceptible/carriers of any meningococcus. Also, the percentage of carriers of the genotype Men B and the rest of genotypes in Spain corresponding to Dec 2011 and Dec 2012.

Year	$t_1 = 2011$	$t_2 = 2012$
Sample size	$n_1 = 3000$	$n_2 = 500$
Susceptible population	2626 (87.53%)	409 (81.8%)
Carrier population of any meningococcus	374 (12.47%)	91 (18.2%)
Genotype Men B	168 (44.92%)	28 (30.77%)
Other genotypes	206 (55.08%)	63 (69.23%)

The transmission dynamics of all the meningococci over the time, t (in months), will be modelled using a discrete SIS type-epidemiological model by identifying the "infected" subpopulation with "carriers", C_t (in percentage). Hereinafter, this model will be referred to as the SCS model. If $\beta > 0$ and $\gamma > 0$ denote the transmission rate of meningicocci and the rate of recovery then, using a classical discrete SCS-type epidemiological model, the dynamics of C_t can described by the following Riccati-type difference equation

$$C_{t+1} = (1 + \beta - \gamma)C_t - \beta C_t^2, \quad \beta, \gamma > 0. \tag{1}$$

Naturally, the percentage of susceptibles derives from the relationship $S_t = 1 - C_t$.

To describe the competition among the genotype of meningococcus bacteria B and the other genotypes of meningococcus, we will consider a fractional Lokta-Volterra model. At this point is important to motivate the use of the fractional derivative instead of the classical derivative in our modelling approach. Indeed we want to take into account the possible memory effects that could appear in the transmission of this disease after its settlement in the human host environment.

In this respect, it is well-known that fractional differential equations may account for these memory effects as these type of derivatives are usually applied to visco-elastic materials and subdiffusive processes [10,13].

In order to motivate the formulation of that fractional model, let us first consider the classical Lotka-Volterra model

$$\begin{cases} X_1'(t) = r_1 X_1(t)(K_1 - X_1(t)) - \alpha_{1,2} X_1(t) X_2(t), \\ \\ X_2'(t) = r_2 X_2(t)(K_2 - X_2(t)) - \alpha_{2,1} X_2(t) X_1(t), \end{cases} \tag{2}$$

where $i = 1$ corresponds to Men B genogroup, $i = 2$ corresponds to non-Men B genogroup, and

- $X_i(t)$, $1 \leq i \leq 2$, denotes the total amount of the genotype i meningoccocus bacteria at the time instant t (in months),
- $r_i > 0$ is the growth rate of the genotype i meningoccocus bacteria,
- $K_i > 0$ is the carrying capacity of the genotype i meningoccocus bacteria,
- $\alpha_{i,j} > 0$ is the effect of the genotype j bacteria on the growth of the genotype i bacteria, $1 \leq j \leq 2$, $j \neq i$.

Since data collected in Table 1 are in percentages, in order to combine the results of this model with those of the SCS model (1), we need scale the model (2). With this aim, we introduce the change of variable $x_i(t) = \frac{X_i(t)}{K_i}$, $i = 1, 2$, that represents the percentage of ecosystem occupied by the i-th-genogroup. Naturally, $x_1(t) + x_2(t) = 1$. This allows us to recast the first equation of model (2) as follows

$$x_1'(t) = N_1 x_1(t)(1 - x_1(t)), \tag{3}$$

where $N_1 = H_1 - M_1$ is defined by $H_1 = r_1 K_1 > 0$, $i = 1$ and $M_1 = \alpha_{1,2} K_2 > 0$. An analogous non-linear differential equation like (3) follows by $x_2(t)$. The classical derivative, $x_1'(t)$ in (3), is substituted by the fractional Caputo derivative, $^C D^\alpha x_1(t) = \frac{1}{\Gamma(1-\alpha)} \int_0^t x_1'(s)(t-s)^{-\alpha} ds$, [14]. This leads to the fractional Lotka-Volterra model

$$^C D^\alpha x_1(t) = N_1 x_1(t)(1 - x_1(t)). \tag{4}$$

Since this continuous model does not admit a closed-form solution, we will solve it numerically using the following discretization [15]

$$(\Delta_*^\alpha x_1)(t) = N_1 x_1(t + \alpha - 1)(1 - x_1(t + \alpha - 1)), \tag{5}$$

being $t \in \mathbb{N}_{1-\alpha} := \{1 - \alpha, 2 - \alpha, \ldots\}$, and Δ_*^α is the Caputo like-delta difference operator [15] defined by

$$(\Delta_*^\alpha x_1)(t) = \frac{1}{\Gamma(1-\alpha)} \sum_{s=0}^{t-n+\alpha} \frac{\Gamma(t-s)}{\Gamma(t-s+1)} (\Delta x_1)(s), \tag{6}$$

where $(\Delta x_1)(t)$ is the discretization of the first derivative of $x_1(s)$ in discrete time, i.e.,

$$(\Delta x_1)(s) = x_1(s) - x_1(s - 1). \tag{7}$$

Putting $x_1(t) = x_t$, the fractional Lotka-Volterra model in discrete time can be rewritten as

$$x_t = x_0 + \frac{N_1}{\Gamma(\alpha)} \sum_{k=1}^{t} \frac{\Gamma(t-k+\alpha)}{\Gamma(t-k+1)} x_{k-1}(1 - x_{k-1}), \tag{8}$$

where N_1 and α are the model parameters to be determined and t is the time in months.

According to the previous development, observe that the proposed combined competition-epidemiological models have four parameters, namely, the infection rate, β; the recovery rate, γ; the balance among the Men B genogroup reproduction and its competition parameter, N_1; and the fractional order, α, of the Caputo derivative that appears in the non-linear discrete equation (8). All these parameters must be quantified from the data collected in Table 1. The following section is addressed to give probabilistic estimation of these model parameters and to construct probabilistic predictions of the target variables.

3   Probabilistic Estimation and Results

In this section we show the results obtained after adjusting the models (1) and (8) using the technical termed Probabilistic Fitting [12].

To treat the available data surveys, we assume that surveys are independent. Hereinafter, $\mathbf{X}^j = (X_1^j, X_2^j)$, $0 \le X_i^j \le n_j$, $i, j = 1, 2$, will denote a random vector whose entries X_1^j and X_2^j are the number of Carriers and Susceptibles, respectively. According to Table 1 the sample size of every survey is $n_1 = 3000$ (Dec 2011) and $n_2 = 500$ (Dec 2012). It is clear that every component of vector $\mathbf{X}^j$ has a binomial distribution of parameters n_j and p_j. Since available data from surveys are limited, we will treat the probability parameter p_j as beta random variable. In this manner, we provide more flexibility to our probabilistic analysis. This leads to assume that X_1^1 has a beta-binomial distribution $X_1^1 \sim$ Bi$(n_1; p_1)$, $p_1 \sim$ Be$(\beta_1^1; \beta_2^1)$. As a consequence of the properties of the binomial and beta distributions $X_2^1 = n_1 - X_1^1 \sim$ Bi$(n_1; 1 - p_1)$, $1 - p_1 \sim$ Be$(\beta_2^1; \beta_1^1)$. We proceed for the second year in a similar manner: $X_1^2 \sim$ Bi$(n_2; p_2)$, $p_2 \sim$ Be$(\beta_1^2; \beta_2^2)$, and $X_2^2 = n_2 - X_1^2 \sim$ Bi$(n_2; 1 - p_2)$, $1 - p_2 \sim$ Be$(\beta_2^2; \beta_1^2)$.

In order to estimate the parameters $\{\beta_1^1, \beta_2^1\}$ and $\{\beta_1^2; \beta_2^2\}$, we have used the maximum likelihood technique (MLE) taking into account the data collected in Table 1. This technique yields

$$\hat{\beta}_1^1 = 6.122 \cdot 10^4; \quad \hat{\beta}_2^1 = 4.2984 \cdot 10^5; \quad \hat{\beta}_1^2 = 3.387 \cdot 10^3; \quad \hat{\beta}_2^2 = 1.5221 \cdot 10^4. \tag{9}$$

Analogously, we introduce the random variables Y_1^j and Y_2^j to denote the Carriers of Meningococcus B and the Carriers of non-Meningococcus B, and again we assume that $Y_1^j \sim$ Bi$(n_j; p_j)$, $p_j \sim$ Be$(\gamma_1^j; \gamma_2^j)$, $j = 1, 2$, and consequently, $Y_2^j \sim$ Bi$(n_j; 1 - p_j)$, $1 - pj \sim$ Be$(\gamma_2^j; \gamma_1^j)$ $j = 1, 2$. In this case, the estimates obtained by MLE technique are

$$\hat{\gamma}_1^1 = 7.7728 \cdot 10^3; \quad \hat{\gamma}_2^1 = 9.5309 \cdot 10^3; \quad \hat{\gamma}_1^2 = 271.1303; \quad \hat{\gamma}_2^2 = 610.0437. \tag{10}$$

In Tables 2 and 3, we show the 95% CI of each component of the random vectors $\mathbf{X}^1 = (X_1^1, X_2^1)$, $\mathbf{X}^2 = (X_1^2, X_2^2)$ and, $\mathbf{Y}^1 = (Y_1^1, Y_2^1)$ and $\mathbf{Y}^2 = (Y_1^2, Y_2^2)$, respectively. These CI have been computed by the binomial-beta probability mass function of every component and then computing the percentiles 2.5 and 97.5.

Table 2. 95% CI of the data surveys for the random variables *the number of Carriers and Susceptibles* in Dec 2011 and Dec 2012.

Dates	Carriers (X_1^j)	Susceptible (X_2^j)
$t_1 = 2011$ $(j = 1)$	$[0.1137\%, 0.1370\%]$	$[0.8630\%, 0.8663\%]$
$t_2 = 2012$ $(j = 2)$	$[0.1480\%, 0.2140\%]$	$[0.7860\%, 0.8520\%]$

Table 3. 95% CI of the data surveys for the random variables *the number of Meningococcus B and non-Meningococcus B* in Dec 2011 and Dec 2012.

Dates	Genotype Men B (Y_1^j)	Genotype non-Men B (Y_2^j)
$t_1 = 2011$ $(j = 1)$	$[0.4343\%, 0.4698\%]$	$[0.5302\%, 0.5657\%]$
$t_2 = 2012$ $(j = 2)$	$[0.2640\%, 0.3440\%]$	$[0.6560\%, 0.7360\%]$

Once we have determined the statistical distributions of the target variables for the two mathematical models (1) and (8), we are able to construct probabilistic predictions of them. To this goal, we have applied the Probabilistic Fitting technique developed in [12]. In Fig. 1, we show our predictions (mean and for 95% CI) $t = 0, 1, \ldots, 119$ (from Dec 2011 until Dec 2021) to the transmission dynamics of the carriers.

The Probabilistic Fitting technique permits to approximate the probability density functions for the model parameters β, γ of the SCS model (2). These functions are shown in Fig. 2. The 95% confidence interval of γ is $[0.1035, 0.7866]$. Since $1/\gamma$ can be interpreted as the average time to clear the meningococcus, then we can conclude, from our results, that the time required to clear any meningococcus bacterium will lie in the interval $[0.1604, 0.9748]$ months with a confidence of 95%.

The above analysis can be also performed for the parameter of fractional Lotka-Volterra (8) model. In Fig. 3 we show the 95% CI and the mean for the Men B from Dec 2011 to Dec 2021. In Fig. 4, we have plotted the probability density functions of both parameters, N_1 and α of that model. It is important to observe that according to the mean value of the probability density function of α, is 0.3190 being $[0.0070, 0.7786]$ its 95% CI. Thus, the fractional order is far from being one. This supports the consideration of a fractional derivative against using its classical version.

The product of the solution stochastic processes, C_t and x_t, of models (1) and (8), respectively, provide the percentage of Men B carriers among all the

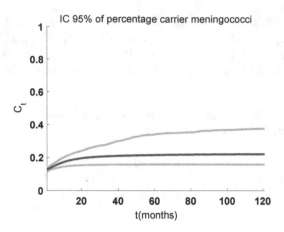

Fig. 1. The green lines represent the 95% CI of the solution stochastic process of the percentage of carriers given by model (1) for each month, t, from Dec 2011 until Dec 2021. The dark line represents the mean of the process. (Color figure online)

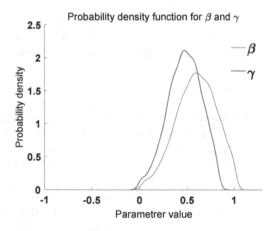

Fig. 2. Probability density functions for the parameters β and γ of the model (1).

population. In a personal communication, Dr. Julio Vázquez, from the Neisserias Reference Laboratory for Meningococcus of the Spanish Institute of Health Carlos III, told us that a percentage of 20%–25% or greater of carriers of Men B in Spain would begin to be concerning. In Table 4, we report the probability that the percentage of carriers of Men B be greater than 20% from Dec 2011 until Dec 2021. According to these figures, the probability of having worrying percentages of carriers of Men B is very low, therefore, an outbreak is not expected.

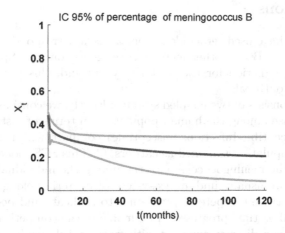

Fig. 3. The green lines represent the 95% CI of the solution stochastic process of the percentage of carriers given by model (8) for each month, t, from Dec 2011 until Dec 2021. The dark line represents the mean of the process. (Color figure online)

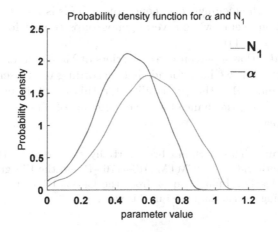

Fig. 4. Probability density functions for the parameters of the Lotka-Volterra model (8).

Table 4. Probability that the percentage of carriers of Men B be greater than 20%.

Date	Probability B $\geq$ 20%
December 2017	0.0029
December 2018	0.0036
December 2019	0.0047
December 2020	0.0062

4   Conclusions

In this work we have used seroepidemiological data corresponding to the prevalence of genogroup B and other meningococci genogroups in Spain to perform a probabilistic prediction for the next few years and, this way, to assess the likelihood of an outbreak.

Our model consists of two coupled systems: Firstly, we consider the transmission of the disease among the human population in terms of a standard version of the SIS (Susceptible-Infectious-Susceptible) compartmental model in which the infectious population is named as carriers, C. This SCS model describes the propagation of the meningococcus bacteria among the individuals. However, in each individual we usually find an ecosystem of bacteria belonging to different genogroups and these compete to each other to replicate and become predominant or marginal as time progresses. In our model the competition is described by a Lotka-Volterra discrete equation with fractional dynamics [10] in terms of a discrete Caputo derivative of index $0 < \alpha < 1$. We find that a better fitting is achieved for $\alpha < 1$ instead of $\alpha = 1$ that would correspond to an ordinary derivative. As fractional derivatives introduce a memory effect involving a contribution from the values at earlier times we conclude that some memory is involved in the competence among the meningococci genogroups. This is not surprising because DNA recombination occurs when several genogroups coexist for a large time in the nasopharynx flora [11].

Experts consider that a percentage of carriers of MenB larger than 20% could unleash a pandemic but we have found that, according to the seroepidemiogical studies of 2011 and 2012, the probability for this event is below a 1% and, consequently, there is little chance for the emergence of MenB as a major threat in the next five years.

Acknowledgements. This work has been partially supported by the Ministerio de Economía y Competitividad grant MTM2013-41765-P and the FIS grant PI13/01459. We also acknowledge Dr. Julio Vázquez from the Carlos III Institute of Health for providing the meningococcal data used in this work.

References

1. Cartwright, K.: Meningococcal carriage and disease. In: Cartwright, K. (ed.) Meningococcal Disease, pp. 71–114. Wiley, Chichester (1995)
2. De Walls, P.: Immunization strategies for the control of serogroup C meningococcal disease in developed countries. Expert. Rev. Vaccines **5**, 269–275 (2006)
3. Pérez-Breva, L., Villanueva, R.J., Villanueva-Oller, J., Acedo, L., Santonja, F., Moraño, J.A., Abad, R., Vázquez, J.A., Díez-Domingo, J.: Optimizing strategies for meningococcal C disease vaccination in Valencia (Spain). BMC Infect. Dis. **14**, 280 (2014)
4. Moore, P.S.: Meningococcal meningitis in sub-Saharan Africa: a model for the epidemic process. Clin. Infect. Dis. **14**(2), 515–525 (1992). doi:10.1093/clinids/14.2.515

5. Valenzuela, M.T., Moreno, G., Vaquero, A., Seoane, M., Hormazábal, J.C., Bertoglia, M.P., Gallegos, D., Sotomayor, V., Díaz, J.: Emergence of W135 meningococcal serogroup in Chile during 2012. Rev. Med. Chil. **141**(8), 959–967 (2013). http://dx.doi.org/10.4067/S0034-98872013000800001

6. Aguilera, J.F., Perrocheau, A., Meffre, C., Hahné, S., W135 Working Group: Outbreak of serogroup W135 meningococcal disease after the Hajj pilgrimage, Europe, 2000. Emerg. Infect. Dis. **8**(8), 761–777 (2002)

7. Paso, C.L., Romeo, A., Pernice, I., Donato, P., Midiri, A., Mancuso, G., Arigò, M., Biondo, C., Galbo, R., Papasergi, S., Felici, F., Teti, G., Beninati, C.: Peptide mimics of the group B meningococcal capsule induce bactericidal and protective antibodies after immunization. J. Immunol. **178**, 4417–4423 (2007). doi:10.4049/jimmunol.178.7.4417

8. Moreno-Pérez, D., García, F.J.Á., Fernández, J.A., Ortega, M.J.C., Raueti, J.M.C., Sánchez, N.G., Merino, A.H., Matos, T.H.-S., Moínao, M.M., del Castillo, L.O., Ruiz-Contreras, J.: Vaccination against meningococcal B disease. public statement of the advisory committee on vaccines of the spanish association of paediatrics (CAV-AEP). Anales de Pediatría **82**(3), 198e.1–198.e9 (2015). doi:10.1016/j.anpede.2014.09.004

9. Montaño, N.M., Sánchez-Yañez, J.M.: Nitrification in tropical soils linked to microbial competition: a model based on Lotka-Volterra theory. Ecosistemas **23**(3), 98–104 (2014). doi:10.7818/ECOS.2014.23-3.13

10. Podlubny, I.: Fractional Differential Equations: An Introduction to Fractional Derivatives, Fractional Differential Equations, to Methods of Their Solution and Some of Their Applications. Academic Press, London (1998)

11. Lamelas, A., Harrisc, S.R., Röltgena, K., Dangya, J.P., Hauser, J., Kingsley, R.A., Connor, T.R., Sied, A., Hodgsone, A., Dougan, G., Parkhill, J., Bentley, S.D., Pluschke, G.: Emergence of a new epidemic Neisseria meningitidis serogroup A clone in the African meningitis belt: high-resolution picture of genomic changes that mediate immune evasion. mBio **5**(5), e01974–e2014 (2014). doi:10.1128/mBio.01974-14

12. Cortés, J.C., Santonja, F.J., Tarazona, A.C., Villanueva, R.J., Villanueva-Oller, J.: A probabilistic estimation and prediction technique for dynamic continuous social science models: the evolution of the attitude of the basque country population towards ETA as a case study. Appl. Math. Comput. **264**, 13–20 (2015). doi:10.1016/j.amc.2015.03.128

13. Yuste, S.B., Acedo, L., Lindenberg, K.: Reaction front in an A+B→C reaction-subdiffusion process. Phys. Rev. E **69**, 036126 (2004)

14. Kilbas, A.A., Srivastava, H.M., Trujillo, J.J.: Theory and Applications of Fractional Differential Equations. Elsevier 633 Science, The Netherlands (2006)

15. Atici, F.M., Eloe, P.W.: Initial value problems in discrete fractional calculus. Proc. Am. Math. Soc. **137**(3), 981–989 (2009)

Smart Sensor and Sensor-Network Architectures

A Novel Wearable for Rehabilitation Using Infrared Sensors: A Preliminary Investigation

Jordan Lui[1](✉), Andrea Ferrone[1,2], Zhi Yih Lim[1], Lorenzo Colace[2], and Carlo Menon[1](✉)

[1] MENRVA Research Group, Simon Fraser University,
Metro Vancouver, Canada
{jdlui,aferrone,zlim,cmenon}@sfu.ca
[2] Roma Tre University, Rome, Italy
{andrea.ferrone,lorenzo.colace}@uniroma3.it

Abstract. In this paper we outline a novel design of a wireless sensor wearable band for tracking patient movements. This technology and design has potential applications for rehabilitation of stroke survivors who suffer from spasticity in their upper extremities. This technology could be used to track patient movement performed in a non-clinical environment, such as inside the comfort of their home. Data on their treatment progress could be transmitted wirelessly both to the clinician and to the patient. This technology could help realize increased monitoring of the patient, quantitative data on patient improvement over time and decreased health care costs. In this paper we demonstrated a preliminary prototype which can track and distinguish classes of movement of a user performing elbow flexion exercises while seated at a table. A study was completed with 6 participants with classification accuracies up to 88%.

Keywords: Rehabilitation · Stroke · Spasticity · Telerehabilitation · Telerehab · Rehabilitation devices · Medical devices · Wearable technology · Biomedical engineering · Machine learning · Physical rehabilitation · Wireless technology · Optical sensing · Sensor fusion · Transfer learning

1 Introduction

Stroke affects more than 15 million people per year worldwide. One third of victims die, and another one third of victims will experience significant long term disability following a stroke [1]. Stroke survivors can suffer from an involuntary muscle tightness condition called spasticity [2]. Spasticity in upper extremities can limit the strength, range of motion, functioning in daily activities, and overall independence of the individual.

1.1 Stroke Rehabilitation

A patient's recovery generally increases with the amount of rehabilitation they receive [3]. While a large degree of a patient's recovery is observed within 6 months, recovery

© Springer International Publishing AG 2017
I. Rojas and F. Ortuño (Eds.): IWBBIO 2017, Part II, LNBI 10209, pp. 573–583, 2017.
DOI: 10.1007/978-3-319-56154-7_51

can continue for as many as 3 years after a stroke. However, patients very rarely receive rehabilitation for a 3-year period. Stroke currently costs the US economy over $65.5 billion per year in medical costs, lost wages, and decreased productivity [4]. Extending the length of a patient's rehabilitation may improve their outcomes by leading to improvements in functional task performance and independence. While there are logistical costs to extend a patient's rehabilitation, the benefits are clear. With improved outcomes of the individual, we would thereby decrease economic costs associated with decreased productivity and lost wages.

1.2   Alternative Rehabilitation

Additional rehabilitation in specialized stroke rehabilitation centres may further improve patient outcomes, however the limited resources and time constraints of rehabilitation centres can present a challenge and fundamentally limit a patient's access to these resources [5].

One alternative solution is tele-rehabilitation. Commute distance from the patient's home to a rehabilitation clinic is a recognized impediment to a patient's adherence and success in a rehabilitation program. Thus, tele-rehabilitation, where phone calls or video conferencing are used to connect a clinician to a patient, have demonstrated improved outcomes on motor activity tests and activities of daily living (ADL) tests for patients [6]. Current tele-rehabilitation requires the therapist to be in direct communication and actively working with the patient during scheduled appointments. Patients, who may be unable to attend therapy sessions at a clinical site may do exercises and therapy in the comfort of their home. However, the patient is only assessed and evaluated during scheduled appointment times. A technology that allows the patient to record and monitor their progress during home practice could aid in the tele-rehabilitation effort and further decrease the requirement for a therapist to be present during all aspects of rehabilitation assessment.

1.3   Rehabilitation Devices and Wearables

Several devices have been proposed for tracking of a stroke patient's movements outside of clinical settings. One such solution is a smart garment which incorporates 4 strain sensors in the torso and arm areas of a shirt to record patient movement [7]. These devices report high accuracies but face some challenges for integration into a clinical regimen due to the varying garment sizes of users. A design that could fit all users would be ideal.

Other studies have investigated use of inertial measurement sensors (IMUs), which allow recording of accelerometer and gyroscope data during movement [8]. IMU data can be used to calculate orientation and position of the body in space through dead reckoning calculations. These methods show promise because they have a potential to be implemented in small, wearable, wirelessly connected electronics. However, several studies illustrate issues in sensor drift occurring from the integration of angular velocity data, resulting in error that must be corrected by repeated recalibrations [9]. An additional limitation in current IMU based systems exists because of the tendency to select professional IMUs which are much larger, and often affixed to the wrist, upper arm, shoulder, and torso of the user [10, 11]. The use of multiple sensor units at different points on the user provides rich sensor data but leads to issues in user acceptance due to weight and complexity in a clinical device. A single integrated hardware solution could eliminate these challenges.

We propose a device that could be used during a patient's regular rehabilitation to accurately measure patient movement during clinical sessions or when the patient performs activities in the comfort of their home. This device can be worn during activities of daily living (ADLs) and may thus have an advantage over traditional therapist tools such as goniometers, which are used for measurements taken during specific, isolated movements. The secure collection of a patient's movement data could provide valuable insights to a clinician who is monitoring the recovery of their patient.

As an initial prototype, we have demonstrated a device which can track the movement of a patient's arm as they perform elbow flexion on the transverse plane and shoulder flexion in the sagittal plane. The goal of this device is to determine proximity of the wrist from the torso during rehabilitation actions, as this can be a differentiator for a patient's stage of recovery from spasticity as based on the Chedoke-McMaster scale [12]. This device could find application in existing rehabilitation programs by monitoring patient activity through the various stages of recovery and providing extended insights to the clinician and patient.

2 Materials and Methods

2.1 Band Design

A wrist-mounted device incorporating passive infrared (PIR) sensors and active IR sensors was built. The device is built to detect motion for a user who performs elbow flexion actions on a table top on the transverse plane (shoulder flexion at approximately 90 degrees, device is measuring elbow flexion). The device contains an inertial measurement unit (IMU) for measurement of acceleration and orientation of movements. Data from the sensors is acquired by a microcontroller unit (MCU), Arduino Uno. The acquired signals are transmitted overall serial connection to a computer, where the data are recorded in LabVIEW (Fig. 1).

2.2 Passive Infrared Sensors

Passive infrared (PIR) sensors are used to measure temperatures in the field of view (FOV) of the detector. Using a silicon lens and a thermopile sensor, the sensor surface is selectively sensitive to thermal emissions that are irradiated by all objects. The sensor surface will generate an output signal voltage that is proportional to thermal radiation received at the sensor from objects in the sensor's FOV. The device also contains an analog circuit and logic circuit required to adjust the measured relative temperature value in relation to the measured ambient temperature. The PIR sensors are arrayed so each sensor can be used to detect multiple temperatures in their FOV. The two PIR sensors cover a combined area of 100° on the transverse (horizontal) plane.

Fig. 1. Label of the image. Hardware prototype

These chosen sensors are responsive to infrared emissions in the Long Wave Infrared (LWIR) range, which ranges from 8–15 μm.

The PIR sensors are mounted with their sensor plane perpendicular to the ground and facing inwards toward the user. When the user sits at a table with their arm in front of them, only the inward PIR sensor will detect thermal emissions coming from the user.

As the user performs an elbow flexion action, the distance from the wrist to body decreases, and the wrist moves towards the centre of the sagittal plane. At this point the outward facing PIR sensor can also detect thermal emissions from the user. This allows us to determine the relative position of the wrist to the body, and to track elbow flexion along the transverse movement plane. Since the PIR sensor is an arrayed sensor, we can gain some resolution on the degree of elbow flexion depending on the signals received at each sensor pixel.

Note that these PIR sensors essentially allow tracking of the orientation of the wrist relative to the body of the user (I.E. elbow flexion), but cannot alone be used to determine distances from wrist to body.

2.3   Active Infrared Sensing

Active infrared IR sensors are employed to determine distance of the wrist from the torso of the person. These sensors actively emit an infrared signal IR in the 900 nm range and use this signal to determine the distance to the nearest object in their FOV. These emissions do not interfere with our chosen PIR sensors. As the user completes an elbow flexion action in a horizontal plane (with the shoulder at 90-degree flexion), the proximity of the wrist from the torso can be determined. Note that the Active IR sensors will detect distance of any object in front of them, person or inanimate. Use of the Active IR sensors for movement classification without the PIR sensors would result in many misclassifications of user movement. The PIR sensors compensate for this active IR sensor vulnerability; this is known as sensor fusion or uncertainty reduction. Sensor fusion is commonly employed to aggregate advantages of various sensors and compensate for the vulnerabilities of individual sensors [13]. The PIR sensor increases system accuracy by confirming the presence and position of the user in the FOV of the Active IR sensors.

2.4   Inertial Measurement Unit (IMU)

An IMU was used to sense acceleration and orientation of the device on the user's wrist. The IMU will record a change in orientation and allow classification of movements in other planes of movement, such as a vertical shoulder flexion.

2.5   Data Analysis and Modelling

Data acquired in LabVIEW is saved and segmented based on movement type. This data is then analyzed with a Support Vector Machine (SVM) Machine Learning model, which trains a model based on a segment of the labeled data. The remaining data segment is used for evaluating the accuracy of the Machine Learning model. A SVM machine learning model was chosen for this dataset due to its ability to produce accurate results in a high dimensional feature space data even with the small initial set of training data points. A radial basis function kernel was employed with highly accurate results. This investigation combines movement readings from multiple participants to train the machine learning model, assuming that there is similarity in participant data when guided by an exercise protocol. While there are inherent variances in exercises performed by different people, the potential benefit of increased training data gives motivation to explore this machine learning technique known as transfer learning.

3   Experimental Protocol

Six healthy participants participated in this study. Each subject completed movement tasks while seated at table. Four classes of movement were performed, with the fourth class serving as the noise baseline for the data analysis. Movements were chosen to simulate motions achievable by stroke survivors who are in various stages of recovery as described by the Chedoke-McMaster Stroke Assessment scale [12]. The stroke assessment scale defines a several stages of recovery in stroke rehabilitation, particularly focusing on a user's recovery from limb spasticity, which inhibits voluntary motion (Table 1).

Table 1. Summary of movement classes

Movement 1: Unaffected elbow flexion	With shoulder flexion at 90°, the participant will perform elbow flexion from 0° to 135° on the horizontal plane. This represents the nominal flexibility and movement of a person who is minimally affected by stroke spasticity, or has recovered to stage 4
Movement 2: Affected elbow flexion	With shoulder flexion at 0°, the participant will perform elbow flexion from 0° to 135° on the horizontal plane, simulating the movement pattern of a stroke affected spastic movement. In this movement, minimal shoulder flexion is present, similar to patients in stage 2 or 3
Movement 3: Upward shoulder flexion and elbow flexion	With shoulder flexion at 0°, the participant will perform upward shoulder flexion from 0° to 90° and elbow flexion from 0° to 135°. This is an action achievable by patients in stage 6
Movement 4: Noise calibration	This fourth movement serves as noise calibration and baseline data for the Machine Learning model. Participants were instructed to rest their arm on the table for 5 recordings, and to hold their arm above the table at 90-degree shoulder flexion for 5 recordings

The participants each completed 10 trials of each of the 4 movement types, for a total of 40 recordings per participant. Data was recorded using the Arduino Uno microcontroller on a 10 Hz sampling rate and wirelessly transmitted to a computer where it was recorded in LabVIEW and saved to a CSV file. A test rig was created to guide user motions and label the motion class during the training data acquisition through use of touch sensors.

This data was input into a Support Vector Machine (SVM) machine learning model which utilized a radial basis function kernel. The data is split into two separate groups for analysis: One portion of data is used to train the model and obtain fitting parameters, and another portion of the data is used for testing the model's ability to predict movement class types on unseen data. The algorithm evaluated several ratios for segmenting data between training and testing, and performed repeated analyses on randomly shuffled data to eliminate outlier results based poor shuffling. Testing and training data was normalized using the mean and standard deviation of the training data.

33 features were analyzed with this machine learning model. This consists of 24 temperature readings from the PIR sensors, 2 distance sensor readings from the Active IR sensor, 3 gyroscope readings from the IMU, and 4 readings from the touch sensors for data labeling. Actual Centigrade temperature values from the PIR sensor were analyzed. Raw values from the Active IR sensors were input, which are voltage signals which are directly proportional to corresponding distance values. Raw gyroscope data from the IMU was analyzed as well, which represents angular velocity in each of the 3 axes. Accelerometer and magnetometer data was not analyzed in this study.

4 Results

Overall analysis of the results from adult male 6 participants suggest that we are indeed able to differentiate between nominal and simulated spasticity-affected movements. Meta-analysis on our algorithm performance indicated that segmenting 10% of data for testing and 90% of data for training yielded optimal results. The algorithm performed 100 random shuffles of the input data and segmented 10% to testing and 90% to training. Average classification accuracy was 79%, with a peak accuracy of 88% in some of the randomly shuffled datasets.

A confusion matrix was constructed (Fig. 2) to show the relative classification accuracies for the four movement types. Each specific column (I.E. column 1 - upward flexion) represents all instances where the model predicted a nominal flexion action. Each row represents the corresponding fraction of data points that were truly class 1, 2, 3, or 4. We can observe that the device delivers strongest classifications of movement types 2 and 3, which are affected elbow flexion and upward shoulder flexion. Classification accuracies were slightly lower for class 1 and the noise baseline movement.

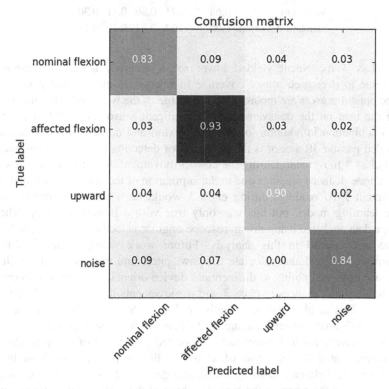

Fig. 2. Confusion matrix showing the test accuracy of movement classification

Considering the lower accuracies of arm motion tracking for class 1 movement compared to class 2, we conclude that arm length of the participant may be a factor which leads to variance to results among participants. This variance is likely more pronounced in the class 1 action, where the user has a larger arc of motion with a shoulder flexion than in class 2. The class 2 action in comparison is completed with a smaller arc of motion, and is perhaps less affected by the differing lengths of patient's arms. Future work should investigate if this is a significant contributor of variance and determine if data normalization increase accuracies and compensate for these variances.

We also examined prediction accuracies across our 100 random shuffles of our data, and obtained the following accuracies and variances, Table 2. We observe that average classification accuracy for class 2 is highest at 94%, and average accuracy is lowest for class 3. Variances are highest for class 3 and 4 movements.

Table 2. Test accuracies across movement classes

		Movement class			
	Overall	1	2	3	4
Mean	0.79	0.81	0.94	0.60	0.83
Maximum	0.88	0.92	1.00	0.94	1.00
Minimum	0.68	0.71	0.80	0.41	0.30
Standard deviation	0.05	0.04	0.04	0.10	0.13

The class 3 movement yielded lower mean accuracy across 100 randomized analyses due to decreased sensor coverage in relation to class 1 and 2 movements. Since the optical sensors are mounted on the inside of the wrist and pointing across the body of the user on the transverse plane, significant sensor FOV is lost as the user performs supination in addition to shoulder flexion and upward arm movement. The wide angled passive IR sensor is the main sensor detecting presence of the individual during a class 3 movement. During the class 3 movement the active IR sensors do not return accurate distance readings due to the supination of the arm. We assumed that the very different IMU readings during class 3 would be a clear differentiator for the machine learning model, but this was only true with a limited accuracy. The IMU gyroscope data analyzed using raw gyroscope angular velocity data; yaw, pitch, or roll data was not utilized in this analysis. Future work should include analysis of accelerometer data and also calculate the yaw, pitch, and roll data as this will likely increase accuracy and ability to differentiate device orientations during movement.

We see higher variances in class 3 and 4 movements than for class 1 and class 2 movements. This is also shown visually in Fig. 3. Standard deviation for class 4 movements is 3 times higher than standard deviation for class 1 and 2 movements.

Class 4 movements had inaccuracy introduced as a result of test procedure. Participants were seated at a table and asked to hold their arm stationary above the table. Some participants held their arms up at different degrees of shoulder flexion, and some participants rested their arm on the table for some of these recordings. We recommend changing this movement class 4 procedure to have the participant rest their arm on the table for this recording to increase consistency in data recordings.

Please note that average accuracies in Fig. 3 may differ from values in Table 2 due to the random shuffling of data during cross validation analysis.

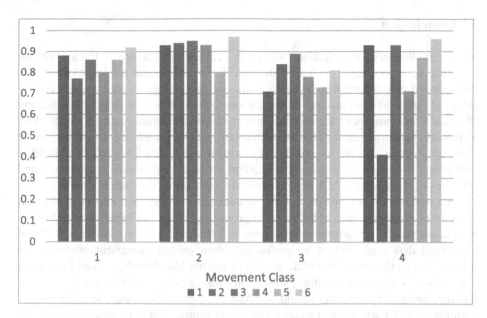

Fig. 3. Intra-class test accuracies for 6 different patients across 4 movement classes

Accuracy levels for individual participants were also examined by repeatedly shuffling the data randomly and analyzing. Higher mean and maximum accuracies were expected on individual participant analysis, but surprisingly the accuracies were only 2% higher. These average accuracies may potentially be higher if more data was obtained for each participant. Analysis on individual participants resulted in analysis using only 10 data files per movement class and could result in higher variances due to a small test set. Minimum accuracy was actually lower than in Table 2, which may also be explained by smaller data test set.

We can observe a notable dichotomy in the participant performance in Table 3. Participants 2, 4, and 5 had lowest mean accuracies and highest variances. Patients 1,3, and 6 had higher mean accuracies and lower variances. It is possible that some participants are simply more consistent in their movements and will yield higher test accuracies, while some patients move with more variance and their motions are harder to classify. However, the hardware setup procedure should also be examined to ensure that the prototype was fastened consistently and securely to each participant; a loosely fitted device could introduce additional movements through slippage and reduce accuracy values.

Table 3. Test accuracies across participants

	Participant						
	Overall	1	2	3	4	5	6
Mean	0.81	0.85	0.69	0.87	0.76	0.78	0.9
Max	0.90	0.91	0.83	0.93	0.91	0.86	0.96
Min	0.62	0.72	0.53	0.74	0.55	0.51	0.64
Standard deviation	0.05	0.03	0.07	0.03	0.07	0.06	0.04

5 Conclusion

In this preliminary investigation, we have demonstrated a novel framework for detecting upper extremity movement through the use of optical sensing technologies incorporated onto a wearable device. Specifically, our design utilizes a novel combination of optical sensors and movement sensors to provide high accuracy classification of a user's arm movement actions when analyzed with a machine learning model. With increasing healthcare costs and an increasing patient preference for home based treatment systems, this device could play a crucial role in emerging tele-rehabilitation technology systems, by providing a method to track rehabilitation exercises and activities of daily living (ADL) exercises in the comfort of the patient's home.

In a trial with 6 participants, a strong test accuracy of 88% was encountered on an analysis of combined data. This is quite promising given the inherent variances in individual patient performance, indicating that transfer learning between different participant data could indeed be performed when guided movement protocols are followed by participants. Overall accuracies of the machine learning model for a multi user system analysis will continue to be somewhat limited by the variance between different patients. Special care should be taken to calibrate the system for each patient, train the patient's actions, and guide a movement routine that is consistent and can be easily followed.

A meta-analysis on feature selection and sensor fusion (such as incorporating yaw, pitch, roll calculations from the IMU data) would likely further increase model accuracy and robustness.

Future studies will increase the number of movement classes to increase versatility and applicability of this device for physical rehabilitation. Conducting a clinical trial with spasticity affected individuals will also provide crucial feedback to the accuracy and suitability of this technology for physical rehabilitation tracking.

References

1. Internet stroke center: Stroke Statistics|Internet Stroke Center. http://www.strokecenter.org/patients/about-stroke/stroke-statistics/
2. Canadian partnership for stroke recovery: Spasticity - StrokeEngine.ca. http://www.strokengine.ca/glossary/spasticity/
3. Teasell, R.: Background concepts in stroke rehabilitation, pp. 1–48 (2016)
4. Di Carlo, A.: Human and economic burden of stroke. Age Ageing 38(1), 4–5 (2009)
5. Bonato, P.: Advances in wearable technology and applications in physical medicine and rehabilitation. J. Neuroeng. Rehabil. 2, 2 (2005)
6. Teasell, R., Foley, N., Mehta, S., Stanines, E.: Upper extremity interventions. In: Teasell, R. (ed.) Evidence-Based Review of Stroke Rehabilitation, vol. 2013, pp. 1–163 (2012)
7. Giorgino, T., Tormene, P., Maggioni, G., Pistarini, C., Quaglini, S.: Wireless support to poststroke rehabilitation: MyHearts neurological rehabilitation concept. IEEE Trans. Inf Technol. Biomed. 13(6), 1012–1018 (2009)
8. Willmann, R.D., Lanfermann, G., Saini, P., Timmermans, A., Te Vrugt, J., Winter, S.: Home stroke rehabilitation for the upper limbs. In: Annual International Conference of the IEEE Engineering in Medicine and Biology, Proceedings, pp. 4015–4018 (2007)

9. Kim, J., Yang, S., Gerla, M.: StrokeTrack: wireless inertial motion tracking of human arms for stroke telerehabilitation. In: Proceedings of First ACM Workshop on Mobile Systems, Applications, and Services for Healthcare, p. 4 (2011)
10. Álvarez, D., Alvarez, J., Gonzalez, R., Lopez, A.: Upper limb joint angle measurement in occupational health. Comput. Methods Biomech. Biomed. Eng. **19**(2), 144–158 (2016)
11. Luinge, H.J., Veltink, P.H., Baten, C.T.M.: Ambulatory measurement of arm orientation. J. Biomech. **40**(1), 78–85 (2007)
12. Gowland, C.K., Griffiths, J., Stratford, P., Barclay-Goddard, R.: Chedoke-Mcmaster stroke assessment. In: Development, Validation and Administration Manual, no. 1995 (2008)
13. Shen, Y., Wu, F., Tseng, K.-S., Ye, D., Raymond, J., Konety, B., Sweet, R.: A motion tracking and sensor fusion module for medical simulation. Stud. Health Technol. Inf. **220**, 363–366 (2016)

Real Time Localization Using Bluetooth Low Energy

Massimo Conti[(✉)]

Dipartimento di Ingegneria dell'Informazione,
Università Politecnica delle Marche, Ancona, Italy
m.conti@univpm.it

Abstract. In recent years the research on real time indoor and outdoor localization is becoming very intensive, due to the possibility offered by new RF technologies and to the wide range of possible applications. Many researches use the Bluetooth Low Energy (BLE) for indoor localization, in particular using the measurement of the Received Signal Strength Indicator (RSSI). This paper presents an inverse model of the distance between two BLE devices depending on the RSSI and on the Packet Error Rate (PER). Furthermore, we derived a model of the accuracy of the estimated distance estimated by the RSSI and PER measurements. Many measurements have been carried out to validate the model.

Keywords: Real time localization · Indoor localization · RSSI · PER · BLE

1 Introduction

In recent years the research on real time indoor and outdoor localization is becoming very intensive. The use of GPS for outdoor positioning and tracking outdoor is widely used. Many solutions are possible in indoor conditions where the GPS is not available, but a compromise between accuracy and cost of the devices must be reached. The technologies used for indoor localization are vision systems, infrared, ultrasound, RFiD, WiFi, ZigBee or Bluetooth or a combination of them. Bluetooth is a widely used and widely studied protocol and many applications have been developed [1–3]. Recently the Bluetooth Low Energy (BLE) extension of the protocol has been developed for low power body area networks. The wide diffusion of BLE allowed the reduction of the cost of BLE devices. The authors in [4] propose a combination of WSN and RFId for the positioning using the RSSI information.

Recently the wide diffusion of RFiD, WiFi, ZigBee or Bluetooth allows the possibility of the position estimation using low cost devices. In particular many researches use the BLE for indoor localization, using the measurement of the Received Signal Strength Indicator (RSSI) [5–9]. In particular papers [5, 7–9] use the measurement of RSSI in a BLE network and different algorithms to improve the accuracy in the triangulation procedure.

Generally, the use the RSSI measurement to estimate the distance between two devices is presented in many research works. In this work we propose, for the estimation of the distance, the measurements of both RSSI and PER and the use more than one transmission power. In addition we propose a model of the error of the distance

© Springer International Publishing AG 2017
I. Rojas and F. Ortuño (Eds.): IWBBIO 2017, Part II, LNBI 10209, pp. 584–595, 2017.
DOI: 10.1007/978-3-319-56154-7_52

estimation as a function of the distance. An experimental characterization of two environments (indoor and open space) has been performed to verify the proposed methodology. Section 2 presents the description of the measurement set up. Section 3 reports the indoor and outdoor propagation model. The model of the accuracy of the distance estimation and experimental results are presented and discussed in Sect. 4.

2 Measuremet Set up Description

The architecture used is the Cypress PSoC-4 BLE demo board with an ARM Cor-texM0 CPU (32bit 48 MHz) and a Bluetooth 4.2 transceiver with RF output power from −18 dBm to +3 dBm, RX sensitivity of −92 dBm and RSSI of 1 dB resolution. The system used consists of 2 Cypress PSoC-4 BLE: one is connected to a PC, it is programmed as server and receives data, it stores the measurement performed and it is placed in a fixed position, the second board is programmed as client and sends data is placed at different distances to the server. The measurement have been carried out in the following way:

- During the advertising phase the client sends continuously advertising packets. The server is listening and tries to create a connection defining the parameters of the connection: connection interval and MTU.
- Once the connection is created the client sends continuously to the server small packets of size 23 MTU for a total amount of 12,000 packets for about 5 s. The server counts the number of packets received, without sending acknowledge to the client. The difference between the number of packets sent and the number of packet received divided by the total number of packets sent, defined as Packet Error Rate (PER), is stored in the PC.
- At the end of the transmission the Received Signal Strength Indicator (RSSI) measured by the server is stored in the PC.

The procedure has been repeated with different conditions:

- indoor in an aisle of the University, at different distances: 1, 3, 5, 7 and 14 m in the same direction, and with different power of the Tx transceiver: 3, 0, −18 dBm. The indoor propagation is affected by the walls and the presence of persons. The packet error rate is influenced by the noise due to electromagnetic interferences such as WiFi or the presence mobile phones.
- outdoor in an empty parking outside the University, at different distances: 1, 3, 5, 7 and 14 m in the same direction, and with different power of the Tx transceiver: 3, 0, −18 dBm. The outdoor propagation is similar to the propagation in free field. The packet error rate is less influenced by the noise and the attenuation is the relevant influencing factor.

Every experiment is repeated for 6 times for a total of: 6×5 (distance) $\times 3$ (power) $\times 2$ (indoor–outdoor) = 180 measurements for the RSSI and 180 for the PER. The measurements have been reported as circles in Figs. 1–2 and 3–4 for the RSSI and PER, respectively. The average value over the 6 measurements is reported as black dots in the same figures. The standard deviation has been calculated, but it can be visually estimated by the spread of the measurements in Figs. 1, 2, 3 and 4.

3 Indoor and Outdoor Propagation Model

The RSSI is defined in IEEE 802.11 standard, as the ratio between the transmitter power and the received power presented by dBm units. A widely used model for the RSSI [4–7] is reported in Eq. (1).

$$RSSI = A - 10nlog_{10}(d+s) \tag{1}$$

where d is the distance between sender and receiver, n is the signal propagation constant, A is the received signal strength at a distance of one meter. The parameters has been introduced in this work as a fitting parameter in order to use Eq. (1) for short and long distances, since the model is used normally for long distance path loss. The RSSI decreases with the distance d, and it used as an indirect measure of the distance between the devices. Equation (2) reports the inverse formula that can be used to estimate the distance from the RSSI measurement, once the parameters n, A and s have been estimated from measurements of the particular device, environment and conditions

$$d_{est,RSSI} = 10^{\left(\frac{RSSI-A}{-10n}\right)} - s \tag{2}$$

The Packet Error Rate (PER) between two devices, that is the percentage of the number of packets lost with respect to the total number of packet sent, depends on the signal to noise ratio and therefore depends on the transmitted power and on the distance between the devices. The PER qualitatively has a sigmoidal relationship with the distance going from 0 when the devices are close each other and 1 when the distance is high. The model used in this work is reported in Eq. (3)

$$PER = \frac{1 + tanh[\alpha(d - \delta)]}{2} + \beta \tag{3}$$

where α is the slope of the sigmoid, δ is the distance at which the PER is ½, β is an additive constant. Equation (4) reports the inverse formula that can be used to calculate the distance from the PER measurement

$$d_{est,PER} = \frac{arctanh[2(PER - \beta) - 1]}{\alpha} + \delta \tag{4}$$

4 Measurements and Model Accuracy

The models in Eqs. (1) and (3) have been used to fit the experimental measurements and the parameters chosen in order to minimize the mean squared error with the data. The results are reported in Figs. 1–2 and 3–4 for the RSSI and PER, respectively.

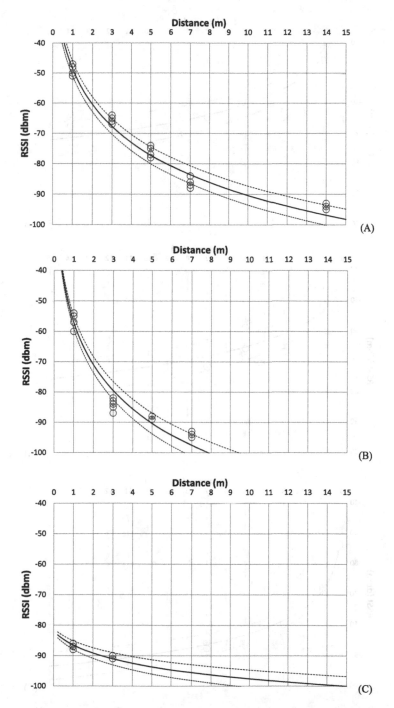

Fig. 1. Model and experimental data of the RSSI as a function of the distance in open space with Tx power = 3 dBm (A), 0 dBm (B) and −18 dBm (C).

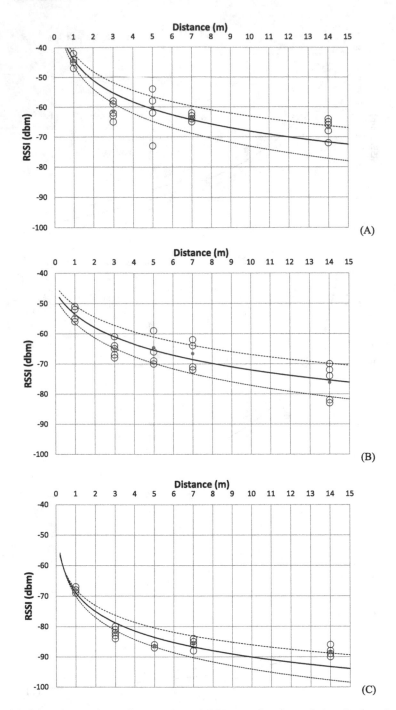

Fig. 2. Model and experimental data of the RSSI as a function of the distance in indoor conditions with Tx power = 3 dBm (A), 0 dBm (B) and −18 dBm (C).

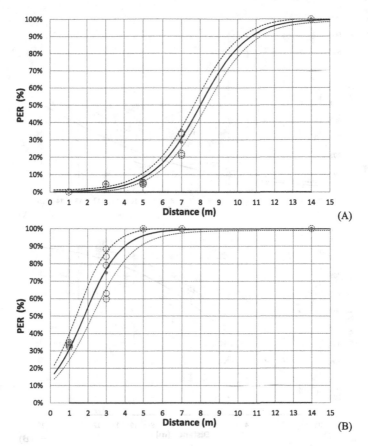

Fig. 3. Model and experimental data of the PER as a function of the distance in open space with Tx power = 3 dBm (A) and −18 dBm (B)

The circles represents the measurements, the black dots the average value. The model is reported in continuous lines.

In Fig. 1 the measurements of RSSI for some distances cannot be seen, in those cases the devices are not able to create a connection.

Some considerations can be drawn with regards to the RSSI measurements and the model. The RSSI model (1) represents better the outdoor measurements with respect to the indoor measurements. The variability of the measurements in outdoor conditions is not relevant (about 4 dBm or less). On the contrary the model error on the experiments is more relevant in indoor conditions and the variability of the measurements is higher (up to 10 dBm for the same conditions). The RSSI is in general higher indoor with respect to outdoor for the same distance.

The variability of the repeated measurements of RSSI and of the PER due to different environmental conditions and the error due to the model inaccuracy depend on the distance.

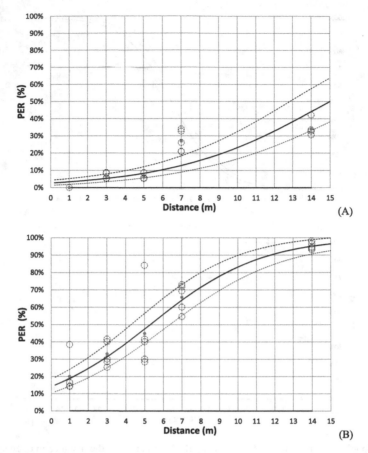

Fig. 4. Model and experimental data of the PER as a function of the distance in indoor conditions with Tx power = 3 dBm (A) and −18 dBm (B)

On the basis of this consideration, we defined a model of this variability considering that the parameters A, n and s of the RSSI may have a variation of ΔA, Δn and Δs as indicated in Eqs. (5–7)

$$A_{min} = A - \Delta A; \quad A_{max} = A + \Delta A; \tag{5}$$

$$n_{min} = n - \Delta n; \quad n_{max} = n + \Delta n; \tag{6}$$

$$s_{min} = s - \Delta s; \quad s_{max} = s + \Delta s; \tag{7}$$

As a consequence the model of the RSSI varies from a minimum value RSSImin to a maximum value RSSImax, as indicated in the Eqs. (8–9)

$$RSSI_{min} = A_{min} - 10n_{max}log_{10}(d + s_{max}) \tag{8}$$

$$RSSI_{max} = A_{max} - 10n_{min}log_{10}(d + s_{min}) \tag{9}$$

We estimated the variations ΔA, Δn and Δs minimizing the error between the variation model and the standard deviation of the experiments, as expressed in Eq. (10).

$$min_{\Delta A, \Delta n, \Delta s} \sum_{i=1}^{K} \left(\sigma_{RSSI,meas}(d_i) - \frac{RSSI_{max}(d_i; \Delta A, \Delta n, \Delta s) - RSSI_{min}(d_i; \Delta A, \Delta n, \Delta s)}{2} \right)^2 \tag{10}$$

where

$$\sigma_{RSSI,meas}(d_i) = \sum_{j=1}^{M} \left(RSSI_{meas,j}(d_i) - \overline{RSSI}_{meas}(d_i) \right)^2 \tag{11}$$

and in our particular measurement set up $M = 6$ is the number of repeated measurements and d_i are the distances expressed in meters

$$d_i(i = 1...K) = \{1, 3, 5, 7, 14\} \tag{12}$$

Finally, the effect of this variation model on the estimated distance is reported in Eqs. (13) and (14).

$$d_{min,RSSI} = 10^{\left(\frac{RSSI-A_{max}}{-10n_{min}}\right)} - s_{min} \tag{13}$$

$$d_{max,RSSI} = 10^{\left(\frac{RSSI-A_{min}}{-10n_{max}}\right)} - s_{max} \tag{14}$$

Once the parameters of the model and of the error model of the RSSI have been estimated from the measurements, for every measurement of the RSSI we derive the estimated distance $d_{est,RSSI}$, $d_{max,RSSI}$ and $d_{min,RSSI}$, as shown in Fig. 5.

The error on the estimated distance $d_{est,RSSI}$ has been defined by the following Eq. (15)

$$d_{error} = \max[(d_{max} - d_{est}); (d_{est} - d_{min})] \tag{15}$$

Figures (1 and 2) report in dashed lines the model of Eqs. (8) and (9).

Similar procedure and model has been used for the PER. Equations (16)–(18), similarly to Eqs. (5)–(7) for the RSSI, report the variations of the PER parameters.

$$\alpha_{min} = \alpha - \Delta\alpha; \quad \alpha_{max} = \alpha + \Delta\alpha; \tag{16}$$

$$\delta_{min} = \delta - \Delta\delta; \quad \delta_{max} = \delta + \Delta\delta; \tag{17}$$

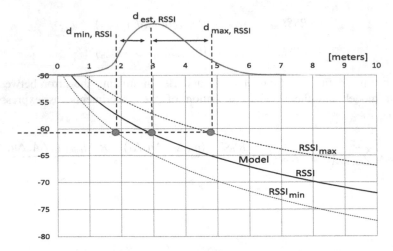

Fig. 5. Error model of the RSSI

$$\beta_{min} = \beta - \Delta\beta; \quad \beta_{max} = \beta + \Delta\beta; \tag{18}$$

Equations (19) and (20) report the model of the variability of the PER, as Eqs. (8) and (9) for the RSSI.

$$PER_{min} = \frac{1 + tanh[\alpha_{min}(d - ax)]}{2} + \beta_{min} \tag{19}$$

$$PER_{max} = \frac{1 + tanh[\alpha_{max}(d - \delta_{min})]}{2} + \beta_{max} \tag{20}$$

Finally, Eqs. (21) and (22) report the variation on the estimation of the distance using the measurements of the PER, similarly to Eqs. (13) and (14) for the RSSI.

$$d_{min,PER} = \frac{arctanh[2(PER - \beta_{max}) - 1]}{\alpha_{max}} + \delta_{min} \tag{21}$$

$$d_{max,PER} = \frac{arctanh[2(PER - \beta_{min}) - 1]}{\alpha_{min}} + \delta_{max} \tag{22}$$

Figures (3 and 4) report in dashed lines the model Eqs. (19) and (20).

Figure 6 reports the error on the distance estimated by the model that uses the RSSI measurements as a function of the distance in indoor conditions with Tx power = 0 dBm and −18 dBm. The figure is obtained using Eq. (15) once the coefficients n, A, s, ΔA, Δn and Δs have been estimated by the measurements. The error of the model that uses the RSSI measurements increases with the distance between the two devices and in general with higher Tx power.

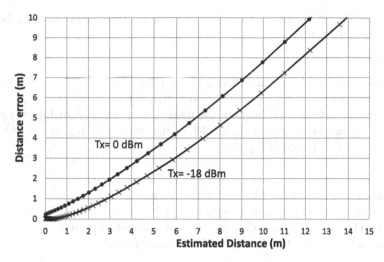

Fig. 6. Error on the distance estimated by the RSSI model as a function of the distance in indoor conditions with Tx power = 0 dBm and −18 dBm

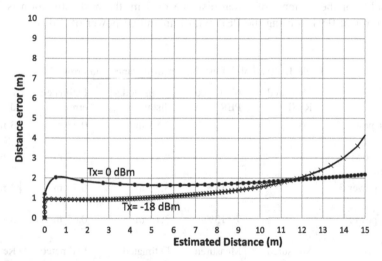

Fig. 7. Error on the distance estimated by the PER model as a funcion of the distance in indoor conditions Tx power = 0 dBm and −18 dBm.

Similarly, Fig. 7 reports the error on the distance estimated by the model that uses the PER measurements as a function of the distance in open space and indoor and with Tx power = 3, −18 dBm.

The error of the model that uses the PER has a minimum for the distance in which the PER changes rapidly from zero to one. On the other hand the error increases when the PER is saturates to 100%. The different shapes of the error of the RSSI model and PER model suggest that we can use one or the other as a function of the estimated distance.

Summarizing, the proposed procedure of the estimation of the distance between two devices is the following:

1. model characterization:

   - measurements of RSSI and PER at defined distances for different values of the transmitted power for the model characterization.
   - estimation of the coefficients of the models n, A, $s\alpha$, δ, β of Eqs. (1) and (3) and of the coefficients of the model of the error ΔA, Δn, Δs $\Delta\alpha$, $\Delta\delta$ and β obtained as reported in Eq. (10).

2. test:

   - measure of RSSI and PER for different transmitted power.
   - estimation of the distances using Eqs. (2) and (4) and of the error using Eqs. (13), (14), (21) and (22) for different transmitted power.
   - use as best choice the distance with the lower error.

As an example we report in Table 1 the measurements of two cases, at a real distance of 3 m and of 7 m. The best estimation is 2.88 m with an error of 0.43 using the RSSI with transmitted power of 3 dBm, for the example of a real distance of 3 m. Conversely, for the example of a real distance of 7 m, the best estimation is 7.91 m with an error of 0.94 m using the PER with transmitted power of 3 dBm.

Table 1. Example of estimation of distance and error

	Measured RSSI	Measured PER	Estimated distance	Estimated error	Real distance
RSSI Tx power 3 dBm	−67		2.88 m	0.43 m	3 m
RSSI Tx power −18 dBm	−90		2.41 m	1.34 m	3 m
PER Tx power 3 dBm		21.18%	3,36 m	0,73 m	3 m
PER Tx power −18 dBm		79.13%	3.13 m	0.65 m	3 m
	Measured RSSI	Measured PER	Estimated distance	Estimated error	Real distance
RSSI Tx power 3 dBm	−84		7.16 m	1.22 m	7 m
RSSI Tx power −18 dBm	No connect		15 m	12 m	7 m
PER Tx power 3 dBm		91.10%	7.91 m	0.94 m	7 m
PER Tx power −18 dBm		No connect	14 m	17 m	7 m

5 Conclusions

We proposed the use of the measurements of both RSSI and PER and the use more than one transmission power, for the estimation of the distance between two BLE devices. In addition we propose a model of the error of the estimated distance as a function of the distance. An experimental characterization of two environments (indoor and open space) has been performed to show the proposed methodology.

The characterization of the particular ambient and environment is fundamental for an accurate estimation. The error on the estimation is in general about 1 m. A light improvement in the accuracy can be obtained using the invers mode of the RSSI or the PER depending on the environment or on the distance between the devices.

The model proposed in this work is used to the estimation of the distance between two BLE devices. The extension to more than two devices and the trilateration will allow a bi- or tri dimensional localization.

References

1. Caldari, M., et al.: SystemC modeling of a Bluetooth transceiver: dynamic management of packet type in a noisy channel. In: Proceedings of the Designers' Forum of the Conference Design Automation and Test in Europe DATE 2003, 3–7 March 2003, Munchen, pp. 214–219 (2003)
2. Scavongelli, C., Conti, M.: A systemC Bluetooth network simulator. In: Proceedings of the 10th Conference on Ph.D. Research Microelectronics and Electronics (PRIME 2014), Grenoble, France, 30 June–3 July 2014, pp. 1–4 (2014)
3. Conti, M., Fedeli, D., Virgulti, M.: B4V2G: Bluetooth for electric vehicle to smart grid connection. In: Proceedings of the 9th International Workshop on Intelligent Solutions in Embedded Systems WISES 2011, Regensburg, Germany, 7–8 June 2011, pp. 13-18 (2011)
4. Xiong, Z., Song, Z., Scalera, A., Ferrera, E., Sottile, F., Brizzi, P., Tomasi, R., Spirito, M.A.: Hybrid WSN and RFID indoor positioning and tracking system. EURASIP J. Embedded Syst. **2013**, 6 (2013)
5. Shan, G., Park, B.H., Nam, S.H., Kim, B., Roh, B.H., Ko, Y.B.: A 3-dimensional triangulation scheme to improve the accuracy of indoor localization for IoT services. In: 2015 IEEE Pacific Rim Conference on Communications, Computers and Signal Processing (PACRIM), pp. 359–363 (2015)
6. Lin, X.Y., Ho, T.W., Fang, C.C., Yen, Z.S., Yang, B.J., Lai, F.: A mobile indoor positioning system based on iBeacon technology. In: 2015 37th Annual International Conference of the IEEE Engineering in Medicine and Biology Society (EMBC), pp. 4970–4973 (2015)
7. Rida, M.E., Liu, F., Jadi, Y., Algawhari, A.A.A., Askourih, A.: Indoor location position based on Bluetooth signal strength. In: 2015 2nd International Conference on Information Science and Control Engineering, Shanghai, 24–26 April 2015, pp. 769–773 (2015)
8. Thaljaoui, A., Val, T., Nasri, N., Brulin, D.: BLE localization using RSSI measurements and iRingLA. In: 2015 IEEE International Conference on Industrial Technology (ICIT), Seville, 17–19 March 2015, pp. 2178–2183 (2015)
9. Jianyong, Z., Haiyong, L., Zili, C., Zhaohui, I.: RSSI based Bluetooth low energy indoor positioning. In: 2014 International Conference on Indoor Positioning and Indoor Navigation (IPIN), pp. 526–533 (2014)

A Sensor Grid for Pressure and Movement Detection Supporting Sleep Phase Analysis

Maksym Gaiduk[1(✉)], Ina Kuhn[1], Ralf Seepold[1],
Juan Antonio Ortega[2], and Natividad Martínez Madrid[3]

[1] HTWG Konstanz, Brauneggerstr. 55, 78462 Konstanz, Germany
{maksym.gaiduk, ralf.seepold}@htwg-konstanz.de,
raina.bertolini@gmail.com
[2] Universidad de Sevilla, Avda. Reina Mercedes s/n, 41012 Seville, Spain
jortega@us.es
[3] Hochschule Reutlingen, Alteburgstraße 150, 72762 Reutlingen, Germany
natividad.martinez@reutlingen-university.de
http://uc-lab.in.htwg-konstanz.de
http://www.us.es
http://iotlab.reutlingen-university.de

Abstract. Sleep quality and in general, behavior in bed can be detected using a sleep state analysis. These results can help a subject to regulate sleep and recognize different sleeping disorders. In this work, a sensor grid for pressure and movement detection supporting sleep phase analysis is proposed. In comparison to the leading standard measuring system, which is Polysomnography (PSG), the system proposed in this project is a non-invasive sleep monitoring device. For continuous analysis or home use, the PSG or wearable Actigraphy devices tends to be uncomfortable. Besides this fact, they are also very expensive. The system represented in this work classifies respiration and body movement with only one type of sensor and also in a non-invasive way. The sensor used is a pressure sensor. This sensor is low cost and can be used for commercial proposes. The system was tested by carrying out an experiment that recorded the sleep process of a subject. These recordings showed the potential for classification of breathing rate and body movements. Although previous researches show the use of pressure sensors in recognizing posture and breathing, they have been mostly used by positioning the sensors between the mattress and bedsheet. This project however, shows an innovative way to position the sensors under the mattress.

Keywords: Sensor grid · Movement detection · Sleep phase · Force resistor sensor

1 Introduction

The average human spends about one third of his or her life sleeping [1]. Depending on how the sleep designated hours are expended, our daily routine can be either positively or negatively influenced. Studies show that the recommended sleep duration varies based on age group. However, the amount of sleep alone does not assure the possibility of a good quality rest [2]. In order to accurately evaluate the quality of sleep, it is necessary to identify the sleep stages and their durations, which present sleep cycles. There exists a

© Springer International Publishing AG 2017
I. Rojas and F. Ortuño (Eds.): IWBBIO 2017, Part II, LNBI 10209, pp. 596–607, 2017.
DOI: 10.1007/978-3-319-56154-7_53

difference between falling asleep and being asleep. As a result, sleep can be categorized into stages, which can be ascertained by the use of various electrophysiological signals during sleep. The electrophysiological signals can be for example Electroencephalography (EEG), Electromyography (EMG) and Electrooculography (EOG), which enable the brain and muscle activity as well as the eye movements to be captured severally [3]. The recorded signals follow the method determined by Rechtschaffen and Kales (R-K) [4] to identify which one of the six sleep stage the body is in.

Rapid Eye Movement (REM) and Non-Rapid Eye Movement (NREM) are the two main categories of sleep. About 25% of the sleep typically occurs in the REM stage, while the remaining 75% occur in the NREM [5]. REM, also known as the dream stage, is the stage where the muscles are shut down with the exception of the eye muscles, with the intention of preventing the physical manifestation of activities or movements being executed in the dream. The eye muscles during this phase are engaged in random movements under the lids, which explicates the name [6]. NREM comprises of four stages of sleep. The first NREM stage, known also as light sleep, is regarded as the transition between being awake and sleep. In other words, it entails the process of falling asleep. The person in this stage of sleep is still a bit conscious of his or her surrounding and can easily be awakened by sounds. This phase usually lasts between 5–10 min [7]. When the second NREM stage is reached, the subject is really sleeping, what means transition from falling asleep to sleep. The person not only becomes less conscious of his or her surrounding, but also breathing and heart rate become more regular and the body temperature drops. People spend approximately 50 percent of their sleep in this stage [5]. Last but not least are the third and fourth sleep stages. The N3-4 is also called deep sleep. Starting in N3 the delta waves or extremely slow waves appear to be switching with some faster waves. In this stage our temperature decreases even more, heart rate and the blood pressure slow down. By N4 stage the brain only produces delta waves. When a subject wake up in the stages N3-4, the first feelings can be groggy and disoriented [8]. These stages form a path that is repeated every 90–120 min [9].

Sleep is not a waste of time. As a matter of fact, research has shown that during the process of sleep, the brain remains active. Sleep also aids and plays a very important role in brain process such as memory consolidation and brain detoxification. During sleep, the brainstem, hippocampus, thalamus and cortex helps to consolidate different kinds of memorial [10]. Furthermore, during sleep, the glymphatic system detoxifies the brain from toxins that are built up while consuming energy during the day [11]. Understanding how we sleep and analyzing the individual sleep patterns, can help to improve life quality. Sleep is of uttermost importance, as a lack thereof can result in so many health issues such as migraine, insomnia and in worst case scenario even death.

Having a good night of sleep is important. Nowadays there are a lot of sleep laboratories where sleep can be analyzed with help of electrophysiological signals. Unfortunately, it is not possible to simulate the sleep environment in such a way that patients feel totally at home This makes it therefore even more difficult to obtain valid results, that are high in accuracy. Not only respiration and heart rate, but also body movements are important in determining sleep behaviour [12]. Besides that, monitoring body movement additionally to breathing during sleep can aid the detection of apnea and myoclonic [13] The aim of this work is to find an efficient way to collect information about movement of a patient while he sleeps without any physical impairments such as wearable sensors. This information should aid the analysis of the quality of sleep.

2 State of the Art

Currently there are different methods used for the sleep stages classification. Polysomnography (PSG) [14] is widely used for measuring sleep patterns. PSG includes data such as EEG, that collects the brain activity. Electrocardiography (ECG) is a method that recognizes and measures the small electrical differences caused by the heart muscle on the skin, which results in the electrical activity of the heart over time. Whereas Electrooculography (EOG) recognizes and measures the standing potential that exists between the back and the front of the human eye. These measurements permit the determination of the sleep stage and the behavior of the eyes like REM and NREM. Muscle activity can be recorded through Electromyogram (EMG).

Apart from PSG, there are other methods used in monitoring sleep. Methods like actigraphy involve the use of worn motion sensors, that measure the body motion [15]. The normal actigraphy method send the read data after some period of time on a computer. New studies have shown a new way to work with wearable devices and read real time data [16]. The demand for the continuous wearing of the devices makes the patient uncomfortable, resulting in the practice of these methods being difficult for long term use. They are in general not long term monitoring devices and can only be put into service in designated environments like sleep laboratories. It also remains uncertain if the patient exhibits the same sleep pattern as exhibited in the laboratory while asleep at home. Combinations of methods are also found in several research papers. An example is the combination of actigraphy and respiratory data [17]. The newest sleep monitoring methods in comparison to PSG cannot provide all data as described by R-K Method, but provide enough to classify the sleep stages and diagnose sleeping disorders.

Apart from the conventional sleep stages, research has also been done to identify new stages such as the pre-wake stage [18]. In this stage, the patient suddenly wakes up. An everyday sleep behavior analysis is in such cases very helpful in detecting any symptoms that might point towards sleep disease syndromes and other health issues. Systems such as PSG when used in such a case for the sleep analysis can be very uncomfortable. For this reason, amongst others, there is a trend of new research using non-invasive sleep analysis systems. Examples of such sensors used for this kind of analyses are; piezoelectric signals, video motion system, optical fiber sensors, radar sensors, load cells, textile recording system, pneumatic method and pressure sensor. Pressure sensor as the name implies, works as a transducer for the pressure [19]. Piezoresistive, capacitive, piezoelectric or optoelectronic technologies are examples of sensors that follow this principle. Another important pressure sensor is the Force Sensing Resistor (FSR) [20], which is also used in several sleep studies. Unlike the piezoresistive sensors that cannot hold up the value under long term use, FSR keeps the value stable. Consisting of polymer thick film technology and interdigitating electrodes, FSR has a resistance value that changes according to the applied force. 0.01 kg to 10 kg is the typical range of applied force recognized by FSR. In this range the deviation of the measured values is not more than ±2%. As a result of the characteristics of FSR, the durability test showed that even under high temperature such as 170 °C and a force test applying 5.44 kg over ca. 1.5 cm^2, using a 3 mm thick 45 shore with a rubber foot, the value of the resistance still remains in the tolerance range of FSR. Most publications have used FSR sensor between the mattress and the bed sheet to analyze the sleeping posture of the subject while asleep. Developing a pressure sensing bed system, that can automatically provide information about a patient to the caregivers so that the risk of bedsore can be avoided, was one of goals of a study about pressure sensing beds [21].

Further research [22] not only classified 2 more positions, giving a total of 8 position, but also has proven a precision of up to 97%. The goal is to analyze the posture of bedbound patients, so that ulcers can be avoided. As a result of the high requirements of the system, this research has a proposal not only to fill in as one of the until now categorized posture detection system. With a low cost pressure mat, built with pressure sensors, the algorithms for continually detecting bed posture has been developed.

In classifying the sleep state, not only the sleep movements, but also respiration and heart rate are important. A research [23] shows that pressure sensors can also be used to analyze sleep states. This research uses the pressure sensor to observe the respiration signal and body movements. For the respiration signal validation, such parameters as Respiratory Rate (RR) was defined, which is the result of the Respiration Per Minute (RPM) and the number of apneas. This number is used to calculate the apnea hypopnea index. RR can also be used in calculating sleep depth and sleep cycles (SC) [24].

Another research [25] also used the FSR sensor to detect respiration rate and posture movements during sleep. The system contains 28 commercial FSR sensors and a wireless network is built with ZigBee. The communication between FSR and computer occurs in real time. The monitoring software for this project was programmed with LabView. The software contains seven views: (1) the value of using color sensors, (2–4) the output signal from the FSR, (6–7) the average value of field 2–4. A real movement or pressure is acknowledged when more than 10% of the value in a FSR sensor has changed. Moreover, this research shows a good result in detecting BM and RR.

3 System Architecture and Movement Detection Method

Knowing that the system is placed in bed to detect movements in order to support sleep analysis, the system does not present any risk to human health. There is no discomfort, inconvenience, molestation and disturbance incurred by the usage of this system. No skin breakage, no contact with mucosa or any internal body cavity beyond a natural or artificial body orifice. These are some of the characteristics included in the definition of a non-invasive system [26]. The furniture chosen is a bed. It is part of the system and consists of a mattress, bed frame, a slatted frame. The bed frame has an open structure, so that changing the sensors and checking the system does not prove any difficulty. This bed frame has a bed surface of 90 × 200 cm. The mattress has a hardness degree between 1 and 2 to ensure, that the exerted pressure goes straight to the slatted frame.

The sensor used to detect the body movement is an inexpensive rugged force sensor. It is flat, flexible and the force range covers the range pressure of the lying body. An array containing points of pressure detection is built with the sensors. The sensor network consists of n sensors. In order to transfer the data acquired in the sensors, a communication channel is necessary. The sensors' only duty is to receive the data resulting from the body. For this reason, a peripheral component is connected to each sensor so that the data can be read and sent to the main component. To control and synchronize the communication between the main component and the peripheral components, a bidirectional communication system is needed. This bidirectional communication ensures that the main component can send a message to the peripheral components requesting data. This data includes important information like physical position and sensor value.

The implementation of it can be done with a master/slave structure. If the implementation is done by one channel, it is important that not all the peripheral components use the channel at the same instant. The system suggests a master/slave communication, but once the channel communication can receive the data information as described, and can be synchronized, then another technology can be used. As already described, each place where the sensor can be plugged has an ID. This information is very important and the current value of the sensor as well. Moreover, to transfer the information to the main component, the peripheral component must implement the communication protocol in order for the data transfer to be accomplished. Furthermore, the sensor has as an already described interface that is adequate for qualitative force. For this reason, the peripheral component needs an analog input channel, so that the sensor value can be read. If some problem exists with the peripheral component, it should be able to send a signal.

The last component to be described is the main component of the system, which has the task of controlling the complete structure, synchronizing the sensor values, saving the data and sending the data to be analyzed. The idea proposed in this work is an open embedded system. The main component has an integrated system with a microelectronic control used for performing complex tasks, but does not have a user interface. This is the primary definition of the main component. The main component is an open embedded system containing at least a microcontroller with 32-bit and multicore systems that permits the execution of more complex tasks. This is in contrast to the depth embedded system that only has an 8-bit microcontroller and can only execute one kind of task. The main component constraints are the following: limited space, compact and small, no user interface, headless structure and diskless. Following characteristics are important: an instant on with a short boot time, fast power off, non-stop $(24/7)$, and long lifetime system. The main component has Integrated Firmware Image (IFWI) which initiates the boot process, the operation system, root file system and application. It is responsible for the administration of all peripheral component and sending the data out for analysis. Furthermore, the main component determines the protocol for the communication channel and requests the data from the peripheral component.

Aimed towards increasing elasticity and snugness, disk-springs are placed upon the bed frame. The smaller sized disk-springs were chosen intentionally as they are flexible and ensure better distribution of the body weight. The disk-springs are arranged next to each other with the maximum distance between the ends not being more than 3 cm.

The chosen sensor to implement was the FSR 406 from Interlink Electronics.

The chosen hardware to connect the FSR sensor into the system was Trinket Pro 5 V [29], however any other hardware that fits the requirements of the system can be used. A maximum of four sensors can be connected to each Trinket Pro.

Figure 1 shows the overview of the system and the concept of the layout. An extra identification wire consisted of a series circuit of resistors was built, to give the Trinket Pro an ID that can categorize the physical position. The resistor value only influences the current value of a circuit. The system's voltage value stays the same and is divided proportionally as long as the resistance has the same value. If the resistors are not equal in value, the voltage will be divided differently. In order to achieve the best measurement results, FSR sensor has to be adjusted with the right series resistance. Its value depends on the working range of the system. FSR sensor decreases in resistance with the increasing in force applied to the sensor-surface. This applied force is converted into the signal. To define the working range, it is important to determine each of the ranges the force applied to the surface of FSR sensor has.

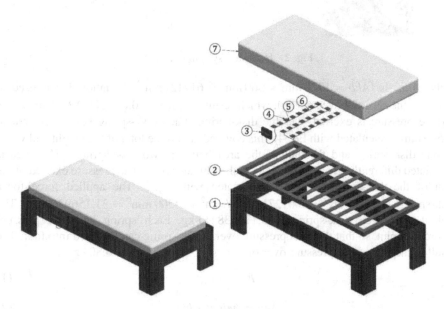

Fig. 1. Overview of the model design system: (1) bed frame, (2) slatted frame, (3) main component, (4) sensors, (5) peripheral component, (6) communication channel, (7) mattress. [28]

Figure 2 shows a hardware overview of the project. It contains 3 main components: Trinket Pro, FSR sensor and Intel Edison. The Intel Edison is the main component of the system. Intel Edison is described as a System-On-Module (SOM), which is used for Internet of Things (IoT) and wearable computing. Intel Edison is based on the dual-core Intel Quark System-on-a-Chip (SoC). A research [30] evaluated some of the important human's body-dimensions. In this research at least 50% of the average case for the dimensions of the human body were put into consideration, a height of 172 cm and a hip area of about 36. The average weight of a human lies between 60–80 kg [31].

To determine the pressure over each FSR sensor, the force applied through the average human (70 kg) and the mattress (10 kg) was calculated using the Eq. 1, so that the total force applied is $F = 80.00\,\text{kg} * 9.81\,\text{m/s}^2 = 784.8\,\text{N}$. To calculate the force applied over an area, the average area was calculated with Eq. 2. The total area of human

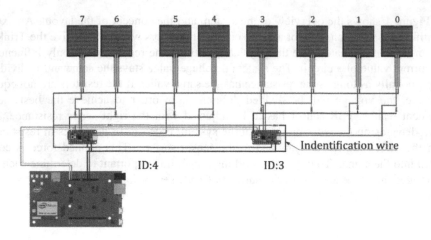

Fig. 2. Hardware system overview

body (Ahb) is $(Ahb = 1720\,\text{mm} * 360\,\text{mm} = 617121\,\text{mm}^2$. Equation 3 was used to calculate the surface pressure (Sp), which results is equal to $0.001271712\,\text{N/mm}^2$. This surface pressure is exerted over the disk-spring. Each disk-spring has cover area of $24500\,\text{mm}^2$ (calculated with Eq. 4), this covered area is the total of the height and width of one disk spring and the half of the area between two disk springs. This area is calculated differently in order to obtain the height and width. The pressure exerted of the area of the disk-spring (Ads), is calculated with Eq. 5. The applied force for a disk-spring Fds, is $Fds = 0.001271712\,\text{N/mm}^2 * 24500\,\text{mm}^2 = 31.15693681\,\text{N}$. The weight over each disk-spring is $3.176038411\,\text{kg}$. Each spring has four pressure recording point, so that the total pressure over a disk-spring can be dived into four. The result is the amount of pressure over one sensor, which is equal $0.80\,\text{kg}$.

$$F = m * g \tag{1}$$

$$Ahb = height * hip \tag{2}$$

$$Sp = \frac{F}{A_{hb}} \tag{3}$$

$$A_{sd} = h * w \tag{4}$$

$$d_s = S_p * A_{ds} \tag{5}$$

As an example based on the graph in Fig. 3, if the system requires a force range of 800–1000 N, the measuring resistor (RM) should be chosen as $10\,\text{k}\Omega$ or $3\,\text{k}\Omega$, because the gradient of this function is the highest in this range.

Figure 4 shows a measured sensor curve with no outlier, where the gradient in the working range is high enough to use it as a calculation base to determine the right RM.

Knowing the Eq. 6 and the Parameters RM, $V+$ and $Vout$ it is now possible to calculate the $RFSR$ curve. The result can be seen in Fig. 5.

$$V_{out} = \frac{R_M * V_+}{R_M + R_{FSR}}$$

(6)

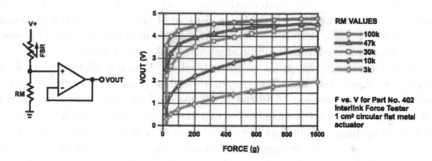

Fig. 3. Voltage divider and FSR curve depended on RM values [32]

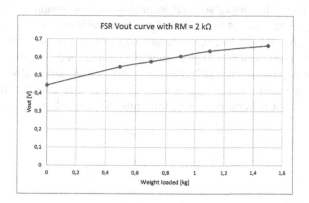

Fig. 4. FSR V_{out} curve with RM = 2 kΩ

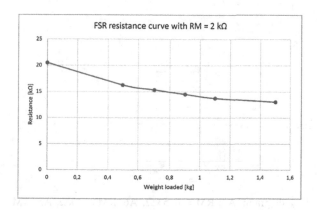

Fig. 5. FSR resistance curve with RM = 2 kΩ

4 Application of System and Results

The sleep test has been carried out with different types of resistance in the Ubiquitous Computing Laboratory. The candidate for the study is a 31-year-old male in good health and with a normal Body-Mass-Index. 20 min of his afternoon nap was recorded in order to be analyzed. Despite the fact that he carries out very little movement during sleep, at least three movements can be identified on the Fig. 6 below.

The first movement happened as he fell asleep. Between 15:44 and 15:45 the candidate scratched his nose. The second movement was very subtle, it was the moment he woke up between 16:59 and 16:00, the third movement shows the time the candidate got out of bed. Compared to the first and second movements where the body does not get up from bed, the last movement showed radical differences in value.

The respiration can be recognized in different times and in different sensors. Previous re-search shows that depending on the subject's position, a particular sensor may be best at recording the subject's respiration movements than others. In this case the left end of the bed was the sensor A7, followed by the A6, A5, A4, A3, A2, A1 and A0 on the end of the bed. All sensors showed recordings of his breathing, but because the subject is lying on his back at the center of the bed, the best rates of breath are in the sensors, A4, A3, A2, A1. To calculate a breathing cycle, it was taken into consideration that an adult breathes between 11–15 times per minute and that inspiration time is shorter than the expiration time [27]. Figure 7 shows four consecutive breathing. By the continuous repetition of the respiration rate cycle, it is assumed that this is the respiration-movement pattern. However, this must be verified by a second device.

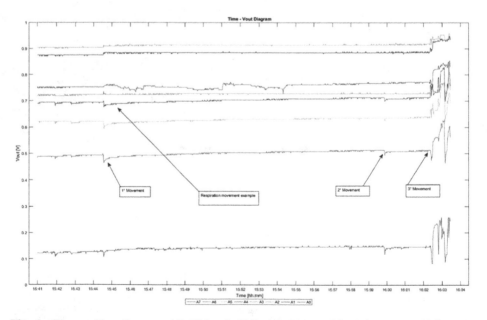

Fig. 6. Time - V_{out} diagram: A7; RM $= 0,5\,k\Omega$; A6; RM $= 1\,k\Omega$; A5; RM $= 1,5\,k\Omega$; A4;\, RM $= 2\,k\Omega$; - without felt gliders. A3; RM $= 0,5\,k\Omega$; A2; RM $= 1\,k\Omega$; A1; RM $= 1,5\,k\Omega$; A0; RM $= 2\,k\Omega$; -with felt gliders

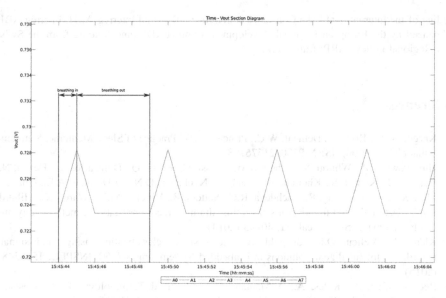

Fig. 7. Time V_{out} section diagram from Fig. 7

5 Conclusion and Future Work

The proposed low-cost sensor grid for pressure and movement detection showed that different parameters for sleep phase analysis can be measured. In addition to that, it proved to be a system that neither come in contact with the subject nor initiates any form of discomfort during sleep. It was demonstrated that FSR sensor is very useful tool in obtaining body movements and respiration signals. This result indicates that the system is well suited for supporting sleep analysis by providing data concerning the following activities carried out during sleep: respiration rate and body movements. The integration of the sensors under the mattress gave a new perspective on how a system is implemented for sleep analysis. At the moment various sleep recordings has been obtained from the same subject and compared using an algorithm. This confirms the extraction of respiration signals and body movements through FSR sensor. Moreover, the system is completely scalable and can be transferred to any bed of the same kind. The designed system shows a promising result with successful validation.

Future work would include a connection to the sleep stage classifier, working with a sleep algorithm [3] that provides a sleep stage classification and a sleep quality analysis. Furthermore, a development with FSR sensor should enable the monitoring of blood pressure and heart rate. To achieve it, further researches of optimal sensor grid topology are necessary. Using the described system as a base for the apnea-recognizing system could be also one of topics for further researches.

Acknowledgement. The project is partly supported by the International University of Lake Constance (IBH). The IBH is a network of 30 universities and colleges located in Austria, Germany, the Principality of Liechtenstein and Switzerland. The IBH budget arises partly from funding of the International Lake Constance conference (IBK) and Interreg ABH. Interreg ABH is financed by the European Regional Development Fund (ERDF) and funding from the Swiss New Regional Policy (NRP) framework.

References

1. Kryger, M.H., Roth, T., Dement, W.C.: Principles and Practice of Sleep Medicine. Saunders, Philadelphia (2000). ISBN 9780721676708
2. Hirshkowitz, M., Whiton, K., Albert, S.M., Alessi, C., Bruni, O., DonCarlos, L., Hazen, N., Herman, J., Katz, E.S., Kheirandish-Gozal, L., Neubauer, D.N., O'Donnell, A.E., Ohayon, M., Peever, J., Rawding, R., Sachdeva, R.C., Setters, B., Vitiello, M.V., Ware, J.C., Hillard, P.J.A.: National sleep foundation's sleep time duration recommendations: methodology and results summary. Sleep Health **1**, 40–43 (2014)
3. Klein, A., Velicu, O.R., Seepold, R.: Sleep stages classification using vital signals recordings. In: Intelligent Solutions in Embedded Systems, pp. 47–50. INSPEC: 15655487 (2015)
4. Rechtschaffen, A., Kales, A.: A Manual of Standardized Terminology, Techniques and Scoring System for Sleep Stages of Human Subjects, pp 1–12. U.S. Department of Health, Education, and Welfare, Washington, D.C. (1968)
5. Sleep Foundation, What Happens When You Sleep? https://sleepfoundation.org/how-sleep-works/what-happens-when-you-sleep. Accessed 05 July 2016
6. Sleep Foundation, How Sleep Affects Brain Function, healthination (2012). https://sleepfoundation.org/video-library. Accessed 02 Sept 2016
7. VeryWell, The Four Stages of Sleep (NREM and REM Sleep Cycles). https://www.verywell.com/the-four-stages-of-sleep-2795920. Accessed 27 Sept 2016
8. National Institute of Neurological Disorders and Stroke, Brain Basics: Understanding Sleep. http://www.ninds.nih.gov/disorders/brain_basics/understanding_sleep.htm. Accessed 06 July 2016
9. Sleep Sync, What normally happens during a typical sleep cycle? http://sleepsync.com/
10. Born, J., Wilhelm, I.: System consolidation of memory during sleep. Psychol. Res. 192–203 (2012). doi:10.1007/s00426-011-0335-6
11. National Institutes of Health, How Sleep Clears the Brain. http://www.nih.gov/news-events/nih-research-matters/how-sleep-clears-brain
12. Muzet, A.: Dynamics of body movements in normal sleep. In: Sleep, pp. 232–234 (1988)
13. Nishida, Y., Hori, T., Sato, T., Hirai, S.: The surrounding sensor approach - application to sleep apnea syndrome diagnosis based on image processing. In: 1999 IEEE International Conference on Systems, Man, and Cybernetics, IEEE SMC 1999 Conference Proceedings, vol. 6, pp. 382–388 (1999). ISBN 0-7803-5731-0
14. Blood, M.L., Sack, R.L., Percy, D.C.: A comparison of sleep detection by wrist actigraphy, behavioural response, and polysomnography. Sleep **20**(6), 388–395 (1997)
15. Hedner, J., Pillar, G., Pittman, S.D., Zou, D., Grote, L., White, D.P.: A novel adaptive wrist actigraphy algorithm for sleep-wake assessment in sleep apnea patients. Sleep **27**(8), 1560–1566 (2004)
16. Velicu, O.R., Martínez Madrid, N., Seepold, R.: Experimental sleep phases monitoring. In: IEEE EMBS International Conference BHI, pp. 625–628 (2016). ISBN 978-1-5090-2455-1
17. Long, X., Fonseca, P., Foussier, J., Haakma, R., Aarts, R.: Sleep and wake classification with actigraphy and respiratory effort using dynamic warping. IEEE J. Biomed. Health Inf. 1272–1284 (2013). ISSN 2168-2194

18. Baran Pouyan, M., Nourani, M., Pompeo, M.: Sleep state classification using pressure sensor mats. In: 2015 37th Annual International Conference of the IEEE Engineering in Medicine and Biology Society (EMBC), pp. 1207–1210, 26 August 2015. ISSN: 1094-687X
19. Wikipedia, the free encyclopedia, Pressure Sensor. https://en.wikipedia.org/wiki/Pressure_ sensor. Accessed 25 Aug 2016
20. Yaniger, S.I.: Force sensing resistors: a review of the technology, 666–668 (1991). doi:10. 1109/ELECTR.1991.718294
21. Hsia, C.C., Liou, K.J., Aung, A.P.W., Foo, V., Huang, W., Biswas, J.: Analysis and comparison of sleeping posture classification methods using pressure sensitive bed system. In: 2009 Annual International Conference of the IEEE Engineering in Medicine and Biology Society, pp. 6131– 6134, 3 September 2009. Print ISSN: 1094-687X
22. Baran Pouyan, M., Ostadabbas, S., Farshbaf, M., Yousefi, R., Nourani, M., Pompeo, M.D. M.: Continuous eight-posture classification for bed-bound patients. In: 6th International Conference on Biomedical Engineering and Informatics, pp. 121–126, December 2013. ISSN 1948-2914
23. Pino, E.J., de la Paz, A.D., Aqueveque, P., Chavez, J.A.P., Moran, A.A.: Contact pressure monitoring device for sleep studies. In: 2013 35th Annual International Conference of the IEEE Engineering in Medicine and Biology Society (EMBC), pp. 4160–4163, 26 September 2013. Electronic ISBN 978-1-4577-0216-7
24. Burioka, N., Cornelissen, G., Halberg, F., Kaplan, D.T., Suyama, H., Sako, T.: Approximate entropy of human respiratory movement during eye-closed waking and different sleep stages. CHEST J. 123(1), 80–86 (2003). doi:10.1378/chest.123.5.1323
25. Lokavee, S., Puntheeranurak, T., Kerdcharoen, T., Watthanwisuth, N., Tuantranont, A.: Sensor pillow and bed sheet system: unconstrained monitoring of respiration rate and posture movements during sleep. In: IEEE International Conference on Systems, Man, and Cybernetics (SMC), Seoul, pp. 1564–1568 (2012). Electronic ISBN 978-1-4673-1714-6
26. Farlex Inc., non-invasive. http://encyclopedia.thefreedictionary.com/Non-invasive. Accessed 06 Sept 2016
27. Wikipedia, the free encyclopaedia, Atmung. https://de.wikipedia.org/wiki/Atmung. Accessed 24 Sept 2016
28. Kuhn, D. (2016)
29. Adafruit: Trinket Pro 5V. https://cdn-learn.adafruit.com/downloads/pdf/introducing-pro-trinket.pdf. Accessed 12 Sept 2016
30. Jürgens, H.W., Matzdorff, I., Windberg, J.: Internationale anthropometrische Daten. Internationale anthropometrische Daten als Voraussetzung für die Gestaltung von Arbeitsplätzen und Maschinen, Wirtschaftsverlag, Nordrhein-Westfalen (1998)
31. Wikimedia: Mensch in Zahlen. https://upload.wikimedia.org/wikibooks/de/archive/f/f6/ 20130102075654!Mensch_in_Zahlen.pdf. Accessed 20 Sept 2016
32. Interlink Electronics: Datasheet, FSR 406. http://www.interlinkelectronics.com/datasheets/ Datasheet_FSR.pdf. Accessed 27 Aug 2016

A Portable Wireless sEMG and Inertial Acquisition System for Human Activity Monitoring

Giorgio Biagetti, Paolo Crippa[✉], Laura Falaschetti, Simone Orcioni, and Claudio Turchetti

DII – Dipartimento di Ingegneria dell'Informazione,
Università Politecnica delle Marche, Via Brecce Bianche 12, 60131 Ancona, Italy
{g.biagetti,p.crippa,l.falaschetti,s.orcioni,c.turchetti}@univpm.it

Abstract. This paper presents a low-cost portable wireless system specifically designed to acquire both surface electromyography (sEMG) and accelerometer signals for healthcare applications, sport, and fitness activities. The system, consists of several ultralight wireless sensing nodes that acquire, amplify, digitize, and transmit the sEMG and accelerometer signals to one or more base stations through a 2.4 GHz radio link using a custom-made communication protocol designed on top of the IEEE 802.15.4 physical layer. Additionally, the system can be easily configured to capture and process many other biological signals such as the electrocardiographic (ECG) signal. Each base station is connected through a USB link to a control PC running a user interface software for viewing, recording, and analysing the data.

Keywords: Surface Electromyography · sEMG · Wireless sensor nodes · Accelerometer · Activity monitoring · ECG · ZigBee · IEEE 802.15.4 · USB

1 Introduction

Human activity monitoring is of paramount importance in several research fields such as pervasive and mobile computing, ambient assisted living, surveillance-based security, sport and fitness activities, healthcare.

Wearable sensors, i.e. sensors that are positioned directly or indirectly on the human body, generate signals (sEMG, accelerometric, ECG, PPG, ...) when the user performs activities. With the progress of the signal processing techniques, more and more information could be derived from such biosignals [1,2]. Therefore they can monitor features that are descriptive of the person's physiological state or movement.

In the recent years, advances in mobile electronics systems, sensor technologies, signal processing, as well as in communication network protocols have launched a new generation of health care systems. Telehealth and fitness monitoring are some examples of an area where integrative research and development

© Springer International Publishing AG 2017
I. Rojas and F. Ortuño (Eds.): IWBBIO 2017, Part II, LNBI 10209, pp. 608–620, 2017.
DOI: 10.1007/978-3-319-56154-7_54

in wearable/portable technology are performed. As a result, high-capacity, low-power, low-cost, tiny and lightweight sensors have been embedded into clothes, shoes, belts, sunglasses, smartwatches and smartphones, or positioned directly on the body in order to collect a large amount of data such as body position and movement, heart rate, muscle fatigue, and skin temperature [3,4].

On the one hand, inertial sensors are probably the most frequently used among wearable sensors for activity monitoring. In particular, they are effective in monitoring activities that involve repetitive body motions, such as running, cycling, lifting weights, walking, climbing stairs [5]. Single inertial sensor systems are often used for activity classification and monitoring and in this case common location choices are the waist, upper arm, wrist, and ankle [6–8]. The waist location has been extensively used in physical activity measurements because it captures major body motions, but algorithms using waist data can underestimate overall expenditure on activities such as bicycling or arm ergometry, where the waist movement is uncorrelated to the movement of the limbs. Similar consideration could be done for recent techniques based on smartwatches and smartphones [9–11]. Therefore several recent studies addressed the problem of detecting human activities using more sensors on the body [12,13].

On the other hand, accurate estimation of biometric parameters recorded from subjects' wrist or waist, when the subjects are performing various physical exercises, is often a challenging problem due to the presence of motion artifacts that corrupt the bioelectric signals, such as the sEMG ones. In order to reduce the motion artifacts, data derived from a triaxial accelerometer have been proven to be very useful [14].

Lightweight wireless sensor devices can be comfortably used during activities of daily living, including sleep for human activity monitoring. In particular, exercise routines and repetitions can be counted in order to track a workout routine as well as determine the energy expenditure of individual movements. Indeed, mobile fitness coaching has involved topics ranging from quality of performing such sports actions to detection of the specific sports activity [15].

Among the biological signals, the surface electromyographic (EMG) signal has been demonstrated to be very useful in monitoring person's body posture, physical performance, and fitness level [4,15–18]. This is due to the fact that it can be obtained using intrinsically noninvasive measurement devices and is relatively easy to acquire. Indeed, this signal originates from the electrical potentials generated by contracting muscles [19,20], and can be collected with electrodes contacting the surface of the skin. Still, its relatively low amplitude requires a carefully designed, high-input-impedance, low-noise amplifier before it can be recorded [21]. The useful bandwidth for EMG signals is typically below 500 Hz, but suffer from possibly severe motion-induced artifacts at frequencies below 5 Hz. These low-frequency artifacts must be rejected by the amplifier, otherwise saturation of the gain stages will occur.

As a result recent works demonstrated as the combination of sEMG and inertial sensors in one device is essential to achieve the results necessary in examining muscle activity, forces, directionality and acceleration that are of paramount importance in sports performance evaluation, injury prevention, rehabilitation, and human activity monitoring in general [22–27].

This work presents an inexpensive and flexible wireless surface EMG system, called WiSE, that is able to acquire both sEMG and motion-related signals using ultralight (23 g) wireless sensor nodes with a software-selectable bandwidth. This selectable bandwidth allows the system to be easily configured, as a possibly useful additional feature, to capture and process further biological signals such as the ECG signal [28].

The sEMG/inertial sensing nodes are able to acquire, amplify, digitize, and transmit the signals to one or more base stations through a 2.4 GHz radio link using a custom-made communication protocol designed on top of the IEEE 802.15.4 physical layer, in order to exploit existing low-cost and low-power transceivers but also to enable the possibility of higher throughput and better synchronization than the standard would have allowed.

The base station can be powered either by an external power supply or by its USB interface, and contains the RF transceiver for the wireless connection to the mobile nodes, a system for simultaneous charging of up to six mobiles nodes, and a 32 bit microcontroller for managing purposes.

Each base station is connected through a USB link to a control PC running a user interface software for viewing, recording, and analysing the data.

The paper is organized as follows. Section 2 gives an overall description of the WiSE system. Section 3 describes the custom-made communication protocol and the main features of the user interface software. Section 4 shows an application example of the system. Finally, conclusions are drawn in Sect. 5.

2   System Implementation

The WiSE system is composed of the following main components:

- Mobile nodes – these are the signal acquisition units with an embedded wireless transceiver.
- Base station – a USB wireless receiver with integrated charger for the mobile nodes, capable of receiving the signals from up to 4 simultaneously active mobile nodes using only a single IEEE 802.15.4 radio channel.
- PC software – a user interface software with system diagnostic, signal liveview, recording and analysis capabilities.

The mobile node is built around an active electromyography sensor coupled to a wireless transmitter, posing as the foundation of the WiSE system. It comprises the active sensor that acquires and amplifies the EMG signal, a 3-axes linear accelerometer, and a microcontroller that digitizes the signals and transmits them to the base station connected through a custom protocol. A distributed software-defined phase-locked-loop (PLL) was designed into the protocol

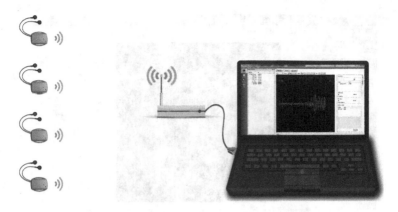

Fig. 1. A WiSE system showing only one base station. Up to 16 base stations can be connected via USB to the computer.

to enable the possibility of synchronizing all the nodes to within a microsecond from the base station clock. All of the node functionalities, as well as the base station's, can be remotely monitored and controlled from the same PC by means of a GPL-licensed software that provides full control of the WiSE system. A picture of the system is shown in Fig. 1.

3   Software

As previously mentioned, WiSE system provides a user interface software that enables real-time acquisition, recording and analysis of EMG data and system diagnosis.

This software, released under a GPL-license, currently runs under Windows 7 and both Linux x86 and x64 flavors, and provides the following functionality:

- supports up to 4 base stations simultaneously connected to the PC and up to 4 active sensors per base station
- configuration of the mobile nodes for real-time signal acquisition (number of nodes, radio channel, gain)
- real-time display of acquired data on the PC monitor
- real-time diagnosis of nodes and base station status
- real-time data storage in EDF/EDF+ format and plain text files for later processing
- off-line display of multiple saved tracks and annotations
- data analysis of stored tracks

It is structured into three main panels, each one containing its own specific features.

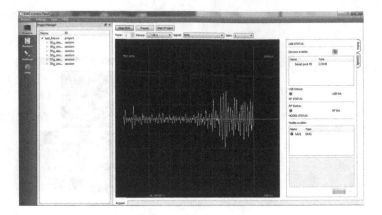

Fig. 2. Screenshot of acquire function of graphical interface.

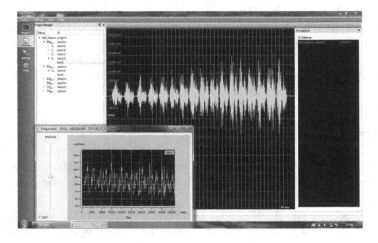

Fig. 3. Screenshot of analyze function of graphical interface.

(1) Acquire: This panel, shown in Fig. 2, allows the automatic start-up of the WiSE system, enabling real-time capture and monitoring of the signal. The PC is responsible for sending, via USB, the commands needed by the base station to wake up the requested nodes, normally in a deep stand-by mode, and to enable them to transmit the acquired data. When the start-up procedure is completed, usually in a few seconds, the measurement starts, and the user can see the real-time trace of the captured signal on the main panel. They can also change the gain, number of traces, scale and type of signal to be displayed on-the-fly, and easily manage stored data and metadata. Different scenarios to categorize different measurement setups can also be created.

(2) Analyze: In this panel, shown in Fig. 3, the analysis of the signals can be performed. The chosen format for data storage is the European Data Format (EDF), which is a simple and flexible format for the exchange and storage of multichannel biological and physical signals, commonly used also by many commercial devices. This functionality is achieved using *EDFbrowser* and the associated library *EDFlib* [29]; together they provide a C/C++ open source, free and multiplatform framework containing functions for reading/writing and displaying data in EDF format.

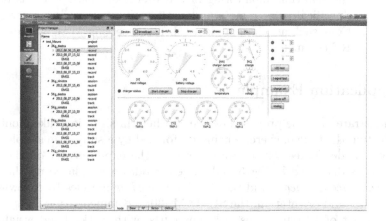

Fig. 4. Screenshot of settings function of graphical interface.

Fig. 5. Photograph of the recording setup with the wireless electromyograph sensors worn on the upper right arm with their electrodes placed on the *biceps brachii*, the *triceps brachii*, and the *deltoideus medius* muscles.

This panel allows the user to open previously saved projects. The signal can then be analyzed with several processing tools (filters, statistical analysis, ...) and the annotations can be edited. In particular, it can be seen in the lower left pane as shown in Fig. 3, a fatigue analysis functionality, which displays the trend of the filtered mean frequency of the EMG spectrogram.

(3) Settings: This panel, shown in Fig. 4, allows a complete diagnostic of the devices that compose the measurement system to be carried out. The Settings panel contains a "nodes" section and a "base stations" section, so the user can easily select the specific device for which the real-time monitoring of the functional parameters is desired. Specifically, for remote nodes, the software provides several functions such as: charging control, reset, setting MAC addresses and RF channel.

4   Application Example

To demonstrate how the proposed device can help in situations like monitoring and detecting the type of exercise being performed by a subject, a simple experiment was made. Three sensors were worn by the experimenter on the upper right arm, as displayed in Fig. 5 and the electrodes were placed on the *biceps brachii*, the *triceps brachii*, and the *deltoideus medius* muscles by following, for their location and orientation, the SENIAM [30] recommendations.

First, a set of 10 repetitions of biceps curls, with a 3 kg hand weight, was performed. Then, another set of 10 repetitions of lateral raises, with the same hand weight, was performed, followed by an isometric contraction held for a few seconds. Finally, a set of 16 repetitions of vertical raises, with the same hand weight, was performed.

The measured signals are shown in Figs. 6, 7, and 8. As can be seen, availability of the accelerometric signals greatly simplifies the task of detecting the various phases of the exercise (curls from about 0 s to 40 s, lateral raises from 40 s to 80 s, vertical raises from 100 s to 140 s), which would be much harder to do on the sEMG signal alone, and can also be used to discriminate between the concentric and eccentric phases, an important but complicated task if the sEMG is to be used to evaluate muscle fatigue [19]. Of course, acceleration is not of much use in evaluating the isometric contraction (from about 80 s to 90 s) which, on the other hand, clearly stands out in the sEMG track relative to the *biceps brachii* muscle being exerted. It is thus apparent how the combination of the two types of sensors helps in achieving a more complete picture of the activity being performed.

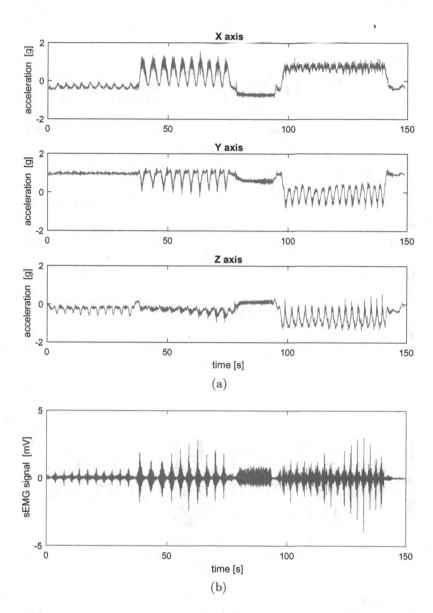

Fig. 6. Three-axis accelerometer signals *(a)* and sEMG signal *(b)* simultaneously acquired from the sensor applied to the *triceps brachii* during the exercise described in Sect. 4, consisting in 10 biceps curls, 10 lateral raises, an isometric contraction, and 16 vertical raises.

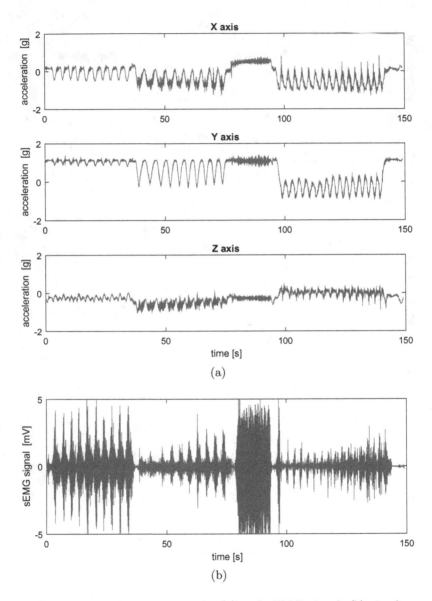

Fig. 7. Three-axis accelerometer signals *(a)* and sEMG signal *(b)* simultaneously acquired from the sensor applied to the *biceps brachii* during the exercise described in Sect. 4, consisting in 10 biceps curls, 10 lateral raises, an isometric contraction, and 16 vertical raises.

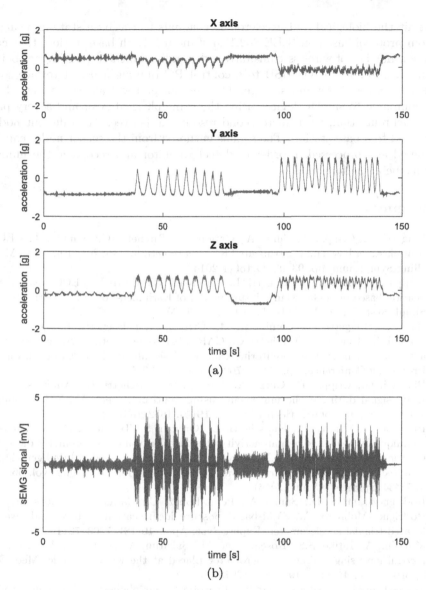

Fig. 8. Three-axis accelerometer signals *(a)* and sEMG signal *(b)* simultaneously acquired from the sensor applied to the *deltoideus medius* during the exercise described in Sect. 4, consisting in 10 biceps curls, 10 lateral raises, an isometric contraction, and 16 vertical raises.

5 Conclusion

In this paper an inexpensive wireless system for sEMG and accelerometer signal acquisition has been presented for healthcare and fitness applications. The system consists of up to four base stations and several sensing nodes that wirelessly transmit the biological and accelerometer signals to the base stations using a custom protocol based on IEEE 802.15.4 standard. Each base station, that can handle a number of wireless transmitters depending on the type of signal being acquired, is connected via USB to a control PC running a user interface software for data analysis and storage. The custom protocol allowed a high data rate compared to similar devices using the same physical layer and a very precise synchronization, with microsecond resolution, between the different nodes connected to a base station. The signals gathered from the sensor nodes can be combined and processed in order to detect, monitor and recognize the human activity being performed.

References

1. Biagetti, G., Crippa, P., Curzi, A., Orcioni, S., Turchetti, C.: A multi-class ECG beat classifier based on the truncated KLT representation. In: 2014 European Modelling Symposium, pp. 93–98, October 2014
2. Crippa, P., Curzi, A., Falaschetti, L., Turchetti, C.: Multi-class ECG beat classification based on a Gaussian mixture model of Karhunen-Loève transform. Int. J. Simul. Syst. Sci. Technol. **16**(1), 2.1–2.10 (2015)
3. Bacà, A., Biagetti, G., Camilletti, M., Crippa, P., Falaschetti, L., Orcioni, S., Rossini, L., Tonelli, D., Turchetti, C.: CARMA: a robust motion artifact reduction algorithm for heart rate monitoring from PPG signals. In: 23rd European Signal Processing Conference, pp. 2696–2700, September 2015
4. Biagetti, G., Crippa, P., Curzi, A., Orcioni, S., Turchetti, C.: Analysis of the EMG signal during cyclic movements using multicomponent AM-FM decomposition. IEEE J. Biomed. Health Inform. **19**(5), 1672–1681 (2015)
5. Biagetti, G., Crippa, P., Falaschetti, L., Orcioni, S., Turchetti, C.: An efficient technique for real-time human activity classification using accelerometer data. In: Czarnowski, I., Caballero, A.M., Howlett, R.J., Jain, L.C. (eds.) Intelligent Decision Technologies 2016. Smart Innovation, Systems and Technologies, vol. 56, pp. 425–434. Springer, Cham (2016)
6. Rodriguez-Martin, D., Samà, A., Perez-Lopez, C., Català, A., Cabestany, J., Rodriguez-Molinero, A.: SVM-based posture identification with a single waist-located triaxial accelerometer. Expert Syst. Appl. **40**(18), 7203–7211 (2013)
7. Mannini, A., Intille, S.S., Rosenberger, M., Sabatini, A.M., Haskell, W.: Activity recognition using a single accelerometer placed at the wrist or ankle. Med. Sci. Sports Exerc. **45**(11), 2193–2203 (2013)
8. Torres-Huitzil, C., Nuno-Maganda, M.: Robust smartphone-based human activity recognition using a tri-axial accelerometer. In: 2015 IEEE 6th Latin American Symposium on Circuits Systems, pp. 1–4, February 2015
9. Anguita, D., Ghio, A., Oneto, L., Parra, X., Reyes-Ortiz, J.L.: Energy efficient smartphone-based activity recognition using fixed-point arithmetic. J. Univ. Comput. Sci. **19**(9), 1295–1314 (2013)
10. Dernbach, S., Das, B., Krishnan, N.C., Thomas, B.L., Cook, D.J.: Simple and complex activity recognition through smart phones. In: 8th International Conference on Intelligent Environments, pp. 214–221, June 2012
11. Khan, A., Lee, Y.K., Lee, S., Kim, T.S.: Human activity recognition via an accelerometer-enabled-smartphone using kernel discriminant analysis. In: 2010 5th International Conference on Future Information Technology, pp. 1–6, May 2010

12. Mannini, A., Sabatini, A.M.: Machine learning methods for classifying human physical activity from on-body accelerometers. Sensors **10**(2), 1154–1175 (2010)
13. Catal, C., Tufekci, S., Pirmit, E., Kocabag, G.: On the use of ensemble of classifiers for accelerometer-based activity recognition. Appl. Soft Comput. **37**, 1018–1022 (2015)
14. Biagetti, G., Crippa, P., Falaschetti, L., Orcioni, S., Turchetti, C.: Artifact reduction in photoplethysmography using Bayesian classification for physical exercise identification. In: Proceedings of the 5th International Conference on Pattern Recognition Applications and Methods, Rome, Italy, pp. 467–474, February 2016
15. Biagetti, G., Crippa, P., Falaschetti, L., Orcioni, S., Turchetti, C.: A rule based framework for smart training using sEMG signal. In: Neves-Silva, R., Jain, L.C., Howlett, R.J. (eds.) Intelligent Decision Technologies. Smart Innovation, Systems and Technologies, vol. 39, pp. 89–99. Springer, Heidelberg (2015)
16. Lee, S.Y., Koo, K.H., Lee, Y., Lee, J.H., Kim, J.H.: Spatiotemporal analysis of EMG signals for muscle rehabilitation monitoring system. In: 2013 IEEE 2nd Global Conference on Consumer Electronics, pp. 1–2, October 2013
17. Chang, K.M., Liu, S.H., Wu, X.H.: A wireless sEMG recording system and its application to muscle fatigue detection. Sensors **12**(1), 489–499 (2012)
18. Pantelopoulos, A., Bourbakis, N.: A survey on wearable biosensor systems for health monitoring. In: 30th Annual International Conference of the IEEE Engineering in Medicine and Biology Society, pp. 4887–4890, August 2008
19. Biagetti, G., Crippa, P., Orcioni, S., Turchetti, C.: Homomorphic deconvolution for MUAP estimation from surface EMG signals. IEEE J. Biomed. Health Inform. **21**(2), 328–338 (2017)
20. Biagetti, G., Crippa, P., Orcioni, S., Turchetti, C.: Surface EMG fatigue analysis by means of homomorphic deconvolution. In: Conti, M., Madrid, N.M., Seepold, R., Orcioni, S. (eds.) Mobile Networks for Biometric Data Analysis. Lecture Notes in Electrical Engineering, vol. 392, pp. 173–188. Springer, Heidelberg (2016)
21. Biagetti, G., Crippa, P., Falaschetti, L., Orcioni, S., Turchetti, C.: Wireless surface electromyograph and electrocardiograph system on 802.15.4. IEEE Trans. Consum. Electron. **62**(3), 258–266 (2016)
22. Nawab, S.H., Roy, S.H., Luca, C.J.D.: Functional activity monitoring from wearable sensor data. In: The 26th Annual International Conference of the IEEE Engineering in Medicine and Biology Society, vol. 1, pp. 979–982, September 2004
23. Spulber, I., Georgiou, P., Eftekhar, A., Toumazou, C., Duffell, L., Bergmann, J., McGregor, A., Mehta, T., Hernandez, M., Burdett, A.: Frequency analysis of wireless accelerometer and EMG sensors data: towards discrimination of normal and asymmetric walking pattern. In: 2012 IEEE International Symposium on Circuits and Systems, pp. 2645–2648, May 2012
24. Ghasemzadeh, H., Jafari, R., Prabhakaran, B.: A body sensor network with electromyogram and inertial sensors: multimodal interpretation of muscular activities. IEEE Trans. Inf. Technol. Biomed. **14**(2), 198–206 (2010)
25. Roy, S.H., Cheng, M.S., Chang, S.S., Moore, J., Luca, G.D., Nawab, S.H., Luca, C.J.D.: A combined sEMG and accelerometer system for monitoring functional activity in stroke. IEEE Trans. Neural Syst. Rehabil. Eng. **17**(6), 585–594 (2009)
26. Wang, Q., Chen, X., Chen, R., Chen, Y., Zhang, X.: Electromyography-based locomotion pattern recognition and personal positioning toward improved context-awareness applications. IEEE Trans. Syst. Man Cybern. Syst. **43**(5), 1216–1227 (2013)

27. Zhang, X., Chen, X., Li, Y., Lantz, V., Wang, K., Yang, J.: A framework for hand gesture recognition based on accelerometer and EMG sensors. IEEE Trans. Syst. Man Cybern. Part A Syst. Hum. **41**(6), 1064–1076 (2011)
28. Biagetti, G., Crippa, P., Orcioni, S., Turchetti, C.: An analog front-end for combined EMG/ECG wireless sensors. In: Conti, M., Madrid, N.M., Seepold, R., Orcioni, S. (eds.) Mobile Networks for Biometric Data Analysis. Lecture Notes in Electrical Engineering, vol. 392, pp. 215–224. Springer, Heidelberg (2016)
29. van Beelen, T.: EDFbrowser. http://www.teuniz.net/edfbrowser/. Accessed 30 June 2014
30. Hermens, H.J., Freriks, B.: European recommendations for surface electromyography [CDROM]. Roessingh Research and Development (1999)

An Automatic and Intelligent System for Integrated Healthcare Processes Management

Virginia Cid-de-la-Paz[✉], Andrés Jiménez-Ramírez,
and M.J. Escalona

Departamento de Lenguajes y Sistemas Informáticos,
Universidad de Sevilla, Seville, Spain
virginia.cid@iwt2.org, {ajramirez,mjescalona}@us.es

Abstract. In this work, an automatic and intelligent system for integrated healthcare processes management is developed on a constraint based system. This project has been carried out in collaboration with a real assisted reproduction clinic. Our goal is to improve the efficiency of the clinic by facilitating the management of the integrated healthcare system. This is very important in an environment in which the healthcare processes present complex temporal and resource constraints.

Keywords: Constraint programming · Clinical guidelines · Integrated healthcare processes · Process management

1 Introduction

In recent years, process management as a mechanism to increase the excellence and the quality of organizations is a fact globally accepted. Healthcare organizations are not an exception. In fact, healthcare process management is essential to (1) ensure an adequate patient care and (2) facilitate the work of health professionals in an area where decision-making based on the best biomedical knowledge which is available is essential [1].

The management of these processes is closely related to the term of integrated healthcare processes (IHPs). IHPs or clinical guidelines are a textual definitions of healthcare processes and a group of clinical decision rules. In certain circumstances, these IHPs support health professionals to make a decision about the health care of their patients [2]. In short, an IHP aims to improve the quality and the safety of patient healthcare and to reduce both the variability in clinical practice and the health costs [2, 3].

However, modeling such guides is a complex task due to the following reasons:

- They are defined in natural language.
- They don´t follow any standard.

The original version of this chapter was revised: The Acknowledgements section was included. The erratum to this chapter is available at https://doi.org/10.1007/978-3-319-56154-7_65

I. Rojas and F. Ortuño (Eds.): IWBBIO 2017, Part II, LNBI 10209, pp. 621–630, 2017.
DOI: 10.1007/978-3-319-56154-7_55

- It is difficult to model the constraints (e.g., time and resources) of a healthcare process in existing standards for process modelling, e.g., Business Process Model and Notation (BPMN).

Since there are multitude of variables and constraints that must be met in a healthcare process, it is difficult to find a healthcare environment that uses a user-centered automatic process management system. This type of systems includes, among other functionalities, the management of appointments, i.e., activities which require the availability of both patients and doctors simultaneously. The management of appointments is another conflicting point in the health systems due to the overload of patients that the sector suffers.

Nowadays, appointment systems can be a source of dissatisfaction for both patients and health care professionals [4]. Patients often complain about the lack of availability, i.e., number of visits allowed per day. However, such availability is directly related to the duration of each appointment. Therefore, the increase of the availability reduces the time which is invested in each patient. In summary, the appointment scheduling systems are at the intersection between efficiency and the time to access to the health services. As a result, it is complex to model an appointment management system.

In addition, for assigning an appointment to a patient, it is necessary to know (1) the IHP and the constraints or relations that exits between their activities, (2) the next activity that should be performed (i.e., the patient's clinical history) and (3) the availability of resources. Given this complexity, health professionals usually have to manage the appointments in a manual way. Therefore, health professionals often are overloaded with administrative tasks instead of taking care of patients' health [5].

In this scenario, this paper focuses on developing a decision support tool based on Constraint Programming (CP) [6] with the aim of enhancing the performance of the health sector i.e., reducing costs and improving the access to health services and quality of the health care.

In addition, decision support systems have already been applied successfully in other service industries, such as airlines, car rental agencies and hotels [7].

For this, the current approach introduces an automatic and intelligent system for process management that:

1. Can be adapted to a defined healthcare processes.
2. Considers the resources which are available in the health center.
3. Manages patient appointments based on the demand and regarding the capacity that the resources offer.
4. Optimizes (1) the use of these resources (2) the waiting times that patients can suffer, direct (i.e., time which the patient waits in the health center) and indirect (i.e., period of time between the day that the user asks for the next activity to be performed until the day of its execution), and (3) the satisfaction of timetabling preferences of both the patients and health professionals.

In addition, the proposed system comprises a user interface to facilitate the management of the IHPs by the patient and health professionals. Therefore, health professionals do not need to allocate and communicate the appointments to their patients manually thus alleviating the workload.

Our proposal is designed for a real assisted reproduction clinic located in Seville. Specifically, to the Fertility Study IHP.

The rest of the paper is structured as follows. Section 2 introduces the background, where necessary terms are defined to improve the understanding of our project. In Sect. 3, we explain the development of our prototype, followed by its evaluation in Sect. 4. Section 5 investigates the existence of proposals similar to ours. Finally, we will conclude with conclusions and future work in Sect. 6.

2 Background

2.1 Planning, Scheduling and Constraint Programming

The area of scheduling includes problems in which it is necessary to determine a schedule for a set of activities related by temporal and resource constraints. A schedule states (1) the start and end times of the activities to be executed and (2) the resource which is assigned to perform each activity. Since different activities may require the same resources, they may compete for limited resources (i.e., resource constraints). In scheduling problems, several objective functions are usually considered to be optimized, in most cases related to temporal measures, or considering the optimal use of resources.

In such context, constraint programming (CP) supplies a suitable framework for dealing with planning and scheduling problems [6]. To solve a problem through CP, it needs to be modelled as a constraint satisfaction problem (CSP, cf. Definition 1).

Definition 1 CSP. A CSP P = (V, D, C) is composed of a set of variables V, a set of domains D which is composed of the domain of values for each variable of V, and a set of constraints C between variables, so that each constraint represents a relation between a subset of variables and specifies the allowed combinations of values for these variables.

3 Our Proposal

Figure 1 show an overview of the architecture which is proposed in the current approach. For this, an automatic and intelligent system was developed (cf. Fig. 1(1)) to execute and manage the activities of an IHP. On the one hand, the system uses the information stored in the database of the health centre (cf. Fig. 1(2)). On the other hand, the system has been integrated with a user interface (cf. Fig. 1(3)). Thus, users of the system (i.e., patients and healthcare professionals) can request the following activity to be per-formed and know their details.

In a first step, after carrying out an exhaustive analysis of the operation of the IHP in its sanitary environment (cf. Fig. 1(a)), an analyst models the IHP as a CSP (cf. Definition 1). For this, is necessary to model the following:

- The phases of the considered IHP including their duration and the type of professional that must perform them.
- The temporal constraints which exist between the different phases.

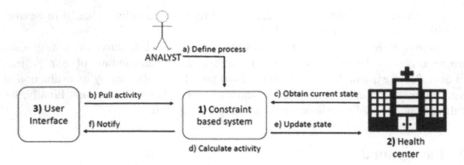

Fig. 1. Overview of the proposal

- The resources of the sanitary environment, contemplating their working hours.
- The state of the health center, taking into account the clinical history of the patients (i.e., the phases of the IHP that the patients have already performed, when they were executed and which resource performed them) and the occupation time of the health professionals.

Once the IHP is modeled in the system, the clients can ask for their next activity to be performed (cf. Fig. 1(b)) through a user interface. This request is sent to the system using a web service architecture.

In order to know the next activity to do by the user, the system must receive the current status of the health center (cf. Fig. 1(c)). Such status includes the patient records and the availability of the different resources between other information.

Thereafter, the proposed system obtains the best schedule for the following activity (cf. Fig. 1(d)) to be carry out by the user. Such calculus considers the clinical history of the user, the information defined in the IHP, the occupation of the resources and both temporal and resources constraints that may present the activity.

Calculated the next activity to be carried out by the user, the system sends the information of this new activity to the system of the health center to update it (cf. Fig. 1 (e)). In a similar way, such activity information is notified to the requesting customer which is shown through the user interface (cf. Fig. 1(f)).

3.1   Implementation Details

The integration of the proposed intelligent system with external systems (i.e., the user interface and a healthcare center) was performed on a scalable web service architecture. In this way, we will favor the ability to connect with other external systems.

In order to model the activities of an IHP we based upon previous proposals [8, 9] where a constraint-based language is defined.

For the development of the intelligent system, we use the constraint-based system (i.e., a system which solves CSP, cf. Definition 1) IBM ILOG CPLEX Optimization Studio (CPLEX) [10] together with the CPLEX CP Optimizer complement [11]. This tool provides for efficient mechanisms to deal with scheduling problems as well as temporal and resources constraints. CPLEX provides a high-level object which can be used to encapsulate activities. Such object is called interval variable. An interval variable has a start, an end and a length. Moreover, an interval variable may be optional, i.e., it may or may not appear in the planning.

Regarding the management of resources, CPLEX includes cumulative function expressions, which can be used to model resource usage functions over time and its limiting capacity.

To model the restrictions that exist between the phases of an IHP, we use high-level constraints on the interval and cumulative function variables. For example, the precedence constraints, which ensure the fulfillment of the precedence and temporal restrictions between the activities of the IHP.

4   Evaluation

In this project the intelligent management system was applied (cf. Fig. 1(1)) for the IHP of Fertility Study (cf. Fig. 2) followed in a real clinic of assisted reproduction since it is a representative process for applying our approach.

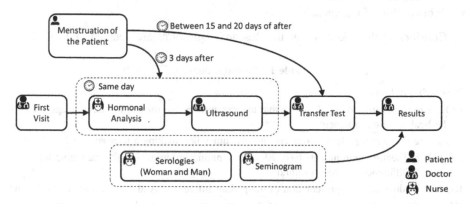

Fig. 2. IHP of Fertility Study

As can be seen, the healthcare process of Fertility Study presents a set of sequential activities. Among some of the activities, there are some temporal relationships (i.e., constraints). These temporal restrictions are represented in the Fig. 2 by a clock. For example, the activities Hormonal Analysis and Ultrasound must be done in the same day and, specifically, three days after the end of the patient's menstruation. In addition, each activity is restricted to being performed by a specific role (i.e., doctor, nurse or patient). These restrictions are an implementation challenge and must be strictly achieved.

Once IHP is analysed, it is modelled as a CSP. Thereafter, it is integrated in the proposed system with the actual clinic, so the system can calculate the following activity that any user requests.

However, it was not possible to carry out a formal validation of the system because only one healthcare process was modeled. Nevertheless, we have test our system with a data base of 25 patient's histories. The system suggested correctly the next activity of the patient that ask for it in the period of one month.

In conclusion, we obtained good impressions in this prototype, so it is presumed to obtain promising results in the continuation of this project.

5   Related Work

In this section, we present the existing works on the intelligent management of PAI in health centers. For this purpose, we did a systematic review of the literature, i.e., a method proposed by Kitchenham [12] that allows us to identify, evaluate and interpret relevant research data in a specific research area. This methodology is highly recommended in software engineering.

The focus was on the following challenging points in the appointments management of health environments:

- Patient waiting time (direct and indirect waiting times)
- Cancellations
- Absence of the patient
- Prioritization of emergencies.

Considering these key points, the Research Questions are detailed in Table 1.

Table 1. Research questions

Research questions
RQ1. Are there currently applications that use dynamic programming to help manage patients in the healthcare environment?
RQ2. Are the user's preferences taken into account when making a user activity?
RQ3. What health environments have adopted dynamic programming for managing user activities? Differences to keep in mind?
RQ4. How does the non-presentation of the patient influence? And the previous cancellation?
RQ5. Is the time between the request of the next activity and the day of the activity (indirect time) taken into account?
RQ6. How to manage emergencies and their priorities?

To respond to such RQs, articles from relevant bibliographic sources (e.g., Google Scholar, IEEE, Scopus and PubMed) are used. In these sources a variety of searches are performed about different keywords (e.g., health, dynamic, appointment, programming, scheduling or review).

Some different combinations or boolean expressions of keywords were made for the search (e.g., health dynamic scheduling, dynamic scheduling software health, health dynamic scheduling review or health dynamic appointment scheduling).

Once the searches were performed, articles were discarded based on defined inclusion/exclusion criteria (e.g., publications since 2006, inclusion of keywords, non-duplicates, relevant number of citations, etc.). Finally, 8 papers [13–20] are included in the review. In spite of obtaining articles of more than five years of antiquity, they are very useful to respond to the RQs. To summarize, the results obtained for the RQs are presented below.

RQ1. Are there currently applications that use dynamic programming to help manage patients in the healthcare environment?

No article has been found about the implementation of this technique in a real sanitary environment.

RQ2. Are the user's preferences taken into account when making a user activity?

Considering the user preferences is a fundamental idea [9] that should not be over-looked. Such preferences directly affect the probability of a user delaying an activity or even cancelling it. This relationship is totally logical, because if the user's preferences are not taken into account, is probably that they lose the interest in their appointment.

However, few articles discuss this subject. Only three articles [9, 15, 19] of the eight contemplated mention the preferences of the patient, two theoretically and one experimental. According to [15], this may be due to the complex mathematical models of arduous computation that we obtain if we consider the needs of patients. I.e., it is difficult to model the user's preferences.

RQ3. What health environments have adopted dynamic programming for managing user activities? Differences to keep in mind?

Practically all studies have been done in all possible health areas: outpatient's clinics, external consultations, nursing, treatment and surgery.

Several articles related to the Ambulatory environment were found, followed closely by the External Consultations. Both domains are very similar. On the contrary, Surgery and Treatment are more hostile environments. For example, for treatments, we include the articles [16, 19], which are intended to manage appointments for the treatment of chemotherapy and radiotherapy respectively. In this scenario, times are essential to ensure the best saving percentage. Moreover, cancellations and delays have a great impact on costs because they are expensive treatments.

About surgery, thanks to one of our basic articles [9] we know the difficulty to model an appointment management system for that environment. The main impediment is the impossibility of generalizing the operating times.

In treatment and surgery environment, managing emergencies is vital, so the waiting time for an emergency must be minimized, without neglecting or disfavouring the other patients.

RQ4. How does the non-presentation of the patient influence? And the previous cancellation?

The experimental study introduced in [16] reinforces the idea of the close relationship between the time from the request of the appointment to the date of the appointment, and the probability of cancellation or absence. However, [16, 17, 19] coincide in the following: if we try to minimize the probability of cancellation, paradoxically the time increases. Therefore, the experimental studies presented are not conclusive because there isn't an efficient technique that attack this problem.

RQ5. Is the time between the request of the next activity and the day of the activity (indirect time) taken into account?

On the one hand, only [16] performs experimental tests trying to minimize the indirect time. On the other hand, [14] claim to continue making experimental tests that contemplating the indirect time and elaborate a study of the reason for such cancellations or absences.

RQ6. How to manage emergencies and their priorities?

Although five of the articles selected [13, 15, 17–20] perform experimental tests attending urgencies, only two of them [17, 18] consider their prioritization.

This studies warn of the complexity of modelling a dynamic appointment management system that accepts urgency. [13, 17–20], save certain time zones to be used for emergencies. However, if these are not used, it is a waste of time, with its consequent influence on costs. In addition, such time reservation increases the time lag between the day that the patient asked for an appointment and the day of the provided appointment which is an undesirable fact. Therefore, it is necessary to keep on researching.

6 Conclusions

In this article, we propose an automatic system of intelligent management of IHPs implemented on a constraint-based system (i.e., CPLEX), which facilitates the modeling of temporal and resource constraints presented in the activities of IHPs.

To the best of our knowledge (cf. Sect. 5) there no exists any direct implementation of intelligent management of IHP systems in health environments.

Our proposal is satisfactory for the sanitary environment of application (cf. Sect. 4) where we can observe the following advantages:

- Healthcare professionals have quick access to the information, thus, avoiding the need to resort to (1) the clinical history of the patient—to know his situation—and (2) the clinical guidelines—which are subjective due to their textual definition—, to know what decision should be taken about the problem of the patient. With our proposal, the health professional quickly knows the situation of the patient and based on it, what activity should be performed.
- Efficient management of the health center. Regarding to the management of IHPs, an intelligent system is in charge of (1) managing efficiently the access to the health service by the patients as well as (2) organizing the schedules of the activities of the IHP in the health center.
- The health sector provides a service that involves a high cost. Therefore, it is important to optimize the management of health centers, not necessarily with the aim of saving, but to provide a better service with the same resources.
  Additionally, optimizing patient waiting times, the efficiency of the sector is enhanced, avoiding possible cancellations and oblivions, and improving health care.
- Healthcare professionals should not be in charge of managing the activities, investing all their working hours to the health care of their patients.

Consequently, the validity and usefulness of these systems in health services is demonstrated.

7 Future Works

Regarding the intelligent management of IHPs, it is important to consider the possibility that a user cancel their activity, is delayed or even does not show up. In these cases, the use of these intelligent systems is very useful since, under a correct design, they can replan the activities. In this way, the intelligent system would improve the efficiency of the sanitary environment. Therefore, we pretend to study its design and implementation in future projects.

In addition, it is interesting to analyse what happens if the health entity has not free resources, i.e., health professionals without free gaps in their schedule of work to attend patients. Therefore, we plan to study how should behave our system in this real situation.

Furthermore, we want to continue modeling other IHPs in our intelligent system in other to generalize the validity of our system.

Acknowledgements. This research has been supported by the Pololas project (TIN2016-76956-C3-2-R) and by the SoftPLM Network (TIN2015-71938-REDT) of the Spanish the Ministry of Economy and Competitiveness.

References

1. de Adana Pérez, R.R.: El blog de Ricardo Ruiz de Adana Pérez, médico de familia sobre actividades preventivas, diagnóstico y tratamiento de problemas de salud, calidad, gestión e investigación en atención primaria. Enero (2011). http://ricardoruizdeadana.blogspot.com.es/2011/01/aprendiendo-disenar-procesos-en-las.html. Accessed Jan 2017
2. de Andalucía, J.: Guía de diseño y mejora continuada de procesos asistenciales (2009)
3. GOC Networking. http://www.gocnetworking.com/ventajas-de-los-procesos-asistenciales-integrados-pai/. Accessed Jan 2017
4. Ballesteros Pérez, A.M., García González, A.L., Alcázar Manzanera, F., Fontcuberta Martíncz, J., Sánchez Rodríguez, F., Pérez-Crespo, C.: La demora en la consulta de atención primaria. Aten. Primaria. **31**, 377–881 (2003)
5. Gestión médica: Seguridad y rentabilidad en medicina privada. Junio (2015). http://gestionmedica.org/gestion-medica-privada/. Accessed Jan 2017
6. Rossi, F., van Beek, P., Walsh, T.: Handbook of Constraint Programming (Foundations of Artificial Intelligence). Elsevier Science Inc., Amsterdam (2006)
7. Talluri, K.T., Van Ryzin, G.J.: The Theory and Practice of Revenue Management. Springer, Heidelberg (2004)
8. Jiménez, A., Barba, I., del Valle, C., Weber, B.: OptBPPlanner: automatic generation of optimized business process enactment plans. In: Linger, H., Fisher, J., Barnden, A., Barry, C., Lang, M., Schneider, C. (eds.) Building Sustainable Information Systems, pp. 429–442. Springer, Boston (2013). doi:10.1007/978-1-4614-7540-8_33
9. Jiménez-Ramirez, A., Weber, B., Barba, I., del Valle, C.: Generating optimized configurable business process models in scenarios subject to uncertainty. Inf. Softw. Technol. **57**, 571–594 (2014)
10. IBM: IBM ILOG CPLEX Optimization Studio (2016). http://www-03.ibm.com/software/products/en/ibmilogcpleoptistud/. Accessed Jan 2017

11. IBM: CPLEX CP Optimizer (2016). http://www-01.ibm.com/software/commerce/optimization/cplex-cp-optimizer. Accessed Jan 2017
12. Kitchenham, B., Charters, S.: Guidelines for performing Systematic Literature Reviews in Software Engineering (2007)
13. Cayirli, T., Veral, E., Rosen, H.: Designing appointment scheduling systems for ambulatory care services. Health Care Manag. Sci. **9**, 47–58 (2006)
14. Gupta, D., Denton, B.: Appointment scheduling in health care: challenges and opportunities. IIE Trans. **40**, 800–819 (2008)
15. Cayirli, T., Veral, E.: Outpatient scheduling in health care: a review of literature. Prod. Oper. Manag. **12**, 519–529 (2009)
16. Liu, N., Ziya, S., Kulkarni, V.G.: Dynamic scheduling of outpatient appointments under patient no-shows and cancellations. Manuf. Serv. Oper. Manag. **12**, 347–364 (2010)
17. Gocgun, Y., Puterman, M.L.: Dynamic scheduling with due dates and time windows: an application to chemotherapy patient appointment booking. Health Care Manag. Sci. **17**, 60–76 (2014)
18. Gupta, D., Lei, W.: Revenue management for a primary-care clinic in the presence of patient choice. Oper. Res. **56**, 576–592 (2008)
19. Laganga, L.R., Lawrence, S.R.: Clinic overbooking to improve patient access and increase provider productivity. Decision Sci. **38**, 251–276 (2007)
20. Sauré, A., Patrick, J., Tyldesley, S., Puterman, M.L.: Dynamic multi-appointment patient scheduling for radiation therapy. Eur. J. Oper. Res. **223**, 573–584 (2012)

A Microcontroller Based System
for Controlling Patient Respiratory Guidelines

Leticia Morales[1(✉)], Manuel Domínguez Morales[2],
Andrés Jiménez-Ramírez[1], and M.J. Escalona[1]

[1] Departamento de Lenguajes y Sistemas Informáticos,
Universidad de Sevilla, Seville, Spain
leticia.morales@iwt2.org,
{ajramirez,mjescalona}@us.es
[2] Departamento de Arquitectura y Tecnología de Computadores,
Universidad de Sevilla, Seville, Spain
mdominguez@atc.us.es

Abstract. The need of making improvements in obtaining (in a non-invasive way) and monitoring the breathing rate parameters in a patient emerges due to (1) the great amount of breathing problems our society suffer, (2) the problems that can be solved, and (3) the methods used so far. Non-specific machines are usually used to carry out these measures or simply calculate the number of inhalations and exhalations within a particular timeframe. These methods lack of effectiveness and precision thus, influencing the capacity of getting a good diagnosis. This proposal focuses on drawing up a technology composed of a mechanism and a user application which allows doctors to obtain the breathing rate parameters in a comfortable and concise way. In addition, such parameters are stored in a database for potential consultation as well as for the medical history of the patients. For this, the current approach takes into account the needs, the capacities, the expectations and the user motivations which have been compiled by means of open interviews, forum discussions, surveys and application uses. In addition, an empirical evaluation has been conducted with a set of volunteers. Results indicate that the proposed technology may reduce cost and improve the reliability of the diagnosis.

Keywords: Respiratory rate · Arduino · Wearable devices

1 Introduction

The process that living things use to introduce oxygen in their body and get rid of carbon dioxide that is not need is what is commonly known as breathing.

According to the World Health Organization (WHO) in 2004 there were 64 million people suffering from respiratory diseases and, at least, 3 million died for this reason [1].

The original version of this chapter was revised: The Acknowledgements section was included. The erratum to this chapter is available at https://doi.org/10.1007/978-3-319-56154-7_65

I. Rojas and F. Ortuño (Eds.): IWBBIO 2017, Part II, LNBI 10209, pp. 631–641, 2017.
DOI: 10.1007/978-3-319-56154-7_56

The Spanish Society of Pneumology and Thoracic Surgery (SEPAR in Spanish) conducted a study which concluded that on January 5, 2016, five of the ten diseases that cause more mortality in the world occur in the respiratory field. Furthermore, according to data published by the National Statistics Institute (INE in Spanish) of Spain in 2014, the diseases that have recorded the highest increase in hospital discharge are related to respiratory pathologies [2].

Early diagnosis may be the key to deal with some of these respiratory diseases [3], among many others. Such diagnosis passes through the recognition of patterns that are abnormal, similar to those formulated when monitoring the respiratory rate.

The respiratory rate corresponds to the number of inhalations and exhalations that a human being performs in a unit of time. The respiratory rate of a persons in resting state is considered as the normal respiratory rate. The normal rate depends on several factors such as age, sex, etc.

Monitoring a normal respiratory rate may reveal abnormal patterns that may indicate many health problems [4]. There are several studies which state the health benefits that the recognition of respiratory rate provides [5–10].

Probably, the small impact that this measure has on the daily clinic practice is due to the lack of technology for obtaining and monitoring the respiratory rate in a non-invasive, easy, comfortable and reliable way.

Nowadays, the methods to measure the respiratory rate in a non-invasive way go through the direct vision, monitors of impedance and with capnographs. However, these methods are limited and have a poor accuracy in getting results.

For all aforementioned facts, this paper presents a design, implementation and test of a low cost and easily usable instrument which obtains the respiratory rate in a non-invasive, precise and accurate way. Furthermore, such instrument connects to a user application so that the breathing rate can be monitored and stored, saving the relevant information in a database along with other data.

The rest of the paper is structured as follows. Section 2 shows the proposal and the details of its development. Section 3 describes how the satisfaction generated by the proposed technology has been measured. Section 4 discusses the proposals that exist and the main differences between them. Finally, Sect. 5 includes a brief summary of the idea, the strengths and weaknesses, and the lines that are open to continue working.

2   Contribution

2.1   Proposed System

The proposed system (cf. Fig. 1) consists of a device for measuring respiratory rate and the software tool for managing the user information and data obtained from the device.

Firstly, the physician considers if it is appropriate to monitor the respiratory rate of a patient (cf. Fig. 1 - Step 1). Then, she has to register the patient in the application by entering the data which is necessary. This data is stored in the database (cf. Fig. 1 - Step 2).

To monitor the respiratory rate, the physician places the device on the patient (cf. Fig. 1 - Step 3). Thereafter, the data acquisition and transfer begins. This is done

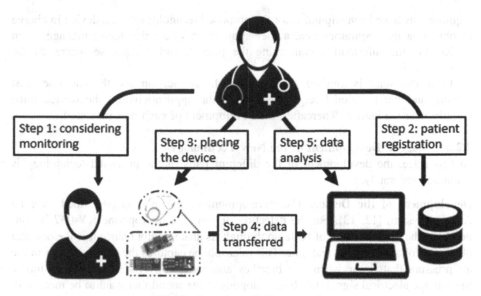

Fig. 1. Overview of the proposal.

through a Bluetooth connection to the application where a graphic is painted according to such data (cf. Fig. 1 - Step 4).

Finally, the health personnel can analyze the data obtained, save and consult the relevant information when necessary (cf. Fig. 1 - Step 5).

2.2 Development Details

For the process of development of the proposed technology, a user-centered methodology is followed. This methodology encompasses a heterogeneous set of methodologies and techniques that share a common goal: to know and understand the needs, objectives, motivations, limitations, behavior and characteristics of the user [11].

In order to clearly understand the development process that has been followed to obtain the technology which proposed in this paper, it is divided into two stages:

2.2.1 Stage 1. Knowing the End Users Thoroughly

In order to learn about the scenario and its users, open interviews, discussion groups and surveys are carried out. This is intended to explain the proposal to users and receive an opinion about it. With the help of these methods, the necessary changes can be made to make the study more fruitful and better adapted to reality.

The end users of this technology are both patients—who need to be monitored— and physicians—who take control of this monitor. Therefore, the technology should fulfill their needs. On the one hand, it should be comfortable for the patient and on the other hand, it should be easy to handle for the doctors, among other things.

In addition, the needs that physicians have when obtaining and storing the information of the different patients have been known. Consequently, a set of functional requirements have been stipulated for the proposed technology, i.e., a device in charge of obtaining the respiratory rate, a user application where the doctor manages such device and the information concerning the patient, and a database where all the information is stored.

Once this stage is finished, the data that the application and the database must contain and the functional requirements that the application and the device must comply with are known. Thereafter, the development of each part may start.

2.2.2   Stage 2. Development of the New Technology

At this stage, the development of the different parts of the proposed technology is explained separately.

Development of the Device. The development of this part is performed over an Arduino system [12, 13]. For the detection of sound, a microphone KY-037 is connected to the hardware part of Arduino through its analog output since it is the one that obtains a voltage signal in real time. This microphone captures the small changes in the air pressure that occur when one breathes and then, converts such changes into a measurable electrical signal which is analogous to the sound that want to be measured.

The hardware part connects to the Arduino software part wirelessly through a Bluetooth device (i.e., Bluetooth JY-MCU). The software part states the COM port where Bluetooth is anchored and the Arduino board model which is used. Additionally, a program based on the language Processing [14] (i.e., Arduino programming language) is developed to indicate the data that want to be obtained.

Development of the Application. For the development of the desktop application, Microsoft Visual Studio 2015 has been used [15] together with the C# programming language.

The application consists of a graphical interface that allows to manage the data of the patients, their clinical histories as well as the data related to the respiratory rate. The graphical interfaces of the application are created through different forms. Each form represents one of the screens that can be accessed in the application. These screens are designed using the graphical components which are provided by Visual Studio.

Figure 2 shows the graphical interface developed for the analysis of the respiratory rate. As shown, a plot is depicted according to data which is being obtained from the device.

When the stipulated time for the test expires, the physician can save the generate plot which is assigned to the patient and contains the date and time when the plot was performed. The peaks of the plot correspond to the expirations that the patient performs throughout the respiratory cycle.

Related to the communication of the application, in the one hand, it maintains a bluetooth communication with the device to obtain the data that the sensor captures and to paint a plot with them. In the other hand, the application communicates with a database developed with Microsoft Access.

To summarize, the development of this software comprises 9 forms with 18 C# classes which contain 4500 lines of code in total. In addition, the prototype was developed by one undergraduate student in 820 h.

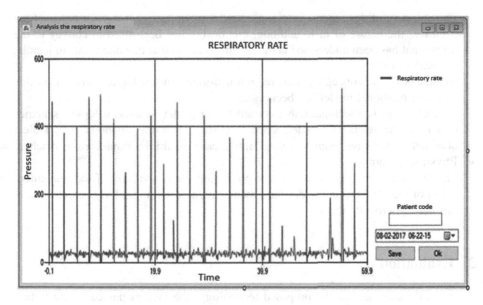

Fig. 2. Part of the graphical interface

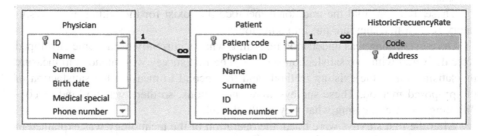

Fig. 3. Medical table design

Design of the Data Base. For the process of designing the database (cf. Fig. 3) the steps proposed in the book "El nuevo PHP. Conceptos avanzados" by Vicente Javier Eslava Muñoz have been followed [16].

The design is divided into three stages:

- **Conceptual Design**
  At this stage an information schema is constructed which is known as a conceptual schema.

  The conceptual schema is constructed using the information that is collected in stage 1 of the development. When constructing this scheme, one understands the meaning of the data that are necessary and the entities, attributes and relationships are found.

The objective of this scheme is to understand the perspective that the user has of the data, i.e., the nature of it. In addition, this design has been used to convey to the users what has been understood about the information that this one wants to handle.

- **Logical Design**
  At this stage the conceptual schema is transformed into a logical schema. In this case, the relational model has been used.
  Normalization is a technique that is used to verify the validity of logical schemes based on the relational model, since it ensures that the relations (i.e., tables) obtained do not have redundant data. In this case the third normal form is reached.
- **Physical Design**
  The physical design is the process of producing the description of the implementation of the database in the database management system that has been chosen, in this case Microsoft Access.

3   Evaluation

After the development of the proposed technology, the system passed to the testing phase. Then, a set of tests were performed to a group of real users, consisting of the use of this technology. Specifically, 20 users participated, of which 10 were doctors and 10 patients.

These users are told the traditional methods that exist for the taking of the respiratory rate so that they are put into situation.

Once everything was understood regarding what is currently used, some enveloped were delivered with two satisfaction surveys. The first survey was intended to measure the satisfaction of the existing methods and the second to measure the satisfaction of the proposed method. These surveys were anonymous, so there was no kind of conditioning when expressing what they really thought.

After the first surveys were filled, the operation of the technology was explained in order to be tested. They had about half an hour to test the proposed technology. During the tests, the group of users were observed and analyzed in order to know how they use the application and the device. As a result, on the one hand, any change in functional requirements was required and, on the other hand, a set of non-functional requirements was detected.

Finally, once the test is finished, users were asked to fill in the second survey, i.e., related to the satisfaction of the proposed method. And once completed, both surveys were inserted in the envelope and delivered.

Surveys were used to measure the reactions of users to this proposal in some way. This is done by looking at user responses and comparing the means of the results. After comparing the means of medical users, interesting results were obtained as shown in Fig. 4. In this figure, Serie 1 corresponds to the means obtained before performing the field work and the Serie 2 corresponds to the means obtained after performing the field work.

Additionally, a comparison of the means of the patient users has been made, which can be seen in Fig. 5. In this case the Serie 1 corresponds to the means obtained before

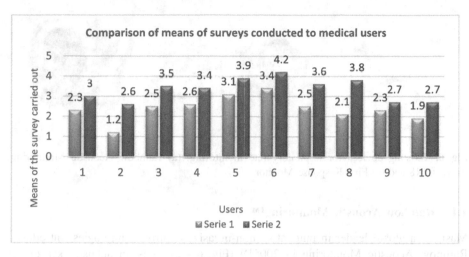

Fig. 4. Comparison of means of the surveys made to medical users.

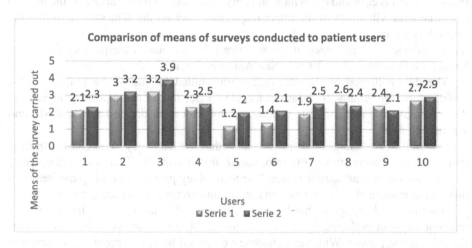

Fig. 5. Comparison of means of the surveys made to patient users.

performing the field work and the Serie 2 corresponds to the means Obtained after the field work.

4 Related Work

As far as we know, there no exist any specialized device to measure and monitor the respiratory rate in a non-invasive way that was reliable in terms of measurement. Although some related research is being conducted (e.g., [17, 18]) only few device are being manufactured (cf. Fig. 6).

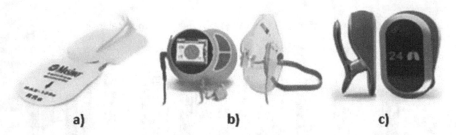

a)   b)   c)

Fig. 6. Existing devices for respiratory rate monitoring. (a) Rainbow Acoustic Monitoring, (b) RespiR8 and (c) First Response Monitor

4.1 Rainbow Acoustic Monitoring™

Masimo, a global leader in innovative non-invasive control technologies, introduced Rainbow Acoustic Monitoring in 2005™ (Fig. 6 - a). It is an acoustic sensor for continuous monitoring of respiratory rate, this sensor has an integrated acoustic transducer and is easy and convenient to apply. It is placed on the surface of the neck to detect acoustic vibrations of the upper respiratory tract on the skin surface during the respiratory cycle.

This device was presented at the American Association of Respiratory Care conference on December 5–7, 2009 in San Antonio, Texas.

The data collected by this sensor are transmitted to one of the platforms also manufactured by Masimo by means of a cable connection. In this platform one can see the values of the respiratory rate collected along with other data if additional Masimo sensors are used.

One of the disadvantages of this product is that it is necessary to purchase a specific monitor from Masimo to see the information that it collects. That is an unnecessary expense for the product carrier center. The technology proposed in this paper removes this inconvenience since it can be installed in any computer that the center owns.

Another disadvantage is that Rainbow Acoustic Monitoring™ requires a cable to transmit the data collected by the sensor to the Masimo monitor, which is uncomfortable for the patient. With the technology proposed here, this problem is overcome by using a bluetooth connection to transmit the sensor information to the computer where the user application is installed.

Finally, unlike Rainbow Acoustic Monitoring™, the technology proposed in this work allows the physician to operate the device through a user application where she can obtain the respiratory rate data of the patient, store data and learn relevant information for diagnosis.

4.2 RespiR8

Anaxsys, a British medical device company that develops and markets respiratory devices, announced on 13 October 2010 the availability of respiR8 (Fig. 6 - b). RespiR8 is a device to continuously monitor respiratory rate through a face mask that uses its own technology through a patented sensor that responds quickly and reliably to differences in the degree of humidity of the inhaled air with respect to exhaled one.

The data collected by the sensor can be read by a screen arranged on a small monitor. The mask and the monitor are connected by cable.

Similarly to Rainbow Acoustic Monitoring™, one needs her own monitor to view the data that the mask collects from the respiratory rate. As stated above, this drawback does not exist with the technology proposed here.

Also, the connection between the monitor and the RespiR8 device is wired which is uncomfortable for patients. As discussed above, the proposed device transmits the data wirelessly.

Another thing worth noting is that the price of the consumables of the proposed technology (i.e., a nasal cannula) is much lower than that of RespiR8.

4.3 First Response Monitor

Finally, Cambridge Design Partnership, a leader in technology and product design, announces on August 6, 2015 that they have created a vital sign monitor for reliable, fast measurement and for real-time display of respiratory rate (Fig. 6 - c). This monitor is shaped like a clip to be placed on the nose. Once placed, the control of the breathing rate begins immediately.

The collected data can be visualized through a screen that is already incorporated or transmitted in real time to a Smartphone or Tablet via Bluetooth.

This technology is not yet in the market so its pros and cons cannot be known accurately. Nonetheless, one of its cons would be that it cannot save relevant information about the patient who the test is being performed to, or store the respiratory rate data information that is obtained.

5 Conclusions and Future Work

This work presents the development of a device from a microcontroller and a respiratory rate detection system with microphone which obtains statistics of the respiratory rate and reflect them in a user application on the computer. The information that is captured by the device is packaged and sent wirelessly using a bluetooth device.

Therefore, the developed technology speeds up obtaining vital parameters and can carry out early diagnosis of diseases in an easier way. To do this, the plots that are generated with the information regarding the respiratory rate that is captured by the sensor are stored the medical equipment. This information includes relevant data of the patient to whom the monitoring of the respiratory rate will be carried out. This information is what has been considered necessary according to the requirements of the doctors.

In short, all the objectives that had been set at the beginning of the work were fulfilled.

Based on the results obtained in the testing part of the system, the development of the technology proposed here is feasible, thus enabling to obtain more information on the evolution of the patient.

Due to the simplicity to the developed system is more comfortable for the patient. In addition the patient will have more confidence in the results of his test, because the proposed technology provides more reliable results compared to the current technology.

Probably, the proposed technology is feasible since the requirements and reactions of end users have been taken into account, i.e., their needs, objectives, motivations, limitations and behavior. Therefore, a greater satisfaction and experience of use is obtained with the minimum effort of the users.

In summary, this technology would achieve, among other things, economic savings, greater reliability in the results and therefore more reliable diagnostics.

Throughout the work have been proposed ideas to improve this technology and the information that is obtained, but lack for time and material, among other things, has not been possible. Some of them correspond to the following:

- Use of sensor with greater accuracy for a better quality and accuracy of the obtained data.
- Extension of the application and graphic interface with the objective of obtaining more information from both the health professional and the patient.
- Implementation of new sensors to be able to measure other vital signs and thus obtaining a more exhaustive examination.
- Optimize the code in order to reduce the software execution time.
- The use of Wi-Fi technology instead of Bluetooth technology in order to reduce the time spent when transferring large files.
- Expansion of the database to store more information.
- Database in the cloud.
- Local storage in the microcontroller.
- Activity alarms.
- For the images that are generated from the respiratory rate, use of the DICOM standard (Digital Imaging and Communication in Medicine) [19].

Acknowledgements. This research has been supported by the Pololas project (TIN2016-76956-C3-2-R) and by the SoftPLM Network (TIN2015-71938-REDT) of the Spanish the Ministry of Economy and Competitiveness.

References

1. OMS|Enfermedad pulmonar obstructiva crónica (EPOC). WHO. http://www.who.int/respiratory/copd/es/. Accessed 09 June 2016
2. SEPAR NP Morbilidad hospitalaria neumología.pdf, Google Docs, 5 ene 2016. https://drive.google.com/file/d/0B3-GelWPMn4dNjdCUThzMS00MkE. Accessed 09 June 2016
3. SEPAR_NP La educacion de los profesionales mejorará el diagnosttico de la FPI.pdf, 2 November 2015, Google Docs. https://drive.google.com/file/d/0B3-GelWPMn4dUi02TVhH b2NPbnM/view?pref=2&pli=1&usp=embed_facebook. Accessed 09 June 2016

4. Frecuencia respiratoria normal|Salud y bienestar. http://lasaludi.info/frecuencia-respiratoria-normal.html. Accessed 21 Jan 2016
5. Fieselmann, J.F., Hendryx, M.S., Helms, C.M., Wakefield, D.S.: Respiratory rate predicts cardiopulmonary arrest for internal medicine inpatients. J. Gen. Intern. Med. **8**(7), 354–360 (1993)
6. Subbe, C.P., Davies, R.G., Williams, E., Rutherford, P., Gemmell, L.: Effect of introducing the modified early warning score on clinical outcomes, cardio-pulmonary arrests and intensive care utilisation in acute medical admissions. Anaesthesia **58**(8), 797–802 (2003)
7. Goldhill, D.R., McNarry, A.F., Mandersloot, G., McGinley, A.: A physiologically-based early warning score for ward patients: the association between score and outcome. Anaesthesia **60**(6), 547–553 (2005)
8. Cretikos, M., et al.: The objective medical emergency team activation criteria: a case-control study. Resuscitation **73**(1), 62–72 (2007)
9. Goldhill, D.R., McNarry, A.F.: Physiological abnormalities in early warning scores are related to mortality in adult inpatients. Br. J. Anaesth. **92**(6), 882–884 (2004)
10. Cook, C.J., Smith, G.B.: Do textbooks of clinical examination contain information regarding the assessment of critically ill patients? Resuscitation **60**(2), 129–136 (2004)
11. Hassan Montero, Y., Ortega Santamaría, S.: Informe APEI sobre usabilidad
12. Arduino - ArduinoBoardNano. https://www.arduino.cc/en/Main/ArduinoBoardNano. Accessed 09 June 2016
13. Arduino - Software. https://www.arduino.cc/en/Main/Software#. Accessed 09 June 2016
14. Processing. MIT Press. https://mitpress.mit.edu/books/processing-0. Accessed 04 Jan 2017
15. Webster, L.: Any Developer, Any App, Any Platform, Visual Studio, 16 November 2016. https://www.visualstudio.com/es/. Accessed 04 Jan 2017
16. Muñoz, V.J.E.: El nuevo PHP. Conceptos avanzados, Vicente Javier Eslava Muñoz (2013)
17. Reyes, B.A., Reljin, N., Chon, K.H.: Tracheal sounds acquisition using smartphones. Sensors **14**(8), 13830–13850 (2014)
18. Nam, Y., Reyes, B.A., Chon, K.H.: Estimation of respiratory rates using the built-in microphone of a smartphone or headset. IEEE J. Biomed. Health Inform. **20**(6), 1493–1501 (2016)
19. DICOM Homepage. http://dicom.nema.org/. Accessed 09 June 2016

Requirements Analysis for User Interfaces in Mobile eHealth Applications

Armando Statti[✉] and Natividad Martinez Madrid

Reutlingen University, Alteburgstr. 150, 72762 Reutlingen, Germany
armando.statti@student.reutlingen-university.de,
natividad.martinez@reutlingen-university.de

Abstract. Medical applications are becoming increasingly important in the current development of health care and therefore a crucial part of the medical industry.

The work focuses on the analysis of requirements and the challenges arisen from designing mobile medical applications in relation to the user interface. The paper describes the current status in the development of mobile medical apps and illustrates the development of e-health market. The author will explain the requirements and will illustrate the hurdles and problems. He refers to the German market which is similar to the European and compares that with the market in the USA.

Keywords: Medical product legislation · eHealth · Mobile medical applications

1 Introduction

The progressive digitization will also take inroads in the healthcare sector. Often referred as the digital medicine, it promises to become one of the most largest market and business areas in the future [1, Sect. 1]. Because of the continuous development of distributed applications and the rapidly growing use of wearables in the health care market, the term e-Health has established itself in the last recent years. Experts are often talking from linked and modernized IT structures in clinical institutions, with a direct networking and connection to the patient.

Patients get the option to receive their medical data over mobile devices and also the possibility for direct participation. In return clinical institutions receive more insight into the everyday life of the patients to support the diagnosis and individual therapy. So you can certainly talk about the digital personalized medicine [2, Sects. 7–8]. The adoption of the e-Health Law in Germany [3] at the end of 2015 shows that the legislator reacts to these circumstances. Under this law a structure for a secure infrastructure should arise. The use of the digital health card and the associated possibilities of electronic health records, medical reports and the digital drug plan, are part of these measures. In addition, also the master data management of patients data will be changed by modern and digital infrastructure. This takes a further reflection in telemedicine, so in future, the platform for medical mobile applications will continue to grow [4, Sect. 27].

© Springer International Publishing AG 2017
I. Rojas and F. Ortuño (Eds.): IWBBIO 2017, Part II, LNBI 10209, pp. 642–652, 2017.
DOI: 10.1007/978-3-319-56154-7_57

Big companies such as Apple and Google are also active in this market with own solutions like the Apple Health Care Center on iOS. The user has already decided by using such services, that a health care digital transformation is desired and required [5].

The market of medical applications, which are more like a medical product than an lifestyle product will grow increasingly. This sort of applications is also known as mHealth applications or Mobile Medical Apps. A major hurdle in the development of medical applications is the target group oriented usability analysis and the interface design to visualize complex and sophisticated medical data as informations for individual assessments of health status from patients. So the amount of information of medical data has to be formatted and embedded sensibly, so that they can be evaluated in an understandable manner for the patient. Actually there are no concepts that deal with a meaningful diagnostic evaluation to provide patients with user-friendly outcomes and informative and visually results. Apple is here a good example with their own application called Health Center. But actually they cover only general health data such as fitness information and basic illness information. Data resulting from therapies or diagnostic forecasts are not considered.

2   mHealth Applications

Mobile medical applications are in addition to the definition a medical device. In this context the developer needs to discriminated accurately between a mobile medical application and a normal mobile application. The Food and Drug Administration (FDA) which is responsible for the rules and laws in USA, has defined mobile medical applications in their guidelines with the following purpose:

> In general, if a mobile app is intended for use in performing a medical device function (i.e. for diagnosis of disease or other conditions, or the cure, mitigation, treatment, or prevention of disease) it is a medical device, regardless of the platform on which it is run. For example, mobile apps intended to run on smart phones to analyze and interpret EKG waveforms to detect heart function irregularities would be considered similar to software running on a desktop computer that serves the same function[...] [6, Sect. 8]

This statement is equal to the comment of the Federal Institute for Drugs and Medical Devices in Germany:

> Standalone software such a smartphone app can be a medical device. For this, the software must be in accordance with 3 No. 1 MPG from manufacturer to be used for people to be diagnosed with at least one of the following purposes:
> – Diagnosis, prevention, monitoring, treatment or alleviation of disease, Diagnosis, monitoring, treatment, alleviation or compensation for an injury or handicap, Investigation, replacement or modification of the anatomy or of a physiological process, Control of conception.

In contrast to pure knowledge deployment in case of a paper-based or electronic book (not a medical device) indicates any form of influence on data or information through the standalone software to a classification as a medical device [7].

The FDA and the Federal Institute in Germany also illustrate this with some examples as well. Applications which are considered as medical devices:

– Apps as an extension of a medical device. The most widely used use case is to present and visualize medical data which are read remotely.
– Apps which perform calculations or analysis for individual patients

Further the FDA defines also conventional applications and applications that are in a grey area:

– Apps which help users to manage everyday life with their illness
– Apps that help patients to understand their disease better
– Apps for communication between doctor and patient
– Apps which have access to clinical information systems

And the FDA defines also applications that are not medical products:

– Apps which are a copy from a book
– Apps which are used for training of hospital personal

All examples are similar to the German Federal Institute, except the spelling and a few deviations [6, Sect. 8] [7].

Medical Applications have advantages for the patient and also disadvantages that are not immediately apparent. The exact determination of the purpose of an application should be used as a help for allocating of the classification.

Actually mobile medical applications are focused on mobile devices such as smartphones, tablets or smart watches. So there is a difference between Health Apps and Medical Apps. Often these two terms are confused and used synonymously. Health Apps are used to the positive and sustainable improvement of physical, mental and social well-being. Medical applications are focused on medical assistance for the purposes of diagnosis, treatment and prevention of disease. Health apps are more general applications. The graph (Fig. 1) shows the exact categorization as an example [8, Sect. 3].

3 Requirement's Analysis

This paper describes the legal requirements to publish a mobile application as a medical device. The focus is on the development process of developing a targeted group-oriented user interface for Mobile Medical Apps. It gives a general overview on the legal status and a detailed explanation for serviceability and usability.

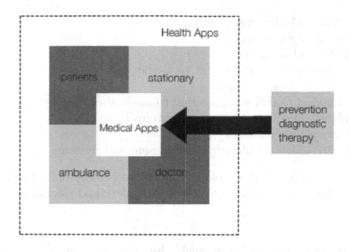

Fig. 1. [8, Sect. 4].

3.1   General Legal Requirements

The work is referred to the two largest markets, the European and the American market. In Europe the Medical Device Directive (MDD) is applied. The implementation of the Medical Devices Directive is a national law in each country and will be implemented by each country itself. In Germany and Austria it is the Medical Device Act.

The parallel tool for the MDD in the USA is the U S Food and Drug Administration (FDA). The main difference is the classification of a medical device based on the risk category. In Europe there are safety classes which could be compared with the Levels of Concern from the FDA. In case of Europe the MDD includes the IEC 62304 Standard for software life cycle processes which defines the following classes:

- Class A: No injury or damage to health is possible
- Class B: None SERIOUS INJURY is possible
- Class C: Death or SERIOUS INJURY is possible [9]

The Level of Concern is defined by the FDA under the Guidance of the "Content of Premarket Submissions for Software Contained in Medical Devices" and has the following types:

- Major: Equal to class C
- Moderate B: Equal to class B
- Minor: Equal to class A [10, Sect. 5]

The following part shows the main differences.

Europe. In Europe is applied the Medical Devices Directive. It specifies requirements covered by international standards. Among them are the following:

- 1 - IEC 62304 - Medical device software
  - Software life cycle processes
- 2 - ISO 14971 - Medical devices
  - Application of risk management to medical devices
- 3 - IEC 62366 - Medical devices
  - Application of usability engineering to medical devices
- 4 - ISO 13485 - Quality management systems
  - Requirements for regulatory purposes
- 5 - IEC 60601 - Medical Electrical Equipment and Systems

If hardware is also included, developers must observe the IEC standard 60601. The 60601 defines the standards for electrical safety and electromagnetic compatibility.

USA. In America the FDA is responsible for classifying and the certification of medical devices. The FDA requirements include the following:

- Quality System Regulations (21 CFR part 820)
- Guidance Document Software Validation
- Guidance Document Contents of Premarket Submissions for Software Contained in Medical Devices
- Guidance Document Cybersecurity

Each point is a kind of guide which provides a framework and some procedures. The scope of content depends on the corresponding risk of the application. In Europe a mobile application needs to be classified in a risk class. In addition to the requirements of the FDA there are further requirements.

FTC: The FTC (Federal Trade Commission) is essentially responsible for regulating competition for law and consumer protection.

FTCs Health Breach Notification Rule: This requirement regulates notifications in the event of incidents, including the misuse of health data.

HIPAA Act: Establishes the standards for data protection and security for the use and exchange of health data.

4   Usability Engineering

This is about the requirements of the usability engineering. The paper is oriented to the European standards and shows the differences from the FDA and the medical device act.

4.1   Requirements Europe

The usability engineering analysis is also regulated by the Medical Device Act. With that, developers have to document the whole process in a medical product file. This file presents the whole documentation and results of the analysis.

Annex 1 from the Guidelines contains the general requirements of products. If it is in use, it must be designed and manufactured under the anticipated conditions and the foreseen purposes. The products do not compromise the clinical condition or the safety of patients or the safety and health of users and third parties [11, Annex 1 General requirements].

The suitability for use depends on the standard DIN EN 62366. Developers are attended by them to comply with certain rules and different requirements. This includes in general:

– Avoid accidental operation
– Acting patient-oriented
– Creating a serviceability act

The product should also satisfy the basic requirements to be successful on the market. These include simple operation and user-friendliness of the product. The usability analysis process includes the following points in chronological order:

– Raise requirements
  • User groups
  • Set intended use
  • Derive usage requirements
– Design Interface
  • Describe usage scenarios
  • Specify user interface
  • Design prototype and check
– Validation: check serviceability [12, Sect. 162]

In addition to the usability analysis there are other standards. So the next logic step would be the processing of the risk analyses, following the quality management and the processes of the software lifecycle. Ideally, all the areas are processed in parallel.

Artifacts of Documentary Serviceability. It should be clarified which requirements have to be met by the standard IEC 62366. The explanation will proceed according to the points of the standard IEC 62366.

Chapter 5.1 - Specification of the application. The intended use and the context in which the product will be used must be explained. The specification of the application occurs by defining the user groups, the Persona and the identification of context scenarios.

Chapter 5.2 - Frequently used functions. The functions which are mainly used by the user must be documented.

The documentation of the most frequently used functions is achieved by identifying the requirements, usage requirements and the core tasks.

Chapter 5.3.1 - Determination of safety-related features. This is a requirement of the risk analysis measures that should be taken to ensure the main functions.

Chapter 5.3.2 - Identifying unknown or foreseeable hazards and hazardous situation. Based on the current knowledge of the user groups, context scenarios and usage scenarios, potential errors and hazards have to be analyzed. This is also a part of the risk analysis.

Chapter 5.4 - Main functions. They consist of the commonly used features, as well as safety-critical functions. A complete editing requires the results of risk analysis.

Chapter 5.5 - Specification of merchantability. This requirement is satisfied with the processing of usage scenarios and the design of the user interface. The result must be used for the validation.

Chapter 5.6 - Validation plan for usability. The test specifications for the validation should be defined. It must be described what procedures are tested and what are the appropriate success and acceptance criteria. The validation must involve all the main functions.

Chapter 5.7 - Design and implementation of the user interface. Design of the user interface in a graphical interface.

Chapter 5.8 - Verification of suitability for use. The suitability for use must be tested according to the standard and by using inspection procedures.

Chapter 5.9 - Validation of serviceability. The serviceability act needs to be checked in compliance with the specification by persons who were not involved in the development process.

4.2   Documentation of the Serviceability Act

The Serviceability Act must have the following documents according to the IEC 62366 standard:

- Specification of the application
    - Frequently used functions
    - Main operations
    - Safety related characteristics
- Identified hazards

- Specification of merchantability
- Validation plan for serviceability
- Shape and design of the user interface product
- Verification and validation of serviceability
- Accompanying documents and training materials

The act includes a large overlap with the risk analysis. Often other documents from standards extend the document of serviceability. The operation and handling of the product is in focus. Therefore, documents for training or with complementary knowledge must be attached.

4.3 Differences to FDA

The FDA requirements are less detailed compared to the European standards. How the serviceability act has to look like is described by FDA in the guide for Applying Human Factors and Usability Engineering to optimize Medical Device Design [13].

The main differences are the prescribed methods. For example, the FDA requirements are using methods for identifying and evaluating relevant hazards already during the development. In the serviceability act from Europe, this must happen only in retrospect. The FDA wants developers to work and test their interface during the design process with mock-ups of the user interfaces. This should help to understand and detect faster problems and errors.

4.4 Compared Requirements

The table (Fig. 2) shows the points which must be fulfilled by the FDA and compared with the Europe act. The table is from the "Applying Human Factors and Usability Engineering to optimize Medical Device Design" [13, p. 34].

Column 2: Descriptions of intended device users, uses, use environments, and training. Will be completed by the European serviceability in his Standard by Chapter 5.2 - met frequently used functions.

Column 3: Description of device user interface. Will be completed by the European serviceability in his Standard by Section 5.5 and 5.8 - met specification verification of serviceability.

Column 4: Summary of known use problems. Is by the European act by Section 5.3.2 - met identifying known or foreseeable hazards and hazardous situations.

Column 5: Analysis of hazards and risks associated with use of the device. This is not covered by the usability analysis. The Europe act requires a risk analysis. The FDA describes that there must be a reference to a risk analysis. In the Medical Device Directive from Europe the risk analysis is directly outsourced to a separate standard and task.

Table A-1. Outline of HFE/UE Report

Sec.	Contents
1	**Conclusion** The <device> has been found to be safe and effective for the intended users, uses and use environments. • Brief summary of HFE/UE processes and results that support this conclusion • Discussion of residual use-related risk
2	**Descriptions of intended device users, uses, use environments, and training** • Intended user population(s) and meaningful differences in capabilities between multiple user populations that could affect user interactions with the device • Intended use and operational contexts of use • Use environments and conditions that could affect user interactions with the device • Training intended for users
3	**Description of device user interface** • Graphical representation of device and its user interface • Description of device user interface • Device labeling • Overview of operational sequence of device and expected user interactions with user interface
4	**Summary of known use problems** • Known use problems with previous models of the subject device • Known use problems with similar devices, predicate devices or devices with similar user interface elements • Design modifications implemented in response to post-market use error problems
5	**Analysis of hazards and risks associated with use of the device** • Potential use errors • Potential harm and severity of harm that could result from each use error • Risk management measures implemented to eliminate or reduce the risk • Evidence of effectiveness of each risk management measure
6	**Summary of preliminary analyses and evaluations** • Evaluation methods used • Key results and design modifications implemented in response • Key findings that informed the human factors validation test protocol
7	**Description and categorization of critical tasks** • Process used to identify critical tasks • List and descriptions of critical tasks • Categorization of critical tasks by severity of potential harm • Descriptions of use scenarios that include critical tasks
8	**Details of human factors validation testing** • Rationale for test type selected (i.e., simulated use, actual use or clinical study) • Test environment and conditions of use • Number and type of test participants • Training provided to test participants and how it corresponded to real-world training levels • Critical tasks and use scenarios included in testing • Definition of successful performance of each test task • Description of data to be collected and methods for documenting observations and interview responses • Test results: Observations of task performance and occurrences of use errors, close calls, and use problems • Test results: Feedback from interviews with test participants regarding device use, critical tasks, use errors, and problems (as applicable) • Description and analysis of all use errors and difficulties that could cause harm, root causes of the problems, and implications for additional risk elimination or reduction

Fig. 2. Anforderungstabelle der FDA [13, Sect. 32]

Column 6: Summary of preliminary analyses and evaluations. Is by the European act met by Section 5.8.

Column 7: Description and categorization of critical tasks. This is not covered by the usability analysis. Here should the risk analysis be used. Thus, there is no additional work because under the serviceability analysis, the risk analysis must be performed.

Column 8: Details of human factors validation testing. Is partially met by the European act in Chapter 5.6 - Validation plan for serviceability and Chap. 5.9 - met validation of serviceability.

5   Conclusion

The development of a medical app lays a heavy financial burden on developers. Therefore the usability should not be developed by a single person. A team of various departments should set out and plan all aspects in close cooperation.

For a full and correct processing of the usability engineering the following work areas should be covered with one person:

- User Experience Designer
- Usability Designer
- Graphic Designer
- Concept Designer

In parallel, additional teams should carry out the processing of the risk analysis. Ideally both teams complement each other with their outcomes. A lot of time and money can be saved when persons take a look into the detailed requirements before the engineering process will start. The usability engineering process from the FDA is not as clearly defined as the standards from Europe. They do not give a straight guidance but they have a detailed vision of methods developers have to use. That means, if developers work with the standards from Europe and analyse the usability with the recommended methods from the FDA and analyse the risk process, they satisfy the standards from the FDA and the European Act, but they have to pay attention to the recommended methods from the FDA.

References

1. Kapitza, T.: Megatrend ehealth mobility, vol. 18 (2015). http://link.springer.com/article/10.1007/s00740-015-0041-x
2. Hnisch, T.: eHealth - eine Begriffsbestimmung, pp. 5–10. Springer Fachmedien Wiesbaden, Wiesbaden (2016). http://dx.doi.org/10.1007/978-3-658-12239-3_2
3. BMG-Bund: Gesetzentwurf e-health.pdf (2015). http://www.bmg.bund.de/fileadmin/dateien/Downloads/E/eHealth/150527_Gesetzentwurf_E-Health.pdf. Accessed 10 Aug 2016
4. Andelfinger, P.V.: eHealth: Grundlagen und Bedeutung für die Gesundheitssysteme heute und morgen, pp. 25–29. Springer Fachmedien Wiesbaden, Wiesbaden (2016). http://dx.doi.org/10.1007/978-3-658-12239-3_5
5. PwC: Weltweiter umsatz mit mobile health (mhealth) bis 2017—statistik, February 2012. http://de.statista.com/statistik/daten/studie/387489/umfrage/weltweiter-umsatz-mit-mobile-health-mhealth/. Accessed 08 Aug 2016

6. FDA: Mobile medical applications. http://www.fda.gov/downloads/Medical Devices/.../UCM263366.pdf. Accessed 11 Aug 2016
7. BfArM: Orientierungshilfe medical apps. http://www.bfarm.de/DE/Medizin produkte/Abgrenzung/medical_apps/_node.html. Accessed 12 Aug 2016
8. Gehring, H., Pramann, O., Imhoff, M., Albrecht, U.-V.: Zukunftstrend "medical apps", vol. 57 (2014). http://link.springer.com/article/10.1007/s00103-014-2061-x
9. D. EN, DIN EN 62304 - Medizingerte-Software - Software-Lebenszyklus-Prozesse, October 2013
10. FDA: Guidance for the content of premarket submissions for software contained in medical devices (2005). http://www.fda.gov/downloads/MedicalDevices/ DeviceRegulationandGuidance/GuidanceDocuments/ucm089593.pdf. Accessed 12 Aug 2016
11. Medizinprodukterichtlinie: Richtlinie 93/42/ewg des rates vom 14. juni 1993 ber medizinprodukte, October 2007. http://eur-lex.europa.eu/LexUriServ/ LexUriServ.do?uri=CONSLEG:1993L0042:20071011:DE:HTML. Accessed 20 Nov 2016
12. Johner, C., Hlzer-Klpfel, M., Wittorf, S.: Basiswissen medizinische Software: Aus- und Weiterbildung zum Certified Professional for Medical Software, 2nd edn. dpunkt, Heidelberg (2015). http://deposit.d-nb.de/cgi-bin/dokserv?id=4784566& prov=M&dok_var=1&dok_ext=htm
13. FDA: Applying human factors and usability engineering to medical devices, February 2016. http://www.fda.gov/downloads/MedicalDevices/.../UCM259760. pdf. Accessed 12 Aug 2016

Mobile Health System for Evaluation of Breast Cancer Patients During Treatment and Recovery Phases

Joaquin Ollero[1]([✉]), Jose Antonio Moral-Munoz[2], Ignacio Rojas[3], and Oresti Banos[4]

[1] City, University of London, London, UK
joaquinollerogarcia@gmail.com
[2] Department of Nursing and Physiotherapy, University of Cadiz, Cádiz, Spain
[3] Research Centre for Information and Communication Technologies,
University of Granada, Granada, Spain
[4] Telemedicine Group, Center for Telematics and Information Technology,
University of Twente, Enschede, Netherlands

Abstract. Breast cancer is the most common tumor in western women and statistically 1 out of 8 women will develop breast cancer over their lifetime. Once overcome it, the stage of rehabilitation that the patient should follow is critical to recover from the suffered disease. In this paper, a system composed of three applications, one for smartwatches, one for smartphones and a web application, is presented. Applications for hand-held devices are directed to the patient who is undergoing rehabilitation and allow to monitor parameters of interest, such as the heart rate, energy expenditure and arm mobility, that will indicate whether the rehabilitation process being followed is improving the health of the patient or not. The web application is directed to a medical expert with the objective of tracking rehabilitation conducted by the patients.

Keywords: Mobile health · mHealth system · Android · Android wear · Smartwatch · Smartphone · Breast cancer · Heart rate sensor · Kinematics sensors · Energy expenditure · Arm mobility · Activity recognition

1 Introduction

The Foundation for the National Institutes of Health defines mobile health or mHealth as the delivery of healthcare services via mobile communications devices [1]. The term is conceived as a conglomeration of hardware and software components, see mobile devices, a software platform that serves basic functionality and applications that provide health services to the user [2]. In the recent past, philosophy and traditions were wrong placed: instead of focusing on healthcare, countries and behaviors were centered in sickcare [3]. This is shifting to a more preventive attitude. For example, health insurance in the United States is drifting to a direction in which companies are tending to reward people who exercise more and take care of their well-being [4].

© Springer International Publishing AG 2017
I. Rojas and F. Ortuño (Eds.): IWBBIO 2017, Part II, LNBI 10209, pp. 653–664, 2017.
DOI: 10.1007/978-3-319-56154-7_58

Breast cancer is one of the most common cancers in the world and by far the most frequent cancer among women. It ranks as the fourth leading cause of death from cancer overall as shown in Fig. 1 [5]. Complications of breast cancer therapies are common, like cardiovascular disease, which is fairly frequent and leads to morbidity, poor quality of life or premature mortality [6]. There has been an appreciable reduction in the breast cancer mortality rate over the past two decades, especially among women younger than 50 years of age (3.1% per year), attributable to improvements in early detection and treatment [7]. After treatment for breast cancer, follow-up care is important to help maintain good a health, manage any side effects from treatment, watch for signs that the cancer has come back after treatment and screen for other types of cancer. A supervised care plan may include regular physical examinations and other medical tests to monitor the recovery during the coming months and years.

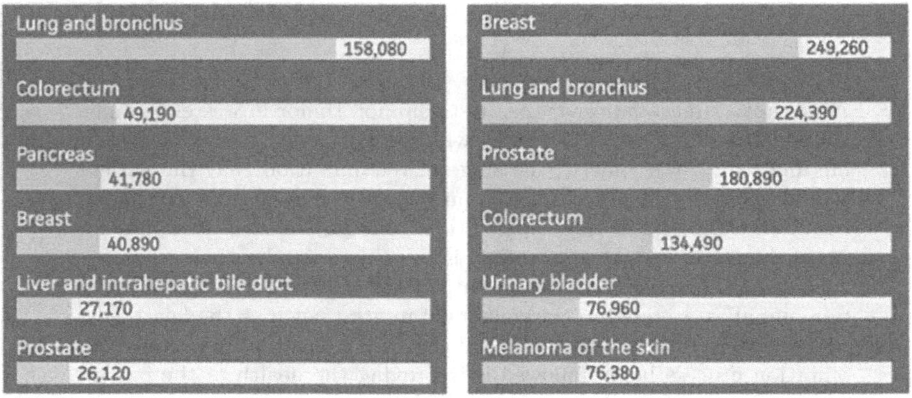

Fig. 1. *(Left)* Estimated deaths and *(right)* new cancer cases in 2016 by cancer type, both sexes combined. American Cancer Society.

Technology can greatly help those women who have had breast cancer and have overcome it. For example, Fitbit, an American company known for products like activity trackers, wireless-enabled wearable technology devices that measure data such as the number of steps walked, heart rate, quality of sleep, steps climbed, and other personal metrics, have announced a collaboration by partnering with the Dana-Farber Cancer Institute for breast cancer research [8]. The study will attempt to figure out if exercise helps prevent breast cancer from recurring.

Smartwatches are one of the most popular and powerful devices in wearable technology. These devices are effectively wearable computers, running applications using a mobile operating system, as Android Wear. The smartwatch industry is fast growing, from USD 1.3 billion in 2014 to expected 117 billion in 2020 [9]. New generations of watches feature continuous measurement of physiological parameters, such as heart rate. Smartwatches connect with accompanying smartphones and receive messages and notifications that are potentially very useful for ubiquitous monitoring applications. These devices are equipped with smart sensors, facilitating one of the main trends in big data science-the quantified self (QS). The QS community is engaged in self tracking or group tracking of physiological, behavioral, and environmental information [10]. The community shares insights, approaches and algorithms. New sensors and systems enable seamless collection of records and integration in databases that can facilitate data mining and new insights. Several companies have created health kit toolsets [11], such as Google Fit or the Apple HealthKit.

The use of wearable sensors has made it possible to have the necessary treatment at home for patients after sudden attacks of diseases such as heart attacks, sleep apnea, Parkinson disease, cancer and so on. Patients after an operation usually go through the recovery process where they follow a strict routine. All the physiological signals as well as physical activities of the patient can be monitored with the help of wearable sensors [12]. During the rehabilitation stage the wearable sensors through media interfaces may provide audio feedback, virtual reality images and other rehabilitative services. The whole activity can be monitored remotely by doctors, nurses or caregivers [13]. Breast cancer rehabilitation can be linked to wearable technology, specifically smartwatches and smartphones, in order to facilitate the recovery from the disease.

In this work, we present a mobile health system for evaluation of breast cancer patients during treatment and recovery phases. The system comprises three different technologies and applications: an application for a smartwatch running on the Android Wear operating system, an application for a smartphone running on the Android operating system and a web server application that can be accessed from any device with an Internet connection. The patient, from the home environment and through the handheld applications, will be able to monitor parameters of importance and send them to the server so the medical expert can visualize them.

The rest of the paper is organized as follows. Section 2 presents the design of the system including the top level view of the system and the description of the theoretical methods used to obtain heart rate, energy expenditure and arm mobility of the patient. Also in this section, a detailed description of the implementation of the system, which consists of by the smartwatch, smartphone and the web applications, is introduced. The global system and the tests conducted to prove its functionality are evaluated in Sect. 3. Finally, achieved aims and future work are presented in the last section of this paper.

2   System

2.1   Design

The objective of this paper is the design and development of an automatic monitoring system for breast cancer survivors, have overcome it and are undergoing rehabilitation. The system obtains useful data of heart rate, energy expenditure and mobility of the affected arm, through which an expert can reach conclusions based on its knowledge of the rehabilitation that the patient is carrying out.

These data of interest are obtained by heart rate, accelerometer and gyroscope sensors embedded into a smartwatch, the LG G Watch R shown in Fig. 2, running on the Android Wear operating system. This information is transmitted to a smartphone running on the Android operating system where data is processed into useful information. Finally, these data are sent to a remote server where they are stored in a database (which also hosts personal patient data available only to the medical expert). Subsequently, the data can be displayed in the form of charts in an accessible web interface to the expert from any mobile device or computer with Internet connection.

Fig. 2. Smartwatch LG G Watch R.

A diagram of the system comprising the three different applications is shown next (Fig. 3):

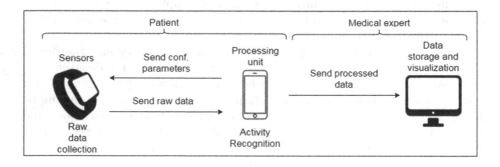

Fig. 3. Top level view of the system.

Each actor of the system has its own set of requirements. Patients need to monitor specific data to provide the doctor with treated information that can easily be interpreted in order to return medical feedback. Due to this reason, information regarding monitoring is displayed to the patient and it is the medical expert the only one who will be able to visualize the performed tests.

Heart Rate. Breast cancer therapy alters autonomic function and contributes to cardiovascular disorders. According to the American College of Cardiology and American Heart Association, oncology patients who have received chemotherapy represent a high-risk for heart failure [14].

Heart rate variability (HRV) is an important noninvasive index of vagal-nerve response and a potential stress marker which can be a useful test for autonomic imbalance. It represents the time differences between beat-to-beat intervals, synonymous with RR variability (RR interval in the electrocardiogram). The analysis of the time differences between successive heartbeats can be accomplished with reference to time i.e. time-domain analysis or frequency i.e. frequency-domain analysis. The parasympathetic and sympathetic nervous systems, which are primarily responsible for changes in the inter beat interval, modify the heart rate and can be quantified by using HRV time- and frequency-domain parameters [7].

The initial objective of the system was to calculate heart rate variability due to its crucial importance but it was not possible to attempt because of the technical limitations of current smartwatches, also the device used here. The LG G Watch R smartwatch, equipped with a PPG heart rate sensor but unable to return the raw signal which is needed to calculate HRV. This device running on the Android Wear operating system only returns the heart rate values expressed in beats per second. Instead, the developed system retrieves heart rate values according to a sampling rate and calculates the average of this set of data while at the same time computes the standard deviation, quantifying the amount of variation or dispersion of this set of data values. These results do not replace the calculation of HRV but they quantify how the heart rate varies over time, which supports the initial objective.

Energy Expenditure. Accelerometers, as the one included on smartwatches, satisfy many of the requirements for physical activity assessment, such as the possibility of measuring physical activities in free-living conditions with minimal discomfort for the subject and in a representative time frame for the average activity level.

The metabolic equivalent of task (MET) is currently the most used indicator for measuring the energy expenditure (EE) of a physical activity (PA) and has become an important measure for determining and supervising the state of health of an individual.

A physical activity can be measured with accelerometers through calculation of counts. Each count is a value that indicates the strength of a movement and can be used in conjunction with other parameters (height, weight, age, gender) to determine the related METs to a PA and thus the EE. The number of METs can be calculated through the number of counts obtained from information from the accelerometers and the aforementioned physiological parameters.

The system has implemented the algorithm expressed in [15], calculating activity counts of a linear accelerometer reading and transform them into METs. First, raw accelerometer values (x, y, z) are collected for a certain period of time with a sampling rate of 15 Hz. Then the linear acceleration of those values is calculated (with low-pass and high-pass filters) and has to be normalized:

$$\sqrt{(linear\_accel_x)^2 + (linear\_accel_y)^2 + (linear\_accel_z)^2}. \quad (1)$$

After this, an integration process is applied to calculate the area under the curve. Using the trapezoidal rule, the sum of the areas returns the total number of counts:

$$\int_a^b f(x)\,\mathrm{d}x = (b-a) \cdot \frac{f(a) + f(b)}{2}. \quad (2)$$

Once obtained the counts, METs are calculated through the following mathematical formula (for all age-groups combined) with the physiological information (height, weight and age) of the patient:

$$METs = 2.7406 + 0.00056 \cdot ACs - 0.008542 \cdot age - 0.01380 \cdot bodymass. \quad (3)$$

where ACs are the activity counts measured in $counts \cdot min - 1$, age measured in years and the body mass index computed as it follows:

$$BMI = \frac{mass(kg)}{height^2(m)}. \quad (4)$$

Arm Mobility. After breast cancer surgery, some women experience numbness, swelling, weakness, or tingling in the arm and shoulder area on the same side of the body on which surgery was conducted. Wearable sensor technology has enabled unobtrusive monitoring of arm movements of different diseases survivors, like stroke survivors, in the home environment, with accelerometry representing the most established approach [16].

Following the method proposed by [17], the system qualitatively assesses functional arm use in the home environment, relying on only a single wrist-worn sensor module, like a smartwatch equipped with both accelerometer and gyroscope sensors. In order to achieve this, first, raw accelerometer and gyroscope values (x, y, z) are obtained for a certain period of time with a sampling rate of 50 Hz. Then, the forearm orientation relative to the earth referential frame is calculated using the gradient descent orientation filter proposed by Madgwick [18], which fuses sensor measurements of gravity and angular rate into an optimal orientation estimate. The filter outputs orientation in a quaternion representation, $q = [q_0, q_1, q_2, q_3]$, which has to be transformed into a 3×3 direction cosine matrix R. To calculate the forearm elevation, the forearm vector expressed in the earth fixed referential is needed:

$$a_e = R^T[1, 0, 0]^T. \quad (5)$$

The elevation Θ of the forearm vector $a_e = [a_{ex}, a_{ey}, a_{ez}]$ can be computed as:

$$\Theta = arctan(\frac{a_{ez}}{\sqrt{a_{ex}^2 + a_{ey}^2}}). \quad (6)$$

and then described in a polar representation. The system calculates the elevation Θ of the forearm activity measuring between -90 and $90°$ on the vertical axis.

2.2 Implementation

The standalone system comprises different devices and technologies. Its design involves three different applications that are connected and synchronized to offer the desired functionality. Through the use of these applications, the patient will be able to monitor automatically valuable information about the rehabilitation that is in current progress and the medical expert will have an adaptive environment to check the data generated by the patient.

Smartwatch Application. Android Wear is a version of the Android operating system of Google designed for smartwatches and other wearables. By pairing with mobile phones running Android version 4.3 or newer (by the accompanying Android Wear app that should be downloaded and configured on the phone), Android Wear integrates mobile notifications and other features. Also, it adds the ability to download applications from the Google Play Store. A differentiating factor over other devices is the large number of sensors included in such devices. For example, a heart rate monitor is a usually included working sensor that can only be found in specific devices prepared to calculate this information. The intrinsic design of smartwatches makes the use of such sensor in these devices ideal.

First, the smartwatch application must receive, from the mobile phone application, the configuration parameters, monitoring time and frequency time, of the desired reading or readings to be performed. Then, it should monitor the desired physiological information (heart rate, as shown in Fig. 4, energy expenditure or arm mobility) of the patient through the integrated sensors of the device and send this data to the mobile phone application. It is essential that it shows a friendly and intuitive user interface (UI), since the potential user of this application may not be experienced with the use of a smartwatch. The UI is divided into three different screens dedicated to display heart rate, energy expenditure and arm mobility respectively.

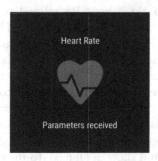

Fig. 4. Smartwatch application ready to monitor heart rate data.

Smartphone Application. Android is a key technology in the system since the Android Wear version is used on the smartwatch and the mobile operating system acts as a gateway between the wrist device and the web server application. The mobile phone application is represented as the processing unit. A compatible Android device that runs Android 2.3 or higher will run the system because it will be able to use the Google Play services APIs to access the Wearable API [19], the Activity Recognition API [20] and the HTTP library Volley [21] which is used to transmit data to the web server application.

This application presents a home screen with a formulary to be filled in with personal information that is needed to calculate and contextualize heart rate, energy expenditure and arm mobility. Then, it is required to send to the smartwatch application the configuration parameters (monitoring time and frequency time) of the heart rate, energy expenditure and arm mobility desired readings. Consequently, the handheld application must receive, from the smartwatch application, all the data generated by the sensors of this device. Then, it processes all the information received from the watch according to each methodology and to build valuable objects containing results of those processings of heart rate, energy expenditure and arm mobility. Additionally, it uses the Activity Recognition API to detect the activity that patients are performing at the same time they are calculating their heart rate, energy expenditure or arm mobility in order to contextualize this information and make it more valuable. The resulting objects of this process are sent to the web server application so the medical expert can graphically visualize this information related to each patient. It presents a configurable and easy-to-use user interface, where the patient can select the configuration parameters and check information about the readings that are in progress. The UI is divided between a home screen with a formulary and three different screens contained in tabs dedicated to both heart rate, energy expenditure and arm mobility respectively.

Web Server Application. A web server application has been implemented in order to give access to the medical expert in a comfortable way so it is possible to check out all the information about its related patients and the tests they have performed from their handheld applications. The web server application has been implemented in Python and Flask programming languages for the back-end. MongoDB has been used to build the databases and the front-end has being built in HTML5, Bootstrap, nvd3 (for charts) and JavaScript.

The main goals of the web server application are to provide a server that can be hosted anywhere (local or cloud) prepared to receive data from the mobile phone application. It must hold a database system to manage the personal information of patients and their related tests performed on the smartwatch and mobile phone applications and show in an easy-to-understand manner the charts of the performed tests in order to obtain conclusions about the rehabilitation. that patients are following. In general terms, it presents a front-end application where the medical expert can log in and manage all the information that is stored in the databases. It shows menus, tabs, formularies, tables and charts to ease the use of the platform.

3   Evaluation

In this section some initial tests are presented to prove the functionality of the system. The tests were performed on healthy patients in a supervised environment while performing daily activities such as being still or walking. The devices used to attempt these tests are the LG G Watch R smartwatch, the Samsung Galaxy A5 smartphone and any device (computer, mobile phone or tablet) with Internet connection. The graphical representation of these tests shows the functionality of the system. The displayed data began being monitored on the smartwatch, passing through the smartphone application for proper processing and ending in the web server application to be stored and represented in charts.

Heart rate tests were performed with a monitoring time of three minutes and a frequency time in intervals of 30 s. Three different charts result from this monitorization: a chart for the average of heart rate as shown in Fig. 5, another one for the standard deviation of the heart rate data and the activity recognition detected while performing the experiment. Tests show that the heart rate relates according to activities carried because it increased when the user was not still (from second 30).

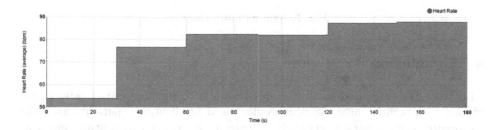

Fig. 5. Heart rate (average) chart.

Energy expenditure tests were performed under the same parameters for the monitoring and frequency times, resulting in a chart for the average of activity counts observed as shown in Fig. 6, another one for the METs that are calculated from the activity counts and finally the activity recognition while performing the readings. The activity counts relate to the activities recognized by the smartphone because first more intense activities were performed, then more relaxed ones.

The arm mobility tests were performed during 1 min of monitorization time, handling all the data generated by the sensors. This results in an arm mobility chart, as shown in Fig. 7 that relates to the real movements of the arm, because it the arm was still at the beginning and then it was moving up and down continuously.

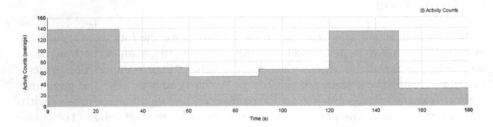

Fig. 6. Activity counts (average) chart.

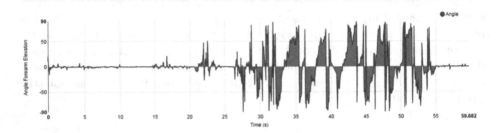

Fig. 7. Arm mobility chart.

4   Conclusion

This work has presented a self contained system to monitor breast cancer survivors. Comprising three different devices and applications, patients are able to monitor crucial parameters such as heart rate, energy expenditure and arm mobility from a smartwatch device running on the Android Wear operating system, send the sensor generated data to a mobile phone application running on the Android operating system so the data is processed and finally send this information to a web server application. A medical expert in charge of these patients is able to access to a multiplatform web server application, to manage patients, visualize performed readings and tests that the patients have performed comfortably in their home environments, without the need of visiting to a medical center.

The basic functionality of the system and all its applications have been achieved by developing a standalone system that allows monitoring of crucial parameters from sensors, such as heart rate, energy expenditure and arm mobility, to supervise patients recovery from breast cancer. Efficient communication has been implemented between the developed applications on different devices and as well as storage mechanisms to save physiological information of patients and tests performed on a remote web server application. All this is presented on user-friendly interfaces to make the use of the applications as easy and comfortable as possible through accessible and interactive elements such as tabs and charts.

Given the huge importance of the wearable technology applied to the health environment, more specifically to the breast cancer and other diseases monitoring, this functional, compact and flexible system could be improved with new functionalities. Some of them are highlighted as it follows. First, it would be important to calculate HRV due to its importance for functional status assessment and to implement the activity recognition on the smartwatch instead of the mobile phone application because it is the smartwatch that generates the sensor data, so the totality of information would come from the same source (available when Android Wear 2.0 is released). Regarding the UIs of the smartwatch and smartphone applications, in order to make the monitoring process completely automatic e.g. handled by the medical expert, they would disappear, so the user does not have to input any parameter making the use of these applications easier and free of obligations. Finally, it has been proposed to conduct experiments with patients in real life to validate the real functionality of the system. This study should be performed with a number of breast cancer survivors and it would thrown interesting and decisive conclusions about the functioning of the system in order to improve it and make it more accessible to the users that use it.

References

1. Merrell, R.C., Doarn, C.R.: m-health. Telemedicine e-Health **20**(2), 99–101 (2014)
2. Liu, C., Zhu, Q., Holroyd, K.A., Seng, E.K.: Status and trends of mobile-health applications for ios devices: a developer's perspective. J. Syst. Softw. **84**(11), 2022–2033 (2011)
3. Sun, M., Burke, L.E., Mao, Z.-H., Chen, Y., Chen, H.-C., Bai, Y., Li, Y., Li, C., Jia, W.: ebutton: a wearable computer for health monitoring and personal assistance. In: Proceedings of the 51st Annual Design Automation Conference on Design Automation Conference, pp. 1–6. ACM (2014)
4. Mekky, S.: Wearable computing and the hype of tracking personal activity (2013)
5. Globocan 2012: Estimated cancer incidence, mortality and prevalence worldwide in 2012. http://globocan.iarc.fr/old/FactSheets/cancers/breast-new.asp. Accessed 23 Jan 2017
6. Arab, C., Dias, D.P.M., de Almeida Barbosa, R.T., de Carvalho, T.D., Valenti, V.E., Crocetta, T.B., Ferreira, M., de Abreu, L.C., Ferreira, C.: Heart rate variability measure in breast cancer patients and survivors: a systematic review. Psychoneuroendocrinology **68**, 57–68 (2016)
7. Caro-Moran, E., Fernandez-Lao, C., Galiano-Castillo, N., Cantarero-Villanueva, I., Arroyo-Morales, M., Rodriguez, L.D.: Heart rate variability in breast cancer survivors after the first year of treatments a case-controlled study. Biol. Res. Nurs. **18**(1), 43–49 (2016)
8. A new breast cancer study enlists fitbit trackers. https://blog.fitbit.com/a-new-breast-cancer-study-enlists-fitbit-trackers/. Accessed 5 July 2016
9. Jovanov, E.: Preliminary analysis of the use of smartwatches for longitudinal health monitoring. In: 37th Annual International Conference of the IEEE on Engineering in Medicine and Biology Society (EMBC 2015), pp. 865–868. IEEE (2015)
10. Banos, O., Amin, M.B., Khan, W.A., Afzal, M., Hussain, M., Kang, B.H., Lee, S.: The mining minds digital health and wellness framework. Biomed. Eng. Online **15**(1), 165–186 (2016)

11. Banos, O., Villalonga, C., Garcia, R., Saez, A., Damas, M., Holgado, J.A., Lee, S., Pomares, H., Rojas, I.: Design, implementation and validation of a novel open framework for agile development of mobile health applications. Biomed. Eng. Online **14**(S2:S6), 1–20 (2015)

12. Banos, O., Villalonga, C., Damas, M., Gloesekoetter, P., Pomares, H., Rojas, I.: Physiodroid: combining wearable health sensors and mobile devices for a ubiquitous, continuous, and personal monitoring. Sci. World J. **2014**(490824), 1–11 (2014)

13. Mukhopadhyay, S.C.: Wearable sensors for human activity monitoring: a review. IEEE Sensors J. **15**(3), 1321–1330 (2015)

14. Bovelli, D., Plataniotis, G., Roila, F., ESMO Guidelines Working Group, et al.: Cardiotoxicity of chemotherapeutic agents and radiotherapy-related heart disease: ESMO clinical practice guidelines. Ann. Oncol. **21**(Suppl. 5), v277–v282 (2010)

15. Ruiz-Zafra, A., Orantes-González, E., Noguera, M., Benghazi, K., Heredia-Jimenez, J.: A comparative study on the suitability of smartphones and IMU for mobile, unsupervised energy expenditure calculi. Sensors **15**(8), 18270–18286 (2015)

16. Patel, S., Park, H., Bonato, P., Chan, L., Rodgers, M.: A review of wearable sensors and systems with application in rehabilitation. J. Neuroeng. Rehabil. **9**(1), 21 (2012)

17. Leuenberger, K., Gonzenbach, R., Wachter, S., Luft, A., Gassert, R.: A method to qualitatively assess arm use in stroke survivors in the home environment. Med. Biol. Eng. Comput. **55**, 141–150 (2016)

18. Madgwick, S.O.H., Harrison, A.J.L., Vaidyanathan, R.: Estimation of IMU and MARG orientation using a gradient descent algorithm. In: 2011 IEEE International Conference on Rehabilitation Robotics, pp. 1–7. IEEE (2011)

19. Wearable API. https://developers.google.com/android/reference/com/google/android/gms/wearable/Wearable. Accessed 23 Jan 2017

20. Activity recognition API, November 2016. https://developers.google.com/android/reference/com/google/android/gms/location/ActivityRecognitionApi. Accessed 23 Jan 2017

21. Transmitting network data using volley. https://developer.android.com/training/volley/index.html. Accessed 23 Jan 2017

Key Factors for Innovative Developments on Health Sensor-Based System

Maria Dolores Peláez[1], Miguel López-Medina[1], Macarena Espinilla[2], and Javier Medina-Quero[2(✉)]

[1] Council of Health for the Andalucian Health Service,
Av. de la Constitución 18, 41071 Sevilla, Spain
{mdolores.pelaez.sspa,miguel.lopez.medina.sspa}@juntadeandalucia.es
[2] University of Jaén, Campus Las Lagunillas s/n, 23071 Jaén, Spain
{mestevez,jmquero}@ujaen.es

Abstract. In the current technological revolution, the proliferation of sensors in smart devices and environments convert users into a real-life data source that ranges from the monitoring of vital signs to the recognition of their lifestyle, behavior and health. In this work, we describe current trends and issues on innovative healthcare systems, which are integrating wearable devices and smart environments into numerous health applications. The report includes a revision of the literature with academic, technical and legal concerns on the development of health solutions.

Keywords: Health technology · Sensor-based systems · Disruptive innovation

1 Introduction

In this work, we present new trends and entrepreneurship opportunities for innovative developments [1] on Health Systems based on the disruptive changes [2] in health systems [3], which is being developed by the deployment of sensor and devices whose accessibility spreads to much of the population of the developed world. The disruptive innovation is key for the next generation of Health System because of being *needed for organizations to survive dynamic and complex markets and uncertain economic situations* [4] and contributing to higher organizational sustainability [5]. Disruptive innovations are usually driven by entrepreneurs and innovative developments due to the initial risk profits and their scarce-resource development [6]. However, when the innovation successes, it achieves a much deeper penetration and impact on individual, functional, company and market levels [7].

The aim of the paper is describing the key factors where the academic world and the entrepreneurship can merge in order to develop the future generation of Health Services. Specifically, we analyze the sensor-based systems for mobile and wearable devices, as well as, within smart environments, emphasizing the challenges and opportunities in these sectors.

© Springer International Publishing AG 2017
I. Rojas and F. Ortuño (Eds.): IWBBIO 2017, Part II, LNBI 10209, pp. 665–675, 2017.
DOI: 10.1007/978-3-319-56154-7_59

The remainder of the paper is structured as follows: in Sect. 2 we describe the issues for innovative developments of health applications for mobile and wearable devices; in Sect. 3, we analyze more complex scenarios of smart environment where the innovation in health services require handling with modelling, storing, privacy an legal issues; finally, in Sect. 4, we present the conclusions of the report.

2   Health Sensors and Devices for Describing the Life of the People

The penetration of mobile devices has enable capabilities of the Ubiquitous Computing [8] in our daily activities. Additionally, sensors embedded at mobile devices and their pervasive communication *enable new applications across a number of sectors but particularly in personal healthcare* [9]. In this way, mobile technologies [10] has been demonstrated to be effective for changing health behaviors [11], for example, increasing exercise time, knowledge of health or weight loss in patients [12].

On a recent stage, wearable devices [13] have emerged as smart portable sensors analyzing user activity by means of reliable measures, such as movement, steps or vital signs, such as hearth rate. In health context, they have provided a new perspective of real-time monitoring for decision support systems and prognosis [14] becoming facilitators of health behavior [15]. We highlight real case studies, which have been carried out by means of remote monitoring on treatments of patients with cardiovascular diseases [16,17].

Moreover, the irruption of these health devices and systems have break down traditional business model translating to an innovative perspective which handle this disruptive process. For that, the report of the European Medical Technology industry [18] has taken a snapshot of development of health devices noting that:

- Prices for implantable medical devices have been reduced between 17% to 34%.
- 25,000 companies are focused on medical technology, where 95% are small and medium-sized enterprises.
- Patents in medical technology represent the largests of all sector (more than 11,000)
- 7.5% of total health expenditure corresponds to medical technologies.
- The Medical Devices Expenditures are centered in USA (40%) and Europe (30%).
- The development life cycle is from 18 to 24 months.

2.1   Ubiquitous and Fog Computing in Mobile Health

Based on the computing capacity of the wearable sensors and health devices, recent paradigms, such as Edge Computing [19] or Fog Computing have translated the computing of data and services within the devices where data are collected [19]. This new perspective displaces the focus from Cloud Computing, with a centralized processing [20], to Internet of Things (IoT) [21,22] with a collaborative network where the Smart objects *interact with each other and cooperate with their neighbors to reach common goals* [23,24].

In these last years, Ubiquitous Computing (UC) [8] has provided a stable development frameworks, to which the capabilities of wearable devices [25] have been integrated into the health applications [15]. These trends have enabled driving remote rehabilitation [26], ubiquitous mobile telemedicine [27,28] or health-monitoring systems [29].

These mobile applications are needing a new perspective to integrate intelligent processes by means of Ubiquitous and Fog Computing. The challenge of mobile health requires locating the information processing of sensors to generate richer and higher-level information [30] in the devices where data are collected in real time. This key issue needs for analysis of raw data, which are summarized and merged with other sensor information by means of context-aware computing [31–33].

On market perspective and enterprise innovation, mobile health applications is currently handling relevant issues, such as: (i) needing for regulatory and funding issues [34], (ii) solving data privacy and security as a major issue on deployments for public health [35], (iii) demonstrating to be cost-effective for handling chronic disease in developing countries [36], (iv) focusing on assistive and monitoring applications, the most frequently used applications [37], (v) integrating collaborative high-quality professional practice protocols [38].

3 Moving Health Care Services to Smart Environments

Smart Environments [39] are interactive spaces where technological devices are adapted to solve daily activities of people. They are developed under networks of physical objects, so Internet of Things (IoT) [22] has recently arisen as a new paradigm where Ambient Intelligence [40] and Ubiquitous Computing [8] converge [41] to provide connected smart things within the Smart Environments.

Among other applications, Smart Environments can provide a successful solution to the ageing population, which is going to raise the percentage of population over 65 up to 15% [42]. In this scenario, the current system, where the health personnel care and supervise patients in an individual way, is untenable. So, smart environments are proposed to help elderly people to stay with the best quality of life as long as possible in their sustainable, healthy and manufacturing homes [43].

The main difference regarding mobile health, is that smart environments analyze a more complete vision of daily-life from users in order to provide an ambient assisted living [44–46]. It is due to the *integration of data from heterogeneous sources* [47], where a wide range of sensors are deployed to collect multiple data from mobile, wearables devices of different users together with ambient devices [48].

Due to this diversity, a key aspect for smart environments is *designing models and structures of knowledge representation*. On semantic health modeling, the development of ontologies have been successfully adapted to human behavior identification [49]. In parallel, other general models have been focused on providing scientist interoperability [50] or the enterprise interoperability [51].

Furthermore, it is necessary to *distribute the information processing of sensors*. The adequate distribution of services in ambient environments is key to provide sensitivity to real time [52] when distributing the information processing in several central processing units [53,54]. In this area, the concept of *middleware* highlight as an infrastructure in which are distributed the sensor streams from ambient and user devices by remote services.

For developing enterprise solutions of smart environments handling this topics, the open source model have resulted as an important initiative making that the disruptive innovation based on open business models [55] had provided high quality tools. We highlight some open tools such as: (i) ZeroC Ice [56], which is an object-oriented distributed computing tool with support for several languages and platforms, (ii) Global Sensor Network (GSN) that applies sliding window management [57] in changing data stream [58], or (iii) W3C Semantic Sensor Networks (SSN) [59,60], an open semantic annotating for sensors.

On the market opportunities, on the first hand, they are focused on translating health care services to smart environments. This ambitious goal aims to solve issues related to dependency and ageing, where we highlight:

- Monitoring chronic diseases, which are suffered by a half of the inhabitants of developed countries [61,62].
- Reducing medication administration errors due to non-compliance with medication instructions, because of (i) arising as cause of approximately 10% of hospitalizations and (ii) producing 23% of hospitalizations of elderly people incomes [63–65].
- Identifying mental disorders by analyzing user activities, such as dementia [66].
- Promoting telenursing, which is demonstrated to *decrease the number of outpatient and emergency room visits, shortening hospital stays, improving health-related quality of life, and decreasing the cost of health care* [67].

On the second hand, these market opportunities can be faced by enterprise innovations for developing *smart health environments*, but in literature several challenges have been noted to be crucial in real deployments:

- Detecting when and where telemedicine is most effective for avoiding ineffective programs [68].
- Training of healthcare professionals in new technology [28].
- Keeping alive the communication between patient and healthcare professionals [69], for example by integrating video conference systems.
- Creating the legal and regulatory infrastructure for telemedicine [68].
- Integrating contact-less technology because of: (i) being now mature to be part of smart environments by means of low-cost and energy-autonomous sensors [70], and (ii) increasing security and minimizing accidents and mistakes [71].

3.1   Big Data, Cloud Services and Security for Health Environments

A special mention is necessary for the services and persistence in smart health environments due to they require handling a vast amount of sensitive data from sensors, which are the key to generate knowledge about patients and illnesses [72]. These huge amount of data is described as Big Data [73], which is on the point of emerging thanks to the proliferation of health sensors and environments [74]. The analysis of these data will be able to provide new researches and to evaluate health care programs using Machine Learning [75]. However it is necessary to solve the key problem of the exponential growth: in 2020, it will only take 73 days for doubling the volume of medical data [76].

For that important reason, the enterprise and innovative solutions, which aim to service a relevant number of environments and patients for health services, needs for handling critical points on health data:

- Including long-term recording of biosignal and sensor streams, by means of recent tools which solve the large-scale of data analysis. For entrepreneurial solutions based on open source, we highlight the great potential of source non-relational, distributed databases, such as Hadoop [77, 78] and Spark [79].
- Including international healthcare informatics interoperability standards, where the most important is Health Level Seven *HL7* [80].
- Integrating the monitoring application in cloud services, making them easily accessible [81].
- Including business intelligence processes. It is a recent and demanding need for *embed analytics into decision-makers across the business gain insight into financial and medical data and become more proactive* [82].
- Adopting the prevention protocols at home by means of the uninterrupted communication with the patients, which allows us to anticipate or predict possible problems, making appropriate decisions at any time. In addition, they optimize the service model itself and reduce the expense of the necessary resources [83].

On the security, privacy an legal issues in electronic health records, the visibility of health data have to be encrypted and restricted to the own users, and in the same way, to be shared with health care personnel from the prior authorization of the owner [84]. These issues are involving relevant works on legal aspects from the real deployments of the innovative health-care systems:

- In [85], the *lack high-quality evidence that supports the adoption of many new technologies and have financial, regulatory, and security hurdles to overcome* is highlighted.
- Base on experiences from England and Australia, the rights and responsibilities of electronic health records need for *moral re-ordering required to transform health care through such means* [86]. In concrete, in data protection of Electronic Health Record (EHR), where *EU and other countries are determined to find solutions, impose policies and standards as to implement EHR at national level and international levels* [87].

Table 1. Technologies, paradigms, applications and issues, which are evolved in sensor-based systems for Health

Technology	Paradigms	Applications	Issues
– Mobile devices	– Ubiquitous Computing	– Access to remote health services and protocols – Hand recording of activity or health status – Knowledge of health process	– Stable mobile application development – Contact-less technology – Low training of health personnel
– Wearable devices – Mobile devices	– Ubiquitous Computing – Fog Computing	– Real time recording of physical activity or health status – Monitoring and physical activity	– Recent development and permanent change – Low familiarity with wearable of health personnel
– Ambient devices – Wearable devices – Mobile devices	– Ubiquitous Computing – Fog Computing – Ambient Intelligence	– Real time recording of daily activity – Telenursering and monitoring rehabilitation at home – Identifying mental disorders at home	– Middleware & Modeling Open tools – Training of healthcare professional
– Big data – Business Intelligence – Ambient devices – Wearable devices – Mobile devices	– Ubiquitous Computing – Fog Computing – Ambient Intelligence	– Evaluation of health care programs – Financial analytics by business intelligence – Identifying mental disorders at home	– Cloud Services – Including interoperability standards – Machine Learning – Permanent changes and uncertain in legal regulation

– In order to provide secure cloud services for health data, it would be necessary to advance in service level agreements for security [88]. These services must include restricted-role access, which permits health personnel to have various view based on the profiles [89].

Once the governments, health institutions and companies agree on a common standardization, the Health Big Data will be a benchmark in health sciences to study and to improve the life of society [90].

4   Conclusions

In this work, we have presented opportunities and key factors on the development of health technology based on mobile, wearable and ambient devices. The most important paradigms, issues and applications have been summarized in Table 1.

In summary, the efforts are focused on two main areas: (i) integrating wearable sensors and mobile applications in daily health activities and (ii) processing huge volumes of data for discovering knowledge and identifying patterns of health problems. Both aim finding answers to current problems of healthcare systems, but to find new questions which improve health of people in the future.

References

1. Abelson, P.H.: A third technological revolution. Science **279**(5359), 2019–2109 (1998)
2. Bower, J.L., Christensen, C.M.: Disruptive technologies: catching the wave, pp. 506–520 (1995). Harvard Business Review Video
3. Schwamm, L.H.: Telehealth: seven strategies to successfully implement disruptive technology and transform health care. Health Aff. **33**(2), 200–206 (2014)
4. Franz, N.K., Cox, R.A.: Time for disruptive innovation. J. Extension **50**(2), 2COM1 (2012)
5. Christensen, C.M., Horn, M.B., Johnson, C.W.: Disrupting Class: How Disruptive Innovation will Change the Way the World Learns, vol. 98. McGraw-Hill, New York (2008)
6. Christensen, C.: The Innovators Dilemma: When New Technologies Cause Great Firms to Fail. Harvard Business Review Press, Boston (2013)
7. Assink, M.: Inhibitors of disruptive innovation capability: a conceptual model. Eur. J. Innov. Manag. **9**(2), 215–233 (2006)
8. Weiser, M.: The computer for the 21st century. Sci. Am. **265**(3), 94–104 (1991)
9. Lane, N.D., Miluzzo, E., Lu, H., Peebles, D., Choudhury, T., Campbell, A.T.: A survey of mobile phone sensing. IEEE Commun. Mag. **48**(9), 140–150 (2010)
10. Malvey, D., Slovensky, D.J.: mHealth: Transforming Healthcare. Springer, New York (2014)
11. Free, C., Phillips, G., Watson, L., Galli, L., Felix, L., Edwards, P., Haines, A.: The effectiveness of mobile-health technologies to improve health care service delivery processes: a systematic review and meta-analysis. PLoS Med. **10**(1), e1001363 (2013)
12. Wantland, D.J., Portillo, C.J., Holzemer, W.L., Slaughter, R., McGhee, E.M.: The effectiveness of Web-based vs. non-Web-based interventions: a meta-analysis of behavioral change outcomes. J. Med. Internet Res. **6**(4), e40 (2004)
13. Lymberis, A., Dittmar, A.: Advanced wearable health systems and applications-research and development efforts in the European Union. IEEE Eng. Med. Biol. Mag. **26**(3), 29–33 (2007)
14. Pantelopoulos, A., Bourbakis, N.G.: A survey on wearable sensor-based systems for health monitoring and prognosis. IEEE Trans. Syst. Man Cybern. Appl. Rev. **40**(1), 1–12 (2010)
15. Patel, M.S., Asch, D.A., Volpp, K.G.: Wearable devices as facilitators, not drivers, of health behavior change. JAMA **313**(5), 459–460 (2015)
16. Szydlo, T., Konieczny, M.: Mobile and wearable devices in an open and universal system for remote patient monitoring. Microprocess. Microsyst. **46**, 44–54 (2016)
17. Albaghli, R., Anderson, K.M.: A vision for heart rate health through wearables. In: Proceedings of the 2016 ACM International Joint Conference on Pervasive, Ubiquitous Computing: Adjunct, pp. 1101–1105. ACM, September 2016
18. MedTech Europe: The European Medical Technology Industry in Figures. MedTech Europe, Brussels (2013)
19. Garcia Lopez, P., Montresor, A., Epema, D., Datta, A., Higashino, T., Iamnitchi, A., Barcellos, M., Felber, P., Riviere, E.: Edge-centric computing: vision and challenges. ACM SIGCOMM Comput. Commun. Rev. **45**(5), 37–42 (2015)
20. Chen, L.W., Ho, Y.F., Kuo, W.T., Tsai, M.F.: Intelligent file transfer for smart handheld devices based on mobile cloud computing. Int. J. Commun. Syst. **30**(1) (2015)

21. Xu, H., Collinge, W.O., Schaefer, L.A., Landis, A.E., Bilec, M.M., Jones, A.K.: Towards a commodity solution for the Internet of Things. Comput. Electr. Eng. **52**, 138–156 (2016)
22. Kopetz, H.: Internet of Things. In: Real-time systems, pp. 307–323. Springer, New York (2011)
23. Atzori, L., Iera, A., Morabito, G.: The Internet of Things: a survey. Comput. Netw. **54**(15), 2787–2805 (2010)
24. Kortuem, G., Kawsar, F., Sundramoorthy, V., Fitton, D.: Smart objects as building blocks for the Internet of Things. IEEE Internet Comput. **14**(1), 44–51 (2010)
25. Lara, O.D., Labrador, M.A.: A survey on human activity recognition using wearable sensors. IEEE Commun. Surv. Tutorials **15**(3), 1192–1209 (2013)
26. Chang, C.-Y., Lange, B., Zhang, M., Koenig, S., Requejo, P., Somboon, N., Sawchuk, A.A., Rizzo, A.A.: Towards pervasive physical rehabilitation using Microsoft Kinect. In: 2012 6th International Conference on Pervasive Computing Technologies for Healthcare (PervasiveHealth) and Workshops, pp. 159–162. IEEE, May 2012
27. Satyanarayanan, M.: Pervasive computing: vision and challenges. IEEE Pers. Commun. **8**(4), 10–17 (2001)
28. Varshney, U.: Pervasive healthcare and wireless health monitoring. Mob. Netw. Appl. **12**(2–3), 113–127 (2007)
29. Custodio, V., Herrera, F.J., Lpez, G., Moreno, J.I.: A review on architectures and communications technologies for wearable health-monitoring systems. Sensors **12**(10), 13907–13946 (2012)
30. Haefner, K.: Evolution of Information Processing Systems: An Interdisciplinary Approach for a New Understanding of Nature and Society. Springer Publishing Company Incorporated, Heidelberg (2011)
31. Emmanouilidis, C., Koutsiamanis, R.A., Tasidou, A.: Mobile guides: taxonomy of architectures, context awareness, technologies and applications. J. Netw. Comput. Appl. **36**(1), 103–125 (2013)
32. Makris, P., Skoutas, D.N., Skianis, C.: A survey on context-aware mobile and wireless networking: on networking and computing environments integration. IEEE Commun. Surv. Tutorials **15**(1), 362–386 (2013)
33. Perera, C., Zaslavsky, A., Christen, P., Georgakopoulos, D.: Context aware computing for the Internet of Things: a survey. IEEE Commun. Surv. Tutorials **16**(1), 414–454 (2014)
34. Dicianno, B.E., Parmanto, B., Fairman, A.D., Crytzer, T.M., Daihua, X.Y., Pramana, G., Coughenour, D., Petrazzi, A.A.: Perspectives on the evolution of mobile (mHealth) technologies and application to rehabilitation. Phys. Ther. **95**(3), 397–405 (2015)
35. Silva, B.M., Rodrigues, J.J., de la Torre Díez, I., López-Coronado, M., Saleem, K.: Mobile-health: a review of current state in 2015. J. Biomed. Inform. **56**, 265–272 (2015)
36. Beratarrechea, A., Lee, A.G., Willner, J.M., Jahangir, E., Ciapponi, A., Rubinstein, A.: The impact of mobile health interventions on chronic disease outcomes in developing countries: a systematic review. Telemed. e-Health **20**(1), 75–82 (2014)
37. Martínez-Pérez, B., De La Torre-Díez, I., López-Coronado, M.: Mobile health applications for the most prevalent conditions by the World Health Organization: review and analysis. J. Med. Internet Res. **15**(6), e120 (2013)
38. Castelnuovo, G., Manzoni, G.M., Pietrabissa, G., Corti, S., Giusti, E.M., Molinari, E., Simpson, S.: Obesity and outpatient rehabilitation using mobile technologies: the potential mHealth approach. Front. Psychol. **5**, 559 (2014)

39. Friess, P.: Internet of Things: Converging Technologies for Smart Environments and Integrated Ecosystems. River Publishers, Denmark (2013)
40. Zelkha, E., Epstein, B., Birrell, S., Dodsworth, C.: From devices to ambient intelligence. In: Digital Living Room Conference, vol. 6, June 1998
41. Marie, P., Desprats, T., Chabridon, S., Sibilla, M.: Extending ambient intelligence to the Internet of Things: new challenges for QoC management. In: Hervás, R., Lee, S., Nugent, C., Bravo, J. (eds.) UCAmI 2014. LNCS, vol. 8867, pp. 224–231. Springer, Cham (2014). doi:10.1007/978-3-319-13102-3_37
42. United Nations, Department of Economic and Social Affairs, Population Division: World Population Ageing (2013). ST/ESA/SER.A/348
43. Branger, J., Pang, Z.: From automated home to sustainable, healthy and manufacturing home: a new story enabled by the Internet-of-Things and Industry 4.0. J. Manag. Anal. 2(4), 314–332 (2014)
44. Yin, J., Tian, G., Feng, Z., Li, J.: Human activity recognition based on multiple order temporal information. Comput. Electr. Eng. 40(5), 1538–1551 (2014)
45. Alam, M.M., Hamida, E.B.: Surveying wearable human assistive technology for life and safety critical applications: standards, challenges and opportunities. Sensors 14(5), 9153–9209 (2014)
46. Van Hoof, J., Wouters, E.J.M., Marston, H.R., Vanrumste, B., Overdiep, R.A.: Ambient assisted living and care in The Netherlands: the voice of the user. In: Pervasive and Ubiquitous Technology Innovations for Ambient Intelligence, Environments, vol. 205 (2012)
47. Bellavista, P., Corradi, A., Fanelli, M., Foschini, L.: A survey of context data distribution for mobile ubiquitous systems. ACM Comput. Surv. (CSUR) 44(4), 24 (2012)
48. Pei, Z., Deng, Z., Yang, B., Cheng, X.: Application-oriented wireless sensor network communication protocols, hardware platforms: a survey. In: IEEE International Conference on Industrial Technology, ICIT, pp. 1–6. IEEE (2008)
49. Villalonga, C., Razzaq, M.A., Khan, W.A., Pomares, H., Rojas, I., Lee, S., Banos, O.: Ontology-based high-level context inference for human behavior identification. Sensors 16(10), 1617 (2016)
50. Nugent, C.D., Finlay, D.D., Davies, R.J., Wang, H.Y., Zheng, H., Hallberg, J., Synnes, K., Mulvenna, M.D.: homeML – an open standard for the exchange of data within smart environments. In: Okadome, T., Yamazaki, T., Makhtari, M. (eds.) ICOST 2007. LNCS, vol. 4541, pp. 121–129. Springer, Heidelberg (2007). doi:10.1007/978-3-540-73035-4_13
51. Weichhart, G., Molina, A., Chen, D., Whitman, L.E., Vernadat, F.: Challenges and current developments for sensing, smart and sustainable enterprise systems. Comput. Ind. 79, 34–46 (2015)
52. Balan, R.K., Satyanarayanan, M., Park, S.Y., Okoshi, T.: Tactics-based remote execution for mobile computing. In: Proceedings of the 1st International Conference on Mobile Systems, Applications and Services, pp. 273–286. ACM, May 2003
53. Verissimo, P., Rodrigues, L.: Distributed Systems for System Architects, vol. 1. Springer Science & Business Media, New York (2012)
54. Liu, G.: Distributing network services and resources in a mobile communications network. U.S. Patent No. 5,825,759. Washington, DC: U.S. Patent and Trademark Office (1998)
55. Chesbrough, H.: Open Business Models: How to Thrive in the New Innovation Landscape. Harvard Business Press, Cambridge (2013)
56. Henning, M.: A new approach to object-oriented middleware. IEEE Internet Comput. 8(1), 66–75 (2004)

57. Salehi, A.: Design and implementation of an efficient data stream processing system, Doctoral dissertation, cole Polytechnique Fdrale de Lausanne (2010)
58. Kifer, D., Ben-David, S., Gehrke, J.: Detecting change in data streams. In: Proceedings of the Thirtieth International Conference on Very Large Data Bases, vol. 30, pp. 180–191. VLDB Endowment, August 2004
59. Compton, M., Barnaghi, P., Bermudez, L., Garcia-Castro, R., Corcho, O., Cox, S., et al.: The SSN ontology of the W3C semantic sensor network incubator group. Web Semant. Sci. Serv. Agents World Wide Web 17, 25–32 (2012)
60. Neuhaus, H., Compton, M.: The semantic sensor network ontology. In: AGILE Workshop on Challenges in Geospatial Data Harmonisation, Hannover, Germany, pp. 1–33 (2009)
61. Ashman, J.J.: Multiple chronic conditions among US adults who visited physician offces: data from the National Ambulatory Medical Care Survey, 2009. Preventing Chronic Dis. 10 (2013)
62. Harbers, M.M., Achterberg, P.W.: Information, Indicators and Data on the Prevalence of Chronic Diseases in the European Union. RIVM, Bilthoven (2012)
63. Veazie, P.J.: An individual-based framework for the study of medical error. Int. J. Qual. Health Care 18(4), 314–319 (2006)
64. Lisby, M., Nielsen, L.P., Mainz, J.: Errors in the medication process: frequency, type, and potential clinical consequences. Int. J. Qual. Health Care 17(1), 15–22 (2005)
65. Lewis, P.J., Dornan, T., Taylor, D., Tully, M.P., Wass, V., Ashcroft, D.M.: Prevalence, incidence and nature of prescribing errors in hospital inpatients. Drug Saf. 32(5), 379–389 (2009)
66. Lotfi, A., Langensiepen, C., Mahmoud, S.M., Akhlaghinia, M.J.: Smart homes for the elderly dementia sufferers: identification and prediction of abnormal behaviour. J. Ambient Intell. Humanized Comput. 3(3), 205–218 (2012)
67. Kamei, T.: Information and communication technology for home care in the future. Jpn. J. Nurs. Sci. 10(2), 154–161 (2013)
68. Kahn, J.M.: Virtual visits - confronting the challenges of telemedicine. N. Engl. J. Med. 372(18), 1684–1685 (2015)
69. Weinstein, R.S., Lopez, A.M., Joseph, B.A., Erps, K.A., Holcomb, M., Barker, G.P., Krupinski, E.A.: Telemedicine, telehealth, and mobile health applications that work: opportunities and barriers. Am. J. Med. 127(3), 183–187 (2014)
70. Amendola, S., Lodato, R., Manzari, S., Occhiuzzi, C., Marrocco, G.: RFID technology for IoT-based personal healthcare in smart spaces. IEEE Internet of Things J. 1(2), 144–152 (2014)
71. Vasquez, A., Huerta, M., Clotet, R., González, R., Rivas, D., Bautista, V.: Using NFC technology for monitoring patients and identification health services. In: Braidot, A., Hadad, A. (eds.) CLAIB 2014. IFMBE Proceedings, vol. 49, pp. 805–808. Springer, Cham (2015)
72. Bates, D.W., Saria, S., Ohno-Machado, L., Shah, A., Escobar, G.: Big data in health care: using analytics to identify and manage high-risk and high-cost patients. Health Aff. 33(7), 1123–1131 (2014)
73. Kuo, M.H., Sahama, T., Kushniruk, A.W., Borycki, E.M., Grunwell, D.K.: Health big data analytics: current perspectives, challenges and potential solutions. Int. J. Big Data Intell. 1(1–2), 114–126 (2014)
74. Chen, M., Ma, Y., Song, J., Lai, C.F., Hu, B.: Smart clothing: connecting human with clouds and big data for sustainable health monitoring. Mob. Netw. Appl. 21(5), 825–845 (2016)

75. Obermeyer, Z., Emanuel, E.J.: Predicting the future - Big Data, machine learning, and clinical medicine. N. Engl. J. Med. **375**(13), 1216–1219 (2016)
76. Bonis, P.: Clinical decision support technology: saving lives. Clin. Serv. J. (2016)
77. Ku, W.Y., Chou, T.Y., Chung, L.K.: The cloud-based sensor data warehouse. In: Proceedings of ISGC 2011 & OGF 31, vol. 75 (2011)
78. Yu, H., Wang, D.: Research and implementation of massive health care data management and analysis based on Hadoop. In: 2012 Fourth International Conference on Computational and Information Sciences (ICCIS), pp. 514–517. IEEE, August 2012
79. Archenaa, J., Anita, E.A.M.: Interactive Big Data management in healthcare using Spark. In: Vijayakumar, V., Neelanarayanan, V. (eds.) ISBCC 2016. SIST, vol. 49, pp. 265–272. Springer, Cham (2016). doi:10.1007/978-3-319-30348-2_21
80. Jin, Z., Chen, Y.: Telemedicine in the Cloud Era: prospects and challenges. IEEE Pervasive Comput. **14**(1), 54–61 (2015)
81. Yadav, S., Chappar, V., Datir, S., Jagtap, P.: An overview of a pervasive and personalized smart health-care system using IoT. Int. Educ. Res. J. **2**(11) (2016)
82. Raghupathi, W., Raghupathi, V.: Big data analytics in healthcare: promise and potential. Health Inf. Sci. Syst. **2**(1), 1 (2014)
83. Lokkerbol, J., Adema, D., Cuijpers, P., Reynolds, C.F., Schulz, R., Weehuizen, R., Smit, F.: Improving the cost-effectiveness of a healthcare system for depressive disorders by implementing telemedicine: a health economic modeling study. Am. J. Geriatr. Psychiatry **22**(3), 253–262 (2014)
84. Fernández-Alemán, J.L., Senor, I.C., Lozoya, P.O., Toval, A.: Security and privacy in electronic health records: a systematic literature review. J. Biomed. Inform. **46**(3), 541–562 (2013)
85. Steinhubl, S.R., Muse, E.D., Topol, E.J.: The emerging field of mobile health. Sci. Transl. Med. **7**(283), 283rv3 (2015)
86. Garrety, K., McLoughlin, I., Zelle, G.: Disruptive innovation in health care: business models, moral orders and electronic records. Soc. Policy Soc. **13**(04), 579–592 (2014)
87. Kaldoudi, E., Drosatos, G., Portokallidis, N., Third, A.: An ontology based scheme for formal care plan meta-description. In: Kyriacou, E., Christofides, S., Pattichis, C.S. (eds.) XIV Mediterranean Conference on Medical and Biological Engineering and Computing 2016. IP, vol. 57, pp. 785–790. Springer, Cham (2016). doi:10.1007/978-3-319-32703-7_153
88. Rong, C., Nguyen, S.T., Jaatun, M.G.: Beyond lightning: a survey on security challenges in cloud computing. Comput. Electr. Eng. **39**(1), 47–54 (2013)
89. Shrestha, N.M., Alsadoon, A., Prasad, P.W.C., Hourany, L., Elchouemi, A.: Enhanced e-health framework for security and privacy in healthcare system. In: 2016 Sixth International Conference on Digital Information Processing and Communications (ICDIPC), pp. 75–79. IEEE, April 2016
90. Murdoch, T.B., Detsky, A.S.: The inevitable application of Big Data to health care. JAMA **309**(13), 1351–1352 (2013)

Time Lapse Experiments and Multivariate Biostatistics

Towards Integration of CFD and Photosynthetic Reaction Kinetics in Modeling of Microalgae Culture Systems

Štěpán Papáček[1(✉)], Jiří Jablonský[1], and Karel Petera[2]

[1] Faculty of Fisheries and Protection of Waters,
South Bohemian Research Center of Aquaculture and Biodiversity of Hydrocenoses,
Institute of Complex Systems, University of South Bohemia in České Budějovice,
Zámek 136, 373 33 Nové Hrady, Czech Republic
spapacek@frov.jcu.cz, jiri.jablonsky@gmail.com
[2] Faculty of Mechanical Engineering, Czech Technical University in Prague,
Technická 4, 166 07 Prague 6, Czech Republic
Karel.Petera@fs.cvut.cz

Abstract. Despite the growing interest in photosynthetic microalgae, e.g., as a source of biofuels, the mass cultivation of microalgae still exhibits many drawbacks and successful industrial applications are scarce. Reliable model integrating all relevant phenomena and enabling to assess the system performance, its in silico scale-up and eventually optimization, is still lacking. Here, we present a unified modeling framework for microalgae culture system. We propose a multidisciplinary modeling framework to bridge biology (cell growth) and physics (hydrodynamics and optics) together. This framework consists of (i) the state system (mass balance equations in form of advection-diffusion-reaction PDEs), (ii) the fluid flow equations (Navier-Stokes equations), and (iii) the irradiance distribution. To validate the method, the Couette-Taylor reactor, which hydrodynamically induces the fluctuating light conditions, was chosen. The results of numerical simulation of microalgae growth in this device show good agreement with experimental data.

Keywords: Microalgae · Mathematical modeling · Photosynthesis · CFD · Microalgae culture systems

1 Introduction

Microalgae and cyanobacteria are incredibly versatile microorganisms attracting a lot of research interest for over a half-century [5,9,14]. These oxygenic unicellular phototrophs utilize the light energy to fix inorganic carbon (CO_2) to synthetize more complex organic molecules in photosynthetic reactions. Moreover, they mitigate carbon dioxide but also consume inorganic nitrogen and phosphorus and thus may participate in waste-water treatment processes [19]. After certain recession, microalgae and cyanobacteria are re-gaining more attention in biotechnology field due to new potential as a biofuel source, for instance modification

© Springer International Publishing AG 2017
I. Rojas and F. Ortuño (Eds.): IWBBIO 2017, Part II, LNBI 10209, pp. 679–690, 2017.
DOI: 10.1007/978-3-319-56154-7_60

for high lipid content [7]. However, reliable methods with a predictive power for *in silico* simulation of microbial growth in microalgae culture systems (MCS), such as photobioreactors (PBR), are rather scarce [4, 16, 20]. The reason resides in the fact that a reliable model of microalgae growth have to encompass very complex problems, e.g., three-dimensional multiphase (gas-liquid-solid) flow dynamics, irradiance distribution inside MCS, multi-dimensional functionality of cellular processes. Moreover, all these parts interact across different timescales. Thus, one has to solve both theoretical (e.g., how to cope with multi-timescale phenomena) and practical (e.g., how to deal with enormous computational requirements associated with an optimization of MCS design and/or operating conditions) issues. As a consequence of the just mentioned problems, one or more interacting parts of the real system are modeled or implemented either in over-simplified or even inadequate way. We argue that the correct integration of a CFD (computational fluid dynamics) code and photosynthetic reaction kinetics, which would cover all spatio-temporal scales, is the cornerstone of successful solution.[1]

While most studies on MCS or PBR performance modeling are focused on specific or partial problems without clear connection to the whole production process, cf. [3, 14] and references within, in this communication we aim to formulate the unified modeling framework (Sect. 2). As a simple case study used for model validation, the numerical simulation of microalgae growth in a Couette-Taylor reactor is provided in Sect. 3. Some final remarks and the discussion of case study results, mainly the comments on the hydrodynamically induced high frequency light-dark cycles regime causing the flashing light enhancement,[2] conclude the work.

2   Model Development

We aim to set up a modeling framework for MCS, i.e., photobioreactor, open pond or raceway, able to bridge biology (microalgae cell growth), biochemistry and physics (hydrodynamics and optics) together. This framework will consist of (i) the state system – mass balance equations in form of advection-diffusion-reaction partial differential equations – PDEs (1–3), (ii) the fluid flow equations,

---

[1] Comparing characteristic times of microalgae growth (it is in order of hours) and that of mixing due to dispersion – turbulent diffusion (it is in order of seconds, similarly that of convective transport), one sees that adopting whatever steady state growth model (e.g., Monod, Haldane), only two alternatives exist: (i) to neglect the details concerning mixing phenomena, e.g., by accepting the hypothesis that the entire cell culture dispersed in medium was homogenized at each calculation step (cf. [12], where the time step Δt for the numerical integration of an underlined model was set to 3600 s), or (ii) to observe the changes due to the hydrodynamic mixing and neglect those of biochemical reaction. Both alternatives completely lose the coupling between transport and reaction phenomena.

[2] Couette-Taylor devices are able to generate the so-called Taylor vortex flow (in laminar flow regime) and subsequently oblige microalgae cells to periodically travel between illuminated wall and the dark side of the bioreactor, cf. [24].

i.e., Navier-Stokes equations (4), and (iii) the irradiance distribution inside MCS, e.g., according to Beer-Lambert's law, cf. [16].

All three part of the MCS model are interconnected, e.g., the material (mass) balance equations for the state variables of interest (e.g., microbial species, nutrients, CO_2, O_2) have to be solved simultaneously with the fluid dynamics (momentum balances, continuity equations). Nevertheless, accordingly to [23], for many biotechnological processes with Newtonian flow behavior, the stationary flow field inside MCS is not affected by mass transfer and reaction. This assumption permits the separation of both state and fluid-dynamic models and consequently the reduction of the computational effort due to the fact that each model can be solved with different numerical method and with different spatial-temporal discretization (on different numerical grids). The separation can be total (for the stationary flow field in a continuous system) or stepwise, e.g., reflecting some sequence of quasi-steady states in a production system operated in batch mode.

2.1   State Model

The material balance equations for the state variables of interest describe the transport and reaction of the reacting species or components [2]:

$$\frac{\partial c_i}{\partial t} + \nabla \cdot (v c_i) - \nabla \cdot (D_e \nabla c_i) = R(c_i) + S(c_i) \ , \ t \in [t_0, T], \ i = 1, ..., m, \quad (1)$$

where $c_i = c_i(x, t)$, is a conservative quantity, e.g., a concentration or cell density, v is the velocity flow field governed by the fluid-dynamic model, cf. (4), and $x \in \Omega \subset R^3$ stands for a position vector in a coordinate system (e.g., cylindrical). The dispersion coefficient $D_e(x)$ is generally a tensor of second order, which corresponds to the diffusion coefficient in microstructure description and becomes mere empirical parameter suitably describing mixing in the system and is influenced by the molecular diffusion and velocity profile, i.e., it is not a material constant.

The reaction kinetics and inlets/outlets of the reacting species are described by the reaction term $R(c_i)$ and source terms $S(c_i)$, respectively. The source term, e.g., the load of nutrients, is usually modeled as a corresponding boundary condition, however, both in order to further simplify the analytic study of the existence of optimal solution, see [1], and respecting that the location of discharge of some material could be inside the domain Ω, we prefer the above form of (1).

The initial condition and boundary condition (impermeability of the domain boundary, e.g., PBR walls) are following:

$$c_{i_0} = c_i(x, t_0), \ x \in \Omega \subset R^3, \ i = 1, ..., m, \quad (2)$$

$$\nabla c_i(x, t) = 0, \ x \in \partial\Omega, \ t \in [t_0, T], \ i = 1, ..., m. \quad (3)$$

2.2   Fluid-Dynamic Model

Generally speaking, the microalgal cells are solid particle and the CO_2 consumed and O_2 evolved are gases, thus we deal with the multiphase flow and transport. Nevertheless, the fluid flow will be treated as a flow of a suspension. The continuous phase is the liquid medium (the gaseous phase is neglected) and the dispersed phase represents the microalgal cells. Although the cell morphology and cell size could be very diverse, we could state that the average diameter of a spherical microalgae is about ten micrometers.

Further, because the mass fraction of cells is low, the suspension is classified as dilute and Newtonian viscosity relationship is supposed. A. Richmond in [19] states that the least ultra-high cell density culture should have the dry weight biomass concentration of about $10\,kg\,m^{-3}$, which represents the 1% mass fraction in the suspension of mass density about $1000\,kg\,m^{-3}$. The volume fraction of cells is similarly low (we assume that the cells are neutrally buoyant). Mass density of the suspension (mixture) is for a general case given by $\rho = \rho_w\,(1-k) + \rho_s\,k$, where ρ_w is the mass density of the medium and ρ_s the cell mass density, and k is the volume fraction. Taking into account a uniform distribution of algal cells and the fact that the characteristic time of algal growth is in order of hours, we assume that mass density of the suspension (mixture) is $\rho = \rho_w$.

Thereafter, by virtue that the inter-particle distances in our case of dilute suspension are sufficiently large, the full flow field over each particle or cell is allowed to be developed. Consequently, the particle velocity differs from the fluid transport velocity only by the fact that the particle could be settling relatively to the fluid velocity in a direction parallel to gravity, i.e. the microalgal cells follow the same trajectories as elementary fluid particles.[3]

Thus, we model only the incompressible liquid phase (suspension of water, nutrients and microalgae), hence the classical system of Navier-Stokes equations and the continuity equation is used as fluid-dynamic model:

$$\frac{\partial v}{\partial t} + (v \cdot \nabla)v = f - \frac{1}{\rho}\nabla p + \nu \nabla^2 v, \quad \nabla \cdot v = 0, \tag{4}$$

in $[t_0, T], \times \Omega$, with suitable boundary conditions on $[t_0, T], \times \partial \Omega$ and initial conditions in Ω, and where v, p, f, ρ and ν denote respectively the fluid velocity, the pressure, the body forces, fluid density and kinematic viscosity.

2.3   Reaction Model and Its Time-Scale Analysis

The photosynthetic reactions depends on some input variables, usually called as substrates. In this work we will discuss only the key variables determining the reaction kinetics, this is the irradiance (the other nutrients are supposed to be present in a sufficient amount, i.e., they do not limit the growth). Having in mind the experimental achievements, we claim that the reliable model of

---

[3] The same assumption was accepted by J. Pruvost *et al.* [17] arguing that the smallest eddy size, given by the Kolmogorov scale for their operating conditions, their annular PBR and their microorganism, is ca. ten times greater than the cell size.

reaction kinetics in MCS has to cover at least three time scales across photosynthetic reactions, including photoinhibition.[4] The following phenomena, which have to be reflected in the reaction term $R(c_i)$ in PDE (1) are: (i) activation of the light harvesting complex (photosynthetic unit – PSU) in light reactions, (ii) production of a certain amount of photoproduct, and (iii) photoinhibition, i.e., damage of a part of PSU by an excessive irradiance. One candidate for the suitable reaction model, namely three-state model of photosynthetic factory (PSF), is presented in Appendix. Here we depict only the key features of the PSF model. We suppose that the microalgal cells exist (with certain probability) in one of three states (*activated – A, inhibited – B* and *rested – R*). Though the fluid-dynamical properties of cells in each of three states are identical, the description of the temporal and spatial dependence of molar concentrations of cells in respective states (similarly to the description of concentration of three different components of one phase), is crucial for further evaluation of microalgal growth, cf. (6).

Let be the concentrations of respective components c_A, c_B, and c_R (with the same units as for the microalgae cell density c_x in whole PBR – generaly 10^6 cell ml^{-1} as in [26]). Then the following relation holds (for $\forall t \in [t_0, t_\infty]$, and $\forall x \in \partial\Omega$):

$$c_A(x, t) + c_B(x, t) + c_R(x, t) = c_x(x, t). \tag{5}$$

The dimensionless scalar values $y_R = c_R/c$, $y_A = c_A/c$, and $y_B = c_B/c$ (molar fractions) are respective states of the PSF model, cf. Appendix.

It is worth to note here, that according to [8, 26], the averaged rate of photosynthetic production (or the specific growth rate $\mu := \dot{c}_x/c_x$) is proportional (there is the constant $\kappa\gamma$) to the activated state fraction (there is also a maintenance term Me, for instance treated as constant):

$$\dot{c}_x = \frac{\kappa\gamma}{T} \int_0^T [y_A(u, t) - Me] \, c_x \, dt. \tag{6}$$

Looking at the above Eq. 6, we realize that while the state y_A of the PSF model is in range of 0 and 1 [-] and is sensitive to the light fluctuations either due to fluid dynamics or the light source (via the irradiance distribution $u(x, t)$ in PSF model, cf. Appendix), the term $\kappa\gamma$ in (6) in fact represent the scale-jump to the hours-days time scale, because $\kappa\gamma \approx 10^{-5}$ [s^{-1}]. By this way, the transition from the time-scale of light fluctuation (time micro-scale) to time-scale of biomass growth (time macro-scale) is effectuated without loss of accuracy.[5]

---

[4] Photoacclimation, the fourth phenomenon, occurs in the largest scale (days), thus can be represented as a steady state process (*via* a reaction model parameter) and no differential equation is needed.

[5] The PSF model has basically three time constant: one, corresponding to the light and dark reactions, is in order of seconds, other, corresponding to the photoinhibition, is in order of minutes, and finally, the last, which corresponds to microalgae growth, is in order of hours.

2.4   Numerical Aspects of PDEs (1–4) Solution

There exist well established methods and CFD (computational fluid dynamics) software packages to solve the above PDEs (1–4). Hence, using a suitable numerical method, PDEs are transformed into a system of linear algebraic equations. In the fluid flow problems, the finite volume approach (FVM) is more popular in obtaining the discretized form of equations than other approaches like FEM or FDM (finite element or finite differences method, respectively), especially with unstructured grids. The system of discretized equations is solved, usually, iteratively to find the values of velocities and other scalar quantities like concentrations in all grid points.

With large number of grid points, the solution domain can be split into several regions (partitions), solving each of them in parallel to decrease the computational time. In CFD parallel computations, it is necessary to exchange some amount of information on the partition interfaces because, for example, when approximating the derivatives on the interface we need to know values from both sides of the interface. This could seriously degrade the parallel speed-up, hence methods with small requirements on the interprocess communication are preferable. One of such methods is the Lattice Boltzmann method (LBM), see [6,21], which provides a superior parallel performance because most of its computational work consists in calculation of a collision operator which is completely local, that is no values from adjacent grid points are necessary in its calculation. The good parallel performance of this method can be advantageously employed in GPU calculations, see [22].

3   Experimental Validation of an Integrated Model

Our idea is to benchmark the experimentally confirmed phenomenon of flashing light enhancement[6] and the results based on our integrated model. The role of experimental device *par excellence* would fulfill a laboratory Couette-Taylor bioreactor (CTBR).

3.1   Photosynthesis in Fluctuating Light

The photosynthetic microorganism growth description is usually based on the photosynthetic steady-state reaction kinetics called *P–I (photosynthesis-irradiance) curve* describing the photosynthetic response in small cultivation systems with a homogeneous light distribution [23]. The MCS models based on *P–I* curves exhibits the loss of sensitivity to the transport term, e.g., hydrodynamic mixing, mixing *via* static mixers. In order to overcome this problems, some authors use the "artificial" interconnection between the steady state kinetic model and the dynamic one, e.g., some corrections or non-linear fit of a model to specific data reflecting the enhancement in photosynthetic efficiency in fluctuating light conditions, see [25]. Aiming to build a predictive model, our proposal

---

[6] For more details, see the well-known flashing light experiments [10,13,25].

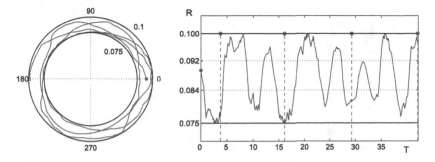

Fig. 1. Result of CFD simulation of one particle trajectory by fluent. The particle trajectory in the CTBR cross-section is shown in the left side. The right-side picture describes the time course of the particle radial position.

should not be based on a specific data, but on some universal principles instead. One of necessary conditions is that the model must cover all time scales of all relevant processes. Nevertheless, even having an adequate dynamic model of microorganism growth (with lumped parameter), another serious difficulty resides in its implementation into a distributed parameter system. The simplistic "Lagrangean" solution is based on decoupling of the fluid dynamics and reaction kinetics. Using either an experimental technique, e.g., CARPT [11], or a CFD code, see Fig. 1, the trajectories of microalgae cells can be described. Then the problem is reduced on a suitable reaction kinetics model in form of an ordinary differential equation with the stochastic input. This input is calculated on the base of "irradiance history" of individual cells. The alternative Eulerian solution is exposed in the next subsection.

3.2 Couette-Taylor Fluid Dynamic Regimes

The choice of CTBR for our validation study resides in the relatively simple integration of CTBR fluid dynamics into microalgae growth model. First of all, the dimensionality of the problem can be significantly reduced because in many operating conditions, the flow field is symmetric. The Couette-Taylor device is mainly composed of two coaxial cylinders. In our laboratory CTBR, the suspension of microalgae in the annular space between cylinders is set in motion by the rotation of the inner cylinder ($r_i = 0.075$ m) along the vertical axis, while the outer cylinder ($r_i = 0.1$ m) is kept at rest. Several hydrodynamic regimes depend on an angular velocity ω, on the geometrical characteristics of the device and on the physical properties of the fluid (kinematic viscosity ν). According to Taylor's results [24], when the so-called Taylor number: Ta $= \frac{(r_e - r_i)^3 r_i}{\nu^2}\omega$, is smaller than a critical value (Ta$_c$), the flow in the system is purely tangential and is called the *Couette flow*. When the Taylor number is superior to this critical value, a transition to a periodic structure is observed. A series of toroidal

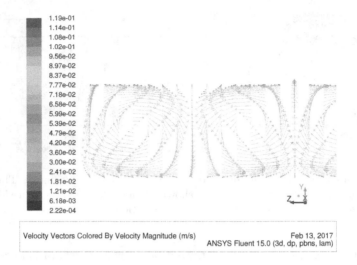

Velocity Vectors Colored By Velocity Magnitude (m/s)   Feb 13, 2017
ANSYS Fluent 15.0 (3d, dp, pbns, lam)

Fig. 2. Fluid flow velocity profile in the axial section of the laboratory CTBR (for the inner cylinder angular frequency $\omega = 2.4$ rad s^{-1}, Re $= 2000$), calculated by CFD code ANSYS Fluent 15.0 (laminar model).

vortices are superposed to the tangential flow, thus the *laminar Taylor vortex flow*, characterized by a laminar cellular vortex motion, occurs,[7] see Fig. 2.

3.3   Biological Performance of Couette-Taylor Bioreactor

For our special case of photosynthetic growth in CTBR, we define a performance index or the objective function J_{CTBR} as the volumetric productivity (usually in grams per liter per day)

$$J_{CTBR} = \frac{\kappa\gamma}{\text{meas}(\Omega)\,T} \int_0^T \int_\Omega [y_A(u,r,t) - Me]\,c_x \,\mathrm{d}r\mathrm{d}t, \tag{7}$$

where r stands for the radial distance measured from the outer CTBR wall. The activated state fraction $y_A(u,r,t)$ in J_{CTBR} has to be determined as the solution of transport-reaction PDE system (1–4).

Based on the Lambert-Beer's law in form $u(r) = u_0\,e^{-k\,r}$, i.e., on exponential decreasing of irradiance level from CTBR wall to the CTBR core,[8] we can

---

[7] According to Taylor's explanation [24], the transition between the two regimes is achieved when the viscous forces do not damp the initial infinitesimal disturbances anymore, and this condition is reached when the Taylor number exceeds the given critical value. A further increase of the Taylor number leads to a sequence of two time-dependent flow regime, the *wavy vortex flow* and the *doubly periodic wavy vortex flow*.

[8] The parameter k represents the attenuation coefficient, unit: m^{-1}, another characteristic parameter describing the attenuation of irradiance by the suspension of microbial cells in the liquid medium is the depth corresponding to the decreasing of the incident irradiance to one half, i.e. $r_{1/2} = \frac{\ln 2}{k}$.

transform the 3D spatial problem into one-dimensional. Consequently, the description of cell motion in direction of light gradient, i.e., perpendicular to CTBR wall, is of most interest. Furthermore, when both the flow velocity profile and the time averaged values of PSF states reach their steady states, i.e., $\frac{\partial c_i}{\partial t} = 0$, then the transport equation (1) can be expressed as follows:

$$\nabla \cdot (vy) - \nabla \cdot (D_e \nabla y) = [\mathcal{A} + u(r,t)\mathcal{B}]y, \tag{8}$$

where y represents the three-state vector and $\mathcal{A}$ and $\mathcal{B}$ stand for the corresponding matrices in PSF model, see (9) in Appendix.

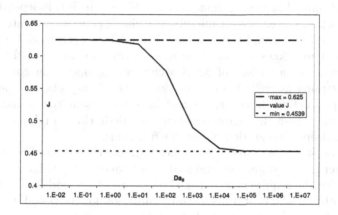

Fig. 3. Performance index J vs. Da_{II} (characteristic number inversely proportional to the angular velocity ω, cf. (7)). For bigger mixing rate, lower Da^{II}, we get better performance J (reaching its maximal theoretical value for $Da_{II} \approx 1$).

Equation (8) can be solved by means of numerical methods, e.g., FDM, FEM, FVM. The numerical process is controlled by the second *Damköhler number* $Da^{II} := k_r \frac{R^2}{D_e(\omega)}$, which submits into relation the characteristic time of dispersion (proportional to the inner cylinder rotation ω) and this of reaction ($\tau_r = 1/k_r$).[9] Generally, when characteristic numbers are either very small or very big, then we can expect some numerical difficulties while solving PDEs, e.g., (8). This could be the case if we would use a steady-state (Monod or Haldane) kinetics instead of a multi time-scale model, being sensitive to the characteristic time of cell transport in direction of light gradient, e.g., PSF model.

The results of biological performance for our Couette-Taylor bioreactor irradiated from outside and operated under constant angular rotation ω are depicted in Fig. 3. It is clearly visible the role of mixing induced by ω, i.e., the flashing light

---

[9] The Damköhler numbers (Da) are widely used in chemical engineering to relate the chemical reaction timescale (reaction rate) to the transport phenomena rate occurring in a system.

effect was reached. One could say: bigger ω (bigger mixing rate, lower Da^{II}) better performance J (reaching actually its maximal theoretical value 0.62). However, the detrimental influence of the hydrodynamically induced shear stress was not taken into account.

4 Conclusion

The crucial point for commercial success of a production plant, e.g., for algae bio-fuels, is the availability of a reliable software tool with predictive capacity enabling simulation and optimization of plant performance. In this work, we proposed the unified modeling framework for MCS, which is independent of the actual production system size or microbial strain. It consists in 3 interconnected parts: (i) the state system, (ii) fluid-dynamic model, and (iii) model of irradiance distribution. This allows for the same model structure to be used to evaluate the performance of a variety of MCS architectures, operating conditions and microalgae strains as well. We do not deal with all aspects and problems of microalgae production, however, we kept in mind the multiscale nature of the problem, mainly the wide separation of characteristic time of relevant processes, and propose deeper integration of cellular functions.

In the case study, we have shown that a simple three-state PSF model well behaves under hydrodynamically induced high frequency light-dark cycles regime and copes with the requirement imposed on the reaction model, i.e., it correctly describes both the steady-state (substrate inhibition kinetics) and dynamic phenomena (flashing light effect and photoinhibition). Furthermore, our state and hydrodynamic model provide an adequate description of microalgae growth in the Couette-Taylor reactor and serve for the optimal control problem formulation. Ongoing modeling efforts will continue in order to develop even more reliable model, e.g., considering the effect of hydrodynamical shear stress on microalgae growth.

Acknowledgment. This work was supported by the Ministry of Education, Youth and Sports of the Czech Republic - projects "CENAKVA" (No. CZ.1.05/2.1.00/01.0024) and "CENAKVA II" (No. LO1205 under the NPU I program).

Appendix

Model of Photosynthetic Factory – PSF Model

Here we describe the multi-timescale three-state model of photosynthetic factory (PSF model) proposed by Eilers and Peeters [8] and further developed by Wu and Merchuk [26]. As follows, PSF model is used for the reaction term $R(c_i)$ derivation, cf. the transport equation (1). PSF model (9) describes the dynamics of three basic phenomena occurring simultaneously in three largely separated time-scales: (i) cell growth (including the shear stress effect), (ii) photoinhibition, and (iii) photosynthetic light and dark reactions. The state vector of the

PSF model is three dimensional, $y = (y_R, y_A, y_B)^\top$, where y_R represents the probability that PSF is in the resting state R, y_A the probability that PSF is in the activated state A, and y_B the probability that PSF is in the inhibited state B.

$$
\begin{bmatrix} \dot{y}_R \\ \dot{y}_A \\ \dot{y}_B \end{bmatrix} = \begin{bmatrix} 0 & \gamma & \delta \\ 0 & -\gamma & 0 \\ 0 & 0 & -\delta \end{bmatrix} \begin{bmatrix} y_R \\ y_A \\ y_B \end{bmatrix} + u(t) \begin{bmatrix} -\alpha & 0 & 0 \\ \alpha & -\beta & 0 \\ 0 & \beta & 0 \end{bmatrix} \begin{bmatrix} y_R \\ y_A \\ y_B \end{bmatrix} \tag{9}
$$

The values of model parameters, i.e., α, β, γ, δ, and κ, cf. 6, are published in Wu and Merchuk (2001) [26]. Concerning the experimental design for parameter estimation and the identifiability study, see [18].

For given input variable, i.e., the irradiance $u(t)$, the ODE system (9) can be solved either by numerical methods or by asymptotic methods. For the special case of the periodic piecewise constant input, the state trajectories were calculated explicitly in [15].

References

1. Alvarez-Vázquez, L., Fernández, F.: Optimal control of a bioreactor. Appl. Math. Comput. **216**, 2559–2575 (2010)
2. Beek, W.J., Muttzall, K.M.K., van Heuven, J.W.: Transport Phenomena. Wiley, Great Britain (2000)
3. Bernard, O., Mairet, F., Chachuat, B.: Modelling of microalgae culture systems with applications to control and optimization. Adv. Biochem. Eng. Biotechnol. **153**, 59–87 (2016)
4. Bernardi, A., Perin, G., Sforza, E., Galvanin, F., Morosinotto, T., Bezzo, F.: An identifiable state model to describe light intensity influence on microalgae growth. Ind. Eng. Chem. Res. **53**, 6738–6749 (2014)
5. Burlew, J.S. (ed.): Algal Culture From Laboratory to Pilot Plant. The Carnegie Institute, Washington, D.C. (1953). Publ. no. 600
6. Chen, S., Doolen, G.D.: Lattice Boltzmann method for fluid flows. Ann. Rev. Fluid Mech. **30**, 329–364 (1998)
7. Chisti, Y.: Biodiesel from microalgae. Biotechnol. Adv. **25**, 294–306 (2007)
8. Eilers, P.H.C., Peeters, J.C.H.: A model for the relationship between light intensity and the rate of photosynthesis in phytoplankton. Ecol. Model. **42**, 199–215 (1988)
9. Grobbelaar, J.U.: Microalgal biomass production: challenges and realities. Photosynth. Res. **106**, 135–144 (2010). doi:10.1007/s11120-010-9573-5
10. Kok, B.: Experiments on photosynthesis by Chlorella in flashing light. In: Burlew, J.S. (ed.) Algal Culture from Laboratory to Pilot Plant, pp. 63–75. The Carnegie Institute, Washington, D.C. (1953). Publ. no. 600
11. Luo, H.-P., Kemoun, A., Al-Dahhan, M.H., Fernandez, J.M., Molina, G.E.: Analysis of photobioreactor for culturing high value microalgae and cyanobacteria via an advanced diagnostic technique: CARPT. Chem. Eng. Sci. **58**(12), 2519–2527 (2003)
12. Muller-Feuga, A., Le Guédes, R., Pruvost, J.: Benefits and limitations of modeling for optimization of porphyridium cruentum cultures in an annular photobioreactor. J. Biotechnol. **103**, 153–163 (2003)

13. Nedbal, L., Tichý, V., Xiong, F., Grobbelaar, J.U.: Microscopic green algae and cyanobacteria in high-frequency intermittent light. J. Appl. Phycol. **8**, 325–333 (1996)
14. Ooms, M.D., Dinh, C.T., Sargent, E.H., Sinton, D.: Photon management for augmented photosynthesis. Nat. Commun. **7**, 12699 (2016). doi:10.1038/ncomms12699
15. Papáček, Š., Čelikovský, S., Štys, D., Ruiz-León, J.: Bilinear system as modelling framework for analysis of microalgal growth. Kybernetika **43**, 1–20 (2007)
16. Park, S., Li, Y.: Integration of biological kinetics and computational fluid dynamics to model the growth of nannochloropsis salina in an open channel raceway. Biotechnol. Bioeng. **112**, 923–933 (2015)
17. Pruvost, J., Legrand, J., Legentilhomme, P., Muller-Feuga, A.: Simulation of microalgae growth in limiting light conditions: flow effect. AIChE J. **48**, 1109–1120 (2002)
18. Rehák, B., Čelikovský, S., Papacek, Š.: Model for photosynthesis and photoinhibition: parameter identification based on the harmonic irradiation O_2 response measurement. In: Joint Special Issue of TAC IEEE and TCAS IEEE, pp. 101–108 (2008)
19. Richmond, A.: Biological principles of mass cultivation. In: Richmond, A. (ed.) Handbook of Microalgal Culture: Biotechnology and Applied Phycology, pp. 125–177. Blackwell Publishing, Hoboken (2004)
20. Sato, T., Yamadab, D., Hirabayashia, S.: Development of virtual photobioreactor for microalgae culture considering turbulent flow and flashing light effect. Energy Convers. Manag. **51**(6), 1196–1201 (2010)
21. Succi, S.: The Lattice Boltzmann Equation. Oxford University Press, Oxford (2001)
22. Štumbauer, V., Petera, K., Štys, D.: The lattice Boltzmann method in bioreactor design and simulation. Math. Comput. Model. **57**, 1913–1918 (2013)
23. Schugerl, K., Bellgardt, K.-H. (eds.): Bioreaction Engineering. Modeling and Control. Springer, Berlin (2000)
24. Taylor, G.I.: Stability of a viscous liquid containing between two rotating cylinders. Phil. Trans. R. Soc. **A223**, 289–343 (1923)
25. Terry, K.L.: Photosynthesis in modulated light: quantitative dependence of photosynthetic enhancement on flashing rate. Biotechnol. Bioeng. **28**, 988–995 (1986)
26. Wu, X., Merchuk, J.C.: A model integrating fluid dynamics in photosynthesis and photoinhibition processes. Chem. Eng. Sci. **56**(11), 3527–3538 (2001)

Kinetic Modelling of Processes Behind S$_{2,3}$-states Deactivation in Photosynthetic Oxygen Evolution

Jiri Jablonsky$^{(\boxtimes)}$ and Stepan Papacek

Faculty of Fisheries and Protection of Waters, South Bohemian Research Center of Aquaculture and Biodiversity of Hydrocenoses, Institute of Complex Systems, University of South Bohemia in Ceske Budejovice, Zamek 136, 373 33 Nove Hrady, Czech Republic
jiri.jablonsky@gmail.com, spapacek@frov.jcu.cz

Abstract. Photosynthetic water splitting, localized in photosystem II, is the source of atmospheric oxygen and possible alternative energy source. It is therefore important to understand the related processes which influence the efficiency of water splitting. We have employed kinetic models of photosystem II to study deactivation processes of higher S-states (redox states of water splitting) in the dark. Our analysis of spinach samples, treated or untreated by electron acceptor phenyl-parabenzoquinone (PPBQ) indicated an unknown mechanism, decay, related to S$_{2,3}$-state deactivation. We concluded that: (1) S$_3$-state decay occurs independently on the PPBQ treatment, i.e., independently on the redox state of the acceptor side of photosystem II, (2) S$_2$-state decay can be fully described by S$_2$-$Q^-_{A,B}$ charge recombination, neglected in previous models, and (3) the mechanism of S$_3$-state decay can be explained by the involvement of slow cooperation within photosystem II dimer between S$_3{}^{PSIIa}$ and S$_{3,2,1}{}^{PSIIb}$ in higher plants. Finally, the slow cooperation is able to explain experimental data both from PPBQ-free and PPBQ treated samples.

Keywords: Kinetic model · Photosynthesis · Photosystem II · Oxygen evolving complex · S-states deactivation

1 Introduction

Over the last decade, we have observed significant advances in Systems biology. This trend is visible especially in the boom of genome-scale models (e.g., [1]) or omics data integration efforts (e.g., [2]). On the other hand, traditional methods such as kinetic modeling are being gradually improved, re-evaluated and made assumption tested for the search of physiologically meaningful interpretation of the data [3,4]. In the case of photosynthesis research, kinetic modeling has very long tradition [5]. One of the hot topics in the photosynthesis research is kinetic modeling of photosystem II (PSII) and related water splitting, a possible renewable energy source. PSII is the prime source of energy for all other processes in

© Springer International Publishing AG 2017
I. Rojas and F. Ortuño (Eds.): IWBBIO 2017, Part II, LNBI 10209, pp. 691–699, 2017.
DOI: 10.1007/978-3-319-56154-7_61

oxygenic photosynthesis but it is threatened by photoinhibition (e.g., [6]), heat degradation (e.g., [7,8]) or reactive oxygen species (e.g., [9]). Further, it was not so long ago that water splitting was viewed as a simple four-step mechanism [10]; now, the influence of relaxation processes on electron transport [11,12] is known, and even deprotonation [13] is considered essential in the kinetic models. In this work, we have developed defragmented kinetic model of PSII in which all the cofactors involved in electron transport, from water splitting to plastoquinone molecule, are considered in one multi-state variable. This improvement was necessary for the analysis of the deactivation processes of $S_{2,3}$-states of oxygen evolving complex occurring in the dark and possible identification of unknown decay mechanism of S_3-state.

2   Methods

The models were developed within SimBiology Toolbox (MathWorks Inc.) in MATLAB (MathWorks Inc.), where all simulations were handled and executed. A detailed description of earlier models, the kinetics used and the handling of simulations are available in Refs. [12,14,15]. The presented simulations originated from seven (see Table 1) versions of PSII kinetic model (see Fig. 1) which differ in the employment of second or first order kinetics and components responsible for deactivation of $S_{2,3}$-states of oxygen evolving complex. The scheme of model$^{C/1}$ is shown in Fig. 1. The significance of model differences is illustrated in the example of S_2-state oxidation by tyrosine Z (Y161-D1), in Fig. 2. The simulations were based on the same conditions as the experiments (number of flashes and initial conditions). Details can be found in the original papers and legends of the figures.

Table 1. A list of the used versions of kinetic model of PSII with the description of differences and with the link to the performed simulations. C stands for cyanobacteria, H stands for higher plants, PPBQ indicates phenylparabenzoquinone. Note: reduction of $S_{2,3}$-states by tyrosine D (Y160-D2) is part of all model versions.

Model	Basic details and modifications	Results
C/1	Three linked multi-state variables [9], see Fig. 2b	Fig. 3a
C/2	C/1 model, with included S_3-Q_B^- charge recombination (ch.r.)	Fig. 3b
C/3	C/2 model but with one multi-state variable, see Fig. 2a	Fig. 3c
H/4	PPBQ treatment: $S_{2,3}$-states decay	Fig. 3d
H/5	PPBQ-free S_3-state artificial decay mechanism, S_3-Q_B^- ch.r	Fig. 4a
H/6	PPBQ-free S_2-$Q_{A,B}^-$ and cooperation between S_3-state and $S_{3,2,1}$-states	Fig. 4b
H/7	PPBQ treatment: cooperation between S_3-state and $S_{3,2,1}$-states	Fig. 3d

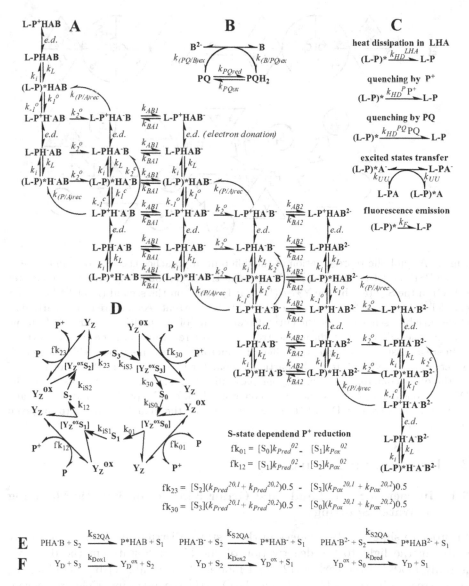

Fig. 1. The kinetic model$^{C/1}$ of PSII designed for cyanobacteria. The meaning of the particular letters in the scheme is as follows: L all light harvesting antennas in PSII, P P680 (special chlorophyll pair), I pheophytin, A Q_A (ubiquinone), B Q_B (plastoquinone), PQ oxidized plastoquinone molecules in the PQ pool, PQH2 reduced and protonated PQ molecules in the PQ pool, Y_Z/Y_{Zox} reduced/oxidized state of tyrosine 161-D1, Y_D/Y_{Dox} reduced/oxidized state of tyrosine 160-D2, Sn (n = 0, 1, 2, 3) the S-states of OEC, [SnYZox] (n = 0, 1, 2, 3) the intermediate S-states of OEC, e.d. in part A means electron donation to P680$^+$ by Y_Z as it is described by part D, asterisk indicates excited state. Description and values of kinetic parameters, as well as explanation of every reaction, is available in our previous work from where this scheme is adopted [12].

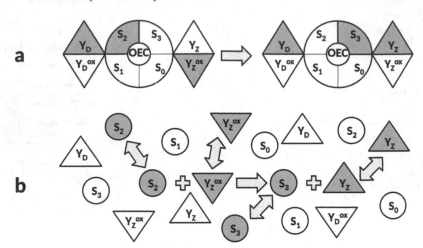

Fig. 2. A symbolic representation of the different approaches used in the donor side of tested PSII models. Schemes **a** and **b** represent the oxidation of oxygen evolution center (OEC) in the S_2-state by the $Y_{Z}ox$ (D1-Y161) based on the second (model$^{C/1}$) or first (model$^{C/3}$) kinetics. The gray color highlights the current redox state of the particular state variable (scheme **b**) or the current redox state of the multistate variable (scheme **a**). The one-side arrows indicate electron transport reactions and the two-side arrows indicate a possible involvement of different cofactors of the same type within the model. Scheme **a**: the redox state of the particular cofactor is connected to the redox state of all the other cofactors representing one particular PSII. Scheme **b**: each S-state of the OEC is considered as the stand alone cofactor which can be oxidized/reduced by any of the tyrosines (red/ox D1-Y161 or D2-Y160) within the model according to the given rules and restrictions.

3   Results and Discussion

3.1   Impact of Second and First Order Kinetics on Kinetic Modeling of Water Splitting

In order to simulate the changes in the S-states distribution from their steady states in the light to the deactivation of the higher S-states in the dark, we initially modified the earlier [12] PSII model$^{C/1}$ (see Table 1) by including the charge recombination between S_3 and the Q_B^- (the second quinone electron acceptor, halftime 380 s, [16]); this reaction is usually omitted for its slowness. The modified model$^{C/2}$ succeeded in the expected stabilization of the S_1-state after several hours in the dark (99.3 %), see Fig. 3a, and matched the experimental deactivation of S_3-state (Fig. 3b). However, model$^{C/2}$ (as well as the original model$^{C/1}$) was, for simplicity, encoded with separated donor and acceptor sides of PSII employing the second order kinetics which could alter the interpretation of the deactivation processes of higher S-states. To exclude any doubts caused by fragmentation of electron cofactors chain, we developed an improved PSII model$^{C/3}$ based on first-order-kinetics for the electron transport within PSII where all cofactors are integrated as one multiplestate variable; the difference

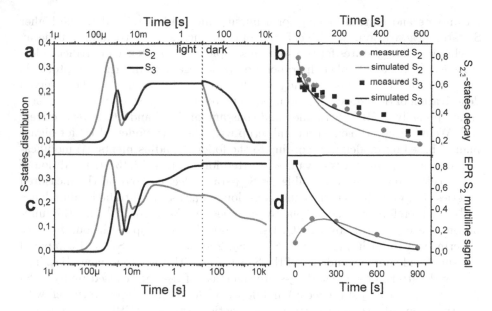

Fig. 3. Changes in the S$_{2,3}$-states distribution induced by continuous light/dark regime or by flashes. Subplots **a** and **c** show the simulated changes of the S$_{2,3}$-states of OEC induced by 10 s of the light with the subsequent 3 h in the dark based on the model[C/1] [12] and model[C/3], respectively. Subplot **b** shows a comparison of the measured (cyanobacterium *Synechococcus elongatus* strain PCC 7942) and simulated deactivation of the S$_{2,3}$-states in the dark based on the model[C/2]. Experimental data for cyanobacterium were redrawn [16] and normalized to theoretical data (the experiments have provided only relative values). Subplot **d** shows comparison of the simulated (model[H/4]) changes of the S$_{2,3}$-states with the measured EPR multiline signal of the S$_2$-state from spinach sample [17]. The sole experimental point of the S$_3$-state in the subplot d means its resultant concentration after the two flashes [17]. The initial conditions for all simulations were same as in the experiments.

between the first and second order kinetics is explained on the example given in Fig. 2. However, the new simulations based on model[C/3] were too far from matching the experiments. The most surprising output was totally missing deactivation of the S$_3$-state and insufficient deactivation of S$_2$-state, see Fig. 3c.

3.2   Decay of Higher S-states in the Presence of Inhibitor PPBQ

Having established the gap between the experiments and our simulations, we were facing three possibilities: (1) to modify the kinetic parameters, especially of S$_3$-Q$_B^-$ charge recombination, (2) to test the possible reversibility of reactions on the donor side of the PSII and (3) to identify a mechanism behind reduction of the higher S-states. The first two options were insufficient to significantly decrease the amount of S$_3$-state with oxidized acceptor side, providing practically identical results as Fig. 3c. We therefore concluded that the model[C/3] was incomplete and started looking for a missing mechanism related to the higher S-states deactivation. We have found our findings similar to unknown decay mechanism of $S_{2,3}$-states for the spinach sample in the presence of artificial electron acceptor PPBQ (phenylparabenzoquinone) maintaining the acceptor side of PSII oxidized and thus disabling any charge recombination [17]. The observed deactivation of $S_{2,3}$-states was assumed to be related to exogenous donor of electrons, possibly PPBQ, with the time constant of 210 s and 260 s, respectively [17]. We mimicked the experimental setup with the new model$^{H/4}$ (H for higher plants) based on model$^{C/3}$ according to the following adjustments: (i) modification of kinetic parameters to values reported for spinach [12,16], (ii) addition of two artificial reactions responsible for $S_{2,3}$-states decay based on the measured kinetics [17]; we note that the exogenous donor was assumed to be in abundance [17] and therefore was not explicitly included in the model$^{H/4}$ and (iii) inhibition of acceptor side by PPBQ. Using model$^{H/4}$, we applied in our *in silico* experiment two flashes (1st flash: $S_1 \rightarrow S_2$, 2nd flash: $S_2 \rightarrow S_3$) and compared the simulated time-dependence of the $S_{2,3}$-states with the experiments [17]. The observed pattern of simulated S_2-state (increase of S_2-state caused by $S_3 \rightarrow S_2$ decay followed by S_2-state decay) and decay of the S_3-state agree very well with the time-dependence of electron paramagnetic resonance (EPR) multiline signal of the S_2-state together with the initial concentration of S_3-state after two flashes [17], see Fig. 3d. The best fit of the experimental data was achieved with the same values of the time constants of the $S_{2,3}$-states relaxation reported in the experiment [17]. As this result supported the early suggested kinetic characterization of unknown decay mechanism [17], we were prepared to test the kinetics in natural conditions (without PPBQ) in the next PSII model (model$^{H/5}$).

3.3   Deactivation of $S_{2,3}$-states in Spinach: Charge Recombination and Cooperation

To explore deactivation of $S_{2,3}$-states in the PPBQ-free spinach sample, we started our analysis with S_2-state decay. In order to advance the majority of oxygen evolving complexes to the S_2-state, one flash is applied which reduces the Q_A and consequently the Q_B and enables a burst of charge recombination between the S_2-state and both Q_A^- and Q_B^-. Our analysis confirmed that S_2-Q_A^- charge recombination (1/k = 1.5 s) together with S_2-Q_B^- charge recombination (1/k = 40 s) in the model$^{H/6}$ are required as the only sources of S_2-state decay; as seen in Fig. 4b. In the case of the S_3-state, its deactivation based on time constant of 260 s [17] was totally insufficient, see Fig. 4a; we note that fast S_3-state deactivation was reported for spinach in natural conditions [16], in contrast to cyanobacteria. The initial match with the experiment (0 s $\rightarrow$ 10 s; Fig. 3a) was caused mainly by a charge recombination and partly by tyrosine D (reduction of $S_{2,3}$-states by Y160-D2 is part of all model versions), which reduced a portion of the S_3-states. Finally, an attempt to modify model$^{H/5}$ by changing the artificial decay time constant, 260 s $\rightarrow$ 24 s, increased the simulated rate of S_3-state deactivation but the exponential decrease was too far from fitting the experimental pattern, see Fig. 4a. We were therefore looking for another process possibly related to decay of $S_{2,3}$-states and the only known candidate was the cooperation between PSIIs [14]. The cooperation occurs between two S-states within the PSII dimer and was suggested for the two sequent S-states (S_1 and S_2) as an assumption necessary for explanation of the higher yield of the oxygen evolution after the second flash in cyanobacteria [14]. Cooperation might be observed even in the long time scale if the formation of the intermediate S-states [12] is reversible or if another pathway of cooperation between two donor sides exists. We have analyzed the most possible scenario based on analysis of the artificial decay rate [17]: besides the S_3-Q_B^- charge recombination, there are one or several pathways of cooperation which might be observed as decay of $S_{2,3}$-states and whose rates might be influenced by PPBQ (to explain the difference between the PPBQ treated/free samples). The following pathways of cooperation were incorporated into the PSII model$^{H/6}$ for a complete description of the S_3-state decay in the PPBQ-free sample; final comparison of simulated and experimental S_3-state decay is shown in Fig. 4b and added reactions with time constants are summarized below:

$$S_3{}^{PSIIa} + S_3{}^{PSIIb} \rightarrow S_4\text{-}S_0{}^{PSIIa} + S_2{}^{PSIIb}, \; k_1^{-1} = 90 \text{ s}$$
$$S_3{}^{PSIIa} + S_2{}^{PSIIb} \rightarrow S_4\text{-}S_0{}^{PSIIa} + S_1{}^{PSIIb}, \; k_2^{-1} = 7.7 \text{ s}$$
$$S_3{}^{PSIIa} + S_1{}^{PSIIb} \rightarrow S_2{}^{PSIIa} + S_2{}^{PSIIb}, \; k_3^{-1} = 11 \text{ s}$$

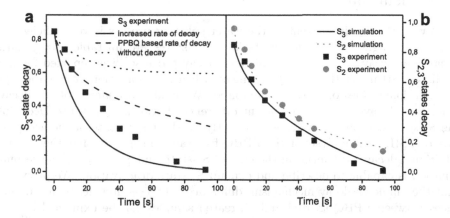

Fig. 4. Analysis of the decay mechanisms in the $S_{2,3}$-states deactivation for higher plants (spinach). Squares show measured S_3-state decay [16], circles show measured S_2-state decay [16]. Subplot **a**: The dashed line represents simulation based on the model$^{H/5}$ where the considered S_3-state decay mechanisms are: (i) S_3-Q_B^- charge recombination and (ii) artificial decay with time constant of 260 s based on the PPBQ experiment [17]. The solid line represents simulation based on the modified model$^{H/5}$ where the relaxation time constant was altered from 260 s to 24 s. The dotted line represents simulation based on the modified model$^{H/5}$ without the artificial decay mechanism, i.e., S_3-Q_B^- charge recombination. Subplot **b**: Simulated $S_{2,3}$-states decay is based on the model$^{H/6}$. The considered S_3-state decay mechanisms are: three slow pathways of cooperation among the $S_{3,2,1}$-states and S_3-Q_B^- charge recombination. The considered S2-state decay mechanisms are: S_2-Q_A^- and $S_2 Q_B^-$ charge recombination.

The slow cooperation between S_3^{PSIIa} and S_3^{PSIIb} was found as the main decay mechanism because the majority of the PSIIs were in the S_3-state after two applied flashes, which also oxidized the majority of Q_B and consequently reduced the chance for S_3-Q_B^- charge recombination ($1/k = 20$ s, [7]). In order to check whether cooperation is sufficient to explain the decay mechanism in PPQB treated sample, the presented pathways of cooperation replaced the reactions for artificial decay of the $S_{2,3}$-states (model$^{H/4}$) within model$^{H/7}$. We have confirmed that the presented pathways of cooperation are fully capable of explaining the relaxation of the $S_{2,3}$-states (data not shown, virtually identical as shown in Fig. 3d). However, the time constants of the pathways of cooperation were several-times increased in order to match the $S_{2,3}$-states decay, probably due to structural changes of D1 protein within the PSII in the presence of PPBQ. Regarding the two pathways of cooperation in which one of the S_3-states is moved to the S_0-state, it was observed that the S_3-state does not always decay through the S_2-state, especially if the artificial electron acceptor is not present [18]. The slow cooperation between S_3^{PSIIa} and $S_{1,2}^{PSIIb}$ was considered partly to improve the match with the experimental S_3-state decay, partly to explain the difference between the PPBQ treated/untreated samples and the initial increase of the S_2-state in the EPR multiline S_2 signal after two flashes, see Fig. 3d.

4 Conclusion

In view of the results presented, the water splitting process is very efficient in appropriate light conditions but its higher S-states are unstable in the dark because of decay processes. An unknown decay processes was reported for the oxidized PSII under the PPBQ treatment, when recombination between acceptor and donor sides of PSII was inhibited. We made an analysis of the decay kinetics and have revealed significant differences between $S_{2,3}$-states. Decay of the S_3-state can be observed independently on the redox state of the acceptor side of PSII. Our analysis of the PPBQ-free sample suggests that there is no additional decay mechanism in the case of S_2-state and S_2-$Q_{A,B}^-$ charge recombination is capable to describe and explain S_2-state deactivation. We proposed that the unknown decay mechanism, differences in $S_{2,3}$-states decay and differences between PPBQ treated and untreated samples can be explained by the involvement of slow pathways of cooperation between S_3^{PSIIa} and $S_{3,2,1}^{PSIIb}$ in higher plants.

Acknowledgment. This work was financially supported by the Ministry of Education, Youth and Sports of the Czech Republic - projects "CENAKVA" (No. CZ.1.05/2.1.00/01.0024) and "CENAKVA II" (No. LO1205 under the NPU I program).

References

1. Knoop, H., Zilliges, Y., Lockau, W., Steuer, R.: The metabolic network of synechocystis sp. PCC 6803: systemic properties of autotrophic growth. Plant Physiol. **154**, 410–422 (2010)

2. Jablonský, J., Papáček, Š., Hagemann, M.: Different strategies of metabolic regulation in cyanobacteria: from transcriptional to biochemical control. Sci. Rep. **6**, 33024 (2016)
3. Nedbal, L., Červený, J., Rascher, U., Schmidt, H.: E-photosynthesis: a comprehensive modeling approach to understand chlorophyll fluorescence transients and other complex dynamic features of photosynthesis in fluctuating light. Photosynth. Res. **93**, 223–234 (2007)
4. Lazár, D., Jablonský, J.: On the approaches applied in formulation of a kinetic model of photosystem II: different approaches lead to different simulations of the chlorophyll alpha fluorescence transients. J. Theor. Biol. **257**, 260–269 (2009)
5. Laisk, A., Nedbal, L., Govindjee: Photosynthesis in silico - Understanding Complexity from Molecules to Ecosystems. Springer, Amsterdam (2009)
6. Lazr, D., Ilk, P., Kruk, J., Strzaka, K., Nau, J.: A theoretical study on effect of the initial redox state of cytochrome b559 on maximal chlorophyll fluorescence level (FM): implications for photoinhibition of photosystem II. J. Theor. Biol. **233**, 287–300 (2005)
7. Kouril, R., Lazr, D., Ilk, P., Skotnica, J., Krchnk, P., Naus, J.: High-temperature induced chlorophyll fluorescence rise in plants at 40–50 degrees C: experimental and theoretical approach. Photosynth. Res. **81**, 49–66 (2004)
8. Lpov, L., Krchnk, P., Komenda, J., Ilk, P.: Heat-induced disassembly and degradation of chlorophyll-containing protein complexes in vivo. Biochim. Biophys. Acta. **1797**, 63–70 (2010)
9. Yamashita, A., Nijo, N., Pospsil, P., Morita, N., Takenaka, D., Aminaka, R., Yamamoto, Y., Yamamoto, Y.: Quality control of photosystem II: reactive oxygen species are responsible for the damage to photosystem II under moderate heat stress. J. Biol. Chem. **283**, 28380–28391 (2008)
10. Kok, B., Forbush, B., McGloin, M.: Cooperation of charges in photosynthetic O2 evolution-I. A linear four step mechanism. Photochem. Photobiol. **11**, 457–475 (1970)
11. Renger, G.: Coupling of electron and proton transfer in oxidative water cleavage in photosynthesis. Biochim. Biophys. Acta BBA - Bioenerg. **1655**, 195–204 (2004)
12. Jablonsky, J., Lazar, D.: Evidence for intermediate S-states as initial phase in the process of oxygen-evolving complex oxidation. Biophys. J. **94**, 2725–2736 (2008)
13. Dau, H., Haumann, M.: Eight steps preceding OO bond formation in oxygenic photosynthesis - a basic reaction cycle of the photosystem II manganese complex. Biochim. Biophys. Acta BBA - Bioenerg. **1767**, 472–483 (2007)
14. Jablonsky, J., Susila, P., Lazar, D.: Impact of dimeric organization of enzyme on its function: the case of photosynthetic water splitting. Bioinforma. Oxf. Engl. **24**, 2755–2759 (2008)
15. Lazr, D.: Chlorophyll a fluorescence rise induced by high light illumination of dark-adapted plant tissue studied by means of a model of photosystem II and considering photosystem II heterogeneity. J. Theor. Biol. **220**, 469–503 (2003)
16. Isgandarova, S., Renger, G., Messinger, J.: Functional differences of photosystem II from Synechococcus elongatus and spinach characterized by flash induced oxygen evolution patterns. Biochemistry (Mosc.). **42**, 8929–8938 (2003)
17. Styring, S., Rutherford, A.W.: Deactivation kinetics and temperature dependence of the S-state transitions in the oxygen-evolving system of photosystem II measured by EPR spectroscopy. Biochim. Biophys. Acta BBA - Bioenerg. **933**, 378–387 (1988)
18. Seibert, M., Lavorel, J.: Oxygen-evolution patterns from spinach photosystem II preparations. Biochim. Biophys. Acta BBA - Bioenerg. **723**, 160–168 (1983)

Observation of Dynamics Inside an Unlabeled Live Cell Using a Bright-Field Photon Microscopy: Evaluation of Organelles' Trajectories

Renata Rychtáriková[✉] and Dalibor Štys

Faculty of Fisheries and Protection of Waters, South Bohemian
Research Center of Aquaculture and Biodiversity of Hydrocenoses,
Institute of Complex Systems, University of South Bohemia in České Budějovice,
Zámek 136, 373 33 Nové Hrady, Czech Republic
{rrychtarikova,stys}@frov.jcu.cz
http://www.frov.jcu.cz/cs/ustav-komplexnich-systemu-uks

Abstract. This article presents an algorithm for the evaluation of organelles' movements inside an unmodified live cell. We used a time-lapse image series obtained using wide-field bright-field photon transmission microscopy as an algorithm input. The benefit of the algorithm is the processing of the primary signal obtained using a color digital camera and the application of the Rényi information entropy, namely a variable called a point information gain, which enables to highlight the borders of the intracellular organelles and to localize the organelles' centers of mass with the precision of one pixel. The algorithm is introduced on the description of intracellular dynamics of a MG-63 cancer cell.

Keywords: Bright-field photon transmission microscopy · Intracellular dynamics · Information entropy · Unlabeled living cell

1 Introduction

The observation of cell biophysics such as organelles' movements plays a pivotal role in the evaluation of physiological state of mammalian cells, e.g. in recognition of the cell pathology. The relevant kinds of commercial software [1–4], freeware [5], and other published software [6–8] were primarily developed for tracing of fluorescently labeled samples. Also, a more advanced label-free plasmonics-based imaging technique was demonstrated for studying the dynamics of organelles (mitochondria) in three dimensions with 5-nm localization precision and 10-ms temporal resolution [9].

Nevertheless, this observation can be also successfully performed using bright-field transmission as a classical light-microscopic method, whose advantage is a possibility of observation of an unlabeled live cell and tissues as well. It also does not require complicated sample preparation. However, the disadvantage of this microscopic technique is a low contrast of the structures observed in

© Springer International Publishing AG 2017
I. Rojas and F. Ortuño (Eds.): IWBBIO 2017, Part II, LNBI 10209, pp. 700–711, 2017.
DOI: 10.1007/978-3-319-56154-7_62

transparent samples due to the interferences of the light which passes the optical path of the microscope. In order to highlight the intracellular structures' borders in digital images, we proposed a novel information-entropic variable derived from the Rényi information entropy – a point information gain. In the image processing, besides other benefits [10–12], this variable enables to work with intensity histograms and to calculate one resulted value of the point information gain for similar intensity values.

In the presented algorithm, this approach is employed to localize centers of mass of dynamical light-diffracting objects in order to track them in time-lapse images of a mammalian cancer cell. These trajectories are further normalized to the scanning time and statistically analyzed.

2 Results and Discussion

2.1 Description of Image Processing Algorithm

The whole time-lapse raw file series scanned using the nanoscope (Sect. 4.2) was processed as follows:

1. Using Matlab® software fortified by an Image Processing Toolbox and Statistics Toolbox (Mathworks, USA), the image of the cell was got rid of the background using a cumulative mask (Fig. 1), where the contribution to the

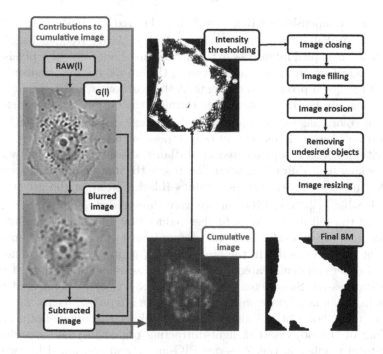

Fig. 1. Algorithm for segmentation of a cell from time-lapse series obtained using wide-field bright-field photon transmission microscopy. (Color figure online)

mask was created from the absolute values of the difference between the green intensities in the original series image and in the blurred version of this image (using a 10-px disk-shaped image filter). The green intensities in the original image were calculated as arithmetic average of each two green pixels in Bayer filter quadruplets [13] which gave a quarter-sized image compared with the original raw data [14].

The obtained mask was further binarized and processed by a standard image-processing methods – by thresholding of the relevant intensities, morphological closing (a 3-px disk-shaped structuring element), filling the holes in the binary image, and morphological erosion to remove the blurred edges of the cell (a 10-px disk-shaped structuring element). The resulted binary mask was two-fold increased to the size of the original raw file and applied to the whole time-lapse series to select the cell of interest at each time step.

Positions of green pixels darker than the intensity mode of the background surrounding the cell in each series image was stored in a .mat file and used for next image processing (see Item 3).

2. For each series segmented raw file, the values of point information gain ($\Gamma_\alpha^{(i)}$, [bit]) were computing using the Image Info Extractor Professional software (Institute of Complex Systems FFPW USB) at omitting black (zero) pixels surrounding the cell of interest as

$$\Gamma_\alpha^{(i)} = \frac{1}{1-\alpha} \log_2 \frac{\sum_{k=1}^{n} (p_k^{(i,j)})^\alpha}{\sum_{k=1}^{n} p_k^\alpha}, \tag{1}$$

where α is a dimensionless Rényi coefficient. For red and blue pixels of the raw data, the probabilities p_j and $p_j^{(i)}$ characterize relative frequencies of occurrences of k occupied intensities in the histogram of relevant color pixels and in the same histogram where one element of the bin i is missing, respectively. In case of the green pixels, two elements at the positions i, j were removed from the histogram. For each color image channel, the output data were stored as an 8-bit color image where black and white pixels correspond to the $\Gamma_\alpha^{(i)}$ with highest and lowest frequencies of occurrences in the raw data.

3. Algorithm in Fig. 2 implemented in Matlab® software (see Sect. 5) was used for tracking light-diffracting organelles inside the images of cell. Through the whole time-lapse series, the zero values (black pixels) in the green channel of each 8-bit $\Gamma_{4.0}^{(i)}$- (i.e., PIG-) image were thresholded. The number of white pixels in the cell interior were further reduced by selection of the intensities darker than original cell's background (see Item 1). In order to segment only relevant large organelles in the binary image, white structures connected to image border and demarcated the cell interior as well as small light structures were suppressed. So-selected organelles (Fig. 3–*right*) as connected components in the binary image were labeled to give a number matrix L.

The main body of the algorithm for tracking the organelles started by labeling of the objects (i.e., light-diffracting organelles) and calculation of their total number in the 1st series PIG-image and continued by two loops. In the higher-order for-loop, each organelle of the 1st label matrix L_1 and the label matrix itself were renamed. Starting the 2nd PIG-image of the time-lapse series, the organelle was traced through the above-mentioned binarized image series while there was one and only overlap of an organelle of interest between two consecutive images (see while-subloop). In this subloop, after thresholding the organelle of interest and finding its centroid in the $(l-1)^{th}$ labeled image, the position of this unique overlap including its number label was found in the lth labeled image. If there was found zero ($k = \emptyset$) or more than one unique overlap, the while-subloop was finished. The cause of disturbing the while-subloop was two-fold:

- If $k = \emptyset$, an organelle did not occur in (vanished from) the $l^{th} + 1$ image.
- If number of elements in k (i.e., if cardinality of the set k) is higher than 1, the organelle decayed in the $l^{th} + 1$ image.

During the while-subloop, the information about the time of saving was read from the names of each pair of original raw files (not illustrated in Fig. 2) which, after the subtraction, gave the scanning time $t_{scan,l}$ of the lth image. After finishing the while-subloop, the centroids saved from the relevant time-lapse images were plotted to obtain a trajectory of the moving organelle (Fig. 4–*left*). The horizontal and vertical components of speed vectors were further computed as ratios of x- and y-components of length of trajectory's step to scanning times $t_{scan,l}$, respectively. Using the Pythagorean Theorem, we obtained the length of trajectory's step d_l and of speed step. From these speed vectors, two resulted speed variables were calculated:

- Average speed characterizes an average path which an organelle travelled per a time unit:

$$AS = \frac{\sum_l (d_l \times px)}{\sum_l t_{scan,l}}, \tag{2}$$

where

$$d_l = \sqrt{(\Delta x)_l^2 + (\Delta y)_l^2}. \tag{3}$$

- Speed resultant describes a total time-normalized path which an organelle travelled:

$$SR = \sum_l \frac{px}{t_{scan,l}} \times (\Delta y)_l. \tag{4}$$

In Eqs. (2)–(4), px is a size of an image pixel ($px = 64\,\text{nm}$). The differences $(\Delta x)_l$ and $(\Delta y)_l$ correspond to the subtraction of a position of the starting and ending point (i.e., pixel) in the lth and $(l+1)^{th}$ image at the 1st and 2nd coordinate.

Finally, these round values of speed vectors underwent the statistical evaluation – namely, plotting the histogram of speed steps (in Sect. 5) and evaluation of the histogram's parameters (Table 1), plotting the trends of organelle's movements (Fig. 4–*right*).

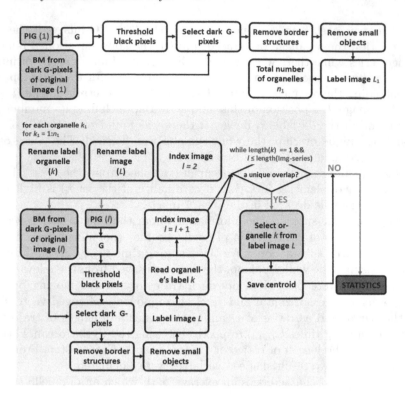

Fig. 2. Algorithm for organelles' tracking and statistical evaluation of movements inside an unlabeled live cell from time-lapse series obtained using wide-field bright-field photon transmission microscopy. *white background* – Segmentation of objects of interests from the 1st PIG-image, *orange background* – for-loop, *gray background* – while-loop, *green boxes* – inputs, *red box* – output. (Color figure online)

2.2 Verification of the Algorithm

The above-described algorithm for analysis of movements of light-diffracting organelles was being verified on an unmodified MG-63 cell line which was scanned in bright-field mode using an inverted optical microscope (Sect. 4). The verification of the algorithm was performed by visual inspection through the time-lapse series (see Sect. 5). The segmented objects (organelles) were labeled by positions of centroids in Img. 1. The results of complete analysis are saved in Supplementary (Sect. 5).

In Img. 1, the algorithm detected 56 relevant objects, from which 46 objects were further found in Img. 2 and analyzed for the cause of disturbing the tracking:

1. Tracking of only one intracellular object (Organelle 4 at the position 095-118) was stopped since there was no overlap with the following image. In other words, the speed of the organelle was faster than requested scanning speed of the camera.

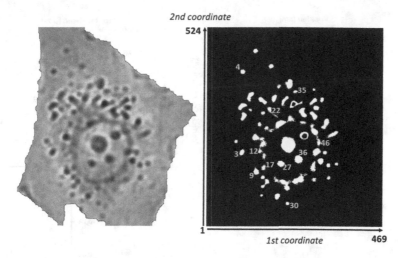

Fig. 3. Results of the segmentation algorithms. *left* – The 1st bright-field photon transmission time-lapse series image of an unmodified live MG-63 human osteosarcoma cell (visualized in 8-bit/c color depth). *right* – A binary image of organelles selected for statistical evaluation of their trajectories and speeds using the algorithm calculating the $\Gamma_{4.0}^{(i)}$. The coordinates describe the positions of the organelles as written in Supplementary. The Nos. of organelles correspond to the Nos. of organelles further in the text, figures, and Supplementary and are ordered according to the position of the organelle in the 1st binary image.

2. Finishing the tracking algorithm due to vanishing of organelle in the following image concerns to 12 organelles where re-appearing of the object in the next-next image occurred in 6 cases. This re-appearing occurred mainly due to the spectral properties of the observed organelle between series images which changes the possibility to detect the organelle using $\Gamma_{\alpha}^{(i)}$ computation.

3. 8 objects were split into more parts and the object in the next-to-last image was overlapped with more objects in the last image. This made the next tracking of the objects difficult. This splitting can be observed if either an intracellular object were divided into two (or more) objects (e.g., in case of division of mitochondria) or two objects were situated each above other and they moved mutually. In case of Organelle 3 (position 093-331), it is difficult to decide if there was its splitting or vanishing in the last image.

4. Only two objects (nucleoli Nos. 27 and 36 at the position 197-361 and 245-350, respectively) were detected through the whole 766-image series.

5. The observation of the rest of the objects was stopped due to joining a neighboring organelle in the next-to-last image followed by re-splitting. Let us note that the joining of the objects shifts the centroid detected in its binary image and can distort the results of the followed statistics which is based on evaluation of trajectory and velocity steps.

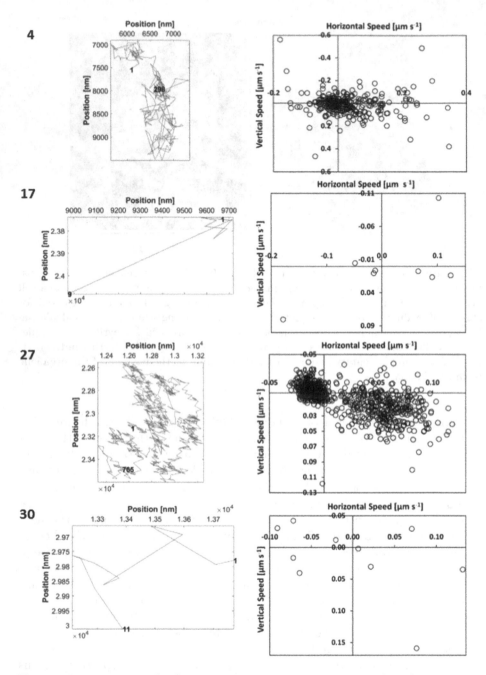

Fig. 4. The trajectories (*left*) and the directions of speed steps (*right*) of moving organelles inside an unmodified live MG-63 human osteosarcoma cell. The Nos. of organelle correspond to the Nos. of organelles in the text, figures, and Supplementary and are ordered according to the position of the organelle in the 1st binary image. The Nos. in the left graphs represent the first and last time steps of tracking.

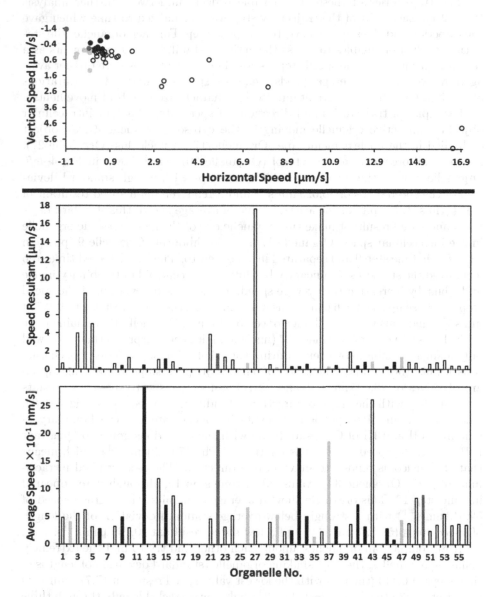

Fig. 5. The directions of speeds (*upper*), the values of speed resultants (*middle*), and the average speeds (*lower*) of moving organelles inside an unmodified live MG-63 human osteosarcoma cell. White, light grey, dark grey, and black bars represent the positions of the speed resultants in Quadrant I, II, III, and IV of the Cartesian plane, respectively. The Nos. of organelle correspond to the Nos. of organelles in the text, figures, and Supplementary and are ordered according to the position of the organelle in the 1st binary image.

The trajectories of these 46 trackable objects underwent further analysis (Sect. 2.1). Each shift of the trajectory step was normalized to time which gave the direction and the size of the vector of speed step. For each organelle, the statistical evaluation enables to assess the activity/stability of the movements as a standard deviation of the speed steps' sizes. These speed steps were further averaged by two ways: by average speeds (i.e., total speeds related to total durations of tracking) and by speed resultants (which characterize trends of movements).

Examples of trajectories and directions of speed steps for four intracellular objects – an active organelle moving in the cytosol (Organelle 4), an object embodied in the nuclear membrane (Organelle 17), a nucleolus (Organelle 27), and a small organelle in the cytosol (Organelle 30) – are shown in Fig. 4–left. Organelle 4 was being tracked for 298 time steps with a high standard deviation of the speed steps of 86.023 nm s^{-1} and a relatively high speed resultant of 8.374 μm s^{-1}. In spite of the average value of average speed, this standard deviation and speed resultant make this organelle one of the most dynamic organelle in the intracellular space. Organelle 17 hit a neighbouring Organelle 9 (position 131-381) in time step 9 and separated in time step 10. The last detected time step 9 changed the statistics significantly by shifting the centroid of the binary image and, thus, by increasing the average speed. The standard deviation of the speed steps increase up to 63.976 μm s^{-1} as well. However, the speed resultant (129.774 nm s^{-1}), and thus the total change of the position of Organelle 17, remained one of the lowest. In case of Organelle 27 (nucleolus), it is not surprised that it showed only a slight shaking movements during the whole time-lapse series. Its characteristics like the standard deviation of lengths of speed steps and average speed are of values of 29.605 nm s^{-1} and 22.471 nm s^{-1}, respectively. These results also correlate with the graph of directions of individual speed steps (Fig. 4–left), where a substantial proportion of points is concentrated at the boundary of Quadrants II and III of Cartesian plane with the deviations going to Quadrant I. However, the speed resultant is relatively high (17.652 μm s^{-1}) which means that the nucleolus tended to move in one direction. The next studied intracellular object – Organelle 30 – vanished in time step 12, although it re-appeared in time step 13. This organelle showed average dynamics (the average speed of 52.513 nm s^{-1}) with a strongly below-average standard deviation of the length of speed steps (50.527 nm s^{-1}) and a low speed resultant (0.169 μm s^{-1}).

If we compare all organelles among each other and try to find the extremely behaving organelles, the least shaking organelle (standard deviation of lengths of time steps of 0.114 μm s^{-1}) with the lowest value speed resultant (7.778 nm s^{-1}) is that of No. 35 (position 236-171), although it was trackable only through three time steps, despite it re-appeared in time step 5. The object with the highest standard deviation of length of time steps (1002.165 nm s^{-1}) is marked by No. 22 (position 171-222) and detected in four time steps. Nevertheless, in the 1st time step, this object seems to be formed by merging of two objects which were disconnected in the 2nd time step and one of these objects was joint to the other organelle. This led to the significant change of the position of the tracked centroid. Organelles 46 (a component of nucleolar envelope) and 12 at the positions 290-282 and

141-327 have the lowest $(7.115 \text{ nm s}^{-1})$ and the highest $(284.381 \text{ nm s}^{-1})$ value of the average speed, respectively. However, in these cases, these resulted values have a low information value. Organelle 46 thus already split in time step 3, whereas Organelle 12 did not too moved during the first two steps but it connected to a neighboring object in time step 3.

As seen in Fig. 5–*upper*, only 7, 3, and 8 intracellular objects tended to move into Quadrant II, III, and IV of the Cartesian plane, respectively. The rest of objects were moving into Quadrant I (southeastern direction) where some attractor can be localized.

Table 1. Statistical evaluation of histograms of speed steps' vectors

	Mean [nm s^{-1}]	Median [nm s^{-1}]	Mode [nm s^{-1}]	STD [nm s^{-1}]
Organelle 4	92.731	66.000	25.000	86.023
Organelle 17	87.875	78.500	14.000	63.976
Organelle 27	37.992	26.000	12.000	29.605
Organelle 30	78.500	76.000	76.000	50.527

3 Conclusions

The presented automated label-free method based on image processing of the bright-field photon microscopic time-lapse series provides an insight into the dynamical processes which can be crucial for understanding of many biological phenomena, including cell growth, mass transportation, signaling transduction and cell migration. The method showed a potential to be used, e.g., in pharmacology and toxicology for testing effects of chemical substances to tissue cultures and is mainly suitable for observation of mitochondria and large endosomes. Nevertheless, the method is believed to be further improved by a possibility to identify the observed objects using calculation of a specific divergence derived from the Rényi information entropy – a point divergence gain – which enables to specify spectral properties of an organelle by finding of pixels of unchanged intensities for two consecutive images in the time-lapse series (if needed, fortified by a zig-zag z-stepping) [12]. The precision of the analysis can be increased by the usage of a camera of a higher scanning frequency and resolution.

4 Methods

4.1 Cell Cultivation

A MG-63 cell line (Serva, cat. No. 86051601) was grown at low optical density overnight at 37°C in a synthetic dropout media with 30% raffinose as the sole carbon source. The nutrient solution for MG-63 cells consists of: 86% EMEM, 10%

newborn-calf serum, 1% antibiotics and antimycotics, 1% L-glutamine, 1% non-essential amino acids, 1% $NaHCO_3$ (all components purchased from the PAA company). During microscopy experiment, cells were cultivated in a Bioptech FCS2 Closed Chamber System.

4.2 Microscopy

Microscopy of a living MG-63 cell culture was performed using a versatile sub-microscope: a nanoscope developed for the Institute of Complex Systems FFPW by the company Optax Ltd. (Czech Republic). The optical path consisted of two Luminus 360 light emitting diodes, the condenser system, a firm sample holder, and an 40× objective system made of two complementary lenses that allow a change of distance between the objective lens and the sample. The UV and IR light emitted by LEDs was blocked by a 450-nm Longpass Filter and 775-nm Shortpass Filter (Edmund Optics), respectively. Next, a projective lens magnified the image to project on a Basler ACA2000-340kc camera chip of resolution of 2048 × 1088 and 12-bit color depth. The size of the original camera pixel using primary magnification was 32×32 nm^2. The combination of the time-lapse scan and zig-zag z-scan with a step of 300 nm was performed automatically with the average time step of (0.396 ± 0.594) fps to give a 766-image series of raw data around the focal plane obtained using a Bayer camera filter.

5   Supplementary

The Supplementary Data are stored at [15].

Acknowledgments. This work was supported by the Ministry of Education, Youth and Sports of the Czech Republic – projects CENAKVA (No. CZ.1.05/2.1.00/01.0024), CENAKVA II (No. LO1205 under the NPU I program), The CENAKVA Centre Development (No. CZ.1.05/2.1.00/19.0380) – and by the CZ-A AKTION programme.

References

1. http://www.bitplane.com. Accessed 8 Feb 2017
2. http://olympusmicro.com/moviegallery/confocal/folumkomito/index.html. Accessed 8 Feb 2017
3. http://www.quantacell.com/applications/cell-analysis/time-lapse-tracking-of-organelles/. Accessed 8 Feb 2017
4. http://phioptics.com/wp-content/uploads/AN02-Cell-Dynamics_small.pdf. Accessed 8 Feb 2017
5. http://cellprofiler.org/tracer/. Accessed 8 Feb 2017
6. Jones, S.A., Shim, S.H., He, J., Zhuang, X.: Fast, three-dimensional super-resolution imaging of live cells. Nat. Methods **8**, 499–508 (2011)
7. Holden, S.J., Uphoff, S., Kapanidis, A.N.: DAOSTORM: an algorithm for high-density super-resolution microscopy. Nat. Methods **8**, 279–280 (2011)

8. Babcock, H., Sigal, Y.M., Zhuang, X.: A high-density 3D localization algorithm for stochastic optical reconstruction microscopy. Opt. Nanoscopy **1**, 6 (2012)
9. Yang, Y., Yu, H., Shan, X., Wang, W., Liu, X., Wang, S., Tao, N.: Label-free tracking of single organelle transportation in cells with nanometer precision using a plasmonic imaging technique. Small **11**(24), 2878–2884 (2015)
10. Rychtáriková, R.: Clustering of multi-image sets using Rényi information entropy. In: Ortuño, F., Rojas, I. (eds.) IWBBIO 2016. LNCS, vol. 9656, pp. 517–526. Springer, Heidelberg (2016). doi:10.1007/978-3-319-31744-1_46
11. Rychtáriková, R., Korbel, J., Macháček, P., Císař, P., Urban, J., Štys, D.: Point information gain and multidimensional data analysis. Entropy, **18**(10) (2016). Article No. 372
12. Rychtáriková, R., Náhlík, T., Shi, K., Malakhova, D., Macháček, P., Smaha, R., Urban, J., Štys, D.: Super-resolved 3-D imaging of live cells organelles from bright-field photon transmission micrographs, under the review in Ultramicroscopy. http://arxiv.org/pdf/1608.05962.pdf
13. Bayer, B.E.: Color imaging array, Patent US3971065 A (1975)
14. Tkačik, G., Garrigan, P., Ratliff, C., Milčinski, G., Klein, J.M., Seyfarth, L.H., Sterling, P., Brainard, D.H., Balasubramanian, V.: Natural images from the birthplace of the human eye. PLoS ONE **6** (2011). Article No. e20409
15. ftp://160.217.215.251:21/Tracking (user: anonymous; password: anonymous)

Automatic Multiparameter Acquisition in Aquaponics Systems

Antonín Bárta[✉], Pavel Souček, Vladyslav Bozhynov,
and Pavla Urbanová

Laboratory of Signal and Image Processing, Faculty of Fisheries and Protection
of Waters, Institute of Complex Systems, South Bohemian Research Center
of Aquaculture and Biodiversity of Hydrocenoses, University of South Bohemia
in České Budějovice, Zámek 136, 37333 Nové Hrady, Czech Republic
abarta@prov.jcu.cz

Abstract. For various purposes from simple water monitoring in maintenance, regulation, control and optimization to behavior models in biometrics, biomonitoring, biophysics and bioinformatics, it is necessary to observe wide field of variables. The precision and accuracy are one of the most important attributes as well as price and operational burden. Online water monitoring is becoming more trendy due to its time saving method and automatic alert included. From that point of view, old manual methods are replaced by these autonomous data obtaining systems, which can simplify the process with data crucial data measurement. Aquaponic systems represents a complex system for many biophysical experiments. In this article, the comparison of basic monitoring solutions is introduced and discussed.

Keywords: Automatic · Multiparameter · Monitoring · Aquaponics · Aquaculture · Hydroponics · Nutrients · Measurement · Control · Systems

1 Introduction

Monitoring of crucial parameters in aquaponics systems is a necessity, which could never be forgotten. It does not matter which system, (commercial, hobby or garden) you are using, because the "heart" of the systems is with small subtleties always the same. Aquaponics is a connection of aquaculture and hydroponics [1]. Both approaches of cultivation are very sensitive to its own parameters. Every aquaponic farmer must think about balancing the right water condition to offer friendly and convenient environment for both, fish and plants. There is a couple of fundamental monitoring solutions, which are suitable for certain sector. Hobby and garden users are mainly using manual measuring device based on chemical reactants. On the other hand, growing aquaponics industry is searching for automatic monitoring solutions, which can transfer online information from measurements of crucial parameters (especially pH, temperature, electric conductivity, dissolved oxygen ammonia, nitrate and iron). From the gardener point of view, first big limitation to set up a commercial aquaponics farm is the lack of the online monitoring system, which is normally very expensive. Laboratory of Signal and Image Processing is now developing smart online water monitoring systems based

I. Rojas and F. Ortuño (Eds.): IWBBIO 2017, Part II, LNBI 10209, pp. 712–725, 2017.
DOI: 10.1007/978-3-319-56154-7_63

on Arduino and RaspberryPi microcomputers to decrease the gap between hobby and commercial farmers and offer the online monitoring solution to general public [2]. The other problem with current commercial solutions is that must be implemented by IT specialist and mostly they are not user friendly. This should be changed to offer "Plug and Play" solution for every farmer who want to jump to the next level of growing (Table 1). These circumstances led us to think about more advanced and user friendly smart online water monitoring solution which help common aquaponics, aquaculture and hydroponics farmers be more precise in their cultivation and offer them the possibility to track the evolution of their growing techniques (Fig. 1).

Fig. 1. Experimental grow bed based aquaponic system

Table 1. General fish and plants water quality tolerances [3]

Organism type	Temp (°C)	pH	Ammonia (mg/litre)	Nitrite (mg/litre)	Nitrate (mg/litre)	DO (mg/litre)
Warm water fish	22–32	6–8.5	<3	<1	<400	4–6
Cold water fish	10–18	6–8.5	<1	<0.1	<400	6–8
Plants	16–30	5.5–7.5	<30	<1	–	>3
Bacteria	14–34	6–8.5	<3	<1	–	4–8

2 State of the Art

2.1 Sera and Other Reactants Based Method of Water Monitoring

The easiest way of water quality monitoring, mainly used by hobby aquarist is based on chemical reactants. Key problem of these solutions is extremely non-accurate results, which could be dangerous for sensitive fish or coral species. In special cases, user has unique combination of fish, plant and organisms and must balance ideal conditions of all living organism in aquarium. The ideal proposition of values (pH, temperature, Calcium, kH, iron and so on) is must be very precise. This approach does not satisfy requirements. There is a big potential to make mistakes by user. Other disadvantage of this solution is usage of dangerous chemicals. For example, during ammonia test, user is working with LiOH and NaClO. These compounds should be hidden from kids.

Here is the another very similar approach from Tetra WaterTest Set Plus procedure (Ammonium/Ammonia):

1. Rinse the test vial with the sample water. Fill the test vial up to 5 ml mark with sample water. Important: The temperature of the sample water must be between 20° and 30 °C.
2. Hold reagent bottle 1 upside down and add 14 drops to the test vial.
3. Close the test vial and shake gently.
4. Open the test vial, hold reagent bottle 2 upsides down and add 7 drops to the test vial.
5. Shake the test vial gently.
6. Hold reagent bottle 3 upsides down and add 7 drops to the test vial.
7. Shake the test vial.
8. Allow 20 min (at room temperature) for the color to develop.
9. Shake the test vial gently and match the shade of the test solution to the color it comes closest to on the color chart. Read the values by holding the test vial approximately 1 cm (a finger's breadth) from the white area on the color chart. After each test, rinse the vial thoroughly with tap water (Fig. 2).

Fig. 2. Tetra water test – ammonia and ammonium comparative chart

The method of calculating total ammonia in water is dependent on pH values. Both forms (ammonium/ammonia) can be brought into equilibrium via the pH value. As a rule, a high pH value (>8.5) represents more toxic ammonia; a low pH value (<7.5)

%	6	6,5	7	7,5	8	8,5	9	9,5	10
0 °C	0,008	0,026	0,082	0,261	0,820	2,55	7,64	20,7	45,3
5 °C	0,125	0,039	0,125	0,394	1,23	3,80	11,1	28,3	55,6
10 °C	0,018	0,059	0,186	0,586	1,83	5,56	15,7	37,1	65,1
15 °C	0,027	0,086	0,273	0,859	2,67	7,97	21,5	46,4	73,3
16 °C	0,029	0,093	0,294	0,925	2,87	8,54	22,8	48,3	74,7
17 °C	0,032	0,101	0,317	0,996	3,08	9,14	24,1	50,2	76,1
18 °C	0,034	0,108	0,342	1,07	3,31	9,78	25,5	52,0	77,4
19 °C	0,037	0,117	0,368	1,15	3,56	10,5	27,0	53,9	78,7
20 °C	0,039	0,125	0,396	1,24	3,82	11,2	28,4	55,7	79,9
21 °C	0,043	0,135	0,425	1,33	4,10	11,9	29,9	57,5	81,0
22 °C	0,045	0,145	0,457	1,43	4,39	12,7	31,5	59,2	82,1
23 °C	0,049	0,156	0,491	1,54	4,70	13,5	33,0	60,9	83,2
24 °C	0,053	0,167	0,527	1,65	5,03	14,4	34,6	62,6	84,1
25 °C	0,057	0,180	0,566	1,77	5,38	15,3	36,3	64,3	85,1
30 °C	0,080	0,254	0,799	2,48	7,46	20,3	44,6	71,8	89,0

Fig. 3. The concentration of undissociated ammonia at temperature and pH of the water.

signifies more harmless ammonium. To make sure your fish are not at risk, check that the pH value is within the acceptable range. Also, keep the ammonia concentration as low as possible.

2.2 Calculate the Concentration of the Toxic Form of Ammonia Nitrogen

For example: The concentration of total ammonia nitrogen (N-NH4$^+$) is 2.5 mg/l. The pH value is 9,5 and temperature is 17 °C.

From Fig. 3 above, we see that at a given temperature and pH is 50,2% in undissociated form.

Concentration N-NH3 = 2.5 * 50.2/100 = 2.5 * 0.502 = 1.255

The toxicity of ammonia is normally written only as NH3 and from this reason we must recalculate N-NH3 value to pure NH3.

NH3 = 1.255 * 1.216 (conversion factor) = 1.526> The concentration of undissociated ammonia **NH3 = 1.53 mg/l.**

At first glance is obvious, that this approach is very time consuming. From the beginning to the final results it took almost 30 min. This method is satisfactory for home aquariums, but inadequate for commercial usage. The chemical reactants methods are very wide area. User can investigate the wide variety of water parameters, from calcium, water hardness to phosphate or iron concentration.

2.3 Portable Sensors and Measuring Devices

Portable sensors offer more flexibility to user's demands. First, it is filling the gap about time consuming. It is very user-friendly solution in comparison with chemical reactants based analysis. Normally, you do not have to recalculate your results to final results value as we could see in ammonia conversion, because the device gives you a final number of concentration in ppm, mg/l or °C. There are three fundamental approaches of portable devices. First is that the device is monitoring only one parameter (for example pH value). The user must buy the other devices with another purpose or it already exists solution which offer the combination of probes in one device. These kombi devices are normally more expensive, but for basic water analysis is a great solution. Of course, there is a lack of online monitoring. The user must always manually put the sensor into the water and rewrite the values from device's screen. The main advantage of this solution per chemical reactants based analysis is time and accuracy.

For example, here are the devices which we are using in our laboratory:

1. **Hanna Instruments HI 9813-6 Portable pH/EC/TDS/Temperature Meter with CAL CHECK**

   The Hanna Instruments HI 9813-6 Portable pH/EC/TDS/Temperature Meter with CAL CHECK Feature is a versatile, water resistant, multiparameter portable instrument specifically designed for agricultural applications such as hydroponics, greenhouses, farming and nurseries. The Hanna Instruments HI9813-6 features an large LCD that clearly displays the parameter being measured as well as calibration instructions. Calibration is fast and easy with knobs located on the front panel of the instrument. The Hanna Instruments HI 98136 includes all the features of the Hanna Instruments HI 9813-5 while incorporating our exclusive CAL CHECK feature. CAL CHECK allows the user to easily check the pH probe calibration status at any time. The Hanna Instruments HI 9813-5 utilizes the HI 1285 series pH/EC/TDS/ temperature probe. This probe features a fiber junction and gel electrolyte making it ideal for fertilizer solutions (Fig. 4).

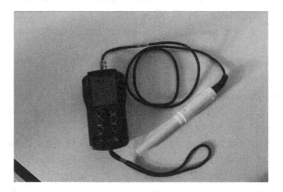

Fig. 4. HI 9813-6 Portable pH/EC/TDS/Temperature Meter

2. **Milwaukee MW600 (0.0–19.9 ppm O2/l) Dissolved Oxygen Meter**

   The Milwaukee MW600 is a compact Portable Dissolved Oxygen meter with Faster Micro Processor. This handy and ergonomically designed portable meter is ideal for anyone working on a low budget and still requires fast and reliable measurements.

This portable meter measures Dissolved Oxygen with a Polarographic probe and is suitable for a wide range of applications, such as Educational and Aquaculture, as well as water and environmental analysis. The Milwaukee MW600 calibrates easily in 2 points (at 100% saturated air and in 0 Oxygen solution) and has Automatic Temperature Compensation which guarantees the highest accuracy. Other features include smaller, ergonomic and lighter case design, 100% larger and easier to read LED Display, low battery warning, easy to replace screw on cap membranes and long battery life. The low battery warning, easy to replace screw on cap membranes make this meter very simple to operate (Fig. 5).

Fig. 5. Milwaukee MW600 (0.0–19.9 ppm O2/l) dissolved oxygen meter

Checker Colorimeters.

The next solution, that we can call portable measurement device is operating on different approach. It is based on colorimetric solution. The Hanna Checker® HC (Handheld Colorimeter) bridges the gap between simple chemical test kits and professional instrumentation. Chemical test kits are not very accurate and only give 5 to 10 points resolution while professional instrumentation can cost hundreds of dollars and can be time consuming to calibrate and maintain. The Checker® is both accurate and affordable. On the other hand, it also takes some time to measure with these devices according to above described sensors. The process of measurement includes combination of reactants and sample water and user must follow the process precisely to avoid mistakes in results. From this point of view, it is very similar to the first chemical reactants base water monitoring described in the beginning of this article. The main advantage is, that the user get the information directly in final concentration. There is no need to process results (Fig. 6).

Seneye.

Next step of water quality monitoring is also based on chemical reactions. In this case was previous method little bit advanced and modified to online water monitoring solution [5]. Seneye works with small stripes, which are working as a pointer for water quality parameters. These stripes are monitored and the color information from stripes

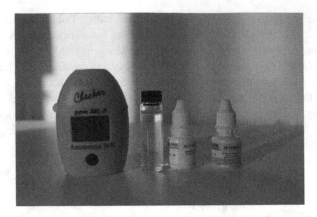

Fig. 6. Hanna instruments – ammonia medium range checker (HC-HI715)

indicates the values of temperature, free ammonia and pH. The seneye device can be used in the following ways. Online will give you constant uploads to seneye.me with alerts. Offline will allow the device to work as a data logger and upload readings when the device is next connected to a PC using SCA (Seneye connect application). Each seneye slide has a unique activation code which must be registered and connected with Seneye device. When the seneye device is connected to computer and the SCA is launched, any stored readings will be automatically uploaded. During this process egg timers, may be shown, and live readings will not show until this process has finished. If the user decided to use the device offline must disconnect the device only when the SCA has finished uploading readings. Graphs and charts for these readings will be produced once they have been received and analyzed. Depending on the number of uploads this process can take a few minutes.

Once the SCA is running you will see that "Current Readings" page is displayed. Your ammonia (NH3), pH and temperature are displayed in the boxes. If one of these parameters is outside of your warning levels, then the box will turn red alerting you. The Seneye device takes reading every 30 min when connected to a PC. The information at the bottom of the page shows when the last reading was completed. This reading will be sent when the Seneye server can be reached (Fig. 7).

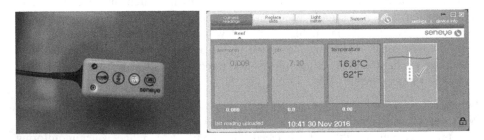

Fig. 7. Seneye - online water monitoring solution

Cooking Hacks – Open Aquarium and Open Garden.

Open aquarium is next level of cheap online water monitoring. Open Aquarium has been designed to help to take care of fish by automating the control and maintenance tasks that take place in the fish tanks and ponds. Open Aquarium consists of two different and complementary kits: Basic and Aquaponics, and many several extra accessories. There are 5 sensors available in the platform to measure the key parameters in the fish tank such as temperature, pH or Conductivity and then some others to control the correct state of the fish tank (water level and leakages). In addition, there are 4 different actuators to automate tasks such as heating or cooling the water, feeding the fish, activating the pumps for water change or medicines administration, and controlling the intensity of the light to simulate the day/night cycles [6] (Fig. 8).

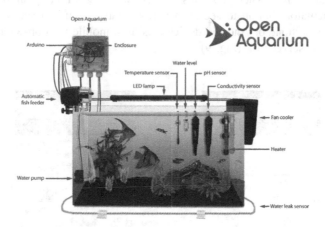

Fig. 8. Open aquarium diagram

Open Aquarium Sensors (The Most Important Sensors).

1. DS18B20 - Water temperature sensor
   This sealed digital temperature probe lets you precisely measure temperatures in wet environments with a simple 1-Wire interface. The DS18B20 sensor provides 9 to 12-bit (configurable) temperature readings over a 1-Wire interface, so that only one wire (and ground) needs to be connected from a central microprocessor.
2. pH sensor
   Measurement range: 0–14 p, temperature of operation: 0–80 °C, zero electric potential: 7 + −0,25p, response time: <1 min, Internal resistance: <=250 MΩ.
3. Dissolved oxygen sensor
   Sensor type: Galvanic cell, Range: 0–20 mg/l, Accuracy: +−2%, Saturation output: 33 mV + −9 mV, Pressure: 0–100 psig (7,5 Bar), Response time: after equilibration, 2 min for 2 mV.

A complete Open Source API is included to easily control the board through Arduino. Solution offers a web application that allows to store in a database the information gathered and visualize it from a browser and from iPhone/Android devices. Open aquarium team released it as Open Source code too, so that you can improve it and make it personal for your own product. Open Aquarium has been designed to work with both 220 V (Europe) & 110 V (US).

Data Visualization.
Open Aquarium includes an application to visualize real data of all your sensors in a web browser. In user field, you should type the user name that you have written in config.php file, and then in password fields, your own password without encryption. Once completed, you should click to connect button at the bottom of the form and you arrive to the main screen when you can see all sensor values. Finally, in the menu, you can click in actuators option to see all actuators states. You can access to the web application from iPhone or Android devices, using a mobile web browser and typing the URL of your server in it (Fig. 9).

Fig. 9. Data visualization – mobile version

Atlas Scientific.
This company convert devices that were originally designed to be used by humans into devices that are specifically designed to be used by robots. The Atlas Scientific company isolated the core functionality of human devices and redesign them to be "robot ready", by doing this Atlas Scientific is empowering the engineer to add the capabilities of many individual devices into a single robot. Atlas Scientific recognizes the importance of high accuracy, repeatability and ease of use [7].

Atlas Scientific does not offer the "Plug and Play" solution. The company is focusing on selling sensors, probes, whole kits and others electronics components. On the other hand, if the user has experience and skills with basic programing of Arduino

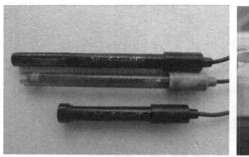

Fig. 10. Atlas scientific – sensors and hardware connection

and RaspberryPi, it is very modifiable solution. There is available their own data logger software called Eniac, but still in BETA version. Atlas Scientific sensors are becoming very popular in open source online water monitoring community (Fig. 10).

3 Big Data

The amount of data in our world has been exploding, and analyzing large data sets—so-called big data—will become a key basis of competition, underpinning new waves of productivity growth and innovation [8]. We are living in an age of "Big data" [8], which is changing all areas of human-kind including science. One of the most important issues in experimental research is the reproducibility of experiments. Achenbach [9] describes the actual situation in the world science, which is measured and driven by the peer-review publishing process. The reproducibility and replicability of biochemistry and biophysics experiments is becoming more and more critical relative to the enormous number of scientific papers published nowadays [10]. The reproducibility is highly connected to the proper description of experimental conditions, which can influence the results of the experiment. The experimental protocol is not only the measurable conditions under which we perform our experiments but it is a complete set of information called experimental metadata. To clarify the concept of experimental metadata, we must start with the general definition of metadata. Source [11, 12] defines metadata as data about data. One of the main challenges in modern science is the amount of data produced by the experimental work; it is difficult to store, organize and share the scientific data and to extract the wealth of knowledge. Experimental method descriptions in scientific publications are often incomplete, which complicates experimental reproducibility. The bioWES system was created in order to address these issues. It provides a solution for management of the experimental data and metadata to support the reproducibility [13] (Fig. 11).

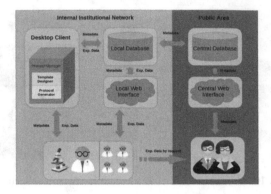

Fig. 11. An overall overview of BioWes system [13]

4 Control

We are living in new age of IoT (Internet of things) [14]. The possibility of connection physical things to the internet is exponentially growing every day. The main advantage of this solution for hydroponics, aquaponics and fishery industry is the reduction of human work in the common tasks connected with water monitoring quality. From the big industry company perspective where hundreds of tanks full of fish are dependent on optimal water parameters (pH, temperature, dissolved oxygen, ammonia level) is critical to be able react very quickly on changes and has general online overview about water quality in fish or hydroponic tanks. Human entity will be always behind the online water monitoring solution. The fundamental idea of the online monitoring system is to save fish and plants welfare and in the same time human work.

Nowadays is becoming very popular mini computers that could change the game of water monitoring industry. Raspberry PI and Arduino are the main players in this game [15]. Arduino is a widely used open-source single-board microcontroller development platform with flexible, easy-to-use hardware and software components. Arduino Uno R3 is based on Atmel Atmega328 microcontroller and has a clock speed of 16 MHz. It has 6 analog inputs and 14 digital I/O pins, so it is possible to connect a number of sensors to a single Arduino board. Arduino-compatible custom sensor expansion board, known as shield, can be developed to directly plug into the standardized pin-headers of the Arduino UNO board. The Raspberry Pi 3 Model measures only about 3.5 by 2.5 inches —small enough to fit in your shirt pocket—and you should be able to reuse any cases or other devices designed for the earlier models. Also, essentially identical are the ports, which offer basic functionality and not a great deal more. You'll need to connect at least a keyboard and a mouse to two of the four USB 2.0 ports, a display to the full-size HDMI port, and a micro USB charger (such as for your cell phone) to power the thing. This time around, though, the Raspberry Pi folks recommend a 2.5-amp adapter "if you want to connect power-hungry USB devices." In any case, there's no Power switch; plug in the charger, and the Pi 3 turns on, or unplug it to turn the system off.

These two excellent mini computers are ready to connect with environmental monitoring probes and sensors. In our case, we are mainly focusing on water monitoring sensors and probes. The most viewed parameters in fishery industry is the "Gold Trinity", pH, temperature and dissolved oxygen. Of course, there are many others parameters (probes) that can be easily added (Ammonia and nitrite etc.), but the "Gold Trinity" is standard which should be always measured because it has the biggest impact on fish welfare.

There are already solutions which offer to customer mini computers based online water monitoring solution. One of them is Slovenia PONNOD company. The company developed software called PONNOD SMART [16]. Unfortunately, their solution is still under development, you can just click add to waiting list.

The advanced software solution has implemented database to make statistics, multivariate analysis, estimation, prediction or event detection. Database will help users to get information or compare information with others. The database system called bioWES developed by Institute of Complex systems with connection of Atlas Scientific sensors and Arduino and RaspberryPi micro computers will definitely satisfy customer's needs for complete online water monitoring solution.

Laboratory of Signal and Image Processing which is focusing on transfer knowhow from cybernetics, mathematics and physics to biology, is now developing online water monitoring system based on mini computers and Atlas Scientific sensors connected with bioWES database solution [17] (Fig. 12).

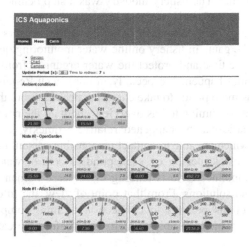

Fig. 12. Data visualization in ICS aquaponics system (not calibrated yet)

5 Results

More than several years was market with Arduino and Raspberry PI transformed to potential inexpensive environmental online monitoring solutions. Today status quo is related with relatively easy process of putting all fundamental parts together. If the user has basic programming skills and he is familiar with essential electronics, he would probably have no troubles with construction of smart online water monitoring solution.

System	pH	T (°C)	EC (µS/cm)	DO (mg/l)	Ammonia (mg/l)	Time of measurement	Price
Atlas scientific (online)	7,425	24,92	243,13	7,65		online	cca 800 USD
Open aquarium (online)	7,15	24,52	236				cca 450 USD
Seneye (online)	8,1	24			0,004	online	cca 160 USD
Hanna-instruments - EGG (manual)	6,9	25,2	210			150 sec	From 50 USD
MW600 (manual)	7,21	25,1		7,54		55 sec	From 350 USD
Reactants based solution (manual)	7		230			300 sec	From 40 USD

Fig. 13. The comparison of different method of measurements

On the other hand, to set up commercial business is much more complicated. Mainly from reasons related with GUI programming. ICS online water monitoring system is actually running but it still need a lot of improvements. Especially GUI (Graphical user interface) and automatic calibration has to be finished. The evaluation process of our measurement devices is summarized in the table below. We can observe the biggest fluctuations in pH values (Fig. 13).

6 Conclusion

We are living in a digital world. We can see the evolution of hundreds market sectors that was changed during last few years. Almost everything is now controlled and connected through internet. The fishery industry was a step behind these IoT (Internet of Things) revolution. Nowadays is raising a big commercial potential for new startup companies. The usage of mini computers, Atlas Scientific sensors and software with database can change the game in fishery online water monitoring quality management. These solutions can save time and protect the water organism from fluctuations in the water quality. Sometimes happened, especially in our Czech Republic region, that the Fishery companies has more ponds to take care than people. From that point of view is very stressful and time consuming to has overview about all the ponds in the company. The catastrophic scenario can be happened relatively quickly because there are no automated indicators which are able to contact responsible person. In medieval age was normal to has a responsible person on each pond to control water quality and fish welfare. But today economic world is tending to decrease human work and increase automated methods and solutions. From that point of view, the future of smart online water monitoring solutions belongs to connection of mini computers, sensors and software with database which are able to alert users before the catastrophic scenario becomes reality.

Acknowledgement. The study was financially supported by the Ministry of Education, Youth and Sports of the Czech Republic - projects 'CENAKVA' (No. CZ.1.05/2.1.00/01.0024), 'CENAKVA II' (No. LO1205 under the NPU I program); and by the GAJU 017/2016/Z.

References

1. Diver, S.: Aquaponics-Integration of hydroponics with aquaculture. Attra, Melbourne (2000)
2. Gertz, E., Di Justo, P.: Environmental Monitoring with Arduino: Building Simple Devices to Collect Data About the World Around Us. O'Reilly Media, Inc., Sebastopol (2012)
3. http://www.heavenandearthaquaponics.com/small-scale-food-production-in-aquaponics-water-quality-part-3/
4. http://web2.mendelu.cz/af_291_projekty2/vseo/print.php?page=3179&typ=html
5. https://www.seneye.me/seneye-userguide.pdf
6. https://www.cooking-hacks.com/documentation/tutorials/open-aquarium-aquaponicsfish-tank-monitoring-arduino/
7. https://www.atlas-scientific.com/
8. Shaw J.: (2014). http://www.harvardmagazine.com/2014/03/why-big-data-is-a-big-deal. Accessed 22 June 2015
9. Achenbach J. (2015) http://www.theguardian.com/science/2015/feb/07/scienctific-research-peer-review-reproducing-data Accessed 22 June 2015
10. Casadevall, A., Fang, F.C.: Reproducible science. Infect. Immun. **78**(12), 4972–4975 (2010). doi:10.1128/IAI.00908-10
11. Cartho W.: Metadata. An overview. (2015) http://www.nla.gov.au/openpublish/index.php/nlasp/article/view/1019/1289. Accessed 22 June 2015
12. Jake, C., Stefano, L. (eds.): Biological Data Mining. Boca Raton. Chapman Hall/ Taylor and Francis, Abingdon-on-Thames (2009)
13. Cisar, P., et al.: BioWes-from design of experiment, through protocol to repository, control, standardization and back-tracking. BioMed. Eng. OnLine, **15**(1), 74 (2016)
14. Kopetz, H.: Internet of things. In: Kopetz, H. (ed.) Real-time Systems, pp. 307–323. Springer US, New York (2011)
15. Ferdoush, S., Li, X.: Wireless sensor network system design using Raspberry Pi and Arduino for environmental monitoring applications. Procedia Comput. Sci. **34**, 103–110 (2014)
16. http://smart.ponnod.com/
17. http://www.frov.jcu.cz/en/institute-complex-systems/lab-signal-image-processing

High Definition Method for Imaging Bacteria in Microconfined Environments on Solid Media

Cesar A. Hernandez[✉], Natalia Lopez-Barbosa, Crhistian C. Segura, and Johann F. Osma

CMUA, Department of Electrical and Electronics Engineering, Universidad de los Andes, Cra. 1E No. 19a-40, Bogota, DC 111711, Colombia
{ca.hernandez11, jf.osma43}@uniandes.edu.co

Abstract. In this paper, the methods and devices used to confine, observe and characterize the growth of bacteria on an agar based nutrient media are described. Selected strain *E. coli* MG1655 [1, 2] was successfully inoculated in a confinement structure composed of a microchannel (7 μm × 60 μm × 6.345 mm) with two culture chambers of 3 mm in diameter on both ends. The microchannel was fabricated on a standard microscope glass slide using a mask-less photolithographic technique and chemical etching. Isolation and manipulation of the confined bacteria was achieved by means of a custom designed 3D printed test cell. Observation was performed on an optical transmission microscope enhanced with a customized automation system. Growth characterization was performed by calculating the surface area colonized by the bacteria through image processing and analysis. The discussion focuses on the comparison of the growth rate within the confinement structure compared to traditional cell counting methods and the description of an observed, but inconsistent, scouting behavior. Finally, we discuss on the possible uses of the reported work and extend a call for introducing this system in current bacteriology research.

1 Introduction

The study of microorganisms in complex and detailed environments is a growing interest in bacteriology [3]. This has increased the use of micro and nanotechnology for the design and fabrication of microenvironments, that allow the monitoring of the behavior of single and multiple bacteria. Fundamental questions regarding their physical and chemical characteristics, such as their shape and morphology at the molecular level under different circumstances, remain unsolved [4]. In addition, the migration of microbiology from a qualitative to a quantitative study has created the challenge to develop reproducible techniques with precisely controlled variables. Thus, microfluidic devices, where bacteria are cultured in confined and controlled environments, represent a suitable fabrication technique for the monitoring of bacteria in the attempt to answer the question on how much of their behavior is because of their genetic material and how much as a response to their environment [5].

© Springer International Publishing AG 2017
I. Rojas and F. Ortuño (Eds.): IWBBIO 2017, Part II, LNBI 10209, pp. 726–736, 2017.
DOI: 10.1007/978-3-319-56154-7_64

The motility and growth of bacteria through a flow in microfluidic devices have been widely studied [5–9], both theoretically and experimentally. This has led to the characterization of the motility of bacteria through different channels and their growth behavior under flow. However, few is known about the behavior of bacteria when growing in static confined devices. Confinement structures and microfluidic devices can be easily developed through photolithography techniques [10], micro-milling machines [11] and micromoulding [12], among others. However, photolithography techniques allow to fabricate micrometric structures and complex designs accurately. This makes it possible to acquire interesting surface characteristics that vary the microenvironment in which bacteria are meant to grow.

For instance, Benhamed et al. [13] showed the effect of the surface characteristics of the skin mucus of fishes in the adhesion of different strains of bacteria. In the same way, Cortés et al. [14] studied the effect of surface roughness in the formation of bacterial biofilms, in which factors as hydrophobicity, surface charge, surface micro-topography and water flow are of special interest. As a consequence, different attempts have been made to introduce metallic [15] and polymeric [16] coatings that prevent or mitigate bacterial growth without the use of antibiotics, as well as microstructures that modified the physical characteristics of the growth surface [3].

Hereby, the methods and devices used to confine, observe and characterize the growth of bacteria on an agar based nutrient media are described. Selected strain E. coli MG1655 [1, 2] was successfully inoculated in a confinement structure composed of a microchannel (7 μm × 60 μm × 6.345 mm) with two culture chambers of 3 mm in diameter on both ends. The microchannel was fabricated on a standard microscope glass slide using a mask-less photolithographic technique and chemical etching. Isolation and manipulation of the confined bacteria was achieved by means of a custom designed 3D printed test cell. Observation was performed on an optical transmission microscope enhanced with a customized automation system. Growth characterization was performed by calculating the surface area colonized by the bacteria through image processing and analysis. The discussion focuses on the comparison of the growth rate within the confinement structure compared to traditional cell counting methods and the description of an observed, but inconsistent, scouting behavior. The fabrication technique places the confinement structure in an easy-to-modify format which can be used for several future growth and motility experiments. Finally, we discuss on the possible uses of the reported work and extend a call for introducing this system in current bacteriology research.

2 Materials and Methods

2.1 Design and Fabrication of the Confinement Structure

Two different types of microstructured devices were fabricated in order to study the growth and motility of E. coli. Figure 1a shows the design used to fabricate the confinement structure. Additional channels were included in a later design in order to study the bacterial behavior in a non-linear structure (Fig. 1b). The main channel is connected to two culture chambers of 3 mm in diameter where it is easy to inoculate by

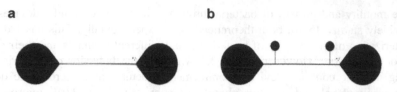

Fig. 1. (a) Design of a linear confinement structure and (b) addition of two additional growth chambers for non-linear studies

means of a 21 G dull needle syringe. Both chambers narrow tangentially with an angle of 39° towards a 60 μm width and 6.345 mm long microchannel. In the case of the non-linear structure, the additional chambers of 300 μm in diameter are connected with the main channel through a 30 μm width and 525 μm long microchannel.

The confinement structure was fabricated by photolithography technique on standard microscope slides (Borosilicate glass, 25.4 × 76.2 × 1.2 mm, Sail Brand, China). Slides were sonicated for 10 min in a CPX2800H sonicator (Branson, USA), while immersed in a 1 M potassium hydroxide (KOH) solution (EK Chem, Germany), rinsed with deionized (DI) water, and dried under a pressurized air stream. Slides were heated at 120 °C for 5 min to avoid excess of humidity.

MICROPOSIT™ SC™ 1827 positive photoresist (MICROPOSIT™, USA) was spun onto the slides by means of a SPIN150 spin coater (SPS Europe B.V., The Netherlands) at 5000 rpm for 1 min, and was pre-cured at 120 °C for 50 s on a hot plate. Spun slides were patterned with a SF-100 maskless lithographysystem (Intelligent micropatterning, FL, USA) system, developed under agitation with MICROPOSIT™ MF™ 319 developer (MICROPOSIT™, USA), and cured on a hot plate at 120 °C for 1 min.

Chemical etching was performed in a solution containing 50 ml of hydrofluoric acid (HF) 40% (Technical grade, PanReac AppliChem, Spain) and 200 mg of sodium sulfite anhydrous (N_2SO_3) (Analytical grade, Merck, Germany) for 5 s. After the etching, glass slides were submerged in DI water. Residual photoresist was removed with Baker PRS-1000 stripper (Avantor, USA), rinsed with DI water, and dried under pressurized air stream.

The height of the channels of the confinement structures were characterized by means of a Dektak 3 profilometer (Veeco, USA), yielding depths for individual structures ranging from 5 to 10 μm.

2.2   Design and Fabrication of the Test Cell

A test cell was designed and fabricated for the isolation and manipulation of the confinement structures and their integration with the experimental setup. Figure 2 shows a schematic of the test cell. The body and cover of the test cell were designed in the CAD software Solid Edge ST (Siemens, Germany), and fabricated in PLA by means of a RepRap Prusa i3 3D printer (Protolab 3D, Colombia).

After placing the confinement structure, the test cell is 40.4 mm wide, 90.2 mm long and 9.5 mm height. The bottom microscope glass slide serves as sealing of the test

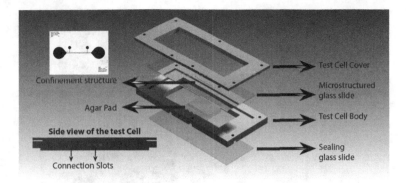

Fig. 2. Schematic of the test cell for the isolation and manipulation of the confinement structure.

cell while allowing the transmission of light from a microscope for observation of the confinement structure. In addition, six connection slots were located at the side of the test cell to facilitate the attachment of sensors or actuators. When assembled, the confinement structure is kept secured by seven 2 mm screws. Parafilm® M (Bemis NA, USA) was used between the confinement structure and the test cell body to obtain a better sealing.

2.3 Bacteria Culturing

E. coli strain MG 1655 was grown for 12 h overnight (ON) at 37 °C. An inoculum of 1 μL was seeded in one of the culture chambers of the confinement structure with a 21G dull needle syringe, and was left to dry for 5 min at room temperature. An agar pad of 20 × 15 × 3 mm, prepared from 2.5% wt of Luria-Bertani (LB) Broth (Miller) (Scharlau, Spain) and 1.5% wt of Agar Bacteriological (Scharlau, Spain), was used to seal the confinement structure. The system was placed and sealed in the test cell and integrated to the monitoring system for observation at 27 °C. Images along the microchannel were recorded every ten minutes by the automated monitoring system.

Growth curves at 27 and 37 °C were recorded by spectrophotometric measurements at 600 nm. 1.5 ml of *E. coli* were inoculated in 250 ml of LB broth, and incubated on a shaker at 300 rpm at both 27 and 37 °C.

2.4 Automated Monitoring System

A light transmission microscope was automated to be able to monitor bacterial growth while controlling time intervals, temperature and position. The hardware of the system was custom made and adapted to an Olympus CX21 microscope (Olympus, Japan). It was designed using the CAD software Solid Edge ST and fabricated in PLA by means of a RepRap Prusa i3 3D printer. A picture of the adapted automation system is shown in Fig. 3.

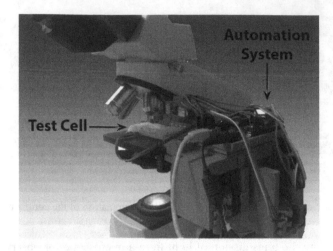

Fig. 3. Adapted automation system on Olympus CX21 microscope. The test cell position is controlled and each image is recorded in a computer

Movement in the X-Y axis was achieved with an Astrosyn SST-024 (Astrosyn International Technology PLC, England) stepper motor coupled to a 3D printed belt-pulley transmission (5:1) which is attached to each mechanical stage control knob. Stepper motors were controlled with an Arduino Mega (Arduino, USA), and A4988 drivers (Allegro MicroSystems LLC, USA) were used as power stage. Similarly, the fine focus knob was moved by a Futaba S33 (Hobbico Inc., USA) servomotor. Serial communication with a C# based PC interface was managed through the Arduino Mega. The program allows the user to visualize the system in real time and acquires images wirelessly with a resolution of 13 pixels per μm by means of a DSC-QX10 (Sony, Japan) camera when using a 10X, 0.25 achromatic objective. X-Y movement and fine focus can also be adjusted from the interface. Users can fix the number and size of position steps along the microchannel. At each step, centering and focusing of the microchannel can be adjusted either manually or automatically. The routine is repeated based on a time period set by the user.

2.5   Image Analysis Software

A software for image analysis and processing was developed in C# under Visual Studio 2015 and uses AForge.NET library [17]. Images recorded with the automated system were loaded to the software and converted to gray scale images.

At a first stage, umbralization was used to determine the section of the image at which the microchannel was located. The relative width in pixels of the microchannel was also measured. Images corresponding to a same section of the microchannel were aligned based on the center point of the microchannel. A second umbralization was performed from a threshold determined based on the percentile method [18] to conduct texture segmentation and differentiate empty sections of the microchannel from those occupied with bacteria. Area of occupation was calculated by counting the number of pixels without bacteria and subtracting it from the total area of the microchannel. The percentage of occupied area of the microchannel section was calculated. The algorithm for calculating the occupied area is based on ELEIMAG™ (Rovira I Virgili University, Spain) [19].

3   Results and Discussion

3.1   Bacteria Culturing

Growth curves of *E.coli* MG 1655 at 27 and 37 °C were recorded by spectrophoto-metric measurements at 600 nm (OD_{600}). Similarly, growth curves in terms of the percentage of occupied area as a function of time were calculated for the confinement structure. Figure 4 compares the results obtained from the OD_{600} and the confinement structure. It was observed that bacterial growth in the culture chamber was comparable with the OD_{600} at 27 °C, while growth in the microchannel was comparable with the OD_{600} at 37 °C.

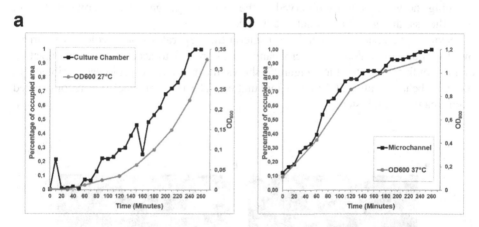

Fig. 4. Growth curves by OD_{600} and recorded in the confinement structure. (a) Shows the results obtained at 27 °C and in the culture chamber. (b) Shows the results obtained at 37 °C and in the microchannel

Growth inside the microchannel can be observed to occur radially from an initial bacterium. Growth rate increased exponentially with time as expected and in accordance with OD_{600}. In addition, since it is possible to observe bacteria as individuals, the growth from a single cell was also observed. Figure 5 shows the division process from a single bacterium at four different sample times (0, 40, 130 and 160 min). It is observed that the time elapsed for the first division to be completed is greater than that needed for the next division to take place.

The monitoring of the confinement structure allowed the observation of the formation of bacterial mono and bilayers. When bacteria reached the walls of the

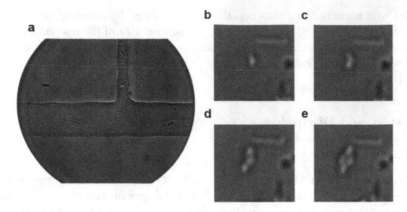

Fig. 5. (a) Division of a single bacteria in the microchannel at times (b) 0 min, (c) 40 min, (d) 130 min, and (e) 160 min

microchannel, a bilayer started growing onto the first layer towards the center of the microchannel at a greater rate than the one of the monolayer. Bilayer formation when touching the walls is easily observed in the recorded images and is shown in Fig. 6. Nonetheless, this growth was not quantified.

Scouting behavior (i.e. individual bacteria that separates from a colony to start its own) was observed along the microchannel as a process to accelerate the colonization of the confinement structure. Figure 7 shows the presence of colonies attributed to scouting behavior after 504 min of inoculation. The microchannel was reconstructed from images at each step.

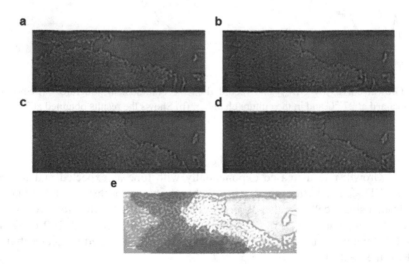

Fig. 6. Bilayer formation in the confinement structure at (a) 110 min, (b) 130 min, (c) 150 min, and (d) 170 min. (e) Shows the compiled expansion profile of the bilayer

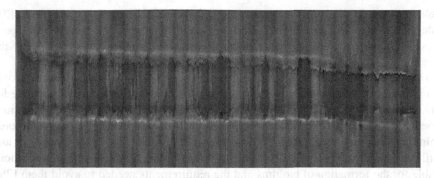

Fig. 7. Reconstruction of the microchannel after 504 min of inoculation. Green areas represent bacteria colonies (Color figure online)

3.2 Discussion

The confinement structure made possible the observation and measurement of bacteria growth in terms of the percentage of occupied area. It was also possible to observe the dynamics of single cells as well as the formation of bilayers. Monolayers showed to be formed from the center of the microchannel toward the walls, while bilayers expanded from the walls toward the center. Also, the rate of expansion of the bilayer was higher than that of the monolayer, which was attributed to the bacterial division in a three dimensional space rather than a surface growth. Growth in the culture chamber had a similar behavior as growth at 27 °C. Since this temperature is not an optimal temperature for *E. coli* growth [20], it takes a longer time for the bacteria to grow, and thus, to reach a plateau. This behavior is similar to the one exhibited in the culture chambers, due to the greater ratio between the growing area and the bacteria size. On the other hand, 37 °C is known as the optimal temperature for *E. coli* growth [20]. Similarly, since the microchannel has a considerably small size, bacteria growth gets saturated as a monolayer and continues growing as a bilayer, which is not reflected as an increase of occupied area.

Growth through the microchannel was also affected by scouting behavior. Individual bacterium seemed to migrate from a main colony to start their own. This seemed to result in a faster rate of colonization of the confinement structure than the one expected in the absence of a scouting behavior.

Although the confined structure was only used for the monitoring of growth and motility of *E. coli*, it can be applied to multiple studies in bacteriology. For instance, the device could be used to analyze the effect of a gradient (e.g. nutrient, pH) along the solid media on the dynamics of the bacterial community. This could give an insight on the incentives that motivate the dynamics of a specific strain. Also, due to the presence of two culture chambers, the interaction of two different strains could be observed without the influence of fluid flow. Under these circumstances, situations as horizontal gene transfer between two strains can be studied. Antibiotic resistance and efficacy can also be studied with more precision than traditional methods due to the ability to observe individual bacteria regardless of their presence in a community. On the other hand, the design of the test cell and the fabrication process of the confined structure, enables the possibility to integrate electrodes for electrochemical measurements [21] inside the channel and their easy connection to an instrumentation circuit. This opens the possibility to fabricate portable electrochemical biosensors with bacteria as the recognition element [22, 23].

Biofilms are structured bacterial communities that are enclosed in a self-grown polymeric matrix and are usually attached to a substrate [24]. They are responsible for many hygiene problems and economic losses in the food industry, as well as infections in medical devices [25]. Due to the way in which the bacteria growth is monitored within the confinement structure, the differentiation in the formation of monolayer and multilayers biofilms can be easily observed. This could be used to study the conditions suitable for the formation of biofilms and the requirements needed to avoid them [26].

4 Conclusions

A confinement structure composed of a microchannel and two culture chambers was successfully designed and fabricated for bacterial growth on solid media. An automation system was adapted to an optical transmission microscope in order to control X-Y movements along the microchannel and record images at defined time intervals. Acquired images from E. coli growth were processed and analyzed to determine the percentage of occupied area as a function of time. Results at the culture chambers and the microchannel were compared with standard growth curves at 27 and 37 °C. It was observed that growth in the culture chambers was comparable with OD_{600} at 27 °C, and growth in the microchannel was comparable with OD_{600} at 37 °C. Scouting behavior, in which individual bacterium left the main colony to start their own, was observed along the microchannel, although it was not quantified. The confinement structure allowed the observation of the formation of a bilayer when the monolayer reached the walls of the microchannel, opening the possibility to use the device for studying biofilms. Finally, we open the discussion on whereas this device could be used for a better characterization of bacterial dynamics and their interaction with different microenvironments.

Acknowledgements. The authors thank the cleanroom facilities from the Department of Electrical and Electronics Engineering at Universidad de los Andes for the financial support. Claudia Camila Barrera Garzón for her help during the bacteria culturing with traditional techniques and important insights during the design stage. The group of Biophysics, specially to David Camilo Durán Chaparro, from the Physics Department of the Universidad de los Andes, for providing us with the E. coli strain, their support and facilities.

Finally, Cesar A. Hernandez thanks Colciencias for their support through doctoral scholarship PDBCNal 567.

Conflict of Interests
In this work, there are no potential conflicts between or related to any author.

References

1. Hayashi, K., Morooka, N., Yamamoto, Y., Fujita, K., Isono, K., Choi, S., Ohtsubo, E., Baba, T., Wanner, B.L., Mori, H.: Highly accurate genome sequences of Escherichia coli K-12 strains MG1655 and W3110. Mol. Syst. Biol. **2** (2006)
2. Sezonov, G., Joseleau-Petit, D., D'Ari, R.: Escherichia coli physiology in Luria-Bertani broth. J. Bacteriol. **189**, 8746–8749 (2007)
3. Hol, F.J.H., Dekker, C.: Zooming in to see the bigger picture: microfluidic and nanofabrication tools to study bacteria. Science **346**, 1251821 (2014). doi:10.1126/science.1251821
4. Weibel, D.B., Diluzio, W.R., Whitesides, G.M.: Microfabrication meets microbiology. Nat. Rev. Microbiol. **5**, 209–218 (2007). doi:10.1038/nrmicro1616
5. Norman, T.M., Lord, N.D., Paulsson, J., Losick, R.: Memory and modularity in cell-fate decision making. Nature **503**, 481–486 (2013)
6. Biondi, S.A., Quinn, J.A., Goldfine, H.: Random motility of swimming bacteria in restricted geometries. AIChE J. **44**, 1923–1929 (1998). doi:10.1002/aic.690440822
7. Frymier, P.D., Ford, R.M.: Analysis of bacterial swimming speed approaching a solid–liquid interface. AIChE J. **43**, 1341–1347 (1997). doi:10.1002/aic.690430523
8. Lauga, E., DiLuzio, W.R., Whitesides, G.M., Stone, H.A.: Swimming in circles: motion of bacteria near solid boundaries. Biophys. J. **90**, 400–412 (2006). doi:10.1529/biophysj.105.069401
9. Ramia, M., Tullock, D.L., Phan-Thien, N.: The role of hydrodynamic interaction in the locomotion of microorganisms. Biophys. J. **65**, 755–778 (1993). doi:10.1016/S0006-3495(93)81129-9
10. Ko, H., Lee, J.S., Jung, C.-H., Choi, J.-H., Kwon, O.-S., Shin, K.: Actuation of digital micro drops by electrowetting on open microfluidic chips fabricated in photolithography. J. Nanosci. Nanotechnol. **14**, 5894–5897 (2014)
11. Gao, J., Manard, B.T., Castro, A., Montoya, D.P., Xu, N., Chamberlin, R.M.: Solid-phase extraction microfluidic devices for matrix removal in trace element assay of actinide materials. Talanta **167**, 8–13 (2017). doi:10.1016/j.talanta.2017.01.080
12. Fukuba, T., Yamamoto, T., Naganuma, T., Fujii, T.: Microfabricated flow-through device for DNA amplification—towards in situ gene analysis. Chem. Eng. J. **101**, 151–156 (2004)
13. Benhamed, S., Guardiola, F.A., Mars, M., Esteban, M.Á.: Pathogen bacteria adhesion to skin mucus of fishes. Vet. Microbiol. **171**, 1–12 (2014). doi:10.1016/j.vetmic.2014.03.008
14. Cortés, M.E., Bonilla, J.C., Sinisterra, R.D.: Biofilm formation, control and novel strategies for eradication. Sci. Against Microb. Pathog. Commun. Curr. Res. Technol. Adv. **2**, 896–905 (2011)
15. Toy, L.W., Macera, L.: Evidence-based review of silver dressing use on chronic wounds. J. Am. Acad. Nurse Pract. **23**, 183–192 (2011)
16. Yue, I.C., Poff, J., Cortés, M.E., Sinisterra, R.D., Faris, C.B., Hildgen, P., Langer, R., Shastri, V.P.: A novel polymeric chlorhexidine delivery device for the treatment of periodontal disease. Biomaterials **25**, 3743–3750 (2004)
17. Kirillov, A.: AForge.NET (2006)
18. van der Valk, F.M., Verweij, S.L., Zwinderman, K.A.H., Strang, A.C., Kaiser, Y., Marquering, H.A., Nederveen, A.J., Stroes, E.S.G., Verberne, H.J., Rudd, J.H.F.: Thresholds for arterial wall inflammation quantified by 18F-FDG PET imaging: implications for vascular interventional studies. JACC Cardiovasc. Imaging **9**, 1198–1207 (2016). doi:10.1016/j.jcmg.2016.04.007

19. Osma, J.F., Toca-Herrera, J.L., Rodríguez-Couto, S.: Environmental, scanning electron and optical microscope image analysis software for determining volume and occupied area of solid-state fermentation fungal cultures. Biotechnol. J. **6**, 45–55 (2011). doi:10.1002/biot. 201000256

20. Noor, R., Islam, Z., Munshi, S.K., Rahman, F.: Influence of temperature on Escherichia coli growth in different culture media. J. Pure Appl. Microbiol. **7**, 899–904 (2013)

21. Lopez-Barbosa, N., Segura, C., Osma, J.F.: Electro-immuno sensors: current developments and future trends. Int. J. Biosens. Bioelectron. **2**, 1–6 (2017). doi:10.15406/ijbsbe.2017.02. 00010

22. Lopez-Barbosa, N., Gamarra, J.D., Osma, J.F.: The future point-of-care detection of disease and its data capture and handling. Anal. Bioanal. Chem. (2016). doi:10.1007/s00216-015-9249-2

23. Lopez-Barbosa, N., Osma, J.F.: Biosensors: migrating from clinical to environmental industries. Biosens. J. (2016). doi:10.4172/2090-4967.100e106

24. Costerton, J.W., Stewart, P.S., Greenberg, E.P.: Bacterial biofilms: a common cause of persistent infections. Science **284**, 1318–1322 (1999)

25. Satpathy, S., Sen, S.K., Pattanaik, S., Raut, S.: Review on bacterial biofilm: an universal cause of contamination. Biocatal. Agr. Biotechnol. **7**, 56–66 (2016). doi:10.1016/j.bcab. 2016.05.002

26. Hernández, C.A., Gaviria, L.N., Segura, S.M., Osma, J.F.: Concept design for a novel confined-bacterial-based biosensor for water quality control. In: 2013 Pan American Health Care Exchanges, pp. 1–3 (2013). doi:10.1109/PAHCE.2013.6568257

Erratum to: Bioinformatics and Biomedical Engineering

Ignacio Rojas$^{(\boxtimes)}$ and Francisco Ortuño

Universidad de Granada, Granada, Spain

Erratum to:
Chapter "An Automatic and Intelligent System for Integrated Healthcare Processes Management" in: I. Rojas and F. Ortuño (Eds.): Bioinformatics and Biomedical Engineering, Part II, LNBI 10209, https://doi.org/10.1007/978-3-319-56154-7_55

The initially published versions of Acknowledgements section were incorrect. This is the correct version:

Acknowledgements. This research has been supported by the Pololas project (TIN2016-76956-C3-2-R) and by the SoftPLM Network (TIN2015-71938-REDT) of the Spanish the Ministry of Economy and Competitiveness.

Erratum to:
Chapter "A Microcontroller Based System for Controlling Patient Respiratory Guidelines" in: I. Rojas and F. Ortuño (Eds.): Bioinformatics and Biomedical Engineering, Part II, LNBI 10209, https://doi.org/10.1007/978-3-319-56154-7_56

The initially published versions of Acknowledgements section were incorrect. This is the correct version:

Acknowledgements. This research has been supported by the Pololas project (TIN2016-76956-C3-2-R) and by the SoftPLM Network (TIN2015-71938-REDT) of the Spanish the Ministry of Economy and Competitiveness.

The updated online version of this book can be found at
https://doi.org/10.1007/978-3-319-56154-7
https://doi.org/10.1007/978-3-319-56154-7_55
https://doi.org/10.1007/978-3-319-56154-7_56

© Springer International Publishing AG 2017
I. Rojas and F. Ortuño (Eds.): IWBBIO 2017, Part II, LNBI 10209, p. E1, 2017.
https://doi.org/10.1007/978-3-319-56154-7_65

Author Index

Printed in the United States
By Bookmasters